MODERN PRACTICE OF GAS CHROMATOGRAPHY

MODERN PRACTICE OF GAS CHROMATOGRAPHY

THIRD EDITION

Edited by

Robert L. Grob, Ph.D.
Professor Emeritus, Analytical Chemistry
Villanova University

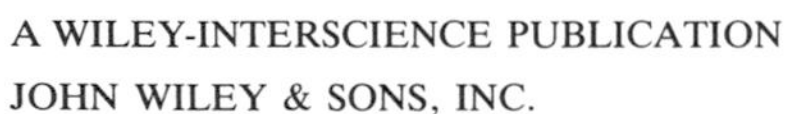
A WILEY-INTERSCIENCE PUBLICATION
JOHN WILEY & SONS, INC.

New York • Chichester • Brisbane • Toronto • Singapore

This text is printed on acid-free paper.

Library of Congress Cataloging in Publication Data:

Modern practice of gas chromatography / edited by Robert L. Grob. -- 3rd ed.

p. cm.

"A Wiley-Interscience publication."

Includes index.

ISBN 0-471-59700-7 (cloth : acid-free paper)

1. Gas chromatography. I. Grob, Robert Lee.

QD79.C45M63 1995

543′.0896--dc20 94-23516

Printed in the United States of America

10 9 8 7 6 5 4 3 2 1

CONTRIBUTORS

Juan G. Alvarez, Department of Obstetrics & Gynecology, Beth Israel Hospital, Harvard Medical School, Boston, MA

Clinton W. Amoss, ARCO Chemical Company, Newtown Square, PA

Eugene F. Barry, Chemistry Department, University of Massachusetts, Lowell, MA

Thomas A. Brettell, New Jersey State Police Forensic Laboratory, Trenton, NJ

Gary W. Caldwell, The R. W. Johnson Pharmaceutical Research Institute, Department of Medicinal Chemistry, Spring House, PA

Frederick J. Debbrecht, Deceased, Landenberg, PA

Cecil R. Dybowski, Chemistry Department, University of Delaware, Newark, DE

Robert L. Grob, Professor Emeritus Analytical Chemistry, Malvern, PA

Lorraine H. Hendrich, ARCO Chemical Company, Newtown Square, PA

Mary A. Kaiser, E.I. DuPont de Nemours & Company, Wilmington, DE

Matthew S. Klee, Hewlett Packard Company, Wilmington, DE

Joseph M. Loeper, Weston Environmental Laboratory, Lionville, PA

John A. Masucci, The R. W. Johnson Pharmaceutical Research Institute, Department of Medicinal Chemistry, Spring House, PA

Donald J. Skahan, ARCO Chemical Company, Newtown Square, PA

Edward F. Smith, Exxon Chemical Americas, Baton Rouge, LA

To
My
Wife and Family

What is written without effort is in general read without pleasure

—Samuel Johnson (1709–1784)
Johnsonian Miscellanies
Vol. ii, p. 309

PREFACE

The third edition of *Modern Practice of Gas Chromatography* represents a number of changes from the first two editions. First, a number of new contributing authors have been involved. These authors were chosen because of their expertise and active participation in the various areas related to gas chromatography (GC). Second, the contents of the various chapters have been changed so as to be all-inclusive. For example, a discussion of the necessary instrumentation has been included in such chapters as Columns and Detectors. Third, the topics from Chapter 9 (2d ed.) have been divided between the chapters on qualitative and quantitative analysis and environmental analysis. Fourth, a separate chapter is dedicated to gas chromatography/mass spectrometry because of its growth in importance since the last edition was published. Another change has been the elimination of several chapters because of their adequate coverage in other texts. The Editor is satisfied that this new edition represents an all-inclusive text that may be used for university courses as well as short courses.

No book will please everyone. Each person has certain ideas concerning what should be covered and how much detail should be given to each topic. The coverage of the theory and basics of GC are what we consider necessary to the beginner for this technique and the nomenclature is that most recently recommended by the IUPAC Commission. The techniques and instrumentation section is greatly detailed and the application chapters cover topics that would be of interest to most people utilizing the gas chromatographic technique.

The Editor thanks the contributing authors for their cooperation and professionalism, thus making this third edition a reality. Most important, the Editor thanks his wife, Marjorie, for her interest, encouragement, and cooperation during these many months of preparation.

ROBERT L. GROB

Malvern, Pennsylvania
June 1995

CONTENTS

MODERN PRACTICE OF GAS CHROMATOGRAPHY

CHAPTER ONE

Introduction

ROBERT L. GROB

Professor Emeritus of Analytical Chemistry, Villanova University

1.1 HISTORY AND DEVELOPMENT OF CHROMATOGRAPHY

Many publications have discussed or detailed the history and development of chromatography (1–3). Rather than duplicate these writings, we present in Table 1.1 a chronological listing of events that we feel are the most relevant in the development of the present state of the field. Since the

Modern Practice of Gas Chromatography, Third Edition, Edited by Robert L. Grob.
ISBN 0-471-59700-7

TABLE 1.1 Development of the Field of Chromatography

Year (Reference)	Scientist(s)	Comments
1834 (4) 1834 (5)	Runge, F. F.	Used unglazed paper and/or pieces of cloth for spot testing dye mixtures and plant extracts
1850 (6)	Runge, F. F.	Separated salt solutions on paper
1868 (7)	Goppelsroeder, F.	Introduced paper strip (capillary analysis) analysis of dyes, hydrocarbons, milk, beer, colloids, drinking and mineral waters, plant and animal pigments
1878 (8)	Schonbein, C.	Developed paper strip analysis of liquid solutions
1897–1903 (9–11)	Day, D. T.	Developed ascending flow of crude petroleum samples through column packed with finely pulverized fuller's earth
1906–1907 (12–14)	Tswett, M.	Separated chloroplast pigment on $CaCO_3$ solid phase and petroleum ether liquid phase
1931 (15)	Kuhn, R. et al.	Introduced liquid–solid chromatography for separating egg yolk xanthophylls
1940 (16)	Tiselius, A.	Earned Nobel Prize in 1948; developed adsorption analyses and electrophoresis
1940 (17)	Wilson, J. N.	Wrote first theoretical paper on chromatography; assumed complete equilibration and linear sorption isotherms; qualitatively defined diffusion, rate of adsorption, and isotherm nonlinearity
1941 (18)	Tiselius, A.	Developed liquid chromatography and pointed out frontal analysis, elution analysis, and displacement development
1941 (19)	Martin, A. J. P., and Synge, R. L. M.	Presented first model that could describe column efficiency; developed liquid–liquid chromatography; received Nobel Prize in 1952
1944 (20)	Consden, R., Gordon, A. H., and Martin, A. J. P.	Developed paper chromatography
1946 (21)	Claesson, S.	Developed liquid–solid chromatography with frontal and displacement development analysis; coworker A. Tiselius

TABLE 1 ***(Continued)***

Year (Reference)	Scientist(s)	Comments
1949 (22)	Martin, A. J. P.	Contributed to relationship between retention and thermodynamic equilibrium constant
1951 (23)	Cremer, E.	Introduced gas–solid chromatography
1952 (24)	Phillips, C. S. G.	Developed liquid–liquid chromatography by frontal technique
1952 (25)	James, A. T., and Martin, A. J. P.	Introduced gas–liquid chromatography
1955 (26)	Glueckauf, E.	Derived first comprehensive equation for the relationship between HEPT and particle size, particle diffusion, and film diffusion ion exchange
1956 (27)	van Deemter, J. J., et al.	Developed rate theory by simplifying work of Lapidus and Ammundson to Gaussian distribution function
1957 (28)	Golay, M.	Reported the development of open tubular columns
1965 (29)	Giddings, J. C.	Reviewed and extended early theories of chromatography

various types of chromatography (liquid, gas, paper, thin-layer, ion exchange, supercritical fluid) have many features in common, they must all be considered in the development of the field. Although the topic of this text, gas chromatography (GC), probably has been the most investigated during the past 30 years, results of these studies have had a great impact on the other types of chromatography, especially modern (high-performance) liquid chromatography (HPLC).

There will, of course, be those who believe that the list of names and events presented in Table 1.1 is incomplete. We simply wish to show a development of an ever-expanding field and to point out some of the important events that were responsible for the expansion. To attempt an account of contemporary leaders of the field could only result in disagreement with some workers, astonishment by others, and a very long listing that would be cumbersome to correlate.

1.2 SEPARATION TECHNIQUES

1.2.1 Various Techniques Used for Separations

Most separation techniques involve the formation of at least two phases, in which the object is to separate and measure the various constituents. There are various ways of describing a phase, that is, gas, liquid, and solid. By proper choice of conditions (temperature and pressure), one is able to convert a solid to a liquid (melting) or a gas (sublimation), a liquid to a solid (freezing) or a gas (distillation), and a gas to a liquid (condensation) or a solid (condensation). When the phase transition(s) are completed, one phase should contain the material of interest and the other(s), materials not of interest. The phases can then be mechanically or physically separated, and the phase containing the component of interest is retained.

Since the component(s) of interest can be in one of three states of matter and these, in turn, can be converted into one of three phase types, many types of separation can be used. The major classifications are shown in Table 1.2. Chromatography is used in four of the nine major types shown.

In our discussion of separations we include not only homogeneous equilibria, but also heterogeneous equilibria and the rates at which these equilibria are obtained. If the equilibrium point and the rate of attainment of said equilibrium are both favorable, the separation can usually be attained in one step. Less favorable systems utilize multistage operations. Multistage separations are both feasible and attractive.

The separations are classified according to mechanical, physical, or chemical processes, as shown in Table 1.2. This is illustrated in Table 1.3. The measurements of the separated components can be made by physical, chemical, or biological means. Several techniques are used within each of these three types of analysis. In the majority of analysis studies most of the discussion relates to an examination of the theoretical background, the experimental limitations, and the applications of the various techniques for making useful measurements. Methods of analysis are usually defined in terms of the final measurement made and thus many give the impression that this stage constitutes the entire subject of analytical chemistry. A more realistic view of analytical chemistry involves decisions regarding what information is needed from a system, how to obtain that information, utilization of one or more separations and measurements, collection and evaluation of the experimental data, and finally drawing some conclusions from the data.

In analysis of materials, any one of the above categories (sample, separation, measurement) may assume more importance than another. It may be more difficult to obtain a representative sample than the separation, or measurement or the separation may be more difficult than the sampling and measurement. Two objectives should be paramount for any analysis: The data must have the *required accuracy and precision* and be produced in the *minimum time*.

The feature that places chromatography in a special category, that is, distinguishes it from other separation techniques, is that one of the phases moves, whereas the other phase is stationary. By combining the states of matter into different pairs, we are able to arrive at a number of different chromatographic techniques (Table 1.4). Viewing chromatography in simple terms, one would reasonably expect these several types of chromatography to have features in common, and this is the case. The principles by which

TABLE 1.2 Classification of Separations[a]

Gas			Liquid			Solid		
Gas	Liquid	Solid	Gas	Liquid	Solid	Gas	Liquid	Solid
Thermal diffusion	GLC, condensation, sorption	GSC, sorption	Volatilization	LLC, distillation, extraction	LSC, precipitation, electrodeposition, crystallization, ring oven	Sublimation	Solution, zone refining	Sieving, magnetic techniques

[a] Abbreviations: GLC, gas–liquid chromatography; GSC, gas–solid chromatography; LLC, liquid–liquid chromatography; LSC, liquid–solid chromatography.

TABLE 1.3 Processes of Separation

Chemical	Mechanical	Physical
Precipitation	Filtration	GLC, GSC, LLC, LSC
Electrodeposition	Centrifugation	Liquid–liquid extraction
Masking	Exclusion chromatography	Distillation
Ion exchange	Dialysis	Sublimation
		Zone electrophoresis
		Zone refining

TABLE 1.4 Types of Chromatography

Mobile Phase	Stationary Phase	Type
Gas	Liquid	GLC
	Solid	GSC
Liquid	Liquid	LLC,[a] PC
	Solid	LSC,[a] TLC, ion exchange

[a] Includes HPLC.

separation is achieved do not vary according to the type of equipment utilized. Experimentally, chromatography is a relatively simple separation technique. Four components are essential in a chromatographic separation: a column, a mobile phase, a sample injector, and a detector.

In this text we are concerned with GC (gas–liquid and gas–solid) and thus discuss only this type. Before discussing the theory of GC per se, let us look at some basic separations and some of the theoretical fundamentals that underlie the technique.

Phase Equilibria. Gas chromatography involves chemical equilibria between phases to bring about a particular separation. Thus a brief discussion of phase equilibria is pertinent at this point. Phase equilibria separations can be understood with the use of the second law of thermodynamics. The phase rule states that if we have a system of C components that are distributed between P phases, the composition of each phase will be completely defined by $C-1$ concentration terms. For definition of the compositions of P phases, therefore, it is necessary to have $P(C-1)$ concentration terms. The temperature and pressure also are variables and are the same for all phases. Assuming that no other forces influence the equilibria, it follows that

$$\text{Total degrees of freedom (variables)} = P(C-1)+2 \tag{1.1}$$

For P phases of C components, we thus have $C(P-1)$ independent equations. It follows that $C(P-1)$ variables are fixed, which leaves

$$[P(C-1)+2]-[C(P-1)] = C-P+2 \tag{1.2}$$

unknown. Hence this number of variables, $C - P + 2$, must be fixed. Stated another way, this is the value of our limiting degrees of freedom. These degrees of freedom may be represented by F. This gives us the recognizable form of the phase rule:

$$F = C - P + 2 \tag{1.3}$$

In its strictest sense the phase rule assumes that the equilibrium between phases is not influenced by gravity, electrical or magnetic forces, or surface action. Thus the only variables are temperature, pressure, and concentration, and if two are fixed, the third is easily determined (another reason for the constant 2 in Equation 1.3).

We need to define exactly the terms in Equation 1.3:

1. By *a phase* we mean any homogeneous or physically distinct part of the system that is separated from other parts of the system by definite bounding surfaces.
2. By a *component* we mean the smallest number of independently variable components from which the composition of each phase can be expressed (directly or in the form of a chemical equation).
3. By a *degree of freedom* we mean the number of variable factors (e.g., temperature, pressure, and concentration) that must be fixed to completely define a system at equilibrium.

We may treat this system in either of two ways: (1) hold temperature constant (isothermal conditions) and vary pressure and concentration or (2) hold pressure constant (isobaric conditions) and vary concentration and temperature. Both situations are shown in Figure 1.1. The plots in Figures 1.1a and 1.1b represent conditions for isothermal and isobaric treatments, respectively.

In this system $C = 2$. If we choose a point that does not fall on the vapor–liquid equilibrium line, all three variables must be known to describe the system. Through selection of a point on the vapor–liquid line phases, however, $P = 2$ and thus degrees of freedom $F = 2 - 2 + 2 = 2$. In other words, only two of the three degrees of freedom (variables) must be known. Referring to Figure 1.1b, if we have a 50/50 mol fraction solution of A and B, the mixture boils at 92°C and the vapor contains 78 mol% of B. In Figure 1.1a the dotted lines indicate the partial pressure of each component; that is, the equation of each line defines Raoult's law:

$$P_{\mathrm{A}} = x_{\mathrm{A}} p_{\mathrm{A}}^{0} \tag{1.4}$$

$$P_{\mathrm{B}} = x_{\mathrm{B}} p_{\mathrm{B}}^{0} \tag{1.5}$$

where P = partial pressure of A or B

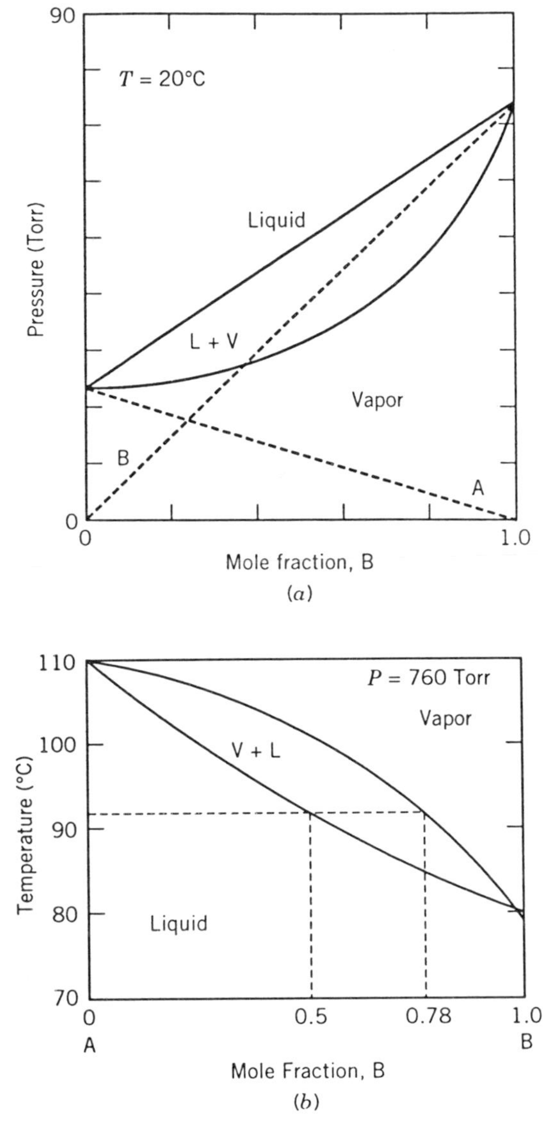

FIGURE 1.1 Phase diagram for two-component system: (a) isothermal conditions; (b) isobaric conditions.

x = mole fraction of A or B in liquid (by *mole fraction* we mean the number of moles of one component divided by the sum of each of the moles of all components present in the mixture)

p^0 = vapor pressure of pure A and B

The dotted lines in Figure 1.1 describe the condition that as the mole percent of either component decreases, so does the partial pressure (since there are fewer molecules at the surface to exert vapor pressure). The upper solid line (boundary between liquid and liquid plus vapor) represents the sum of the two dotted lines. The equation for this line defines Dalton's law:

$$P_{\mathrm{A}} = Y_{\mathrm{A}} p \tag{1.6}$$

$$P_{\mathrm{B}} = Y_{\mathrm{B}} p \tag{1.7}$$

where P = partial pressure of A or B
Y = mole fraction of A or B in the vapor
p = total pressure of the system

In the preceding discussion regarding liquid–vapor equilibria we assumed that our representative systems were ideal, that is, that there are no differences in attractions between molecules of different types (intra- and intermolecular attractions). Few systems are ideal, and most show some deviation from ideality and do not follow Raoult's law. Deviations from Raoult's law may be positive or negative. Positive deviations (for binary mixtures) occur when the attraction of like molecules, such as A–A or B–B, are stronger than unlike molecules, such as A–B (total pressure greater than that computed for ideality). Negative deviations result from the opposite effects (total pressure lower than that computed for ideality). A mixture of two liquids can exhibit nonideal behavior by forming an azeotropic mixture (a constant boiling mixture).

Raoult's law assumes that the liquid phase is idea, that is, that the partial pressure of the component A is equal to the mole fraction of A, in the liquid, times the vapor pressure of pure A. The same could be said of component B, and so on. Mathematically we write Raoult's law as shown in Equations 1.4 and 1.5; therefore, the sum of the partial pressures equals the total pressure of the system, namely, Dalton's law:

$$p = P_{\mathrm{A}} + P_{\mathrm{B}} \tag{1.8}$$

This relationship is represented by the plot in Figure 1.2. Raoult's law is usually followed when x_{A} is a large value. In some systems it does hold for low values of x_{A}. When x_{A} is a small value, the system is said to follow Henry's law, which states that the partial pressure of a component is equal to the mole fraction in the liquid, multiplied by a constant:

$$P_{\mathrm{A}} = H_{\mathrm{A}} x_{\mathrm{A}} \tag{1.9}$$

where H_{A} = Henry's constant. If the system being studied is ideal, Raoult's and Henry's laws are identical; that is, the H_{A} term is the vapor pressure of the pure component, p_{A}^{0}.

A short discussion of thermodynamics is necessary to place the topic of

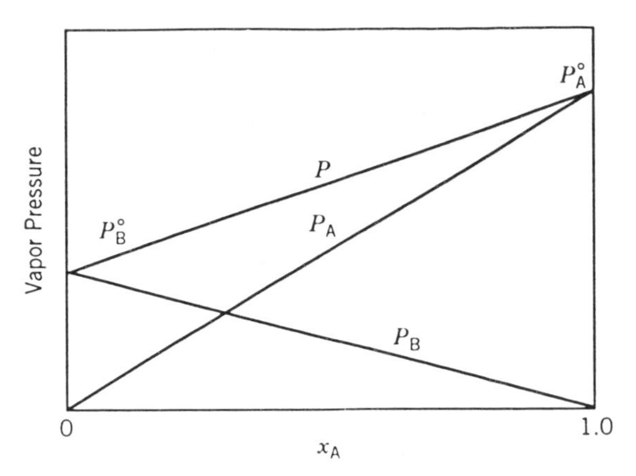

FIGURE 1.2 Plot of Raoult's law for two-component system.

equilibrium into proper perspective. From the viewpoint of thermodynamics a system is in equilibrium when the free energy G is equal to zero. Free energy is the energy available to do work. The free energy of a system depends on the enthalpy (heat content) H and the entropy (disorder or randomness of the molecules) S:

$$G = H - TS \tag{1.10}$$

and for the isothermal process the free energy change is dependent on the changes in both enthalpy and entropy:

$$\Delta G = \Delta H - T\Delta S \tag{1.11}$$

Adsorption. Solids have a residual surface force field, and there is a tendency for the free energy of this surface to decrease. This phenomenon is responsible for adsorption. Temperature and pressure are the two main variables that affect the process of adsorption. Decrease in the temperature or increase in the pressure results in increase in adsorption. At low temperatures, adsorption of gases increases very rapidly with small changes in pressure. Increase in the temperature causes a decrease in adsorption, which implies evolution of heat in the adsorption process. Curves that show the variation in pressure P with temperature T are referred to as *isoteres*. Plotting of the log P against the reciprocal of temperature $1/T$ gives a straight line, indicating that ΔH is independent of temperature. Van 't Hoff's equation represents the variation of the equilibrium constant K with the temperature T for a reaction involving gases in terms of the change in heat content (heat of reaction H at constant pressure):

$$\frac{d \ln K}{dT} = \frac{\Delta H}{RT^2} \tag{1.12}$$

The adsorption capacity of various solids for a specific gas depends primarily on the effective area of solids. For a series of gases, the order of increasing adsorption is the same for all solid adsorbents. These similarities hold as long as there are no chemical bonding factors intervening in the adsorption process. Gases that are the most easily liquified are the most readily adsorbed at a solid surface. Table 1.5 depicts data for the adsorption of a number of gases on 1 g of activated charcoal at 15°C.

Adsorption phenomena are divided into two main categories: physical (van der Waals or dispersion forces) and chemical (analogous to valence bonding). The physical type results in small heats of adsorption (same order of magnitude as heats of vaporization), in which the equilibrium of the gas is reversible and easily attained with changes in temperature or pressure. All gases exhibit van der Waals adsorption, and some also exhibit chemisorption. This second type shows heats of adsorption of the order of chemical combinations. Adsorption equilibria are usually presented as an adsorption isotherm (quantity adsorbed as a function of pressure with temperature constant).

Diffusion. From the gas laws we know that M/V is a measure of the density (ρ) of a gas. Therefore, we can arrive at an equation for the speed of

TABLE 1.5 Correlation of Adsorption of Gases With Critical Temperature

Gas	Critical Temperature (K)[a]	Volume Adsorbed (mL)
H_2	33	4.7
N_2	126	8.0
O_2	154	8.2
CO	134	9.3
CH_4	190	16.2
CO_2	304	48.0
N_2O	310	54.0
HC	324	72.0
H_2S	373	99.0
NH_3	406	181.0
Cl_2	417	235.0
SO_2	430	380.0

[a] The critical temperature (maximum temperature at which a gas can be liquified) is related to the boiling point.

gaseous molecules:

$$(\bar{C}^2)^{1/2} = \left(\frac{3P}{\rho}\right)^{1/2} \tag{1.13}$$

where $\bar{C}^2$ is the mean square velocity for all the molecules in a gas. Equation 1.13 tells us that the velocity of gaseous molecules is inversely proportional to the square root of the density. Thus it follows that the rate of diffusion is inversely proportional to the square root of the molecular weight M of a gas.

Diffusion processes occur in all systems where concentration differences exist. Diffusion is the main mechanism that aids in the elimination of concentration gradients. Fick's first law of diffusion defines this phenomenon by correlating mass flow with concentration gradient:

$$\frac{\partial N}{\partial t} = -D \frac{\partial n}{\partial l} \tag{1.14}$$

where N = number of molecules passing through a unit surface
t = time for molecules to pass unit surface
D = diffusion coefficient (weight diffusing across a plane 1 cm^2 in unit time under a concentration gradient of unity)
n = concentration of gas molecules
l = distance that molecules diffuse
$\partial n/\partial l$ = concentration gradient

The right side of Equation 1.14 carries a negative sign to indicate that diffusion is taking place in the direction of lower concentration. Stated otherwise, Equation 1.14 illustrates that the amount of material diffusing through the unit surface in unit time is proportional to the concentration gradient in the direction of the diffusion.

If diffusion is taking place in a system, there must be a conservation of mass in the process. In a gas chromatographic column we are concerned primarily with the longitudinal diffusion, and this can be described by Fick's second law:

$$\frac{\partial n}{\partial t} = D \frac{\partial^2 n}{\partial l^2} \tag{1.15}$$

These diffusion processes are random, and one may express these in terms of a statistical distribution. Equation 1.15 can be stated

$$n = \frac{A}{t^{1/2}} e^{-l^2/4Dt} \tag{1.16}$$

where A is a constant. If the system can be given in terms of the total quantity of diffusing material m, we can solve for the constant A:

$$A = \frac{m}{2(\pi D)^{1/2}} \tag{1.17}$$

Solution of the differential Equation 1.16 and substitution of Equation 1.17 for A gives

$$n = \frac{m}{2(\pi Dt)^{1/2}} e^{-l^2/4Dt} \tag{1.18}$$

Equation 1.18 is an equation of a Gaussian curve. This curve is described by its maximum and width. We can represent the Gaussian curve in terms of Equation 1.18, as shown in Figurc 1.3. Thc basc width of the curve is $2(2DT)^{1/2}$.

One may study zone broadening in GC by observing the shape of the elution peak, which is Gaussian in ideal systems. The base width of thc Gaussian curve is measured in standard deviation units; therefore,

$$\sigma = (2Dt)^{1/2} \tag{1.19}$$

Squaring of the standard deviation term gives the variance, a term that is used in the rate theory description for the gas chromatographic process (see Section 2.3.2):

$$\sigma^2 = 2Dt \tag{1.20}$$

As Equations 1.19 and 1.20 illustrate, the diffusion curve is determined by time (time elapsed from beginning to end of separation) and the value of D (diffusion coefficient, which is different for each gas). One may obtain the value of D from the kinetic theory of gases.

Thus far we have discussed diffusion in terms of molecular or free diffusion, where the diffusion rate is determined by molecular collisions and the particle voids, which are larger than the mean free path. In packed gas

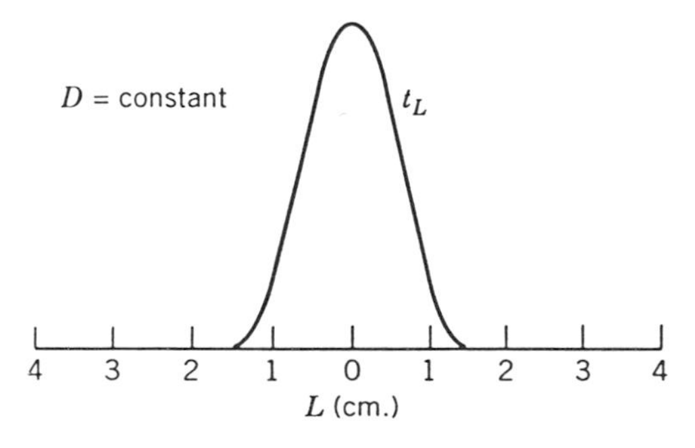

FIGURE 1.3 Gaussian curve in terms of Fick's second law.

chromatographic columns the diffusion process follows other laws. Under these conditions we can encounter four types of diffusion.

1. *Surface or Volmer Diffusion*. When pores are small, the adsorbed molecules diffuse from the pore walls toward the less densely coated areas. The D_g values are less than $10^{-3}\,cm^2/sec$.
2. *Knudsen Diffusion*. When the pores have diameters <0.1 μm, the collisions with the walls are more frequent than intermolecular collisions (also referred to as *capillary diffusion*). The D_g values are approximately $10^{-2}\,cm^2/sec$.
3. *Free or Molecular Diffusion*. This type of diffusion is used to explain the preceding model. It takes place between particles and in pores of diameters greater than 0.1 μm. Here we encounter the largest values for D_g, that is, $10^{-1} - 1.0\,cm^2/sec$.
4. *Solid Diffusion*. This takes place in pore diameters of about 0.001 μm (10A) (D_g values of order of $10^{-5}\,cm^2/sec$).

The first three types of diffusion occur in gas–solid chromatography on microporous adsorbents (activated charcoal, molecular sieves). Knudsen diffusion may also occur in gas–liquid chromatographic supports. Since most diffusion types (Volmer, Knudsen, and solid) are orders of magnitude smaller than molecular diffusion, they contribute little to the longitudinal diffusion process in gas chromatographic columns. They can greatly affect mass-transfer rate, however, and this effect is evident in broadening of the zones.

Since flow in capillary columns follows a parabolic flow profile, it is usually laminar (see Section 2.3.2 and Figure 2.22). Tube diameter and mobile phase velocity u increase axial dispersion; however, increase in diffusivity of solute molecules results in decreased bandspreading (30).

Diffusion in liquids is four to five orders of magnitude less than that found in gases. For this reason we may neglect longitudinal diffusion effects in the liquid phase in zone broadening. However, longitudinal diffusion must be considered in equilibrium effects because it may determine the rate of mass transfer.

Distillation. Separations that take place in a chromatographic column are very similar to other types of separation; thus we discuss several of the more important separation techniques. A chromatographic column may be likened to a distillation column in many respects; in fact, some of the terminology in chromatography is taken from distillation theory.

The distillation technique is used not for separation of complex mixtures, but more for preparation of large quantities of pure substances or the separation of complex mixtures into fractions. The technique depends on the distribution of constituents between the liquid mixture and the component vapors in equilibrium with the mixture; two phases exist because of partial evaporation of the liquids. How effective the distillation becomes depends on the type of equipment employed, the method of distillation, and the properties of the mixture components. In distillation all components are volatile, whereas in evaporation volatile and nonvolatile components are separated from each other. An example of distillation would be the separation of ethyl alcohol and benzene. An evaporative separation would be the separation of water from an aqueous solution of some inorganic salt, for example, sodium sulfate.

We can construct a distillation column that has separate steps or plates (i.e., a bubble-cap-type distillation column). Each plate would correspond to an evaporation–condensation step. The phase diagram for such a system is illustrated in Figure 1.1b. This type of column is a useful example for describing the concept of plates.

Efficient laboratory distillation systems use a fractionating column that is a packed column rather than a column with separative plates. Here we cannot refer to the plate or step where the evaporation–condensation step occurs. Thus we refer to a "theoretical plate" that will produce a liquid with a particular composition. The number of these theoretical plates in a fractionation column is given the symbol n. This number of theoretical plates may be determined experimentally by distilling a binary mixture and comparing the data obtained with the phase diagram.

Efficiency of the distillation column is measured by the *height equivalent to a theoretical plate*, abbreviated HEPT or simply H. The length of the column is L; thus $H = L/n$. The H value is independent of L, whereas the n value is dependent on L.

This HEPT has been applied to chromatographic separations. The reader should keep in mind that actual *plates* do not exist in a chromatographic column, either.* The n value is a measure of efficiency for the column. The number of theoretical plates for a chromatographic column is a relative measure of the zone broadening that occurred during the passage of the sample through the column. A direct comparison of efficiencies between the distillation technique and the chromatographic technique is not possible. A given separation would require more theoretical plates by the chromatographic technique than by the distillation technique.

Liquid–Liquid Extraction. In the previous section we were concerned with the phase transitions between liquid and vapor and discussed the various techniques for effecting such changes. In this section consider transferring solute components from one liquid phase to a second liquid phase. This technique is referred to as *liquid–liquid extraction* (LLE). The main

* We visualize the chromatographic column as if it were divided into a number of regions called *theoretical plates*. We further assume that equilibrium exists between the solute in the mobile and stationary phases.

restriction on this separation technique is that the two phases must be immiscible. By *immiscible liquids* we mean two liquids that are completely insoluble in each other. A little reflection will reveal that it is very difficult to have two liquids that are mutually insoluble. If such a system were achievable, the total pressure P of the system would be defined by

$$P = p_A^0 + p_B^0 \tag{1.21}$$

where p_A^0 and p_B^0 are the respective vapor pressures of liquids A and B in the pure state. The composition of the vapor, in terms of moles of each liquid (n_A, nB), would be

$$\frac{n_A}{n_B} = \frac{p_A^0}{p_B^0} \tag{1.22}$$

As with any liquid system, this system would boil when the total vapor pressure P equals atmospheric pressure. However, the boiling point of a mixture of two completely immiscible liquids is lower than that of either constituent because the pressure of the mixture is higher at all temperatures. This total vapor pressure is independent of the amounts of the two phases, and thus the boiling point remains constant as long as the two layers are present.

Once a quantity of a third substance (solute) is added to a system of two immiscible liquids, it will distribute or divide between the layers in definite proportions. If we apply the phase rule to such a system, we find that we have a system of three components C and two phases P. Thus the system has three degrees of freedom F: pressure, temperature, and concentration.

The distribution of the third substance (solute) between the two phases is governed to a first approximation by its solubility in each of the two phases. At a definite temperature, therefore, the ratio of the concentration in each phase is a constant. This is the basis of the distribution law first stated by Berthelot and later by Nernst. Simply stated, it is

$$K = \frac{C_{II}}{C_I} \tag{1.23}$$

where C_I and C_{II} are the concentrations of the solute in liquids I and II, respectively.

Equilibrium of Separation. If we allow a system of two immiscible liquids, containing a solute (i) to come to equilibrium, we can express our equilibrium distribution coefficient $\tilde{K}_c$ as

$$(i)_1 \rightleftharpoons (i)_2$$

$$\tilde{K}_c = \frac{[i]_2}{[i]_1} \tag{1.24}$$

$$K^0 = \frac{(ai)_2}{(ai)_1} = \frac{[i]_2(\gamma)_2}{[i]_1(\gamma)_1}$$

where γ is the activity coefficient. Our distribution coefficient can also be expressed in terms of weight and volume:

$$\tilde{K}_c = \frac{(W_i)_2/MW_i}{V_2} \div \frac{(W_i)/MW_i}{V_1} \tag{1.25}$$

where W_i is the weight of solute (g), MW_i represents the molecular weight of solute, and V_1 and V_2 represent volumes of the two phases.

Equation 1.25 becomes

$$\tilde{K}_c = \frac{(W_i)_2}{(W_i)_1} \cdot \frac{V_1}{V_2} \tag{1.26}$$

The ratio of the total amount of solute in phase 2 to the total amount of solute in phase 1 is known as the *capacity factor* and is given the symbol k ; thus

$$k = \frac{(W_i)_2}{(W_i)_1} = \frac{(C_i)_2V_2}{(C_i)_1V_1} \tag{1.27}$$

We can introduce another term that is used a great many times in chromatography, the phase ratio β. It is the ratio of the volumes of the two phases:

$$\beta = \frac{V_1}{V_2} \tag{1.28}$$

Incorporation of Equations 1.27 and 1.28 into Equation 1.26 gives us

$$\tilde{K}_c = k\beta \tag{1.29}$$

Another way to consider the efficiency of an extraction system is to present the data in terms of the fraction of solute extracted and the fraction unextracted. The fraction of total solute in a given phase ϕ can be written

$$\phi_2 = \frac{C_2V_2}{C_1V_1 + C_2V_2} \tag{1.30}$$

where C_1 and C_2 are concentrations of solute in two phases and V_1 and V_2 represent volumes of the two phases.

Since $\tilde{K}_c = C_2/C_1$, it follows that

$$\phi_2 = \frac{\tilde{K}_cV_2}{\tilde{K}_cV_2 + V_1} = \frac{\tilde{K}_c(V_2/V_1)}{\tilde{K}_c(V_2/V_1) + 1} \tag{1.31}$$

Equation 1.27 may be written as

$$\frac{V_1}{V_2}k = \frac{C_2}{C_1} = \tilde{K}_c \tag{1.32}$$

Combining Equations 1.31 and 1.32, we obtain

$$\phi_2 = \frac{(V_1/V_2)k(V_2/V_1)}{(V_1/V_2)k(V_2/V_1) + 1} = \frac{k}{k+1} \tag{1.33}$$

By a similar substitution manipulation, we obtain

$$1 - \phi_2 = \frac{1}{1+k} \tag{1.34}$$

We may now derive equations for the fraction of solute extracted ϕ_2 and the fraction unextracted $(1 - \phi_2)$. Therefore,

$$\phi_2 + (1 - \phi_2) = 1 \tag{1.35}$$

$$\phi_2 = 1 - (1 - \phi_2) \tag{1.36}$$

Also, the ratio of fraction extracted to that of unextracted would be the same as the capacity factor k:

$$k = \frac{\phi_2}{1 - \phi_2} \tag{1.37}$$

From Equation 1.29, we would have

$$\tilde{K}_c = \frac{\phi_2}{1 - \phi_2} \cdot \frac{V_1}{V_2} \tag{1.38}$$

giving

$$\phi_2 = \frac{\tilde{K}_c V_1}{V_2 + \tilde{K}_c V_1} \tag{1.39}$$

Division by V_2 gives

$$\phi_2 = \frac{\tilde{K}_c(1/\beta)}{1 + \tilde{K}_c(1/\beta)} = \frac{k}{1+k} \tag{1.40}$$

Thus Equations 1.33 and 1.40 are similar.

For fraction unextracted,

$$\begin{aligned} 1 - \phi_2 &= \frac{1 - \tilde{K}_c V_1}{V_2 + \tilde{K}_c V_1} = \frac{V_2 + \tilde{K}_c V_1 - \tilde{K}_c V_1}{V_2 + \tilde{K}_c V_1} \\ &= \frac{V_2}{V_2 + \tilde{K}_c V_1} \end{aligned} \tag{1.41}$$

Dividing by V_2, we obtain

$$1 - \phi_2 = \frac{1}{1 + \tilde{K}_c(1/\beta)} = \frac{1}{1+k} \tag{1.42}$$

If $V_1 = V_2$, the $\beta = 1$ and Equations 1.39 and 1.41 become

$$\phi_2 = \frac{\tilde{K}_c}{1 + \tilde{K}_c} \tag{1.43}$$

and

$$1 - \phi_2 = \frac{1}{1 + \tilde{K}_c} \tag{1.44}$$

Carrying this one step further, if we are performing multiple extractions, after n extractions, we would have

$$(1 - \phi_2)_n = \frac{V_0}{(\tilde{K}_c V_0 + V_w)^n} = \left(\frac{V_0}{\tilde{K}_c V_0 + V_w}\right)^n \tag{1.45}$$

and

$$(\phi_2)_n = 1 - (1 - \phi_2)_n = 1 - \left(\frac{V_0}{\tilde{K}_c V_0 + V_w}\right)^n \tag{1.46}$$

Liquid–Gas Extraction. Liquid–gas extraction, known as *vapor equilibrium technique* (31) or *headspace sampling* (32, 33), is an often overlooked extraction technique. It has been used mainly for blood alcohol content analyses and petroleum source identification after spills. It is being used more and more for the analysis of environmental samples (see Chapters 13 and 14) by those active in this area.

One may define the distribution constant K_D as

$$K_D = \frac{W_L/V_L}{W_G/V_G} = \frac{W_L/V_G}{V_L W_G} \tag{1.47}$$

where W_L and W_G are the mass of solute in condensed (liquid) and gaseous phases, respectively and V_L and V_G are the volumes of the two phases.

A mass balance must be maintained between the two phases, and if we have a component i distributed between the gas and the liquid phases, then

$$W_i = W_{iG} + W_{iL} \tag{1.48}$$

The distribution constant K_D for a volatile component may be calculated by using equal volumes of liquid phase V_L and headspace gas V_G. Inject a known volume of headspace gas into gas chromatograph and measure the peak area or peak height. Relate this measurement to concentration. Remove the remaining gas phase and equilibrate the liquid phase with another equal volume of headspace gas. Inject same known volume of headspace gas into the gas chromatographic column, and relate the area or peak height to concentration. Repeat the process several times. Plot log C_i against the equilibration number. The resulting plot will be linear, have a negative slope, and be described by the following equation:

$$\log(i_G)n = -n \log(K_D + 1) + \log K_D(i_L)_1 \tag{1.49}$$

where $(i_G)n$ is the amount of component i in the gas phase after n equilibrations and $(i_L)_1$ represents the initial quantity of i in the liquid phase. The slope of the line is $(-)\ \log(K_D + 1)$, and its intercept is $\log K_D(i_L)_1$. Selection of any two points on this plot allows one to divide the larger value by the smaller value and subtract one from the quotient to obtain $\log K_D$. Division of the intercept $\log K_D(i_L)_1$ by $\log K_D$ yields $(i_L)_1$.

Maximum Efficiency in Extraction. One may determine the maximum possible efficiency of an extraction by considering the fraction of solute that remains in the original phase (raffinate) following equilibrations. The symbol for this fraction is ϕ_R; thus

$$\phi_R = \frac{V_1}{\tilde{K}_c V_2 + V_1} \tag{1.50}$$

where V_1 is the volume of the original phase containing all the solute before extraction and V_2, the volume of the extracting phase.

If n extractions are carried out on V_1, we have

$$\phi_R^n = \left[\frac{V_1}{\tilde{K}_c(\Sigma V_2/n) + V_1}\right]^n \tag{1.51}$$

where V_2 represents the fixed total volume of extractant. The limit of this expression as n approaches infinity would be

$$\phi_R^\infty = e^{-\tilde{K}_c \Sigma V_2/V_1} \tag{1.52}$$

By plotting the percent efficiency of extraction $[100(\phi_R^n)]$ against the number of extractions n, one would obtain a plot like that in Figure 1.4. The plot shows that a limit value is approached when $V_2 = 5V_1$ (assuming $\tilde{K}_c = 1$).

When the value of k is near to 0.5 or when equal amounts of solute are present in each phase, we evidence the most sensitive changes in $\tilde{K}_c$ or V_2/V_1. Also, the fractional amount of total solute ϕ, in a given phase asymptotically approaches one or zero for large or small values, respectively, of the capacity factor k. What this says is that when k is large or small, little effect is noted for ϕ when a large change occurs in k. Under the preceding conditions it becomes very difficult to remove the last traces of a component from either phase. It is for this reason that more than five

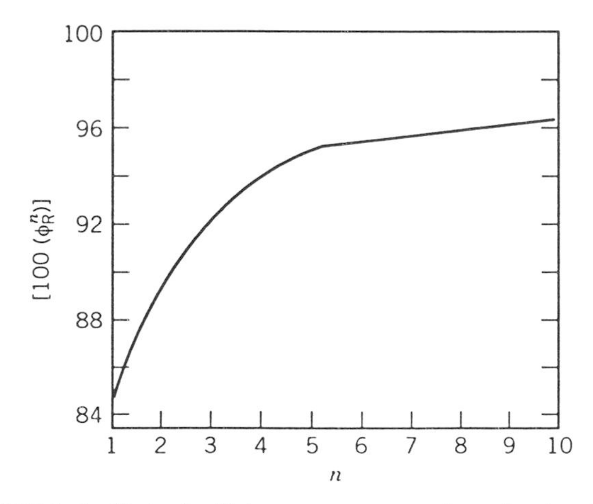

FIGURE 1.4 Plot of efficiency versus number of extractions.

extractions accomplishes little in regard to quantitative separation. Thus, when $(1 - \phi) \leq 0.01$, we assume complete separation of solute from one phase to another phase.

Countercurrent Extraction. We have discussed extractions from the viewpoint of one substance being transferred from one phase to another or the separation of two solutes by selective extraction. When we have a system in which the distribution constants K or the equilibrium distribution coefficients $\tilde{K}_c$ differ by 10^3, we can recover the extracted solute in only about 97% purity. Continued extractions will increase yield but not purity. Good separations, with high purity, of two or more solutes can be achieved when there is a difference in the thermodynamic behavior of the various solutes, that is, a difference in the distribution constants K or coefficients $\tilde{K}_c$. A measure of this degree of separation is the separation factor α for pairs of solutes, which is defined as

$$a = \frac{K_A}{K_B} = \frac{\tilde{K}_{cA}}{\tilde{K}_{cB}} = \frac{(C_A)_2/(C_A)_1}{(C_B)_2/(C_B)_1} = \frac{[(C_A)_2(C_B)_1]}{[(C_A)_1(C_B)_2]} \tag{1.53}$$

If there are other equilibria (interphase) involved, the K values are replaced by the distribution ratio D. For extractions, it makes little difference which solute appears in the numerator because the term is usually defined in such a manner that it will have a numerical value equal to or greater than one. Separation factor applied to chromatographic systems is expressed as a ratio of retention data (see Section 1.3):

$$\alpha = \frac{(V_R')_A}{(V_R')_B} \tag{1.54}$$

where V_R' is the adjusted retention volume. The component that is more retained (the greater value of V_R') is placed in the numerator.

Before proceeding, the author wishes to point out that the separation factor α as defined above is the term used by most people working in chromatography and the term recommended by the IUPAC. Readers will encounter statements to the contrary in some references (5, 7).

Countercurrent extraction was applied most widely before the popularity of gas and liquid chromatography, when one needed to separate two components that had distribution ratio values D that were very similar. The extraction is performed with a series of tubes, each containing a similar volume of the solvent in which the sample is dissolved. A layer of extracting solvent is equilibrated in the first tube and then moved to second tube. A fresh layer of extracting solvent is placed in the first tube and the process continued until a sufficient number of transfers have been completed to separate the sample components.

If θ is the fraction extracted and $1-\theta$ is the amount unextracted, the concentration in the various tubes is given by the binomial expansion:

$$[\theta + (1-\theta)]^n = 1 \tag{1.55}$$

where n is the number of transfers of solvent along the tubes.

If a large number of extractions were performed, one could calculate the fraction of solute present in the rth tube after n extractions by Equation 1.56:

$$f_{n,r} = \frac{n!}{r!(n-r)!}\frac{(DV)^r}{(1+DV)^n} \tag{1.56}$$

where $V = V_0/V_w$.

Field Flow Fractionation. Field flow fractionation (FFF) is a technique similar to chromatography and is based on the simultaneous influence of external field forces and fluid flow nonuniformities on macromolecules or particulate species undergoing separation. Giddings (34) published the first paper on this expanding technique. The separation is brought about inside a narrow channel composed of two highly polished plane-parallel surfaces that have a spacer between them. The fluid passes along this channel and the sample is injected on the inlet side and emerges at the outlet into a detector. During the flow, a physical or chemical field is directed across the channel walls. This field interacts with the molecules or particles of the solute, which are compressed against one wall. A concentration gradient is formed and causes a diffusion flux to be induced in the reverse direction.

The solute macromolecules or particles move in the direction of the longitudinal axis of the channel at different velocities, depending on their distance from the walls. The ratio of the average longitudinal velocities of a retained solute v_S and that of the solvent v is represented by R:

$$\begin{aligned} R &= \frac{v_S}{v} = 6[\coth(2)^{-1} - 2] \\ &= l/w \end{aligned} \tag{1.57}$$

where l is the mean layer thickness and w is the distance between channel walls. The number of components that can be resolved (peak capacity) increases with the field strength. Peak capacity variation with retention volume for a given efficiency is greater in FFF than in chromatography. The best resolution in FFF is for those solutes retained the most; thus flow programming (gradual increase in flowrate) makes it possible to work in the range of optimum parameters and to decrease the time of analysis.

The separation processes in FFF take place in the fluid phase (where no active stationary phase of large surface area is present), making the technique of fundamental importance for fractionation of materials of biological origin. The total surface area of an FFF channel (plays no active part in separation process) is lower than the active surface area of a chromatographic column with similar separation capacity. The range of retentions in FFF is variable because the field strength can be varied continuously over a wide range. The application possibilities of the FFF method is greatly extended because of the ability to

- Program the field strength
- Program the flow velocity
- Vary geometry of the channel and thus form the shape of velocity profile
- Apply both gases and liquids as fluids in the channel

The various FFF techniques differ in respect to the character of the field applied. The following are representative techniques of FFF:

1. *Electrical FFF*, in which the field is induced by an electrical current across the channel walls (two semipermeable membranes that permit passage of small ions).
2. *Thermal FFF*, in which the upper block is heated electrically and the lower block cooled by water. Solute accumulates near the cold wall.
3. *Flow FFF*, in which the flow of solvent is perpendicular to the flow of the basic medium in the channel (two plane-parallel semipermeable membranes).
4. *Sedimentation FFF*, in which natural gravitational or centrifugal forces serve as the effective field. Programming of the field strength can be effected by gradually decreasing the number of revolutions of centrifuge and programming the solvent density.
5. *Magnetic FFF*, in which the channel walls cause a magnetic gradient field.
6. *Steric FFF*, in which field strength causes the smaller molecules (or particles) to move closer to channel walls and thus elute later.

The reader is referred to several excellent reviews on this technique (35–40) for more detailed information.

Solid-Phase Extraction. See Chapter 14 for details for this technique.

Supercritical-Fluid Extraction. See Chapter 14 for details of this technique.

1.2.2 Unifying Theory of Separations

Giddings (41) presented a classificatory system as a basic approach to separations methods. His essay is an attempt to point out a system whereby all separation methods may be viewed considering driving forces and flow. Justification for this approach is viable when one considers the different phases, fields, flows, columns, packing and stage geometries, procedures, apparatus size, sample size, time scales, and overall objectives of the many separation methods. Separation systems are viewed from the aspects of analysis time, resolution, and adaptability to identification by the analytical chemist, whereas the engineer is more concerned about the sample throughput, continuous processing, and energy consumption.

One can discuss the various classes of methodologies by discussing

- Types of external fields and types of nonmiscible phases
- Multiple stages
- High pressures
- Chemical reactions
- Columns
- Large sample capacity
- Presence or absence of flow
- Continuous throughput
- Discrete zones
- Elution
- Porous support
- Energy efficiency

but this does not exist in one comprehensive system.

Separation systems require the use of a polythetic classification (similarities and differences of many properties) as opposed to monothetic classification (use of only one property or characteristic). Gidding's system proposes to describe the structure and relationship of different separation methods. His basic observation is that a separation method relies on the spatial disengagement of components of a mixture. The displacement processes are defined by two categories:

1. *Bulk (Flow) Displacement.* The displacement is not selective, and the mixture component follows the movement of its surrounding mass. Thus displacement is achieved by flow and mechanical relocation.
2. *Relative Displacement (Selective).* An effective force (external fields, gradients, or affinity of mixture component for the chemical phase employed) drives the component through its surrounding medium. This force may be *continuous* (*c*), with external fields or gradients, or *discontinuous* (*d*), that is, forces at membrane surfaces.

For the separation to occur, a *flow F* may or may not be used (*S*, static nonflow system) to complete the separation. Since *flow* is a vector quantity,

TABLE 1.6 Categories of Separation Methods

Category	Separation Methods
Sc	Electrophoresis, isoelectric focusing, dielectrophoresis, electrostatic precipitation, sedimentation, and isopycnic centrifugation
Sd	Single-stage extraction, adsorption, crystallization, sublimation, and dialysis
$F(=)c$	Countercurent electrophoresis and elutriation
$F(=)d$	Filtration, ultrafiltration, reverse osmosis, pressure dialysis, and zone melting
$F(+)c$	Field flow fractionation, thermogravimetric methods, and electrophoresis–convection
$F(+)d$	Chromatography, distillation, countercurrent distribution, adsorption, crystallization, foam fractionation, and mineral floatation

it can be either parallel or perpendicular to the *force*. Giddings defined two symbols:

$$F(=) = \text{flow and force parallel}$$

$$F(+) = \text{flow and force perpendicular}$$

Not included in the flow category, are parasitic flows (electroosomotic flow), passive flows (curtain electrophoresis), and stirring flows.

Most separation methods can be placed into six basic categories (see Table 1.6) employing the flow [$F(=)$ or $F(+)$] and force (c or d) symbolism.

It should be noted that the term "driving forces" used by Giddings is different from the term used by Strain et al. (42).

1.3 DEFINITIONS AND NOMENCLATURE

The definitions given in this section are a combination of those used widely and those recommended by the International Union of Pure and Applied Chemistry (IUPAC) (43). The recommended IUPAC symbol appears in parentheses if it differs from the widely used symbol.

Adjusted retention time, t'_R. The solute total elution time minus the retention time for an unretained peak (hold-up time):

$$t'_R = t_R - t_M$$

Adjusted retention volume, V'_R. The solute total elution volume minus the retention volume for an unretained peak (holdup volume):

$$V'_R = V_R - V_M$$

Adsorbent. An active granular solid used as the column packing or a wall coating in gas–solid chromatography that retains sample components by adsorptive forces.

Adsorption chromatography. This term is synonymous with gas–solid chromatography.

Adsorption column. A column is used in gas–solid chromatography, consisting of an active granular solid and a metal or glass column.

Air peak. The air peak results from a sample component nonretained by the column. This peak can be used to measure the time necessary for the carrier gas to travel from the point of injection to the detector.

Absolute temperature, K. The temperature stated in terms of the Kelvin scale:

$$K = °C + 273.15°$$

$$0°C = 273.15\ K$$

Analysis time, t_{ne}. The minimum time required for a separation:

$$t_{ne} = 16R_s^2 \frac{H}{\bar{u}} \left(\frac{\alpha}{\alpha - 1}\right)^2 \frac{(1+k)^3}{k^2}$$

Area normalization (*raw area normalization*). The peak areas of each peak are summed; each peak area is then expressed as a percentage of the total:

$$A_1 + A_2 + A_3 + A_4 = \Sigma A\ ; \qquad \%A_1 = \frac{A_1}{\Sigma A}\text{, etc.}$$

Area normalization with response factor, ANRF. The area percentages are corrected for the detector characteristics by determining response factors. This requires preparation and analysis of standard mixtures.

Attentuator. An electrical component made up of a series of resistances that is used to reduce the input voltage to the recorder by a particular ratio.

Band. Synonymous with zone. It is the volume occupied by the sample component during passage and separation through the column.

Band area. Synonymous with the peak area *A*. It is the area of peak on the chromatogram.

Baseline. The portion of a detector record resulting from only eluant or carrier gas emerging from the column.

Bed volume. Synonymous with the volume of a packed column.

Bonded phase. A stationary phase that is covalently bonded to the support particles or to the inside wall of the column tubing. The phase may be immobilized only by in situ polymerization (cross-linking) after coating.

Capacity factor, $k(D_m)$. See *Mass distribution ratio*. (In GSC, $V_A > V_L$; thus smaller β values and k values occur.) It is a measure of the ability of the column to retain a sample component:

$$k = \frac{t_R - t_M}{t_M}$$

Capillary column. Synonymous with open tubular column (OTC). This column has small-diameter tubing (0.25–1.0 mm i.d.) in which the inner walls are used to support the stationary phase (liquid or solid).

Carrier gas. Synonymous with mobile or moving phase. This is the phase that transports the sample through the column.

Chromatogram. A plot of the detector response (which uses effluent concentration or another quantity used to measure the sample component) versus effluent volume or time.

Chromatograph (*verb*). A transitive verb meaning to separate sample components by chromatography.

Chromatograph (*noun*). The specific instrument employed to carry out a chromatographic separation.

Chromatography. A physical method of separation of sample components in which these components distribute themselves between two phases, one stationary and the other mobile. The stationary phase may be a solid or a liquid supported on a solid.

Column. A metal, plastic, or glass tube packed or internally coated with the column material through which the sample components and mobile phase (carrier-gas) flow and in which the chromatographic separation takes place.

Column bleed. The loss of liquid phase that coats the support or walls within the column.

Column efficiency, N. See *Theoretical plate number*.

Column material. The material in the column used to effect the separation. An adsorbent is used in adsorption chromatography; in partition chromatography, the material is a stationary phase distributed over an inert support or coated on the inner walls of the column.

Column oven. A thermostatted section of the chromatographic system containing the column, the temperature of which can be varied over a wide range.

Column volume, V_c. The total volume of column that contains the stationary phase. (The IUPAC recommends the column dimensions be given as the inner diameter (i.d.) and the height or length L of the column occupied by the stationary phase under the specific chromatographic conditions.) Dimensions should be given in meters, millimeters, feet, or centimeters.

Component. A compound in the sample mixture.

Concentration distribution ratio, D_C. The ratio of the analytical concentration of a component in the stationary phase to its analytical concentration in the mobile phase:

$$D_C = \frac{\text{Amount component/mL stationary phase}}{\text{Amount component/mL mobile phase}} = \frac{C_S}{C_M}$$

Corrected retention time, t_R^0. The total retention time corrected for pressure gradient across the column:

$$t_R^0 = jt_R$$

Corrected retention volume, V_R^0. The total retention volume corrected for the pressure gradient across the column:

$$V_R^0 = jV_R$$

Cross-sectional area of the column. The cross-sectional area of the empty tube:

$$A_c = r_c^2\pi = \frac{d_c^2}{4}\pi$$

Dead time, t_M. See *Holdup time.*

Dead volume, V_M. See *Holdup volume.* This is the volume between the injection point and the detection point, minus the column volume V_c. This is the volume needed to transport an unretained component through the column.

Derivatization. Components with active groups such as hydroxyl, amine, carboxyl, olefin, and others can be identified by a combination of chemical reactions and GC. For example, the sample can be shaken with bromine water and then chromatographed. Peaks due to olefinic compounds will have disappeared. Similarly, potassium borohydride reacts with carbonyl compounds to form the corresponding alcohols. Comparison of before and after chromatograms will show that one or more peaks have vanished whereas others have appeared somewhere else on the chromatogram. Compounds are often derivatized to make them more volatile or less polar (e.g., by silylation, acetylation, methylation) and consequently suitable for analysis by GC.

Detection. A process by which a chromatographic band is recognized.

Detector. A device that signals the presence of a component eluted from a chromatographic column.

Detector linearity. The concentration range over which the detector response is linear. Over its linear range the response factor of a detector (peak area units per weight of sample) is constant. The linear range is characteristic of the detector.

Detector minimum detectable level, MDL. The sample level, usually given in weight units, at which the signal-to-noise (S/N) ratio is 2.

Detector response. The detector signal produced by the sample. It varies with the nature of the sample.

Detector selectivity. A selective detector responds only to certain types of compound (FID, N-FID, ECD, PID, etc.). The thermal conductivity detector is universal in response.

Detector sensitivity. Detector sensitivity is the slope of the detector response for a number of sample sizes. A detector may be flow sensitive or mass sensitive.

Detector volume. The volume of carrier gas (mobile phase) required to fill the detector at the operating temperature.

Differential detector. This detector responds to the instantaneous difference in composition between the column effluent and the carrier gas (mobile phase).

Direct injection. A term used for the introduction of samples directly onto open tubular columns (OTC) through a flash vaporizer without splitting (should not be confused with on-column injection).

Discrimination effect. This occurs with the split injection technique for capillary columns. It refers to a problem encountered in quantification with split injection onto capillary columns in which a nonrepresentative sample goes onto the capillary column as a result of the difference in rate of vaporization of the components in the mixture from the needle.

Displacement chromatography. An elution procedure in which the eluant contains a compound more effectively retained than the components of the sample under examination.

Distribution coefficient, D_g. The amount of a component in a specified amount of stationary phase, or in an amount of stationary phase of specified surface area, divided by the analytical concentration in the mobile phase. The distribution coefficient in adsorption chromatography with adsorbents of unknown surface area is expressed as

$$D_g = \frac{\text{Amount component/g dry stationary phase}}{\text{Amount component/mL mobile phase}}$$

The distribution coefficient in adsorption chromatography with well-characterized adsorbent of known surface area is expressed as

$$D_s = \frac{\text{Amount component/m}^2\text{ surface}}{\text{Amount component/mL mobile phase}}$$

The distribution coefficient when it is not practicable to determine the weight of the solid phase is expressed as

$$D_v = \frac{\text{Amount component stationary phase/mL bed volume}}{\text{Amount component/mL mobile phase}}$$

Distribution constant, $K(K_D)$. The ratio of the concentration of a sample component in a single definite form in the stationary phase to its concentration in the mobile phase. IUPAC recommends this term rather than the partition coefficient:

$$K = \frac{C_S}{C_G}$$

Efficiency of column. This is usually measured by column theoretical plate number. It relates to peak sharpness or column performance.

Effective theoretical plate number, $N_{eff}(N)$. A number relating to column performance when resolution R_S is taken into account:

$$N_{eff} = \frac{16\,R_S^2}{(1-\alpha)^2} = 16\left(\frac{t'_R}{w}\right)^2$$

Effective plate number is related to theoretical plate number by

$$N_{eff} = N\left(\frac{k}{k+1}\right)^2$$

Electron capture detector, *ECD*. A detector utilizing low-energy electrons (furnished by a tritium or ^{63}Ni source) that ionize the carrier gas (usually argon) and collect the free electrons produced. An electron-capturing solute will capture these electrons and cause a decrease in the detector current.

Eluant. The gas (mobile phase) used to effect a separation by elution.

Elution. The process of transporting a sample component through and out of the column by use of the carrier gas (mobile phase).

Elution chromatography. A chromatographic separation in which an eluant is passed through a column during or after injection of a sample.

External standardization technique, *EST*. This method requires the preparation of calibration standards. The standard and the sample are run as separate injections at different times. The calibrating standard contains only the materials (components) to be analyzed. An accurately measured amount of this standard is injected. *Calculation steps for standard*: (1) For each peak to be calculated, calculate the amount of component injected from the volume injected and the known composition of the standard; then (2) divide the peak area by the corresponding component weight to obtain the absolute response factor (ARF):

$$\text{ARF} = \frac{A_1}{W_1}$$

Calculation step for sample: For each peak, divide the measured area by the absolute response factor to obtain the absolute amount of that component injected:

$$\frac{A_i}{\text{ARF}} = W_i$$

Filament element. A fine tungsten or similar wire that is used as the variable-resistance sensing element in the thermal conductivity cell chamber.

Flame ionization detector, FID. This detector utilizes the increased current at a collector electrode obtained from the burning of a sample component from the column effluent in a hydrogen and air jet flame.

Flame photometric detector, FPD. A flame ionization detector (utilizing a hydrogen-rich flame) that is monitored by a photocell. It can be specific for halogen-, sulfur-, or phosphorous-containing compounds.

Flash vaporizer. A device used in GC where the liquid sample is introduced into the carrier gas stream with simultaneous evaporation and mixing with the carrier gas prior to entering the column.

Flow controller. A device used to regulate the flow of mobile phase through the column.

Flow programming. In this procedure the rate of flow of the mobile phase is systematically increased during a part or all of the separation of higher boiling components.

Flowrate, F_c. The volumetric flowrate of the mobile phase, in milliliters per minute, is measured at the column temperature and outlet pressure:

$$F_c = \frac{\pi r^2 L}{t_M}$$

Frontal chromatography. A type of chromatographic separation in which the sample is fed continuously onto the column.

Fronting. Asymmetry of a peak such that, relative to the baseline, the front of the peak is less sharp than the rear portion.

Gas chromatograph. A collective noun for those chromatographic modules of equipment in which gas chromatographic separations can be realized.

Gas chromatography, GC. A collective noun for those chromatographic methods in which the moving phase is a gas.

Gas–liquid chromatography, GLC. A chromatographic method in which the stationary phase is a liquid distributed on an inert support or coated on the column wall and the mobile phase is a gas. The separation occurs by the partitioning (differences in solubilities) of the sample components between the two phases.

Gas sampling valve. A bypass injector permitting the introduction of a gaseous sample of a given volume into a gas chromatograph.

Gas–solid chromatography, GSC. A chromatographic method in which the stationary phase is an active granular solid (adsorbent). The separation is performed by selective adsorption on an active solid.

Heart cutting. This technique utilizes a precolumn (usually packed) and a capillary column. With this technique only the region of interest is transferred to the main column; all other materials are backflushed to the vent.

Height equivalent to an effective plate, H_{eff}. The number obtained by dividing the column length by the effective plate number:

$$H_{eff} = \frac{L}{N_{eff}}$$

Height equivalent to a theoretical plate, H. The number obtained by dividing the column length by the theoretical plate number:

$$H = \frac{L}{N} = \text{HETP}$$

$$= \frac{H}{d}$$

where d is the particle diameter in a packed column or the tube diameter in a capillary column.

Holdup time, t_M. The time necessary for the carrier gas to travel from the point of injection to the detector. This is characteristic of the instrument, the *mobile-phase* flowrate, and the column in use.

Holdup volume, V_M. The volume of mobile phase from the point of injection to the point of detection. In GC it is measured at the column outlet temperature and pressure and is a measure of the volume of carrier gas required to elute an unretained component (including injector and detector volumes):

$$V_M = t_M F_c$$

Initial and final temperatures, T_1 *and* T_2. This temperature range is used for a separation in temperature-programmed chromatography.

Injection point, t_0. The starting point of the chromatogram, which corresponds to the point in time when the sample was introduced into the chromatographic system.

Injection port. Consists of a closure column on one side and a septum inlet on the other through which the sample is introduced (through a syringe) into the system.

Injection temperature. The temperature of the chromatographic system at the injection point.

Injector volume. The volume of carrier gas (mobile phase) required to fill the injection port of the chromatograph.

Integral detector. This detector is dependent on the total amount of a sample component passing through it.

Integrator. An electrical or mechanical device employed for a continuous summation of the detector output with respect to time. The result is a measure of the area of a chromatographic peak (band).

Internal standard. A pure compound added to a sample in known concentration for the purpose of eliminating the need to measure the sample size in quantitative analysis and for correction of instrument variation.

Internal standardization technique, IST. A technique that combines the sample and standard into one injection. A calibration mixture is prepared containing known amounts of each component to be analyzed, plus an added compound that is not present in the analytical sample. *Calculation steps for calibration standard:*

1. For each peak, divide the measured area by the amount of that component to obtain a response factor:

$$(\mathrm{RF})_1 = \frac{A_1}{W_1}, \text{ etc.}$$

2. Divide each response factor by that of the internal standard to obtain relative response factors, RRF:

$$\mathrm{RRF}_1 = \frac{(\mathrm{RF})_1}{(\mathrm{RF})_i}$$

Calculation steps for sample:

1. For each peak, divide the measured area by the proper relative response factor to obtain the corrected area:

$$(\mathrm{CA})_1 = \frac{A_1}{\mathrm{RRF}_1}$$

2. Divide each corrected area by that of the internal standard to obtain the amount of each component relative to the internal standard:

$$(\mathrm{RW})_1 = \frac{(\mathrm{CA})_1}{(\mathrm{CA})_i}$$

3. Multiply each relative amount by the actual amount of the internal

standard to obtain the actual amounts of each component:

$$(\mathrm{RW})_1 W_i = W_1$$

Interstitial fraction, ε_{I}. The interstitial volume per unit of packed column:

$$\varepsilon_{\mathrm{I}} = \frac{V_{\mathrm{I}}}{X}$$

Interstitial velocity of carrier gas, u. The linear velocity of the carrier gas inside a packed column calculated as the average over the entire cross section. Under idealized conditions it can be calculated as

$$u = F_c \varepsilon_{\mathrm{I}}$$

Interstitial volume, $V_{\mathrm{G}}(V_{\mathrm{I}})$. The volume occupied by the mobile phase (carrier gas) in a packed column. This volume does not include the volumes external to the packed section, that is, the volume of the sample injector and the volume of the detector. In GC it corresponds to the volume that would be occupied by the carrier gas at atmospheric pressure and zero flowrate in the packed section of the column.

Ionization detector. A chromatographic detector in which the sample measurement is derived from the current produced by the ionization of sample molecules. This ionization may be induced by thermal, radioactive, or other excitation sources.

Isothermal mode. A condition wherein the column oven is maintained at a constant temperature during the separation process.

Katharometer. This term is synonymous with thermal conductivity cell; it is sometimes spelled "catharometer."

Linear flowrate, (F_{c}). The volumetric flowrate of the carrier gas (mobile phase) measured at column outlet and corrected to column temperature, and F_{a} is volumetric flowrate measured at column outlet and ambient temperature:

$$F_{\mathrm{c}} = F_{\mathrm{a}} \left(\frac{T_{\mathrm{c}}}{T_{\mathrm{a}}} \right) \frac{P_{\mathrm{a}} - P_{\mathrm{w}}}{P_{\mathrm{a}}}$$

where T_{c} is column temperature (K), T_{a} is ambient temperature (K), P_{a} is ambient pressure, and P_{w} is partial pressure of water at ambient temperature.

Linear velocity, u. The linear flowrate F_{c}, divided by the cross-sectional area of the column tubing available to the mobile phase:

$$u = \frac{F_{\mathrm{c}}}{A_{\mathrm{c}}} = \frac{F_{\mathrm{c}}}{r_{\mathrm{c}}^2 \pi} = \frac{L}{t_{\mathrm{M}}}$$

where A_{c} is the cross-sectional area of the column tubing, r_{c} is the tubing radius, and π is a constant. The equation given above is applicable for capillary columns but not for packed columns; for packed columns, the equation becomes

$$u = \frac{F_c}{\varepsilon_I r_c^2 \pi}$$

Thus one must account for the interstitial fraction of the packed column.

Liquid phase. Synonymous with stationary phase or liquid substrate. It is a relatively nonvolatile liquid (at operating conditions) that is either sorbed on the solid support or coated on the walls of OTCs, where it acts as a solvent for the sample. The separation results from differences in solubility of the various sample components.

Liquid substrate. Synonymous with stationary phase.

Marker. A reference component that is chromatgraphed with the sample to aid in the measurement of hold-up time or volume for the identification of sample components.

Mass distribution ratio, $k(D_m)$. The fraction $(1 - R)$ of a component in the stationary phase divided by the fraction R in the mobile phase. The IUPAC recommends this term in preference to capacity factor k:

$$k(D_m) = \frac{1 - R}{R} = \frac{K}{\beta} = \frac{C_L V_L}{C_G V_G} = K\left(\frac{V_L}{V_G}\right)$$

Mean interstitial velocity of the carrier gas, $\bar{u}$. The interstitial velocity of the carrier gas multiplied by the pressure-gradient correction factor:

$$\bar{u} = \frac{F_c j}{\varepsilon_I}$$

Mobile phase. Synonymous with carrier gas or gas phase.

Moving phase. See *Mobile phase*.

Net retention volume, V_N. The adjusted retention volume multiplied by the pressure-gradient correction factor:

$$V_N = jV'_R$$

Nitrogen–phosphorus detector, NPD. This detector is selective for monitoring nitrogen or phosphorus.

On-column injection. Refers to the method wherein the syringe needle is inserted directly into the column and the sample is deposited within the column walls rather than a flash evaporator. On-column injection differs from direct injection in that the sample is usually introduced directly onto the column without passing through a heated zone. The column temperature is usually reduced, although not as low as with splitless injections ("cool" on-column injections).

Open-tubular columns, OTC. Synonymous with capillary columns.

Packed column. A column packed with either a solid adsorbent or solid support coated with a liquid phase.

Packing material. An active granular solid or stationary phase plus solid support that is in the column. The term "packing material" refers to the conditions existing when the chromatographic separation is started, whereas the term "stationary phase" refers to the conditions during the chromatographic separation.

Partition chromatography. Synonymous with gas–liquid chromatography.

Partition coefficient. Synonymous with the distribution constant.

Peak. The portion of a differential chromatogram recording the detector response or eluate concentration when a compound emerges from the column. If the separation is incomplete, two or more components may appear as one peak (unresolved peak).

Peak area. Synonymous with band area. The area enclosed between the peak and peak base.

Peak base. In differential chromatography, this is the baseline between the base extremities of the peak.

Peak height, h. The distance between the peak (band) maximum and the peak base, measured in a direction parallel to the detector response axis and perpendicular to the time axis.

Peak maximum. The point of maximum detector response when a sample component elutes from the chromatographic column.

Peak resolution, R_S. The separation of two peaks in terms of their average peak widths:

$$R_S = \frac{2\Delta t_R}{w_a + w_b} = \frac{2\Delta t'_R}{w_a + w_b}$$

Peak width, w_b. The bar segment of the peak base intercepted by tangents to the inflection points on either side of the peak and projected on to the axis representing time or volume.

Peak width at half-height, w_h. The length of the line parallel to the peak base, which bisects the peak height and terminates at the intersections with the two limbs of the peak, projected onto the axis representing time or volume.

Performance Index, PI. This is used with open tubular columns; it is a number (in poise) that provides a relationship between elution time of a component and pressure drop. It is expressed as

$$\mathrm{PI} = 30.7H^2\left(\frac{u}{K}\right)\frac{1+k}{k+\frac{1}{16}}$$

Phase ratio, β. The ratio of the volume of the mobile phase to the stationary phase on a partition column:

$$\beta = \frac{V_I}{V_S} = \frac{V_G}{V_A} = \frac{V_0}{V_S}$$

Photoionization detector, PID. A detector in which detector photons of suitable energy cause complete ionization of solutes in the inert mobile phase. Ultraviolet radiation is the most common source of these photons. Ionization of the solute produces an increase in current from the detector, and this is amplified and passed onto the recorder.

PLOT. An acronym for porous-layer open tubular column, which is an open tubular column with fine layers of some adsorbent deposited on the inside wall. This type of column has a larger surface area than a wall-coated open tubular column (WCOT).

Polarity. Sample components are classified according to their polarity (measuring in a certain way the affinity of compounds for liquid phases), for example: nonpolar—hydrocarbons; medium polar—ethers, ketones, aldehydes; and polar—alcohols, acids, amines.

Potentiometric recorder. A continuously recording device whose deflection is proportional to the voltage output of the chromatographic detector.

Precolumn sampling (OTC). Synonymous to selective sampling with open tubular columns.

Pressure, P. Pressure is measured in pounds per square inch at the entrance valve to the gas chromatograph (psi = pounds per square inch = $lb/in.^2$; psia = pounds per square inch absolute = ata; psig = pounds per square inch gauged = atii; 1 psi = 0.069 bar).

Pressure-gradient correction coefficient, j. This factor corrects for the compressibility of the mobile phase in a homogeneously filled column of uniform diameter:

$$j = 3/2 \left[\frac{(p_i/p_0)^2 - 1}{(p_i/p_0)^3 - 1} \right]$$

Programmed-temperature chromatography. A procedure in which the temperature of the column is changed systematically during a part or the whole of the separation.

Purged splitless injection. This term is given to a splitless injection (see *Splitless injection*) wherein the vent is open to allow the large volume of carrier gas to pass through the injector to remove any volatile materials that may be left on the column. Most splitless injections are purged splitless injections.

Pyrogram. The chromatogram resulting from the sensing of the fragments of a pyrolyzed sample.

Pyrolysis. A technique by which nonvolatile samples are decomposed in the inlet system and the volatile products are separated on the chromatographic column.

Pyrolysis gas chromatography. A process that involves the induction of molecular fragmentation to a chromatographic sample by means of heat.

Pyrometer. An instrument for measuring temperature by the change in electrical current.

Qualitative analysis. A method of chemical identification of sample components.

Quantitative analysis. This involves the estimation or measurement of either the concentration or the absolute weight of one or more components of the sample.

Relative retention, $r_{a/b}$. The adjusted retention volume of a substance related to that of a reference compound obtained under identical conditions:

$$r_{a/b} = \frac{(V_g)_a}{(V_g)_b} = \frac{(V_N)_a}{(V_N)_b} = \frac{(V'_R)_a}{(V'_R)_b} \neq \frac{(V_R)_a}{(V_R)_b}$$

Required plate number, n_{ne}. The number of plates necessary for the separation of two components based on resolution R_S of 1.5:

$$n_{ne} = 16R_S^2 \left(\frac{\alpha}{\alpha - 1}\right)^2 \left(\frac{1+k}{k}\right)^2$$

Resolution, R_S. Synonymous with peak resolution; it is an indication of the degree of separation between two peaks.

Retention index, I. A number relating the adjusted retention volume of a compound A to the adjusted retention volume of normal paraffins. Each n-paraffin is arbitrarily alloted, by definition, an index of 100 times its carbon number. The index number of component A is obtained by logarithmic interpolation:

$$I = 100N + 100\frac{[\log V'_R(A) - \log(V'_R)(N)]}{[\log V'_R(n) - \log V'_R(N)]}$$

where N and n are the smaller and larger n-paraffin, respectively, that bracket substance A.

Retention time (absolute), t_R. The amount of time that elapsed from injection of the sample to the recording of the peak maximum of the component band (peak).

Retention volume (absolute), V_R. The product of the retention time of the sample component and the volumetric flowrate of the carrier gas (mobile phase). The IUPAC recommends that it be called total retention volume because it is a term used when the sample is injected into a flowing stream of the mobile phase. Thus it includes any volume contributed by the sample injector and the detector.

Sample. The gas or liquid mixture injected into the chromatographic system for separation and analysis.

Sample injector. A device used for introducing liquid or gas samples into the chromatograph. The sample is introduced directly into the carrier-gas stream (e.g., by syringe) or into a chamber temporarily isolated from the system by valves that can be changed so as to instantaneously switch the gas steam through the chamber (gas sampling valve).

SCOT. An acronym for support-coated open tubular column. These are capillary columns in which the liquid substrate is on a solid support that coats the walls of the capillary column.

Selective sampling. Refers to the transportation of a portion of a mixture onto the capillary column after it has passed through another chromatographic column, either packed or open tubular.

Separation. The time elapsed between elution of two successive components, measured on the chromatogram as the distance between the recorded bands.

Separation efficiency, N/L. A measure of the "goodness" of a column. It is usually given in terms of the number of theoretical plates per column length, that is, plates per meter for open tubular columns.

Separation factor, $\alpha_{a/b}$. The ratio of the distribution ratios or coefficients for substances A and B measured under identical conditions. By convention the separation factor is usually greater than unity:

$$\alpha_{a/b} = \frac{K_{D_a}}{K_{D_b}} = \frac{D_a}{D_b} = \frac{K_a}{K_b}$$

Separation number, n_{sep} *or SN*. The possible number of peaks between two n-paraffin peaks resulting from components of consecutive carbon numbers:

$$n_{sep} = \frac{(t_{R_2} - t_{R_1})}{(w_h)_1 + (w_h)_2} - 1 = \mathrm{SN}$$

See *Trennzahl number*.

Separation temperature. The temperature of the chromatographic column.

Septum bleed. Refers to the detector signal created by the vaporization of small quantities of volatile materials trapped in the septum. It is greatly reduced by allowing a small quantity of carrier gas to constantly sweep by the septum to vent.

Solid support. The solid packing material on which the liquid phase is coated and which does not contribute to the separation process.

Solute. A synonymous term for components in a sample.

Solvent. Synonymous with liquid phase (stationary phase or substrate).

Solvent effect (OTC). An effect noted in splitless injections for concentrating higher boilers at the head of the column so that the peak band will reflect the efficiency of the column and not the volume of the injection port liner. For this effect to occur, the oven temperature must be close to the boiling point of the major solvent component in the system so that it condenses at the head of the column and acts as a barrier for the solute.

Solvent efficiency, α. Synonymous with separation factor.

Solvent venting (OTC). Refers to the elimination of the solvent or major ingredient in a mixture by heart-cutting and flushing the solvent through the vent.

Span of the recorder. The number of millivolts required to produce a change in the deflection of the recorder pen from 0 to 100% on the chart scale.

Specific retention volume, V_g. The net retention volume per gram of stationary phase corrected to 0°C:

$$V_g = \frac{273 V_N}{T W_L} = \frac{j V'_R}{T W_L}$$

Specific surface area. The area of a solid granular adsorbent expressed as square meter per unit weight (gram) or square meter per milliliter.

Split injection (OTC). The term given to the classical method of injecting samples into a capillary system wherein the sample is introduced into a flash vaporizer and the splitter reduces the amount of sample going onto the column by the use of restrictors so that the majority of the sample goes into the vent and not onto the capillary column. Typical split ratios are 100:1 and 200:1, where the lower number refers to the quantity going onto the column.

Splitless injection (OTC). The term applied to a flash vaporization technique wherein the solvent is evaporated in the injection port and condenses on the head of the column. After a suitable time (usually 0.5 min) the splitter is opened and any of the remaining material in the flash vaporizer is vented. The solvent that will have condensed at the head of the column is then slowly vaporized through column temperature programming. Splitless injection is used to concentrate small quantities of solute in a large injection (2–3 μL) onto a capillary column. The solute should have a higher boiling point than the condensed solvent so that its relative retention time is at least 1.5 and its retention index is greater than 600.

Splitter. A fitting attached to the injection port or column exit to divert a portion of the flow. It is used on the inlet side to permit the introduction of very small samples to a capillary column and on the outlet side to permit introduction of a very small sample of the column effluent to the detector, to permit introduction of effluent to two detectors simultaneously or collect part of a peak from a destructive detector.

Stationary phase. Synonymous with liquid phase, distributed on a solid, in gas–liquid chromatography or the granular solid adsorbent in gas–solid chromatography. The liquid may be chemically bonded to the solid.

Stationary phase fraction, ε_S. The volume of the stationary phase per unit volume of the packed column:

$$\varepsilon_S = \frac{V_S}{X}$$

Stationary phase volume, $V_L(V_S)$. The total volume of stationary-phase liquid on the support material in a particular column:

$$V_L = \frac{w_L}{\text{Density}_L}$$

Surface area. The area of a solid granular adsorbent A.

Tailing. In this condition the asymmetry of a peak is such that, relative to the baseline, the front is steeper than the rear.

Temperature programming. In this procedure the temperature of the column is changed systematically during part or all of the separation process.

Theoretical plate number, N. This number defines the efficiency of the column or sharpness of peaks:

$$N = 16\left(\frac{\text{Peak retention time}}{\text{Peak width}}\right)^2$$

$$= 16\left(\frac{t_R}{w}\right)$$

Thermal conductivity. A physical property of a substance, serving as an index of its ability to conduct heat from a warmer to a cooler surface.

Thermal conductivity detector, *TCD*. A chamber in which an electrically heated element will reflect changes in thermal conductivity within the chamber atmosphere. The measurement is possible because of the change in resistance of the element.

Thermistor bead element. A thermal conductivity detection device in which a small glass-coated semiconductor sphere is used as the variable resistive element in the cell chamber.

Trennzahl number, Tz. This term is comparable with separation number and is calculated from the resolution between two consecutive members of a homologous hydrocarbon series. It is usually considered as the number of peaks that could be placed between those two members of the series. It is used predominantly in capillary column work and is expressed as

$$Tz = \left[\frac{t_{R_2} - t_{R_1}}{(w_h)_1 + (w_h)_2} \right] - 1$$

True adsorbent volume, V_A. The weight of the adsorbent packing is divided by the adsorbent density:

$$V_A = \frac{W_A}{D_A}$$

van Deemter equation. This equation expresses the extent a component band spreads as it passes through the column in terms of physical constants and the velocity of the mobile phase:

$$\mathrm{HEPT}(H) = A + \frac{B}{u} + Cu$$

where HEPT = height equivalent to a theoretical plate
u = linear velocity of carrier gas (mobile phase)
A = a constant that accounts for the effects of "eddy" diffusion in the column
B = a constant that accounts for the effect of molecular diffusion of the vapor in the direction of the column axis
C = a constant proportional to the resistance of the column packing to mass transfer of solute through it

Velocity of mobile phase, u. Synonymous with linear velocity.

WCOT. An acronym for wall-coated open tubular column. It is a capillary column in which the inside wall is coated with the stationary phase.

Weight of stationary liquid phase, W_L. The weight of liquid phase in the column.

WWCOT. A whisker-wall-coated open tubular column. It is a WCOT in which the walls have been etched before the stationary phase is deposited.

WWPLOT. An acronym for whisker-wall porous-layer open tubular column. It is a PLOT column in which the walls have been etched before the depositing of the support.

WWSCOT. An acronym for whisker-wall-support-coated open tubular column. It is a SCOT column in which the walls have been etched before depositing of the support.

Zone. The position and spread of a solute within the column, the region in the chromatographic bed where one or more components of the sample are located. See *Band*.

1.4 LITERATURE ON GAS CHROMATOGRAPHY

We do not attempt to present an all-inclusive listing of the literature available in which one may find chromatographic information. Rather, we list categories of source material and a few examples of each category. The expansion of the chromatographic field has made the reading of all literature an impossible task. One must choose a particular aspect of the field and follow it closely while at the same time scanning the topics and abstracts of the other aspects of chromatography.

Novices to the field of chromatography should become aware of the basic texts that are available. Once familiar with these general sources, they can then proceed to specific areas of chromatography and sample types or topics within an area. With this background, they can then pursue the scanning of journal contents an abstracts of the chromatographic literature.

Gas chromatography is one of the most active areas of analytical chemistry, but many references in GC are found in sources other than just literature on chromatography or analytical chemistry. Thus literature searches should take one to the journals on topics where GC may be utilized, such as journals of biochemistry, organic chemistry, physical chemistry, catalysis, environmental studies, drug analysis, forensic chemistry, petroleum chemistry, and inorganic chemistry.

A compilation of the journals may be found in the *Chemical Abstracts* listing of periodicals abstracted by the *Chemical Abstracts* of the American Chemical Society, Washington, DC, 20036. Familiarity with this journal is a must for anyone involved in any type of chemical research. In addition to these abstracts, there are other abstract services that are concerned solely with chromatographic references. Typical of these would be the following:

1. *Gas and Liquid Chromatography Abstracts*. These are published under the auspices of the Chromatography Discussion Group of London, England. They are published quarterly and may be obtained from the Applied Science Publishers Ltd., Ripple Road, Barking, Essex, England. The fourth-quarter volume lists an author and a detailed subject index. These abstracts are a worthwhile investment for those working in GC or liquid chromatography (LC).

2. *Gas–Liquid Chromatography Abstracts*. These are published in separate volumes: GC abstracts are published monthly and LC abstracts, bimonthly. They are published by the Preston Technical Abstracts Services, Niles, IL 60648. The last issue of the year contains an author and detailed subject index.
3. *CA Selects*. These are published bimonthly by computer search of the Chemical Abstracts Service Information Base by Chemical Abstracts Service, Columbus, OH 43210. There is a separate publication for GC.

Several journals are dedicated solely to chromatographic articles. These include the *Journal of Chromatography*, *Journal of Chromatographic Science* (formerly *Journal of Gas Chromatography*), *Chromatographia*, *Journal of High Resolution Chromatography and Chromatography Communications*, and *Separation Science*. Other prominent journals containing a large fraction of the remaining literature are *Analytical Chemistry*, *Nature*, *Zeitschrift fur Analytische Chemie*, *Analyst*, *Journal of the American Chemical Society*, *Angewandte Chemie*, *Talanta*, *Brennstoff–Chemie Bulletin de la Society Chemique de France*, and *Revista Italiana delle Sostanze Grasse*.

Two excellent publications that review the various techniques and their applications are (1) *Annual Fundamental Reviews in Analytical Chemistry*, published biannually in even-numbered years by the *Analytical Chemistry Journal* of the American Chemical Society, Washington, DC 20036, and (2) *Annual Applied Reviews in Analytical Chemistry*, published biannually in odd-numbered years by the *Analytical Chemistry Journal* of the American Chemical Society, Washington, DC 20036.

Additional abstract sources are *Chemisches Zentralblatt Analytical Abstracts* and the Bibliography Section of the *Journal of Chromatography*. As pointed out previously, any research journal may contain an article in which GC was used. In this case the researcher should refer to the journal for the specific area or topic, for example, *Analytical Biochemistry*, *Food Technology*, *Journal of Agricultural and Food Chemistry*, *Journal of the American Oil Chemist's Society*, *Clinical Chemistry*, or *Journal of Chemical Physics*.

Finally, there are many good books that deal with GC specifically or with chromatography in general:

Suggested Reading on GC

S. Dal Nogare and R. S. Juvet, *Gas–Liquid Chromatography, Theory and Practice*, Interscience, New York, 1962.

J. C. Giddings, *Dynamics of Chromatography*, Part 1, *Principles and Theory*, Dekker, New York, 1965.

W. E. Harris, *Programmed Temperature Gas Chromatography*, Wiley, New York, 1966.

L. S. Ettre and A. Zlatkis (Eds.), *The Practice of Gas Chromatography*, Interscience, New York, 1967.

J. Tranchant (Ed.), *Practical Manual of Gas Chromatography*, Elsevier, Amsterdam, 1969.

J. Sevcik, *Detectors in Gas Chromatography*, Elsevier, Amsterdam, 1975.

R. L. Grob (Ed.), *Chromatographic Analysis of the Environment*, 2d ed., Dekker, New York, 1983.

C. F. Poole and S. K. Poole, *Chromatography Today*, Elsevier, New York, 1991.

E. Heftmann (Ed.), *Chromatography*, 5th ed., Part A and B, Elsevier, New York, 1992.

R. L. Grob and M. A. Kaiser, *Environmental Problem Solving Using Gas and Liquid Chromatography*, Elsevier, New York, 1982.

1.5 COMMERCIAL INSTRUMENTATION

All leading instrument manufacturers produce and market gas chromatographs. In addition, many smaller speciality companies also manufacture and market GC units. Which instrument should be considered depends on the use to which they are to be utilized, and this ultimately establishes the criteria for purchase. GC units come in a variety of makes and models, from simple student instructional types (e.g., Gow-Mac Instrument Co.) up to deluxe multicolumn, interchangeable detector types (e.g., Hewlett-Packard Company). We refer the reader to the "Lab Guide" issue of the *Journal of Analytical Chemistry* (44), rather than to one particular company, for a listing of the instrument manufacturers.

REFERENCES

1. V. Heines, *Chem. Technol.*, **1**, 280–285 (1971).
2. L. S. Ettre, *Anal. Chem.*, **43**(14), 20A–31A (1971).
3. G. Zweig and J. Sherma, *J. Chromatogr. Sci.*, **11**, 279–283 (1973).
4. F. F. Runge, *Farbenchemie, I and II* (1834, 1843).
5. F. F. Runge, *Ann. Phys. Chem.*, *XVII*, **31**, 65 (1834); *XVIII*, **32**, 78 (1834).
6. F. F. Runge, *Farbenchemie*, *III*, 1850.
7. F. Goppelsroeder, *Zeit. Anal. Chem.*, **7**, 195 (1868).
8. C. Schönbein, *J. Chem. Soc.*, **33**, 304–306 (1878).
9. D. T. Day, *Proc. Am. Philos. Soc.*, **36**, 112 (1897).
10. D. T. Day, *Congr. Intern. Pétrole Paris*, **1**, 53 (1900).
11. D. T. Day, *Science*, **17**, 1007 (1903).
12. M. Twsett, *Ber. Deut. Bot. Ges.*, *XXIV*, 316 (1906).

13. M. Twsett, *Ber. Deut. Bot. Ges.*, *XXIV*, 384 (1906).
14. M. Twsett, *Ber. Deut. Bot. Ges.*, *XXV*, 71–74 (1907).
15. R. Kuhn, A. Wunterstein, and E. Lederer, *Hoppe-Seyler's Z. Physiol. Chem.*, **197**, 141–160 (1931).
16. A. Tiselius, *Ark. Kemi. Mineral. Geol.*, **14B**(22) (1940).
17. J. N. Wilson, *J. Am. Chem. Soc.*, **62**, 1583–1591 (1940).
18. A. Tiselius, *Ark. Kemi. Mineral. Geol.*, **15B**(6) (1941).
19. A. J. P. Martin and R. L. M. Synge, *Biochem. J.* (London), **35**, 1358 (1941).
20. R. Consden, A. H. Gordon, and A. J. P. Martin, *Biochem. J.*, **38**, 224–232 (1944).
21. S. Claesson, *Arkiv. Kemi. Mineral. Geol.*, **23A**(1) (1946).
22. A. J. P. Martin, *Biochem. Soc. Symp.*, **3**, 4–15 (1949).
23. E. Cremer and F. Prior, *Z. Elektrochem.*, **55**, 66 (1951); E. Cremer and R. Muller, ibid., **55**, 217 (1951).
24. C. S. G. Phillips, J. Griffiths, and D. H. Jones, *Analyst*, **77**, 897 (1952).
25. A. T. James and A. J. P. Martin, *Biochem. J.*, **50**, 679–690 (1952).
26. E. Glueckauf, in *Ion Exchange and Its Applications*, Society of the Chemical Industry, London, 1955, pp. 34–36.
27. J. J. van Deemter, F. J. Zuiderweg, and A. Klinkenberg, *Chem. Eng. Sci.*, **5**, 271–289 (1956).
28. M. J. E. Golay, in *Gas Chromatography*, V. J. Coates, H. J. Noebels, and I. S. Fagerson (Eds.), Academic, New York, 1958, pp. 1–13.
29. J. C. Giddings, *Dynamics of Chromatography*, Part I, *Principles and Theory*, Dekker, New York, 1965, pp. 13–26.
30. C. Horvath and W. R. Melander, "Theory of Chromatography," in *Chromatography, Fundamentals and Applications of Chromatographic and Electrophoretic Methods*, Part 1, A. E. Heftmann (Ed.), Elsevier, Amsterdam, 1983.
31. C. McAuliffe, *Chem. Technol.*, **1**, 46 (1971).
32. J. Drozd and J. Novak, *J. Chromatogr.*, **165**, 141 (1979).
33. G. R. Umbreit and R. L. Grob, *J. Environ. Sci. Eng.*, Part A, **A15**(5), 429 (1980).
34. J. C. Giddings, *Sep. Sci.*, **1**, 123 (1966).
35. J. C. Giddings, M. N. Meyers, F. J. F. Yang, and L. K. Smith, in *Colloid and Intersurface Science*, Vol. 4, M. Kerker (Ed.), Academic, New York, 1976.
36. J. C. Giddings, S. R. Fisher, and M. N. Meyers, *Am. Lab.*, **10**, 15 (1978).
37. J. C. Giddings, *Pure Appl. Chem.*, **51**, 1459 (1979).
38. J. C. Giddings, M. N. Meyers, K. D. Caldwell, and S. R. Fisher, in *Methods of Biochemical Analysis*, Vol. 26, D. Glick (Ed.), Interscience, New York, 1980, p. 79.
39. J. C. Giddings, M. N. Meyers, and K. D. Caldwell, *Sep. Sci. Technol.*, **16**, 549 (1981).
40. J. C. Giddings, *Anal. Chem.*, **53**, 1170A (1981).

41. J. C. Giddings, *Sep. Sci. Technol.*, **73**(1), 3–24 (1978).
42. H. H. Strain, T. R. Sato, and J. Engelke, *Anal. Chem.*, **26**, 90 (1954).
43. *Unified Nomenclature for Chromatography*, IUPAC, *J. Pure Appl. Chem.*, **65**(4), 819–872 (1993).
44. "1993 Lab Guide," *Anal. Chem.*, **65**(16) (1993).

PART I

THEORY AND BASICS

Science moves, but slowly slowly, creeping on from point to point.

—Alfred, Lord Tennyson (1809–1892)

Locksley Hall, line 134

CHAPTER TWO

Theory of Gas Chromatography

ROBERT L. GROB

Professor Emeritus of Analytical Chemistry, Villanova University

Modern Practice of Gas Chromatography, Third Edition. Edited by Robert L. Grob.
ISBN 0-471-59700-7

2.1 CHROMATOGRAPHIC METHODS

2.1.1 Classification of Methods

In the strictest sense, the term "chromatography" is a misnomer. Most of the materials chromatographed today are either colorless, or, if they were colored, one would not be able to perceive them in most instances. A number of workers in the field have offered contemporary definitions of the term, but not all practitioners of the technique use these terms or even agree with them. In the paragraph that follows we present our own definition but do not declare it to be unique or more representative of the process.

Chromatography encompasses a series of techniques that have in common the separation of components of a mixture by a series of equilibrium operations that result in separation of the entities as a result of their partitioning (differential sorption) between two different phases, one stationary with a large surface and the other a moving phase in contact with the first. Chromatography is not restricted to analytical separations. It may be used in the preparation of pure substances, the study of the kinetics of reactions, structural investigations on the molecular scale, and the determination of physicochemical constants, including stability constants of complexes, enthalpy, entropy, and free energy (see Chapter 9).

Using the definition given in the preceding paragraph (or any other definition of chromatography), one can tabulate numerous variations of the technique (see Figure 2.1). Our specific concern is the gas chromatographic technique. For this technique we have available different types of column that may be used to perform the separation. More details are found in Chapters 3 and 4.

2.1.2 General Aspects

The mixture to be separated and analyzed may be either a gas, a liquid, or a solid in some instances. All that is required is that the sample components

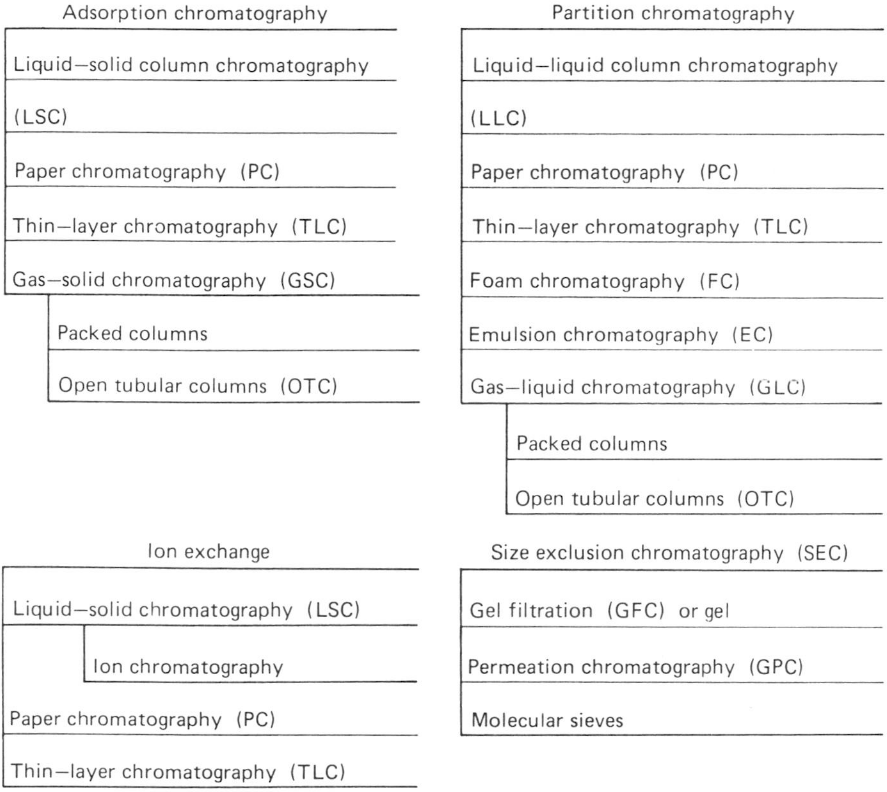

Adsorption chromatography

- Liquid–solid column chromatography (LSC)
- Paper chromatography (PC)
- Thin–layer chromatography (TLC)
- Gas–solid chromatography (GSC)
  - Packed columns
  - Open tubular columns (OTC)

Partition chromatography

- Liquid–liquid column chromatography (LLC)
- Paper chromatography (PC)
- Thin–layer chromatography (TLC)
- Foam chromatography (FC)
- Emulsion chromatography (EC)
- Gas–liquid chromatography (GLC)
  - Packed columns
  - Open tubular columns (OTC)

Ion exchange

- Liquid–solid chromatography (LSC)
  - Ion chromatography
- Paper chromatography (PC)
- Thin–layer chromatography (TLC)

Size exclusion chromatography (SEC)

- Gel filtration (GFC) or gel Permeation chromatography (GPC)
- Molecular sieves

FIGURE 2.1 Various chromatographic techniques.

be stable, have a vapor pressure of approximately 0.1 Torr at the operating temperature, and interact with the column material (either a solid adsorbent or a liquid stationary phase) and the mobile phase (carrier gas). The result of this interaction is the differing distribution of the sample components between the two phases, resulting in the separation of the sample component into zones or bands. The principle that governs the chromatographic separation is the foundation of most physical methods of separation, for example, distillation and liquid–liquid extraction.

The separation of the sample components may be achieved by one of three techniques: frontal analysis, displacement development, or elution development.

2.1.3 Frontal Analysis

The liquid or gas mixture is fed into a column containing a solid packing. The mixture acts as its own mobile phase or carrier, and the separation depends on the ability of each component in the mixture to become a sorbate (see Figure 2.2). Once the column packing has been saturated (i.e.,

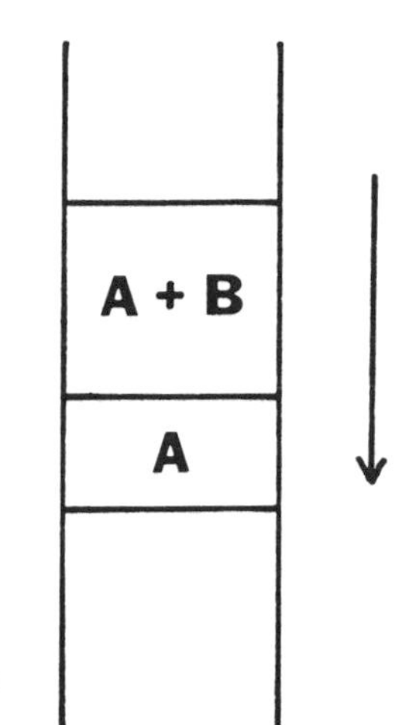

FIGURE 2.2 Frontal analysis. Component B is more sorbed than component A.

when it is no longer able to sorb more components), the mixture then flows through with its original composition. The early use of this technique involved measurement of the change in concentration of the front leaving the column; hence the name "frontal analysis." The least-sorbed component breaks through first and is the only component to be obtained in a pure form. Figure 2.3 illustrates the integral-type recording for this type of system. In this figure we illustrate the recording of the fronts from a four-component sample.

Frontal analysis requires that the system have convex isotherms (see Section 2.1.6). This results in the peaks having sharp fronts and well-formed steps. An inspection of Figure 2.3 reflects the problem of analytical frontal analysis—it is difficult to calculate initial concentrations in the sample. One

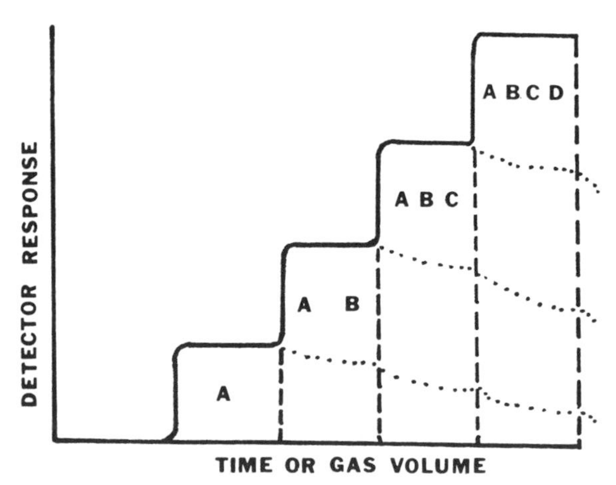

FIGURE 2.3 Integral-type chromatogram from frontal analysis. Component A is the least sorbed of four components.

can, however, determine the number of components present in the sample. If the isotherms are linear, the zones may be diffuse. This may be caused by three important processes: inhomogeneity of the packing, large diffusion effects, and nonattainment of sorption equilibrium.

2.1.4 Displacement Development

In this technique the developer is contained in the moving phase, which may be a liquid or a gas (Figure 2.4). One necessary requirement is that this moving phase be more sorbed than any sample components. One always obtains a single pure band of the first component in the sample. In addition, there is always an overlap zone for each succeeding component, which is an advantage of this technique over frontal analysis. The disadvantage, from the analytical viewpoint, is that the component bands are not separated by a region of pure mobile phase. The result of this displacement mechanism (Figure 2.5) assumes a three-component mixture and a developer or displacing agent. The step height is utilized for qualitative identification of components, whereas the step length is proportional to the amount of the component.

As with frontal analysis, displacement analysis requires convex isotherms. Once equilibrium conditions have been attained, an increase in column length serves no useful purpose in this technique because the separation is more dependent on equilibrium conditions than on the size of column.

2.1.5 Elution Development

In this technique, components A and B travel through the column at rates determined by their retention on the solid packing (Figure 2.6). If the differences in sorption are sufficient or the column is long enough, a

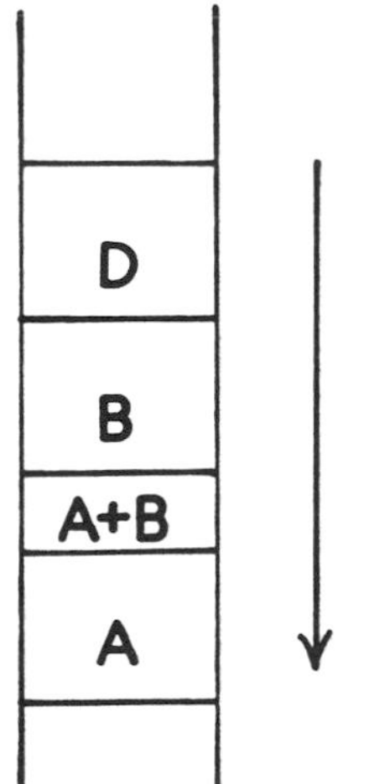

FIGURE 2.4 Displacement development (D = displacer). D is more sorbed than B, which is more sorbed than A.

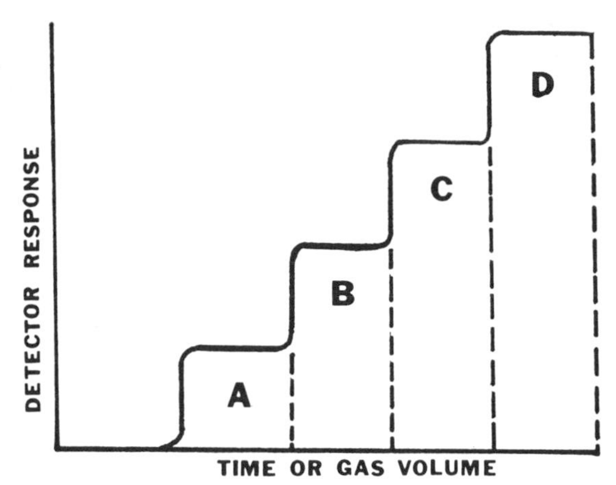

FIGURE 2.5 Integral-type chromatogram from displacement development. Order of sorption: D > C > B > A.

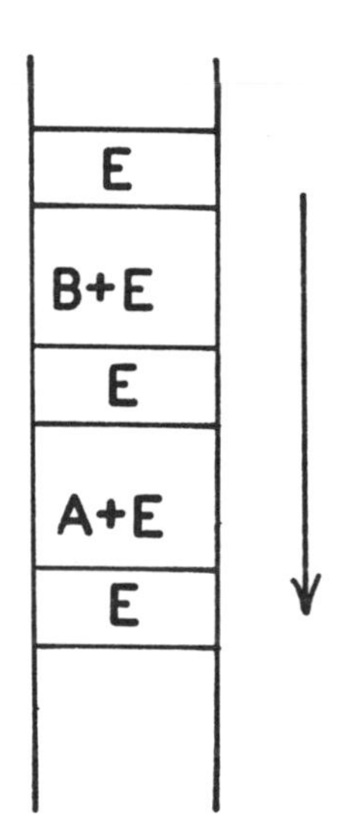

FIGURE 2.6 Elution development (E = eluant). B is more sorbed than A.

complete separation of A and B is possible. Continued addition of eluant causes the emergence of separated bands or zones from the column. A disadvantage of this technique is the very long time interval required to remove a highly sorbed component. This can be overcome by increasing the column temperature during the separation process. Figure 2.7 depicts a typical chromatogram for this technique. The position of peak maximum on the abscissa qualitatively identifies the component, and the peak area is a measure of the amount of each component.

Summary. The *frontal* technique does not lend itself to many analytical applications because of the overlap of the bands and the requirement of a

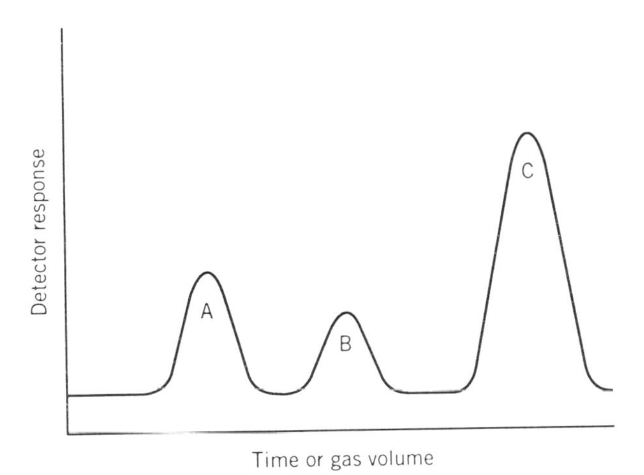

FIGURE 2.7 Differential chromatogram from elution development. Order of retention: C > B > A.

large amount of sample. However, it may be used to study phase equilibria (isotherms) and for preparative separations. (Many of the industrial chromatographic techniques use frontal analysis.) *Displacement* development has applications for analytical liquid chromatography (LC). (For instance, it may be used as an initial concentrating step in GC for trace analysis.) This technique may also be used in preparative work. The outstanding disadvantage of both of these techniques is that the column still contains sample or displacer at the conclusion of the separation; thus the column must be regenerated before it can be used again.

It is in this regard that *elution* chromatography offers the greatest advantage—at the end of a separation, only eluant remains in the column. Thus the bulk of the discussion in the subsequent chapters is concerned with elution GC. The isotherms and chromatograms of elution chromatography are discussed in Sections 2.1.6, 2.1.7, 2.1.8, and 2.1.9.

2.1.6 Isotherms

An isotherm is a graphical presentation of the interaction of an adsorbent and a sorbate in solution (gas or liquid solvent) at a specified temperature. The isotherm is a graphical representation of the partition coefficient or distribution constant K:

$$K = \frac{C_S}{C_G} \tag{2.1}$$

where C_S is the concentration of sorbate in stationary phase or at the solid surface, and C_G is the concentration of sorbate in the gas phase. The

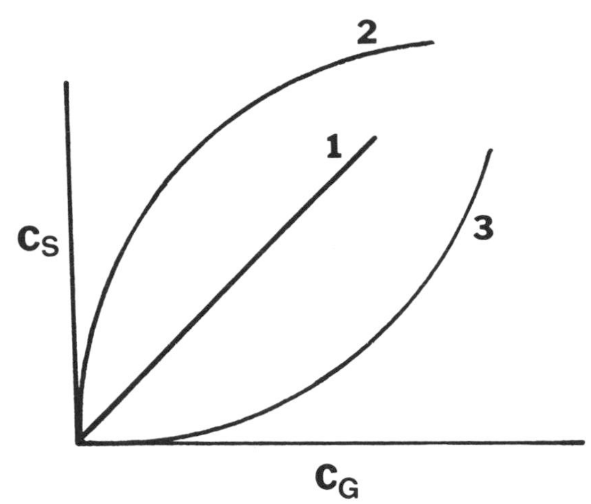

FIGURE 2.8 Isotherms. C_S = concentration at solid surface or in a stationary phase; C_G = concentration in solution at equilibrium. 1, Linear isotherm; 2, convex isotherm; 3, concave isotherm.

concentration of the substance sorbed per unit mass of sorbent is plotted against the concentration of the substance in equilibrium with the phase present at the interface. Three types of isotherm are obtainable: one linear and two curved. We describe the nonlinear isotherms as either concave (curved away from the abscissa) or convex (curved toward the abscissa). Figure 2.8 depicts these three isotherms.

The *linear isotherm* is obtained when the ratio of the concentration of substance sorbed per unit mass and the concentration of the substance in solution remains constant. This means that the partition coefficient or distribution constant K (see Section 1.3) is constant over all working concentration ranges. Thus the frontal and rear boundaries of the band or zone will be symmetrical.

The *convex isotherm* demonstrates that the K value is changing to a higher ratio as concentration increases. This results in movement of the component through the column at a faster rate, thus causing the front boundary to be self-sharpening and the rear boundary to be diffuse.

The *concave isotherm* results from the opposite effect (where the K value changes to a lower value), and the peak will have a diffuse front boundary and a self-sharpening rear boundary. In other words, the solute increasingly favors the surface of the stationary phase as the solution concentration increases. These effects are depicted in Figure 2.9. When the isotherms curve in either direction (convex or concave) as concentration is varied, one obtains complex chromatograms. Changing the sample concentration or physical conditions (temperature, flowrate, pressure, etc.) can help in converting the rear and front boundaries to Gaussian shape.

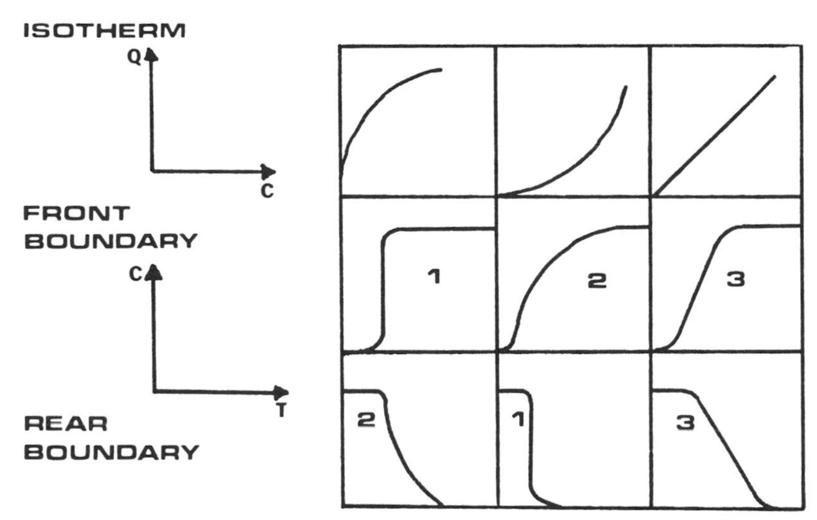

FIGURE 2.9 Dependence of boundary profile on form of partition isotherm. C = concentration (mL/mol) of solute in gas phase; Q = concentration in liquid or adsorbed phase; T = time for band to emerge from the column. 1, Self-sharpening profile; 2, diffuse profile; 3, Gaussian profile. (Courtesy of Wiley-Interscience Publishers.)

The most frequently applied isotherm equations are those due to Freundlich and Langmuir (1).

1. *Freundlich Equation.* This equation represents the variation of adsorption with pressure over a limited range, at constant temperature:

$$\frac{x}{m} = kp^{1/n} \tag{2.2}$$

where x = mass of adsorbed gas
m = grams of adsorbing material
p = pressure
k,n = constants

The exponent $1/n$ is usually less than one, indicating that the amount of adsorbed gas does not increase in proportion to the pressure. If the exponent $1/n$ were unity, the Freundlich equation would be equivalent to the distribution law. Converting Equation 2.2 to log form, we obtain

$$\log \frac{x}{m} = \log k + \left(\frac{1}{n}\right) \log p \tag{2.3}$$

which is an equation of a straight line; thus the log x/m–log p relationship is linear (linear isotherm). If a value of $1/n$ being unity gives a linear isotherm, a value of $1/n > 1$ gives a concave isotherm. When $1/n < 1$, a convex isotherm results.

2. *Langmuir Equation*. It is probable that adsorbed layers have a thickness of a single molecule because of the rapid decrease in intermolecular forces with distance. The Langmuir adsorption isotherm equation is

$$\frac{x}{m} = \frac{k_1 k_2 p}{1 + k_1 p} \tag{2.4}$$

where k_1, k_2 are constants for a given system and p is the gas pressure, which may be written as

$$\frac{p}{x/m} = \frac{1}{k_1 k_2} + \frac{p}{k_2} \tag{2.5}$$

A plot of $p(x/m)$ versus p produces a straight line with slope of $1/k_2$ and an intercept of $1/k_1k_2$. Deviations from linearity are attributed to nonuniformity, leading to various types of adsorption on the same surface, that is, non-monomolecular adsorption on a homogeneous surface.

2.1.7 Process Types in Chromatography

The process of chromatographic separation can be defined by two conditions:

1. The distribution isotherms (representation of the partition coefficient or distribution constant K) may be either linear or nonlinear (see Section 2.1.6).
2. The chromatographic system is either *ideal* or *nonideal*. *Ideal* chromatography infers that the exchange between the two phases is thermodynamically reversible. In addition, the equilibrium between the solid granular particles or liquid-coated particles and the gas phase is immediate; that is, the mass transfer is very high, and longitudinal and other diffusion processes are small enough to be ignored. In *nonideal* chromatography these assumptions cannot be made.

Using these two sets of conditions, we can then describe four chromatographic systems: (1) linear–ideal chromatography, (2) linear–nonideal chromatography, (3) nonlinear–ideal chromatography, and (4) nonlinear–nonideal chromatography.

2.1.8 Linear–Ideal Chromatography

This is the most direct and simple theory of chromatography. The transport of the solute down the column will depend on the distribution constant (partition coefficient) K and the ratio of the amounts of the two phases in the column. Band (zone) shape does not change during this movement through the column.* The system can be visualized as illustrated in Figure 2.10.

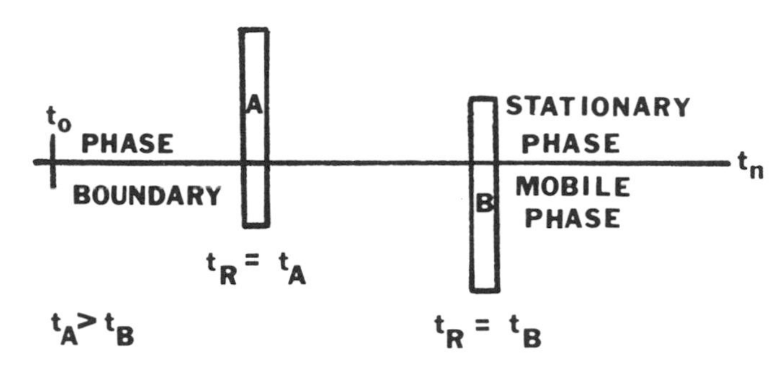

FIGURE 2.10 Linear–ideal chromatography. t_0 = start of separation (point of sample injection); t_A = retention time of component A; t_B = retention time of component B; t_n = time for emergence of mobile phase from t_0.

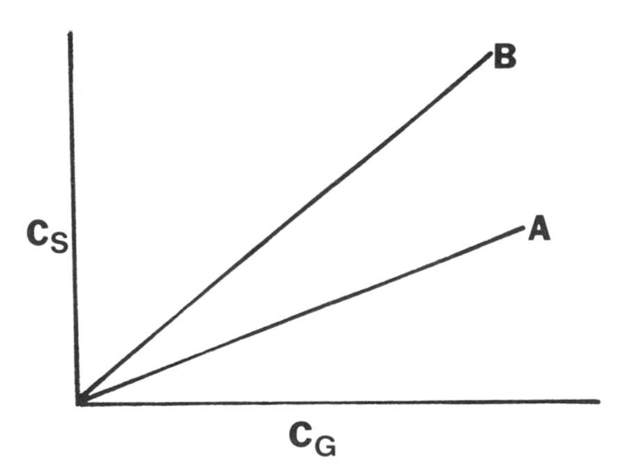

FIGURE 2.11 Isotherms for linear–ideal chromatography. C_S = concentration at surface or in stationary phase; C_G = concentration in solution at equilibrium.

* This type of chromatography would be the best of all worlds—that is, there are no diffusion effects, and mass transfer between phases is instantaneous (see Section 2.3.2). The isotherms that result from this system would be linear (see Figure 2.11).

2.1.9 Linear–Nonideal Chromatography

In this system the bands (zones) broaden because of diffusion effects and nonequilibrium. This broadening mechanism is fairly symmetrical, and the resulting elution bands approach the shape of a Gaussian curve. This system best explains liquid or gas partition chromatography. The system may be viewed in two ways:

1. *Plate Theory*. Envision the chromatographic system as a discontinuous process functioning the same as a distillation or extraction system, that is, one consisting of a large number of equivalent plates.
2. *Rate Theory*. Consider the chromatographic system as a continuous medium where one accounts for mass transfer and diffusion phenomena.

These two points of view usually are used to discuss gas chromatographic theory. Linear–nonideal chromatography may be visualized by the relationships shown in Figures 2.12 and 2.13.

2.1.10 Nonlinear–Ideal Chromatography

Liquid–solid chromatography is representative of this system type because nonlinearity effects are usually appreciable. Mass transfer is fast, and longitudinal diffusion effects may be ignored in describing the system. The net result is that the bands (zones) develop self-sharpening fronts and diffuse

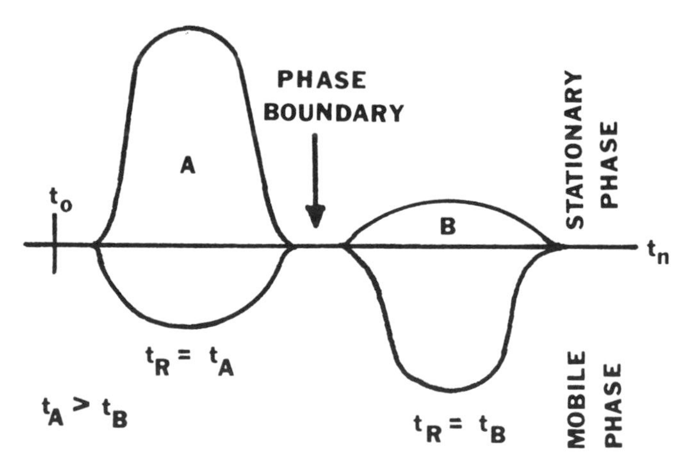

FIGURE 2.12 Linear nonideal chromatography. t_0 = time at start of separation (point of sample injection); t_A = retention time of component A; t_B = retention time of component B; t_n = time for emergence of mobile phase from t_0.

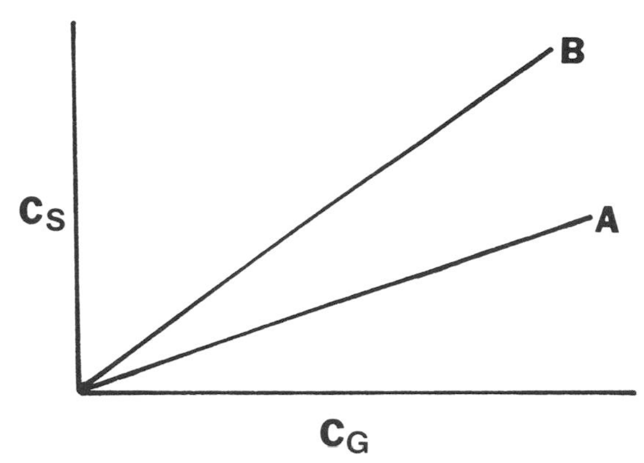

FIGURE 2.13 Isotherms for linear–nonideal chromatography. C_S = concentration at surface or in stationary phase; C_G = concentration in solution (mobile phase) at equilibrium.

rear boundaries. Because of this tailing, this technique is unsuitable for elution analysis. This system is represented by Figures 2.14 and 2.15.

2.1.11 Nonlinear–Nonideal Chromatography

Gas–solid chromatography is best described by this theory. Here one finds diffuse front and rear boundaries with definite tailing of the rear boundary. Mathematical descriptions of systems of this type can become very complex; however, with proper assumptions mathematical treatments do fairly repre-

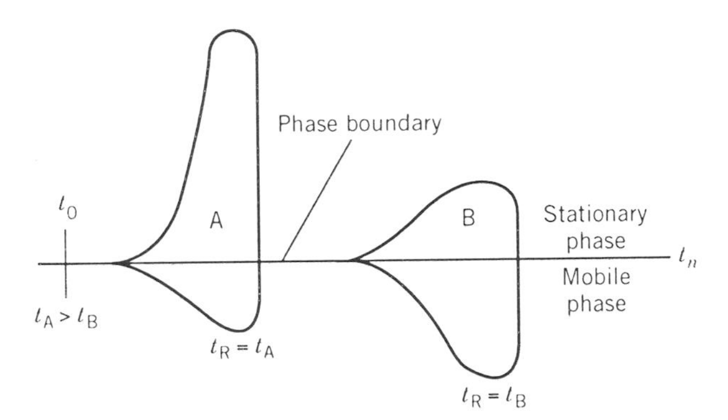

FIGURE 2.14 Nonlinear–ideal chromatography. t_0 = start of separation (point of sample injection); t_A = retention time of component A; t_B = retention time of component B; t_n = time of emergence of mobile phase from t_0.

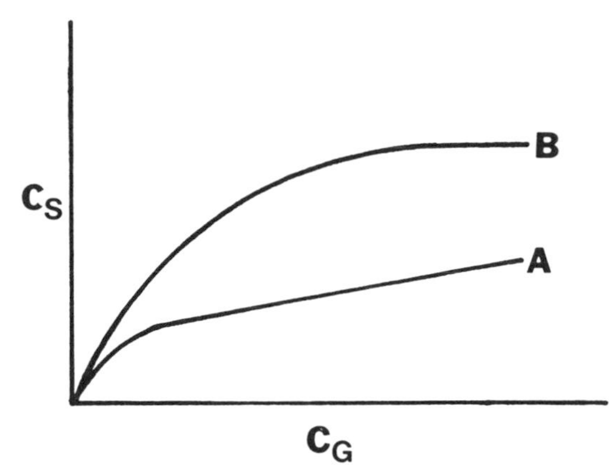

FIGURE 2.15 Isotherms for nonlinear ideal chromatography. C_S = concentration at surface or in stationary phase; C_G = concentration in solution (mobile phase) at equilibrium.

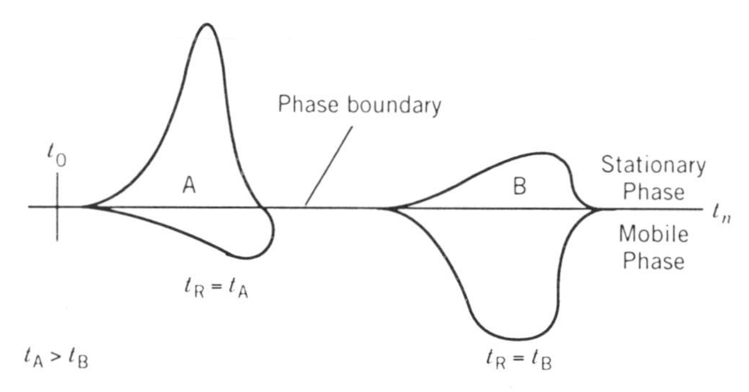

FIGURE 2.16 Nonlinear–nonideal chromatography. t_0 = start of separation (point of sample injection); t_A = retention time of component A; t_B = retention time of component B; t_n = time of emergence of mobile phase from t_0.

sent the experimental data. The bands (zones) are similar to those shown in Figures 2.16 and 2.17.

2.2 GENERAL ASPECTS OF GAS CHROMATOGRAPHY

2.2.1 Applications of Gas Chromatography

Gas chromatography is a unique and versatile technique. In its initial stages of development it was applied to the analysis of gases and vapors from very volatile components. The work of Martin and Synge (2) and then James and Martin (3) in gas–liquid chromatography (GLC) opened the door for an analytical technique that has revolutionized chemical separations and

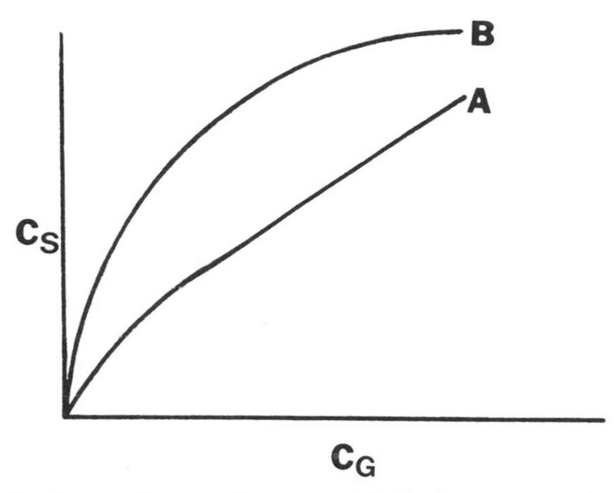

FIGURE 2.17 Isotherms for nonlinear–nonideal chromatography. C_S = concentration at surface or in stationary phase; C_G = concentration in solution (mobile phase) at equilibrium.

analyses. As an analytical tool, GC can be used for the direct separation and analysis of gaseous samples, liquid solutions, and volatile solids.

If the sample to be analyzed is nonvolatile, the techniques of derivatization or pyrolysis GC can be utilized. This latter technique is a modification wherein a nonvolatile sample is pyrolyzed prior to its entry into the column. Decomposition products are separated in the gas chromatographic column, after which they are qualitatively and quantitatively determined. Analytical results are obtained from the *pyrogram* (a chromatogram resulting from the detection of pyrolysis products). This technique can be compared to mass spectrometry, a technique in which analysis is based on the nature and distribution of molecular fragments that result from the bombardment of the sample component with high-speed electrons. In pyrolysis GC the fragments result from chemical decomposition by heat. If the component to be pyrolyzed is very complex, complete identification of all the fragments may not be possible. In a case of this type the resulting pyrogram may be used as a set of "fingerprints" for subsequent study.

Pyrolysis may be defined as the thermal transformation of a compound (single entity) into a other compound or compounds, usually in the absence of oxygen. In modern pyrolysis the sample decomposition is rigidly controlled. One should keep in mind that pyrolysis gas chromatography (PGC) is an indirect method of analysis in which heat is used to change a compound into a series of volatile products that should be characteristic of the original compound and the experimental conditions.

Gas chromatography is the analytical technique used for product identification (under very controlled conditions) and must be directly coupled to a mass spectrometer when information other than a comparative fingerprint (pyrogram) is required, such as positive identification of peaks on the chromatogram.

Ettre and Zlatkis (6) classified pyrolysis types according to extent of degradation of the sample compound:

1. *Thermal Degradation.* Usually occurs in the temperature range of 100 to 300°C but may occur as high as 500°C. This type may be carried out in the injection port of the instrument. Rupture of carbon–carbon bonds is minimal.
2. *Mild Pyrolysis.* Occurs between 300 and 500°C and carbon–carbon bond breakage occurs to some extent.
3. *Normal Pyrolysis.* Occurs between 500 and 800°C and involves cleavage of carbon–carbon bonds. Very useful for characterizing polymers and copolymers (see Chapter 11).
4. *Vigorous Pyrolysis.* Occurs at temperatures between 800 and 1100°C. The end results is the breaking of carbon–carbon bonds and cleaving organic molecules into smaller fragments.

The pyrolysis process may be performed by three different methods:

1. *Continuous-Mode Method.* May involve tube furnaces or microreactors. In this mode the heated wall of the reactor is at a higher temperature than the sample and secondary reactions of pyrolysis products will most likely occur.
2. *Pulse-Mode Pyrolysis.* Sample is in direct contact with a hot wire, thus minimizing secondary reactions. Although the temperature profile is reproducible, the exact pyrolysis temperature cannot be measured. Another disadvantage is that the sample weight cannot be known accurately. This is also known as Curie-point pyrolysis (see Chapter 11).
3. *Laser-Mode Pyrolysis.* Directs very high energies to the sample, which usually result in ionization and the formation of plasma plumes. Thus, laser pyrolysis results in fewer and sometimes different products than thermal pyrolysis.

To a first approximation, good interlaboratory reproducibility of the pyrolysis profile is obtainable; however, intralaboratory matchings have been disappointing. Several major parameters influence pyrolysis reproducibility:

1. Type of pyrolyzer

2. Temperature
3. Sample size and homogeneity
4. Gas chromatographic conditions and column(s) used
5. Interface between the pyrolyzer and the gas chromatograph.

Therefore, optimization of the pyrolyzer by use of reference standards is important. Thermal gradients across the sample may be avoided by use of thin samples. For good results in PGC one must have rapid transfer of the pyrolysis products to the column, minimization of secondary reaction products, and elimination of poor sample injection profiles.

When employing PGC for qualitative and quantitative analysis of complex unknown samples it is essential to use *pure* samples of suspected sample components as a reference. One should never base identification of unknown pyrolyzate peaks on the retention time of pyrolyzate product peaks obtained from the standard (7). A peak in the chromatogram from the pyrolysis of the unknown may be from the matrix and not the suspected component. The use of selective detectors (i.e., a NPD with a FID or a FID with an ECD) will furnish element information but not molecular or structural information about the component peak. The matrix components (in the absence of the suspected analyte) may yield the same peak at the same retention time.

Another important variable in PGC is temperature control. Small changes in temperature may have pronounced effects on the resulting chromatogram. The effects may be manifested several ways:

1. Increased number of peaks
2. Decreased number of peaks
3. Partial resolution of overlapping peaks
4. Increase or decrease in the peak areas for same sample size of unknown (indicating different pyrolysis mechanism)
5. Changes in peak shape of pyrolysis products

Thus, caution must be used when identifying a peak on a pyrogram for an unknown. This means that a reliable identification should not be based on retention time data. The two best techniques for identifying unknown peaks are infrared spectroscopy (IRS) and mass spectrometry (MS). Mass spectrometry is the better of the two techniques because one obtains a mass number that may be matched with a mass number in a library of mass spectra of known compounds. *All* the ions from a known compound must be present for positive identification. Infrared spectroscopy will validate the presence of functional groups in the molecule. If the peak is single entity, one may match the spectrum (IR) obtained with a spectrum of a standard compound.

In addition to analysis, GC may be used to study structure of chemical compounds, determine the mechanisms and kinetics of chemical reactions, and measure isotherms, heats of solution, heats of adsorption, free energy of solution and/or adsorption, activity coefficients, and diffusion constants (see Chapter 9). Another significant application of GC is in the area of the preparation of pure substances or narrow fractions as standards for further investigations. Gas chromatography is also utilized on an industrial scale for process monitoring. In adsorption studies it can be used to determine specific surface areas (4,5). A novel use is its utilization for elemental analyses of organic components (8–10). Distillation curves may also be plotted from gas chromatographic data.

Gas chromatography can be applied to the solution of many problems in various fields. A few examples are enumerated:

1. *Drugs and Pharmaceuticals*. Gas chromatography is used not only in the quality control of products of this field, but also in the analysis of new products and the monitoring of metabolites in biological systems.

2. *Environmental Studies*. A review of the contemporary field of air pollution analyses by GC was published in the first volume of *Contemporary Topics in Analytical and Clinical Chemistry* (11). A book by Grob and Kaiser (12) discussed the use of LC and GC for this type of analyses. Many chronic respiratory diseases (asthma, lung cancer, emphysema, and bronchitis) could result from air pollution or be directly influenced by air pollution. Air samples can be very complex mixtures, and GC is easily adapted to the separation and analysis of such mixtures. Two publications concerned with the adaptation of cryogenic GC to analyses of air samples are Refs. 13 and 14. Chapter 14 covers the application of GC in the environmental area.

3. *Petroleum Industry*. The petroleum companies were among the first to make widespread use of GC. The technique was successfully used to separate and determine the many components in petroleum products. One of the earlier publications concerning the response of thermal conductivity detectors to concentration resulted from research in the petroleum field (15). The application of GC to the petroleum field is discussed in Chapter 10.

4. *Clinical Chemistry*. Gas chromatography is adaptable to such samples as blood, urine, and other biological fluids (see Chapter 12). Compounds such as proteins, carbohydrates, amino acids, fatty acids, steroids, triglycerides, vitamins, and barbiturates are handled by this technique directly or after preparation of appropriate volatile derivatives.

5. *Pesticides and Their Residues*. Gas chromatography in combination with selective detectors such as electron capture, phosphorus, and electrolytic conductivity detectors (see Chapter 5) have made the detection of such components and their measurement relatively simple. Detailed information in this area may be found in a monograph by Grob (16).

6. *Foods*. The determination of antioxidants and food preservatives is an active part of the gas chromatographic field. Adaptations and sample types are almost limitless, and include analysis of fruit juices, wines, beers, syrups, cheeses, beverages, food aromas, oils, dairy products, decomposition products, contaminants, and adulterants.

2.2.2 Types of Detection

The various detectors employed in GC are discussed in Chapter 5. Our purpose here is only to categorize the detection system according to whether they are an integral-type system or a differential-type system. This classification is an old one; any detection system can be made integral or differential simply by a modification of the detector electronics. A more modern categorization would be *instantaneous* (differential) and *cumulative* (integral). Chromatograms that result from this classification of detectors are shown in Figure 2.18.

2.2.3 Advantages and Limitations

From the limited discussion so far one can visualize the versatility of the gas chromatographic technique. There are so many reasons for this, and we shall enumerate some of the advantages. It should be stressed that what one person considers a disadvantage may be an advantage to someone else. Additionally, a current disadvantage, may be an advantage several years from now.

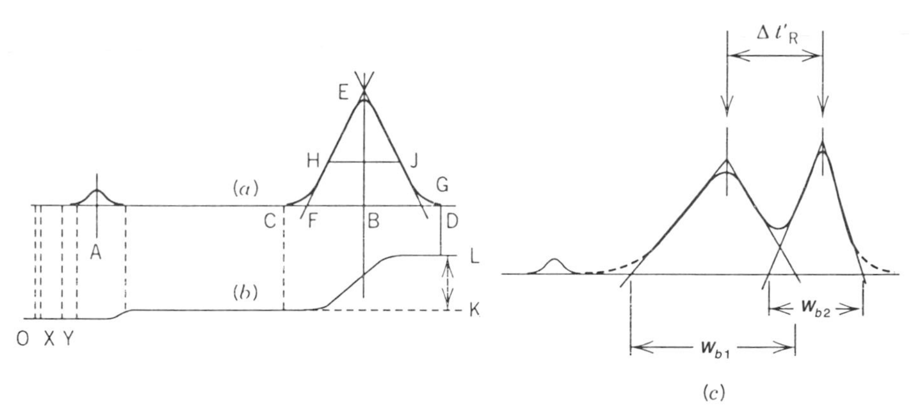

FIGURE 2.18 Types of chromatogram: a, differential chromatogram; b, integral chromatogram; c, peak resolution; *O*, injection point; *OX*, injector volume; *OY*, detector volume; *OA*, holdup volume V_M; *OB*, total retention volume V_R; *AB*, adjusted retention volume $V_R - V_M$; *CD*, peak base; *FG*, peak width w_b; *HJ*, peak width at half-height w_h; *BE*, peak height; *E*, peak maximum; *CHEJD*, peak area (space incorporated within these letters); *KL*, step height of integral chromatogram.

A few broad comments regarding GC would include the following:

1. *An Analytical Technique*. This is used not only for the qualitative identification of components in a sample, but also for quantitative measurements.

2. *A Physical Research Technique*. This may be used to investigate various parameters of a system, such as determination of partition coefficients, thermodynamic functions, and adsorption isotherms (see Chapter 9).

3. *A Preparative Technique*. Once the analytical conditions have been determined, the system may be scaled up to separate and collect gram amounts of components. This approach is summarized in Section 2.3.7.

4. *An On-Line Monitoring Probe*. A gas chromatograph can be locked into a process line so that the process stream may be monitored on a 24-hour basis.

5. *An Automated System*. A gas chromatograph may be interfaced to a computer with an automatic sampler so that routine analyses can be run overnight.

Following are some overall advantages of GC that should be stressed:

1. *Resolution*. The technique is applicable to systems containing components with very similar boiling points. By choosing a selective liquid phase or the proper adsorbent, one can separate molecules that are very similar physically and chemically. Components that form azeotropic mixtures in ordinary distillation techniques may be separated by GC.

2. *Sensitivity*. This property of the gas chromatographic system largely accounts for its extensive use. The simplest thermal conductivity detector cells can detect a few parts per million; with an electron capture detector or phosphorous detector, parts per billion or picograms of solute can easily be measured. This level of sensitivity is more impressive when one considers that the sample size used is of the order of 1 μL or less.

3. *Analysis Time*. Separation of all the components in a sample may take from several seconds up to 30 min. Analyses that routinely take an hour or more may be reduced to a matter of minutes, because of the high diffusion rate in the gas phase and the rapid equilibrium between the moving and stationary phases.

4. *Convenience*. The operation of GC is a relatively straightforward procedure. It is not difficult to train nontechnical personnel to carry out routine separations.

5. *Costs*. Compared with many analytical instruments available today, gas chromatographs represent an excellent value.

6. *Versatility*. Gas chromatography is easily adapted for analysis of samples of permanent gases as well as high-boiling liquids or volatile solids.

7. *High Separating Power*. Since the mobile phase has a low degree of viscosity, very long columns with excellent separating power can be employed.

8. *Assortment of Sensitive Detecting Systems*. Gas chromatographic detectors (see Chapter 5) are relatively simple and highly sensitive, and possess rapid response rates.

9. *Ease of Recording Data*. Detector output from gas chromatographs can be conveniently interfaced with recording potentiometers, integrating systems, computers, and a wide variety of automatic data storing modules (see Chapter 5).

10. *Automation*. Gas chromatographs may be used to monitor automatically various chemical processes in which samples may be periodically taken and injected onto a column for separation and detection.

2.3 GAS CHROMATOGRAPHY

It was pointed out in Section 2.1 that chromatographic separations can be evaluated by the shape of the peaks from a particular system. The shapes of the peaks depend on the isotherms that describe the relationship between concentration of solute in stationary phase to the solute concentration in the carrier gas. If the isotherms are linear, the peaks are Gaussian in shape and the separations proceed with little or no problems. If the isotherms are nonlinear, the peaks become asymmetric. Some isotherms are linear over a limited range and as long as we work in this limited range few problems are encountered. If the isotherm is concave to the gas-phase concentration axis (so that the distribution ratio decreases with the increase in solute concentration in the mobile phase), the band will have a sharp front and a long tail. If, on the other hand, the isotherm is convex to the gas-phase concentration axis (so that the distribution ratio increase increases with the increase in solute in mobile phase), the band will have a leading front and a sharp rear edge.

If chromatographic theory is explained on the basis of a discontinuous model, several assumptions are made: (1) equilibrium between solute concentration in the two phases is reached instantaneously, (2) diffusion of solute, in mobile phase, along column axis is minimal; and (3) the column is packed uniformly or wall-coated uniformly. All these conditions are not present in all chromatographic separations.

If the rate constants for the sorption–desorption processes are small, equilibrium between phases need not be achieved instantaneously. This effect is often called *resistance-to-mass transfer*, and thus transport of solute from one phase to another can be assumed to be diffusional in nature. As the solute migrates through the column, it is sorbed from the mobile phase into the stationary phase. Flow is through the void volume of the solid particles, with the result that the solute molecules diffuse through the interstices to reach the surface of the stationary phase. Likewise, the solute must diffuse from the interior of the stationary phase to get back into the mobile phase.

When the term "longitudinal diffusion" is applied to chromatographic band we include the true longitudinal molecular diffusion (Section 1.2.1) and apparent longitudinal diffusion or eddy diffusion. True longitudinal diffusion occurs because of concentration gradients within the mobile phase, but eddy diffusion results from uneven velocity profiles because of unequal lengths and widths of the large number of zigzag paths. As a result of these diffusion effects, some solute molecules move ahead, whereas others lag behind the center of the zone (band). The widening of the band as it moves down the column is of paramount importance in GC. The extent to which the band spreads (peak sharpness) determines the column efficiency, N (theoretical plate number).

2.3.1 Plate Theory

From the equilibrium shown in Equation 1.50 it follows that

$$\alpha = \frac{k_A}{k_B} \tag{2.6}$$

and that optimum separation occurs when $k_A k_B = 1$. If $k_A = 100$ and $k_B = 0.01$, then $\alpha = 10^4$ and $k_A k_B = 1$. This would indicate a good separation because 1% of A would remain unextracted and 1% of B would be extracted. However, if $k_A = 1.0$ and $k_B = 10^{-4}$, we still obtain an α of 10^4 but $k_A k_B = 10^{-4}$, meaning that 50% of A is in each phase and 0.01% of B is extracted. Solution B has been significantly unextracted but not separated from A. This second example of a separation lends itself to countercurrent distribution (extraction) (see Chapter 1). L. C. Craig (17–20) can be credited with the refinement of this technique. This extraction technique can be used to partially explain what occurs in a chromatographic column. It also is illustrative for explaining zone broadening in multistage processes. What one assumes is that the system is made of individual, discontinuous steps (theoretical plates) and that the system comes to equilibrium as solute passes from one step (plate) to the next. Thus it is referred to as the "plate" model. This model and the "rate" model (discussed in Section 2.3.2) may both be used to describe the theory of chromatography. Both models arrive at same basic conclusion that the zone broadening is proportional to the square root of the column length and that the zone shape follows the normal distribution law. Figure 2.19 illustrates the similarity between the countercurrent extraction (CCE) process and the chromatographic process.

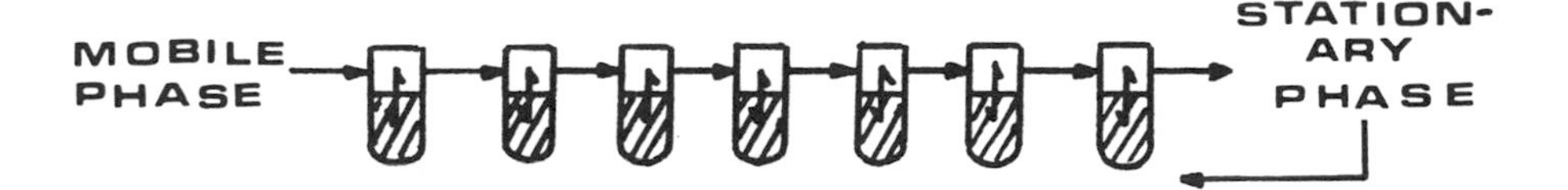

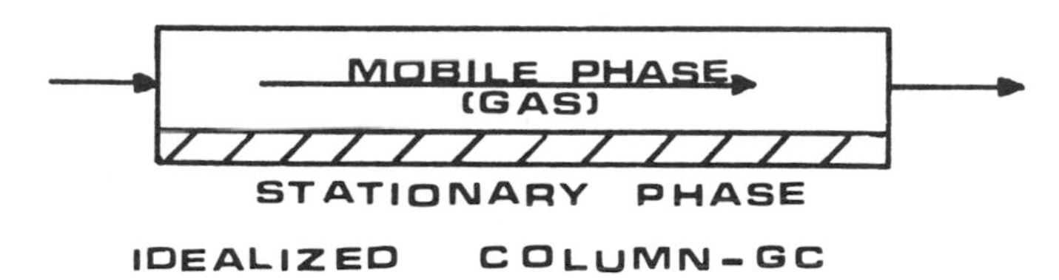

FIGURE 2.19 Comparison of countercurrent extraction and the chromatographic process.

The concept of plate theory was originally proposed for the performance of distillation columns (21). However, Martin and Synge (22) first applied the plate theory to partition chromatography. The theory assumes that the column is divided into a number of zones called *theoretical plates*. One determines the zone thickness or height equivalent to a theoretical plate (HEPT) by assuming that there is perfect equilibrium between the gas and liquid phases within each plate. The resulting behavior of the plate column is calculated on the assumption that the distribution coefficient remains unaffected by the presence of other solutes and that the distribution isotherm is linear. The diffusion of solute in the mobile phase from one plate to another is also neglected.

Martin and Synge (22) derived an expression for the total quantity q_n of solute in plate n and the volume of mobile phase (carrier gas) that passes through the column:

$$q_n = \frac{1}{(2\pi N)^{1/2}} \exp\left\{\frac{-[(V/v_R) - N]^2}{2N}\right\} \tag{2.7}$$

where v_R is the retention volume per plate and V is the total retention volume.

If $q_n = v_R C_n$, where C_n is solute concentration in mobile phase of plate n, then

$$C_n = \frac{1}{v_R(2\pi N)^{1/2}} \exp\left\{\frac{-[(V/v_R) - N]^2}{2N}\right\}$$

or (2.8)

$$C_n = \frac{1}{(v_R\sqrt{N})\sqrt{2\pi}} \exp\left[\frac{-(V - Nv_R)^2}{2(v_R\sqrt{N})^{1/2}}\right]$$

Equation 2.8 has the form of the normal error curve, and from the geometric properties of the curve we can show that

$$N = 16\left(\frac{V_R}{w_b}\right)^2 \tag{2.9}$$

where w_b is the base width of the peak. Equation 2.9 is a measure of the efficiency of a gas chromatographic column. Sometimes the number of plates is measured at the bandwidth at half-height w_h. From statistics

$$W = \left(\frac{2}{\ln 2}\right)^{1/2} w_h \tag{2.10}$$

Equation 2.9 may be expressed as Equation 2.11 in terms of w_h:

$$\begin{aligned} N &= 8 \ln 2\left(\frac{V_R}{w_h}\right)^2 \\ &= 8(2.30 \log 2)\left(\frac{V_R}{w_h}\right)^2 \\ &= 5.54\left(\frac{V_R}{w_h}\right)^2 \end{aligned} \tag{2.11}$$

If the number of theoretical plates in a column is known, it is possible to calculate the maximum sample size (i.e., the volume of sample that will not cause more than 10% peak broadening) that can be injected:

$$V_{max} = a\frac{V_m^0 + KV_L}{\sqrt{N}} \tag{2.12}$$

where $V_m^0 = jV_m$, and a is a constant. In terms of the column internal diameter d_c, we obtain

$$V_{max} = a\left(\frac{\pi}{4}\right) d_c^2 L + \frac{KV_L}{\sqrt{N}} \tag{2.13}$$

Equation 2.13 states that the maximum sample size is inversely related to the number of theoretical plates. This equation can be written more precisely as

$$V_{max} = a(V_R^0/\sqrt{N}) \tag{2.14}$$

which relates sample size to corrected retention volume. The constant a is determined experimentally by the successive injection of smaller samples until no more improvement in resolution is seen.

As long as the sample occupies less than $0.5(N)^{1/2}$ theoretical plates, there will be no band broadening because of sample size. As the total number of theoretical plates increases within a column, the maximum space (in terms of theoretical plates) that should be occupied by the sample will also increase. However, the percentage of column length available for sample will decrease (because the number of theoretical plates per column length increases), as shown in Table 2.1.

Having calculated the number of theoretical plates and knowing the length of the column, one may determine the HETP:

$$H = \text{HETP} = \frac{L}{N} = \frac{L}{16}\left(\frac{w_b}{V_R}\right)^2 = \frac{L}{16}\left(\frac{w_b}{t_R}\right)^2 \tag{2.15}$$

Plate theory disregards the kinetics of mass transfer; therefore, it reveals little about the factors influencing HETP values. Plate theory tells us that HETP becomes smaller with increasing flowrate; however, experimental evidence show that a plot of HETP versus flowrate always goes through a minimum.

The "theoretical plate" defined in GC is not the same as that in distillation or other countercurrent mass-transfer operations. In the latter, the number of theoretical plates represents the number of equilibrium stages on the equilibrium curve of a binary mixture that causes a given concentration change. In other words, HETP is the length of column producing a concentration change that corresponds to one equilibrium stage. In GC, the number of theoretical plates is a measure of peak broadening for a single component during the lifetime of the column. For a given column of constant length, therefore, the HETP represents the peak broadening as a function of retention time. In a gas chromatographic column, each component will yield different N and HETP values. Those solutes with high retention (high K values) will result in greater numbers of theoretical plates

TABLE 2.1 Sample Space in Terms of Theoretical Plates in Column

	Maximum Space Available for Sample	
Number of Plates in Column	In Terms of Theoretical Plates[a]	In Terms of % Column Length[b]
4	1	25
100	5	5
400	10	2.5
10,000	50	0.5

[a] $0.5\sqrt{N}$.
[b] $(0.5\sqrt{N}/N)100$.

and thus lower HETP values. It is generally found that the necessary number of theoretical plates for packed gas chromatographic columns is 10 times greater than in distillation for a similar separation.

Fritz and Scott (23) derived simple statistical expressions for calculating the mean and variance of chromatographic peaks that are still on a column (called *position peaks*) and these same peaks as they emerge from the column (called *exit peaks*). The classical plate theory is derived by use of simple concepts from probability theory and statistics. In this treatment, each sample chemical substance molecule is examined separately, whereas its movement through the column is described as a stochastic process. Equations are given for a discrete- and continuous-flow model. They are derived by calculating the mean and variance of a chromatographic peak as a function of the capacity factor k.

Using this statistical approach, Fritz and Scott studied two classical models falling into the category of plate theory: the discrete-flow model and the continuous-flow model. According to peak theory, the chromatographic column is considered to be divided into "plates" or "disjoint segments." In their discussion, those authors refer to these "disjoint segments" as *theoretical segments* (TS). Therefore, the sample molecules move from one TS to the next until they reach the last segment from which they elute from the column.

1. *Discrete-Flow Model.* This model requires several assumptions: (a) all the mobile phase moves from one segment to the next segment at the end of a discrete interval; and (b) the sample molecules are always in equilibrium with the mobile and stationary phases. On the basis of these assumptions, the equilibrium condition expresses the probability p that the molecule is in the mobile phase:

$$p = \frac{1}{1+k} \tag{2.16}$$

and the probability $(1-p)$ that the molecule is in the stationary phase:

$$1 - p = \frac{k}{1+k} \tag{2.17}$$

2. *Continuous-Flow Model.* The assumptions in this model are that (a) the mobile and stationary phases remain in equilibrium throughout the separation, (b) the mobile phase flows from one segment to the next segment at a constant rate, and (c) perfect mixing takes place in all segments.

Theoretical plate number N and effective theoretical plate number N_{eff} may then be calculated for both the discrete- and continuous-flow models:

Discrete-Flow Model

$$N = \frac{[E(T)]^2}{\text{var}(T)} = \frac{[r(1+k)]^2}{rk(1+k)} = r\left(\frac{1+k}{k}\right) \tag{2.18}$$

$$N_{\text{eff}} = \frac{[E(T) - t_0]^2}{\text{var}(T)} = \frac{(rk)^2}{rk(1+k)} = r\left(\frac{k}{1+k}\right) \tag{2.19}$$

Continuous-Flow Model

$$N = \frac{[E(T)]^2}{\text{var}(T)} = \frac{[r(1+k)]^2}{r(1+k)^2} = r \tag{2.20}$$

$$N_{\text{eff}} = \frac{[E(T) - t_0]^2}{\text{var}(T)} = \frac{(rk)^2}{r(1+k)^2} = r\left(\frac{k}{1+k}\right)^2 \tag{2.21}$$

where $E(T)$ = expected exit time of sample substance
$\text{var}(T)$ = variance of the time
r = number of theoretical segments in column
k = capacity factor
t_0 = exit time for nonsorbed substance

Therefore, only the plate numbers N for the continuous-flow model are independent of the capacity factor k (Equation 2.20).

A number of chromatographic systems from the literature were examined by Fritz and Scott. In all cases they demonstrated the applicability of the actual data to their system.

2.3.2 Rate Theory

Although HETP is a useful concept, it is an empirical factor. Since plate theory does not explain the mechanism that determines these factors, we must use a more sophisticated approach, the rate theory, to explain chromatographic behavior. Rate theory is based on such parameters as rate of mass transfer between stationary and mobile phases, diffusion rate of solute along the column, carrier-gas flowrate, and the hydrodynamics of the mobile phase.

Glueckauf (24) studied the effect of four factors on the chromatographic process: (1) diffusion in the mobile phase normal to the direction of flow, (2) longitudinal diffusion in the mobile phase, (3) diffusion into the particle, and (4) size of the particle.

The interpretation of the resulting chromatogram will indicate how well a separation has been performed. This interpretation can be viewed from two points: (1) how well the centers of the solute zones have been disengaged and (2) how compact the resulting zones are. Many chromatographic separations accomplish the first point but not the second, which results in the two zones spreading into each other.

We consider the three variables that cause zone spreading: ordinary diffusion, eddy diffusion, and local nonequilibrium. We approach this discussion from the random walk theory, since the progress of solute molecules through a column may be viewed as a random process.

First we define these three types of diffusion:

Ordinary Diffusion. This process results when there exists a region of high concentration and a region of low concentration. The migration is from the higher to the lower concentration region in the axial direction of the column. Diffusion occurs on the molecular level, resulting from movement of molecules after collision. Once the sample has been placed at the top of the column (in the minimum number of theoretical plates), these gradient regions exist.

Eddy Diffusion. Visualize a column packed with marbles of equal diameter. The void space along the column is essentially uniform (74% of column volume is occupied with the marbles and 26% is open or void volume). As the size of the marbles (particles) decreases, it becomes increasingly difficult to control uniformity in size and to prevent crushing or fractionation of the particle. This is especially true for column support materials that are easily fractionated if excessive vibrating or tapping is used in the packing procedure. With the particle size used in analytical gas chromatographic columns, 60/80 mesh (0.25–0.17 mm) for 0.25-in.-i.d. columns and 110/120 mesh (0.13–0.12 mm) for 0.125-in.-i.d. columns, it is very difficult to have all the particles of the same diameter, and some of these particles might fit into void spaces between particles. The overall effect is that the spaces along the column are not uniform. When a sample migrates down the column, therefore, each molecule "sees" different paths and each path is of a different length. Some molecules take the longer paths and others take the shorter paths. There are also variations in the velocities of the mobile phase within these pathways. The overall result is that some molecules lag behind the center of the zone, whereas others move ahead of the zone. Therefore, the eddy diffusion process results from flow along randomly spaced variable-size particles in the column.

Local Nonequilibrium. As the zone of solute molecules migrates through the column (approximating a Gaussian curve), there exists a variable concentration profile from the leading edge through the center to the trailing edge. As this zone continues to migrate down the column, it is constantly bringing an ever-changing concentration profile in contact with the next part of the column. This effect results in different rates of equilibration along the column. Thus each section (theoretical plate) in the column is constantly attempting to equilibrate with a variable concentration zone in the mobile phase. At one time the zone attempts to equilibrate with a low concentration in the mobile phase, and then at another time with a high concentration. If no flow were present, equilibration would proceed; however, we are in a dynamic system and there is always flow. These overall processes result in nonequilibrium at each theoretical plate. The overall process is determined by kinetic rate processes that account for transfer of the solute molecules between the two phases in the column; that is, the mass-transfer rate from mobile phase to stationary phase is different from the mass-transfer rate from stationary phase to mobile phase.

Viewing the zone migration as discussed previously, we can conclude that increasing the mobile phase velocity will increase the nonequilibrium effect, providing for more rapid exchange of solute molecules between the mobile and stationary phases and thus decreasing the nonequilibrium effect. Theory tells us that horizontal displacement (perpendicular to flow) is constant throughout the zone, proportional to velocity of flow, but inversely proportional to rate of restoring equilibrium. On the other hand, vertical displacement (parallel to flow) is proportional to the concentration gradient.

Since the three processes discussed earlier are all random diffusion processes, we can evaluate the zone broadening from the viewpoint of a random walk. If a process results from the random back and forth motion of solute molecules, we have a concentration profile that is Gaussian in shape. (That is, there is an equal number of molecules preceding the zone center as there are trailing the zone center.) The extent of spreading for normal Gaussian-distributed molecules is described by the standard deviation σ. This bandspreading σ is defined in random walk model by the number of steps taken n and the length of each step l:

$$\sigma = l n^{1/2} \tag{2.22}$$

Equation 2.22 states that zone spreading is proportional to step length but not to the number of steps. For instance, movement is random; it takes 16 steps to give a displacement four times the average length of each step.

We know from statistical treatments that standard deviations are not additive. However, variances, the square of the standard deviation, are additive. In terms of the chromatographic process, three diffusive process variables contribute to zone spreading. Thus we can sum these variables in terms of variances to give the overall contribution of zone spreading. The combined effect may be shown as

$$\sigma^2 = \Sigma\, \sigma_i^2 \tag{2.23}$$

where the $\Sigma\, \sigma_i$ term is a sum of each of the three processes: σ_{D} for ordinary diffusion, σ_{E} for eddy diffusion, and σ_{K} for nonequilibrium diffusion effects.

The ordinary diffusion process term is defined by the Einstein diffusion equation:

$$\sigma_{\mathrm{D}}^2 = 2Dt \tag{2.24}$$

where D is the coefficient of diffusion, and t represents the time molecules spend in the mobile phase from the start of the random process. The term t also can be expressed in terms of the distance the zone has moved L and the velocity of the mobile phase u; thus

$$t = \frac{L}{u} \tag{2.25}$$

and Equation 2.24 becomes

$$\sigma_{\mathrm{D}}^2 = \frac{2DL}{u} \tag{2.26}$$

The reader should keep in mind when developing a theory of zone spreading that we must have a point of reference to show how the spreading develops. This point of reference is the zone center.

The eddy diffusion term σ_{E} describes the change in pathway and velocity of solute molecules in reference to the zone center. If the molecules are in a "fast" channel, they can migrate ahead of the zone center; if in a "slow" channel, they can lag behind the zone center. To quantify the eddy diffusion term, we must describe the step length and the number of steps taken in a specified period of time. The void or channel volume between particles would be expected to be in the order of one particle diameter d_{p}. As molecules move from one channel to another, their velocity will be of the order of $+d_{\mathrm{p}}$ or $-d_{\mathrm{p}}$ (in respect to the zone center). So on average, the molecules will take an equivalent step of d_{p}.

We can determine the number of steps in terms of the total column length L and the equivalent length of the step. Therefore,

$$n = \frac{L}{d_{\mathrm{p}}} \tag{2.27}$$

On reflection it is apparent that channels cannot be regarded as either "fast" or "slow." Rather, there will be a range of velocities with some average value for the entire column length. Also, the column voids or channels will not be exactly equal to d_{p}, but will vary from larger than d_{p} to smaller than d_{p}, with an overall average of d_{p}. In light of the preceding description we can equate d_{p} for length of step l and L/d_{p} for number of steps. Substitution into Equation 2.22 gives

$$\sigma_{\mathrm{E}} = d_{\mathrm{p}}\left(\frac{L}{d_{\mathrm{p}}}\right)^{1/2} = (Ld_{\mathrm{p}})^{1/2} \tag{2.28}$$

This equation states that eddy diffusional effects on zone spreading increase with the square root of zone displacement and particle size.

Equations 2.24 and 2.28 account for the effect of ordinary and eddy diffusion in the zone broadening process. Now we need to express nonequilibrium effects that are concerned with the time the solute molecules spend in the two phases. Let us define a few more terms in order to set up some mathematical relationships:

k_1 = transition rate of the molecule from mobile phase to stationary phase

$1/k_1$ = average time required for one sorption to occur

k_2 = transition rate of molecules from stationary phase to mobile phase

$1/k_2$ = time required for one desorption to occur

A molecule in the mobile phase is moving faster than the center of the zone. The velocity of the zone is Ru, where R is the fraction of solute molecules in mobile phase and u is the mobile phase velocity. Therefore, $1 - R$ is the fraction of solute molecules in the stationary phase with a velocity of zero. Now, molecules move back and forth with respect to the zone center as each phase transfer occurs. In terms of random walk, n is the number of transfers our molecules take between the two phases. In terms of sorptions—desorptions, n is twice the number of desorptions (one desorption occurs for each sorption), and the time needed for the solute zone to move through the column (distance $= L$) at its velocity Ru is

$$t = \frac{L}{Ru} \tag{2.29}$$

During this time t, the molecules will spend the fraction R in mobile phase and the fraction $(1 - R)$ in the stationary phase. So the time that the fraction of molecules $(1 - R)$ spend in the stationary phase will be

$$t = \frac{(1 - R)L}{Ru} \tag{2.30}$$

The number of desorptions is the time spent by the molecules in the stationary phase (Equation 2.30) divided by $1/k_2$:

$$n_{\text{des}} = \frac{(1 - R)L/Ru}{1/k_2} = \frac{k_2(1 - R)L}{Ru} \tag{2.31}$$

Since there are twice as many phase transfers as there are desorption processes, the number of steps n is equal to two times Equation 2.31, or

$$n = \frac{2k_2(1 - R)L}{Ru} \tag{2.32}$$

To obtain a value for the distance a molecule moves back with respect to the zone center l, we need to consider $1/k_2$, the lifetime of a molecule in the stationary phase. The center of the zone moves forward $[Ru \times (1/k_2)]$ or (Ru/k_2) during the time the molecule is in the stationary phase; thus our step length also is Ru/k_2. By similar reasoning we arrive at the same value for the forward movement of molecules ahead of the zone center.

We now can describe an equation for the effect of nonequilibrium on zone spreading viewed as a random walk. Substituting Ru/k_2 for l and $2k_2(1-R)L/Ru$ for n in Equation 2.22, we have

$$\sigma_K = \frac{Ru}{k_2}\left[\frac{2k_2(1-R)L}{Ru}\right]^{1/2}$$

$$= \left[\frac{2R(1-R)Lu}{k_2}\right]^{1/2} \tag{2.33}$$

Equation 2.33 indicates that an increase in flow velocity causes an increase in nonequilibrium effects. Provision for rapid exchange of solute molecules between phases decreases these effects.

We may now return to Equation 2.23 and make the appropriate substitutions from 2.24, 2.28, and 2.33:

$$\sigma^2 = 2Dt + Ld_{\mathrm{p}} + \frac{2R(1-R)Lu}{k_2} \tag{2.34}$$

$$t = \frac{L}{u}\ (\text{Equation 2.25})\ ;\quad \therefore$$

$$\sigma^2 = \frac{2DL}{u} + Ld_{\mathrm{p}} + \frac{2R(1-R)Lu}{k_2} \tag{2.35}$$

$$\sigma^2 = L\left[\frac{2D}{u} + d_{\mathrm{p}} + \frac{2R(1-R)u}{k_2}\right] \tag{2.36}$$

Martin and Synge (2) introduced height equivalent to a theoretical plate H as a measure of zone spreading:

$$H = \frac{\sigma^2}{L} \tag{2.37}$$

So Equation 2.36 may be written as

$$H = \frac{2D}{u} + d_{\mathrm{p}} + \frac{2R(1-R)u}{k_2} \tag{2.38}$$

Rearrangement of Equation 2.38 yields

$$H = d_p + \frac{2D}{u} + \frac{2R(1-R)u}{k_2} \qquad (2.39)$$

To find the correct flow velocity u, which gives the minimum plate height H_{min}, we take the first derivative of Equation 2.38 and set dH/du equal to zero. This results in

$$u = \left[\frac{k_2 D}{R(1-R)}\right]^{1/2} \qquad (2.40)$$

The van Deemter equation (25) is used for describing the gas chromatographic process. This equation was evolved from the earlier work (26) and was later extended with Glueckauf's theory. The equation was derived from consideration of the resistance to mass transfer between the two phases as arising from diffusion:

$$H = \frac{2D_c}{u} + \left(\frac{8}{\pi^2}\right)\frac{k}{(1+k)^2}\left(\frac{d_f^2}{D_l}\right)u \qquad (2.41)$$

where D_c = overall longitudinal diffusivity of solute in gas phase
k = capacity factor
d_f = effective film thickness of liquid phase
D_l = diffusivity of solute in liquid phase
u = apparent linear flowrate of gas phase

The first term in Equation 2.41 is the contribution due to overall longitudinal diffusion, and the second term is contribution due to resistance to mass transfer in the liquid phase.

The overall longitudinal diffusivity D_c is the sum of apparent longitudinal diffusivity D_a and true molecular diffusivity D_g:

$$D_c = D_a + \gamma D_g \qquad (2.42)$$

The factor γ is used to account for irregular diffusion patterns and usually is less than unity because molecular diffusivity is smaller in packed columns than in open tubes.

Klinkenberg and Sjenitjer (27) showed statistically that

$$D_a = \lambda u d_p \qquad (2.43)$$

where λ is a dimensionless constant characteristic of packing. This equation is indicative of how poor or effective the packing homogeneity is in the column; for regular packings, $\lambda < 1$; for nonuniform packed columns with channels, $\lambda > 1$. The term d_p is the particle diameter in centimeters.

Uneven distribution of the stationary phase liquid on the solid support particles causes band dispersion. This may be rationalized if one considers that molecules entering a thin part of the liquid film permeate faster than in a thicker part of the liquid film. This effect causes some molecules to spend more time in the liquid phase than other molecules. This slow movement through the column results in spreading of the band.

Considering all the effects discussed above and combining Equations 2.41 and 2.43, we have the expression

$$H = 2\lambda d_{\mathrm{p}} + \frac{2\gamma D_{\mathrm{g}}}{u} + \frac{8}{\pi^2}\frac{k}{(1+k)^2}\left(\frac{d_{\mathrm{f}}^2}{D_l}\right)u \qquad (2.44)$$

This equation predicts that for maximum column performance, we must minimize the contribution of each term while maintaining a constant linear flowrate. The first term accounts for the geometry of the packing, the second for longitudinal diffusion in the gas phase, and the third for resistance to mass-transfer processes.

The general form for the van Deemter equation is

$$H = A + \frac{B}{u} + Cu \qquad (2.45)$$

where $A = 2\lambda d_{\mathrm{p}}$ = eddy diffusion term
$B = 2\gamma D_{\mathrm{g}}$ = longitudinal or ordinary diffusion term
$C = (8/\pi^2)[k/(1+k)^2](d_{\mathrm{f}}^2/D_l)$
= nonequilibrium or resistance to mass-transfer term

A representation of this equation is given in Figure 2.20, which shows the effect of H with changes in linear gas velocity. Equation 2.45 represents a hyperbola that has a minimum at velocity $u = (B/C)^{1/2}$ and a minimum H

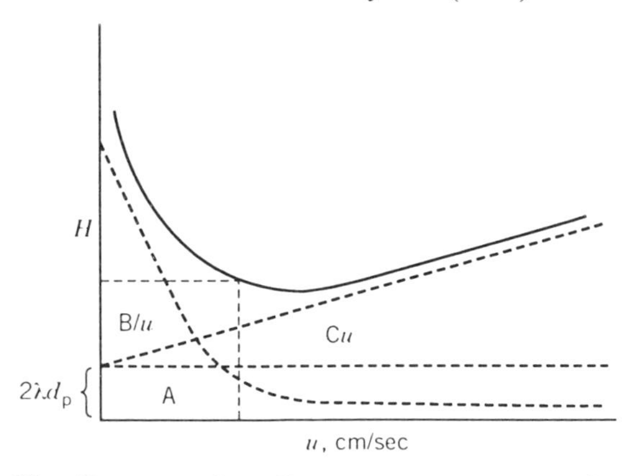

FIGURE 2.20 Van Deemter plot. Change in H versus linear gas velocity u: $H_{\mathrm{min}} = A + (2BC)^{1/2}$; $u_{\mathrm{opt}} = (BC)^{1/2}$.

value (H_{min}) at $A + 2(BC)^{1/2}$. The constants may be graphically calculated from an experimental plot of H versus linear gas velocity as shown in Figure 2.20.

It is also useful to plot H against $1/u$ (Figure 2.21). In both presentations (Figures 2.20, 2.21) the intercept of the linear portion of the H plot will equal $2\lambda d_p$. Thus, if the particle size d_p is known, λ can be calculated and a measure of packing regularity obtained. From the slope of the linear part of the H–u curve one also can estimate the film thickness d_f, if D_l and k are known (resistance to mass transfer in liquid-phase term).

The constants A, B, C can also be determined by the method of least squares. A gradual approximation of B may be calculated from a plot of H–Cu versus $1/u$, and C can be approximated from a plot of H–Bu versus u.

Let us take a better look at the effect of the terms of Equation 2.44 on plate height. The contribution of the $2\lambda d_p$ term can be decreased by reducing the particle size. As the particle size becomes smaller, however, the pressure drop through the column increases. The value of λ usually increases as d_p decreases. Of the three terms in Equation 2.44, only this first one is independent of linear flowrate.

The second term, $2\gamma D_g/u$, is a measure of the effect of molecular diffusion on zone spreading. This term may be decreased by reducing the molecular diffusivity D_g. We know from the kinetic theory of gases that the D_g value depends on the nature of the vapor and the temperature and the pressure of the system. Diffusion in low-molecular-weight gases (H_2 and He) is high compared to that in higher-molecular-weight gases (N_2 or CO_2). If this were the only criterion for choice of carrier gas, one would choose N_2 or CO_2 rather than He. This is evidenced by the fact that optimum gas velocity is governed by $(B/C)^{1/2}$. One obtains a value for $(B/C)^{1/2}$ by differentiating Equation 2.45 with respect to u and then setting $dH/du = 0$;

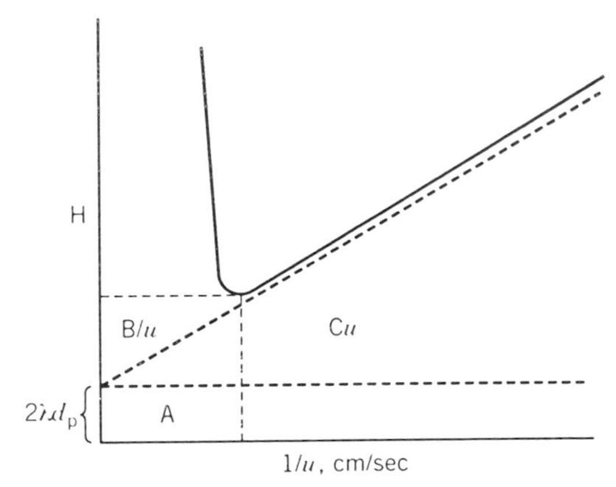

FIGURE 2.21 Rate theory equation plotted as H versus $1/u$: $H_{min} = A + (2BC)^{1/2}$; $u_{opt} = (BC)^{1/2}$.

$u_{opt}(H_{min})$ is then equal to $(B/C)^{1/2}$. However, other factors affect the choice of a carrier gas, such as the effect of the sensitivity of the detector employed. If in a particular system the C term is small and high flowrates are allowable (thus reducing term $2\gamma D_g/u$), the nature of the carrier is not too important. For columns of low permeability, a low-molecular-weight (less viscous) gas might be the best choice (e.g., He). If the C term is large and low flowrates are used, the term $2\gamma D_g/u$ becomes important and the carrier gas can exert influence on the HETP. In this case, high-molecular-weight gases (e.g., N_2 or CO_2) would be the preferred because solute diffusion coefficients would be small.

The third term of Equation 2.44 accounts for resistance to mass transfer in liquid phase. An obvious way of reducing this term is to reduce the liquid film thickness d_f. This causes a reduction in k and an increase in the term $k/(1+k)^2$. However, the use of thinly coated column packings increases the probability of adsorption of solute molecules on the surface of support material, which might result in tailing of peaks.

The k term is temperature dependent, so we increase k and decrease $k(1+k)^2$ by lowering of temperature. Lowering of temperature increases viscosity and thus decreases D_l. Therefore, the effects of the factors $k/(1+k)^2$ and $1/D_l$ counteract each other.

Thus it can be seen that the observed HETP is not only a function of column packing but depends on operating conditions and the properties of the solute. This is why different values of HETP (or different numbers of theoretical plates per unit column length) are obtained for various solutes.

Many modifications to the original van Deemter plate height equation have appeared in the literature (28–32). Some account for mass transfer in the gas phase (28, 29), and other modifications have been made for velocity distribution because of flow retardation of interfacial resistance (30, 31). Improvements were attempted, usually stochastic theories based on random walk theory (32). However, we elaborated on the work of Giddings (33), who described plate height contributions as a function of the diffusional character of zone broadening by accounting for local nonequilibrium.

Modifications of the van Deemter Equation. If one accounts for the fact that resistance to mass transfer can occur in the stationary phase as well as in the mobile phase, Equation 2.45 may be written as

$$H = A + \frac{B}{u} + C_l u + C_g u \tag{2.46}$$

The last term accounts for the resistance to mass transfer in the gas phase. Low-loaded liquid coatings cause the C_g term to be significant. Equation 2.46 was further extended to account for velocity distributions due to gas flow retardation in the layers C_1 and the interaction of the two types of gas resistance C_2:

$$H = A + \frac{B}{u} + C_l u + C_g u + C_1 u + C_2 u \tag{2.47}$$

The term $C_g u$ may be defined as

$$C_g u = c_a \frac{k^2}{(1+k)^2} \left(\frac{d_g^2}{D_g} \right) u \tag{2.48}$$

where c_a is a proportionality constant, d_g is the gas diffusional pathlength, and D_g is the diffusion coefficient of solute molecules in gas phase.

The C_1 term becomes significant with rapidly eluted but poorly sorbed components. The value of C_1 depends on the particle size of the packing:

$$C_1 u = \left(\frac{c_b d_p^2}{D_g} \right) u \tag{2.49}$$

where c_b is a proportionality factor approximately equal to unity. Giddings and Robinson (34) realized that the processes in the gas phase cannot be considered independent with respect to their effect on H. Thus they stated that the term A (flow characteristic) and the effect of resistance to mass transfer in the gas phase must be treated dependently. So Equation 2.46 becomes

$$H = \frac{1}{1/A + 1/C_1 u} + \frac{B}{u} + C_l u + C_g u + H_e \tag{2.50}$$

The term

$$\frac{1}{1/A + 1/C_1 u}$$

results from the merging of the eddy diffusion term and the velocity distribution term ($C_1 u$) of Equations 2.47 and 2.49. The term H_e is introduced to account for the characteristics of the equipment used in the system. The first term (Equation 2.50) is not simple (35, 36). Depending on the nature of the packing and the flow, five possible mechanisms can take place, so our term becomes a summation term:

$$H = \sum_{i=1}^{5} \frac{1}{(1/A) + 1/C_1 u} + \frac{B}{u} + C_l u + C_g u + H_e \tag{2.51}$$

The five possible mechanisms of band broadening occur because of flow (1) through channels between particles, (2) through particles, (3) resulting from uneven flow channels, (4) between inhomogeneous regions, and (5) throughout the entire column length.

All preceding discussion has assumed no compressibility of the gas stream. With columns where the pressure drop is large, the change in gas velocity should be considered. (Gas expansion also causes zone spreading.) DeFord et al. (37) demonstrated the importance of a pressure correction and after considering the A term to be negligible, developed the following equation:

$$H = \frac{B^0}{p_0 u_0} + (C_g^0 + C_1^0) p_0 u_0 f + C_l^0 u_0 j \tag{2.52}$$

where B^0, C_g^0, C_1^0, C_l^0 = coefficients determined by measuring H for various outlet pressures and outlet velocities

p^0 = outlet pressure

u_0 = outlet gas velocity

j = James–Martin pressure correction factor (compressibility factor)

$$= \frac{3}{2}\left[\frac{(p_i/p_0)^2 - 1}{(p_i/p_0)^3 - 1}\right]$$

f = pressure correction

$= p_i(p_0 + 1)j^2/2$

f = usually unity and can be neglected except in accurate theoretical work

Flow. The rate at which zones migrate down the column is dependent on equilibrium conditions and mobile phase velocity; on the other hand, how the zone broadens depends on flow conditions in the column, longitudinal diffusion, and the rate of mass transfer. Since various types of columns are used in GC—namely, open tubular columns, support coated open tubular columns, packed capillary columns, and analytical packed columns—we should study the conditions of flow in a gas chromatographic column. Our discussion of flow is restricted to Newtonian fluids, that is, those in which the viscosity remains constant at a given temperature.

Flow through an open tube is characterized by the dimensionless Reynolds number,

$$\mathrm{Re} = \frac{\rho d u}{\eta} = \frac{dG}{\mu} \tag{2.53}$$

where ρ = fluid density (g/mL)

d = tube diameter (cm)

u = fluid velocity (cm/sec)

η = fluid viscosity (poise)

G = mass velocity (g/cm^2 sec)

μ = absolute viscosity (g/cm sec)

Inertial forces of the fluid increase with density and the square of velocity (ρu^2), whereas viscous forces decreases with increasing diameter of tube ($\eta u/d$) and increase with viscosity and velocity. High Reynolds numbers ($\mathrm{Re} > 4000$) result in turbulent flow; with low Reynolds number ($\mathrm{Re} < 2000$), the flow is laminar. Laminar flow results from formation of layers of fluid with different velocities after a certain flow distance, as illustrated in Figure 2.22, segment A. Flow at the walls is zero and increases on approach to the center of the tubes. The laminar flow pattern results from layers of mobile phase with different velocities traveling parallel to each other. The maximum flow at the center is twice the average flow velocity of fluid. Molecules in the field can exchange between fluid layers by molecular diffusion. Most open tubular columns operate under laminar flow conditions.

Turbulent flow results because of the radical mixing of layers to equalize flowrates. The mixing of the layers is due to the increased eddies, and mass

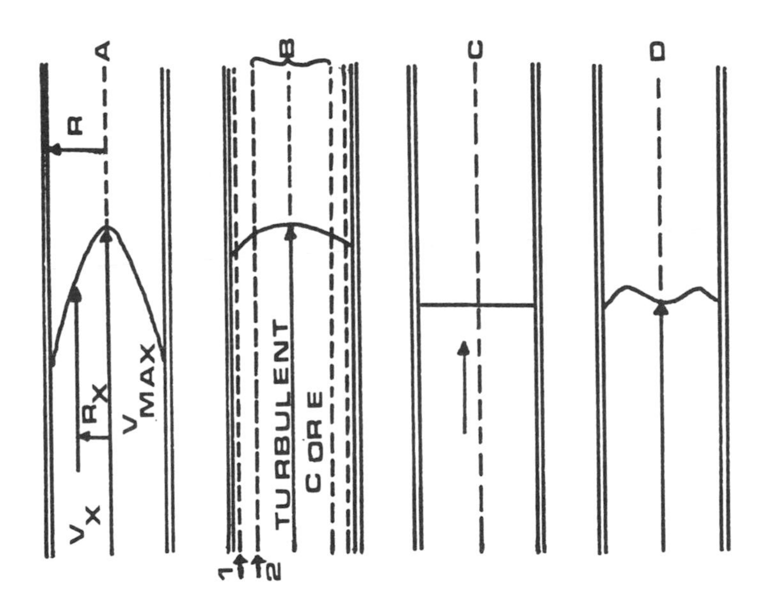

FIGURE 2.22 Flow profiles in tubes and packed columns. Segment A, laminar flow: r = tube radius, V_x = stream path velocity at radial position r_x, V_{max} = maximum flow velocity at tube center. Most open tubular columns operate with this profile. Segment B, turbulent flow: 1 = laminar sublayer, 2 = buffer layer. Segment C, plug flow. Segment D, flow in a packed column. Effect is more pronounced with smaller tube diameter:particle size ratios.

transfer occurs by eddy diffusivity. Turbulent diffusivity increases in proportion to mean flow velocity, as depicted in Figure 2.22, segment B. Figure 2.22, segment C represents plug flow, which is unattainable in practice but does suggest a model from which other flows may be considered. The flow usually attained in packed columns is illustrated in Figure 2.22, segment D.

A considerable difference exists between flow through an open column and a packed column, as illustrated in Figure 2.23. Darcy's law, which governs flow through packed columns, states that flow velocity is proportional to the pressure gradient:

$$u_0 = \frac{B^2}{\eta} \frac{p_o - p_i}{L} \tag{2.54}$$

True average fluid velocity may be expressed as

$$u = \frac{B^0}{\epsilon\eta} \frac{p_o - p_i}{L} \tag{2.55}$$

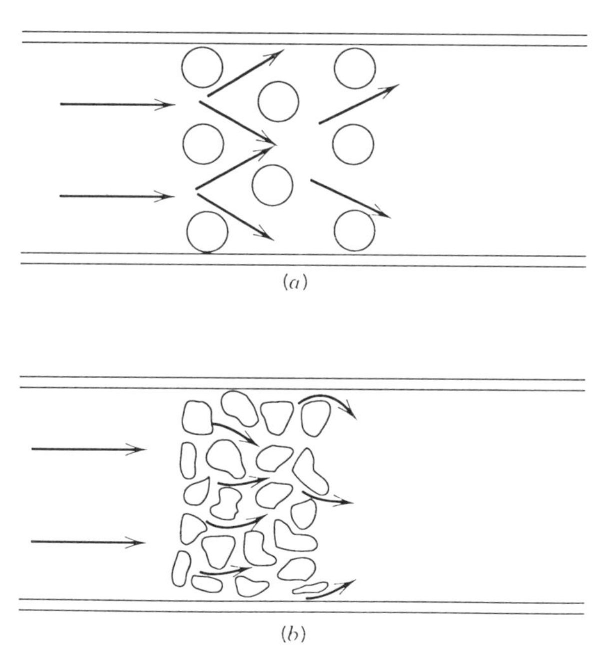

FIGURE 2.23 Representative flow through a packed column. (a) Simplified diagram of column with "uniform" particles. (b) Representative diagram of column with experimental particles. The degree of tortuosity of the path becomes dependent on the particle packing structure. Plug flow usually results.

and mean velocity of a fluid is represented by

$$u \text{ (mean velocity)} = \frac{(p_i - p_o)r^2}{8\eta L} \tag{2.56}$$

Combination of Equations 2.54 and 2.56 gives the specific permeability coefficient B^0:

$$B^0 = \frac{r^2}{8} \tag{2.57}$$

If we express the free cross section of the column bed by the interparticle porosity (knowing that the total porosity of packed beds with porous particles is larger because of intraparticle space), we can obtain the true average fluid velocity u:

$$u_0 = \frac{B^0(p_o - p_i)}{\epsilon \eta L} \tag{2.58}$$

where u_0 = superficial velocity (average velocity without packing)
B^0 = specific permeability coefficient (1 darcy = 10^{-8} cm^2)
ϵ = interparticle porosity (0.4 ± 0.03)
η = mobile phase viscosity
L = column length
p_o = outlet pressure
p_i = inlet pressure

By combining the cross-sectional area of the tube $r^2\pi$ and the mean velocity (Equation 2.56), therefore, we can come up with an expression for the volumetric flowrate F_c:

$$F_c = \frac{(p_i - p_o)r^4\pi}{8\eta L} \tag{2.59}$$

Flow in packed columns may be expressed in terms of modified Reynolds numbers (Re)m, which take into account a geometric factor for the diameter of the particle rather than the diameter of the column (see Equation 2.53):

$$(\text{Re})\text{m} = \frac{\rho u dp}{\eta} = \frac{dpG}{\mu} \tag{2.60}$$

For laminar flow (Re)m values are less than 10 and with turbulent flow (Re)m > 200. Packed gas chromatographic columns normally operate with a (Re)m of <10, so they may be considered to operate with laminar flow.

Assuming a column of diameter $d = 0.3$ cm and particles of diameter $d_p = 0.02$ cm, Table 2.2 shows the change in Reynolds number with flowrate. In gas chromatographic procedures the carrier-gas flow usually is

TABLE 2.2 Change in (Re)m With Mobile Phase Flowrate

Flowrate	(Re)m for H_2	(Re)m for N_2
25 mL/min	0.11	0.78
50 mL/min	0.22	1.56

measured after the column by a soap-bubble flowmeter. To obtain the average flowrate $\bar{F}_c$ in the column, one must account for three factors:

1. Compressibility correction j.
2. Correction for flow being measured at room temperature T_0, rather than column temperature T, that is, T/T_0.
3. Correction factor for vapor pressure of water p_w when using flowmeter $(p_0 - p_w)/p_0$, where p_0 is atmospheric pressure.

The initial flow F_0 into the column in terms of the measured flow F_c is

$$F_0 = F_c \left(\frac{T}{T_0}\right) \frac{p_0 - p_w}{p_0} \tag{2.61}$$

The average flowrate in the column then is determined by

$$\bar{F}_c = jF_c \left(\frac{T}{T_0}\right) \frac{p_o - p_w}{p_o} \tag{2.62}$$

This average flowrate term should be used to measure precise retention volumes.

2.3.3 The Solid Support

One should keep two things in mind when choosing a support: (1) structure and (2) surface characteristics. Structure contributes to the efficiency of the support, whereas the surface characteristics govern the support's participation in the resulting separations. The perfect column material would be chemically inert toward all types of sample. It would have a large surface area so that the liquid phase could be spread in a thin film and the structure of the surface would be such that it would properly retain the liquid film. However, large surface area is not a guarantee of an efficient column.

The most commonly used column support materials are made from diatomite. Other materials include sand, Teflon, inorganic salts, glass beads, porous-layer beads, porous polymers, and carbon blacks. We discuss the diatomite supports in some detail, and additional information may be obtained in Chapter 3.

Basically, two types of support are made from diatomite. One is pink and derived from firebrick, and the other is white and derived from filter aid.

German diatomite firebrick is known as *Sterchmal*. Diatomite itself is a diatomaceous earth, as is the German kieselguhr. Diatomite is composed of diatom skeletons or single-celled algae that have accumulated in very large beds in numerous parts of the world. The skeletons consist of a hydrated microamorphorous silica with some minor impurities (e.g., metallic oxides). The various species of diatoms number well over 10,000 from both freshwater and saltwater sources. Many levels of pore structure in the diatom cell wall cause these diatomites to have large surface areas ($20\ m^2/g$). The basic chemical differences between the pink and white diatomite may be summarized as follows:

1. The white diatomite or filter aid is prepared by mixing it with a small amount of flux (e.g., sodium carbonate), and calcining (burning) at temperatures greater than 900°C. This process converts the original light-gray diatomite to white diatomite. The change in color is believed to be the result of converting the iron oxide to a colorless sodium iron silicate.
2. The pink or brick diatomite has been crushed, blended, and pressed into bricks, which are calcined (burned) at temperatures greater than 900°C. During the process the mineral impurities form complex oxides and silicates. It is the oxide of iron that is credited for the pink color.

It has been well established that the surface of the diatomites are covered with silanol (Si–OH) and siloxane (Si–O–Si) groups. The pink diatomite is more adsorptive than the white; this difference is due to the greater surface area per unit volume rather than in any fundamental surface characteristic. The pink diatomite is slightly acidic (pH 6–7), whereas the white diatomite is slightly basic (pH 8–10). Both types of diatomites have two sites for adsorption: (1) van der Waals sites and (2) hydrogen-bonding sites. Hydrogen-bonding sites are more important, and there are two different types for hydrogen bonding: *silanol groups*, which act as a proton donor, and the *siloxane group*, where the group acts as a proton acceptor. Thus samples containing hydrogen bonds (e.g., water, alcohol, and amines) may show considerable tailing, whereas those compounds that hydrogen-bond to a lesser degree (e.g., ketones, esters) do not tail as much.

The elimination of adsorption sites (i.e., deactivation of surface) can be performed in several ways:

1. Removal by acid washing
2. Removal by reaction of silanol group
3. Saturation of surface with a liquid substrate phase
4. Coating with solid material

It is not entirely clear what is accomplished by acid treatment. It is generally believed that some species, perhaps iron, is removed from the support. Regarding reaction at silanol groups, it has been suggested (38) that during treatment with dimethyldichlorosilane (DMCS) there is a reaction between the surface hydroxy groups:

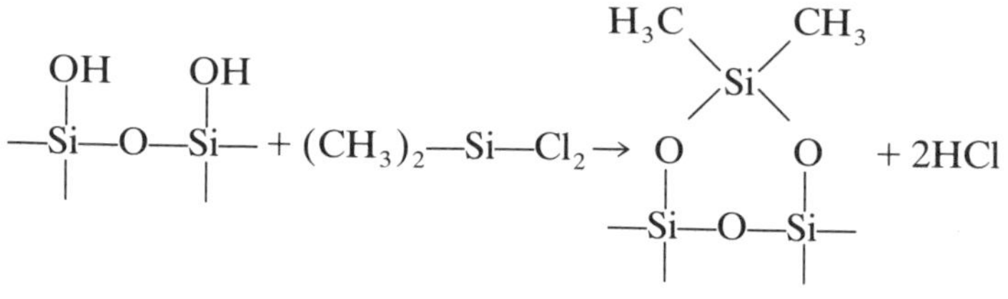

$$\text{—Si(OH)—O—Si(OH)—} + (CH_3)_2\text{—Si—}Cl_2 \rightarrow \text{—Si—O—Si—}\ (\text{both O bridged by } Si(CH_3)_2,\ H_3C\text{—Si—}CH_3) + 2HCl$$

Silyl ethers

If only one hydroxy group is available, the reaction may be

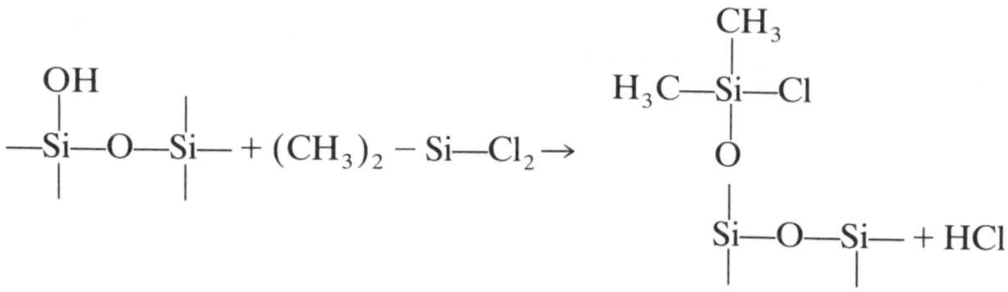

$$\text{—Si(OH)—O—Si—} + (CH_3)_2 - \text{Si—}Cl_2 \rightarrow H_3C\text{—Si(}CH_3\text{)(Cl)—O—Si—O—Si—} + HCl$$

Silyl ethers

In the case of hexamethyldisilazane (HMDS) the following reaction has been proposed:

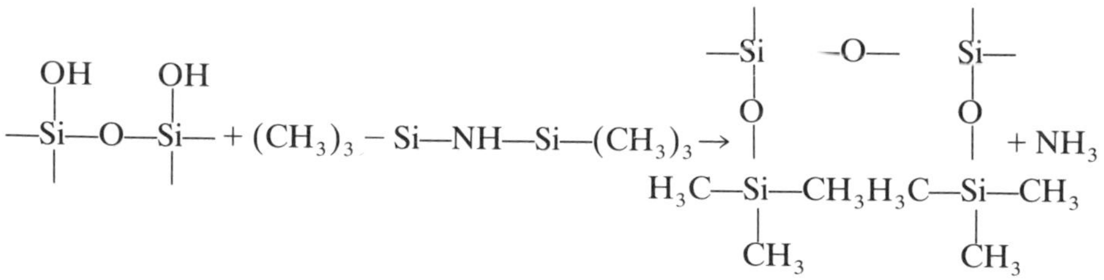

$$\text{—Si(OH)—O—Si(OH)—} + (CH_3)_3 - \text{Si—NH—Si—}(CH_3)_3 \rightarrow \text{—Si—O—Si—}\ (\text{each Si—O—Si}(CH_3)_3:\ H_3C\text{—Si(}CH_3\text{)—}CH_3\ \ H_3C\text{—Si(}CH_3\text{)—}CH_3) + NH_3$$

Silyl ethers

It should be noted that silanization reduces the surface area of the support. Thus one generally should not use more than 10% stationary phase loading.

Small particles should be used in gas chromatographic columns since the HETP is directly proportional to particle diameter. However, column permeability is proportional (and pressure drop inversely proportional) to the square of the particle diameter. If particles are too small, therefore, pressure requirements increase tremendously.

It has become the practice to refer to chromatographic supports in terms of the mesh range. For sieving of particles for chromatographic columns, both the Tyler Standard Screens and the U.S. Standard Series are frequently used. Tyler screens are identified by the actual number of meshes per linear inch. The U.S. sieves are identified by either micrometer (micron) designations or arbitrary numbers. Thus a material referred to as 60/80 mesh means particles that will pass through a 60-mesh screen but not an 80-mesh screen. You may also see this written as −60 + 80 mesh. Particle size is much better expressed in micrometers (microns); therefore, 60/80 mesh would correspond to 250–177 μm (μ) particle size range. Table 2.3 shows the conversion of column-packing particle sizes and also the relationship between mesh size, micrometers, millimeters, and inches. Table 2.4 shows the relationship between particle size and sieve size.

Lack of the proper amount of packing in a gas chromatographic column often is the source of a poor separation. How can one tell when a column is properly packed? The answer is twofold: by column performance (efficiency) and by peak symmetry. Many factors affect column performance, but one of the easiest ways to check is the amount of packing per foot of

TABLE 2.3 Conversion Table of Column Packing Particles

Mesh Size	Micrometers	Millimeters	Inches
4	4760	4.76	0.185
6	3360	3.36	0.131
8	2380	2.38	0.093
12	1680	1.68	0.065
16	1190	1.19	0.046
20	840	0.84	0.0328
30	590	0.59	0.0232
40	420	0.42	0.0164
50	297	0.29	0.0116
60	250	0.25	0.0097
70	210	0.21	0.0082
80	177	0.17	0.0069
100	149	0.14	0.0058
140	105	0.10	0.0041
200	74	0.07	0.0029
230	62	0.06	0.0024
270	53	0.05	0.0021
325	44	0.04	0.0017
400	37	0.03	0.0015
625	20	0.02	0.0008
1250	10	0.01	0.0004
2500	5	0.005	0.0002

TABLE 2.4 Relationship Between Particle Size and Screen Openings

Sieve Size	Top Screen Openings (μm)	Bottom Screen Openings (μm)	Micrometer Spread
10/30	2000	590	1410
30/60	590	250	340
35/80	500	177	323
45/60	350	250	100
60/80	250	177	73
80/100	177	149	28
100/120	149	125	24
120/140	125	105	20
100/140	147	105	42

column length (grams per foot). If the amount of packing varies greatly from its optimum value, poor separations can result. Knowledge of the number of grams per foot will allow one to predict column performance.

Loosely packed columns generally are inefficient. A column that is too tightly packed gives excessive pressures drop or may even become completely plugged because the support particles have been broken and fines are present. Applied Science Laboratories, Inc. has prepared a useful column-packing guide, which is reproduced in Table 2.5. A column packed within ±10% of the values shown in the table will provide satisfactory efficiency.

2.3.4 Mobile Phase

Sample components are transported through the column by means of a proper carrier gas. Use of a gas as the mobile phase enables rapid equilibrations between moving and stationary phases, resulting in the high performance of the gas chromatographic technique. Gaseous mobile phases have low flow resistance, allowing long columns with high separating power. Detection of the emerging gaseous sample is simple with the use of highly sensitive detectors (see Chapter 5).

The mobile phase is an important component of the overall gas chromatographic system. Constant and reproducible flow conditions should be maintained at all times. To achieve this, the mobile phase must have the proper auxiliary components.

1. *Gas Cylinder or Generator*. In most cases the supply of mobile phase is from commercial cylinders. These are connected to the carrier gas system of the chromatograph by means of a reducing valve. Hydrogen necessary for a flame-ionization detector may be supplied from a commercial tank or electrolytically generated. Purity of gases used has a profound effect on the results that one may expect from their system. Helium should be 99.995% pure, and hydrogen should be 99.95% pure.

TABLE 2.5 Column Packing Guide Information (g/ft)[a]

			Supports Used (grams packing/ft column length)				
Mesh Size of Support	Type of Metal Tubing	O.d. (inches) of Tubing	Gas Chrom S, A, P, Z, Q	Gas Chrom R, RA, RP, Rz Chromosorb P	Chromsorb W	Porous Chromsorb G	Polymer Beads[b]
45/60	SS	$\frac{1}{8}$	0.45[c]	0.55	0.5	—	—
		$\frac{3}{16}$	—	2.1	—	—	—
		$\frac{1}{4}$	—	4.0[d]	2.8[c]	4.6	—
60/80	SS	$\frac{1}{8}$	0.4	0.6	0.4	0.75	0.5
		$\frac{1}{4}$	2.4	3.5	2.8	4.6	2.5
80/100	SS	$\frac{1}{8}$	0.45	0.6	0.45	0.8	0.5
		$\frac{1}{4}$	2.7	3.7	2.9	4.7	2.6
100/120	SS	$\frac{1}{8}$	0.5	0.6	0.5	0.8	0.5
		$\frac{1}{4}$	2.7	3.7	2.9	4.7	2.8[c]
45/60	Al	$\frac{1}{8}$	—	0.4[d]	0.3[d]	—	—
60/80	Al	$\frac{1}{8}$	0.25[c]	0.4	0.25[c]	—	—
		$\frac{1}{4}$	1.4	2.8[c]	—	3.2	—
		$\frac{3}{8}$	5.1	6.8[d]	—	10.3	—
80/100	Al	$\frac{1}{8}$	0.3	0.4	0.3[c]	—	—
		$\frac{1}{4}$	1.7	—	—	—	—
		$\frac{3}{8}$	5.3[c]	—	—	10.3	—
100/120	Al	$\frac{1}{8}$	0.3	0.45[b]	—	—	—
		$\frac{1}{4}$	1.7[c]	—	—	—	1.6[d]
		$\frac{3}{8}$	5.3[c]	—	—	—	5.0
45/60	Cu	$\frac{1}{4}$	—	3.4	—	—	—
60/80	Cu	$\frac{1}{8}$	0.3	0.35[c]	—	—	—
		$\frac{1}{4}$	1.6	3.2[d]	—	—	—
80/100	Cu	$\frac{1}{8}$	0.3	0.45[c]	0.3	—	—

[a] Reproduced from Applied Science Laboratories, Inc., *GAS-CHROM Newsletter*, Jan./Feb. 1970, *Technical Bulletin No. 7*, p. 4.

[b] Porous Polymer Beads must be packed tightly; figures are thought to be minimum acceptable value.

[c] Estimated.

[d] Limited data.

2. *Purifier*. Commercially prepared tank gases are usually technical grade, which can contain oil vapors, oxygen, and water. The impurities present in these gases reduce column activity and life as well as interfere with the proper functioning of the detector(s). Oxygen (~5 ppm) may be removed by passing carrier gas through a catalyst (copper or nickel) at 100–105°C. For removal of CO and hydrocarbon (1–100 ppm) impurities, the same catalysts may be operated in excess of 600°C. Palladium is very efficient for removing oxygen from hydrogen streams. Carbon dioxide may

be removed by passing the gas through solid soda–lime or soda–asbestos. Oil vapors and other heavier contaminants are best removed with activated charcoal. Molecular sieves serve well for removing water. Purification of air for the flame ionization detector is accomplished by combustion over quartz at 800°C.

3. *Pressure and/or Flow Controls.* Gas valve regulators supplied by most gas suppliers are adequate for controlling pressure and flow. Maintenance of constant flowrate is accomplished with a large pressure outside of column (40–60 psi).

4. *Pressure-Measuring Device.* Accurate pressure measurements are made with a manometer if needed. In most instruments the carrier gas, hydrogen and air (for flame ionization detector), and argon–methane (for electron capture detector) are provided with separate gauges.

5. *Flowmeter.* A variety of devices are available for monitoring of flow of carrier gas, such as differential capillary, thermal conductivity, ionization, rotameters, and calibrated soap-film tubes. Measurement of the flow may be either continuous or intermittent, and the flowmeter may be placed either in front of the column or at the carrier-gas outlet. The soap-film type is most commonly used (at exit of column) because of its economy and ease of operation.

6. *Preheater.* By means of an optional piece of equipment, the carrier gas is heated before it enters the sample injector. This prevents condensation of high boiling components and subsequent blocking of outlets. If a thermal conductivity detector is used, another advantage with the use of a preheater is that it ensures that identical thermal conditions are in effect with the reference and measuring sides of the cell.

Any gas may be used as the carrier as long as it does not react with the sample or the stationary phase. However, other properties must be consid ered, depending on the type of detector employed. With a thermal conductivity detector, a gas with high heat conductivity is used because thermal conductivity of a gas is inversely proportional to the square root of the molecular weight. Thus very low molecular weight gases are optimum. Helium is generally preferred, since hydrogen is reactive and inflammable. Argon is used with β-radiation ionization detectors and at times with the gas density balance detector.

As was shown in Section 2.3.2, the efficiency of a column is not only a function of linear flow velocity, but also the type of carrier gas. Helium generally is the best compromise as a carrier gas. However, Lloyd et al. (29) have shown that nitrogen permits the highest column efficiency, but not the greatest separation. Nitrogen is the choice for highly loaded columns, whereas hydrogen is best for lower loaded columns. Optimum flowrate and retention time are both dependent on the cross-sectional area of the column;

therefore, a narrower bore column gives a more efficient separation in the same time.

2.3.5 Stationary Phase

One reason for the wide acceptance of GLC is that there exists such a variety of liquid phases with different properties. Because of this large number of liquid phases a great amount of work has been done to clarify the interaction between the liquid phase and the solute molecules. There have been attempts to find some theoretical basis for choosing a liquid phase to accomplish a particular separation, and lately there has been an effort to decrease the number of liquid phases used. We now wish to discuss in general terms the role of the liquid phase and describe some of the criteria needed to discuss its role in a chromatographic separation.

In choosing a liquid phase some fundamental criteria must be considered:

1. Is the liquid phase selective toward the components to be separated?
2. Will there be any irreversible reactions between the liquid phase and the components of the mixture to be separated?
3. Does the liquid phase have a low vapor pressure at the operating temperature? Is it thermally stable?

Let us consider some of the information available to answer these questions, although we are not now attempting to develop a pattern for selection of the proper liquid substrate; this is discussed to some extent in Chapter 3 and to a greater extent in Chapter 4. The vapor pressure of the liquid phase should be less than 0.1 Torr at the operating temperature of the column. This value can change depending on the detector used (see Chapter 5), since bleed from the liquid phase will cause noise and elevate background signal and thus decrease sensitivity. Information from a plot of vapor pressure versus temperature is not always completely informative because adsorption of the liquid phase on the solid support results in decrease in the actual vapor pressure of the liquid phase. Other than its effect on the detector noise, liquid-phase bleed may interfere with analytical results and determine the life of the column. Also, some supports may have a catalytic effect to decompose the liquid phase, thus reducing its life in the column. Contaminants in the carrier gas (e.g., O_2) also may interfere with the stability of a liquid phase.

Two other properties of the liquid phase to be considered are viscosity and wetting ability. Ideally, liquid phases should have low viscosity and high wetting ability (ability to form a uniform film on the solid support or column wall).

It is uninformative to refer to a liquid phase as being selective, since all liquid phases are selective to varying degrees. "Selectivity" refers to the relative retention of two components and gives no information regarding the mechanism of separation. Most separations depend on boiling point difference, variations in molecular weights of the components, and the structure of the components.

The relative volatility or separation factor α depends on the interactions of the solute and the liquid phase, that is, van der Waals cohesive forces. These cohesive forces may be divided into three types:

1. *London Dispersion Forces*. These are due to the attraction of dipoles that arise from the arrangement of the elementary charges. Dispersion forces act between all molecular types and especially in the separation of nonpolar substances (e.g., saturated hydrocarbons).
2. *Debye Induction Forces*. These forces result from interaction between permanent and induced dipoles.
3. *Keesom Orientation Forces*. These forces result from the interaction of two permanent dipoles, of which the hydrogen bond is the most important. Hydrogen bonds are stronger than dispersion or inductive forces.

If the two components have the same vapor pressure, separation can be achieved on the basis of several properties. These properties are (in the order of their ease of separation) (1) difference in the functional groups, (2) isomers with polar functional groups, and (3) isomers with no functional groups.

Polarity is another property that has been used to tabulate liquid phases. By "polarity" we mean the electrical field effect in the immediate vicinity of the molecule, which depends on the number, nature, and arrangement of the atoms and on the type of bond and the groups. Rohrschneider (39) introduced a polarity scale p^*, which ranks solvents according to their polarity:

$$p^* = a\left(\log \frac{{}_2V_{\mathrm{R}.p}}{{}_1V_{\mathrm{R}.p}} - \log \frac{{}_2V_{\mathrm{R}.u}}{{}_1V_{\mathrm{R}.u}}\right) \tag{2.63}$$

where a = constant
subscripts 1 and 2 = butane and butadiene, respectively
p = a phase with polarity p^*
u = a nonpolar phase, squalane (standard).

On the scale, $p^* = 100$ for β,β'-oxydipropionitrile (polar liquid) and $p^* = 0$ for squalane (nonpolar liquid).

All other liquids fall between these two limits. This scale has two very good features: (1) it permits a rapid selection of a liquid by minimizing the number of different liquid phases since many have equivalent p^* values, and (2) it allows us to pick solvents on a general basis of polarity when we have many components to separate. Details of this polarity scale are given in Chapter 3.

2.3.6 Open Tubular (Capillary) Gas Chromatography

The use of open tubular columns (OTC) or capillary columns was initially suggested by Martin (3). However, it wasn't until several years later than Golay (40) published the theoretical and practical results for their use. Over the years a number of books have been written about these unique columns; perhaps one of the better and up-to-date is that by Freeman (41).

The inside diameters (i.d.) of these columns range from 0.2 to 1.0 mm. The wall-coated open tubular (WCOT) column has the smallest inside diameter (0.2–0.35 mm). These are referred to as *narrow-bore WCOT columns*. The wide-bore WCOT columns vary from 0.5 to 0.75 mm i.d. If the inside surface of the capillary column has been extended by means of macro-elongated crystal deposits, they are referred to as *porous-layer open tabular* (PLOT) columns. Support-coated open tabular (SCOT) columns result from the coating of the column wall with a mixture of a finely divided solid support and a liquid phase. The later two types of capillary column are usually 0.50–0.75 mm i.d.

Equation 2.45 takes a different form for capillary columns:

$$H = \frac{B}{u} + C_{G}u \tag{2.64}$$

The A term is nonexistent because there is only one flow path and no packing material. The resistance-to-mass-transfer term C has the greatest effect on band broadening, and its effect in capillary columns is controlled by the mass transfer in the gas phase C_G.

Mobile phase velocity (cm/sec) is dependent on column length, carrier gas viscosity, pressure drop through the column, and the column permeability. Column permeability is expressed by the specific permeability coefficient B_0:

$$B_0 = \frac{2\eta\epsilon L p_0 u_0}{p_i^2 - p_o^2} = \text{cm}^2 \tag{2.65}$$

where p_i = inlet pressure (dynes/cm^2)
p_o = outlet pressure (dynes/cm^2)
η = viscosity of carrier gas (dyne-sec/cm^2)
L = length of column (cm),
$u_0 = F_c L/V_c = \bar{u}a/j$
V_c = tube volume (cm^3)
$\bar{u}$ = average linear velocity of air peak
a = packing porosity
ϵ = interparticle porosity

If a is known, B_0 can be calculated from the retention time of the air peak.

The permeability of open tubular columns can be calculated by use of the Hagen–Poiseuille equation:

$$B_0 = \frac{d^2}{32} \tag{2.66}$$

where d is the tube diameter. Permeability of open tubular columns is 10–100 times greater than that of packed columns. Because of this, open tubular columns can be longer than packed columns by the same factors.

More information about capillary columns may be found in Chapters 3, 4.

2.3.7 Preparative Gas Chromatography

Gas chromatography can be utilized for preparative-scale separations as well as for analytical-type separations. For the qualitative and quantitative analyses of a small sample (see Chapter 7), microliter or microgram sample sizes are used. After formulating the conditions for such an analytical separation, one may wish to isolate larger amounts of one or more components of a mixture. Use of GC for this purpose is referred to as *preparative gas chromatography*.

Preparation of milligram quantities of substances can readily be performed with an analytical gas chromatographic column system by repetitive injection and collection. Larger sample quantities require modifications to an analytical apparatus but are more easily obtained with the use of a special preparative unit. It has been postulated that the sample size approximately increases with the fourth power of the column diameter ($\leq$1 g).

Samples larger than 1 g necessitate gas chromatographic equipment different from that for analytical systems. Preparative GC involves the consideration of several possibilities.

1. *Automatic Sample Injection and Fraction Collection*. The manual injection of repetitive analytical size samples is one of the simplest and most obvious approaches to preparative GC. This may be improved by automatically injecting the samples and collecting fractions with a mechanical system.

2. *Rotating Columns and Moving Bed Equipment*. Preparative GC can be carried out where the sample is continuously introduced into a set of columns that may move in a transverse direction to the mobile phase flow and sample injection point. In such an arrangement, the elution point on the cylindrical base x_i is

$$x_i = 2\pi rat_{Ri} \tag{2.67}$$

where i = component and t_{Ri} = retention time of component i
a = number of revolutions of cylinder per unit time
r = cylinder radius

Maximum efficiency is realized when the nth component is eluted adjacent to the first component. The frequency of revolutions must be

$$a = \frac{1}{t_{Rn} - t_{Ri}} \quad (2.68)$$

where t_{Rn} represents the retention time of the nth component. Presumably, this system would have the efficiency of a single-column arrangement. However, optimum efficiencies are not realized because of the column-to-column variations and the tendency to use excess fed volumes.

3. *Large-Diameter Columns (5–20 cm).* This has been one of the primary approaches to preparative GC, with the assumption that sample size can be increased in proportion to cross-sectional area of the column. Actually this is not strictly the case. The basic problem with large-diameter columns is that lateral mixing becomes inefficient with increased column diameter (42). There are several ways to overcome this deficiency: (a) placing rings in the column at various intervals (e.g., 10 cm for a 100-cm column)—these act as baffles, thus causing the mobile phase at the wall to be forced back to the column center (43)—and (b) constructing the preparative columns from two concentric tubes with the column packing between the two walls. This eliminates the center of the column packing and is reputed to provide a more uniform flow pattern (44).

4. *Small-Diameter Columns in Parallel (15–20 mm).* This is an alternative to increased column diameter. Such a system maintains the inherent efficient manifold system for dispersing the sample uniformly into a matched cross-sectional area. The disadvantage of this system is the development of an inefficient manifold system for disbursing the sample uniformly into matched columns to obtain identical retention characteristics (45).

5. *Long, Small-Diameter Columns (8–12 mm).* This setup does permit useful separations; however, separation efficiency deteriorates quickly since the column is overloaded with sample. The justification for such a system is that preparative-size sample capacity is proportional to the amount of solid support, mesh size, and column dimensions (46).

6. *Sample Recirculation.* This technique involves the recirculation of the solute mixture between two columns until a satisfactory separation is obtained. One is able to provide a large number of theoretical plates from two short columns; however, this system is restricted to two neighboring components. If the mixture contains other components with longer retention times, they complicate the separation unless removed from the system (47).

In preparative GC one is primarily concerned not with resolution or theoretical plates, but with the high separation efficiency and high speed of this technique.

Partitioning is much faster in gas-phase systems than liquid–liquid or liquid–solid systems. This results from the high surface area relative to the liquid volume, which provides shorter equilibrations than one could achieve in, say, preparative distillation. Distillations also are limited in terms of high-quality separations of two-component systems and the large sample volume required for efficient columns.

In preparative GC, long columns can be more of a disadvantage than an advantage. Longer length gives increased retention times and more time for unwanted diffusion effects to become pronounced. A practical maximum for column length in preparative GC is 20 ft. If good resolution cannot be obtained on an analytical column with proper adjustment of flow, temperature, and sample size, increasing column length generally will not improve the situation.

Verzele (48) studied variables in preparative GC, including column length, particle size, liquid substrate loading, and sample size. Studies showed that column length can be increased if mesh size of particles is decreased (larger particles). This permits reasonable flows of carrier gas with modest inlet pressures.

Essential Components of a Preparative Gas Chromatographic System. The most important components for the system are (1) the sample inlet system, (2) the column, and (3) the collection system. It is just as important in preparative GC as in analytical GC to introduce the sample as a "plug" to maintain the resolution. As noted previously, the longer the component spends in the column, the more it diffuses, with resulting peak broadening. Unfortunately, it is not easy to inject a "plug" of sample in preparative GC. Injection of a large liquid sample results in a large amount of vapor and relatively large pressures in the injection port. As a result there exists a large backpressure against the syringe, causing a longer time for sample injection and possibly a safety hazard. One of the best ways to circumvent this problem is to inject sample as a liquid (low temperature) and then begin to temperature program. Another trick is to inject liquid at a rate so as to create vapor at the same rate as the carrier gas flowrate. Purnell (49) has calculated that if the sample bandwidth is less than 25% of the average peak bandwidth, the effects of injection have a minimal influence on the final peak (band) width.

We now focus on the various component parts of a preparative gas chromatographic column: the column tube, the solid support, the liquid substrate, the carrier gas, and the column length and diameter. In preparative GC it is particularly important that solvent efficiency α be maximized as in analytical GC. A liquid substrate should be used that will give the maximum separation of peaks (high selectivity). The higher the term, the fewer theoretical plates N are required and the larger samples that can be injected.

The number of required theoretical plates may be calculated by Equation 2.86. As the value of k increases beyond $k = 5$, the second term of Equation 2.86 becomes less effective and can be ignored. Thus at $k > 5$ the number of theoretical plates is especially dependent on the α term. Generally, the maximum sample size increases in proportion to column diameter (d^4), and if column diameter remains constant, the volume of sample that can be injected is in proportion to column length. There is also a relationship between maximum sample size and liquid loading. However, this effect is significant only when small sample sizes are used (46). Horvath (50) states a range of 10–300 mm as the optimum inside diameter and a liquid loading greater than 20% w/w for a preparative column.

Bayer et al. (43) studied the efficiency of preparative columns and showed that when using the van Deemter equation (Section 2.3.2), an additional term had to be introduced to account for the nonuniform flow distribution. This can be represented by

$$H_{\mathrm{p}} = \frac{Ed^{0.58}}{u^{1.806}} \tag{2.69}$$

where H_{p} = preparative term to be added to van Deemter equation
E = correlation factor
d = column diameter
u = linear gas velocity

Increased column diameter may cause a reduction in the homogeneity of the packing, resulting in flowrate distortion of the component zones across the column cross section and higher H (HETP) values. Diminution of column efficiency also can be related to (1) the nonuniformity of liquid coating, which, in turn, affects mass transfer, (2) the radial temperature gradient that exists within the large-diameter column, and (3) variation in the carrier gas between the wall and the center of column (this is especially true in large coiled columns). These variables generally increase in effect as the column diameter increases.

One of the greatest problems with packing preparative columns results from the radial separation of particles; that is, large particles tend toward the column walls, whereas fines remain more to the center. Since the use of larger-diameter columns tends to introduce more problems, the chromatographer is advised to purchase prepacked columns (see Chapters 3 and 4).

Detectors. Most of the work in preparative GC has utilized either thermal conductivity detectors or flame ionization detectors. With thermal conductivity detectors, one normally passes the entire sample through the detector.

However, bypass designs are necessary in work with the more sensitive thermal conductivity detectors now on the market. A bypass arrangement must be used with flame ionization detectors, allowing about 0.1–1.0% of the column effluent to be burned for signal purposes. Detectors that have slow response time and a wide linear range are recommended in preparative GC. The reader is referred to Chapter 5 for details concerning detectors.

2.3.8 Gas–Solid Chromatography

All previous discussion regarding the theory of GC has been concerned with gas–liquid chromatography. By and large the majority of gas chromatographic investigations have dealt with gas–liquid chromatography rather than gas–solid chromatography. In recent years, however, more attention has been given to gas–solid chromatography, a unique and versatile technique. Following are some of the reasons for this interest:

1. Higher temperatures are possible (500°C).
2. Solid surfaces that are stable and do not undergo chemical reactions (e.g., oxidation) are available.
3. High column efficiencies are possible as there is no liquid phase contribution to band spreading.
4. There is great specificity of surfaces for solute molecule configurations.
5. Liquid substrate bleed effects are eliminated.
6. This technique is applicable in studying the physicochemical measurements possible at solid surfaces (e.g., isotherms and heats and entropies of adsorptions); see Chapter 9.
7. Catalysts, kinetics of catalytic processes, and catalyst reactions in general can be evaluated.

There are some modifications of the definitions and terms used in gas–liquid chromatography compared to those used in gas–solid chromatography. First, the phase ratio is described as

$$\beta = \frac{V_{\mathrm{G}}}{V_{\mathrm{A}}} \tag{2.70}$$

where V_{A} is the true adsorbent volume (weight of adsorbent/density $= V_{\mathrm{A}}$).

In the rate theory of gas–solid chromatography, the equation for H has essentially the same terms except that C_k replaces C_l; C_k is a term characteristic of adsorption kinetics. Equation 2.51 then becomes

$$H = \sum_{i=1}^{5} \frac{1}{1/A + 1/C_l u} + \frac{B}{u} + C_k u + C_{\mathrm{g}} + H_{\mathrm{e}} \tag{2.71}$$

Theoretical considerations indicate that, on homogeneous surfaces, C_k is smaller than C_l, and this implies highly efficient applications of gas–solid chromatography.

One considerable obstacle still existing in gas–solid chromatography is the lack of adequate adsorbent structure descriptions, as well as the distribution and dimensions of the pores. Another is the lack of reproducibility of adsorbents, not only among manufacturers (different products, presumably the same), but within the same manufacturer (different lots).

Adsorbent Properties. In the use of a solid adsorbent for gas–solid chromatography, its properties must be considered from several points of view. First, the specific surface area (m^2/g) is important. The greater the surface area, the higher probability of some sorption process occurring. Also, the more active sites per unit area, the more reactive will be the sorbate molecules with the sorbent surface. Second, the chemical composition of the surface layer and its crystal structure is of interest. Information of this type allows one to speculate more correctly regarding wanted or unwanted reactions that may take place at the gas–solid interface. Finally, a knowledge of the porous structure is helpful in identifying molecules that may selectively be trapped or sorbed on the surface. In spite of the fact that much work has been done to determine the surface area of solids, primary information regarding adsorption phenomena comes from the analysis of adsorption isotherms.

Adsorption of Gases at Solid Surface. Regardless of the process at the gas–solid surface, physical adsorption of the gas or vapor on the solid surface is part of the mechanism. The distance between sorbed molecules is shorter than the distances found between molecules of a real gas, but these distances are larger than those encountered in chemical interactions. The interaction energy between molecules and the surface of a sorbent may be estimated by initially assuming that the molecules are spherical in shape and located in a field of infinite sorbent and then considering the temperature to be sufficient to reduce the effect of molecular interactions to a negligible value. Thus, by introducing N gas molecules at a temperature T and a pressure P into a container holding a sorbent with a uniform surface, one can calculate the "apparent volume" V_a:

$$V_a = \frac{NkT}{P} \tag{2.72}$$

If the "void" volume V_m represents the sorbent uniform surface, we can designate V_0 as the limiting value of V_a as P goes to zero; thus

$$\frac{1}{V_a} = \frac{1}{V_0} = \frac{PV_aC}{kTV_0^3} \tag{2.73}$$

and

$$V_0 = V_m + \int V_m \exp\left(\frac{-E}{kT}\right) dV \tag{2.74}$$

where E is the potential energy of gas molecules in sorbent field and $1/V_0$ is the intercept of the linear plot of $1/V_a$ versus PV_a. The value of E may then be calculated from the integration of Equation 2.74.

The carrier gas may have a significant effect on the separation process. Adsorbents with high specific surfaces adsorb the carrier gas, thus decreasing some of the active sites (adsorption centers). This results in a change of component adsorption, which is in proportion to the adsorption capacity of the carrier gas. A change in carrier gas from hydrogen or helium to nitrogen may produce sharper peaks because of the higher adsorption capacity for nitrogen (51).

Sorbates (solute molecules) may be grouped according to their intermolecular interactions. These groupings are based on electronic configurations, electron density, and functional groups in the molecule.

1. *Group A Molecules*. These molecules have spherically symmetrical electron shells. Examples are noble gases and the saturated hydrocarbons, which have only sigma bonds between the carbon atoms. Molecules of this type will interact nonspecifically, through dispersion forces resulting from concordant electronic motion in the interacting molecules.

2. *Group B Molecules*. These molecules have a concentrated electron density (negative charge) (e.g., unsaturated and aromatic hydrocarbons) and or π electron bonds (e.g., N_2, H_2O, ROH, ROR, RCOR, NH_3, NHR_2, NR_3, RSH, RCN).

3. *Group C Molecules*. These include molecules with locally concentrated positive charges within small radius linkages, but these should not be adjacent groups with concentrated electron densities (e.g., –OH or =NH groups). Organometallic compounds exemplify this group. This type of compound interacts specifically with Group B molecules but nonspecifically with Group A molecules.

4. *Group D Molecules*. These molecules are with adjacent bonds, one with positive charge and one with electron density on the periphery of the other. Molecules with –OH and =NH functional groups, for example, H_2O, ROH, and 1° and 2° amines, constitute this group. Group D molecules may interact specifically with Group B and C molecules and with each other; however, they interact nonspecifically with Group A molecules.

Adsorbents are classified as either specific or nonspecific. The specificity results from the type of molecules or functional groups attached to the absorbent surface. Three classifications result:

1. *Nonspecific Adsorbents*. There are no functional groups or exchange ions on the surface of these adsorbents. Adsorbents of this are carbon black, boron nitride, and polymeric saturated hydrocarbons (e.g., polyethylene).

2. *Specific Adsorbents With Positive Surface Charges*. These adsorbents have acidic hydroxyl groups (hydroxylated acid oxides such as silica), aprotic acid centers, or small-radius cations (zeolites) on the surface. Adsorbents of this type will interact with molecules that have locally concentrated electron densities, that is, Group B and Group D molecules.

3. *Specific Adsorbents With Electron Densities on Surface*. Graphitized carbon blacks with dense monolayers of Group B molecules or macromolecules on surface are found in this classification. Adsorbents with a functional group, for example, cyano, nitrile, or carbonyl, would also be included in this category.

The adsorbents may also be classified according to their structure. These classifications are summarized in Table 2.6.

In gas–solid chromatography, retention of sorbate components is determined by (1) the chemical nature and geometric pore structure of the sorbent, (2) molecular weight of sorbate molecules and their geometric and electronic structures, and (3) temperature of the column.

Column separating power depends on selectivity of the sorbent and diffuseness or spreading of chromatographic bands moving through the sorbent. Thus any chromatographic column will be most effective when bands are less diffuse. Band spreading is caused by thermodynamic, kinetic, injection, and diffusion effects. These may be summarized as follows:

1. Nonsymmetrical band spreading may be attributed to nonlinearity of the equilibrium adsorption isotherm (i.e., deviation of isotherm from

TABLE 2.6 Adsorbents, Classified by Type

Type	Description and Classification
I—Nonporous	Nonporous mono- and polycrystalline sorbents (e.g., graphitized carbon black, NaCl) Porous amorphous sorbents (e.g., Aerosil and thermal blacks)
II—Uniform wide pores	Large-pore glasses, wide-pore Xerogels, and compressed powders made from nonporous particles (>100 Å in size and specific surface areas <300 m^2/g)
III—Uniform fine pores	Amorphous fine-pore Xerogels, fine-pore glasses, many activated charcoals, and porous crystals (type A and X zeolites)
IV—Nonuniform pores	Chalklike silica gels obtained from hydrolysis of salts from strong acids in a silicate solution

Henry's law). This causes the sorbate to move through the column at different rates (rate dependent on sorbate concentration).

2. Peak diffuseness can be attributed to the various processes occurring during the transport of the sorbate through the column. The diffusion processes are very complex; this complexity is due to (a) ordinary diffusion in the gas phase; (b) band movement through particle layers of different size and shape and packed in various ways, causing diffuseness related to a nonuniform distribution of gas flowrates over each cross-sectional area; and (c) difference in local flow velocities from the average velocity through the column. Columns can exhibit what is referred to as the *wall effect*, which means that flow at the walls is higher than the average of the column because resistance at the wall is less. The great effectiveness of capillary columns is due mainly to the absence of specific diffusion processes caused by particles. However, one does observe diffuseness in capillary or unpacked columns resulting from the parabolic velocity distribution over the column cross section. The gas velocity is higher at the center and lower near the walls than the average velocity of the band (see Figure 2.22).

3. Peak diffuseness can be a result of the kinetics of the sorption–desorption process (i.e., slow mass transfer or exchange at sorbent surfaces). Peak diffusion in this case is usually nonsymmetrical because the rates of sorption and desorption are not the same. The band spreading due to the final rate of mass exchange is closely related to the diffusion phenomena. Physical adsorption, for all practical purposes, is instantaneous. The overall process of sorption, however, consists of several parts: (1) movement of sorbate molecules toward sorbent surface, resulting from integrain diffusion (outer diffusion), (b) movement of sorbate molecules to inside of pores (i.e., internal diffusion of the sorbate molecules in the pores and surface diffusion in the pores), and (c) the sorption process in general.

4. Apparent diffusion may be influenced by the time it takes for sample injection.

Following are some of the more common adsorbents used for gas–solid chromatography:

1. *Carbon*. These are nonspecific-type adsorbents because of the lack of functional groups, ions, or unsaturated bonds. Most interactions are due to dispersion forces.

2. *Metal Oxides*. These include (a) silica gel, which is a specific type of adsorbent because there are free hydroxyl groups on surface—polar molecules are easily separated, and wide-pore silicas with homogeneous surfaces are used for analytical gas–solid chromatography; and (b) alumina, which is used more in liquid–solid than gas–solid chromatography. In gas–solid chromatography most peaks are symmetrical in shape. It is sometimes coated for use in gas–liquid chromatography.

3. *Zeolites*. Zeolites are useful in gas–solid chromatography because of their porous structure and good adsorption properties. They are a specific type of adsorbent, with cavities allowing sieving action for low molecular weight molecules able to enter "holes" (windows).

4. *Inorganic Adsorbents*. These have two general classifications: (a) inorganic salts (e.g., alkali metal nitrates and halides (52), alkaline earth halides (53), vanadium, manganese, and cobalt chlorides (54), and barium salts (55) and (b) inorganic salts, coated on surfaces of silica, alumina, carbon, and so on.

5. *Organic Adsorbents*. These include (a) organic crystalline compounds (e.g., benzophenone on firebrick, anthraquinone on graphitized carbon black, phthalic anhydride, and/or phthalic acid isomers on Chromosorb G), (b) liquid phases below their melting point (e.g., SE-30 on Chromosorbs and Carbowax 20M on Chromosorbs), (c) organic clay derivatives (e.g., Bentone 34), and (d) porous polymers (e.g., Porapaks, Chromosorb 101 and 102, Polysorb, and Synachrom).

Thermodynamics of Processes at Gas–Solid Interface. The processes taking place at the gas–solid interface may be interpreted by the use of adsorption thermodynamics. A term much used in chromatography is the *retention volume*, that is, the volume of carrier gas needed to transport the sample molecules through the column and into the detector system. In gas–solid chromatography we refer to the retention volume in the same way, but the sample molecule is more specifically referred to as the *sorbate*. Nomenclature between gas–solid and gas–liquid chromatography does not necessarily agree. In gas–liquid chromatography, retention volume is represented by V_R (meaning the total retention volume or the absolute retention volume). The term V_R has been used in gas–solid chromatography to represent corrected retention volume (56), whereas corrected retention volume in gas–liquid chromatography is given the symbol V_R^0. Specific retention volume in gas–solid chromatography is V_s^T, but in gas–liquid chromatography it is V_g.

The determination of the adsorbent parameter V_g at three temperatures permits the calculation of the free energy, enthalpy, and entropy of adsorption. A plot of log V_g versus $1/T$ has a slope of $-H_{ads}/2.3R$. ΔG_{ads} is obtained by

$$-\Delta G_{ads} = RT \ln K = RT \ln V_g \tag{2.75}$$

The value of ΔS_{ads} is then obtained with proper substitution in Equation 2.76:

$$\log V_g = \frac{-\Delta H_{ads}}{2.3RT} + \frac{\Delta S_{ads}}{2.3R} \tag{2.76}$$

These calculations may be performed manually or by use of an appropriate computer program. The reader may obtain more details in Ref. 57 and/or Chapter 9.

2.3.9 Evaluation of Column Operation

Several parameters can be used to evaluate the operation of a column and to obtain information about a specific system. If we plot the concentration of solute (in percent) against volume of mobile phase or number of plate volumes for the tenth, twentieth, and fiftieth plate in the column, we will obtain a plot as shown in Figure 2.24. Improved separation of component peaks is possible for columns that have a larger plate number. Similar information is obtained if we plot concentration of solute (percent of total) versus plate number. Figure 2.25 shows the band positions after 50, 100, and 200 equilibrations with the mobile phase.

A good gas chromatographic column is considered to have high separation power, high speed of operation, and high capacity. One of these factors can be improved usually at the expense of another. Thus a number of column parameters must be discussed to enable us to arrive at an efficient

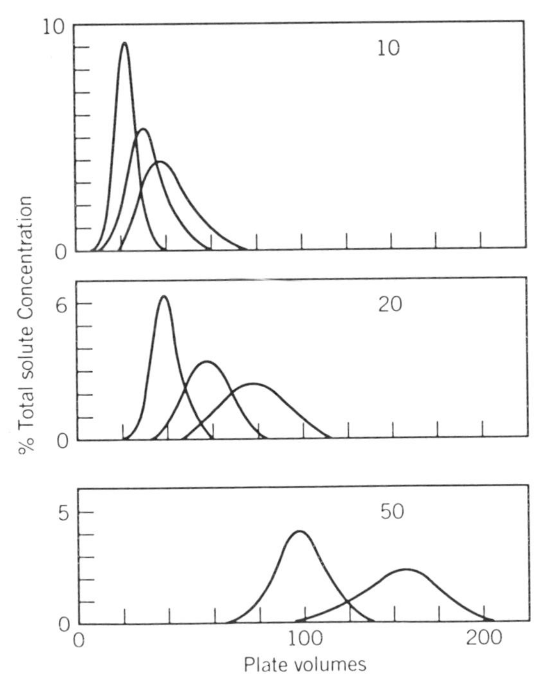

FIGURE 2.24 Elution peaks for three solutes from various plate columns. *Top*: 10 plates; *middle*: 20 plates; *bottom*: 50 plates.

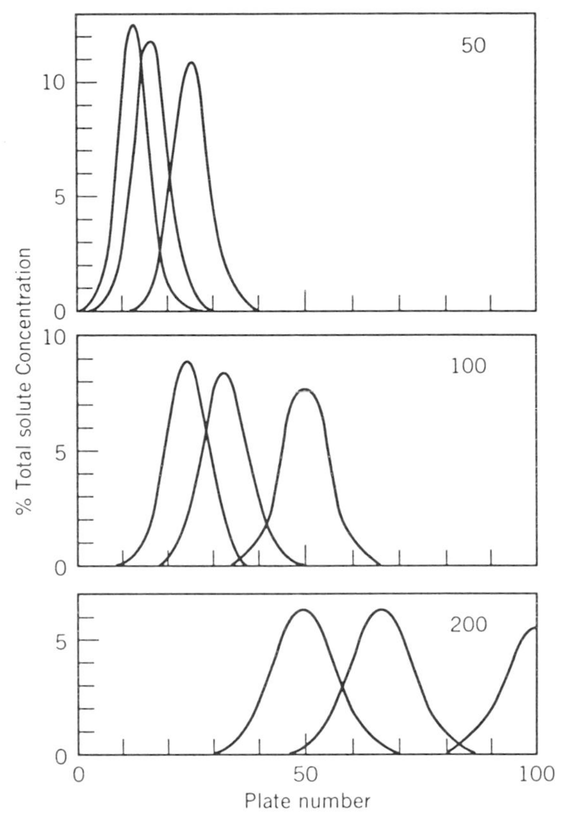

FIGURE 2.25 Plate position of components after variable number of equilibrations. *Top*: 50 equilibrations; *middle*: 100 equilibrations; *bottom*: 200 equilibrations.

operation of a column. We now consider several of these parameters and illustrate with appropriate relationships.

Column Efficiency. Two methods are available for expressing the efficiency of a column in terms of HETP: measurement of the peak (Figure 2.18) width at (1) the baseline (Equation 2.9) and (2) half-height (Equation 2.11).

In determining N, we assume that the detector signal changes linearly with concentration. If it does not, N cannot measure column efficiency precisely. If Equations 2.9 or 2.11 are used to evaluate peaks that are not symmetrical, positive deviations of 10–20% may result. Since N depends on column operating conditions, these should be stated when efficiency is determined.

There are several ways by which one may calculate column efficiency other than the two equations shown (Equations 2.9 and 2.11). Figure 2.26

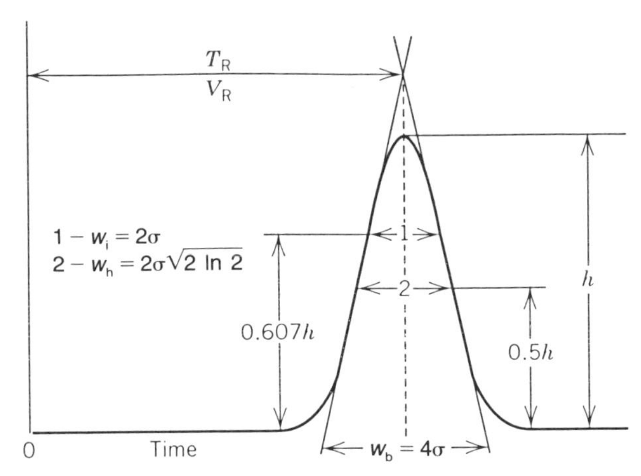

FIGURE 2.26 Characteristic data of the peak for calculation of column efficiency.

and Table 2.7 illustrate other ways in which the information may be obtained.

Effective Number of Plates. Desty et al. (58) introduced the term "effective number of plates" N_{eff} to characterize open tubular columns. In this relationship adjusted retention volume V'_R, in lieu of total retention volume V_R, is used to determine the number of plates. This equation is identical to Purnell's separation factor discussed below (Equation 2.91):

$$N_{eff} = 16\left(\frac{V'_R}{w_b}\right)^2 = 16\left(\frac{t'_R}{w_b}\right)^2 \tag{2.77}$$

This N_{eff} value is useful for comparing a packed and an open tubular column or two similar columns when both are used for the same separation. Open tubular columns generally have a larger number of theoretical plates. One can translate regular number of plates N to effective number of plates N_{eff}

TABLE 2.7 Calculation of Column Efficiency From Chromatograms

Standard Deviation Terms	Measurements	Plate Number $N =$
$A/h(2\pi)^{1/2}$	t_R and band area A and height h	$2\pi(t_R h/A)^2$
$W_i/2$	t_R and width at inflection points $(0.607h)w_i$	$4(t_R/w_i)^2$
$w_h/(8 \ln 2)^{1/2}$	t_R and width at half-height w_h	$5.55(t_R/w_h)^2$
$w_b/4$	t_R and baseline width w_b	$16(t_R/w_h)^2$

by the expression

$$N_{\text{eff}} = N\left(\frac{k}{1+k}\right)^2 \tag{2.78}$$

as well as the plate height to the effective plate height:

$$H_{\text{eff}} = H\left(\frac{1+k}{k}\right)^2 \tag{2.79}$$

Similarly, the number of theoretical plates per unit time can be calculated:

$$\frac{N}{t_R} = \frac{\bar{u}(k)^2}{t_R(1+k)^2} \tag{2.80}$$

where $\bar{u}$ is the average linear gas velocity. This relationship accounts for characteristic column parameters, thus offering a way to compare different-type columns.

Resolution. The separation of two components as the peaks appear on the chromatogram (see Figure 2.18) is characterized by

$$R_s = \frac{2\Delta t'_R}{w_{b1} + w_{b2}} \tag{2.81}$$

where $\Delta t'_R = t'_{R_2} - t'_{R_1}$. If the peak widths are equal, that is, $w_{b1} = w_{b2}$, Equation 2.81 may be rewritten

$$R_s = \frac{\Delta t'_R}{w_b} \tag{2.82}$$

The two peaks will touch at the baseline when $\Delta t'_R$ is equal to $4\,\sigma$:

$$t'_{R_2} - t'_{R_1} = \Delta t'_R \tag{2.83}$$

If two peaks are separated by a distance 4σ, therefore, $R_s = 1$. If the peaks are separated by a 6σ, then $R_s = 1.5$.

Resolution also may be expressed in terms of Kovat's retention index (see Chapter 3):

$$R_s = \frac{I_2 - I_1}{w_h f} \tag{2.84}$$

where f is the correction factor (1.699) because $4\sigma = w_b = 1.699 w_h$.

A more useable expression for resolution is

$$R_s = \frac{1}{4}(N)^{1/2}\left(\frac{\alpha - 1}{\alpha}\right)\left(\frac{k}{1+k}\right) \tag{2.85}$$

where N and k refer to the later eluted compound of the pair. Since α and k are constant for a given column (under isothermal conditions), resolution will be dependent on the number of theoretical plates N. The k term generally increases with a temperature decrease as does α, but to a lesser extent. The result is that at low temperatures one finds that fewer theoretical plates or a shorter column are required for the same separation.

Required Plate Number. If the capacity factor k and the separation factor α are known, the required number of plates (n_{ne}) can be calculated for the separation of two components. (The k value refers to the more readily sorbed component.) Thus

$$n_{ne} = 16R_s^2\left(\frac{\alpha}{\alpha - 1}\right)^2\left(\frac{1+k}{k}\right)^2 \tag{2.86}$$

The R_s value is set at the 6σ level or 1.5. In terms of the required effective number of plates, Equation 2.86 would be

$$N_{eff} = 16R_s^2\left(\frac{\alpha}{\alpha - 1}\right)^2 \tag{2.87}$$

Taking into account the phase β ratio, we can write Equation 2.86 as

$$n_{ne} = 16R_s^2\left(\frac{\alpha}{\alpha - 1}\right)^2\left(\frac{\beta}{k_2} + 1\right)^2 \tag{2.88}$$

Equations 2.86 and 2.88 illustrate that the required number of plates will depend on the partition characteristics of the column and the relative volatility of the two components, that is, on K and β. Table 2.8 gives the values of the last term of Equation 2.86 for various values of k. These data suggest a few interesting conclusions: if $k < 5$, the plate numbers are controlled mainly by column parameters; if $k > 5$, the plate numbers are controlled by relative volatility of components. The data also illustrate that k values greater than 20 cause the theoretical number of plates N and effective number of plates N_{eff} to be the same order of magnitude; that is,

$$N \simeq N_{eff} \tag{2.89}$$

TABLE 2.8 Values for Last Term of Equation 2.86

k:	0.25	0.5	1.0	5.0	10	20	50	100
$(1 + k/k)^2$:	25	9	4	1.44	1.21	1.11	1.04	1.02

The relationship in Equation 2.86 also can be used to determine the length of column necessary for a separation L_{ne}. We know that $N = L/H$; thus

$$L_{ne} = 16R_s^2 H \left(\frac{\alpha}{\alpha - 1}\right)^2 \left(\frac{1+k}{k}\right)^2 \tag{2.90}$$

Unfortunately, Equation 2.90 is of little practical importance because the H value for the more readily sorbed component must be known but is not readily available from independent data.

Let us give some examples from the use of Equation 2.86. Table 2.9 gives the number of theoretical plates for various values of α and k, assuming R_s to be at 6σ (1.5). Using data in Table 2.9 and Equation 2.86, we can make an approximate comparison between packed and open tubular columns. As a first approximation, β values of packed columns are 5–30 and for open tubular columns, 100–1000—thus a ten- to hundredfold difference in k. Examination of the data in Table 2.9 shows that when $\alpha = 1.05$ and $k = 5.0$ we would need 22,861 plates in a packed column, which would correspond to an open tubular column with $k = 0.5$ having 142,884 plates. Although a greater number of plates is predicted for the open tubular column, this is relatively easy to attain because longer columns of this type have high permeability and smaller pressure drop than the packed columns.

Separation Factor. The reader will recall that the separation factor α in Section 1.2.1 is the same as the relative volatility term used in distillation theory. In 1959 Purnell (59, 60) introduced another separation factor term (S) to describe the efficiency of a column. It can be used very conveniently to describe efficiency of open tubular columns:

$$S = 16\left(\frac{V'_R}{w_b}\right)^2 = 16\left(\frac{t'_R}{w_b}\right)^2 \tag{2.91}$$

where V'_R and t'_R = adjusted retention volume and adjusted retention time, respectively. Equation 2.91 may be written as a thermodynamic quantity

TABLE 2.9 Number of Theoretical Plates for Values of α and k (R_s at $6\sigma = 1.5$)

k	α: 1.05	1.10	1.50	2.00	3.00
0.1	1,920,996	527,076	39,204	17,424	9,801
0.2	571,536	156,816	11,664	5,184	2,916
0.5	142,884	39,204	2,916	1,296	729
1.0	63,504	17,424	1,296	576	324
2.0	35,519	9,801	729	324	182
5.0	22,861	6,273	467	207	117
8.0	20,004	5,489	408	181	102
10.0	19,210	5,271	392	173	97

that is characteristic of the separation but independent of the column. In this form we assume resolution R_s at the 6σ level or having a value of 1.5. Therefore, from Equation 2.87,

$$S = 36\left(\frac{\alpha}{\alpha - 1}\right)^2 \tag{2.92}$$

Separation Number. As an extension of the term separation factor S discussed above, we also can calculate a separation number SN as another way of describing column efficiency (61). By "separation number" we mean the number of possible peaks that appear between two n-paraffin peaks with consecutive carbon numbers. It may be calculated by

$$\mathrm{SN} = \left[\frac{t_{\mathrm{R}_2} - t_{\mathrm{R}_1}}{(w_{\mathrm{h}})_1 + (w_{\mathrm{h}})_2}\right] - 1 \tag{2.93}$$

This equation may be used to characterize capillary columns or for application of programmed pressure or temperature conditions for packed columns.

Analysis Time. If possible, we like to perform the chromatographic separation in a minimum time. Time is important in analysis but it is particularly important in process chromatography. Analysis time is based on the solute component that is more readily sorbed. Using the equation for determination of retention time, we obtain

$$t = \frac{L(1 + k)}{\bar{u}} = \frac{NH}{\bar{u}}(1 + k) \tag{2.94}$$

and substituting the value for the required number of plates, n_{ne} for n (Equation 2.86), we arrive at an equation for the minimum analysis time t_{ne}:

$$t_{\mathrm{ne}} = 16R_s^2\frac{H}{\bar{u}}\left(\frac{\alpha}{\alpha - 1}\right)^2\frac{(1 + k)^3}{(k)^2} \tag{2.95}$$

the term $H/\bar{u}$ can be expressed in terms of the modified van Deemter equation (Section 2.3.2, Equation 2.45).

$$\frac{H}{\bar{u}} = \frac{A}{\bar{u}} + \frac{B}{\bar{u}^2} + C_l + C_{\mathrm{g}} \tag{2.96}$$

For minimum analysis time, high linear gas velocities are used; thus the first two terms on right side of Equation 2.96 may be neglected. Therefore,

$$\frac{H}{\bar{u}} = C_l + C_{\mathrm{g}} \tag{2.97}$$

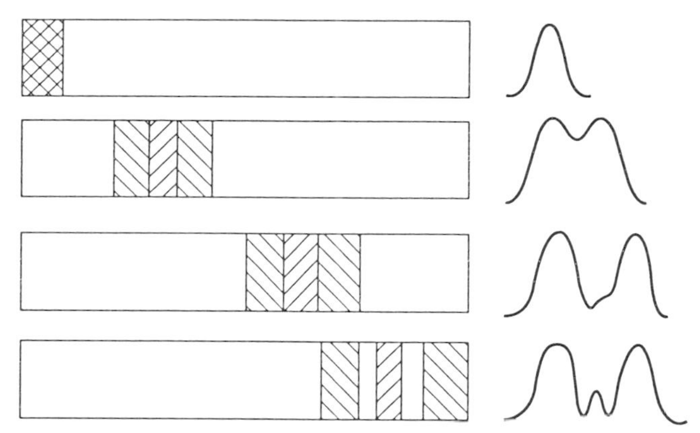

FIGURE 2.27 Idealized separation process in a gas chromatograph with three components, two major and one minor.

Substituting Equation 2.86 and 2.97 into Equation 2.95 we obtain

$$t_{ne} = n_{ne}(C_l + C_g)(1 + k) \tag{2.98}$$

This equation indicates that minimal separation time depends on plate numbers, capacity factor, and resistance to mass transfer. It should be pointed out that the analysis times calculated from Equation 2.98 also depend on the desired resolution. Our example calculations were made on the basis of resolution $R_s = 1.5$. For a resolution of 1.00, even shorter analysis times can be achieved.

Figure 2.27 gives a representation of an idealized separation of component zones and the corresponding chromatographic peaks for a three-component system. With columns of increasing number of plates, we see better resolution as column efficiency increases.

REFERENCES

1. S. Brunauer, *The Adsorption of Gases and Vapors*, Vol. 1, Princeton University Press, Princeton, NJ, 1945.
2. A. J. P. Martin and R. L. M. Synge. *Biochem. J.* (London), **35**, 1358 (1941).
3. A. T. James and A. J. P. Martin, *Biochem. J.* (London), **50**, 679 (1952).
4. R. L. Grob, M. A. Kaiser, and M. J. O'Brien, *Am. Lab.*, **7**(6), 13–25 (1975).
5. R. L. Grob, M. A. Kaiser, and M. J. O'Brien, *Am. Lab.*, **7**(8), 33–41 (1975).
6. L. S. Ettre and A. Zlatkis, *The Practice of Gas Chromatography*, Interscience, New York, 1967, p. 474.
7. T. P. Wampler and E. J. Levy, *J. Anal. Appl. Pyrolysis*, **12**, 75 (1987).
8. P. W. Rulon, *Organic Microanalysis by Gas Chromatography*, Thesis, Villanova University, May 1976.

9. J. F. Sullivan, R. L. Grob, and P. W. Rulon, *J. Chromatogr.*, **261**, 265–272 (1983).
10. J. F. Sullivan and R. L. Grob, *J. Chromatogr.*, **268**, 219–227 (1983).
11. R. L. Grob, "Analysis of the Environment," in *Contemporary Topics in Analytical and Clinical Chemistry*, Vol. 1, D. M. Hercules, M. A. Evenson, G. M. Hieftje, and L. R. Snyder (Eds.), Plenum, New York, 1977.
12. R. L. Grob and M. A. Kaiser, *Environmental Problem Solving Using Gas and Liquid Chromatography*, Elsevier, Amsterdam, 1982.
13. J. A. Giannovario, R. J. Gondek, and R. L. Grob, *J. Chromatogr.*, **89**, 1 (1974).
14. J. A. Giannovario, R. L. Grob, and P. W. Rulon, *J. Chromatogr.*, **121**, 285 (1976).
15. D. M. Rosie and R. L. Grob, *Anal. Chem.*, **29**, 1263 (1957).
16. R. L. Grob (Ed.), *Chromatographic Analysis of the Environment*, Dekker, New York, 1st ed., 1975; 2d ed., 1983.
17. L. C. Craig, *J. Biol. Chem.*, **155**, 519 (1944).
18. B. Williamson and L. C. Craig, *J. Biol. Chem.*, **168**, 687 (1947).
19. L. C. Craig and O. Post, *Anal. Chem.*, **21**, 500 (1949).
20. L. C. Craig, *Anal. Chem.*, **22**, 1346 (1950).
21. W. A. Peters, *Ind. Eng. Chem.*, **14**, 476 (1922).
22. A. J. P. Martin and R. L. M. Synge, *Biochem. J.*, **35**, 1359 (1941).
23. J. S. Fritz and D. M. Scott, *J. Chromatogr.*, **271**, 193–212 (1983).
24. E. Glueckauf, *Disc. Faraday Soc.*, No. 7, 199–213 (1949).
25. J. J. van Deemter, F. J. Zuiderweg, and A. Klinkenberg, *Chem. Eng. Sci.*, **5**, 271 (1956).
26. L. Lapidus and N. R. Amundson, *J. Phys. Chem.*, **56**, 984 (1952).
27. A. Klinkenberg and F. Sjenitzer, *Chem. Eng. Sci.*, **5**, 258 (1956).
28. E. Glueckauf, M. J. E. Goley, and J. H. Purnell, *Ann. N. Y. Acad. Sci.*, **72**, 612 (1956).
29. R. J. Loyd, B. O. Ayers, and F. W. Karasek, *Anal. Chem.*, **32**, 689 (1960).
30. W. L. Jones, *Anal. Chem.*, **33**, 829 (1961).
31. R. Kieselbach, *Anal. Chem.*, **33**, 806 (1961).
32. J. H. Beynon, S. Clough, D. A. Crooks, and G. R. Lester, *Trans. Farad. Soc.*, **54**, 705 (1958).
33. J. C. Giddings, *Dynamics of Chromatography*, Part I, Vol. 1, Chromatographic Science Series, Dekker, New York, 1965.
34. J. C. Giddings and R. A. Robinson, *Anal. Chem.*, **34**, 885 (1962).
35. J. C. Giddings, *Anal. Chem.*, **35**, 2215 (1963).
36. J. C. Giddings, in *Gas Chromatography, 1964*, A. Goldup (Ed.), Butterworths, London, 1964, p. 3.
37. D. D. DeFord, R. J. Loyd, and B. O. Ayers, *Anal. Chem.*, **35**, 426 (1963).
38. J. Bohemen, S. H. Langer, R. H. Perrett, and J. H. Purnell, *J. Chem. Soc.*, 2444 (1960).
39. L Rohrschneider, in *Advances in Chromatography*, Vol. 4, J. C. Giddings and R. A. Keller (Eds.), Dekker, New York, 1968.

40. M. J. E. Golay, *Nature*, **180**, 435 (1957).
41. R. R. Freeman, *High Resolution Gas Chromatography*, 2d ed., Hewlett-Packard Company, Avondale, PA 1981.
42. J. H. Purnell, *Ann. N. Y. Acad. Sci.*, **72**, 614 (1959).
43. E. Bayer, K. P. Hupe, and H. Mack, *Anal. Chem.*, **35**, 492 (1963).
44. J. C. Giddings, *Anal. Chem.*, **34**, 37 (1962).
45. J. H. Purnell, *J. Roy. Inst. Chem.*, **82**, 586 (1958).
46. M. Verzele, J. Bouche, A. DeBruyne, and M. Verstappe, *J. Chromatogr.*, **18**, 253 (1965).
47. A. J. P. Martin, in *Gas Chromatography*, V. J. Coates, H. Noebels and I. Fagerson (Eds.), Academic, New York, 1958, p. 237.
48. M. Verzele, *J. Gas Chromatogr.*, **3**, 186 (1965).
49. J. H. Purnell, *Gas Chromatography*, Wiley, New York, 1962, pp. 113–116.
50. C. S. Horvath, in *The Practice of Gas Chromatography*, L. S. Ettre and A. Zlatkis (Eds.), Interscience, New York, 1967.
51. A. C. Locke and W. W. Brandt, in *Gas Chromatography*, *1962*, 4th Symposium, Hamburg, M. van Swaay (Ed.), Butterworths, London, 1962, p. 66.
52. R. L. Grob, G. W. Weinert, and J. W. Drelich, *J. Chromatogr.*, **30**, 305–324 (1967).
53. R. L. Grob, R. J. Gondek, and T. A. Scales, *J. Chromatogr.*, **53**, 477–486 (1970).
54. R. L. Grob and E. J. McGongile, *J. Chromatogr.*, **59**, 13–20 (1971); **101**, 39–50 (1974).
55. L. D. Belyakova, A. V. Kiselev, and G. A. Soloyan, *Chromatographia*, **3**, 254–259 (1970).
56. D. T. Sawyer and D. J. Brookman, *Anal. Chem.*, **40**, 1847 (1968).
57. R. L. Grob, in *Progress in Analytical Chemistry*, Vol. 8, I. V. Simmons and G. W. Ewing (Eds.), Plenum, New York, 1976, pp. 151–194.
58. D. H. Desty, A. Goldup, and W. T. Swanton, in *Lectures on Gas Chromatography—1962*, H. A. Szymanski (Ed.), Plenum, New York, 1963, p. 105.
59. J. H. Purnell, *Nature*, **184**, 2009 (1959).
60. J. H. Purnell, *J. Chem. Soc.*, **54**, 1268 (1960).
61. R. Kaiser, *Z. Anal. Chem.*, **189**, 11 (1962).

CHAPTER THREE

Columns: Packed and Capillary/Column Selection in Gas Chromatography

EUGENE F. BARRY

University of Massachusetts Lowell

Modern Practice of Gas Chromatography, Third Edition. Edited by Robert L. Grob.
ISBN 0-471-59700-7

Part 1 Overview

3.1 CENTRAL ROLE PLAYED BY COLUMN

The gas chromatographic column is the central item in a gas chromatograph. Over the last three decades the nature and design of the column has changed considerably from one containing either a solid adsorbent or a liquid deposited on an inert solid support packed into a length of tubing to one containing an immobilized or cross-linked stationary phase bound to the inner surface of a much longer length of fused silica tubing. With respect to packing materials, solid adsorbents such as silica gel and alumina have been replaced by porous polymeric adsorbents, while the the vast array of stationary liquid phases in the 1960s were greatly reduced in number by the next decade to a smaller number of phases of greater thermal stability. These became the precursors of the chemically bonded or cross-linked phases of today. Column tubing fabricated from copper, aluminum, glass, and stainless steel served the early analytical needs of gas chromatographers. Presently, fused silica capillary columns having a length of 10–60 m and an inner diameter of 0.20–0.53 mm are in widespread use.

Although gas chromatography (GC) may be viewed in general as a mature analytical technique, improvements in column technology, injection, and detector design steadily appear nonetheless. Innovations and advancements in gas chromatography of the last decade have been made with the merits of the fused-silica column as the focal point and have been driven primarily by the environmental and toxicological fields.

3.2 JUSTIFICATION FOR COLUMN SELECTION AND CARE

The cost of a gas chromatograph can range from \$6000 to over \$100,000 depending on the type and number of detectors, injection systems, and peripheral devices such as data system, headspace, and thermal desorption units, pyrolyzers, autosamplers, etc. When one also factors in the purchase of high-purity gases on a regular basis required for operation of the chromatograph, it quickly becomes apparent that a sizeable investment has been made in capital equipment. For example, cost-effectiveness and good chromatographic practice dictate that users of capillary columns should give careful consideration to column selection. The dimensions and type of capillary column should be chosen with the injection system and detectors in mind, considerations that are virtually nonissues with packed columns. Careful attention should be also paid to properly implemented connections of the column to the injector and detector and the presence of high boilers, particulate matter in samples, etc.

The price of a column ($200–$800) may be viewed as relatively small compared to the initial, routine, and preventative maintenance costs of the instrument. In fact, a laboratory may find that the cost of a set of air and hydrogen gas cylinders of research grade purity for FID operation is far greater than the price of a single conventional capillary column! Consequently, to derive maximum performance from a gas chromatographic system the column should be carefully selected for an application, handled with care following the suggestions of its manufacturer, and installed as recommended in the user's instrument manual.

3.3 SURVEYS OF PACKED AND CAPILLARY-COLUMN USAGE

The introduction of inert fused-silica capillary columns in 1979 markedly changed the practice of gas chromatography, enabling high-resolution separations to be performed in most laboratories. Previously such separations were achieved with reactive stainless steel columns and with glass columns. After 1979 the use of packed columns began to decline. A further decrease in the use of packed columns occurred in 1983 with the arrival of the megabore capillary column of 0.53 mm inner diameter, which serves as a direct replacement for the packed column. These developments, in conjunction with the emergence of immobilized or cross-linked stationary phases specifically tailored for fused-silica capillary columns and overall improvements in column technology, have been responsible for the greater acceptance for capillary GC.

The results of a survey of 12 leading experts in gas chromatography appeared in 1989 and outlined their thoughts on projected trends in gas chromatographic column technology, including the future of packed columns versus capillary columns (1). Following are some responses of that panel:

1. Packed columns are used for approximately 20% of gas chromatographic analyses.

2. Packed columns are employed primarily for preparative applications, for fixed gas analysis, for simple separations, and for those separations where high resolution is not required or not always desirable (PCBs).
3. Packed columns will continue to be used for gas chromatographic methods that were validated on packed columns where time and cost of revalidation on capillary columns would be prohibitive.
4. Capillary columns will not replace the packed column in the near future although few applications require packed columns.

In 1990 Majors summarized the results of a more detailed survey on column usage in gas chromatography, this one, however, soliciting response from *LC*GC* readership (2). Some conclusions drawn from this survey include the following:

1. Likewise, nearly 80% of the respondents use capillary columns.
2. Capillary columns of 0.25 and 0.53 mm i.d. are the most popular, as are columns lengths of 10–30 m.
3. The methyl silicones and polyethylene glycol stationary phases are the most preferred for capillary separations.
4. Packed columns are used mostly for gas–solid chromatographic separations such as gas analyses.
5. The majority of respondents indicated the need for stationary phases of higher thermal stability.

Column manufacturers rely on the current literature and results of surveys to keep abreast of the needs of practicing gas chromatographers. The fused silica capillary column has clearly emerged as the column of choice for most gas chromatographic applications. A market research report (3) showed that $100 million were spent on capillary columns worldwide in 1993 alone and, at an estimated average cost of $400 for a column, this figure represents about 250,000 columns (4). Despite the maturity of capillary GC, research continues to improve instrument performance and enhance column technology. Chromatographers can expect to see continued efforts by capillary-column manufacturers on producing columns having lower residual activity and being capable of withstanding higher column temperature operation with reduced column bleeding. There should also be increased availability of capillary columns exhibiting stationary phase-selectivity tuned for specific applications (4).

3.4 LITERATURE ON GAS CHROMATOGRAPHIC COLUMNS

The primary journals where developments in column technology and applications are published include *Analytical Chemistry*, *Journal of Chroma-*

tography, *Journal of Chromatographic Science*, *Journal of High Resolution Chromatography*, and *LC*GC Magazine*. The biennial review issue of *Analytical Chemistry—Fundamental Reviews* (published in even-numbered years), contains concise summaries of developments in gas chromatography. An abundance of gas chromatographic applications may be found in the companion issue, *Application Reviews* (published in odd-numbered years), covering the areas of polymers, geological materials, petroleum and coal, coatings, pesticides, forensic science, clinical chemistry, environmental analysis, air pollution, and water analysis.

Most laboratories have access to literature-searching through one of a number of on-line computerized-database services. Although the location of articles on gas chromatography in primary journals is relatively easy, finding publications of interest in lesser known periodicals can be a challenge and may prove to be tedious at times. *CA Selects* and *Current Contents* are convenient alternatives. The biweekly *CA Selects*, gas chromatography topical edition, available from Chemical Abstracts Service is a condensation of information reported throughout the world. *Current Contents* in diskette format provides weekly coverage of current research in the life sciences, clinical medicine, the physical, chemical and earth sciences, as well as agricultural, biology, and environmental sciences.

The periodic commercial literature and annual catalog of column manufacturers describing applications for their columns also contains more and more useful technical information of a generic nature with each passing year. In addition, this author strongly recommends *LC*GC* magazine as a valuable resource in which appear not only timely technical articles but also sections devoted to "Column Watch" and troubleshooting for GC.

Part 2 Packed-Column Gas Chromatography

Packed columns are still utilized for a variety of applications in gas chromatography. A packed column consists of three basic components: (1) tubing in which packing material is placed, (2) packing retainers (such as glass wool plugs) inserted into the ends of the tubing to keep the packing in place, and (3) the packing material itself. In Part 2 the role and properties of solid support materials, adsorbents, commonly used stationary phases, and procedures for the preparation of packed columns are described. Factors affecting packed-column performance are also presented.

3.1 SOLID SUPPORTS AND ADSORBENTS

3.5.1 Supports for GLC: Diatomaceous Types, Halocarbons

The purpose and role of the solid support is the accommodation of a uniform deposition of stationary phase on the surface of the support. The most commonly used support materials are primarily diatomite supports and graphitized carbon (which is also an adsorbent for GSC), and to a lesser extent, Teflon. There is no perfect support material because each has limitations. Pertinent physical properties of a support for packed-column GC are particle size, porosity, surface area, and packing density. Particle size impacts column efficiency via the *A* term or eddy diffusion contribution in the van Deemter expression (Equation 2.4). The surface area of a support is governed by its porosity, the more porous supports requiring greater amounts of stationary phase for surface coverage. A photomicrograph of Chromosorb W HP of 80/100 mesh appears in Figure 3.1, where the complex pore network is clearly evident.

A support should have sufficient surface area so that the chosen amount of stationary phase can be deposited uniformly and not leave an exposure of active sites on its surface. Conversely, if excessive phase (above the upper coating limit of the support) is deposited on the support, phase may have a tendency to "puddle or pool" on a support particle and can even spread over to an adjacent particle resulting in a decrease in column efficiency due to unfavorable mass transfer.

Diatomite Supports. These supports can be categorized into three groups:

1. Firebrick-type supports, obtained after sequentially heating, pulverizing, and sieving bricks derived from diatomaceous earth.

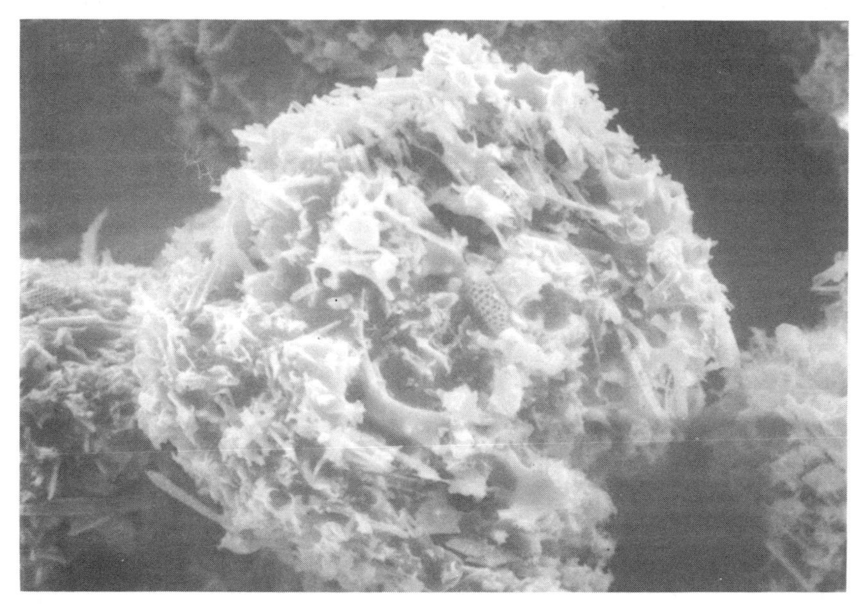

FIGURE 3.1 Scanning electron micrograph of 80/100-mesh Chromosorb W. (From Reference 5.)

2. Filter-aid supports, prepared by adding sodium carbonate to diatomaceous earth and heating the product at elevated temperatures, which results in the formation of aggregates of the individual skeletons.
3. Chromosorb G, manufactured in a similar fashion to filter-aid supports, but considerably more durable.

The pink-colored firebrick supports such as Chromosorb P and Gas Chrom R are very strong particulates that provide higher column plate numbers than most supports. Because of their high specific surface area, these supports can accommodate up to 30% loading of liquid phase and their use is reserved for the analysis of nonpolar species such as hydrocarbons. They must be deactivated, however, when employed for the analysis of polar compounds such as alcohols and amines. As a result, the white-colored filter-aid supports of lower surface area (Chromosorb W, Gas Chrom Q, and Supelcoport, to name several), are preferable, although they are more fragile and permit a slightly lower maximum percent loading of about 25% by weight of liquid phase. The harder and improved support, Chromosorb G, is denser than the Chromosorb W but also exhibits a lower surface area and is used for the analysis of polar compounds. Its maximum loading is 5% by weight.

A support should ideally be inert and not interact with sample components in any way, otherwise a component may decompose on the column resulting in peak-tailing or even disappearance of the peak in a chromatogram. The presence of active silanol groups (Si–OH functionalities) and metal ions constitute two types of active adsorptive sites on support materials. Polar analytes, acting as Lewis bases, can participate in hydrogen bonding with silanol sites and display peak-tailing. The degree of tailing increases in the sequence hydrocarbons, ethers, esters, alcohols, carboxylic acids, etc., and also increases with decreasing concentration of a polar analyte. Treatment of the support with the silylating reagent dimethyldichlorosilane (DMDCS) converts silanol sites into silyl ether functionalities and generates a deactivated surface texture. Since this procedure is both critical and difficult (HCl is a product of the reaction), it is advisable to purchase DMDCS-deactivated support materials or column packings prepared with this chemically modified support material from a column manufacturer. A word of caution: The presence of moisture in the chromatographic system due to either impure carrier gas or water content in injected samples can hydrolyze silanized supports, reactive them, and initiate degradation of many liquid phases.

Metal ions such as Fe^{3+} present on a diatomite support surface can likewise cause decomposition of both sample and stationary liquid phase. These ions can be considered Lewis acids, which can also induce peak-tailing of electron-dense analytes, e.g., aromatics. The ions can be leached from the support surface by washing with hydrochloric acid followed by thorough rinsing to neutrality with deionized water of high quality. A Chromosorb support subjected to this treatment has the suffix -AW after its name; the untreated or non-acid-washed version of the same support is designated by the abbreviation -NAW. A support that is both acid-washed and deactivated with DMDCS is represented as -AW-DMDCS. The designation -HP is used for the classification of a support as high-performance grade, namely, the best available quality. A cross-reference of Chromosorb supports and the popular Gas Chrom series of supports is outlined as a function of type of diatomite and treatment in Table 3.1 and pertinent support properties are displayed in Table 3.2.

Small quantities of acids and bases may also be added to the stationary to cover or neutralize active sites on support. They usually have the same

TABLE 3.1 Cross-Reference of Solid Supports

Source	Acid-Washed, DMDCS-Treated	Non-Acid-Washed	Acid-Washed
Firebrick	Chromosorb P AW-DMDCS	Chromosorb P NAW	Chromosorb P AW
	Gas Chrom RZ	Gas Chrom R	Gas Chrom RA
Celite filter aid	Chromosorb W AW-DMDCS	Chromosorb W NAW	Chromosorb W AW
	Chromosorb W HP[a]		
	Chromosorb G AW-DMDCS	Chromosorb G NAW	Chromosorb G AW
	Supelcoport[a]		
Other filter aid	Gas Chrom QII[a]		
	Gas Chrom Q (also base washed, then silanized)		
	Gas Chrom Z	Gas Chrom S	Gas Chrom A

Note. Chromosorb, Gas Chrom, and Supelcoport are trademarks of Johns–Manville, Alltech, and Supelco, respectively.

[a] High-performance support or best available grade of support.

TABLE 3.2 Properties of Selected Diatomaceous Earth Supports

Support	Packing Density (g/mL)	Surface Area (m^2/g)	Pore Volume (mL/g)	Maximum Liquid Phase Loading %
Chromosorb P NAW	0.32–0.38	4–6	1.60	30
Chromosorb P AW	0.32–0.38	4–6		
Chromosorb P AW-DMDCS	0.32–0.38	4–6		
Chromosorb W AW	0.21–0.27	1.0–3.5	3.56	15
Chromosorb W HP	0.23	0.6–1.3		
Chromosorb G NAW	0.49	0.5	0.92	5
Chromosorb G AW-DMDCS	0.49	0.5		
Chromosorb G HP	0.49	0.4		

Source. Data obtained from References 6 and 7.

acid–base properties of the species being analyzed and are referred to as "tail reducers." Phosphoric acid-modified packings are effective for analyzing fatty acids and phenols; potassium hydroxide has been used with success for amines and other basic compounds.

An often overlooked parameter in the selection of a packed column is the packing density of the support material. Packing density can have a rather pronounced effect on retention data. The stationary phase is coated on a support on a weight percent basis, whereas the packing material is placed in the column on a volume basis. If the packing density of a support increases, then the total amount of stationary phase in the column increases, even if the loading percentage is constant. Packing density varies among support materials (Table 3.2) and may even vary from batch to batch for a given type of support. Consider the following scenario. Two packings are prepared—10% Carbowax 20M on Chromosorb G HP and the other 10% Carbowax 20M on Chromosorb W HP—and each is subsequently packed into glass columns of identical dimensions. The column containing the impregnated Chromosorb G HP will contain approximately twice as much stationary phase as the other column. Therefore, careful adherence should be paid to the nature and properties of a support in order to generate meaningful retention data and compare separations.

Teflon Supports. Although diatomite supports are widely used support materials, analysis of corrosive or very polar substances require even more inertness from the support. Halocarbon supports offer enhanced inertness and a variety have been tried, including Fluoropak-80, Kel-F, Teflon, and other fluorocarbon materials. However, Chromosorb T made from Teflon 6 powder is perhaps the best material available, because high column efficiencies can be obtained when it is coated with a stationary phase having high surface area, such as polyethylene glycols. Chromosorb T has a surface area of 7–8 m^2/g, a packing density of 0.42 g/mL, an upper coating limit of 20% and a rather low upper temperature limit of 250°C. Applications where this type of support is recommended are the analyses of water, acids, amines, HF, HCl, chlorosilanes, sulfur dioxide, and hydrazine. Difficulties in coating Chromosorb T and packing columns may be encountered as the material tends to develop static charges. This situation is minimized by (1) using plasticware in place of glass beakers, funnels, etc., (2) chilling the support to 10°C prior to coating, and (3) also chilling the column before packing. Consultation of References 8–12 yields further information for successful results with this support. However, preparation of columns containing Teflon-coated stationary phases is best performed by the column manufacturers.

The interested reader desiring further details about solid supports is urged to consult the comprehensive reviews of Ottenstein (8, 9) and the benchmark book, *The Packed Column in Gas Chromatography*, written by Supina (10).

3.5.2 Adsorbents for GSC: Porous Polymers, Molecular Sieves, Carbonaceous Materials

Surface adsorption is the prevailing separation mechanism in gas–solid chromatography (GSC), while great care is taken to avoid this effect in gas–liquid chromatography (GLC). In GSC an uncoated adsorbent serves as the column packing, although special effects in selectivity by a mixed retention mechanism can be obtained by coating the adsorbent with a stationary phase. The latter case is an illustration of gas–liquid–solid chromatography (GLSC). Permanent gases and very volatile organic compounds can be analyzed by GSC as their volatility is problematic in GLC because their volatility causes rapid elution.

Porous Polymers. Porous polymcrs are the adsorbents of choice for most applications focusing on the analysis of gases, organics of low carbon number, acids, amines, and water (15, 16). The presence of water is detrimental to gas–liquid chromatographic packings. Because water is eluted with symmetrical band profiles on a number of porous polymers, these adsorbents may be employed for the analyses of aqueous solutions and the determination of water in organic matrices. There are three separate product lines of commercially available porous polymers: the Porapaks (Waters Associates), the Chromosorb Century Series (Johns-Manville), and HayeSep (Hayes Separation) polymers. Within each of the product lines, there are several members, each differing in chemical composition and, therefore, exhibiting unique selectivity, as may be observed in Table 3.3. On the other hand, some adsorbents are quite similar, such as is the case with Porapak Q-Chromosorb 102 (both styrene–divinylbenzene copolymers) and HayeSep C-Chromosorb 104 (both acrylonitrile–divinylbenzene copolymers). In the future we should expect to see new polymers addressing old separation problems, as was the case with the arrival of HayeSep A, which can resolve a mixture of nitrogen, oxygen, argon, and carbon monoxide at room temperature (17).

Molecular Sieves. These sorbents are also referred to as zeolites, which are synthetic alkali or alkaline earth metal aluminum silicates and are utilized for the separation of hydrogen, oxygen, nitrogen, methane, and carbon monoxide. These substances are separated on molecular sieves because the pore size of the sieve matches their molecular diameter. Two popular types of molecular sieves are used in GSC: Molecular Sieve 5A (pore size of 5 Å with calcium as the primary cation) and Molecular Sieve 13X (pore size of 13 Å with sodium as the primary cation). At normal column temperatures, molecular sieves permanently adsorb carbon dioxide, which gradually degrades the O_2–N_2 resolution. The use of a silica gel precolumn, which adsorbs carbon dioxide, eliminates this problem. Molecular sieve columns must be conditioned at 300°C to remove residual moisture

TABLE 3.3 Porous Polymeric Adsorbents for GSC

Adsorbent	Polymeric Composition or Polar Monomer (PM)	Maximum Temperature (°C)	Applications
HayeSep A	DVB-EGDMA	165	Permanent gases, including hydrogen, nitrogen, oxygen, argon, CO, and NO at ambient temperature; can separate C2 hydrocarbons, hydrogen sulfide and water at elevated temperature.
HayeSep B	DVB-PEI	190	C1 and C2 amines; trace amounts of ammonia and water.
HayeSep C	ACN-DVB	250	Analysis of polar gases (HCN, ammonia, hydrogen sulfide) and water.
HayeSep D	High-purity DVB	290	Separation of CO and carbon dioxide from room air at ambient temperature; elutes acetylene before other C2 hydrocarbons; analyses of water and hydrogen sulfide.
Porapak N HayeSep N	DVB-EVB-EGDMA EGDMA (copolymer)	190 165	Separation of ammonia, carbon dioxide, water, and separation of acetylene from other C2 hydrocarbons.
Porapak P HayeSep P	Styrene-DVB Styrene-DVB	250 250	Separation of a wide variety of alcohols, glycols, and carbonyl analytes
Porapak Q HayeSep Q	EVB-DVB copolymer DVB polymer	250 275	Most widely used; separation of hydrocarbons, organic analytes in water, and oxides of nitrogen.
Porapak R HayeSep R	Vinyl pyrollidone (PM) DVB-4-vinyl-pyrollidone	250 250	Separation of ethers and esters; separation of water from chlorine and HCl.
Porapak S HayeSep S	Vinyl pyridine (PM) DVB-4-vinyl-pyridine	250 250	Separation of normal and branched chain alcohols.
Porapak T HayeSep T	EGDMA (PM) EGDMA polymer	190 165	Highest polarity Porapak and offers greatest water retention; determination of formaldehyde in water.
Chromosorb			
101	Styrene-DVB	275	Separation of fatty acids, alcohols, glycols, esters, ketones, aldehydes, ethers, and hydrocarbons.
102	Styrene-DVB	250	Separation of volatile organics and permanent gases, no peak-tailing for water and alcohols.
103	Cross-linked PS	275	Separation of basic compounds, such as amines, and ammonia; useful for separation of amides, hydrazines, alcohols, aldehydes, and ketones.

TABLE 3.3. *(Continued)*

Adsorbent	Polymeric Composition or Polar Monomer (PM)	Maximum Temperature (°C)	Applications
104	ACN-DVB	250	Nitriles, nitroparaffins, hydrogen sulfide, ammonia, sulfur dioxide, carbon dioxide, vinylidene chloride, vinyl chloride, trace water content in solvents.
105	Cross-linked polyaromatic	250	Separation of aqueous solutions of formaldehyde, separation of acetylene from lower hydrocarbons and various classes of organics with boiling points up to 200°C.
106	Cross-linked PS	225	Separation of C2 to C5 alcohols; separation of C2 to C5 fatty acids from corresponding alcohols.
107	Cross-linked acrylic ester	225	Analysis of formaldehyde, sulfur gases, and various classes of compounds.
108	Cross-linked acrylic	225	Separation of gases and polar species, such as water, alcohols, aldehydes, ketones, and glycols

Note. DVB, divinyl benzene; EGDMA, ethylene glycol dimethacrylate; PEI, polyethyleneimine; ACN, acrylonitrile; EVB, ethylvinyl benzene.

Source. Data obtained from References 6, 13, and 14.

from the packing; otherwise the permanent gases elute too quickly, with little or no resolution, and coelution or reversal in elution order for CO–methane may occur (18).

Carbonaceous Materials. Adsorbents containing carbon are commercially available in two forms: carbon molecular sieves and graphitized carbon blacks. The use of carbon molecular sieves as packings for GSC was first reported by Kaiser (19). They behave similarly to molecular sieves because their pore network is also in the angstrom range. Permanent gases and C1–C3 hydrocarbons may be separated on carbonaceous sieves such as Carbosphere and Carboxen.

Graphitized carbons play a dual role in GC. They are a nonspecific adsorbent in GSC having a surface area in the range of 10–1200 m^2/g. Adsorbents such as Carbopacks and Graphpacs may also serve as a support in GLC and in GLSC where unique selectivity is acquired and a separation is based on molecular geometry and polarizability considerations. Coated graphitized carbons can tolerate aqueous samples and have been used for the determination of water in glycols, acids, and amines by DiCorcia and coworkers (20–22). In the latter roles, since graphitized carbon has a nonpolar surface texture, it must be coated with a stationary phase for deactivation of its surface. The resulting packing reflects a separation that is a hybrid of gas–solid and gas–liquid mechanisms. Frequently, the packing is further modified by the addition of H_3PO_4 or KOH to reduce peak-tailing for acidic and basic compounds, respectively. Separations of alcohols and amines are displayed in Figure 3.2.

3.6 STATIONARY PHASES

3.6.1 Requirements of a Stationary Phase

An ideal stationary liquid phase for GLC should exhibit selectivity and differential solubility of components to be separated and a wide operating temperature range. A phase should be chemically stable and have a low vapor pressure at elevated column temperatures. A minimum temperature limit near ambient temperature, where the liquid phase still exists as a liquid and not as a solid, is desirable for separations at or near room temperature and eliminates a gas–solid adsorption mechanism prevailing, e.g., as with Carbowax 20M below 60°C. In the selection of a stationary phase a compromise between theory and practice must be reconciled. For example, theory dictates that a stationary phase of low viscosity or fluid in texture is preferable over a chemically equivalent, more viscous gum phase, as may be ascertained from the contribution of D_l, the solute diffusivity in the stationary phase, appearing in Equation 2.44. However, this same fluid possessing a lower molecular weight or weight distribution if polymeric in nature will typically have poorer thermal stability and a lower maximum operating temperature. Although unfavorable from the viewpoint of mass transfer in the van Deemter expression, practical considerations may favor the gum for separations requiring high column temperatures. Equation 2.44 indicates that a higher column efficiency is obtained with a column containing a low percent loading of stationary phase compared to a same column packed with a higher phase loading. But in practice, the deposition of a thin coating of stationary phase on a support may yield insufficient coverage of the active sites on the surface of the support, resulting in peak-tailing, and reestablish a need for a higher percentage of stationary phase loading. Note in Figure 3.3B the peak-tailing of the *n*-alkanes on a lightly loaded packing (less than 3% OV-101 on Chromosorb W HP) and the elimination of tailing with a heavier coating of stationary phase (Figure 3.3A).

Separations in GLC are the result of selective solute–stationary phase interactions and differences in the vapor pressure of solutes. The main forces that are responsible for solute interaction with a stationary phase are dispersion, induction, orientation, and donor–acceptor interactions (23–25), the sum of which serves as a measure of the "polarity" of the stationary toward the solute. Selectivity, on the other hand, may be viewed in terms of

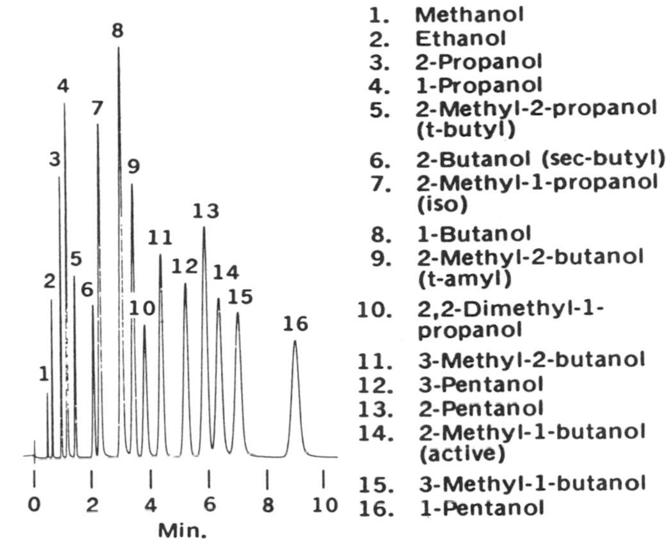

80/100 Carbopack C/0.2% Carbowax 1500, 6′ x 2mm ID glass, Col. Temp.: 135°C, Flow Rate: 20mL/min., N_2, Det.: FID, Sample Size: 0.02μL.

(*a*)

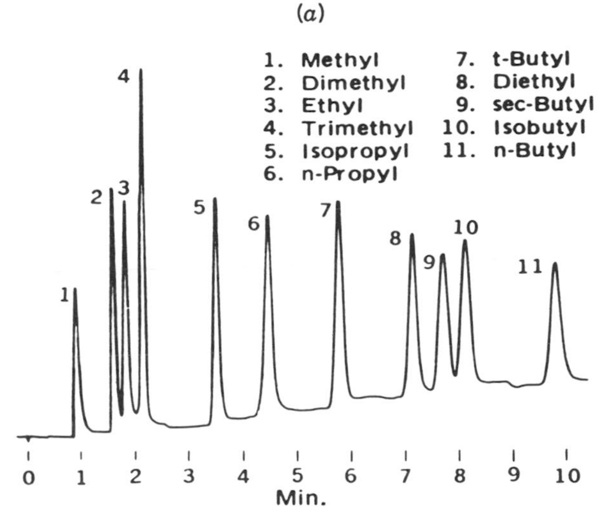

60/80 Carbopack B/4% Carbowax 20M/0.8% KOH, 6′ x 2mm ID glass, Col. Temp.: 90°C to 150°C @ 4°C/min., Flow Rate: 20mL/min., N_2, Det.: FID, Sens.: 4 x 10^{-11} AFS, Sample: 0.5μL, 100ppm each amine in water.

(*b*)

FIGURE 3.2 Separation of C_1–C_5 alcohols (A) and aliphatic amines (B) on graphitized carbon. Reproduced from reference (18): W. A. Supina, in *Modern Practice of Gas Chromatography*, R. L. Grob (Ed.) second edition, Copyright 1985, John Wiley & Sons, Inc. Reprinted by permission of John Wiley & Sons, Inc.

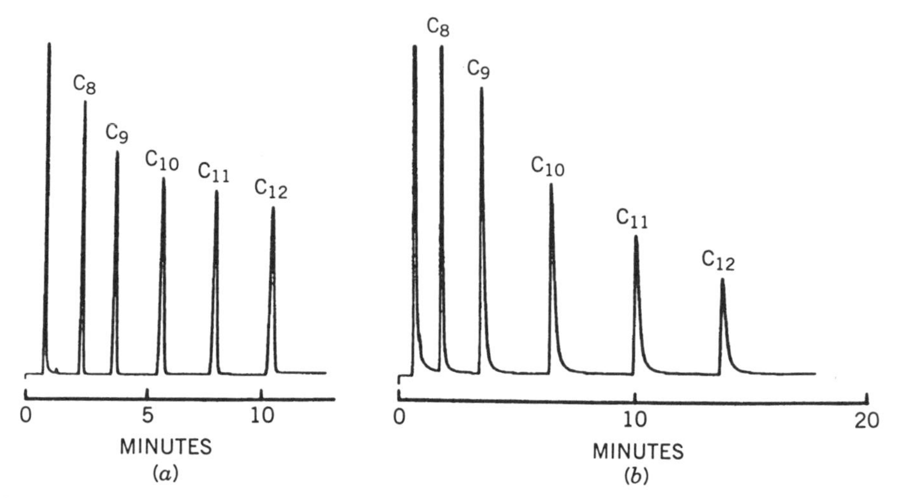

FIGURE 3.3 Chromatograms of *n*-alkanes on (A) a 6-ft glass column, 2 mm i.d., containing 20% OV-101 on 80/100-mesh Chromosorb W HP, column conditions: 100–175°C at 6°C/min; (B) same as in (A) but with 3% OV-101 on same support; column conditions: 50–120°C at 4°C/min. Flowrate: 25 mL/min He, Det: FID. (From Reference 5.)

the magnitude of the individual energies of interaction. In GLC, the selectivity of a column governs band spacing or the degree to which peak maxima are separated. The following parameters influence selectivity:

1. The nature of the stationary phase
2. The concentration of the stationary phase
3. Column temperature
4. The choice and pretreatment of solid support or adsorbent

Differences in selectivity are significant because they permit the separation of solutes of similar or even the same polarity by a selective stationary phase.

In the early practice of gas chromatography, the concept of polarity and even the requirements for a stationary phase were not clearly understood. There was a proliferation of liquid phases encompassing (1) those that had marginal gas chromatographic properties, such as nujol, glycerol, diglycerol, and Tide, the laundry detergent; (2) those that were industrial grade lubricants of variable composition, such as the Apiezon greases and Ucon oils; and (3) an abundance of phases that simply duplicated the chromatographic behavior of others. In retrospect, the vast array of stationary phases can probably be attributed to the compensation for the inefficiency of a packed column by achieving some acceptable degree of selectivity for the resolution of two solutes (Equation 2.85). Conversely, the high efficiency of a capillary column necessitates the availability of a relatively few stationary phases, each differing in selectivity to achieve any required resolution.

The stationary phase requirements of selectivity and higher thermal stability then become more clearly defined; the process of stationary phase selection and classification became logical after the studies of McReynolds (26) and Rohrschneider (27, 28) were published, both of which were based on the Kovats Retention Index (29). The Kovats Retention Index procedure and McReynolds constants are discussed in detail in the following section. Kovats retention indices remain today a widely used technique for reporting retention data, while every stationary phase developed for packed and capillary GC has been characterized by generation of its McReynolds constants.

3.6.2 Kovats Retention Indices

This universal approach solved the problems pertaining to the use, comparison, and characterization of gas chromatographic retention data. The reporting of retention data as absolute retention time t_R is meaningless because virtually every chromatographic parameter and any related experimental fluctuation affect a retention time measurement. The use of relative retention data ($\alpha = t'_{R2}/t'_{R1}$) offered some improvement, but the lack of a universal standard suitable for wide temperature range on stationary phases of different polarities has discouraged its utilization. In the Kovats approach, the retention index I of an n-alkane is assigned a value equal to 100 times its carbon number. Thus, for example, the I values of n-octane, n-decane, and n-dodecane are equal to 800, 1000, and 1200, respectively, by definition and are applicable on any column, packed or capillary, and any liquid phase and are independent of every chromatographic condition, including column temperature. For all other compounds, the chromatographic conditions, such as the stationary phase, its concentration, support, and column temperature for packed columns, must be specified. Since retention indices are also the preferred method for reporting retention data with capillary columns, the stationary phase, film thickness, and column temperature likewise have to be specified for compounds other than n-alkanes; otherwise the I values are meaningless.

An I value of a component can be determined by spiking a mixture of n-alkanes with the component(s) of interest and chromatographing the resulting mixture under the specified conditions. A plot of log-adjusted retention time versus retention index is generated and the retention index of the solute under consideration is determined by extrapolation, as depicted in Figure 3.4 for isoamyl acetate. The selectivity of a particular stationary phase can be established by comparing the I values of a solute on a nonpolar phase such as squalane or OV-101 ($I = 872$) with the corresponding value of I of 1128 associated with a more polar column containing Carbowax 20M,

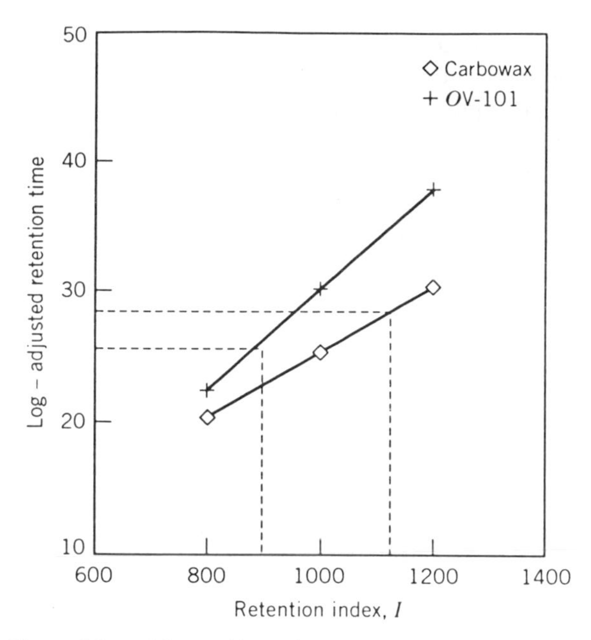

FIGURE 3.4 Plot of logarithm adjusted retention time versus Kovats retention index: isoamyl acetate at 120°C.

for example. This difference of 256 units indicates the greater retention produced by the Carbowax 20M column. More specifically, isoamyl acetate elutes between *n*-C11 and *n*-C12 on a Carbowax 20M, but more rapidly on OV-101 where it elutes after *n*-octane.

Alternatively, the retention index of an analyte at an isothermal column temperature can be calculated from the following equation:

$$I_x = 100Z + \frac{100(\log t'_{R,x} - \log t'_{R,z})}{\log t'_{R,z+1} - \log t'_{R,z}} \tag{3.1}$$

where $t'_{R,x}$ is the adjusted retention time of the component under consideration, $t'_{R,z}$ is the adjusted retention time of the *n*-alkane eluting before it, $t'_{R,z+1}$ is the adjusted retention time of the *n*-alkane eluting after it, and Z is the carbon number of the *n*-alkane having retention $t'_{R,z}$. For temperature programmed runs, the adjusted retention times in Equation 3.1 are replaced by the appropriate elution temperatures in °K. An I value computed by Equation 3.1 is strongly recommended because it is inherently more accurate than that obtained by the graphical approach.

Retention indices normalize instrumental variables in gas chromatographs, allowing retention data generated on different systems to be compared. For example,, isoamyl acetate with a retention index of 1128 will elute between *n*-C11 and *n*-C12 under the same chromatographic conditions. Retention indices are also very helpful in comparing relative elution orders of series of analytes on a specific column at a given temperature and for comparing selective behavior of two or more columns.

McReynolds has tabulated retention indices for a large number of compounds on various liquid phases (30); an excellent review of the retention index system has been prepared by Ettre (31). The postrun calculation of retention indices is greatly facilitated by using a reporting integrator or a date acquisition system. Consistent with the growing trend of computer assistance in gas chromatography is the availability of retention index libraries for drugs and pharmaceuticals, organic volatiles, pesticides, herbicides and PCBs of environmental significance, methyl esters of fatty acids, food and flavor volatiles, solvents, and chemicals (32).

3.6.3 McReynolds Classification of Stationary Phases

The most widely used system of classifying liquid phases is the McReynolds system (26), which has been employed to characterize virtually every stationary phase. McReynolds selected 10 probe solutes of different functionality, each designated to measure a specific interaction with a liquid phase. He analyzed these probe solutes and measured their I values on over 200 phases, including squalane, which served as a reference liquid phase under the same chromatographic conditions. A similar approach was previously implemented by Rohrschneider (27, 28) with five probes. In Table 3.4 the probes used in both approaches and their function are listed. McReynolds calculated for each probe a ΔI value, where

$$\Delta I = I_{\text{liquid phase}} - I_{\text{squalane}}$$

As the difference in the retention index for a probe on a given liquid phase and squalane increases, the degree of specific interaction associated with that probe increases. The cumulative effect, when summed for each of the 10 probes, is a measure of overall "polarity" of the stationary phase. In a tabulation of McReynolds constants the first five probes usually appear and are represented by the symbols X', Y', Z', U', and S'. Each probe is assigned a value of zero with squalane as reference liquid phase.

Several significant consequences resulted from these classification procedures. Phases that have identical chromatographic behavior also have identical constants. In this case the selection of a stationary phase could be based on a consideration such as thermal stability, lower viscosity, cost, or availability. McReynolds constants of the more popular stationary phases for packed-column GC are listed in Table 3.5. Note that the DC-200 (a silicone oil of low viscosity) and OV-101 or SE-30 (a dimethylpolysiloxane) have nearly identical constants, but also observe that these two polysiloxanes have a more favorable higher temperature limit. Comparisons of this type curtailed the proliferation of phases, eliminated the duplication of

TABLE 3.4 Probes Used in the McReynolds and Rohrschneider Classifications of Liquid Phases

Symbol	McReynolds Probe	Rohrschneider Probe	Measured Interaction
X′	Benzene	Benzene	Electron density for aromatic and olefinic hydrocarbons
Y′	*n*-Butanol	Ethanol	Proton donor and proton acceptor capabilities (alcohols, nitriles)
Z′	2-Pentanone	2-Butanone	Proton acceptor interaction (ketones, ethers, aldehydes, esters)
U′	Nitropropane	Nitromethane	Dipole interactions
S′	Pyridine	Pyridine	Strong proton acceptor interaction
H′	2-Methyl-2-pentanol	—	Substituted alcohol interaction similar to *n*-butanol
J′	Iodobutane	—	Polar alkane interactions
K′	2-Octyne	—	Unsaturated hydrocarbon interaction similar to benzene
L′	1,4-Dioxane	—	Proton acceptor interaction
M′	*cis*-Hydrindane	—	Dispersion–interaction

phases, and simplified column selection. Many phases quickly became obsolete and were replaced by a phase having identical constants but of higher thermal stability, such as a polysiloxane phase. Polysiloxane phases are the most commonly used stationary phases today for both packed (and capillary-column) separations because they exhibit excellent thermal stability, have favorable solute diffusivities, and are available in a wide range of polarities. They are discussed in greater detail in Part 3.

There was also an impetus to consolidate the number of stationary phases in use during the mid 1970s. In 1973 Leary et al. (34) reported the application of a statistical nearest-neighbor technique to the 226 stationary phases in the McReynolds study and suggested that just 12 phases could replace the 226. The majority of these 12 phases appear in Table 3.5. Delley and Friedrick found that four phases—OV-101, OV-17, OV-225, and Carbowax 20M—could provide satisfactory gas chromatographic analysis for 80% of a wide variety of organic compounds (35). Hawkes et al. (36) reported the findings of a committee effort on this subject and recommended a condensed list of 6 preferred stationary phases on which almost all gas–liquid chromatographic analysis can be performed: (1) a dimethylpolysiloxane (e.g., OV-101, SE-30, SP-2100), (2) a 50% phenylpolysiloxane (OV-17, SP-2250), (3) polyethylene glycol of mol wt > 4000 (Carbowax...), (4) DEGS, (5) a 3-cyanopropylpolysiloxane (Silar 10C, SP-2340), and (6) a trifluoropropylpolysiloxane (OV-210, SP-2401).

Another feature of the McReynolds constants is the guidance in the selection of a column that will separate compounds with different functional groups, such as ketones from alcohols, ethers from olefins, and esters from nitriles. If an analyst wishes a column to elute an ester after an alcohol, the stationary phase should have a larger Z' value with respect to its Y' value. In the same fashion, a stationary phase should exhibit a larger Y' value with respect to Z', if an ether is to elute before an alcohol. The appendices in Reference 10 list McReynolds constants in order of increasing ΔI for each probe in successive tables that are handy and greatly facilitate the column-selection process.

3.6.4 Optimization of Packed-Column Separations

Examination of the parameters in the van Deemter expression (Equation 2.44), term by term, provides a basis for optimizing a packed column separation. The plate height h of a packed column may be represented as the sum of the eddy diffusion, molecular diffusion, and mass transfer effects. Thus, to attain maximum column efficiency, each term in the plate height equation should be minimized.

Eddy Diffusion ($2\lambda d_p$). Also referred to as the multiple-path effect, eddy diffusion is minimized by using small particles of support materials. A support of 100/120 mesh produces a more efficient column than 60/80 mesh particles and should be used wherever possible. A support of lower mesh, e.g. 80/100 or 60/80 should be selected to avoid a high pressure drop within a long column. This term is also independent of linear velocity or flow-rate.

Molecular Diffusion ($2\gamma D_g/\mu$). This term becomes significant at very low flow-rates. This contribution may be minimized by using a carrier gas of high molecular weight (nitrogen, carbon dioxide or argon) because their diffusion coefficient, D_g, is lower than that of a lower molecular weight carrier gas (helium or hydrogen) yielding a lower minimum in a H vs μ profile (Figure 3.5). Factors affecting the choice of helium vs hydrogen will be discussed later. However, other factors can override carrier gas selection such as detector compatibility. Thus, helium is preferred over nitrogen as carrier gas with a thermal conductivity detector.

Mass Transfer Contribution, $[8kd_f^2/\pi^2(1+k)^2D_l]\mu$. This term requires compromises to be made. The magnitude of this term can be clearly

TABLE 3.5 McReynolds Constants and Cross-Reference of Commonly Used Stationary Phases

Phase	Min/Max Temperature (°C)	Chemical Nature	X′	Y′	Z′	U′	S′	Phases of Similar Structure
Squalane	20/100	Cycloparaffin	0	0	0	0	0	
Polysiloxanes								
DC 200	0/200	Dimethylsilicone	16	57	45	66	43	SP-2100, SE-30, OV-101, OV-1
DC-710	5/250	Phenylmethylsilicone	107	149	153	228	190	OV-11
SE-30	50/300	Dimethyl	15	53	44	64	41	SP-2100, OV-101, OV-1
SE-54	50/300	5% Phenyl, 1% vinyl	33	72	66	99	67	
OV-1	100/350	Dimethyl (gum)	16	55	44	65	42	SP-2100
OV-3	0/350	10% Phenyl-phenylmethyldimethyl	44	86	81	124	88	
OV-7	0/350	20% Phenyl-phenylmethyldimethyl	69	113	111	171	128	
OV-11	0/350	35% Phenyl-phenylmethyldimethyl	102	142	145	219	178	DC-710
OV-17	0/375	50% Phenyl-50% methyl	119	158	162	243	202	SP-2250
OV-22	0/350	65% Phenyl-phenylmethyldiphenyl	160	188	191	283	253	
OV-25	0/350	75% Phenyl-phenylmethyldiphenyl	178	204	208	305	280	
OV-61	0/350	33% Phenyl-diphenyldimethyl	101	143	142	213	174	
OV-73	0/325	5.5% Phenyl-diphenyldimethyl (gum)	40	86	76	114	85	
OV-101	0/350	Dimethyl (fluid)	17	57	45	67	43	SP-2100, SE-30, OV-1
OV-105	0/275	Cyanopropylmethyl-dimethyl	36	108	93	139	86	
OV-202	0/275	Trifluoropropyl-methyl (fluid)	146	238	358	468	310	

OV-210	0/275	Trifluoropropyl-methyl (fluid)	146	238	358	468	310	SP-2401
OV-215	0/275	Trifluoropropyl-methyl (gum)	149	240	363	478	315	
OV-225	0/265	Cyanopropylmethyl-phenylmethyl	228	369	338	492	386	SP-2300, Silar 5 CP
OV-275	25/275	Dicyanoallyl	629	872	763	1106	849	SP-2340
OV-330	0/250	Phenyl silicone-Carbowax copolymer	222	391	273	417	368	
OV-351	50/270	Carbowax-nitroterephthalic acid polymer	335	552	382	583	540	SP-1000
OV-1701	0/250	14% Cyanopropylphenyl	67	170	153	228	171	
Silar 5 CP	0/250	50% Cyanopropyl–50% phenyl	319	495	446	637	531	SP-2300, OV-225
Silar 10 CP	0/250	100% Cyanopropyl	520	757	660	942	800	SP-2340
SP-2100	0/350	Methyl	17	57	45	67	43	SE-30, OV-101, OV-1
SP-2250	0/375	50% Phenyl	119	158	162	243	202	OV-17
SP-2300	20/275	50% Cyanopropyl	316	495	446	637	530	OV-225
SP-2310	25/275	55% Cyanopropyl	440	637	605	840	670	
SP-2330	25/275	90% Cyanopropyl	490	725	630	913	778	
SP-2340	25/275	100% Cyanopropyl	520	757	659	942	800	Silar 10 CP
SP-2401	0/275	Trifluoropropyl	146	238	358	468	310	OV-210
NONSILICONE PHASES								
Apiezon L	50/300	Hydrocarbon grease	32	22	15	32	42	
Carbowax 20M	60/225	Polyethyleneglycol	322	536	368	572	510	Superox 4, Superox 20M
DEGS	20/200	Diethylene glycol succinate	496	746	590	837	835	
TCEP	0/175	1,2,3-Tris(2-cyanoethoxy) propane	594	857	759	1031	917	
FFAP	50/250	Free fatty acid phase	340	580	397	602	627	OV-351

Source. Data obtained from References 6, 26, 33.

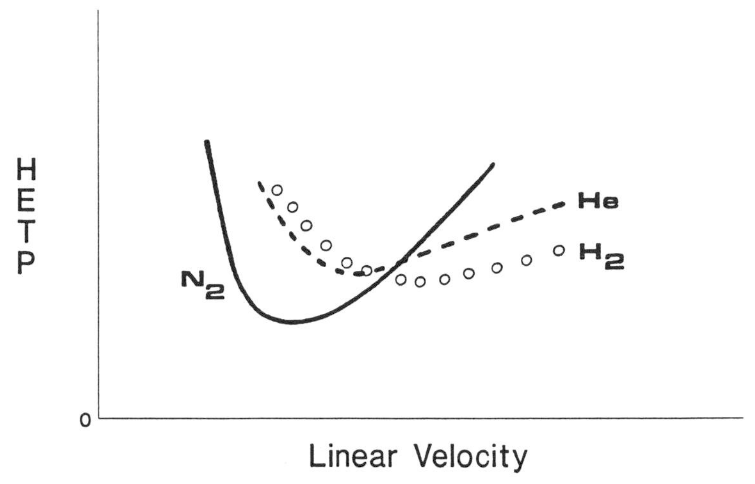

FIGURE 3.5 Plot of HETP versus linear velocity.

minimized by decreasing film thickness (by using a packing have a lower loading of stationary phase). Therefore, column efficiency increases (and time of analysis decreases) with a decrease in stationary phase loading, as may be seen in Figure 3.6. If a support is too thinly coated, the exposure of active sites on the support may cause adsorption of solutes. A decrease in column temperature lowers the magnitude of the k' term; however, lowering of the column temperature also decreases D_l by increasing the viscosity of the stationary phase. Effects of the various changes in chromatographic parameters on resolution that can be implemented are schematically illustrated in Figure 3.7.

A relationship between stationary phase concentration and column temperature is depicted in Figure 3.8. Decreasing column temperature increases time of analysis; to have the same analysis time on a heavier loaded packing in an identical column at the same flowrate requires a higher column temperature.

3.7 COLUMN PREPARATION

Most laboratories today purchase columns of designated dimensions (length, o.d., and i.d.) containing a specified packing, e.g., untreated or treated support coated with a given liquid phase loading, directly from a chromatography vendor. In-house preparation of packings and filling columns can be time-consuming and is false economy; more importantly, vendors can do the job better. Some supports are difficult to coat uniformly, while some packings present a problem when filling a column. Nevertheless, presented

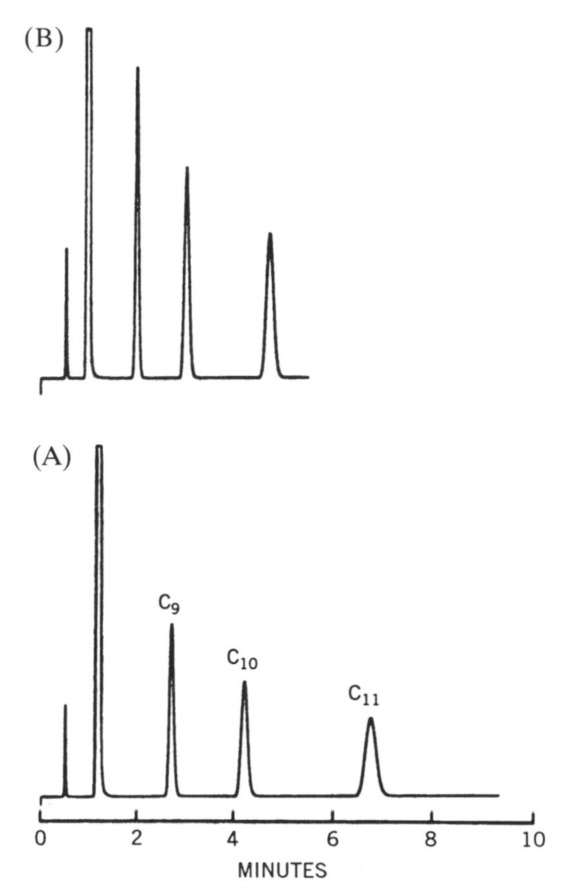

FIGURE 3.6 Chromatograms of *n*-alkanes on (A) a 6-ft glass column, 2 mm i.d., containing 20% OV-101 on 80/100-mesh Chromosorb W HP, column temperature: 140°C; (B) same as in (A) but with thinner film of chemically bonded OV-101 on same support. Flowrate: 25 mL/min He, Det: FID. (From Reference 5.)

below are guidelines and concise descriptions of recommended procedures to be employed for preparing a packed column.

3.7.1 Description of Coating Methods

The techniques of solvent evaporation and solution coating are the most commonly used procedures for deposition of a liquid phase on a support. Solvent evaporation is employed for coating supports with high concentrations (>15%) of viscous phases, while the solution-coating method produces a more uniform phase deposition and is more widely utilized.

In solvent evaporation, a known amount of stationary phase is dissolved in an appropriate solvent. A weighed amount of support is added to the

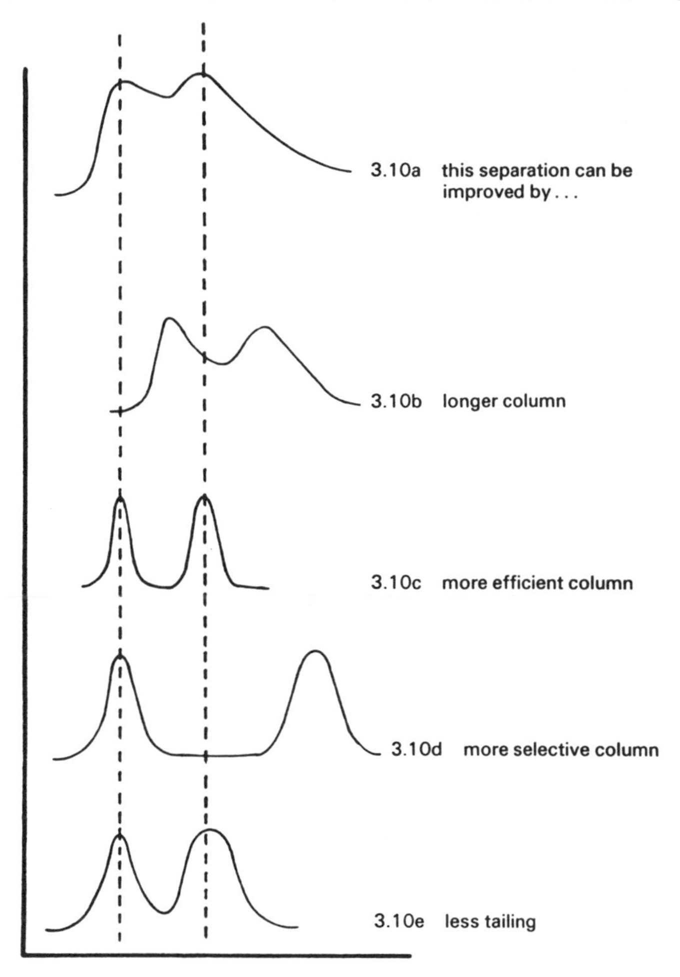

FIGURE 3.7 Effect of selected column changes on resolution. Reproduced from reference (18): W. A. Supina, in *Modern Practice of Gas Chromatography*, R. L. Grob (Ed.) second edition, Copyright 1985, John Wiley & Sons, Inc. Reprinted by permission of John Wiley & Sons, Inc.

solution and the solvent is allowed to slowly evaporate from the slurry. Since all stationary phase is deposited on the support, stirring or thorough mixing is a necessity, otherwise nonuniform deposition of phase will result. In Figure 3.9 a series of scanning electron micrographs of 20% Carbowax 20M on 80–100-mesh Chromosorb W HP are illustrated. A photomicrograph of a nonhomogeneous deposition of phase is shown in Figure 3.9A, where a large amount of polymer distributed between two particles is visible in the left-hand portion of the photograph. This packing ultimately yielded a column of low efficiency because of unfavorable mass transfer, as opposed to the higher column efficiency associated with a column packed with a more uniformly coated support (Figure 3B).

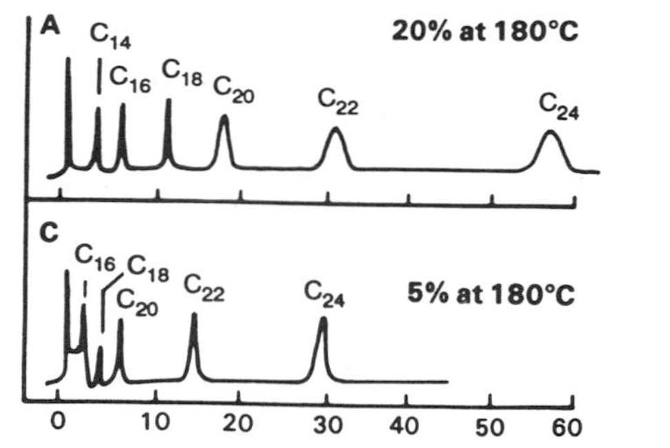

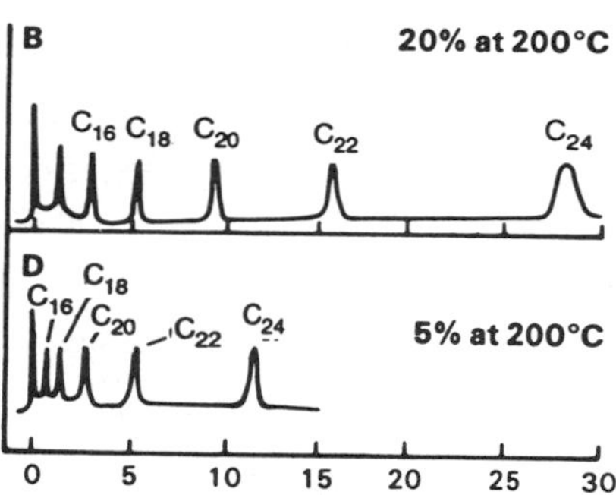

FIGURE 3.8 Effect of concentration of stationary phase and column temperature on sample resolution (methyl esters of fatty acids). Reproduced from reference (18): W. A. Supina, in *Modern Practice of Gas Chromatography*, R. L. Grob (Ed.) second edition, Copyright 1985, John Wiley & Sons, Inc. Reprinted by permission of John Wiley & Sons, Inc.

The technique of solution coating consists of the following steps:

1. A solution of known concentration of liquid phase in its recommended solvent is prepared.
2. To a known volume of this solution is added a weighed amount of solid support.
3. The resulting slurry is transferred into a Buchner funnel, and the remaining solvent is removed by vacuum.
4. The volume of filtrate is measured.
5. After suction is completed the wet packing is allowed to air-dry on a tray in a hood to remove residual solvent. Do not place damp packing into a laboratory drying oven!
6. The mass of liquid phase retained on the support is computed since the concentration of liquid phase in the solution is known.

This technique produces a uniform coating of a support, minimum generation of fines from the support particles, and minimum oxidation of the stationary phase. Further details about these procedures are described in References 10 and 18.

3.7.2 Tubing Materials and Dimensions

The nature and reactivity of the sample will govern the choice of tubing for packed-column GC. Of the available materials, glass is the most inert and the best material for most applications, although somewhat fragile. Overtightening a fitting attached to a glass column can cause the dreaded "ping" sound of broken glass. Utilization of a special torque wrench, which breaks apart itself instead of breaking the glass column when a specific torque level is exceeded, is a good investment for a gas chromatographic laboratory.

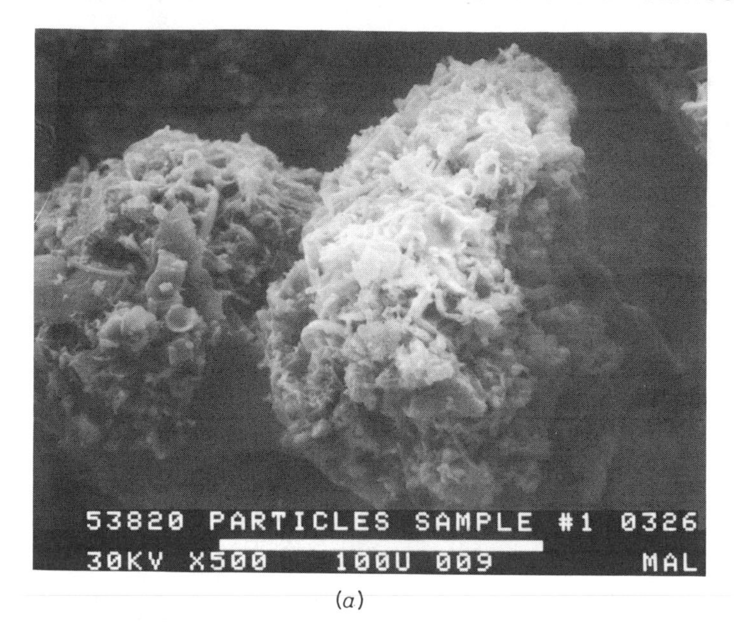

(*a*)

(*b*)

FIGURE 3.9 Scanning electron photomicrographs of (A) nonuniform coating of 20% Carbowax 20M on Chromosorb W HP, 80/100 mesh (note the stationary phase "pooling" in left-hand portion of photograph) and (B) a more uniform coating of Carbowax 20M. (From reference 5.)

Empty glass columns need deactivation or silylation prior to packing. Usually this is accomplished by filling a thoroughly cleaned empty column with a 5% solution of DMDCS in toluene. After standing for 30 min, the column is rinsed successively with toluene and methanol, then purged with dry nitrogen, after which it is ready to be packed. Moreover, as opposed to a metal column, the use of glass permits direct visualization of how well a column is packed after filling and also after the column has been used for separations. Teflon tubing, also inert, is used for the analysis of sulfur gases, halogens, HF, and HCl, but has a temperature limitation of 250°C.

Nickel tubing offers the attractive combination of the durability and strength of metal tubing with the favorable chemical inertness of glass. Stainless steel is the next least reactive material and is utilized for analysis of hydrocarbons, permanent gases, and solvents. As is the case with all metal tubing, stainless steel columns should be rinsed with nonpolar and polar solvents to remove residual oil and greases. When used for the analysis of polar species, a higher grade of stainless steel tubing with a polished inner surface is recommended. Copper and aluminum tubing have been employed for noncritical separations, but their use is not recommended and should be restricted to the plumbing of cylinder-instrument gas lines. Oxide formation can occur on the inner surface of these materials, resulting in adsorptive tailing and/or catalytic problems under chromatographic conditions.

3.7.3 Glass Wool Plugs and Column Fittings

Chromatographic packings are retained within a column by a wad or plug of glass wool. Since the chemical nature of the wool closely resembles that of the glass column, it should be deactivated by the same procedure used for glass columns. It is advisable to further soak the wool in a dilute solution of H_3PO_4 for the analysis of acidic analytes such as phenols and fatty acids. Untreated or improperly treated glass wool exhibits an active surface and can cause peak-tailing. Alternatives to glass wool are stainless steel frits and screens for gas chromatographic purposes, available from vendors of chromatographic supplies.

The ferrules and metal retaining nuts are used to form a leak-tight seal of a column in a gas chromatograph. Criteria for selection of the proper ferrule material are column diameter, column tubing material, maximum column temperature, and whether the connection is designated for a single use or for multiple connections and disconnections. Ferrules fabricated from various materials for metal-to-metal, glass-to-metal, and glass-to-glass connections are commercially available for use with $\frac{1}{16}$-, $\frac{1}{8}$-, and $\frac{1}{4}$-in. o.d. packed columns. The properties and characteristics of common types are presented in Table 3.6.

3.7.4 Filling the Column

The following procedure for packing columns, with practice, can produce the desired goal of a tight packing bed with minimum particle fracturing.

TABLE 3.6 Ferrule Materials for Packed Columns

Material	Temperature Limit (°C)	Properties
Metal		
Brass	250	Permanent connection on metal columns.
Stainless steel	450	Permanent connection on metal columns.
Teflon	250	Low upper temperature limit and cold-flow properties render this material unsuitable for temperature programming and elevated temperature operation, reusable to some extent.
Ceramic-filled Teflon	250	Isothermal use only; conforms easily to glass. Used for connections to mass spectrometers.
Graphite (G)	450	High temperature limit with no bleed or decomposition; soft and easily deformed upon compression; may be resealed only a limited number of times.
Vespel		
100% polymide (PI)	350	Good reusability factor; can be used with glass, metal, and Teflon columns; may seize on metal and glass columns with use at elevated temperatures.
Vespel/graphite		
85% PI, 15% G 60% PI, 40% G	400	Excellent reusability; will not seize to glass or metal; performs better than graphite and Vespel alone; 60% PI composite seals with lesser torque and has added lubricity.

First, a metal column is precoiled for easy attachment to the injector and detector of the instrument in which it is to be installed, or a precoiled glass column configured for a specific instrument is procured from a vendor. Insert a large wad of glass wool partially into one end of the column, align the excess wool along the outside of the tubing, overlap the excess wool with vacuum tubing, and attach the other end of the vacuum tubing to a faucet aspirator or pump (10). After securing a small funnel to the other end of the column, add packing material in small incremental amounts into the funnel and gently tap the packing bed while suction is applied. After the column is completely packed, insert a small piece of silanized glass wool into the inlet end of the column, disconnect the vacuum, remove the vacuum tubing and the large wad of glass wool, replacing it with a smaller plug of silanized wool. This approach eliminates the exasperating sight of your packing material zipping out of the column during filling if a insufficient tight wad of silanized glass wool was initially inserted into the outlet end of the column.

3.7.5 Conditioning the Column and Column Care

Before a column is used for analyses, it must be thermally conditioned. This is done by heating the column overnight at an oven temperature below the upper limit of the stationary phase with a normal flowrate of carrier gas. The column should not be connected to the detector during conditioning. The purpose of conditioning is the removal from the column of residual volatiles and low boiling species present in the stationary phase that otherwise would produce an unsteady baseline at elevated column temperatures, commonly referred to as column bleed, and contaminate the detector. Conditioning a column also helps in the redistribution of the liquid phase on the solid support. The degree of conditioning is dependent on the nature and amount of liquid phase in the column; usually heating a column overnight at an appropriate elevated temperature produces a steady baseline under chromatographic conditions the following day. Analyses using the more sensitive detectors (ECD, NPD, MS) may require an even longer column conditioning period.

The following guidelines can prolong the life-time of a column:

1. Any gas chromatographic column, new or conditioned, packed or capillary, should be purged with dry carrier gas for 15–30 min before heating to a final elevated temperature to remove the detrimental presence of air.
2. A column should not be rapidly or ballistically heated to an elevated temperature, but should be heated by slow to moderate temperature programming to the desired final temperature.
3. Excessively high conditioning and operating temperatures reduce the life-time of any gas chromatographic column.
4. Use "dry" carrier gas or install a moisture trap in the carrier gas line. Do not inject aqueous sample on a column containing a stationary phase intolerant of water.
5. The accumulation of high boiling compounds from repetitive sample injections occurs at the inlet end of column and results in discoloration of the packing. It is a simple matter to remove the discolored segment of packing and replace it with fresh packing material. This action prolongs the column lifetime.
6. Do not thermally shock a column by disconnecting a column while hot. Allow the column to cool to ambient temperature prior to disconnection. Packings are susceptible to oxidation when hot.
7. Cap the ends of a column for storage to prevent air and dust particles from entering the column. Save the box in which a glass column was shipped for safe storage of the column.

Part 3 Capillary-Column Gas Chromatography

3.8 INTRODUCTION

The capillary column, also referred to as an open tubular column because of its open flow path, offers a number of advantages over the packed column. These merits include vastly improved separations with higher resolution, reduced time of analysis, smaller sample size requirements, and often higher sensitivities. The arrival of fused silica as a capillary-column material had a major impact on capillary gas chromatography and, in fact, markedly changed the practice of gas chromatography. In 1979 the number of sales of gas chromatographs with capillary capability was less than 10%; this figure increased to 60% by 1989 (37) and is consistent with the results of the user surveys discussed in Section 3.3. It is a safe prediction that the number of capillary-column users will continue to increase as further developments in capillary-column technology and instrumentation are made. Here in Part 3, theoretical and practical considerations of the capillary column are discussed.

3.8.1 Significance and Impact of Capillary GC

Marcel Golay is credited with the discovery of the capillary column. In 1957, Golay presented the first theoretical treatment of capillary-column performance when he illustrated that a long length of capillary tubing having a thin coating of stationary phase coating the narrow inner diameter of the tube offered a tremendous improvement in resolving power compared to a conventional packed column (38). Such a column is also often referred to as a wall-coated open tubular column (WCOT). The high permeability or low resistance to carrier-gas flow of capillary column enables a very lengthy column to generate a large number of theoretical plates.

In contemporary practice, separations of high resolution are attainable by capillary GC, as illustrated in the chromatogram in Figure 3.10, which was generated with a conventional fused-silica capillary column. An exploded view of this chromatogram of a gasoline-contaminated diesel fuel with a data acquisition system indicates the presence of over 400 chromatographic peaks. Because of its separation power, capillary gas chromatography has become synonymous with the term "high-resolution gas chromatography."

3.8.2 Chronology of Achievements in Capillary GC

The first column materials employed in the developmental stage of the technique were fabricated from plastic materials (Tygon and nylon) and metal (aluminum, nickel, copper, stainless steel, and gold). Plastic capillaries, being thermoplastic in nature, had temperature limitations, whereas metallic capillary columns had the disadvantage of catalytic activity. Rug-

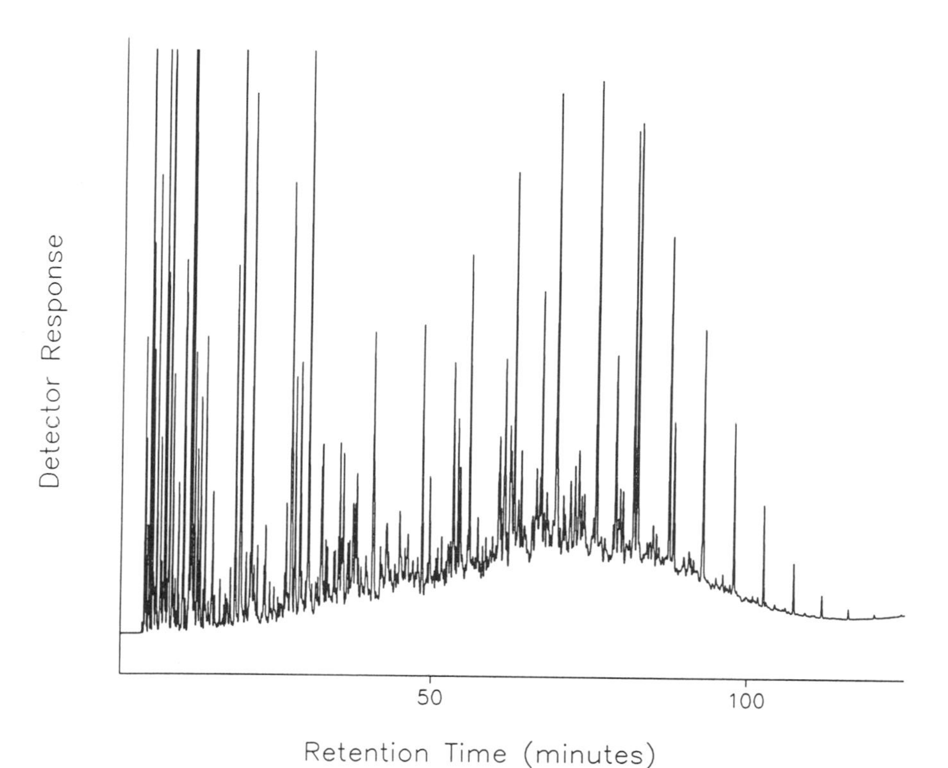

FIGURE 3.10 Chromatogram of a sample of diesel fuel contaminated with gasoline on 30-m × 0.25-mm-i.d. HP-1 (0.25-μm film). Column Temperature conditions: 30°C (5 min), 2°C/min to 250°C; split injection (100:1), Det: FID.

ged, flexible stainless steel columns rapidly became state of the art, and were widely used for many applications, mainly for petroleum analyses. The reactive metallic surface proved to be unfavorable in the analysis of polar and catalytically sensitive species. In addition, only the split-injection mode was available for quite some time for the introduction of the small quantities of sample dictated by a thin film of stationary phase within the column. As the surface chemistry of glass was gradually studied and understood, capillaries made of borosilicate and soda–lime glass became popular in the 1970s and replaced metal capillary columns (39). Here the metal oxide content and presence of silanol groups on the glass surface necessitated carefully controlled deactivation and coating procedures, but separations obtained on glass capillaries were clearly superior to those obtained with metal capillaries. The fragility of glass often proved to be problematic, requiring restraightening of a capillary end upon breakage with a minitorch followed by a recoating of the straightened portion with a solution of stationary phase to deactivate the straightened segment.

Today equivalent or superior separations with a fused-silica capillary column can be generated with the additional feature of ease of use; the flexibility of fused silica is illustrated in Figure 3.11. The most significant advancement in capillary gas chromatography occurred in 1979 when Dandeneau and Zerenner of Hewlett–Packard introduced fused silica as a column material (40, 41). The subsequent emergence of fused silica as the column material of choice for high-resolution gas chromatography is responsible for the widespread use of the technique and has greatly extended the range of gas chromatography. In the next decade, there were several other major developments in capillary gas chromatography. Instrument manufacturers responded to the impact of fused-silica columns by designing chromatographs with injection and detector systems optimized in performance for fused silica columns. There were also concurrent advances in the area of microprocessors. Reporting integrators and fast data acquisition systems with increased sampling rates became available to be compatible with the narrow bandwidths of capillary peaks. The stature of the capillary column was further enhanced by development of the cross-linked

FIGURE 3.11 Photograph illustrating the flexibility of fused silica. (Photograph courtesy of the Hewlett–Packard Company.)

or bonded stationary phase within the column. A column containing a cross-linked phase has an extended lifetime because it has high thermal stability and can tolerate large injection aliquots of solution without redistribution of the stationary phase. Inlet discrimination was addressed with the development of on-column injection, the programmed temperature vaporizer, and, more recently, electronic pressure-controlled injection.

Since Golay's proposal of the use of the capillary column, capillary gas chromatography has exhibited spectacular growth, maturing into a powerful analytical technique. Some of the more notable achievements in capillary gas chromatography are listed in Table 3.7.

3.8.3 Comparison of Packed and Capillary Columns

Three stages in the evolution of the capillary-column technology are presented in Figure 3.12: a packed-column separation and two separations with a stainless steel and glass capillary column. Better resolution is evident with the capillary chromatograms because more peaks are separated and smaller peaks can be detected. The superior performance of the glass capillary column is clearly apparent.

In addition to providing a separation where peaks have narrower bandwidths compared to a packed-column counterpart, a properly prepared

TABLE 3.7 Advancements in Capillary Gas Chromatography

Year	Achievement
1958	Theory of capillary column performance, GC Symposium in Amsterdam
1959	Sample inlet splitter
1959	Patent on Capillary Columns by Perkin–Elmer
1960	Glass drawing machine developed by Desty
1965	Efficient glass capillary columns
1975	First capillary column symposium
1978	Splitless injection
1979	Cold on-column injection
1979	Fused silica introduced by Hewlett–Packard
1981–1984	Deactivation procedures and cross-linked stationary phases
1983	Megabore column introduced as an alternative to the packed column
1981–1988	Interfacing capillary columns with spectroscopic detectors (MS, FTIR, AED)
1981–1992	Programmed temperature vaporizer and electronic pressure controlled sample inlet systems; more affordable bench-top GC-MS systems

Source: Data courtesy of the Americas Training Center, Hewlett–Packard Co., Atlanta, GA.

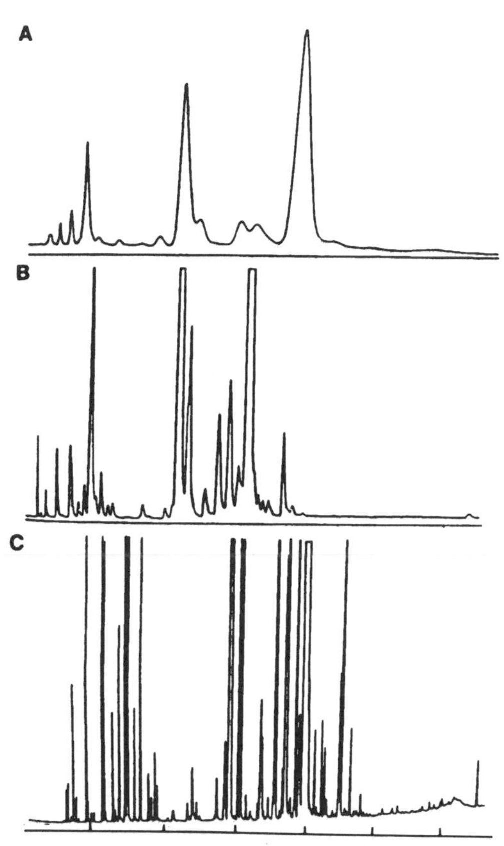

FIGURE 3.12 Optimized separations of peppermint oil on (A) a 6-ft × 0.25-in.-i.d. packed column, (B) a 500-ft × 0.03-in.-i.d. stainless steel capillary column, and (C) a 50-m × 0.25-mm-i.d. glass capillary column. Stationary phase on each column was Carbowax 20M; reference: W. Jennings, *J. Chromatogr. Sci.* 17, 637 (1979). Reproduced from the *Journal of Chromatographic Science* by permission of Preston Publications, a Division of Preston Industries, Inc.

fused-silica capillary column, which has an inert surface (less potential for adverse adsorptive effects toward polar species), yields better peak shapes; i.e., bands are sharper with less peak tailing, which facilitates trace analysis as well as provides more reliable quantitative and qualitative analyses. Sharp, narrow bands of the trace components present in a capillary chromatogram such as that in Figure 3.10 have increased peak height relative to the peak of the same component at identical concentration in a packed-column chromatogram where the peak may be unresolved or disappear in the baseline noise. Moreover, due to the low carrier gas flowrate, greater detector sensitivity, stability, and signal-to-noise levels are possible with a capillary column. One drawback of the capillary column, though, is its limited sample capacity, which requires dedicated inlet systems to introduce small quantities of sample commensurate with low amount of stationary phase. Operational parameters of packed and capillary columns are further contrasted in Table 3.8.

The superior performance of a capillary column can be further viewed in the following manner. Because of the geometry and flow of a gas through a packed bed, molecules of the same solute can take a variety of paths through the column enroute to the detector (via eddy diffusion), whereas in a capillary column all flow paths have nearly equal length. The open geometry of a capillary column causes a lower pressure drop, allowing longer columns to be used. Since a packed column contains much more stationary phase, often thickly coated on an inert solid support, there are locations in the packing matrix where the stationary phase spans or spreads over to adjacent particles (Figure 3.9A). Some molecules of the same component encounter thinner regions of stationary phase, whereas other molecules have increased residence times in these thicker pools of phase, all of which create band broadening. On the other hand, a capillary column contains a relatively thin film of stationary uniformly coated on the inner wall of the tubing. These factors, collectively considered, are responsible for the sharp band definition and narrow retention time distribution of molecules of a component eluting from a capillary column.

At higher column oven temperatures with increased linear velocity of carrier gas, capillary separations can be achieved that mimic those on a packed column but with a shortened time of analysis. The reduced amount of stationary phase in a capillary column imparts another advantage to the chromatographer, namely, one observes less bleed of stationary phase from the column at elevated temperatures and this means less detector contamination. Theoretical considerations of the capillary column are discussed in Section 3.10.

TABLE 3.8 Comparison of Wall-Coated Capillary Columns With Packed Columns

	Packed	Capillary
Length (m)	1–5	5–60
Inner diameter (mm)	2–4	0.10–0.53
Plates per meter	1000	5000
Total plates	5000	300,000
Resolution	Low	High
Flowrate (mL/min)	10–60	0.5–15
Permeability (10^7 cm^2)	1–10	10–1000
Capacity	10 μg/peak	<100 ng/peak
Liquid film thickness (μm)	10	0.1–1

Source. Data obtained in References 42 and 43.

3.9 CAPILLARY-COLUMN TECHNOLOGY

3.9.1 Capillary-Column Materials

After Desty developed a glass drawing and coiling apparatus (44), focus then shifted away from metal capillary columns to fabrication of columns from more inert borosilicate and soda–lime glass. Glass is inexpensive and readily available, and glass columns could be conveniently drawn in-house with dimensions (length and inner diameter) tailored to individual needs. Investigators quickly realized that this increase in the column inertness of glass was at the expense in flexibility. With fused silica, a column could be fabricated from a material having the flexibility of stainless steel with an inner surface texture more inert than glass. Thus, fused silica quickly replaced glass as the capillary-column material of choice.

Fused Silica and Other Glasses. To cultivate an understanding for the widespread use of used silica as a column material for capillary GC, it is helpful to examine the chemical structures of glasses in Figure 3.13 that have been used as column materials and the corresponding metal oxide con-

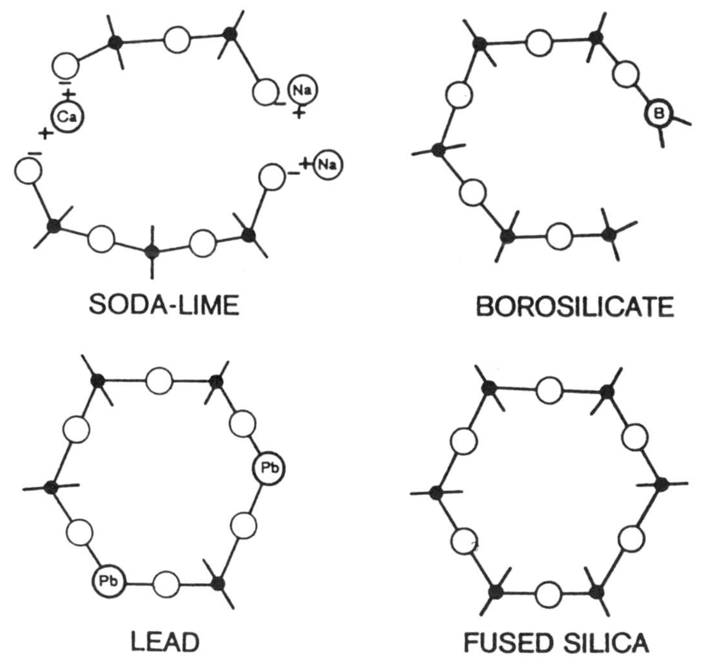

FIGURE 3.13 Schematic representations of the structures of different glasses. (Reproduced from Reference 46 and reprinted with permission of Dr. Alfred Huethig Publishers.)

TABLE 3.9 Approximate Glass Composition (%)

Glass	SiO_2	Al_2O_3	Na_2O	K_2O	CaO	MgO	B_2O_3	PbO	BaO
Soda-lime (Kimble R6)	68	3	15	—	6	4	2	—	2
Borosilicate (Pyrex 7740)	81	2	4	—	—	—	13	—	—
Potash soda lead (Corning 120)	56	2	4	9	—	—	—	29	—
Fused silica	100	Less than 1 ppm total metals							

Source. Data abstracted from References 42 and 46.

centration data presented in Table 3.9. Column activity may be attributed to exposed silanol groups and metal ions on the surface of a glass capillary. While soda–lime borosilicate glasses, for example, have percentage levels of metal oxides, the metal content of synthetic fused silica is less than 1 ppm. Although quartz tubing is commercially available, its metal oxide content (10–100 ppm) is considered to be too great for use in capillary gas chromatography (45). Metal oxides are considered to be Lewis acids and can serve as adsorptive sites for electron–donor species such as ketones and amines and as an active site for species with π bonding capability (aromatics and olefins). Boron impurities in glass also act as Lewis acid sites capable of chemisorbing electron donors (46). The absence of these adsorptive sites in fused silica is responsible for its remarkable inertness and is a direct result of the synthesis of this material. However, the hydroxyl groups attached to tetravalent silicon atoms are of paramount significance because they can contribute residual column activity.

Synthetic fused silica is formed by introducing pure silicon tetrachloride into a high-temperature flame followed by reaction with the water vapor generated in the combustion (46). The process can be described by the reaction

$$SiCl_4 + 2H_2O \rightarrow SiO_2 + 4HCl \tag{3.2}$$

There are three distinct categories of silanol groups present on the surface of fused silica shown in Figure 3.14. First, there are free silanol groups, which are acidic adsorptive sites with $K_a = 1.6 \times 10^{-7}$. The surface concentration of free silanols on fused silica has been calculated to be 6.2 μmol/m^2. This type of silanol group has a direct bearing on column behavior. Geminal silanols, the situation where two hydroxyl groups are attached to the same silicon atom, are also present at a concentration of 1.6 μmol/m^2 (47). Third, there are vicinal silanol functionalities characterized by hydroxyl groups attached to adjacent silicon atoms. Here steric effects become important. For instance, vicinal silanols represent a rather weak adsorptive

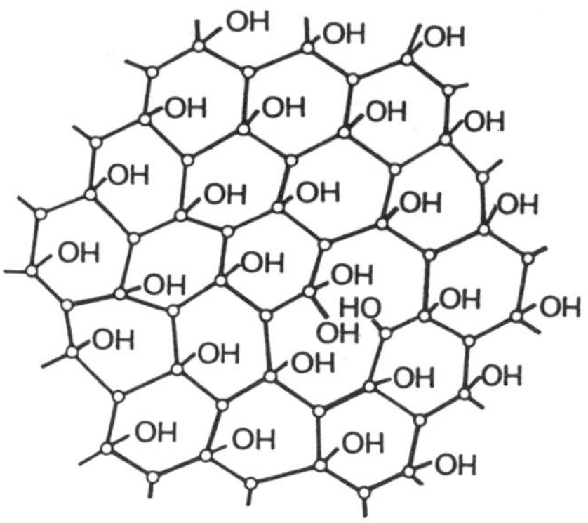

FIGURE 3.14 Probable structure of fused silica. (Reproduced from Reference 46 and reprinted with permission of Dr. Alfred Huethig Publishers.)

site but in the presence of water can be rendered active (46):

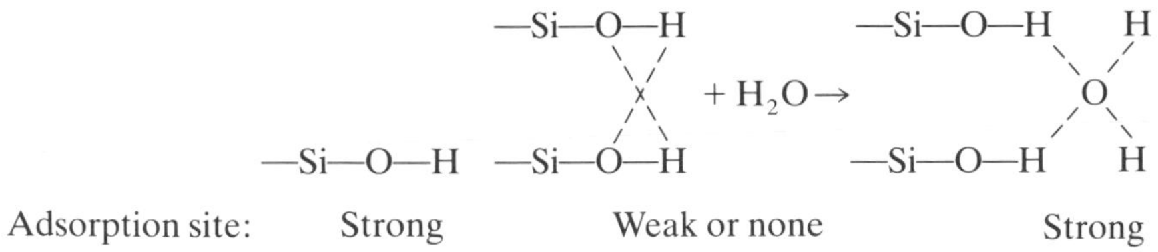

```
                      —Si—O—H                 —Si—O—H    H
                          \ /                         \  /
                           X     + H2O→                 O
                          / \                         /  \
—Si—O—H               —Si—O—H                 —Si—O—H    H
```

Adsorption site:	Strong	Weak or none	Strong

(3.3)

If the interatomic distance of neighboring oxygen atoms is between 2.4 and 2.8 Å, the groups are hydrogen bonded, if this distance exceeds 3.1 Å, hydrogen bonding does not occur and the free silanol behavior dominates (48). Bound silanol groups can dehydrate producing siloxane bridges under certain conditions. More detailed information on the complexities of silica surface chemistry may be found in the books by Jennings (46) and Lee et al. (42).

Extrusion of a Fused-Silica Capillary Column. Three steps are involved in the preparation of a fused silica column: (1) the high-temperature extrusion of the blank capillary tubing from a preform, where the capillary receives a protective outer coating in the same process; (2) the deactivation of the inner surface of the column; and (3) the uniform deposition of a stationary phase of a desired film thickness on the deactivated inner surface. In this section the extrusion of fused silica will be described; the procedures employed for the deactivation and coating of fused silica capillaries are presented in the following section.

Prior to extrusion the fused silica preform is usually treated with dilute hydrofluoric acid to remove any imperfections and deformations present on the inner and outer surfaces and then rinsed with distilled water and followed by annealing (49). In a clean-room atmosphere, the preform is vertically drawn through a furnace maintained at approximately 2000°C. Guidance and careful control of the drawing process is achieved by focusing an infrared laser beam down the middle of the capillary in conjunction with feedback control electronic circuity in order to maintain uniformity in the specifications of the inner and outer diameters in the final product.

Fused silica drawn in this manner exhibits a very high tensile strength and has excellent flexibility due to the thin wall of the capillary. However, the thin wall of the capillary is subject to corrosion upon exposure to atmospheric conditions and is extremely fragile. To eliminate degradation and increase its durability, the fused-silica tubing receives a protective outer coating, usually of polyimide, although other coating materials have been used, including silicones, gold, vitreous carbon, and polyamides. Polyimide, which also serves as a water barrier, is most widely used because it offers temperature stability to 400°C. The color of polyimide seems to vary slightly from one column manufacturer to another, with no effect, however, on column performance.

Aluminum-Clad Fused-Silica Capillary Columns. There are number of application areas requiring columns to be operated at or above 400°C, such as the analysis of waxes, crude oils, and triglycerides. These have driven efforts to replace the polyimide outer coating with a thin (20 μm) layer of aluminum and extend the temperature range of capillary gas chromatography, as illustrated in the chromatograms of high-temperature capillary separations in Figure 3.15. An Al-clad capillary column has excellent heat transfer while maintaining the same flexibility and inertness of the fused-silica surface as the polyimide-coated columns. Trestianu and coworkers showed that for an alkane of high carbon number the elution temperature on a high-temperature column is 100°C lower than with a corresponding packed column (50). It must be emphasized here that to obtain optimum column performance, the injection mode is critical. Cold on-column or programmed temperature vaporizer injectors are recommended to avoid inlet discrimination problems for the analysis of solutes of high molecular weight with this type of capillary column.

Fused-Silica-Lined Stainless Steel Capillary Columns. A third type of protective outer coating, stainless steel, for fused silica offers an alternative to aluminum-clad fused silica for elevated column temperatures. This technology is the inverse of that for polyimide-clad fused-silica capillary where a layer of fused silica is deposited on the inner surface of a stainless capillary. In Figure 3.16 scanning electron micrographs are displayed to compare the rough surface of stainless steel with the smooth surface of untreated fused silica and the surface of stainless steel after a micron layer of deactivated fused silica is bonded to its interior wall, termed Silcosteel. In addition to high thermal stability, a distinguishing feature of a fused-silica-lined, thin-walled stainless steel capillary column is that it can be coiled in a diameter less than 4 in. (compared to larger diameters with polyimide-clad fused silica) without breakage, making it a very favorable column material

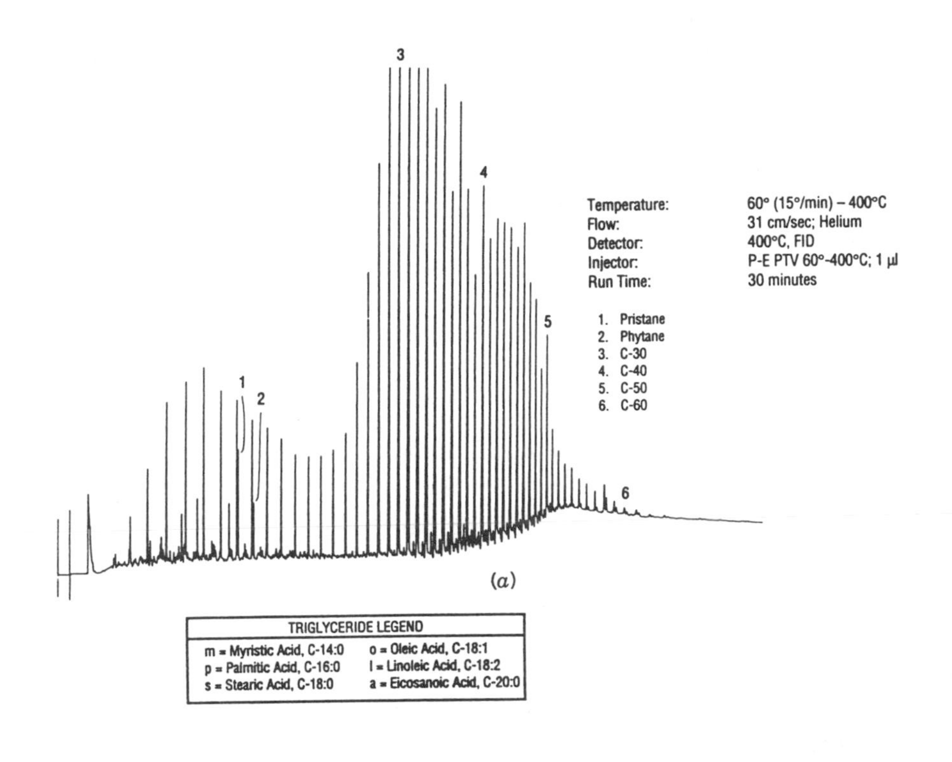

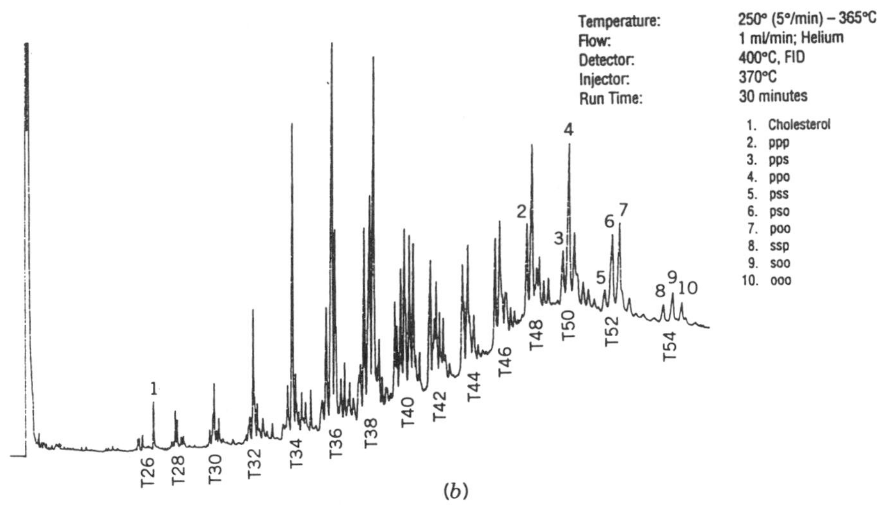

FIGURE 3.15 Chromatogram of separation of (A) Canadian wax on a 15-m × 0.25-mm-i.d. aluminum-clad capillary column (0.1-μm film) and (B) triglycerides on a 25-m × 0.25-mm-i.d. aluminum-clad capillary column (0.1-μm film). (Chromatograms courtesy of the Quadrex Corporation.)

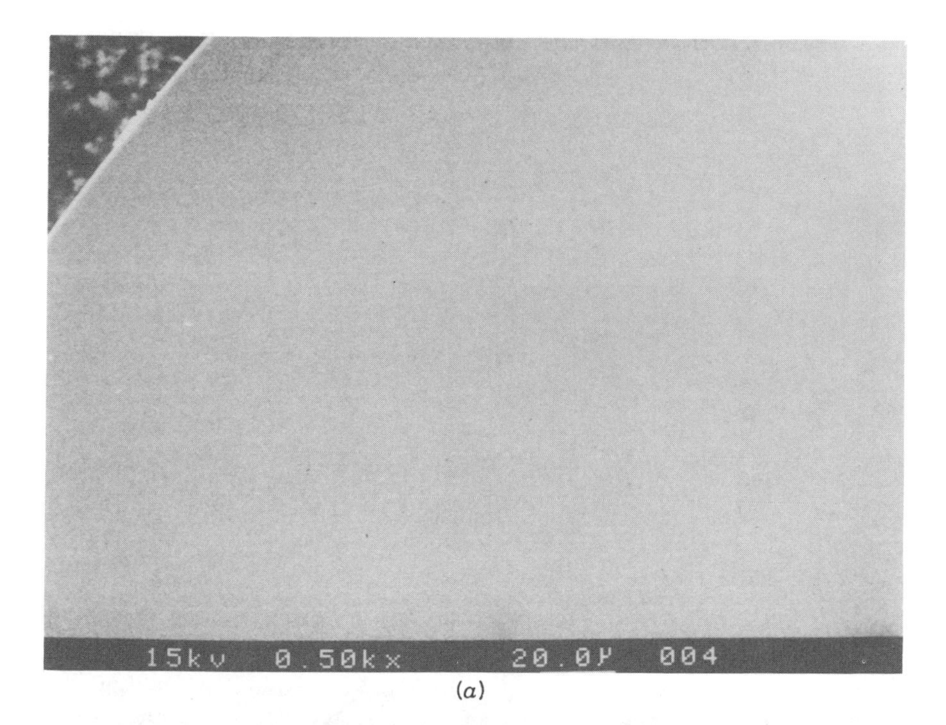

(*a*)

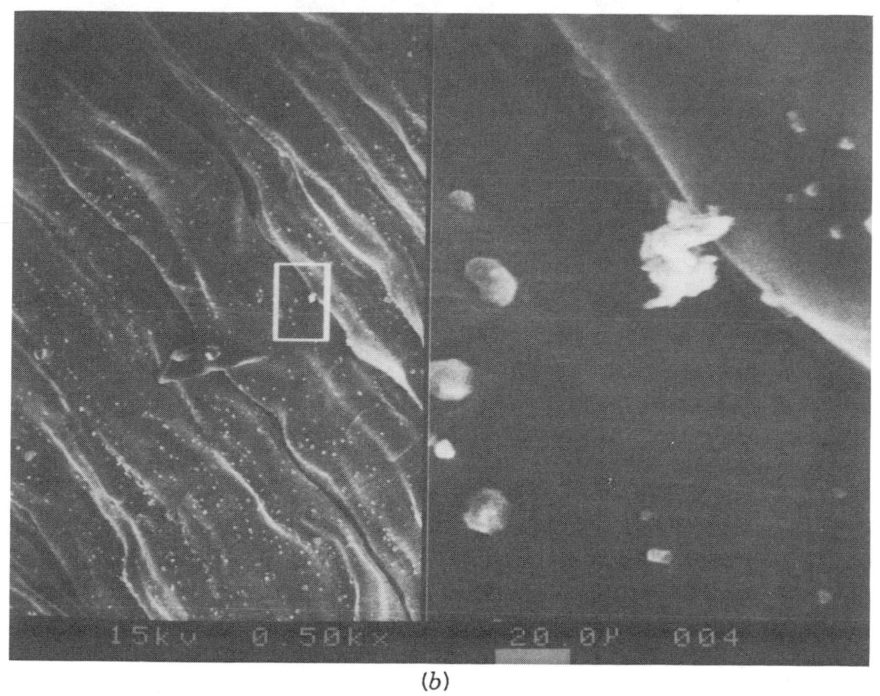

(*b*)

FIGURE 3.16 Scanning electron micrographs of (A) untreated fused silica, (B) rough inner surface of stainless steel capillary tubing, and (C) the smoother inner surface of the stainless steel capillary tubing after deposition of thin layer of fused silica. (C) also illustrates regions where fused silica lining was selectively removed to expose the untreated stainless steel surface below. (Scanning electron micrographs courtesy of the Restek Corporation.)

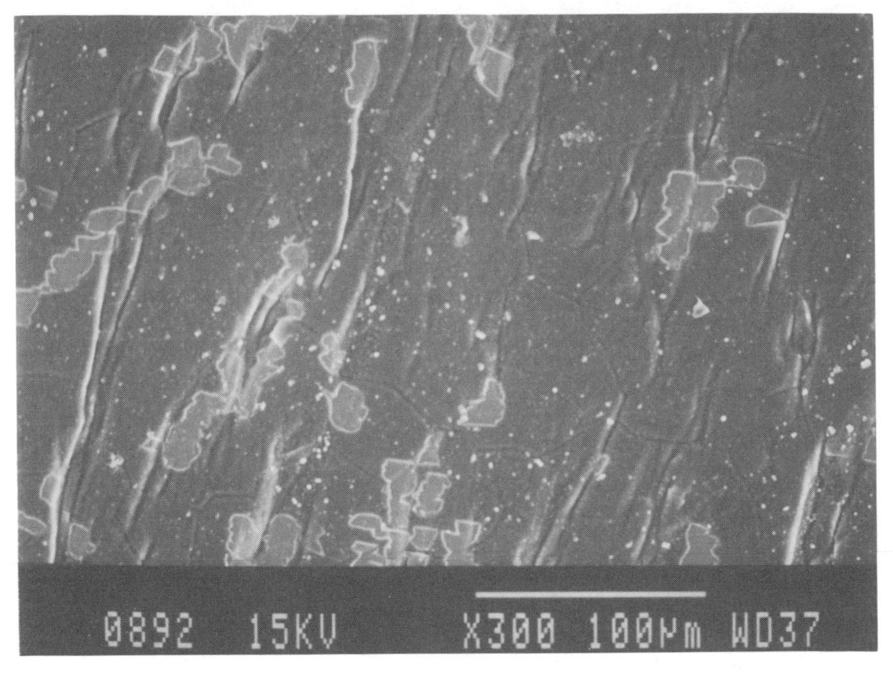

(*c*)

FIGURE 3.16. (*Continued*)

for process control and portable gas chromatographs where size of a column oven, shock resistance, and ruggedness become limiting factors. In Figure 3.17 photographs of Al-clad and fused-silica-lined stainless steel capillaries are illustrated.

3.9.2 Preparation of a Fused-Silica Capillary Column

Most users of a modern capillary column regard it as a high-precision and sophisticated device and elect to purchase columns from a vendor (and correctly so). Few give any thought to the steps involved in column preparation. Their number one priority is understandably the end result of accurate and reproducible chromatographic data that the column can provide. In this section, an overview of deactivation and coating of a fused-silica column with stationary phase is discussed.

Silanol Deactivation. For maximum column performance, blank or raw fused-silica tubing must receive pretreatment prior to the final coating with stationary phase. The purpose of pretreatment is twofold: to cover up or deactivate active surface sites and to create a surface more wettable by the phase. The details of the procedure are dependent on the stationary phase to be subsequently coated, but deactivation is essential for producing a column having a uniform film deposition along the inner wall of the capillary.

Although metal ions are not a factor with fused silica, the presence of silanol groups still must be addressed, otherwise the column has residual surface activity. Column activity can be demonstrated in several ways. The chromatographic peak of a given solute can completely disappear, the peak can partially disappear, being diminished in size, or a peak can exhibit tailing. A chromatogram of a test mixture showing the activity of an uncoated fused-silica column is displayed in Figure 3.18A; the inherent acidity associated with surface silanol groups is responsible for the complete disappearance of the basic probe solute, 2,6-dimethylaniline. When this column is deactivated with a precoating of Carbowax 20M, the residual surface column is considerably reduced (Figure 3.18B).

A variety of agents and procedures have been explored for deactivation purposes (51–65). For subsequent coating with nonpolar and moderately polar stationary phases such as polysiloxanes, fused silica has been deactivated by silylation at elevated temperatures, thermal degradation of poly-

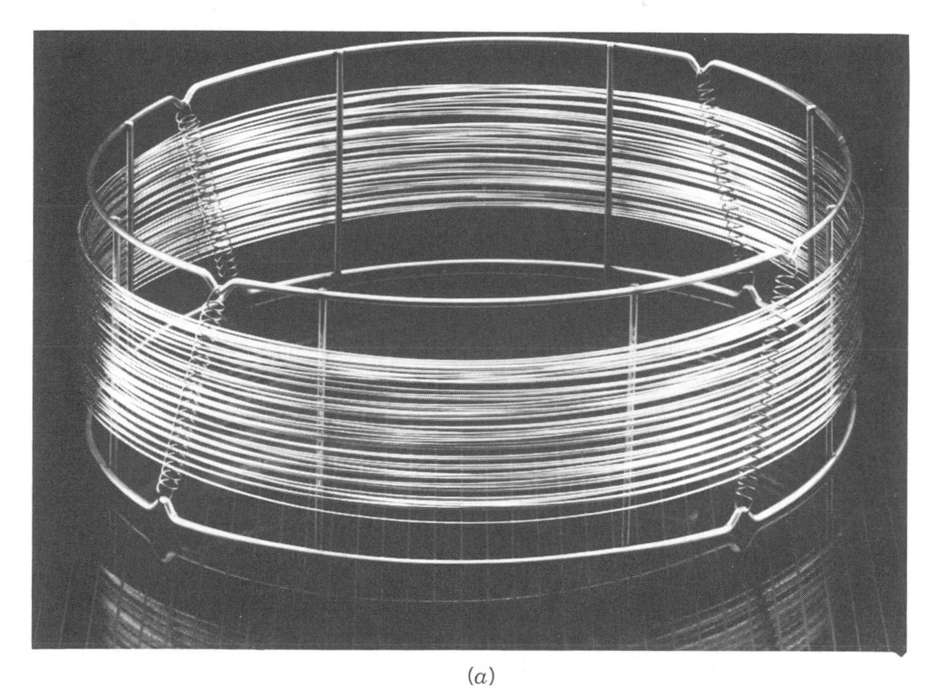

(*a*)

FIGURE 3.17 Photographs of metal-clad capillary columns: (A) aluminum-clad capillary column (photograph courtesy of the Quadrex Corporation); (B) fused silica-lined stainless steel capillary column (lower) and polyimide-clad fused silica capillary columns (upper) (photograph courtesy of the Restek Corporation).

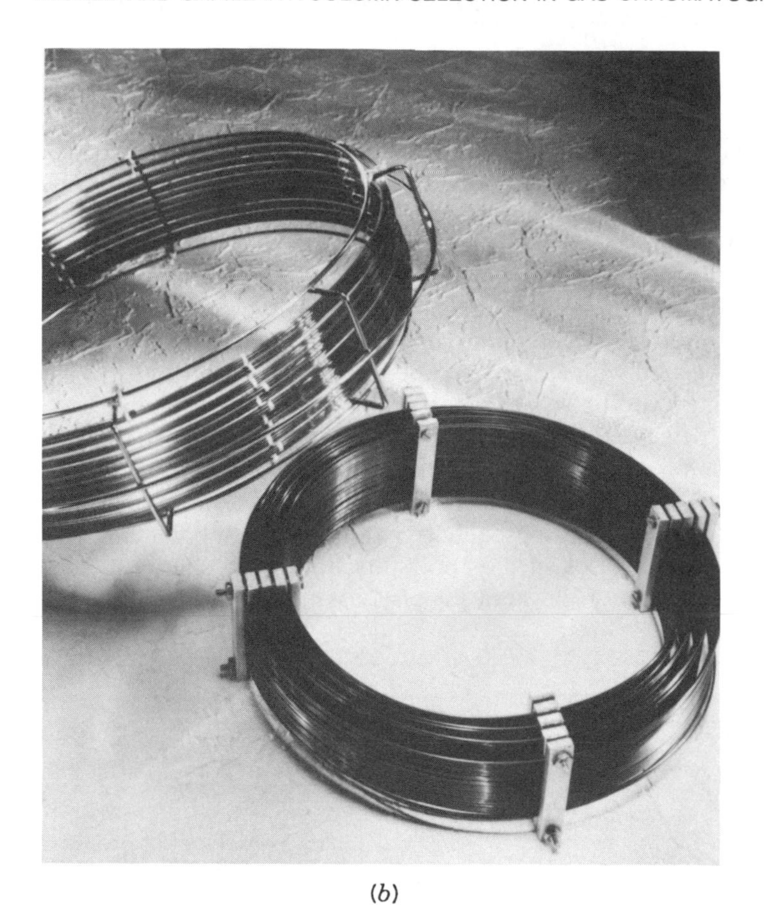

(*b*)

FIGURE 3.17. (*Continued*)

siloxanes and polyethylene glycols, and the dehydrocondensation of silicon hydride polysiloxanes (62, 66–70).

Blomberg has published a comprehensive review of deactivating methods using polysiloxanes (71). One approach has been suggested by Schomburg et al. (68) who prepared columns having excellent thermal stability with polysiloxane liquid phases as deactivators and proposed that the decomposition products formed at the elevated temperatures chemically bond to surface silanols. Surface-stationary phase compatibility has also been achieved with cyclic siloxanes having the same side functional groups as the silicone stationary phase. Octamethylcyclotetrasiloxane (D_4) has been decomposed at 400°C by Stark et al. (72), who postulated that the process involved opening the D_4 ring to form a 1,4-hydroxyoctamethyltetrasiloxane. They indicate that a terminal hydroxyl group interacts with a protruding silanol group eliminating water, and in a secondary reaction the other hydroxyl reacts with another silanol or even a tetrasiloxane. Well-deacti-

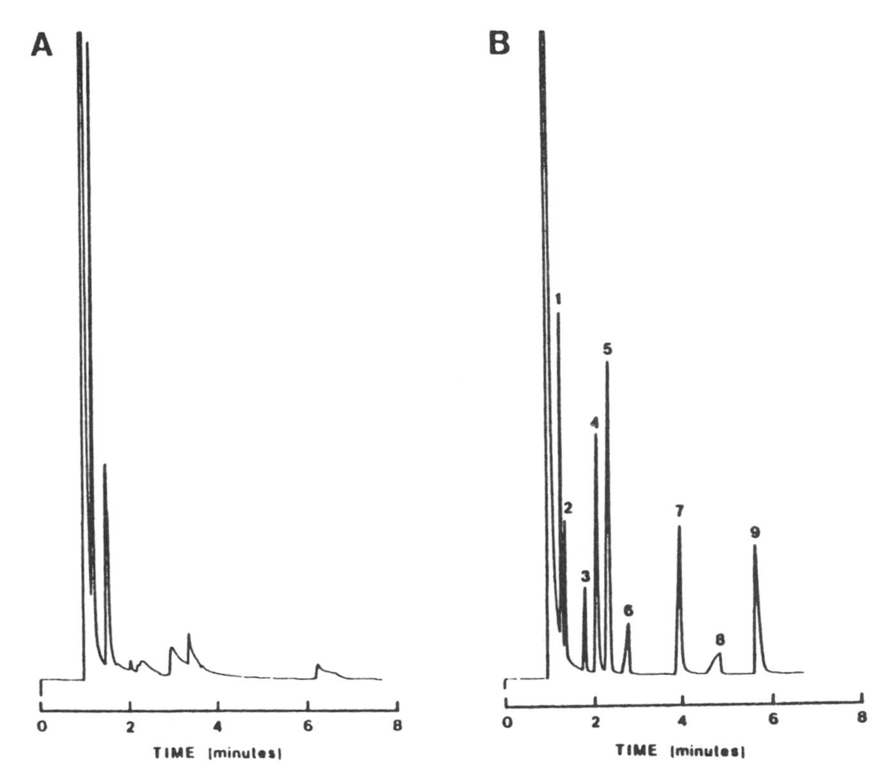

FIGURE 3.18 Chromatograms of an activity mixture on a 15-m × 0.25-mm (A) uncoated fused silica and (B) a fused silica capillary column after deactivation with Carbowax 20M. Column temperature: 70°C; 25 cm/sec He; split injection (100:1). Peaks: 1, *n*-dodecane; 2, *n*-tridecane; 3, 5-nonanone; 4, *n*-tetradecane; 5, *n*-pentadecane; 6, 1-octanol; 7, napthalene; 8, 2,6-dimethylaniline; and 9, 2,6-dimethylphenol. (From Reference 5.)

vated capillary columns can be prepared by this technique (73). In Figure 3.19 the effectiveness of the D_4 deactivation procedure is demonstrated for both acidic and basic test mixtures where the components have excellent band profiles. Woolley et al. outlined an easily implemented deactivation procedure employing the thermal degradation of polyhydrosiloxane at about 260°C, where the silyl hydride groups undergo reaction with surface silanols to form rather stable Si–O–Si bonds and also hydrogen gas (69). This method has the merits of a reaction time less than a hour, a relatively low reaction temperature, and a high degree of reproducibility. A representation of selected deactivated surface textures is displayed in Figure 3.20.

Carbowax 20M has also been successfully used to deactivate column surfaces (75, 76). After coating a thin film of Carbowax 20M, for example, on the column wall, the column is heated to 280°C, then exhaustively extracted with solvent, leaving a nonextractable film of Carbowax 20M on

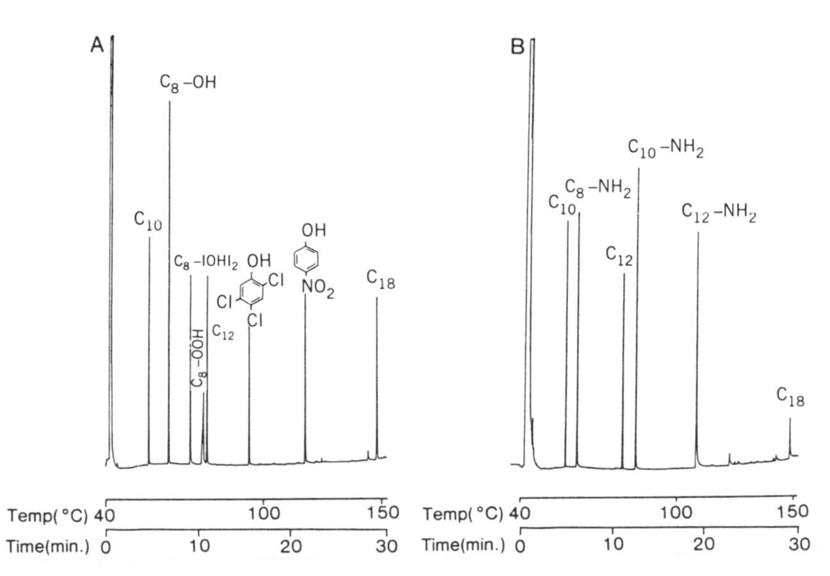

FIGURE 3.19 Chromatogram illustrating the high degree of inertness possible on a D_4-deactivated SE-54 (0.25-μm) fused silica column with (A) an acidic test mixture and (B) a basic test mixture. Temperature programmed from 40°C at 4°C/min after 2 min isothermal hold; H_2 carrier gas at 45 cm/sec. (Reproduced from Reference 73 and reprinted with permission of Elsevier Scientific Publishing Company.)

the surface. Both apolar and polar stationary phases, including Carbowax, can then be coated on capillaries subjected to this pretreatment (77). Dandeneau and Zerenner used this procedure to deactivate their first fused-silica columns (40). Other polyethylene glycols used for deactivating purposes have been Carbowax 400 (78), Carbowax 1000 (79), and Superox-4 (80). Moreover, when a polar polymer is used for deactivation, it may alter the polarity of the stationary phase and this effect becomes particularly problematic with a thin film of a nonpolar phase where the resulting phase has retentions of a mixed phase. Furthermore, silazanes and cyclic silazanes, as deactivating agents, ultimately yield a basic final column texture, whereas chlorosilanes, alkoxysilanes hydrosilanes, hydrosiloxanes, siloxanes, and Carbowax produce an acidic column (4). An alternative procedure is the utilization of OH-terminated stationary phases where deactivation and immobilization of the phase occurs in a single-step process (Section 3.11.5).

Static Coating of Capillary Columns. The goal in coating a capillary column is the uniform deposition of a thin film, ranging from 0.1 to 8 μm in thickness, on the inner wall of a length of clean, deactivated fused silica tubing. Jennings (81) has reviewed the various methods for coating station-

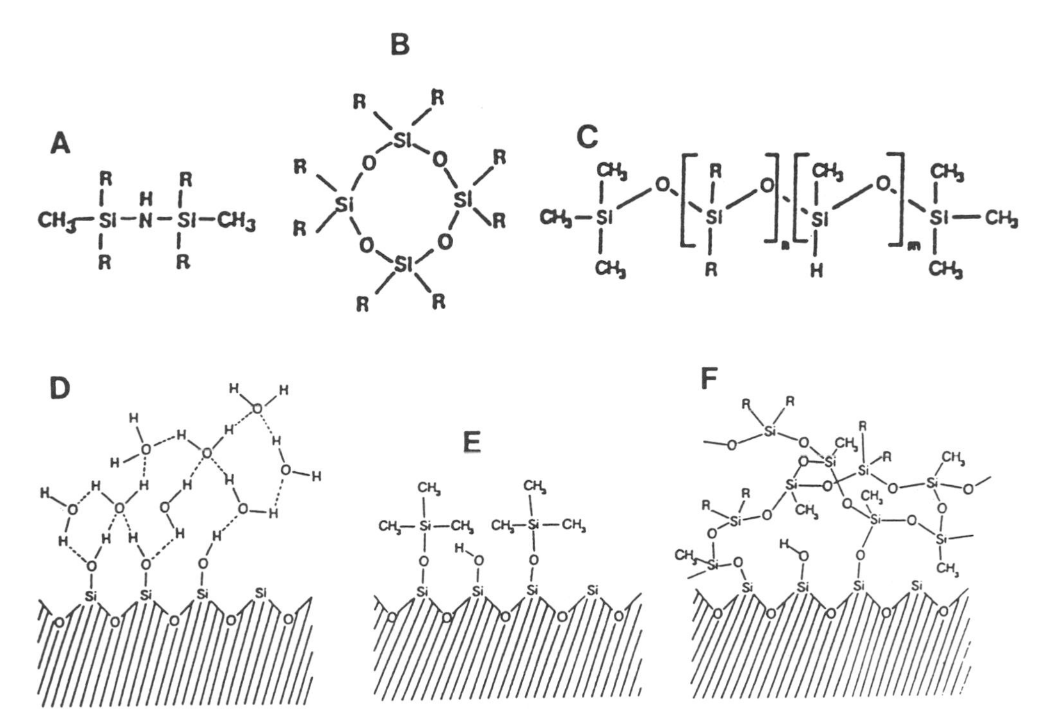

FIGURE 3.20 Selected reagents used for deactivation of silanol groups: (A) disilazanes, (B) cyclic siloxanes, and (C) silicon hydride polysiloxanes. Lower portion is a view of fused silica surface with (D) adsorbed water, (E) after deactivation with a trimethylsilylating reagent, and (F) after treatment with a silicon hydride polysiloxane. (Reproduced from Reference 74 and reprinted with permission from Elsevier Scientific Publishing Company.)

ary phases. The static method of coating is discussed here because it is most widely used today by column manufacturers.

This procedure was first described by Bouche and Verzele (82), where the column is initially completely filled with a solution of known concentration of stationary phase. One end of the column is sealed and the other is attached to a vacuum source. As the solvent evaporates, a uniform film is deposited on the column wall. The column must be maintained at constant temperature for uniform film deposition. The coating solution should be free of microparticulates and dust, be degassed so no bumping occurs during solvent evaporation, and there should be no bubbles in the column. Pentane is the recommended solvent because of its high volatility and should be used wherever stationary phase solubility permits. Evaporation time is approximately half of that required to evaporate methylene chloride. The static coating technique offers the advantage of an accurate determination of the phase ratio (Section 3.10.3) from which the film thickness of the stationary phase can be calculated.

Capillary Cages. Since the ends of flexible fused silica capillary tubing are inherently straight, columns must be coiled and confined on a circular frame, also called a cage (Figure 3.17). The capillary column can then be mounted securely in the column oven of a gas chromatograph. Fused-silica capillary columns of 0.10–0.32 mm i.d. are wound around a 6- or 7-in.-diameter cage whereas an 8-in. cage is used with megabore columns (0.53 mm i.d.). Installation of a capillary column is greatly facilitated, since the ends of a fused-silica column can easily be inserted at the appropriate recommended lengths into sample inlets and detector systems. The ultimate in gas chromatographic system inertness is attainable with on-column injection, where a sample encounters only fused silica from the point of injection to the tip of a FID flame jet.

Test Mixtures for Monitoring Column Performance. The performance of a capillary column can be evaluated with a test mixture whose components and resulting peak shapes serve as monitors of column efficiency and diagnostic probes for adverse adsorptive effects and the acid/base character of a column. These mixtures are used by column manufactures in the quality control of their columns and are likewise recommended for the chromatographic laboratory.

A chromatogram of a test mix and a report are usually supplied with a commercially prepared column. Using the same indicated chromatographic conditions, the separation should be duplicated by the user prior to running samples with column. In the test report evaluating the performance of the column, chromatographic data are listed. These may include retention times of the components in the text mix, corresponding Kovats retention indices of several, if not all, of the solutes, the number of theoretical plates N and/or the effective plate number N_{eff}, and the acid/base inertness ratio (the peak-height ratio of the acidic and basic probes in the test mixture). The values of two additional chromatographic parameters, separation number (Trennzahl number) and coating efficiency, may also be included in the report; the significance of TZ is discussed in Section 3.10.5 and coating efficiency is treated in 3.10.6.

The first chromatogram obtained on a new column may be viewed as the "birth certificate" of a column and defines column performance at time $t = 0$ in the laboratory; a test mix should also be analyzed periodically to determine any changes in column behavior occurring with age and use. For example, a column may acquire a pronouncedly basic character if it has been employed routinely for amine analyses. Another important but often overlooked aspect is that a test mixture serves to monitor the performance of the *total* chromatographic system, not just the performance of the column. If separations gradually deteriorate over time, the problem may not

TABLE 3.10 Test Mixture Components and Role

Probe	Function
n-Alkanes, typically C_{10}–C_{15}	Column efficiency; Trennzahl number (TZ)
Methyl esters of fatty acids, usually C_9–$C1_2$ (E9–E12)	Separation number; column efficiency
1-Octanol (ol)	Detection of hydrogen-bonding sites, silanol groups
2,3-Butanediol (D)	More rigorous test of silanol detection
2-Octanone	Detection of activity associated with Lewis acids
Nonanal (al)	Aldehyde adsorption other than via hydrogen-bonding
2,6-Dimethylphenol (P)	Acid/base character
2,6-Dimethylaniline (A)	Acid/base character
4-Chlorophenol	Acid/base character
n-Decylamine	Acid/base character
2-Ethylhexanoic acid (S)	More stringent measure of irreversible adsorption
Dicyclohexylamine (am)	More stringent measure of irreversible adsorption

Note. Abbreviations of the components in the comprehensive Grob mix indicated in parentheses.

always be column-related but could be due to extracolumn effects, such as a contaminated or activated inlet liner. Commonly used components, their accepted abbreviations, and functions are listed in Table 3.10.

An ideal capillary column should be well deactivated, have excellent thermal stability and high separation efficiency. The extent of deactivation is usually manifested by the amount of peak-tailing for polar compounds. The most comprehensive and exacting test mixture is the solution reported by Grob et al. (83) and is more sensitive to residual surface activity than other polarity mixes. Adsorption may cause (1) broadened peaks of Gaussian shapes, (2) a tailing peak of more or less the correct peak area, (3) a reasonably shaped peak with reduced area, and (4) a skewed peak of correct area but having an increased retention time. Furthermore, irreversible adsorption cannot always be detected by peak shape. In the Grob procedure one measures peak heights as a percentage of that expected for complete and undistorted elution. The technique encompasses all types of peak deformations (broadening, tailing, and irreversible adsorption). A solution whose components are present at specific concentrations is analyzed under recommended column temperature programming conditions.

In practice, the percentage of the peak height is determined by drawing a line (the 100% line) connecting the peak maxima of the nonadsorbing peaks (*n*-alkanes and methyl esters), as shown in Figure 3.21. Alcohols are more sensitive than the other probes to adsorption caused by hydrogen bonding to exposed silanols. The acid and base properties are ascertained with probe solutes such as 2,6-dimethylaniline and 2,6-dimethylphenol, respectively. However, most column manufacturers recommend a modification of the Grob scheme to circumvent the lengthy time involved and, instead, tailor the composition of the mix and column temperature conditions to be

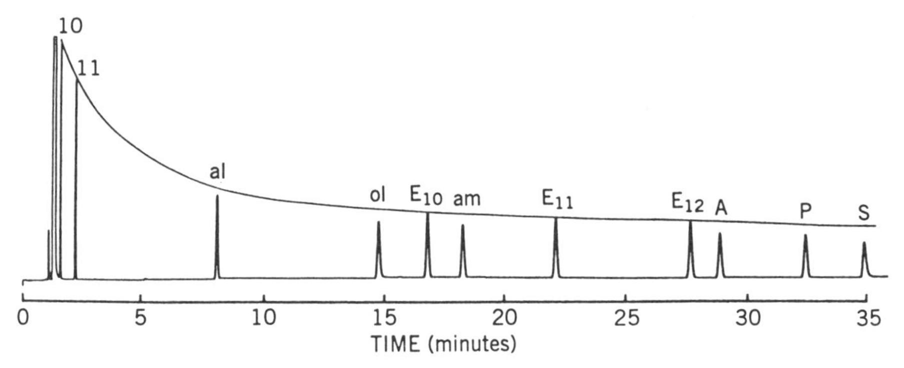

FIGURE 3.21 Chromatogram of a comprehensive Grob mixture on a 15-m × 0.32-mm-i.d. Carbowax 20M capillary column. Column conditions: 75–150°C at 1.7°C/min; 28 cm/sec He. Designation of solutes appears in Table 3.10. (From Reference 5.)

commensurate with the particular deactivation procedure and stationary phase under consideration.

Guthrie and Harland (4) have commented that the effects of deactivation, the chemistry of the stationary phase and its cross-linking, as well as the effect of any postprocess treatment all appear in the final version of a column. An example of this situation is the separation of the Grob mixture (Figure 3.22A) performed on a 15-m × 0.25-mm-i.d. fused silica capillary column deactivated with Carbowax 20M, after which the column received a recoat of the polymer. After cross-linking of the stationary phase (Figure 3.22B) column behavior changed markedly. The 2,3-butanediol peak (D), absent in Figure 3.22A, is present on the cross-linked phase which has acquired increased acidity in the cross-linking process. Note the decreased peak height of the dicyclohexylamine probe (am) and the increased peak height of 2-ethylhexanoic acid (S). Thus, any change or a minor modification in column preparation can affect the final column performance.

3.10 CHROMATOGRAPHIC PERFORMANCE OF CAPILLARY COLUMNS

3.10.1 Golay Equation Versus the van Deemter Expression

The fundamental equation underlying the performance of a gas chromatographic column is the van Deemter expression (Equation 2.44; further discussed in Section 3.6.4), which may be expressed as

$$H = A + \frac{B}{\bar{u}} + C\bar{u} \tag{3.4}$$

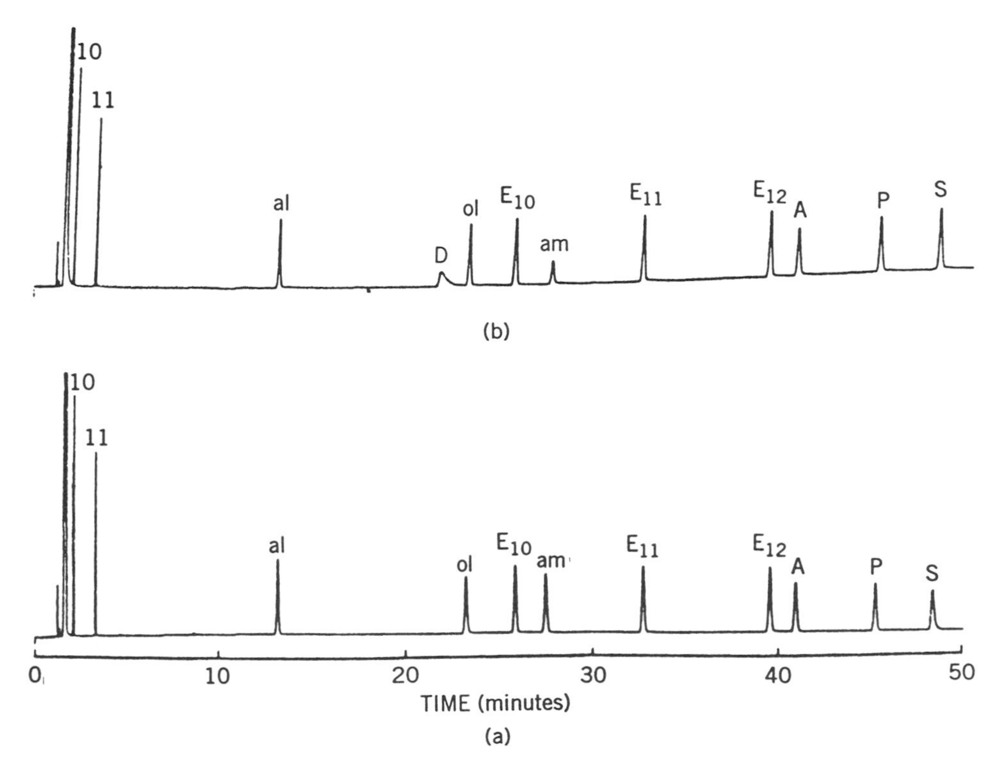

FIGURE 3.22 Chromatogram of a comprehensive Grob mixture on a 15-m × 0.32-mm-i.d. Carbowax 20M capillary column (A) after coating and (B) after cross-linking the stationary phase. Column conditions: 75–150°C at 2°C/min; 28 cm/sec He. Designation of solutes appears in Table 3.10. (From Reference 5.)

where H = the height equivalent to a theoretical plate
A = the eddy diffusion or multiple path term
B = the longitudinal diffusion contribution
C = the resistance to mass transfer term
$\bar{u}$ = the average linear velocity of the carrier gas

In the case of a capillary column, the A term is equal to zero because there is no packing material. Thus, the above equation simplifies to

$$H = \frac{B}{\bar{u}} + C\bar{u} \tag{3.5}$$

This abbreviated expression is often referred to as the Golay equation (38). The B term may be expressed as $2D_g/\bar{u}$, where D_g is the binary diffusion coefficient of the solute in the carrier gas. Peak broadening due to longitudinal diffusion is a consequence of the residence time of the solute within the column and the nature of the carrier gas. This effect becomes pertinent only at low linear velocities or flowrates and is less pronounced at high velocities (Figures 2.20 and 2.21).

However, the major contributing factor contributing to band broadening is the C term, in which the resistance to mass transfer can be represented as the composite of the resistance to mass transfer in the mobile phase C_g and that in the stationary phase C_l:

$$C = C_g + C_l \tag{3.6}$$

where

$$C_g = \frac{r^2(1 + 6k + 11k^2)}{D_g 24(1 + k)^2} \tag{3.37}$$

$$C_l = \frac{2kd_f^2 \bar{u}}{3(1 + k)^2 D_l} \tag{3.8}$$

D_l is the diffusion coefficient of the solute in the stationary phase, k is the capacity factor the solute, d_f is the film thickness of stationary phase, and r is the radius of the capillary column. With capillary columns, C_s is quite small compared to the magnitude of C_g because of the thin film of stationary phase. (The reverse is true with packed columns where $C_s \gg C_g$.) The Golay equation may then be rewritten as

$$H = \frac{B}{\bar{u}} + C_g \bar{u} = \frac{2D_g}{\bar{u}} + \frac{r^2(1 + 6k + 11k^2)\bar{u}}{D_g 24(1 + k)^2} \tag{3.9}$$

The optimum linear velocity corresponding to the minimum in a plot of H versus u (Figure 2.20) can be obtained by setting $dH/du = 0$ and solving for $\bar{u}$:

$$\frac{dH}{du} = 0 = -\frac{B}{\bar{u}^2} + C_g \tag{3.10}$$

Thus, $u_{opt} = (B/C_g)^{1/2}$ and the value of H corresponding to this optimum linear velocity, H_{min}, is

$$H_{min} = r\sqrt{\frac{1 + 6k + 11k^2}{3(1 + k)^2}} \tag{3.11}$$

Consequently, as the diameter of a capillary column decreases, both maximum column efficiency N and maximum effective efficiency N_{eff} increase and are also dependent on the particular solute retention k. Retention in capillary GC is usually expressed as k, where $k = (t_R - t_M)/t_M$.

In Table 3.11 the effect of column inner diameter on maximum attainable column efficiency N is presented as a function of capacity factor. For a capillary of a given inner diameter, one can see that there is an increase in plate height with increasing k, with a corresponding decrease in plate number and an increase in effective plate count. As the inner diameter of a capillary column increases, column efficiency N drops markedly, while the effective plate number N_{eff} increases. For separations requiring high resolution, columns of small inner diameter are recommended. Expressing efficiency in terms of plates per meter allows the efficiency of columns of unequal lengths to be compared. Also included in Table 3.11 are data for 0.53 mm i.d., the diameter of the "megabore" capillary column, which has been termed the alternative to the packed column. The merits and features of this particular type of column are discussed in Section 3.10.4.

TABLE 3.11 Column Efficiency as a Function of Inner Diameter and Capacity Factor

Inner Diameter (mm)	k	h_{min}	Maximum Plates per Meter, N	Effective plates per Meter, N_{eff}
	1	0.061	16,393	4,098
	2	0.073	13,697	6,027
	5	0.084	11,905	6,667
0.10	10	0.090	11,111	9,222
	20	0.093	10,752	9,784
	50	0.095	10,526	10,105
	1	0.153	6,536	1,634
	2	0.182	5,495	2,442
	5	0.210	4,762	3,307
0.25	10	0.224	4,464	3,689
	20	0.231	4,329	3,925
	50	0.236	4,237	4,073
	1	0.196	5,102	1,276
	2	0.232	4,310	1,896
	5	0.269	3,717	2,082
0.32	10	0.286	3,497	2,903
	20	0.296	3,378	3,074
	50	0.302	3,311	3,179
	1	0.325	3,076	769
	2	0.384	2,604	1,146
	5	0.445	2,247	1,258
0.53	10	0.474	2,110	1,751
	20	0.490	2,041	1,857
	50	0.500	2,000	1,920

3.10.2 Choice of Carrier Gas

Capillary-column efficiency is dependent on the carrier gas used, the length and inner diameter of the column, the capacity factor of the particular solute selected for the calculation of the number of theoretical plates, and the film thickness of stationary phase. Profiles of H versus u for three carrier gases with a thin-film capillary column are displayed in Figure 3.23. Although the lowest minimum and, therefore, the greatest efficiency are obtained with nitrogen, speed of analysis must be sacrificed. The increasing portion of the curve is steeper for nitrogen, which necessitates working at or near u_{opt}; otherwise, loss in efficiency (and resolution) quickly results. On the other hand, if one is willing to accept a slight loss in the number of theoretical plates, a more favorable analysis time is possible with helium and hydrogen as carrier gases, because u_{opt} occurs at a higher linear velocity. Moreover, the mass transfer contribution or rising portion of a curve is less steep with helium or hydrogen, which permits working over a wider range of linear velocities without substantial sacrifice in resolution. This advantage becomes evident in comparing the capillary separation of the components in calmus oil with nitrogen and hydrogen as carrier gases in Figure 3.24.

In comparing these carrier gases, another benefit becomes apparent at linear velocities corresponding to equal values of plate height. With the lighter carrier gases solutes can be eluted at lower column temperatures during temperature programming with narrower band profiles, since higher linear velocities can be used. Thus, either helium or hydrogen is rec-

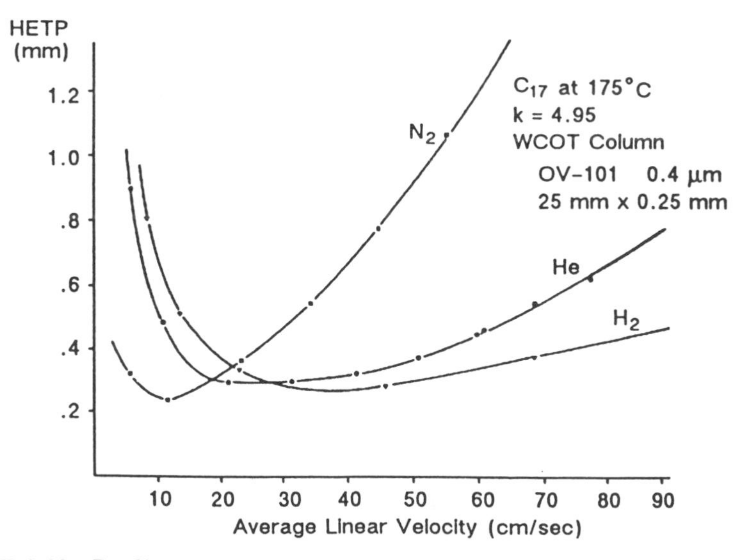

FIGURE 3.23 Profiles of HETP versus linear velocity for the carrier gases helium, hydrogen, and nitrogen. (Courtesy of Americas Training Center, Hewlett–Packard Co., Atlanta, GA.)

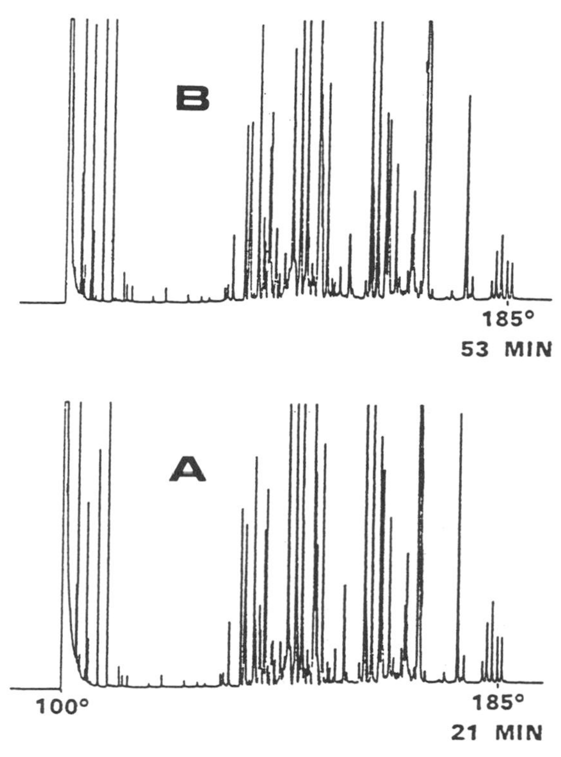

FIGURE 3.24 Chromatograms of the separation of calmus oil using (A) hydrogen as carrier gas, 4.2 mL/min at a programming rate of 4.0°C/min, and (B) nitrogen as carrier gas, 2.0 mL/min programming rate of 1.6°C/min on a 40-m × 0.3-mm i.d. capillary column (0.12-μm film). (Reproduced from Reference 89 and reprinted with permission from Dr. Alfred Huethig Publishers.)

ommended over nitrogen and indeed these gases are used today as carrier gases for capillary gas chromatography. One advantage of using hydrogen is that plate number varies less for hydrogen than for helium as linear velocity increases. The use of hydrogen for any application in the laboratory always requires safety precautions in the event of a leak. Precautionary measures should be taken for the safe discharge of hydrogen from the split vent in the split-injection mode.

Measurement of Linear Velocity. The flowrate through a capillary column whose inner diameter is less than 0.53 mm is difficult to measure accurately and reproducibly by a conventional soap-bubble meter. Instead, the flow of carrier gas through a capillary column is usually expressed as linear velocity rather than as a volumetric flowrate. Linear velocity may be calculated by injecting a volatile, nonretained solute and noting its retention time t_M (sec).

For a capillary column of length L in centimeters,

$$u \ (\text{cm/sec}) = \frac{L}{t_M} \tag{3.12}$$

For example, the linear velocity of carrier gas through a 30-m column where methane has a retention time of 2 min is 3000 cm/120 sec or 25 cm/sec. If desired, the volumetric flowrate F (mL/min), can be computed from the relationship

$$F \ (\text{mL/min}) = 60\pi r^2 u \tag{3.13}$$

where r is the radius of the column in centimeters. An injection of methane is convenient to use with a FID to determine t_M and or a headspace injection of methylene chloride and acetonitrile can be made with an ECD and NPD, respectively. Recommended linear velocities and flowrates of helium and hydrogen for capillary columns of various inner diameters are listed in Table 3.12.

Effect of Carrier-Gas Velocity on Linear Velocity. Chromatographic separations using capillary columns are achieved under constant pressure conditions, as opposed to packed columns, which are usually operated in a flow-controlled mode. The magnitude of the pressure drop across a capillary column necessary to produce a given linear velocity is a function of the particular carrier gas and length/inner diameter of the column. The relationship between viscosity and temperature for any gas is linear, as shown in Figure 3.25 for helium, hydrogen, and nitrogen. In gas chromatography, as column temperature increases, linear velocity decreases because of increased viscosity of the carrier gas. Thus, initially higher linear velocities are established for temperature programmed analyses than for isothermal separations. If we compare columns of identical dimensions and operate them at the same inlet pressure and temperature, the linear velocity will be

TABLE 3.12 Recommended Linear Velocities and Flowrates With Helium and Hydrogen

	Linear Velocity (cm/sec)		Flowrate (mL/min)	
Inner Diameter (mm)	Helium	Hydrogen	Helium	Hydrogen
0.18	20–45	40–60	0.3–0.7	0.6–0.9
0.25	20–45	40–60	0.7–1.3	1.2–2.0
0.32	20–45	40–60	1.2–2.2	2.2–3.0
0.53	20–45	40–60	4.0–8.0	6.0–9.0

Note. 30-m column length.

Source. Data abstracted from 1994–95 J&W Scientific Catalog.

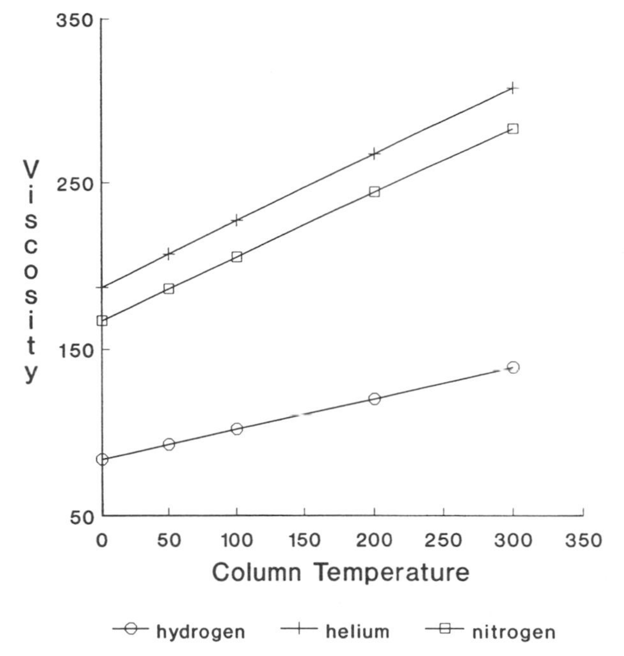

FIGURE 3.25 Effect of temperature on carrier gas viscosity. (Data for curves generated from viscosity–temperature relationships in Reference 74.)

highest for hydrogen and lowest for helium. Therefore, whenever a change in the type of carrier gas is made in the laboratory, linear velocities should actually be measured and one should not reconnect the pressure regulator using the same delivery pressure.

3.10.3 Phase Ratio

In addition to the nature of the carrier gas, column efficiency and, ultimately, resolution and sample capacity of a capillary column are affected by the physical nature of the column, namely, the inner diameter and film thickness of stationary phase. An examination of the distribution coefficient K_D as a function of chromatographic parameters is helpful here. K_D is constant for a given solute-stationary phase pair and is dependent only on column temperature. K_D may be defined as

$$K_D = \frac{\text{Concentration of solute in stationary phase}}{\text{Concentration of solute in carrier gas}} \tag{3.14}$$

or

$$K_D = \frac{\text{Amount of solute in stationary phase}}{\text{Amount of solute in mobile phase}} \times \frac{\text{Volume of carrier gas}}{\text{Volume of stationary phase in the column}} \quad (3.15)$$

K_D can now be expressed as

$$K_D = k\beta' = k\left(\frac{r}{2d_f}\right) \quad (3.16)$$

where β is the phase ratio and is equal to $r/2d_f$, r is the radius of the column, and d_f is the film thickness of stationary phase. At a given column temperature, retention increases as the phase ratio of the column decreases, which can be manipulated either by decreasing the diameter of the column or increasing the film thickness of the stationary phase; likewise a decrease in retention is noted with an increase in β. Since K_D is a constant at a given column temperature, film thickness and column diameter play key roles in determining separation power and sample capacity. In selecting a capillary column, the phase ratio should be considered. The variation in β as a function of stationary phase film thickness is illustrated in Figure 3.26 for columns of 0.18, 0.25, and 0.32 mm i.d.

As the film thickness decreases, k or retention also decreases at constant temperature, column length, and i.d. Conversely, with an increase in film thickness in a series of columns having the same dimensions, retention increases under the same temperature conditions. This effect of film thickness on separation is demonstrated in the series of parallel chromatograms appearing in Figure 3.27. Column diameter limits the maximum amount of stationary phase that can be coated on its inner wall. Small-diameter columns usually contain thinner films of stationary phase, while thicker films can be coated on wider-bore columns. The concept of phase ratio allows two columns of equal length to be compared in terms of sample capacity and resolution.

As depicted in Table 3.11, column efficiency increases as column diameter decreases. Sharper peaks yield improved detection limits. However, as column diameter decreases, so does sample capacity. Column temperature conditions and linear velocity of the carrier gas can usually be adjusted to have a more favorable time of analysis. In Figure 3.28 these parameters are placed into perspective in a pyramidal format as a function of the inner diameter of a capillary column.

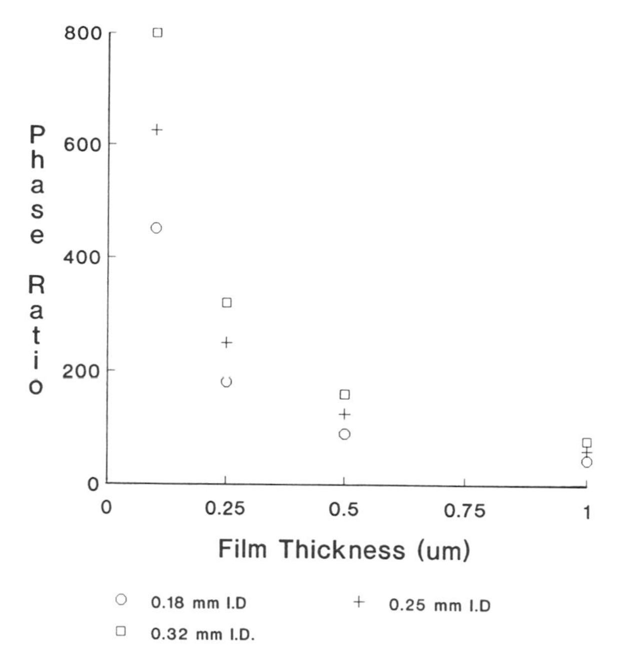

FIGURE 3.26 The effect of film thickness on phase ratio for capillary columns of several inner diameters.

3.10.4 Practical Considerations of Column Diameter and Film Thickness

Guidelines for the selection of column diameter, film thickness of stationary phase, and length will now be established based on practical gas chromatographic considerations.

Column Diameter

1. Sample capacity increases as column diameter increases. Samples having components present in the same concentration range can be analyzed on a column of any diameter. The choice is dependent on resolution required.
2. For complex samples, select a column having the smallest diameter and sample capacity compatible with the concentration range of the sample components.
3. Samples whose components differ widely in concentration should be analyzed on a column of larger i.d. (>0.25 mm) to avoid overload of the column by solutes of higher concentration.
4. The selection of column i.d. may be based on the type of sample inlet

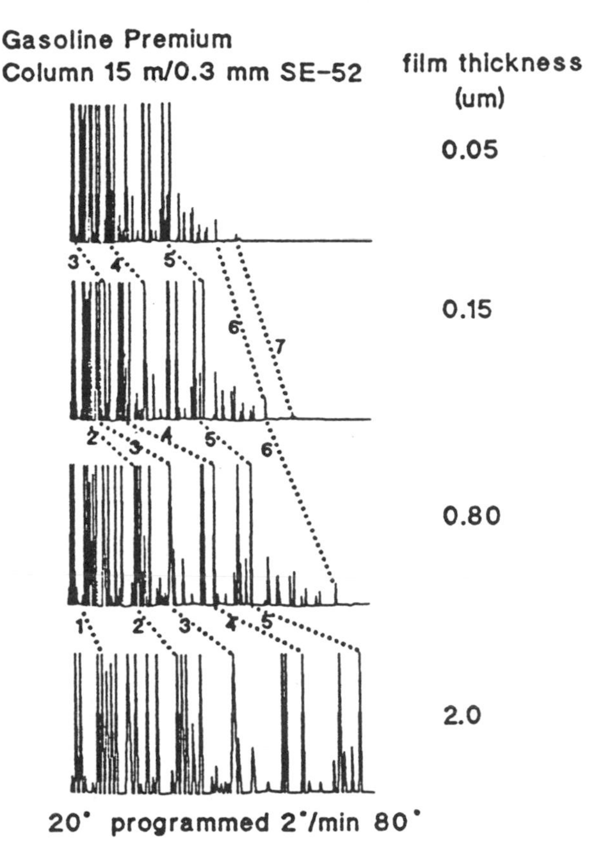

FIGURE 3.27 Chromatograms of gasoline on capillary columns with varying film thickness of stationary phase SE-52. (Reproduced from Reference 89 and reprinted with permission of Dr. Alfred Huethig Publishers.)

system. Generally, a 0.25- or 0.32-mm-i.d. column may be used for split and splitless injections, 0.32 mm i.d. for splitless and on-column injections, and 0.53 mm i.d. for direct injections.

5. Capillary columns of 0.18 and 0.25 mm i.d. should be used for GC-MS systems, because the lower flowrates with these columns will not exceed the limitations of the vacuum system.

Film Thickness of Stationary Phase

1. Retention and sample capacity increase with increasing film thickness with a concurrent decrease in column efficiency.
2. Thin-film columns provide higher resolution of high boiling solutes but lower resolution of more volatile components under any set of column temperature conditions.

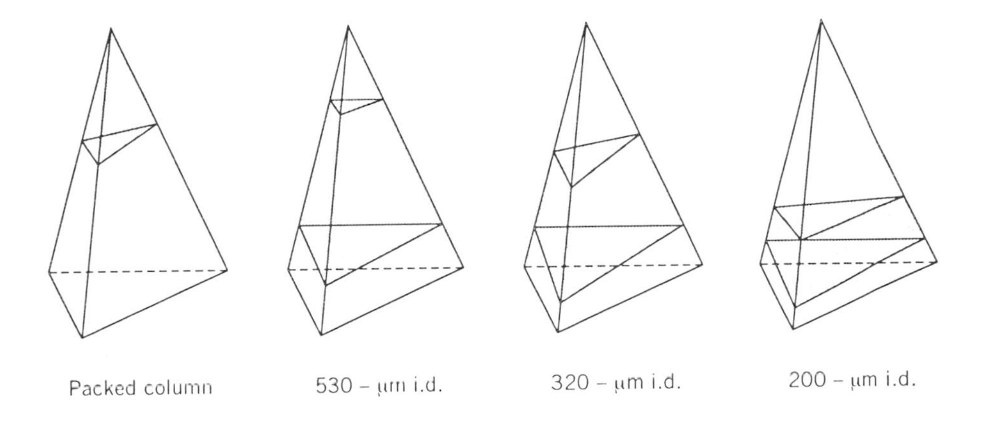

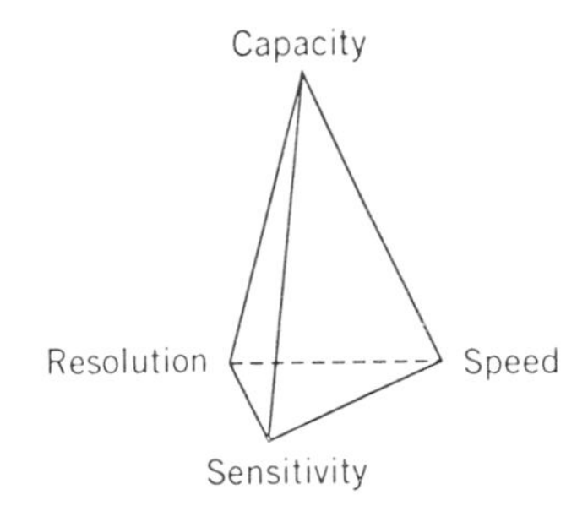

FIGURE 3.28 The "chromatographic pyramids" for packed and capillary columns of varying inner diameter. (Courtesy of Americas Training Center, Hewlett Packard Co., Atlanta, GA.)

3. The sample capacity of thin-film columns may be inadequate and require cryogenic temperature control of the column oven.
4. Film thicknesses of $<0.2\ \mu m$ permit the use of longer columns for complex samples.
5. A solute will exhibit a lower elution temperature as film thickness decreases; thus, thin-film columns are ideal for high-boiling petroleum fractions, triglycerides, etc.
6. Thicker films of stationary phase ($>1\ \mu m$) should be used for analysis of more volatile solutes. Very thick films ($>5\ \mu m$) should be selected for analyses to be performed at room temperature.
7. Thicker film columns necessitate higher elution temperatures, but incomplete elution of all sample components may result.
8. Higher elution temperatures for prolonged periods of time mean a reduced column lifetime and more column bleed.
9. A capillary column, 30 m or longer, with a thick film of stationary

phase, offers an alternative to cryogenic oven temperature control for solute-focusing purposes, especially attractive with auxiliary sample introduction techniques of purge and trap and thermal desorption.

A summary of sample capacities for capillary columns of several inner diameters with different film thicknesses is listed in Table 3.13.

Capillary Columns of 0.53 mm i.d. (The Megabore Column). Many applications previously performed on a packed column can now be done with a megabore column, a capillary column of 0.53 mm i.d. A megabore column with a fairly thick film of stationary phase has a low phase ratio and exhibits retention characteristics and sample capacities similar to those of a packed column. A 0.53-mm-i.d. column offers the best of both worlds, because it combines the attributes of a fused silica capillary column with the high sample capacity and ease of use of a packed column. Analytical methods developed with a packed column can be easily transferred for many applications to a megabore column with the appropriate stationary phase.

Peaks generated with a megabore column typically are sharper and exhibit less tailing compared to those with a packed column. Redistribution of the stationary phase can occur at the inlet of packed column with large injections of solvent and leave an exposure of silanol sites on a diatomaceous earth support. With a cross-linked phase in a megabore column this problem is eliminated. Lewis acid sites, which are a problem with supports, are likewise absent in this larger-diameter capillary column. Analysis time is also shorter as a rule (Figure 3.29). On the other hand, long megabore

TABLE 3.13 Column Capacity as a Function of Inner Diameter and Film Thickness

Inner Diameter (mm)	Film Thickness (μm)	Capacity[a] (ng/component)
0.25	0.15	60–70
	0.25	100–150
	0.50	200–250
	1.0	350–400
0.32	0.25	150–200
	0.5	250–300
	1.0	400–450
	3.0	1200–1500
0.53	1.0	1000–1200
	1.5	1400–1600
	3.0	3000–3500
	5.0	5000–6000

Source. Data abstracted from 1994–95 J&W Scientific Catalog.

[a] Capacity is defined as the amount of component where peak asymmetry occurs at 10% at half-height.

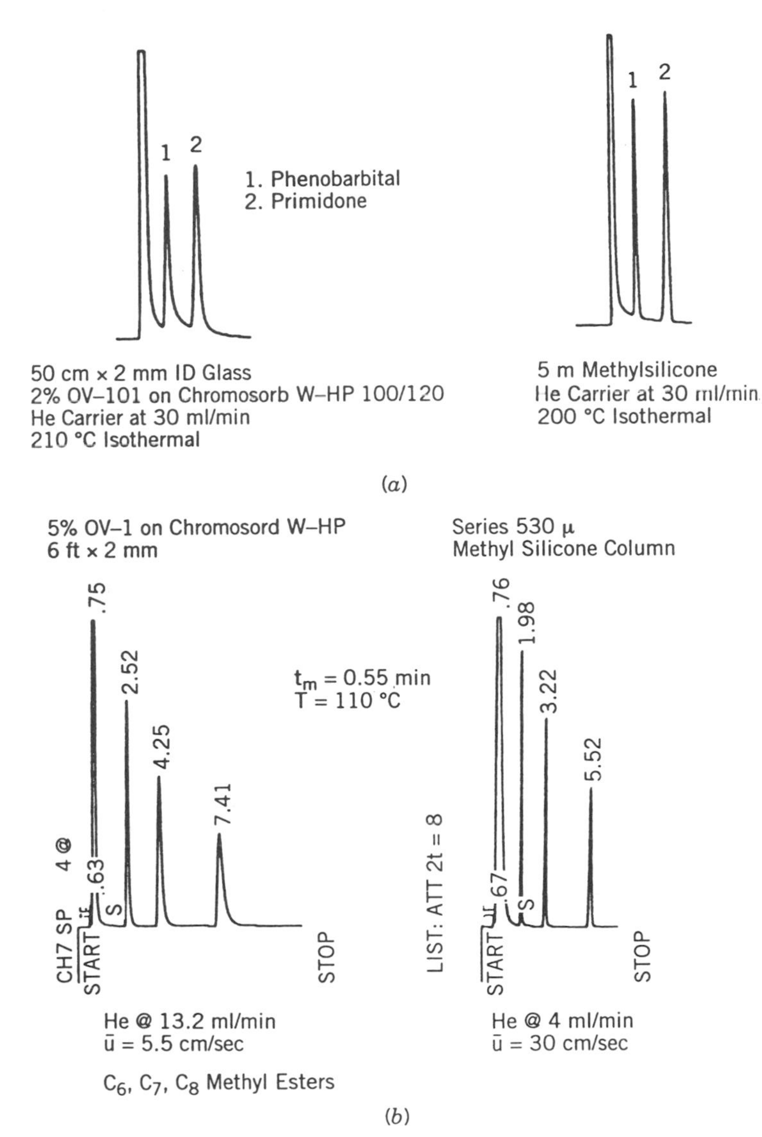

FIGURE 3.29 Chromatograms comparing (A) the effect of the inertness of a 0.53-mm-i.d. column to the more active surfaces within a packed column and (B) the retention characteristics of a packed and 0.53-mm-i.d. column. (Courtesy of Americas Training Center, Hewlett–Packard Co., Atlanta, GA.)

columns can be used for the analysis of more complex samples, such as the separation of the reference standard for EPA Method 502.2 (VOCs in drinking water) presented in Figure 3.30.

Column Length. Resolution is a function of the square root of the number

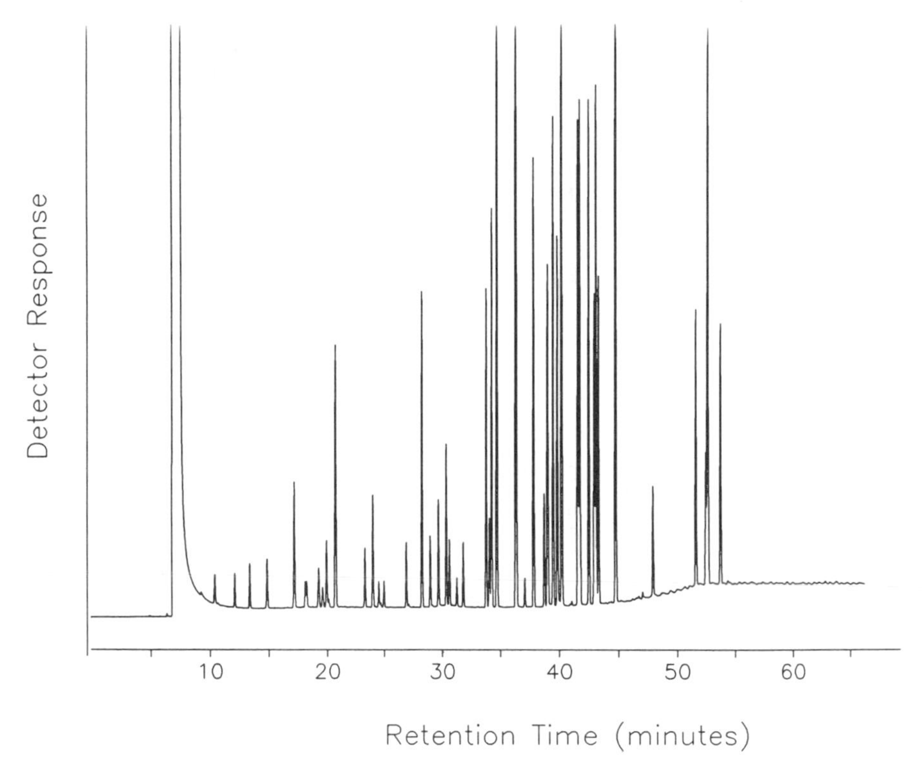

FIGURE 3.30 Chromatogram of the separation of reference standard for EPA Method 502.2 on a 75-m × 0.53-mm-i.d. methyphenyl-cyano propylsilicone capillary column (2.5-μm film). Column conditions: 35°C at 4°C/min to 280°C after 10 min isothermal hold; 10 mL/min He; splitless injection and FID.

of theoretical plates or column length. One must consider the trade-off of the increase in overall resolution in a separation by augmenting column length with the simultaneous increase in analysis time under isothermal conditions. Increasing the length of a capillary column from 15 to 30 m, for example, results in an improvement by a factor of 1.4 (the square root of 2) in resolution, but analysis time also doubles (Equation 2.95), which may limit sample throughput in a laboratory. To double resolution between two adjacent peaks, one needs a fourfold increase in column length. If one is already using a 30-m column, increasing the column length to 120 m is unreasonable. Here a column of the same initial length (or shorter sometimes) having another stationary phase will have a different selectivity and solve the problem. The situation is slightly different under temperature programming conditions where a large improvement in resolution can sometimes be obtained with only a moderate increase in analysis time.

The best approach is the selection of a 25- or 30-m column for general analytical separations (Figure 3.31) and for fingerprinting chromatograms

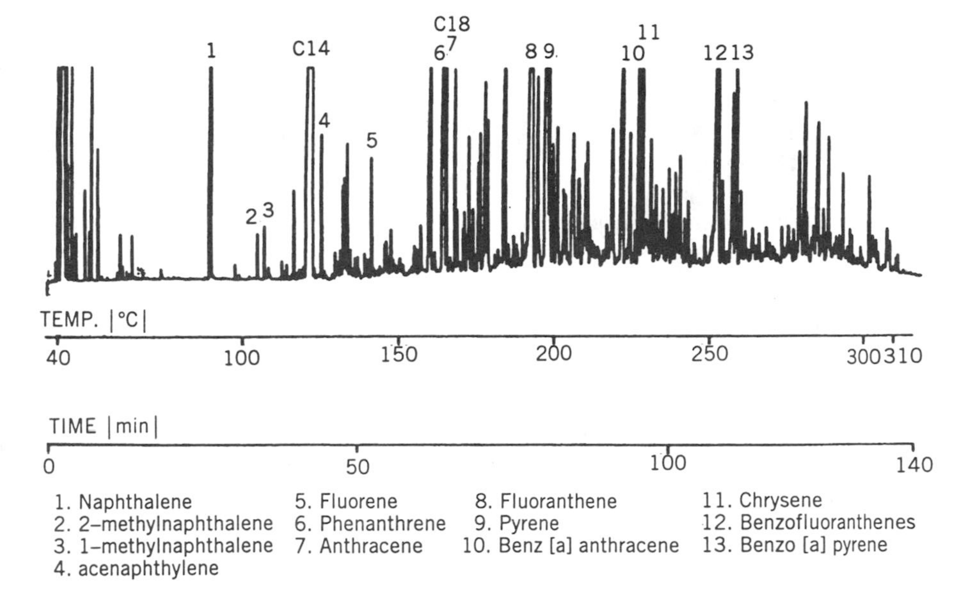

FIGURE 3.31 Chromatogram of separation of a polynuclear aromatic hydrocarbon fraction isolated from a creosote sample on a 30-m × 0.25-mm-i.d. DB-5 column (0.25-μm film). Temperature conditions: 70°C at 2°C/min to 300°C for 10 min; split injection (100:1), Det: FID, 29 cm/sec He. (Reproduced from Reference 90.)

generated under the same chromatographic conditions for comparison of samples (Figure 3.32). A shorter length of column may be employed for rapid screening or simple mixtures or a 60-m column for very complex samples (a longer column also generates more column bleed). Temperature programming ramp profiles can be adjusted to optimize resolution. A number of studies dealing with computer simulation based on optimization of column temperature have been reported (84–88).

3.10.5 Separation Number (Trennzahl)

An alternative measure of column efficiency is provided by the separation or Trennzahl number (TZ), which may be computed as follows:

$$TZ = \frac{t_{R(N+1)} - t_{R(N)}}{w_{0.5(N+1)} + w_{0.5(N)}} - 1 \qquad (3.17)$$

TZ represents the number of peaks that can be inserted between the peaks of two consecutive homologs in a chromatogram. A TZ value of 15, for example, calculated from peak data for the separation of C_9–C_{10} fatty acid methyl esters suggests that 15 peaks can be placed between the peaks of the

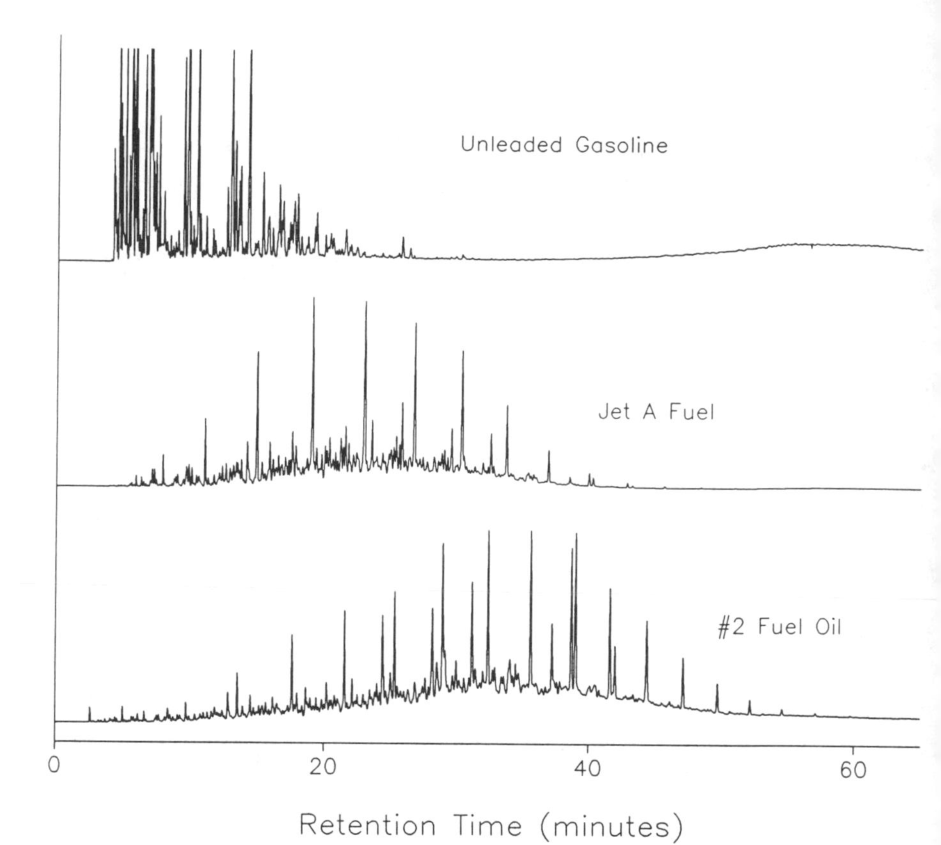

FIGURE 3.32 Chromatograms showing the separation of an unleaded gasoline, Jet A fuel, and No. 2 fuel oil under the same column conditions. Column: 30-m × 0.25-mm-i.d. HP-1 (0.25-μm film); temperature conditions: 35°C (2 min) at 4°C/min to 260°C, Det: FID, 25 cm/sec He.

two esters. TZ is related to resolution

$$\mathrm{TZ} = \frac{R}{1.18} - 1 \tag{3.18}$$

The concept is very useful in temperature-programmed separations where TZ is the sole measure of column efficiency.

3.10.6 Coating Efficiency

This parameter, also called the utilization of theoretical efficiency (UTE), is the ratio of the actual efficiency of a capillary column to its theoretical maximum possible efficiency. Coating efficiency or UTE is expressed as

$$\% \text{ Coating efficiency} = \frac{H_{min}}{H} \times 100 \tag{3.19}$$

where H_{min} is defined in Equation 3.11. Coating efficiency is a measure of how well a column is coated with stationary phase. Coating efficiencies of nonpolar columns range from 90 to 100%. Polar columns have somewhat lower coating efficiencies, 60–80%, because a polar stationary phase is more difficult to coat uniformly. Also, larger-diameter columns tend to have higher coating efficiencies. This parameter is listed on a test report shipped with a new column and is important to column manufacturers for monitoring the quality of their product. It is usually of no concern to most capillary column users.

3.11 STATIONARY PHASE SELECTION FOR CAPILLARY GAS CHROMATOGRAPHY

3.11.1 Requirements

The use of packed columns for gas chromatographic separations requires having an assortment of columns available with different stationary phases in order to compensate for column inefficiency by a commensurate gain in selectivity. With a capillary column, the demands on selectivity, although important, are not as stringent, because of the high plate count possible with a capillary column. However, with the transition from the era of the packed to the capillary column, a gradual redefinement in the requirements of the stationary phase took place.

Many liquid phases for packed-column purposes were unacceptable for capillary GC. Although they offered selectivity, overriding factors responsible for their disfavor were overall lack of thermal stability and the instability of the stationary phase as a thin film at elevated temperatures and during temperature programming. In the latter processes, it is crucial that the phase remain a thin uniform film; otherwise, loss of both inertness and column efficiency result. Today, these problems have been solved and the refinements are reflected in the high performance of commercial columns.

Factors influencing the choice of inner diameter, film thickness of stationary phase, and column length have been discussed in Section 3.10.4. Let us focus now on the selection of the stationary phase, the most important aspect in column selection.

In choosing a stationary phase for capillary separations, remember the adage "like dissolves like." As a starting point, try to match the functional groups present in the solutes under consideration with those in a stationary phase. In the analysis of polar species, for instance, select a polar stationary

TABLE 3.14 Cross-reference of Columns From Manufacturers

Stationery Phase	Manufacturer								
	Alltech	Chrompack	HP	J&W	Perkin–Elmer	Quadrex	Restek	SGE	Supelco
				Polysiloxanes					
100% Dimethyl	AT-1 SE-30	CP Sil 5 CB	HP-1, Ultra-1 HP-101, PONA	DB-1, DB-1HT DB-Petro DB-Petro 100 DB-2887	PE-1	007-1	Rtx-1, MXT-1 Rtx-2887, MXT-2887	BP-1	SPB-1, SP-2100 SPB-1 sulfur Petrocol DH 50.2 Petrocol 2887 Petrocol EX, Petrocol 3710
5% Phenyl– 95% dimethyl	AT-5 SE-54	CP Sil 8 CB	HP-5, Ultra-2 PAS-5	DB-5, DB-5 HT DB-5 MS DB-5.625 SE-54	PE-2	007-2	Rtx-5, MXT-5 XTI-5	BP-5	SPB-5, PTE-5 SE-54
6% Cyanopropyl-phenyl– 94% dimethyl	—	—	HP-1301	DB-1301 DB-624	—	007-502	Rtx-1301 Rtx-624 MXT-624	—	—
20% Diphenyl– 80% dimethyl	—	—	—	—	PE-7	007-7	Rtx-20 MXT-20	—	SPB-20 VOCOL
35% Diphenyl– 65% dimethyl	AT-35	—	HP-35	—	PE-11	007-11	Rtx-35	—	SPB-35 SPB-608
14% Cyanopropyl-phenyl– 86% dimethyl	AT-1701	CP Sil 19CB	PAS-1701	DB-1701	PE-1701	007-1701	Rtx-1701 MXT-1701	BP-10	SPB-1701
Trifluoropropyl	AT-210	—	—	DB-210	—	—	Rtx-200	—	—
50% Methyl– 50% phenyl	AT-50	—	HP-17 HP-50+	DB-17 DB-17 HT DB-608	PE-17	007-17	Rtx-50	—	SP-2250 SPB-50
65% Diphenyl- 35% methyl	—	TAP-CB	—	—	—	400 -65TG	Rtx-65 Rtx-65TG	—	—
50% Cyanopropyl methyl– 50% phenylmethyl	AT-225	CP Sil 43CB	HP-225	DB-225	PE-225	007-225	Rtx-225	—	—

Carbowax	AT-Wax	CP Wax 52 CB	HP-20M InnoWax	DB-Wax	PE-CW	007-CW	Stabilwax MXT-Wax	BP-20	Supelcowax-10 Omegawax Carbowax PEG 20M
Carbowax (basic)	—	—	—	CAM	—	—	Stabilwax	—	Carbowax-Amine
Carbowax (acidic)	AT-1000 FFAP	CP Wax -58 CB	HP-FFAP	DB-FFAP OV-351	PE-FFAP	007-FFAP	Stabilwax-DA	BP-21	Nukol SP1000
90% Biscyanopropy–10% phenylcyanopropyl	AT-Silar	—	—	—	—	—	Rtx-2330 Rt-2330	—	SP2330 SP2331 SP2380 SP2560
100% Biscyanopropyl	—	CP Sil 88	—	—	—	—	Rtx-2340		SP2340
Specialty Phases									
EPA VOCs	AT- 624	CP Sil 13 CB	—	DB-624	PE- 502	007- 624	Rtx-Volatiles	—	VOCOL
EPA Method 502.2 524.2	—	—	HP-VOC	DB-624 DB-VRX	PE- 502	007-502	RTx-502.2	—	VOCOL
EPA Method 504	—	—	—	DX-3	—	—	Rtx-Volatiles	—	—
EPA Method 608	AT-5	CP Sil 8	HP-5,PAS-5	DB-5	PE-2	007-2	Rtx-5	BP-5	SPB5
	AT-35	—	HP-608	—	PE-608	007-608	Rtx-35	—	SPB608
	AT-1701	CP Sil 19	PAS-1701	DB-1701	PE-1701	007-1701	Rtx-1701	BP-10	SPB1701
	AT-50	—	HP-50	DB-17 DB-608	PE-17	007-17	Rtx-50	—	SPB608

Chiral Phases

Chirasil-Val (Alltech), Cyclodex-B (J&W), Rtx-βDEXm (Restek), Cyclex-B (SGE), B-DEX (Supelco)

PLOT Columns

Molesieve 5A (Chrompack), MoleSiev 5A (HP), GS-Molesiev (J&W), PLT-5A (Quadrex)
PoraPlot Q (Chrompack), GS-Q (J&W), Poraplot Q (HP)
GS-Alumina (J&W), Al_2O_3/KCL (Chrompack), Aluminum oxide/KCl (HP), Aluminum oxide/Na_2SO_4(HP)

Source. Column designations obtained from Reference 32.

phase. Fine-tune this choice, if necessary, by examining McReynolds constants of specific interactions of a particular solute with a stationary phase. However, for reasons that will be elucidated throughout the remainder of Section 3.11, a polar phase, when compared to a nonpolar one, tends to exhibit slightly less column efficiency, has a lower maximum temperature limit, and will have a shorter lifetime if operated for a prolonged period of time at an elevated temperature. The effect of the lower thermal stability of polar phases can be alleviated by selecting a thinner film of stationary phase and a shorter column length for more favorable elution temperatures. Thus, it is common in capillary GC to choose a phase slightly less polar in nature than that which would ordinarily be selected for a corresponding packed-column separation. In addition, a nonpolar phase is more "forgiving," because it is more resistant to oxidation and hydrolysis than a polar phase.

The list of different stationary phases available for capillary separations (Table 3.14) is not an extensive one and includes basically polysiloxanes and polyethylene glycol phases, which are suitable for most applications. The majority of analyses can be performed on columns containing 100% dimethyl polysiloxane or 5% phenyl–95% methylpolysiloxane, a cyano-polysiloxane, and a polyethylene glycol. Additional selectivity in a separation can be achieved by using a trifluoropropylpolysiloxane or phases of varying cyano and phenyl content. Separation of permanent gases and light hydrocarbons can now be performed on a capillary column containing an adsorbent (a porous polymer, alumina, molecular sieves), which serves as a direct substitute of the packed-column version.

History. The coating of a glass capillary column was achieved by roughening its inner surface prior to coating for enhanced wettability by stationary phases having a wide range of polarities and viscosities, but this option is unavailable with fused silica. The wettability of fused silica proved to be more challenging because its thin wall does not permit aggressive surface modification. Consequently, fewer phases initially could be coated on fused silica compared to glass capillaries. Although polar phases could be deposited successfully on glass capillaries, fused-silica columns coated with polar phases were especially inferior in terms of efficiency and thermal stability.

Viscosity of the film of stationary phase after deposition under the thermal conditions of GC proved to be an important consideration. Wright and coworkers (91) correlated viscosity of a stationary phase with coating efficiency and stability of the coated phase. The results of their study supported the experimental success of viscous gum phases, which yielded higher coating efficiencies and had greater thermal stability than corresponding nonviscous counterparts. The popularity of the nonpolar polysiloxane phases is in part due to the fact that their viscosity is nearly independent of temperature (92). However, the introduction of phenyl and more polar functionalities on the polysiloxane backbone causes a decrease in viscosity of a polysiloxane at elevated temperatures, resulting in thermal instability. Three areas have enhanced the quality of stationary phases for capillary GC: (1) in situ free radical cross-linking of stationary phases coated on fused silica, (2) the synthesis or commercial availability of a wide array of highly viscous gum phases, and (3) the use of OH-terminated polysiloxanes. These are discussed in Sections 3.11.3 and 3.11.4.

3.11.2 Cross-Reference of Columns From Manufacturers

A cross-reference of capillary columns offered by the major column manufacturers (listed in alphabetical order) is presented in Table 3.14 and serves as a handy reference of stationary phases and their chemical composition when comparing columns and chromatographic methods. This author hastens to add that the information presented here is the best available at the time. Perusal of this table indicates several trends in the bewildering array of column designations, which in itself is testimony to the widespread use of capillary GC. Each manufacturer has its own alpha or numeric designation for is product line, e.g., J & W (DB-) and Quadrex (007-). In many instances, the numerical suffix corresponds to the numerical suffix of the appropriate OV (Ohio Valley) phase appearing in Table 3.5 and is representative of the percentage of the polar functional group or modifier in a given polysiloxane. For example, HP-5 is listed as being *chemically similar* to DB-5 in terms of its chromatographic properties, including its selectivity and retention characteristics. For a given polysiloxane, likewise, inference should not be drawn that the phases Rtx-50 and SP-2250 are identical, only that they are chemically similar and behave similarly under chromatographic conditions. Since each manufacturer has optimized column preparation for a specific stationary phase, column dimensions, and, in some cases, an intended application of the column, slight differences in chromatographic behavior are to be expected. A manufacturer considers the steps involved in column preparation to be proprietary information.

Two types of stationary phases are most popular: the polysiloxanes and polyethylene glycol phases. Both types of phases may be characterized as having the necessary high viscosity and the capability for cross-linking and/or chemical bonding with fused silica. One should note the presence of more recent additions to the capillary column family, namely, specialty columns designed for selected EPA Methods, chiral separations, and gas–solid chromatographic separations. These specialty phases are considered in Section 3.11.6.

3.11.3 Polysiloxanes

Polysiloxanes are the most widely used stationary phases for packed and capillary-column GC. They offer high solute diffusivities coupled with excellent chemical and thermal stability. The thorough review of polysiloxane phases by Haken (93) and the overview of stationary phases for capillary GC by Blomberg (94, 95) are strongly recommended readings.

One measure of the polarity of a stationary phase is the cumulative value of its McReynolds constants, as discussed in Section 3.6.3. Because a variety of functional groups can be incorporated into the structure, polysiloxanes exhibit a wide range of polarities. Since many polysiloxanes are viscous gums and, as such, coat well on fused silica and can be cross-linked, they are ideally suited for capillary GC. The basic structure of 100% dimethylpolysiloxane can be illustrated as

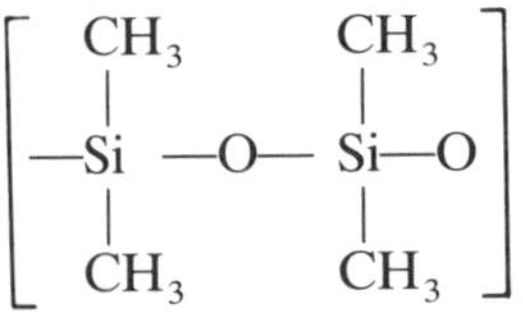

$$\left[-\underset{CH_3}{\overset{CH_3}{Si}} - O - \underset{CH_3}{\overset{CH_3}{Si}} - O \right]$$

Replacement of the methyl groups with another functionality enables polarity to be imparted to the polymer. The structure of substituted polysiloxanes in Tables 3.5 and 3.14 can be depicted by the following general representation:

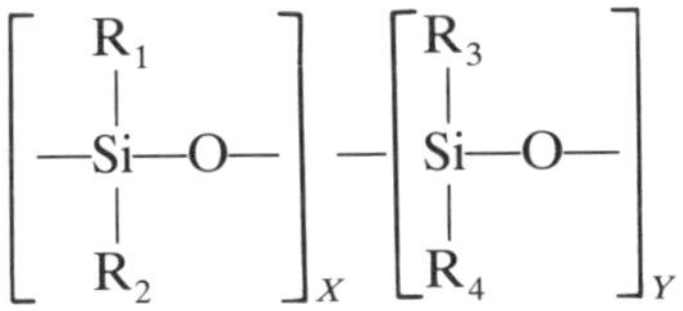

$$\left[-\underset{R_2}{\overset{R_1}{Si}} - O - \right]_X - \left[\underset{R_4}{\overset{R_3}{Si}} - O - \right]_Y$$

where the R groups can be CH_3, phenyl, $CH_2CH_2CF_3$, or $CH_2CH_2CH_2CN$, and X and Y indicate the percentage of an aggregate in the overall polymeric stationary phase composition. In the case of the phase, DB-1301 or one of its chemically equivalents (6% cyanopropylphenyl–94% dimethylpolysiloxane, $R_1 = CH_2CH_2CH_2CN$, $R_2 =$ phenyl, R_3 and R_4 are methyl groups; X and Y have the values of 6 and 94%, respectively. For phases equivalent to 50% phenyl–50% methylpolysiloxane, R_1 and R_2 are methyl groups, while R_3 and R_4 are aromatic rings; X and Y are each equal to 50%.

Phase selectivity, which also impacts resolution in a chromatographic separation, is governed by solute–stationary phase interactions, such as dispersion, dipole, acid/base, and hydrogen bond donors/acceptors. A column containing a polar stationary phase can display greater retention for a solute having a given polar functional group compared to other solutes of different functionality, while on a less polar stationary phase elution order may be reversed or altered to varying degrees under the same chromato-

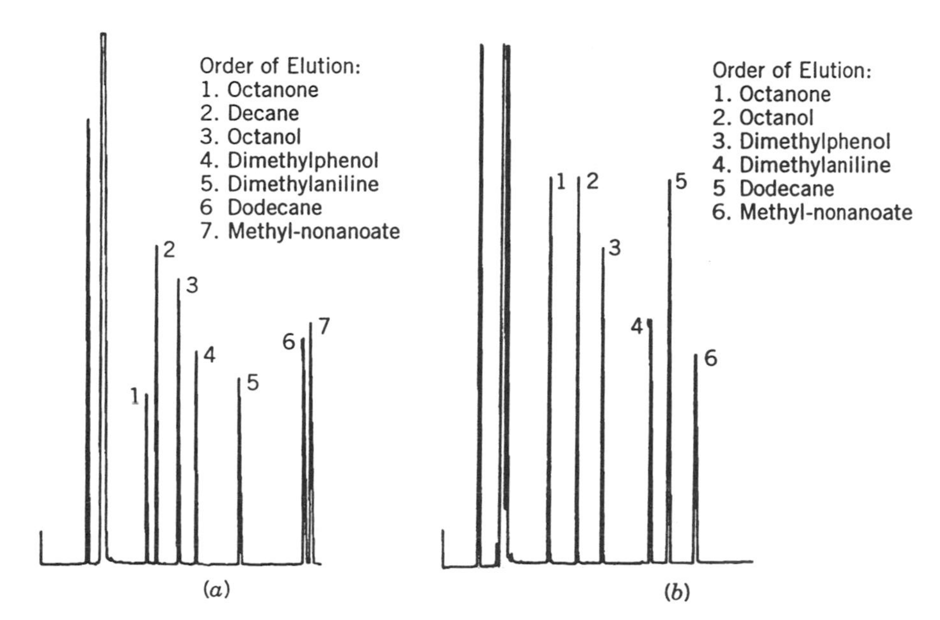

FIGURE 3.33 Chromatograms of an activity mix on immobilized (A) polydimethylsiloxane and (B) 5% phenyl–95% methylpolysiloxane; column: 25 m × 0.25 mm i.d. (0.25-μm film); column temperature: 110°C, 25 cm/sec He, Det: FID.

graphic conditions. Several illustrations of how elution order is affected by stationary phase selectivity are presented in Figures 3.33 and 3.34.

3.11.4 Polyethylene Glycol Phases

The most widely used non-silicon-containing stationary phases are the polyethylene glycols. They are commercially available in a wide range of molecular weights under several designations, such as Carbowax 20M and Superox-4. The general structure of a polyethylene glycol may be described as

$$HO—CH_2—CH_2—(—O—CH_2—CH_2—)_n—O—CH_2—CH_2—OH$$

The popularity of polyethylene glycols stems from their unique selectivity and high polarity as a liquid phase. Unfortunately, they do have some limitations. Characteristic of Carbowax 20M, for example, is its rather low upper temperature limit of approximately 225°C and a minimum operating temperature of 60°C. In addition, trace levels of oxygen and water have adverse effects on most liquid phases, but particularly so with Carbowax 20M, where they accelerate the degradation process of the phase. Verzele and coworkers have attempted to counteract these drawbacks by preparing a

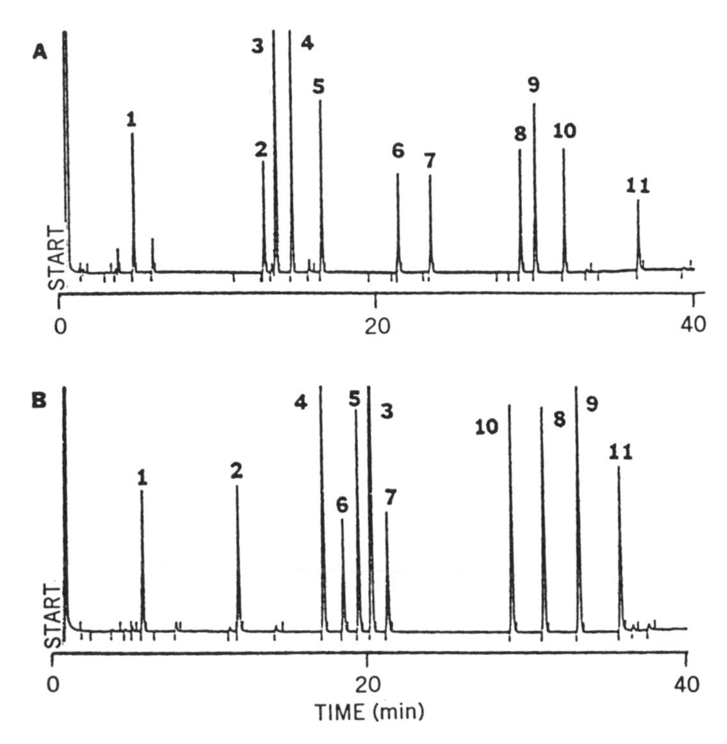

FIGURE 3.34 Chromatograms of the separation of nitro-containing compounds on 10 m × 0.53 mm with (A) CP Sil 19 CB and (B) CP Sil 8 CB as stationary phase. Column conditions: 50°C (3 min) at 6°C/min to 250°C, Det: FID, 24 cm/sec He. Solutes: 1, nitropropane; 2, 2-nitro-2-methyl-1-propanol; 3, *n*-dodecane; 4, nitrobenzene; 5, *o*-nitrotoluene; 6, 2-nitro-2-methyl-1,3-propanediol; 7, 2-nitro-2-ethyl-1,3-propanediol; 8, *p*-nitroaniline; 9, *p*-nitrobenzyl alcohol; 10, *o*-nitrodiphenyl; and 11, 4-nitrophthalimide. (From Reference 146.)

very high molecular weight polyethylene glycol, Superox-4 (96). Other successful attempts include free radical cross-linking and bonding, which is discussed in the next section.

3.11.5 Cross-Linked Versus Chemically Bonded Phases

The practice of capillary GC has been enriched by the advances made in the immobilization of a thin film of a viscous stationary phase coated uniformly on the inner wall of fused silica tubing. At present, two pathways are employed for the immobilization of a stationary phase: free radical cross-linking and chemical bonding. By immobilizing a stationary phase by either approach, the film is stabilized and is not disrupted at elevated column temperatures or during temperature programming. Thus, less column bleed and higher operating temperatures can be expected with a phase of this nature, a consideration especially important in GC-MS. A column containing an immobilized stationary phase is also recommended for on-column injection where large aliquots of solvent are injected without dissolution of the stationary phase. Likewise, a column having an immobilized phase can be backflushed to rinse contamination from the column without disturbing the stationary phase.

Cross-Linking of a Stationary Phase. The ability of a polymer to cross-link is highly dependent on its structure. The overall effect of cross-linking is that the molecular weight of the polymer steadily increases with the degree of cross-linking, leading to branched chains until eventually a three-dimensional rigid network is formed. Since the resultant polymer is rigid, little opportunity exists for the polymer chains to slide past one another, thereby increasing the viscosity of the polymer. Upon treatment of a cross-linked polymer with solvent, the polymer does not dissolve, but rather a swollen gel remains behind after decantation of the solvent. Under the same conditions, an uncross-linked polymer of the same structure would dissolve completely. In summary, the dimensional stability, viscosity, and solvent-resistance of a polymer are increased as a result of cross-linking. The mechanism for cross-linking 100% dimethyl polysiloxane is described below where R· and γ (gamma radiation) are free radical initiators:

$$
\begin{array}{ccccccc}
\begin{array}{c} CH_3 \\ | \\ -Si-O- \\ | \\ CH_3 \end{array} & & \begin{array}{c} CH_3 \\ | \\ -Si-O- \\ | \\ \cdot CH_2 \end{array} & & \begin{array}{c} CH_3 \\ | \\ -Si-O- \\ | \\ CH_2 \\ | \\ CH_2 \\ | \\ -Si-O- \\ | \\ CH_3 \end{array} & \begin{array}{c} +\,2RH \\ \\ \text{or} \\ \\ H_2 \end{array} & (3.20) \\
 & \begin{array}{c} R\cdot \\ +\ \text{or} \rightarrow \\ \gamma \end{array} & & \rightarrow & & & \\
\begin{array}{c} CH_3 \\ | \\ -Si-O- \\ | \\ CH_3 \end{array} & & \begin{array}{c} CH_3 \\ | \\ -Si-O- \\ | \\ \cdot CH_2 \end{array} & & & &
\end{array}
$$

The series of photographs presented in Figure 3.35 permits visualization of the cross-linking process for two stationary phases, 100% dimethylpolysiloxane (SE-30) and trifluoropropylmethylpolysiloxane, as a function of increasing degree of cross-linking by gamma radiation (97).

Madani et al. provided the first detailed description of capillary columns where polysiloxanes were immobilized by hydrolysis of dimethyl and

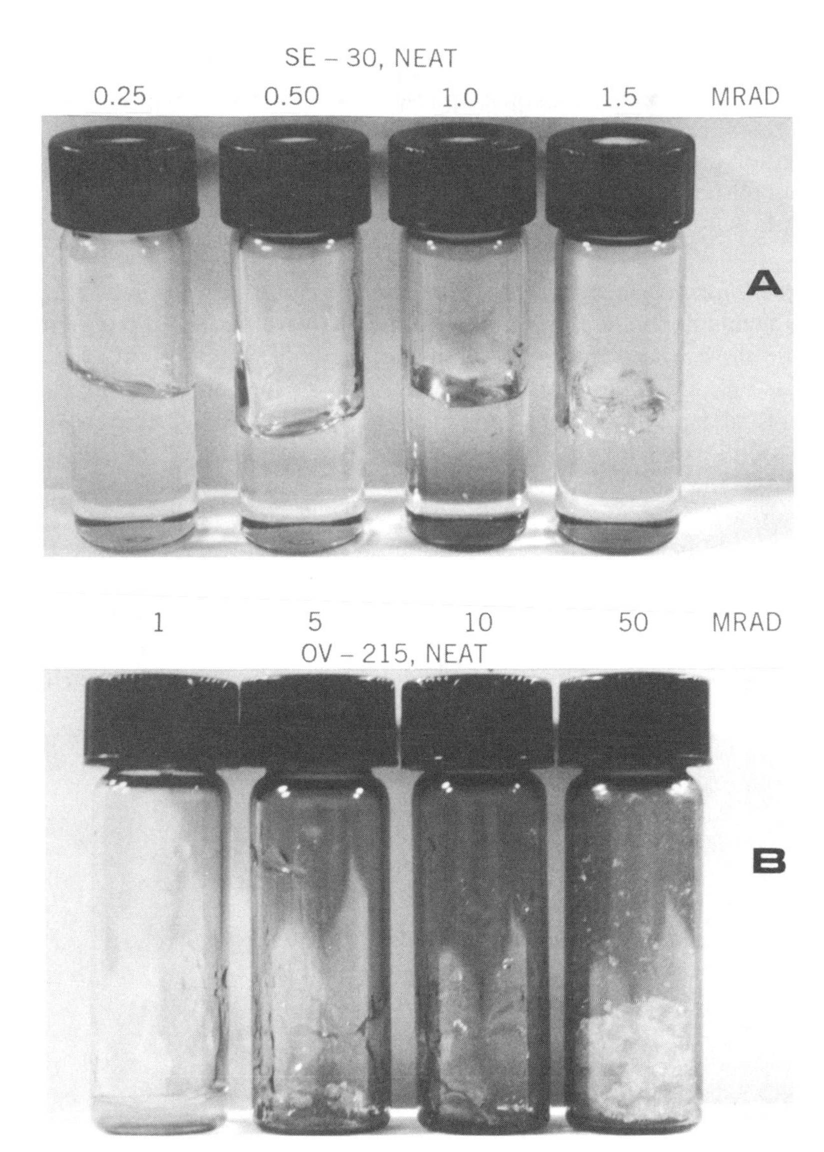

FIGURE 3.35 Effect of gamma radiation on degree of cross-linking: (A) SE-30 (polydimethylsiloxane) and (B) OV-215 (trifluoropropylmethylpolysiloxane); reference 97. Reproduced from the *Journal of Chromatographic Science* by permission of Preston Publications, a Division of Preston Industries, Inc.

diphenylchlorosilanes (98, 99). Interest increased when Grob found that the formation of cross-linked polysiloxanes resulted in enhanced film stability (100). Blomberg et al. illustrated in situ synthesis of polysiloxanes with silicon tetrachloride as a precursor, followed by polysiloxane solution

(101, 102). Since then, various approaches for cross-linking have been investigated. These include chemical additives such as organic peroxides (103–111), azo compounds (73, 112, 113), ozone (114), and gamma radiation (68, 115–118). Several different peroxides have been evaluated, with dicumyl peroxide being the most popular. However, peroxides can generate polar decomposition products that remain in the immobilized film of stationary phase. Moreover, oxidation may also occur, which increases the polarity and decreases the thermal stability of a column. These adverse effects are eliminated with azo species, as free radical initiators. Lee et al. have cross-linked a wide range of stationary phases, from nonpolar to polar, in their studies using azo-*tert.*-butane and other azo species (73, 112, 113). If an azo compound or a peroxide exists as a solid at room temperature, the agent is spiked directly into the solution of the stationary phase used for coating of the column. On the other hand, for free radical initiators that are liquid at ambient temperature, the column is first coated with stationary phase, then saturated with vapors of the reagent (42, 73).

Gamma radiation from a cobalt-60 source has also been used by Schomburg et al. (68), Bertsch et al. (115), and Hubball and coworkers (97, 116–118) as an effective technique for cross-linking polysiloxanes. In a comparative study of gamma radiation with peroxides, Schomburg et al. (68) noted that each approach immobilized polysiloxanes, but that the formation of polar decomposition products is avoided with radiation. Radiation offers the additional advantages of the cross-linking reaction occurring at room temperature and columns can be tested both before and after the immobilization of the stationary phase.

Not all polysiloxanes can be directly or readily cross-linked. The presence of methyl groups facilitates cross-linking. Consequently, the nonpolar siloxanes exhibit very high efficiencies and high thermal stability. However, as the population of methyl groups on a polysiloxane phase decreases and as these groups are replaced by phenyl or more polar functionalities, cross-linking of a polymer becomes more difficult. The incorporation of vinyl or tolyl groups into the synthesis of a polymer tailored for use as a stationary phase for capillary GC overcomes this problem. Lee (109, 112) and Blomberg (110, 111) have successfully synthesized and cross-linked stationary phases of high phenyl and high cyanopropyl content that also contain varying amounts of these free radical initiators. Colloidal particles have also been utilized to stabilize cyanoalkyl stationary phase films for capillary GC (119). Recently, favorable thermal stability and column inertness were obtained by a binary cross-linking reagent, a mixture of dicumyl peroxide and tetra(methylvinyl)cyclotetrasiloxane (120).

Developments in the cross-linking of polyethylene glycols have been slower in forthcoming, although successes have been reported. Immobilization of this phase by the following procedures increases its thermal stability and its compatibility and tolerance for aqueous solutions. DeNijs and de Zeeuw (121) and Buijten et al. (122) immobilized a polyethylene glycol in situ, the latter group with dicumyl peroxide and methyl(vinyl) cyclopentasiloxane as additives. Etler and Vigh (123) used a combination of gamma radiation with organic peroxides to achieve immobilization of this polymer, while Bystricky selected a 40% solution of dicumyl peroxide (126). George (5) and Hubball (124) have successfully cross-linked polyethylene glycol using radiation. The chromatographic separation of a cologne in Figure 3.36 was generated on a capillary column containing Carbowax 20M cross-linked by gamma radiation and indicates acceptable thermal stability to 280°C. Horka and colleagues described a procedure for cross-linking Carbowax 20M with pluriisocyanate reagents (125). Recently, thermally bondable polyethylene glycols and polyethyleneimines have been popular phases and yield chromatographic selectivity similar to the traditional polyethylene glycols. Despite these efforts, the upper temperature limit of polyethylene glycol columns generally remains below 300°C.

Chemical Bonding. In the last several years interest in chemically bonding a stationary phase of the inner wall of a fused-silica capillary appears to be intensifying among column manufacturers. As the term suggests, an actual chemical bond is formed between fused silica and the stationary phase. The foundation of this procedure was first reported by Lipsky and McMurray (127) in their investigation of hydroxy-terminated polymethylsilicones and was later refined by the work of Blum et al. (128–132), who employed OH-terminated phases for the preparation of inert, high-temperature stationary phases of varying polarities. The performance of hydroxy-terminated phases has also been evaluated by Schmid and Mueller (133) and Welsch and Teichmann (134). Other published studies include the behavior

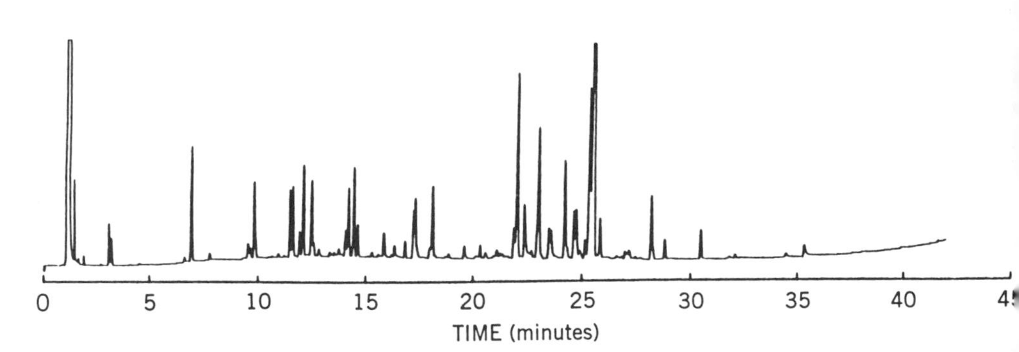

FIGURE 3.36 Chromatogram illustrating the separation of a cologne sample on a capillary column (15 m × 0.32 mm i.d., 0.25-μm film) containing Carbowax 20M cross-linked by gamma radiation. Column conditions: 40°C (2 min) at 6°C/min to 280°C, Det: FID, 25 cm/sec He.

of hydroxyl phases of high cyanopropyl content by David et al. (135) and trifluoropropylmethylpolysiloxane phases by Aichholz (136).

In the chemical bonding approach to stationary phase immobilization, a capillary column is coated in the conventional fashion with an OH-terminated polysiloxane and then temperature programmed to an elevated temperature, during which time a condensation reaction occurs between the surface silanols residing on the fused-silica surface and those of the phase. It is important to note here that both deactivation and coating are accomplished in a single-step process and result in the formation of a Si–O–Si bond more thermally stabile than the Si–C–C–Si bond created via cross-linking. Cross-linking of the stationary phase is not a necessary requirement. But, if a stationary phase contains a vinyl group (or another free radical initiator), cross-linking can simultaneously occur. Phases that cannot be cross-linked during the bonding process can be cross-linked afterward with an azo compound, for example. Grob, after observing the increased inertness and thermostability of OH-terminated phases, commented that they might reflect a "revolution in column technology" (137). A blend of both cross-linking and chemically bonding is utilized by column manufacturers today.

Since the mid 1980s, immobilization of polysiloxanes and polyethylene glycols is no longer a subject of rapid advancements reported in the literature. The fixation of these stationary phases via cross-linking and chemical bonding for capillary GC is now a well-defined and a matured technology. A capillary column containing such a stationary phase is the resultant of elegant efforts of people such as M. L. Lee, L. Blomberg, K. Grob and his family members, G. Schomburg and their colleagues, and many, many others too numerous to mention here. Attention has now been shifted to such areas as stationary phase selectivity tuned or optimized for specific applications, multidimensional chromatographic techniques, and immobilization of chiral stationary phases, which are discussed in Section 3.11.6.

3.11.6 Specialty Columns

EPA Methods. Column manufacturers have responded to the increasing analytical and environmental demands for capillary columns for use in EPA 500 Series, EPA 600 Series, and EPA 8000 Series methods. To simplify the column selection for a given EPA method by a user, a column is so designated, e.g., the alpha or numeric tradename of the manufacturer—the EPA method number, as shown in Table 3.14. These columns have been configured in length, inner diameter, film thickness, and stationary phase composition for optimized separation of the targeted compounds under the chromatographic conditions specified in a particular method. Another factor relating to column dimensions considered by manufacturers is the compatibility of thicker film columns with methods that stipulate purge and trap sampling by eliminating the need for cryogenic solute focusing prior to chromatographic separation.

Chiral Separations. Capillary columns having a chiral stationary phase are used for the separation of optically active isomers or enantiomers, namely, species that have the same physical and chemical properties with the exception of the direction in which they rotate plane-polarized light. Enantiomers may also have different biological activity and, therefore, enantiomeric separations are important in the food, flavor, and pharmaceutical areas. A chiral stationary phase can recognize differences in the optical activity of solutes to varying extent.

Most of the earlier chiral phases have limited thermal stability. By chemically bonding a chiral stationary phase to a polysiloxane, the upper temperature limit can be extended. Chirasil-Val is perhaps the most famous stationary phase in this category. Recent work has employed β-cyclodextrin as the key chiral recognition component in stationary phases. The mechanism of separation (or enantiomeric selectivity) is based on the formation of solute–β-cyclodextrin complexes occurring in the barreled-shaped opening of the cyclodextrin and can be manipulated by varying the size of the opening of the cyclodextrin ring as well as by the substitution pattern on the ring. Although cyclodextrins can be used as a stationary phase (138, 139), the current practice is to either place them in solution with a polysiloxane (140) or immobilize them by bonding to a polysiloxane (141). The chromatographic principles underlying chiral separations by GC have recently been reviewed by Hinshaw (142).

Gas–Solid Adsorption Capillary Columns. These columns are also referred to as PLOT or porous layer open tubular columns. A PLOT column consists of fused-silica capillary tubing in which a layer of an adsorbent lines the inner wall in place of a liquid phase. An early use of a PLOT column was reported by de Nijs (143) who prepared a fused-silica column coated with submicron particles of aluminum oxide for the analysis of light hydrocarbons. De Zeeuw et al. (144) subsequently prepared PLOT columns with a 10- to 30-μm layer of a porous polymer of the Porapak Q type (styrene–divinylbenzene) by in situ polymerization of a coating solution. The selection of adsorbents currently available in a PLOT column configuration includes aluminum oxide/KCl, aluminum oxide/sodium sulfate, molecular sieves, alumina, graphitized carbon, and porous polymers. These columns are intended to be direct replacements for a packed column containing the same adsorbent and feature faster regeneration of the adsorbent.

3.11.7 Capillary-Column Care and First Aid

Ferrule Materials and Fittings. Ferrules for capillary columns are usually fabricated from graphite and Vespel/graphite composites. The characteristics of these materials are presented in Table 3.6. It is important to select the proper ferrule i.d. to be compatible with the o.d. of the capillary column or a carrier gas leak will result after installation. Ferrules having an i.d. of 0.4 mm are recommended for 0.25-mm-i.d. columns, 0.5-mm-i.d. ferrules are recommended for 0.32-mm-i.d. columns, and 0.8-mm-i.d. ferrules are recommended for 0.53-mm-i.d. capillary columns.

Outlined below are guidelines for the preparation of a capillary column for installation.

1. Slide the retaining fitting or nut over the end of a new column, then the ferrule and position them at least 6 in. away from the column end. It will be necessary to cut several inches from the end of the column because ferrule particles may have entered the column and can cause tailing and adverse adsorptive effects.
2. With a scoring tool, gently scribe the surface of the column several inches away from the end. While holding the column on each side of the scoring point, break the end at the scoring point at a slight downward angle. Any loose chips of fused silica or polyimide will fall away and will not enter the column, as may happen if it is broken in a completely horizontally configuration. This procedure eliminates the possibility of chips of fused silica or polyimide from residing in the end of the column. Alternatively, an excellent cut can be made with a ceramic wafer.
3. Examine the end of the column with a 10–20 × magnifier or an inexpensive light microscope. The importance of a properly made cut cannot be overstated. An improperly cut column, as illustrated in Figure 3.37, where a jagged edge of fused silica is evident, can generate active sites and may cause peak-tailing, peak-splitting, or solute adsorption.
4. A supply of spare ferrules and related tools (Figures 3.38 and 3.39) are convenient to have in the laboratory when removal or changing columns is required. Ferrules, ferrule-handling accessories, tool kits, and other handy gadgets are available from many column manufacturers.

Column Installation. Define the injector and detector ends of the column. Align the end to be inserted into the injector with a ruler and mark the recommended distance of insertion as specified in the instrument manual with typewriter correction fluid. Then slide the ferrule and nut closer to the point of application of the correction fluid and mount the column cage on the hanger in the column oven. Alleviate any stress, sharp bends, or contact with sharp objects along the ends of the column. Insert the measured end into the injector and tighten the fitting. If the column is to be conditioned, leave the detector end of the column disconnected; otherwise, insert this

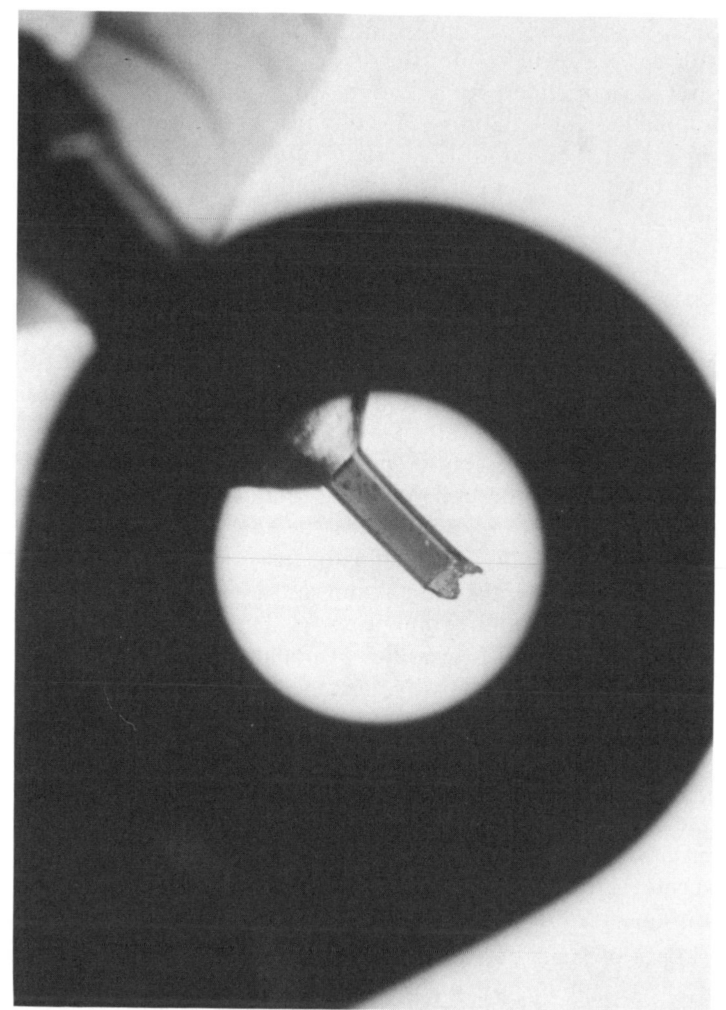

FIGURE 3.37 Examination of a poorly terminated fused silica column. (Photograph courtesy of the Hewlett–Packard Company.)

end into the jet tip of the FID at the specified recommended distance, usually 2 mm down from the top of the jet. In Figure 3.40 a photograph of the Quick-Connect fitting is shown. This device facilitates column installation in most gas chromatographs without the use of wrenches and extends ferrule lifetime.

Column Conditioning. Conditioning of a capillary column removes residual volatiles from the column. There are three essential rules for

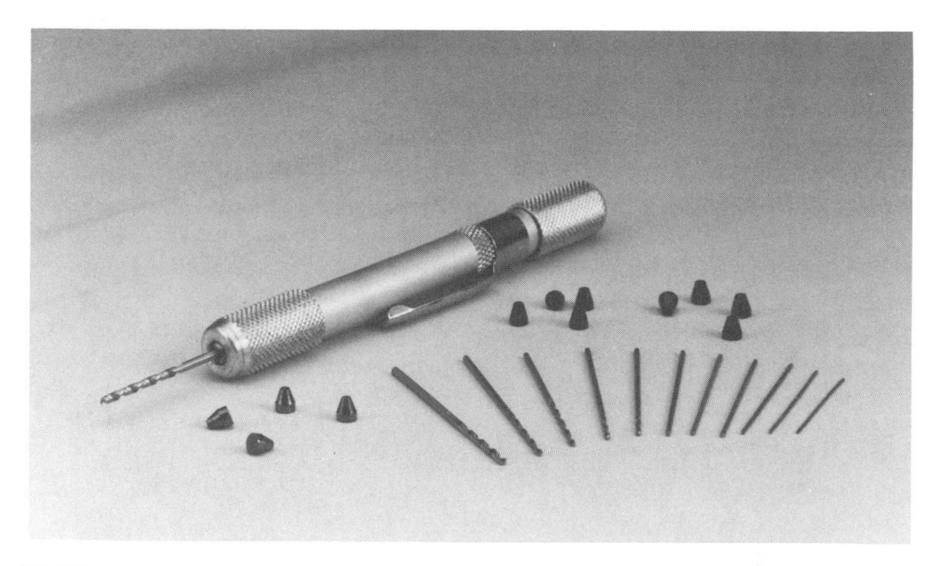

FIGURE 3.38 Photograph of a capillary column ferrule kit containing an assortment of ferrules and a pin vise drill with bits for drilling or enlarging the bore of ferrules. (Reprinted with permission of Supelco, Inc., Bellefonte, PA 16823, USA.)

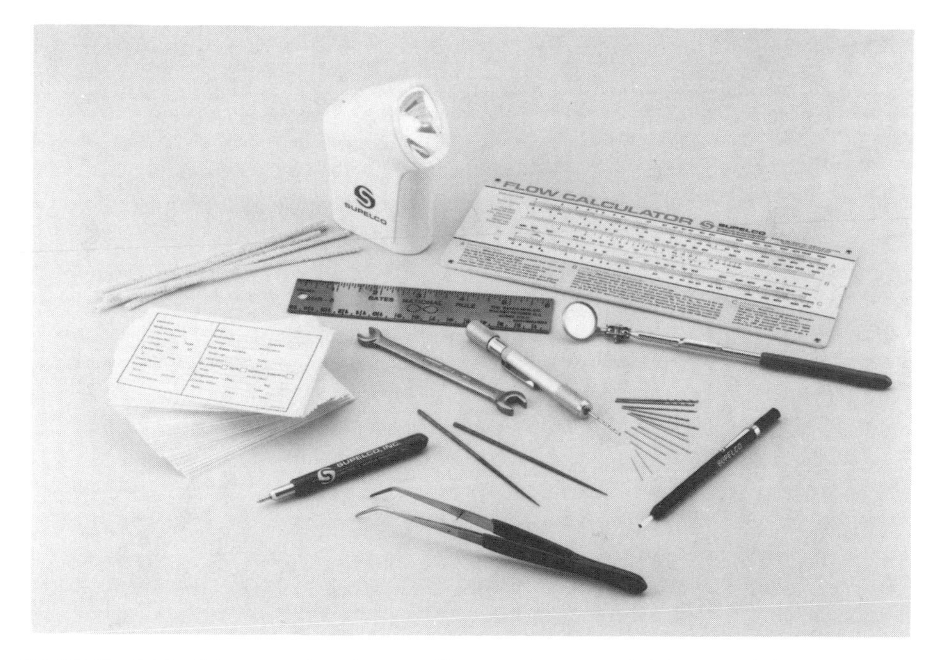

FIGURE 3.39 Photograph of a capillary column tool kit containing tweezers, needle files, scoring tool, pin vise drill kit, flow calculator, pocket mirror, mini flashlight, labels, septum puller, stainless steel ruler, and pipe cleaners. (Reprinted with permission of Supelco, Inc., Bellefonte, PA 16823, USA.)

(a)

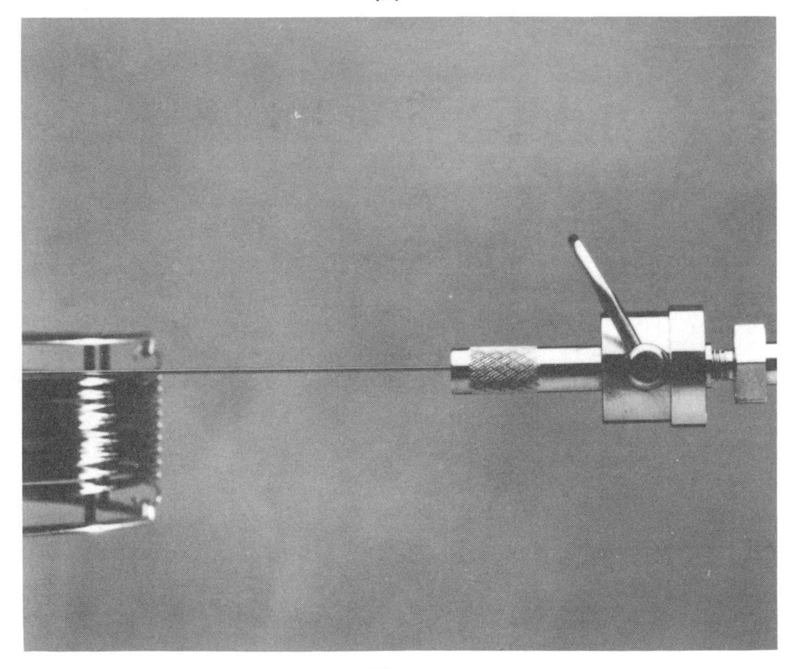

(b)

FIGURE 3.40 Illustration of the Quick-Connect fitting for installation of capillary columns: (A) internal sealing mechanism and (B) sealing mechanism inserted into fitting and locked into place. (Photographs courtesy of the Quadrex Corporation.)

conditioning a capillary column, the first two of which also apply to the conditioning of a packed column (Section 3.7.5):

1. Carrier-gas flow must be maintained *at all times* when the column temperature is above ambient temperature and there should be no gas leaks.
2. Do not exceed the maximum allowable temperature limit of the stationary phase or permanent damage to the column can result.
3. As opposed to the conditioning of a packed column, overnight conditioning of a capillary column is usually unnecessary. Instead, purge the column with normal carrier-gas flow for 30 min at room temperature, then temperature program the column oven at 4°C/min to a temperature 20°C above the anticipated highest temperature at which the column will be subjected without exceeding the maximum allowable temperature limit. Usually after a column has been maintained at this elevated temperature for several hours, a steady baseline is obtained and the column is ready for analyses to be conducted. Use of high-purity carrier gas and a leak-free chromatographic system will greatly extend the lifetime of any column, packed or capillary. Other details pertaining to conditioning and column care can be found in Section 3.7.5.

Column Bleed. Column bleed is a phrase used to describe the rise in baseline during a blank temperature programming run and is the inevitable consequence of increasing thermal degradation of a polymer with an increase in column temperature. One should expect some degree of bleeding with every column; some phases just generate more bleed than others. For example, a nonpolar phase bleeds less than a polar phase because the former typically has a higher temperature limit and thus is more thermally stable. Moreover, in comparing capillary columns of different dimensions, the level of bleed will increase with increasing amount of stationary phase in the column. Therefore, longer and wider diameter columns yield more bleed than shorter or narrower columns. Likewise, column bleed increases with increasing film thickness of the stationary phase.

The rate of temperature programming or ramp rate can influence the bleed profile from a column. As the rate of temperature programming increases, column bleed also increases. Finally, the more sensitive element-specific detectors, e.g., an ECD or NPD, will generate a more pronounced bleed profile if the stationary phase contains a heteroatom or functional group (–CN or –F) to which a detector responds in a sensitive fashion.

Retention Gap and Guard Columns. A 0.5- to 5.0-m length of deactivated fused-silica tubing installed between the injector and analytical column is often referred to as a retention gap or guard column (Figure

3.41). The term, retention gap, is used to describe this segment for on-column injection where the condensed solvent resides after injection, but both solvent and solutes are not retained once vaporization occurs via temperature programming. As a guard column, this short length of deactivated tubing preserves the lifetime of an analytical column by collecting nonvolatile components and particulate matter in dirty samples that would otherwise accumulate at the inlet of the analytical column. As such, its latter role in capillary GC parallels the function of the guard column in HPLC. A guard column is considered to be a consumable item, requiring replacement from time to time, usually when the detector response of active analytes begins to decrease substantially. It eliminates the need for the repetitive

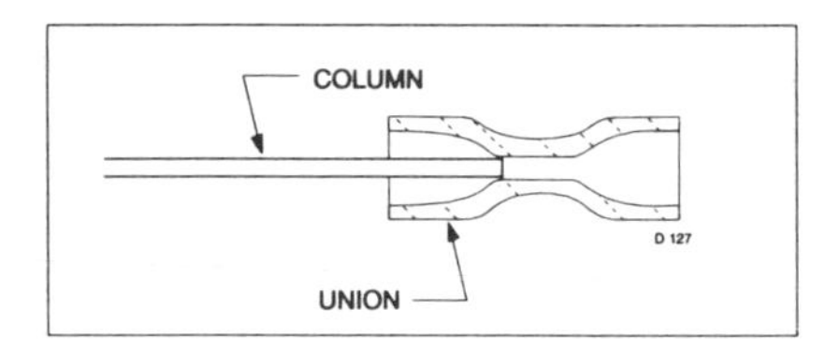

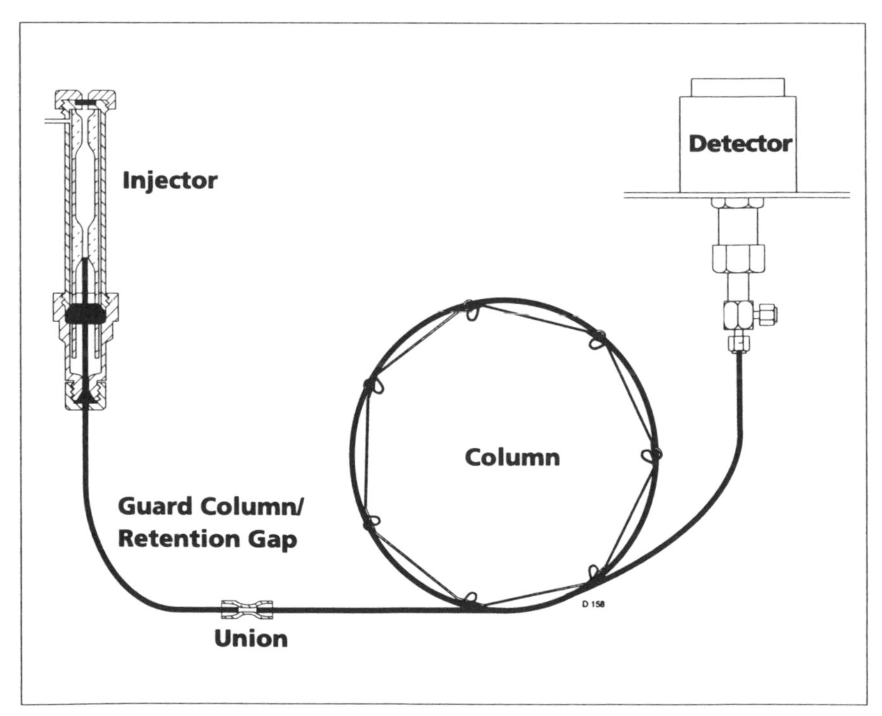

FIGURE 3.41 Schematic diagram of a guard column/retention gap. (Illustration courtesy of J&W Scientific, Division of Curtin Matheson Scientific, Inc.)

removal of small sections at the inlet end of an analytical column with the buildup of contamination.

Proper implementation of the connection between the guard and analytical columns is essential for the preservation of the chromatographic integrity of the system. The generation of active sites within the fitting can cause adsorptive losses and peak-tailing. Commercially available fittings for this purpose include the metal butt connector of low dead volume, Press-Tight connectors, and a capillary Vu-Union. An illustration of the primary and secondary sealing mechanisms in the capillary Vu-Union is shown in Figure 3.42, where the two column ends are positioned into ferrules located inside a deactivated tapered glass insert. This type of connector combines the benefits of a low dead volume connection with the sturdiness of a ferrule seal. Furthermore, the glass window permits visual confirmation of the connection.

Column Fatigue and Regeneration. Deterioration in column performance can occur by the contamination of the column with the accumulation of nonvolatiles and particulate matter, usually in the injector liner and column inlet. Column contamination is manifested by adsorption and peak-tailing of active analytes, excessively high column bleed levels, and changes in the retention characteristics of the column. Rejuvenation of the column can be attempted by several paths. First, remove one or two meters of column from the inlet end. If the column still exhibits poor chromatographic perform-

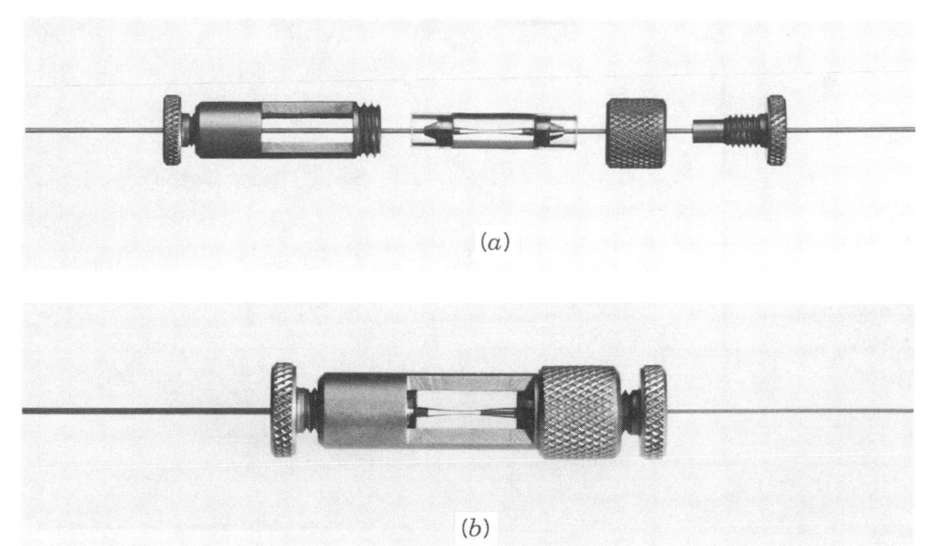

FIGURE 3.42 Photographs of a capillary/microbore Vu-Union (A) disassembled, showing the primary and secondary sealing mechanisms and (B) assembled. (Photographs courtesy of the Restek Corporation.)

ance, try turning the column around and reconditioning it overnight disconnected from the detector. A third approach is solvent rinsing, an extreme measure that should be attempted only with cross-linked or chemically bonded phases. Solvent-rinse kits, such as the one schematically described in Figure 3.43, are commercially available and enable a column to be backrinsed of contamination by slowly introducing 10–30 mL of an appropriate solvent into the detector end of the column by nitrogen gas pressure. This approach is worthwhile for heavily contaminated columns, but in all cases the recommendations for rinsing outlined by the column manufacturer should be followed.

3.11.8 Applications

Numerous applications of capillary separations appear throughout the remaining chapters of this book and provide the reader with an appreciation for the vast scope and power of capillary GC. A condensed summary of some general application areas and a corresponding suggested stationary phase is presented in Table 3.15. This summary is obviously intended not to be all-inclusive, but rather to serve as a starting point for column selection.

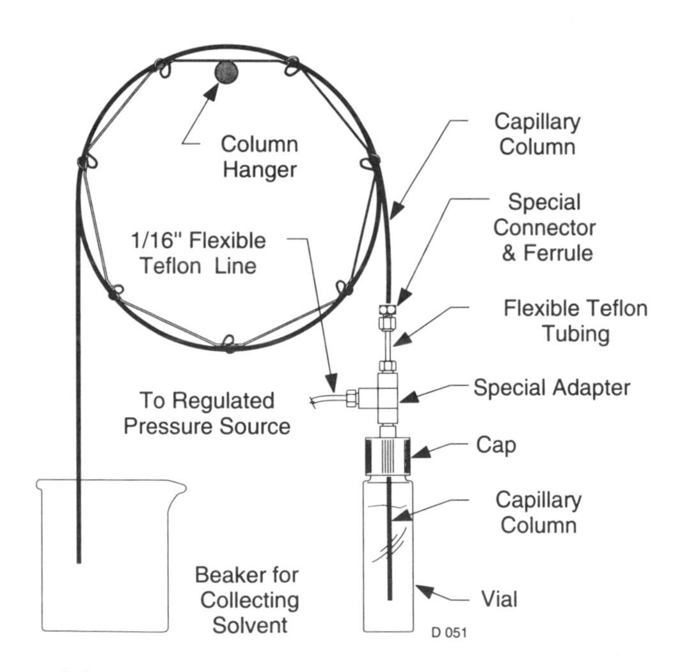

FIGURE 3.43 Schematic diagram of a solvent rinse kit. (Illustration courtesy of J&W Scientific, Division of Curtin Matheson Scientific, Inc.)

TABLE 3.15 Recommended Stationary Phases for Selected Applications

Stationary Phase Polysiloxane	Application
100% Methyl	Alkaloids, amines, drugs, FAME, hydrocarbons, petroleum products, phenols, solvents, waxes, general purposes
5% Phenyl–95% dimethyl	Alcohols, alkaloids, aromatic hydrocarbons, flavors,fuels, halogenates, herbicides, pesticides, petroleum products, solvents, waxes, general purposes
50% Methyl–50% phenyl	Alcohols, drugs, herbicides, pesticides, phenols, steroids, sugars
14% Cyanopropylphenyl–86% dimethyl	Alcohols, aroclors, alcohol acetates, drugs, fragrances, pesticides
50% Cyanopropylmethyl–50% phenyl	Carbohydrates, FAME
Trifluoropropyl	Drugs, environmental samples, ketones, nitroaromatics
Polyethylene glycol	Alcohols, flavors, fragrances, FAME, amines, acids

Part 4 Column Oven Temperature Control

3.12 THERMAL PERFORMANCE VARIABLES AND ELECTRONIC CONSIDERATIONS

Gas chromatographic columns are installed in a column oven where the temperature must be accurately and precisely controlled, because column temperature has a pronounced influence on retention time. Any fluctuation in column temperature will yield an impact on the measurement of retention data and retention indices. Present oven geometries and electronic temperature control components are capable of thermostatting a column oven to ±0.1°C.

There are several additional requirements that a column oven must satisfy. A column oven should be thermally insulated from heated injector and detector components, a requirement that becomes more demanding as the selected column oven temperature approaches ambient temperature. Ideally, the temperature of a column oven should remain constant and independent of environmental changes in the laboratory and any line-voltage fluctuations. Versatility in the operating temperature capability is also necessary to achieve column temperatures ranging from subambient temperature to elevated temperatures above 400°C (for separations with metal-clad capillary columns). With recent advances in adsorbents and PLOT columns, the need for cryogenic cooling of a column oven for the subambient separations of permanent gases and light hydrocarbons is no longer required, but has been replaced by the need of cryogenic capability for solute focusing purposes with on-column injection and auxiliary sampling techniques, such as thermal desorption and purge and trap.

Current gas chromatographic oven design is also a product of the age of miniaturization. Early column ovens were relatively large in volume (up to $3500\,in^3$) to accommodate U-shaped glass columns for biomedical and preparative separations. Thermal gradients were common in these huge vertically configured rectangular ovens. On the other hand, the typically smaller oven geometry of today (i.e., $30 \times 27 \times 15$ cm or approximately 12 L) can comfortably accommodate two capillary columns, a packed and a capillary column, or two packed columns. Forced air convection is the most popular type of gas chromatographic oven, because it provides a uniform temperature in the column oven. Modern oven designs also permit fast cooldown rates after temperature programming, an important consideration because it governs sample throughput in a laboratory.

In modern gas chromatographs the temperature controller of a column oven is a microprocessor incorporated into a feedback loop, allowing both temperature programming ramp profiles and isothermal heating to be accomplished accurately and reproducibly. Under microprocessor control, a flap or door movement permits the blending of the proper amount of ambient lab air with oven air in the control of oven temperature. In addition, a cryogenic value can be opened by a microprocessor for delivery of carbon dioxide or liquid nitrogen in the column oven.

3.13 ADVANTAGES OF TEMPERATURE PROGRAMMING OVER ISOTHERMAL OPERATION

Isothermal operation of a chromatographic column has a number of drawbacks, as illustrated in the scenario depicted for the separations in Figure 3.44. If the selected isothermal column temperature is too low, the early eluting peaks will be closely spaced, while the more strongly retained components will be broad and low-lying (Figure 3.44C). These strongly retained components can be more quickly eluted by selecting a higher isothermal temperature, which will also improve their detectability (Figure 3.44A). However, in doing so, rapid coelution of components, peaks too closely spaced, and an overall loss in resolution result in the beginning of a chromatogram. This situation, which prevails in all practiced versions of elution chromatography, is often called the "general elution problem"; it is solved in GC by temperature programming, where the column oven temperature is gradually increased at a linear rate during an analysis (Figure 3.44B).

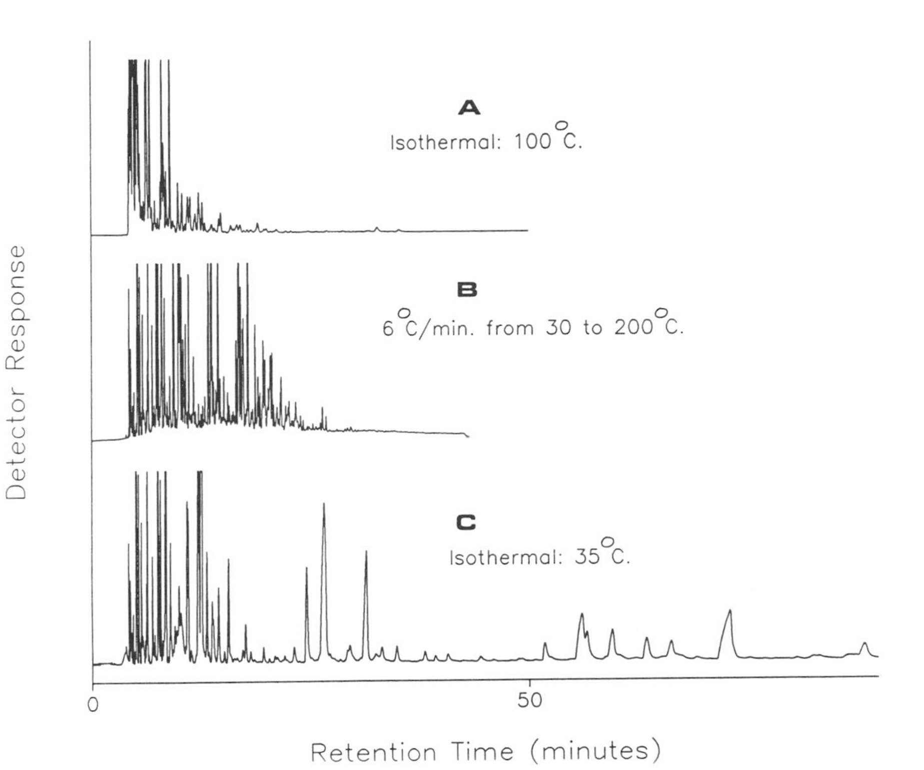

FIGURE 3.44 Demonstration of the general elution problem in gas chromatography with a gasoline sample. Chromatographic conditions as noted in figure; otherwise, the same as in Figure 3.33.

Temperature programming offers several attractive features. One can expect a reduced time of analysis and improved overall detectability of components (peaks are sharper and have nearly equal bandwidths throughout the chromatogram). In the case of unknown samples or samples of high complexity, high-boiling components, which might not elute or be detected under isothermal conditions, can exhibit a more favorable retention time. Temperature programming also helps "clean-out" a column of remnant high-boiling species from previous injections. The interested reader is urged to consult the classic book *Programmed Temperature Gas Chromatography* by Harris and Habgood (145) for a detailed treatment of the subject.

3.14 OVEN TEMPERATURE PROFILE FOR PROGRAMMED TEMPERATURE GC

Three basic types of temperature programming profiles are used in GC: ballistic, linear, and multilinear. Ballistic programming occurs when an oven maintained at a given isothermal temperature is rapidly changed to a much higher isothermal temperature (Figure 3.45A) and is sometimes used for fast conditioning of a gas–solid chromatographic column after it has been unused for a period of time. More commonly, programming of this type is incorporated into peripheral sampling methods to quickly drive off solutes from an adsorbent. An example is the purge-and-trap procedure for the determination of volatiles in aqueous solution, where collected solutes are

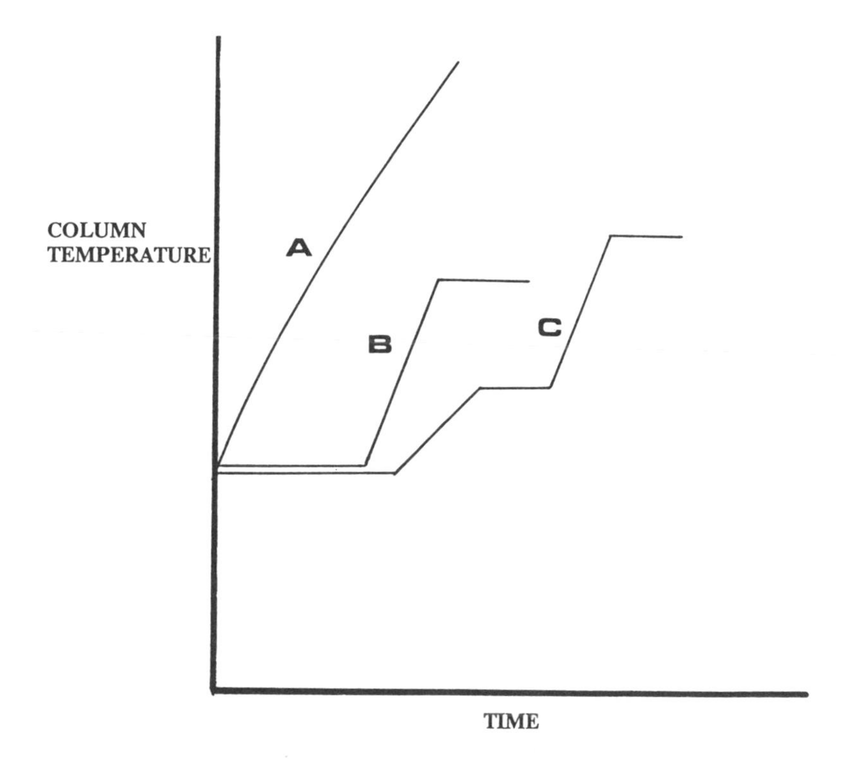

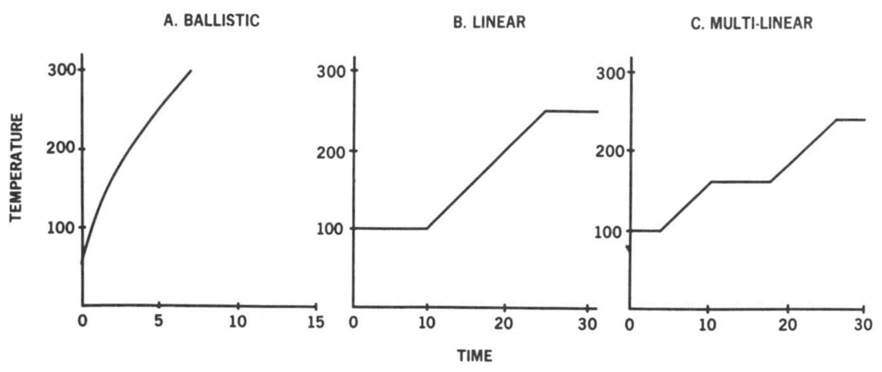

FIGURE 3.45 Types of temperature programming: (A) ballistic, (B) linear, and (C) multilinear.

thermally desorbed by ballistic programming (and also by rapid controlled linear temperature programming) from a silica gel/charcoal/Tenax trap so that they migrate as a narrow zone to the inlet of a capillary column where they are focused. A ballistic ramp may also be used with cryogenic solute focusing to elevate the column temperature quickly to above ambient temperature. However, a chromatographic column maintained at an elevated temperature, then ballistically programmed, can suffer severe damage due to disruption of the stationary phase film caused by this thermal shock.

The most widely used temperature program is the linear profile, as described in Figure 3.45B. Here the run begins at a low initial temperature, which may be maintained for a certain number of minutes (an isothermal hold), after which the column oven temperature is raised at a linear rate to the selected final temperature where it can also be maintained for a specific time interval. The initial temperature and hold period are usually determined from a scouting run made while noting elution temperatures; proper selection of the initial conditions will permit the separation of the low boilers in the separation, while the final temperature chosen should be sufficient for the elution of the more strongly retained components in the sample (keeping in mind the upper temperature limit of the stationary phase). Multilinear profiles (Figure 3.45C) may be employed in some instances to fine-tune or enhance the resolution in a separation, but are more commonly used in conjunction with on-column injection. In this injection mode, a low column temperature is maintained during sample introduction into the retention gap, then initiation of the first and usually faster ramp rate induces the solvent and components to start moving in the analytical column and the final ramp is implemented for elution of the components.

The ramp rate governs the trade-off between analysis time and resolution. The compromise between resolution and time of analysis is contrasted in Figure 3.46 for parallel capillary column separations of lemon oil generated at two different programming rates, 2 and 12°C/min. Relying on experience and intuition in establishing optimum column temperature conditions can be time-consuming and inefficient. An alternative route is the use of computer simulation for method development (84–88).

3.15 CAPILLARY CAGE DESIGN

A capillary column is placed on a cage after it is prepared by a column manufacturer. A cage serves several purposes. Since the ends of flexible fused-silica capillary tubing are inherently straight, the column must be coiled and retained securely on the cage. To alleviate stress on coiled fused silica, the diameter of the cage must be compatible with the inner diameter on the column. In other words, the column cannot be coiled too tightly or fracture of the tubing can occur. Megabore columns are coiled on 8-in.

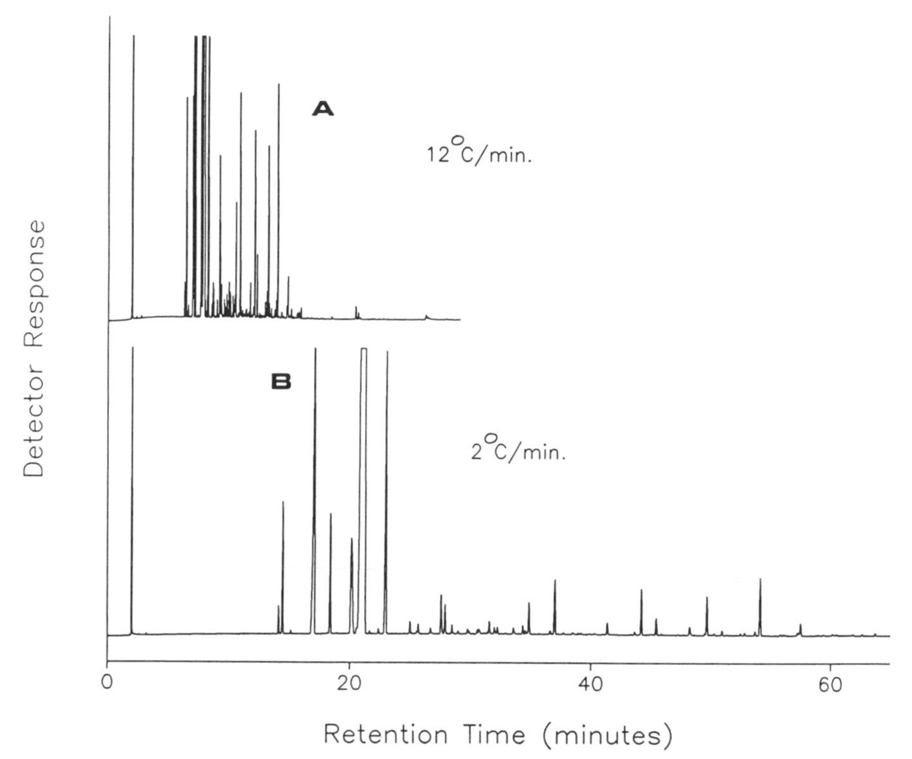

FIGURE 3.46 Effect of rate of temperature programming on resolution and analysis time; column: 30 m × 0.25 mm i.d. DB-1 (0.25-μm film); split injection (100:1), 27 cm/sec He, Det: FID. (E. F. Barry and S. Leepipatpiboon, unpublished results.)

cages, whereas 0.10- to 0.32-mm-i.d. columns are wound on 6- or 7-in. frames.

The cage also confines the column such that it will not flop around in the column oven under the influence of forced air convection currents. Before a column cage is mounted on the hanging bracket in the oven, the column should be examined to see if the coils are evenly spaced or distributed around the width of the cage with minimum overlap, a pertinent consideration for uniform heating of the column. It is also important that ends of the column extending from the cage for connection to the injector and detector do not contact any metallic parts of the oven where abrasion can erode the protective outer layer of polyimide on the column surface.

3.16 SUBAMBIENT OVEN TEMPERATURE CONTROL

Most gas chromatographs have the capability to operate the column oven at subambient temperatures. An accessory kit is available for either liquid nitrogen (−99°C) or carbon dioxide (−40°C) as a coolant and includes a cryogenic valve that is microprocessor controlled. The valve opens and closes, depending on the demand for coolant. In the open position, coolant is sprayed into the oven where it chills the oven down with assistance from forced-air convection.

REFERENCES

1. R. E. Majors, *LC*GC*, **7**, 888 (1989).
2. R. E. Majors, *LC*GC*, **8**, 444 (1990).
3. *Analytical Instrument Industry Report*, **10**(14), 4 (1993).
4. E. J. Guthrie and J. J. Harland, *LC*GC*, **12**, 80 (1994).
5. W. A. George, Thesis, University of Lowell, 1986.
6. 1994 Catalog of Supelco Chromatography Products, Supelco, Inc., Bellefonte, PA.
7. *Diatomite Supports for Gas–Liquid Chromatography*, Johns–Manville Corp., Ken-Caryl Ranch, Denver, CO, 1981.
8. D. M. Ottenstein, *J. Chromatogr. Sci.*, **11**, 136 (1973).
9. D. M. Ottenstein, *Advances in Chromatography*, Vol. 3, J. C. Giddings and R. A. Keller (Eds.), Dekker, New York, 1966, p. 137.
10. W. R. Supina, *The Packed Column in Gas Chromatography*, Supelco, Inc., Bellefonte, PA (1974).
11. J. J. Kirkland, *Anal. Chem.*, **35**, 2003 (1963).
12. Publication FF-202A, Johns–Manville Corp., Ken-Caryl Ranch, Denver, CO, 1980.
13. *Chromosorb Century Series Porous Polymer Supports*, Johns–Manville Corp., Ken-Caryl Ranch, Denver, CO, 1980.
14. Foxboro/Analabs Catalog of Chromatography Chemicals and Accessories, K-23B.
15. O. L. Hollis, *Anal. Chem.*, **38**, 309 (1966).
16. O. L. Hollis and W. V. Hayes, *J. Gas Chromatogr.*, **4**, 235 (1966).
17. G. E. Pollack, D. O'Hara, and O. L. Hollis, *J. Chromatogr. Sci.*, **22**, 343 (1984).
18. W. A. Supina, *Modern Practice of Gas Chromatography*, R. L. Grob (Ed.) 2d ed., Wiley, New York, 1985, p. 124.
19. R. Kaiser, *Chromatographia*, **1**, 199 (1968).
20. A. DiCorcia and R. Samperi, *Anal. Chem.*, **46**, 140 (1974).
21. A. DiCorcia and R. Samperi, *Anal. Chem.*, **51**, 776 (1979).
22. A. DiCorcia, R. Samperi, and C. Severini, *J. Chromatogr.*, **170**, 325 (1980).
23. R. V. Golovnya and T. A. Misharina, *J. High Resolut. Chromatogr.*, **3**, 4 (1980).
24. C. F. Poole and S. K. Poole, *Chem. Rev.*, **89**, 377 (1989).
25. H. Lamparczyk, *Chromatographia*, **20**, 283 (1985).

26. W. O. McReynolds, *J. Chromatogr. Sci.*, **8**, 685 (1970).
27. L. Rohrschneider, *J. Chromatogr.*, **22**, 6 (1966).
28. L. Rohrschneider, *Advances in Chromatography*, Vol. 4, J. C. Giddings and R. A. Keller (Eds.), Dekker, New York, 1967, p. 333.
29. E. Kovats, *Helv. Chim. Acta*, **41**, 1915 (1958).
30. W. O. McReynolds, *Gas Chromatographic Retention Data*, Preston Technical Abstracts Co., Evanston, IL, 1966.
31. L. Ettre, *Anal. Chem.*, **36**, 31A (1964).
32. 1994/95 Catalog of Chromatography Products, Restek Corporation, Bellefonte, PA.
33. J. K. Haken, *J. Chromatogr.*, **300**, 1 (1984).
34. J. J. Leary, J. B. Justice, S. Tsuge, S. R. Lowry, and T. L. Isenhour, *J. Chromatogr. Sci.*, **11**, 201 (1973).
35. R. Delley and K. Friedrich, *Chromatographia*, **10**, 593 (1977).
36. S. Hawkes, D. Grossman, A. Hartkopf, T. Isenhour, J. Leary, and J. Parcher, *J. Chromatogr. Sci.* **13**, 115 (1975).
37. K.J. Hyver, (Ed.), *High Resolution Gas Chromatography*, 3rd ed., Hewlett–Packard, Wilmington, DE, 1989, Chapter 2.
38. M. J. E. Golay, in *Gas Chromatography 1957* (East Lansing Symposium), V. J. Coates, H. J. Noebels, and I. S. Fagerson, Eds., Academic, New York, 1958, pp. 1–13.
39. M. L. Lee and B. W. Wright, *J. Chromatogr.*, **184**, 234 (1980).
40. R. Dandeneau and E. H. Zerenner, *J. High Resolut. Chromatogr., Chromatogr. Commun.*, **2**, 351 (1979).
41. R. Dandeneau and E. H. Zerenner, *Proceedings of the Third International Symposium on Glass Capillary Gas Chromatography*, Hindelang, 1979, pp. 81–97.
42. M. L. Lee, F. J. Yang, and K. D. Bartle, *Open Tubular Gas Chromatography: Theory and Practice*, Wiley, New York, 1984, p. 4.
43. R. R. Freeman, *High Resolution Gas Chromatography*, 2nd ed., Hewlett–Packard, Wilmington, DE, 1989, p. 3.
44. D. H. Desty, J. N. Haresnape, and B. H. Whyman, *Anal. Chem.*, **32**, 302 (1960).
45. S. R. Lipsky and W. J. McMurray, *J. Chromatogr.*, **217**, 3 (1981).
46. W. G. Jennings, *Comparisons of Fused Silica and Other Glasses Columns in Gas Chromatography*, Heuthig, Heidelberg, 1986, p. 12.
47. L. Borksanyi, O. Liardon, and E. Kovats, *Adv. Colloid Interface Sci.*, **6**, 95 (1976).
48. E. R. Lippincott and R. Schroeder, *J. Phys. Chem.*, **23**, 1099 (1955).
49. S. R. Lipsky, W. J. McMurray, M. Hernandez, J. E. Purcell, and K. E. Billeb, *J. Chromatogr. Sci.*, **18**, 1 (1980).
50. S. Trestianu and G. Gilioli, *J. High Resolut. Chromatogr., Chromatogr. Commun.*, **8**, 771 (1985).

51. J. Buijten, L. Blomberg, K. Markides, and T. Wannman, *J. Chromatogr.*, **237**, 465 (1982).
52. G. Rutten, J. de Haan, L. van de Ven, A. van de Ven, H. van Cruchten, and J. Rijks, *J. High Resolut. Chromatogr., Chromatogr. Commun.*, **8** 664 (1985).
53. C. L. Woolley, K. E. Markides, M. L. Lee, and K. D. Bartle, *J. High Resolut. Chromatogr., Chromatogr. Commun.*, **9**, 506 (1986).
54. C. L. Woolley, R. C. Kong, B. E. Richter, and M. L. Lee, *J. High Resolut. Chromatogr., Chromatogr. Commun.*, **7**, 329 (1984).
55. K. E. Markides, B. J. Tarbet, C. M. Schregenberger, J. S. Bradshaw, M. L. Lee, and K. D. Bartle, *J. High Resolut. Chromatogr., Chromatogr. Commun.*, **8**, 741 (1985).
56. H. Traitler, *J. High Resolut. Chromatogr., Chromatogr. Commun.*, **6**, 60 (1983).
57. W. Blum, *J. High Resolut. Chromatogr., Chromatogr. Commun.*, **9**, 120 (1986).
58. L. J. M. van de Ven, G. Rutten, J. A. Rijks, and J. W. de Haan, *J. High Resolut. Chromatogr., Chromatogr. Commun.*, **9**, 741 (1986).
59. M. A. Moseley and E. D. Pellizari, *J. High Resolut. Chromatogr., Chromatogr. Commun.*, **5**, 472 (1982).
60. A. Venema and J. T. Sukkei, *J. High Resolut. Chromatogr., Chromatogr. Commun.*, **8**, 705 (1985).
61. V. Pretorius and D. H. Desty, *J. High Resolut. Chromatogr., Chromatogr. Commun.*, **4**, 38 (1981).
62. B. Xu and N. P. E. Vermulen, *J. High Resolut. Chromatogr., Chromatogr. Commun.*, **8**, 181 (1985).
63. R. C. Kong, C. L. Woolley, S. M. Fields, and M. L. Lee, *Chromatographia*, **18**, 362 (1984).
64. T. Welsch and H. Frank, *J. High Resolut. Chromatogr., Chromatogr. Commun.*, **8**, 709 (1985).
65. K. E. Markides, B. J. Tarbet, C. L. Woolley, C. M. Schrengenberger, J. S. Bradshaw, M. L. Lee, and K. D. Bartle, *J. High Resolut. Chromatogr., Chromatogr. Commun.*, **8**, 378 (1985).
66. K. Grob, *Making and Manipulating Capillary Columns for Gas Chromatography*, Heuthig, Heidelberg, 1986.
67. T. Welsch, *J. High Resolut. Chromatogr.*, **11**, 471 (1988).
68. G. Schomburg, H. Husmann, S. Ruthe, and T. Herraiz, *Chromatographia*, **15**, 599 (1982).
69. C. L. Woolley, K. E. Markides, and M. L. Lee, *J. Chromatogr.*, **367**, 9 (1986).
70. M. Hetem, G. Rutten, B. Vermeer, J. Rijks, L. van de Ven, J. de Haan, and C. Cramers, *J. Chromatogr.*, **477**, 3 (1989).
71. L. Blomberg and K. E. Markides, *J. High Resolut. Chromatogr., Chromatogr. Commun.*, **8**, 632 (1985).
72. T. T. Stark, R. D. Dandeneau and L. Mering, *1980 Pittsburgh Conference*, Abstract 002, Atlantic City, NJ.

73. B. W. Wright, P. A. Peaden, M. L. Lee, and T. Stark, *J. Chromatogr.*, **248**, 17 (1982).
74. C. F. Poole and S. K. Poole, *Chromatography Today*, Elsevier, New York, 1991, p. 147.
75. K. Grob and G. Grob, *J. Chromatogr.*, **125**, 471 (1976).
76. D. A. Cronin, *J. Chromatogr.*, **97**, 263 (1974).
77. L. Blomberg and T. Wannman, *J. Chromatogr.*, **148**, 379 (1978).
78. J. L. Marshall and D. A. Parker, *J. Chromatogr.*, **122**, 425 (1976).
79. K. Grob, G. Grob, and K. Grob, *J. High Resolut. Chromatogr., Chromatogr. Commun.*, **1**, 149 (1978).
80. R. F. Arrendale, R. F. Severson, and O. T. Chortyk, *J. Chromatogr.*, **208**, 209 (1981).
81. W. Jennings, *Gas Chromatography with Glass Capillary Columns*, 2d ed., Academic, New York, 1980.
82. J. Bouche and M. Verzele, *J. Gas Chromatogr.*, **6**, 501 (1968).
83. K. Grob, G. Grob, and K. Grob, *J. Chromatogr.*, **156**, 1 (1976).
84. G. N. Abbay, E. F. Barry, S. Leepipatpiboon, T. Ramsted, M. C. Roman, R. W. Siergiej, L. R. Snyder, and W. Winniford, *LC*GC*, **9**, 100 (1991).
85. R. L. Grob, E. F. Barry, S. Leepipatpiboon, J. M. Ombaba, and L. A. Colon, *J. Chromatogr. Sci.*, **30**, 177 (1992).
86. D. E. Bautz, J. W. Dolan, W. D. Raddatz, and L. R. Snyder, *Anal. Chem.*, **62**, 1560 (1990).
87. D. E. Bautz, J. W. Dolan, and L. R. Snyder, *J. Chromatogr.*, **541**, 1 (1991).
88. J. W. Dolan, L. R. Snyder, and D. E. Bautz, *J. Chromatogr.*, **541**, 21 (1991).
89. K. Grob and G. Grob, *J. High Resolut. Chromatogr., Chromatogr. Commun.*, **2**, 109 (1979).
90. Stephen J. MacDonald, Thesis, University of Lowell, 1987.
91. B. W. Wright, P. A. Peaden, and M. L. Lee, *J. High Resolut. Chromatogr., Chromatogr. Commun.*, **5**, 413 (1982).
92. W. Noll, *Chemistry and Technology of Silicones*, Academic, New York, 1968, p. 464.
93. J. K. Haken, *J. Chromatogr.*, **300**, 1 (1984).
94. L. Blomberg, *J. High Resolut. Chromatogr., Chromatogr. Commun.*, **5**, 520 (1982).
95. L. Blomberg, *J. High Resolut. Chromatogr., Chromatogr. Commun.*, **7**, 232 (1984).
96. M. Verzele and P. Sandra, *J. Chromatogr.*, **158**, 211 (1978).
97. J. A. Hubball, P. R. DiMauro, E. F. Barry, E. A. Lyons, and W. A. George, *J. Chromatogr. Sci.*, **22**, 185 (1984).
98. C. Madani, E. M. Chambaz, M. Rigaud, J. Durand, and P. Chebroux, *J. Chromatogr.*, **126**, 161 (1976).
99. M. Rigaud, P. Chebroux, J. Durand, J. Maclouf, and C. Mandini, *Tetrahedron Lett.*, **44**, 3935 (1976).

100. K. Grob, *Chromatographia*, **10**, 625 (1977).
101. L. Blomberg, J. Buijten, J. Gawdzik, and T. Wannman, *Chromatographia*, **11**, 521 (1978).
102. L. Blomberg and T. Wannman, *J. Chromatogr.*, **168**, 81 (1979).
103. K. Grob and G. Grob, *J. Chromatogr.*, **213**, 211 (1981).
104. K. Grob, G. Grob, and K. Grob, Jr., *J. Chromatogr.*, **211**, 243 (1981).
105. K. Grob and G. Grob, *J. High Resolut. Chromatogr., Chromatogr. Commun.*, **4**, 491 (1981).
106. K. Grob and G. Grob, *J. High Resolut. Chromatogr., Chromatogr. Commun.*, **5**, 13 (1982).
107. P. Sandra, G. Redant, E. Schacht, and M. Verzele, *J. High Resolut. Chromatogr., Chromatogr. Commun.*, **4**, 411 (1981).
108. L. Blomberg, J. Buijten, K. Markides, and T. Wannman, *J. High Resolut. Chromatogr., Chromatogr. Commun.*, **4**, 578 (1981).
109. P. A. Peaden, B. W. Wright, and M. L. Lee, *Chromatographia*, **15**, 335 (1982).
110. J. Buijten, L. Blomberg, K. Markides, and T. Wannman, *Chromatographia*, **16**, 183 (1982).
111. K. Markides, L. Blomberg, J. Buijten, and T. Wannman, *J. Chromatogr.*, **254**, 53 (1983).
112. B. E. Richter, J. C. Kuei, J. I. Shelton, L. W. Castle, J. S. Bradshaw, and M. L. Lee, *J. Chromatogr.*, **279**, 21 (1983).
113. B. E. Richter, J. C. Kuei, N. J. Park, S. J. Crowley, J. S. Bradshaw, and M. L. Lee, *J. High Resolut. Chromatogr., Chromatogr. Commun.*, **6**, 371 (1983).
114. J. Buijten, L. Blomberg, S. Hoffman, K. Markides, and T. Wannman, *J. Chromatogr.*, **289**, 143 (1984).
115. W. Bertsch, V. Pretorius, M. Pearce, J. C. Thompson, and N. G. Schnautz, *J. High Resolut. Chromatogr., Chromatogr. Commun.*, **5**, 432 (1982).
116. J. A. Hubball, P. DiMauro, E. F. Barry, and G. E. Chabot, *J. High Resolut. Chromatogr., Chromatogr. Commun.*, **6**, 241 (1983).
117. E. F. Barry, G. E. Chabot, P. Ferioli, J. A. Hubball, and E. M. Rand, *J. High Resolut. Chromatogr., Chromatogr. Commun.*, **6**, 300 (1983).
118. E. F. Barry, J. A. Hubball, P. R. DiMauro, and G. E. Chabot, *Am. Lab.*, **15**, 84 (1983).
119. G. Alexander, *J. High Resolut. Chromatogr.*, **13**, 65 (1990).
120. H. Liu, A. Zhang, Y. Jin, and R. Fu, *J. High Resolut. Chromatogr.*, **12**, 537 (1989).
121. R. C. M. de Nijs and J. de Zeeuw, *J. High Resolut. Chromatogr., Chromatogr. Commun.*, **5**, 501 (1982).
122. J. Buijten, L. Blomberg, K. E. Markides, and T. Wannman, *J. Chromatogr.*, **268**, 387 (1983).
123. O. Etler and G. Vigh, *J. High Resolut. Chromatogr., Chromatogr. Commun.*, **8**, 42 (1985).
124. J. A. Hubball, Thesis, University of Connecticut, 1987.

125. M. Horka, V. Kahle, K. Janak, and K. Tesarik, *Chromatographia*, **21**, 454 (1985).
126. L. Bystricky, *J. High Resolut. Chromatogr., Chromatogr. Commun.*, **9**, 240 (1986).
127. S. R. Lipsky and W. J. McMurray, *J. Chromatogr.*, **289**, 129 (1984).
128. W. Blum, *J. High Resolut. Chromatogr., Chromatogr. Commun.*, **8**, 718 (1985).
129. W. Blum, *J. High Resolut. Chromatogr., Chromatogr. Commun.*, **9**, 350 (1986).
130. W. Blum, *J. High Resolut. Chromatogr., Chromatogr. Commun.*, **9**, 120 (1986).
131. W. Blum and L. Damasceno, *J. High Resolut. Chromatogr., Chromatogr. Commun.*, **10**, 472 (1987).
132. W. Blum and G. Eglinton, *J. High Resolut. Chromatogr., Chromatogr. Commun.*, **12**, 290 (1989).
133. P. Schmid and M. D. Mueller, *J. High Resolut. Chromatogr., Chromatogr. Commun.*, **10**, 548 (1987).
134. T. Welsch and O. Teichmann, *J. High Resolut. Chromatogr.*, **14**, 153 (1991).
135. F. David, P. Sandra, and G. Diricks, *J. High Resolut. Chromatogr., Chromatogr. Commun.*, **11**, 256 (1988).
136. R. Aichholz, *J. High Resolut. Chromatogr.*, **13**, 71 (1990).
137. K. Grob and G. Grob, *J. Chromatogr.*, **347**, 351 (1985).
138. W. A. Konig, S. Lutz, P. Mischnick-Lubecke, R. Brassat, and G. Wenz, *J. Chromatogr.*, **441**, 471 (1988).
139. D. W. Armstrong and W. Y. Li, *Anal. Chem.*, **62**, 217 (1990).
140. V. Schurig and H. Nowotny, *J. Chromatogr.*, **441**, 155 (1988).
141. V. Schurig, D. Schmalzing, U. Muhleck, M. Jung, M. Schleimer, P. Mussche, C. Duvekot, and J. C. Buyten, *J. High Resolut. Chromatogr.*, **13**, 713 (1990).
142. J. V. Hinshaw, *LC*GC*, **11**, 644 (1993).
143. R. C. M. de Nijs, *J. High Resolut. Chromatogr., Chromatogr. Commun.*, **4**, 612 (1981).
144. J. de Zeeuw, R. C. M. de Nijs, J. C. Buijten, J. A. Peene, and M. Mohnke, *Am. Lab.*, **19**, 84 (1987).
145. W. E. Harris and H. W. Habgood, *Programmed Temperature Gas Chromatography*, Wiley, New York, 1966.
146. R. S. Brophy, Thesis, University of Lowell, 1987.

CHAPTER FOUR

Optimizing Separations in Gas Chromatography

MATTHEW S. KLEE

Hewlett-Packard Company

Modern Practice of Gas Chromotography, Third Edition, Edited by Robert L. Grob.
ISBN 0-471-59700-7

4.1 OVERVIEW OF THE COMPROMISES IN OPTIMIZATIONS

4.1.1 Triangle of Compromise: Speed, Resolution, Capacity and Sensitivity

The key to optimizing separations in gas chromatography is to understand the compromises that must be made and their interrelationships. Figure 4.1 is a pictorial representation of the major attributes one strives to optimize in any separation. Each apex of the triangle corresponds to one pure attribute. For example, the top represents maximum capacity (and sensitivity), but minimum speed and resolution. If one desired to maximize capacity to

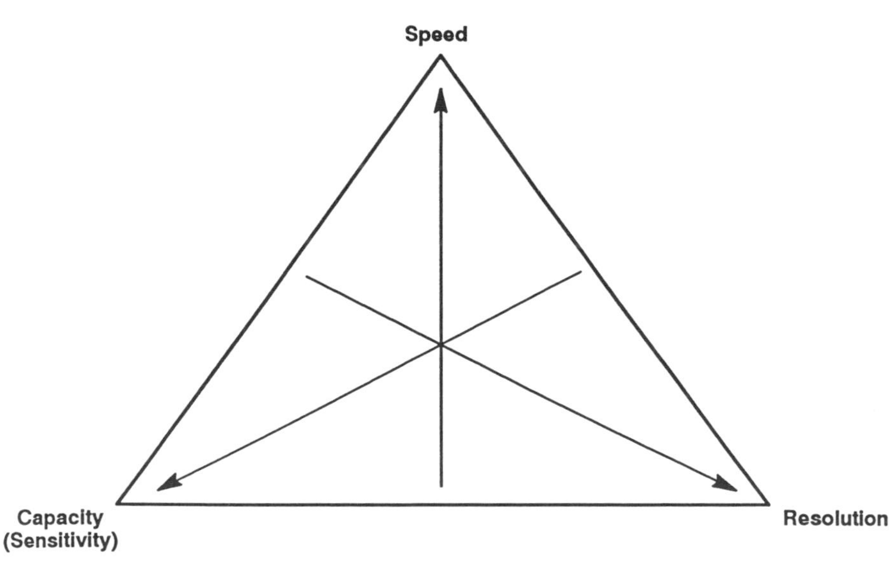

FIGURE 4.1 Chromatographic triangle of compromise. Each corner is a pure attribute. Inside the triangle corresponds to mixtures of the three attributes (e.g., the middle corresponds to equal parts of each corner).

TABLE 4.1 General Approach to Optimizing Gas Chromatographic Analyses

1.	Define goals and objectives of the separation.
2.	Make initial selection of inlet, detector, column, stationary phase, and carrier gas.
3.	Screen sample with fast temperature program and flowrates.
4.	Evaluate peak shape, relative concentrations, and elution temperatures.
5.	Adjust conditions as necessary to facilitate evaluation and approach analysis goals.
6.	Change instrument choices or column if not encouraging; repeat steps 3–7.
7.	Repeat screen with widely different temperature program or flowrate.
8.	Evaluate sensitivity to temperature/flow, peak shifts, and elution order changes.
9.	Fine tune temperature program rate, flowrate to meet goals.

analyze both major and minor components in the same sample or to be able to detect trace components in dilute solutions, one would select chromatographic conditions to allow large injection volumes. However, this would be accompanied by low resolution and long analysis times. An analogous situation exists at the other apexes: Maximum resolution coincides with lowest speed and capacity, and maximum speed occurs with lowest resolution and capacity.

Clearly, one rarely has the luxury of being able to maximize one variable at the expense of the others. A finite column capacity, analysis time, and resolution are always required. So, the task of optimizing a given separation centers around the specific needs of the analysis and the capabilities of the available gas chromatograph and columns.

A general approach to optimizing separations in gas chromatography is listed in Table 4.1 and will be the basis for the remainder of this chapter.

4.1.2 Setting Goals

The most important task in optimizing separations is to define the goals. Common requirements of analytical separations include

- Specific limit of quantification
- Analysis time limit
- Minimum resolution of peaks of interest from each other and interferences
- Linearity of detector response
- Finite cost per analysis
- Compatibility with current chromatographic instrumentation
- Use of standard columns and consumables

- Minimum sample consumption
- Ruggedness
- Minimum sample degradation
- Minimum sample handling

Goals are usually prioritized by musts and wants; often they have finite limits attached. For example, the analysis must quantify 16 specific organochlorine pesticides at concentrations above 1 ppb, with linear detector response of at least two orders of magnitude, and the results must be confirmed using a second column with different stationary phase. Once goals such as these are defined, the process of narrowing in on an initial instrument configuration and starting conditions can commence.

4.2 OVERVIEW OF VARIABLES

4.2.1 Summary Table

Table 4.2 summarizes the magnitude and direction of the influence that gas chromatographic variables have on capacity (and sensitivity), resolution, and speed of analysis. Each is explained in detail later in this chapter. For Table 4.2, it is assumed that only one variable at a time is changed. The resulting changes in speed, resolution, and capacity are stated only relative to the original results. It is assumed for a change that results in a 1.4 decrease in resolution that the resulting separation is still adequate, or the twofold decrease in time would be meaningless.

The relationship between capacity and sensitivity is not exactly 1:1, but it is close enough and will be grouped together with it herein. For example, sensitivity is a function of both the amount of sample that can be injected on the column (capacity) and the efficiency of the column (narrowness of the resulting peak). Since efficiency decreases with larger amounts of stationary phase, peak width will increase slightly as column capacity increases. However, the gain in sensitivity due to larger injection volumes is larger than the loss in sensitivity due to loss in efficiency. So, references to the influences of experimental variables on capacity will be assumed to influence sensitivity in generally the same manner.

4.2.2 General Importance of Instrumental Variables in Optimizations

Proper selection of inlets, detectors, injection parameters, and column format (capillary vs. packed) are critical to meeting analysis goals, and each is optimized individually based on criteria presented in other chapters. The choice of column dimensions and stationary phase can affect all aspects of a separation and are difficult to change once chosen, but may be worth taking the time to optimize for high-volume “standard” methods. Therefore, these

TABLE 4.2 Influence of Chromatographic Variables on Speed, Capacity, and Resolution

Variable	Resolution	Capacity	Analysis Time		
Column dimensions					
2*X* diameter	1/4	>4	1		
2*X* length	1.4	2	2		
2*X* particle size	0.7	1	>1		
Packed to capillary	>1	0.002	1/2	(same length)	
Stationary phase (fixed column dimensions, optimum flowrate)					
2*X* beta	2	1/2	1/2		
2*X* film thickness	<1	2	2		
Change liquid phase type	+++/−−−	+/−	+/−		
GSC to GLC	+/−	++	−− (faster)		
Carrier-gas type and flow				PSI	mL/min out
N2 @ optimum	1	NI	12	12	0.244
N2, 3*X* optimum	0.5	NI	4	39	1.30
He, optimum	1	NI	5	36	1.0
He, 3*X* optimum	0.5	NI	1.5	119	7.74
H2, optimum	1	NI	3	27	1.4
H2, 3*X* optimum	0.5	NI	1	89	9.7
Column temperature (same column, low k' pair, high k' pair)					
Low (*X*°C)	2, 8	NI	4		
Med (*X* + 30°C)	1.4, 6	NI	2		
High (*X* + 60°C)	1, 4	NI	1		
Program (*X* to *X* + 60°C)	2, 6	NI	2		

Note. +, Potential for positive influence; ++, potential for large influence; +++, potential for very large influence; NI, no influence; −, potential for negative influence; −−, potential for large negative influence; −−−, potential for very large negative influence.

require considerable thought before selection is made and optimization is started. Column temperature and carrier-gas flowrate are more easily adjusted and are used to fine-tune separations if scouting runs look promising.

4.3 SELECTING THE RIGHT INLET AND CONDITIONS

In contrast to the continuous and predictable impact of the variables listed in Table 4.2, choice of inlet, injection parameters, column installation, and detector have the potential to seriously degrade ultimate system performance. These topics are covered in more detail in other chapters, and should be understood before initiating optimizations.

Improper selection of inlet or operating conditions will negatively affect (degrade) a separation. One must always keep in mind the goals of the separation and the limits imposed by the sample when selecting inlet and conditions. It is always wise to use the most moderate conditions to minimize sample degradation, inlet overload, contamination, ghosting, and column degradation. Once reasonable separations have been achieved, inlet conditions can be modified to identify any dependencies on conditions.

4.3.1 Relative Merits of Inlets for GC

Chapter 8 describes the relative merits of inlets for gas chromatography and considerations for their use, which are summarized in Table 8.2. Selection of an inlet is sometimes a matter of practicality, since we rarely have access to all possible configurations. However, the table may point out conflicts between available hardware and analysis goals.

4.4 SELECTING THE RIGHT DETECTOR AND ENSURING APPROPRIATE OPERATION

Proper selection of detector is also important to ensure that the goals of the separation can be met. Characteristics of common gas chromatographic detectors are summarized in Table 4.3 relative to typical analysis goals.

4.4.1 Summary of Guidelines

Issues of sensitivity, selectivity, universality, and ruggedness should be considered when choosing a detector. Refer to Chapter 5 for more information on individual detectors. For any given detector, manufacturers' instructions should be followed to ensure the highest system performance. This includes setting of gas flows and temperatures, proper column installation, cleaning, and maintenance.

TABLE 4.3 Detector Attributes Compared to Analysis Goals

Analysis Goal	Detectors
Universal response	FID, TCD, MS, IR
Selective response	MS, FPD, NPD, ECD, ELCD, PID, AED
Wide dynamic range	FID, MS
Low detection limits	ECD, SIM-MS, ELCD, (PID, AED)
Ruggedness	FID, TCD
Ease of use	TCD, FID (ECD, FPD)
Molecular information	MS, IR
Atomic information	AED (ELCD)
Low cost of operation	TCD, FID, ECD

4.4.2 Selective Detection

Good separation of sample components is usually the best way to guarantee reliable data. When optimized chromatography fails to separate the components of interest from each other or interferences, a detector that selectivity responds to the component(s) of interest can be very useful. Even though the FID is considered one of the most universal detectors, it does not respond to water or components of air. Even it, therefore, can simplify analyses such as organics in water.

Selectivity can be achieved through several physical and chemical means, and the reader is directed to Chapter 5 for details on individual detectors. The key is to select a detector that responds only to the component(s) of interest. This results in lower limits of quantification, reduced analysis time, higher accuracy, and higher reproducibility. Most of these benefits stem from the decreased dependency on the quality of the separation. The detector is, in effect, separating the components in another dimension from the chromatography.

4.5 SELECTING CARRIER GAS AND FLOWRATE

4.5.1 Influence of Carrier-Gas Choice

The maximum efficiency and resolution occur when the carrier gas flowrate is at its "optimum." This corresponds to the minimum in the van Deemter curve as explained in Chapter 2, and is illustrated in Figure 4.2. The optimum flowrate for gases depends on viscosity and diffusion rates. This is a function of the mass of the gas molecule, so optimum flowrates follow the order of $H_2 > He >> N_2 > Ar$. In addition to faster optimum flowrates, the shape of the van Deemter curve is much flatter for the lighter gases, so there is less loss of efficiency at higher than optimum flowrates than with the

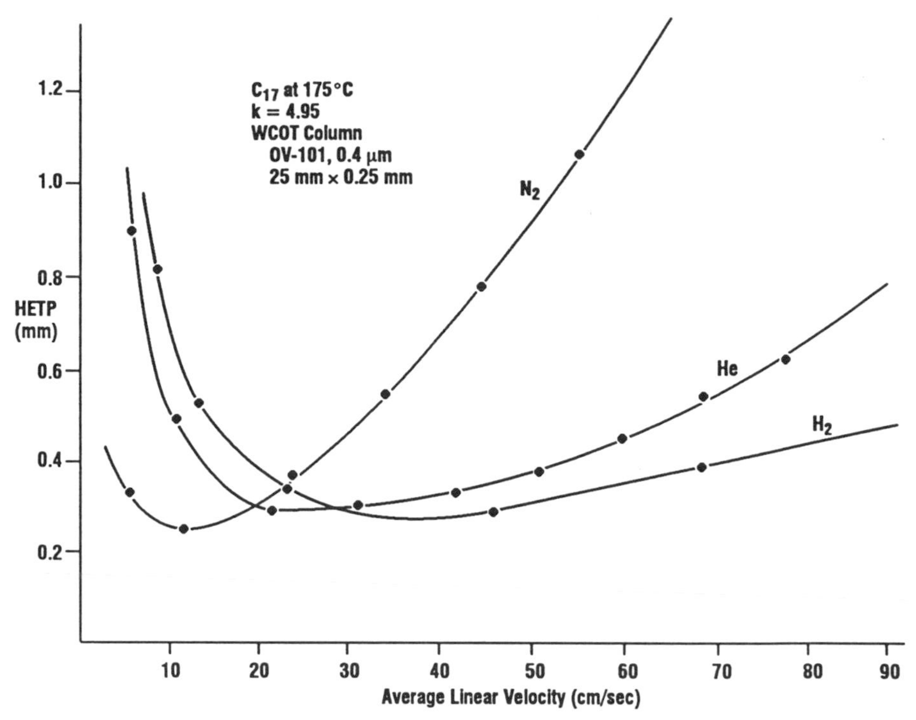

FIGURE 4.2 Van Deemter curves for common carrier gases. Highest efficiencies are at the minimum of the curves. (Reprinted from Reference 28 with permission.)

heavier carrier gases. To maximize speed of analysis, hydrogen is the carrier gas of choice.

4.5.2 Optimum Practical Gas Velocity

Doing separations at the optimum flowrate yields the highest efficiencies and theoretically allows one to minimize the length of column required for a given separation. Usually, however, one does not cut columns to shorter lengths, so this advantage is realized only for well characterized and dedicated analyses when development of specialized columns can be justified.

Figure 4.3 illustrates that at a given head pressure, linear velocity will decrease as column temperature is increased. This plot is for open tubular columns, but the general trend occurs with packed columns also. This is because gases increase in viscosity with increased temperature—the opposite of liquids. If the head pressure is set to the optimum for a given column and temperature, and the temperature is then increased during the run, not only does the run time increase due to decreased flow, but column efficiency will significantly degrade. Changes in flow are avoided by using mass flow

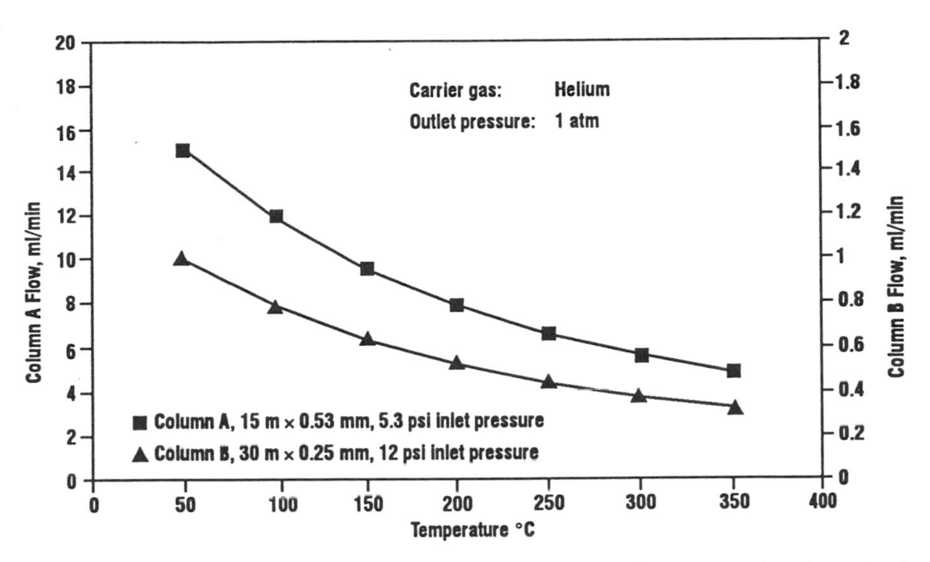

FIGURE 4.3 Decrease in open tubular column flow rate (at constant head pressure) as a function of column temperature. (Reprinted from Reference 28 with permission.)

controllers with packed columns, and by electronic pressure control with capillary columns (Chapter 8).

In capillary chromatography, setting the carrier-gas flowrate to the optimum practical gas velocity (OPGV) ensures maximum efficiency per unit time and prevents loss in efficiency in constant-pressure systems as column temperature is increased. The OPGV (illustrated in Figure 4.4) is defined as the point in the van Deemter curve where HETP versus *m* first becomes linear (1). The benefit of operating at the OPGV is illustrated in

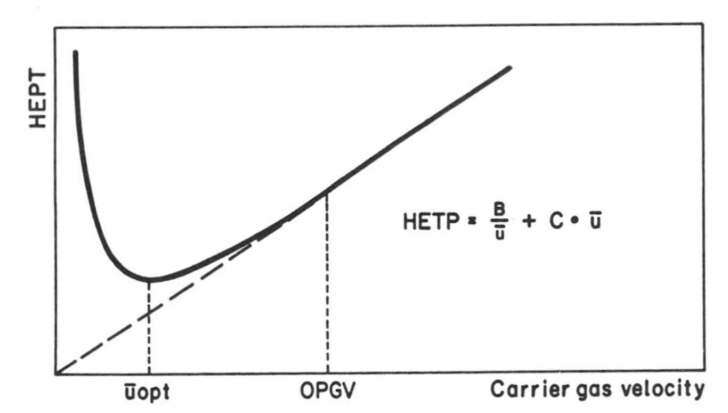

FIGURE 4.4 Typical location of the optimum practical gas velocity on the van Deemter curve. Running an open tubular column at its OPGV ensures maximum number of plates/time. (Reprinted from Reference 2 with permission.)

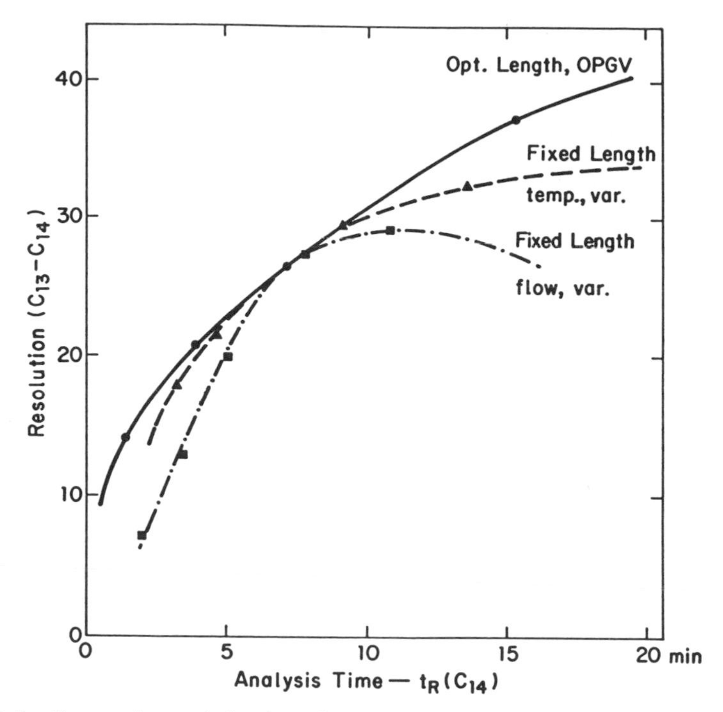

FIGURE 4.5 Comparison of the benefits achievable by varying column temperature, flowrate, or column length (at OPGV) to achieve a given resolution. On average, using a longer column length at high flowrates will yield more resolution in less time than trying to have many columns, each at optimal flowrate and temperature. If the longer column yields excess resolution, the flowrate can be increased to decrease analysis time further. (Reprinted from Reference 2 with permission.)

Figure 4.5. Using longer columns at the OPGV will yield higher resolution in less time than will changing column temperature or flowrate with a given column. Maximum resolution between two peaks is achieved when the later peak elutes around $k' = 6$ for capillary columns operated at the OPGV (2). Table 4.4 lists approximate optimum linear and OPGV velocities for common carrier gases and open tubular columns.

TABLE 4.4 Approximate Optimum and Optimum Practical Linear Velocities

Carrier Gas	Optimum (cm/sec)	OPGV (cm/sec)
H_2	40	100
He	25	65
N_2	10	25

4.6 SELECTING THE RIGHT STATIONARY PHASE

The most significant determinant of successful separations in gas chromatography is the choice of stationary phase; the mobile phase (gas) is inert. Stationary phases are selected based on the volatility and polarity of sample components and on inherent selectivity of the phase (see Chapter 3). Optimizations usually involve screening with an initial phase and if it is not promising, switching to one with very different selectivity to maximise the probability of success in the shortest time.

4.6.1 Selecting Packed Versus Capillary Columns

Capillary columns are preferred over packed columns for inertness and absolute number of plates (longer columns are possible because of lower pressure drops). The need for drawers full of packed columns with different stationary phases has widely been replaced by a few long capillary columns. In contrast to when capillary columns were inconsistently fabricated from glass, the question has become "When am I prevented from using a capillary column?" instead of "When am I forced to use a capillary column?" In fact, the few cases where packed columns were the only choice—for molecular sieves, alumina, and porous polymer phases—can now be handled by PLOT columns. The PLOT column formats have the same selectivities and characteristics as the packed-column versions, with the added benefits of higher efficiencies (longer column lengths) and faster optimum linear velocities. A comparison of typical packed and capillary column characteristics is given in Table 4.5.

TABLE 4.5 Typical Packed- and Capillary-Column Characteristics

	Packed	Capillary
Length	2 m	30 m
Diameter	2 mm (100/120 mesh)	250 μm
Liquid phase film thickness	5% wt/wt	0.25 μm
Phase ratio	36	250
Capacity	15 mg	50 ng
Plates/meter	2500	3500
Total plates	5000	105,000
Optimum average linear velocity (He)	13 cm/sec	25 cm/sec
Void time	15 sec	120 sec
Plates/sec ($k = 1.5$)	133	350
Resolution, $\alpha = 1.1$	3.1 ($k = 14$)	13.8 ($k = 2$)
Peak capacity (20 min analysis)	60	275
Cost	$100	$350

Packed columns are preferred for ruggedness, capacity, compatibility with gas sampling valves, higher selection of polar phases, and low cost. For these reasons, packed columns have been extremely effective for analyses requiring specialized stationary phases, in process (on-line) analysis, and for operators with limited training. Solid adsorbent phases (e.g., porous polymers) are preferred over coated (liquid) stationary phases for process analysis, especially of very volatile components, because they are more stable and do not bleed as much.

Some process analyses presently done using packed columns could be improved by using capillary columns. The packings in packed columns can be disturbed by the rapid pressure changes accompanying valve switching, resulting in loss of efficiency and split peaks. Capillary columns are open tubes, and the stationary phase is not disturbed by sudden pressure changes. In addition, most capillary columns are now made of flexible fused-silica tubing, facilitating installation, and they have cross-linked stationary phases, which increases lifetime and decreases stationary phase bleed. So, careful attention to the goals of the analysis will help determine whether packed columns or capillary columns are most appropriate.

4.6.2 Selecting Stationary Phase

Stationary phases are chosen primarily for their ability to separate the component(s) of interest, but factors such as stability, availability, ease of use, and cost must also be considered. Stationary phases retain solutes through a combination of nonpolar (dispersive) and polar interactions. Nonspecific interactions can account for 1–10 kcal/mol and hydrogen bonding interactions can contribute 0–8 kcal/mol (3).

Dispersive interactions are nonselective. Stationary phases whose major interactions are dispersive (methyl silicones, styrene/divinyl benzene copolymers) elute compounds based on their molar volumes, or boiling points. This is extremely useful in the optimization process to determine the elution temperature range and complexity of a sample. These phases also tend to be the most stable and have the widest usable temperature ranges.

Polar stationary phases are often required for analysis of polar compounds because of their column deactivating properties and miscibility with solutes. Solutes that are tailed or irreversibly adsorbed on nonpolar columns are often eluted with good peak shapes when a polar stationary phase is used. Polarity may also add the selectivity necessary to separate components that coelute on nonpolar phases.

A major consideration in selecting what type of stationary phase to use is the state that the sample is in at room temperature. In general, use

- Solid adsorbents for gases
- Liquid stationary phases for liquids or solids

- Polar phases for polar solutes
- Nonpolar phases for nonpolar solutes

Liquid phases act as solvents, and the solubilities of gases are small in common liquid stationary phases. Most light gases, therefore, are not retained long enough by gas–liquid partition to be separated unless very small phase ratios are used and/or the column is held at subambient (cryogenic) temperatures. For this reason, the stronger interactions afforded by gas–solid adsorption are most often used. For solutes that are liquids or solids at room temperature, gas–liquid partition is most successful. Table 4.6 summarizes guidelines for initial selection of stationary phase type.

For both adsorbents and liquid phases, the polarity of the stationary phase should be matched with the polarity of the solute for best peak shape and capacity. Several techniques have been used to define "polarity" of stationarity phases and they rely on relative retention behavior of polar probes relative to each other and to nonpolar solutes. All of these approaches have pointed out that most liquid phases have similar polarities and selectivities, as do most porous polymers.

Increased polarity of stationary phases helps to improve separations of polar solutes in two general ways: (1) by retaining solutes with polar functional groups longer than nonpolar solutes and (2) by selectively retaining solutes by specific polar interactions. Selectivity has been difficult to define for most stationary phases. McReynolds constants, ΔI values (4), are not easy to interpret, since all the constants tend to increase monotonically with increasing "polarity" of the stationary phase. This indicates only that polar solutes are being retained longer with respect to *n*-alkanes than on phases with lower ΔI values. A selective phase would have a single ΔI value that is considerably larger than the others, indicating a preferential polar interaction.

Table 4.7 represents a traditional way of describing characteristics of

TABLE 4.6 Usual Choices of Stationary Phases Based on Sample State

Compond Class	Types of Solutes	Stationary Phase
Inert and permanent gases	H_2, O_2, HN_2, CH_4, CO, He, Ne, Ar(+CO_2)	Molecular sieves or carbosieve
Light gases	CH_4, CO_2, light polar, and non-polar compounds	Porous polymers
	Non-polar hydrocarbons	Alumina, silica gel
Nonpolar liquids, solids	Hydrocarbons, aromatics, ethers, esters, silyl derivatives	Nonpolar liquid phases (methyl, substituted methyl silicones)
Polar liquids, solids	Alcohols, diols, acids, amines	Polar phases (Carbowax, 50% phenyl, trifluoropropyl silicones)

TABLE 4.7 Application and Properties of Porous Polymer Adsorbents[a]

Chromosorb Century Series 101–108							
				Retention Indices[d]			
Chromosorb	Application	T_{max}[b,c] (°C)	Polymer Type	Benzene	*t*-Butanol	2-Butanone	Acetonitrile
101	Fast, efficient separation of free fatty acids, glycols, alcohols, alkanes, esters, ketones, aldehydes, hydrocarbons, and ethers	275	Styrenedivinyl-benzene	745	565	645	580
102	No tailing of water, alcohols, and most other oxygenated compounds; useful for light and permanent gases and low-molecular-weight compounds; high surface area	250	Styrenedivinyl-benzene	650	525	570	460
103	Fast, efficient separation of basic compounds and amines; also for separation of alcohols, aldehydes, hydrazines, amides, and ketones; glycols are adsorbed	275	Cross-linked polystyrene	720	575	640	565
104	Separation of nitriles, nitroparaffins, aqueous, H_2S, xylenols, NH_3, and oxides of nitrogen, sulfur, and carbon; trace water in benzene; different selectivity from 101, 102, and 103	250	Acrylonitrile di-vinylbenzene	845	735	860	885

105	Separations of aqueous solutions containing CH_2O, separation of acetylene from lower hydrocarbons, separation of most gases and organic compounds in boiling range up to 200°C; less polar than 104, but with different selectivity, i.e., acetic acid before butanol (the reverse order of elution is observed on 105)	250	Polyaromatic	635	545	580	480
106	Retains benzene and nonpolar organic compounds; separation of $C_2 - C_5$ fatty acids from corresponding alcohols	250	Cross-linked polystyrene	605	505	540	405
107	Intermediate polarity, providing efficient separation of various classes of compounds in general and of CH_2O in particular	250	Cross-linked acrylic ester	660	620	650	550
108	Separation of gases and polar materials, such as water alcohols, aldehydes, ketones, glycols, etc.; retention characteristics differ from other Chromosorbs	250	Cross-linked acrylic ester	710	645	675	605

TABLE 4.7. *(Continued)*

Porapaks							
				Retention Indices[d]			
Porapak	Application	T^{max}(°C)	Polar Monomer	Benzene	*t*-Butanol	2-Butanone	Acetonitrile
P	Least polar; separates a wide variety of carbonyl compounds, glycols, and alcohols	250	–[e]	765	560	650	590
P-S	Surface-silanized version of P that minimizes tailing; separation of aldehydes and ketones	250	–	NA	NA	NA	NA
Q	Most widely used; particularly effective for hydrocarbons, organic compounds in water, and oxides of nitrogen	250	–[f]	630	538	580	450
Q-S	Surface-silanized version of Q that eliminates tailing; separates organic acids and other polar compounds with minimum tailing	250	–	625	525	565	445
R	Moderate polarity; long retention and good resolution observed for ethers; separation of H_2O from Cl_2 and HCl	250	Vinylpyrolidone	645	545	580	455

S	Separation of normal and branched-chain alcohols	250	Vinylpyridine	645	550	575	465
N	Separation of CO_2, NH_3, and H_2O and of acetylene from other C_2 hydrocarbons; high water retention	190	Vinylpyrolidone	735	605	705	595
T	Highest polarity and greatest water retention; determination of CH_2O in aqueous solutions	190	Ethylene glycol	–	675	700	635

[a] Reprinted from Ref. 31 with permission.
[b] Maximum temperature in programmed temperature operation could be about 25°C higher for short duration.
[c] Might discolor after column conditioning and after extended use. This in no way affects performance.
[d] Data reported S. Dave, *J. Chromatogr.*, 7, 389–399 (1969).
[e] Prepared by polymerizing styrene-divinylbenzene mixtures.
[f] Prepared by polymerizing ethylvinylbenzene and divinylbenzene.

porous polymers and their compositions, which includes some McReynolds constants. Several newer porous polymers have been developed and often exhibit higher efficiencies and inertness than the Poropaks and Chromosorb 100 series polymers, but the retention and selectivity characteristics are nearly the same. Qualitative approaches such as that in Table 4.7 are useful but do not provide a quantitative means to compare available phases and are difficult to use for predicting relative retention behavior.

A complementary approach to comparing the selectivity of polar stationary phases relied on the relative contributions of hydrogen bond donor, hydrogen bond acceptor, and dipole/induced dipole interactions to retention of polar probes (5, 6). The approach focuses on three probes (ethanol, 1,4-dioxane, and nitromethane) to measure hydrogen bonding and dipole interactions. Selectivity parameters $x_{e,d,n}$ are calculated as the relative McReynolds constants of the three probes:

$$x_{\mathrm{e,d,n}} = \frac{\Delta I_{\mathrm{e,d,n}}}{\Sigma\, \Delta I} \qquad (4.1)$$

where x_{e} is the selectivity parameter for proton acceptor interactions, x_{d} is for proton donor interactions, x_{n} is for dipole interactions, and $\Delta I_{\mathrm{e,d,n}}$ are McReynolds constants of the probes on the stationary phase of interest. Plotting the selectivity parameters on trilinear coordinates (a triangle) helps to visualize differences between phases.

Figure 4.6 shows the position of several stationary phases on the selectivity triangle. The selectivity triangle indicates the relative differences in polar selectivity between phases. Most commercial phases have polar selectivities near the center of the triangle, outline by the box in Figure 4.7, indicating that these phases do not have unique polar selectivity. To exhibit unique selectivity, a phase would be located more toward a corner of the triangle.

Figure 4.8 adds a third dimension to the selectivity triangle, $\Sigma\Delta I$. This is

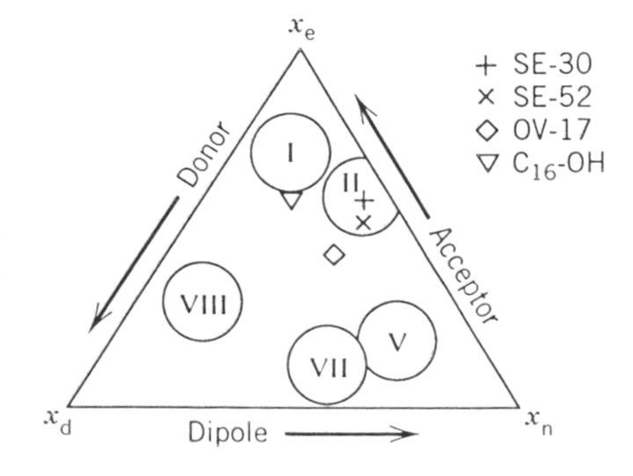

FIGURE 4.6 Expanded selectivity triangle showing the location of several stationary phases. Dipole and proton acceptor axes 0.2–0.7, proton donor axis 0.1–0.6. Alcohols, group I; ethers, group II; chlorinated hydrocarbons, group V; aromatics, group VII; weak acids, group VIII. (Group locations from Reference 29, reprinted from Reference 5 with permission.)

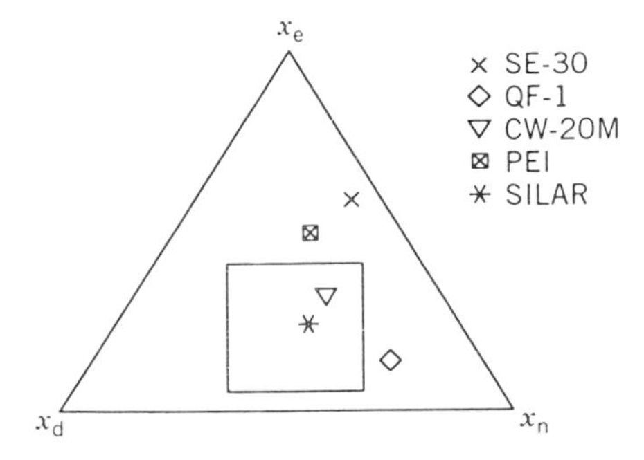

FIGURE 4.7 Expanded selectivity triangle showing location of five common liquid phases. Most commercially available phases have selectivities in the boxed-in portion. The "polar" phases, Carbowax 20M and Silar 10CP, do not exhibit unique polar selectivities. (Reprinted from Reference 5 with permission.)

similar to the $\Sigma\Delta I$ quoted with McReynolds constants, except that in this case it is the sum of only three constants (for ethanol, 1,4-dioxane, and nitomethane). The $\Sigma\Delta I$ reflects the magnitude of the polar interactions of a phase relative to its non-polar interactions. For phases with a small $\Sigma\Delta I$, such as SE-30 (polydimethyl silicone), the overwhelming majority of its interaction potential is nonpolar, so any polar selectivity that may be present is small and will not contribute significantly to the overall retention of solutes.

Phases with larger $\Sigma\Delta I$ will preferentially retain solutes with polar functional groups relative to nonpolar solutes. If a phase with large $\Sigma\Delta I$ is also selective (is located toward a corner of the selectivity triangle), it will very significantly exhibit this selectively. For example, a phase with a $\Sigma\Delta I$ of

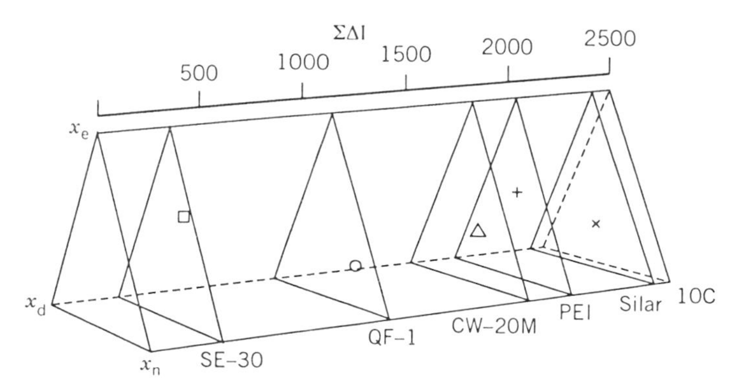

FIGURE 4.8 The same selectivity triangle as in Figure 4.7, plotted against $\Sigma\Delta I$. The phases with higher $\Sigma\Delta I$, such as Silar 10CP and Carbowax 20M, have a higher proportion of polar interaction versus nonpolar interaction compared to other phases, and are most prone to exhibit any inherent polar selectivity. (Reprinted from Reference 5 with permission.)

2000 and a selectivity in the upper portion of the triangle (proton acceptor) selectively retains compounds with proton donor functional groups (acid protons). In addition, all polar solutes would be retained longer than nonpolar solutes of similar molar volume or boiling point.

Unfortunately, few phases have both large $\Sigma\Delta I$ and unique selectivities. This is especially the case for proton donor phases. The few polar phases that do fit these criteria are trifluoropropyl silicones (e.g., OV-210, QF-1, SP-2401), which are selective toward dipoles; polypropylencimine and polyethyleneimine, which are proton acceptors; and CW-20M and 50% cyanopropyl silicones (e.g., Silar 10CP), which have the largest $\Sigma\Delta I$ values and retain polar solutes significantly longer than nonpolar solutes of similar boiling points, but which do not have particularly unique selectivities. Of these phases, the imines are not stable and the 50% cyanopropyl phases yield very low coating efficiencies and stability.

For general analysis, then, three liquid stationary phases are recommended to be stocked:

1. Polydimethyl silicone
2. Trifluorpropyl methyl silicone
3. Carbowax 20M

These cover the widest chromatographic selectivity space and are available in most column formats. For maximum speed of separation, inertness, and efficiencies, these phases should be coated on fused-silica capillary columns. For maximum stability and lowest bleed, they should also be cross-linked or immobilized.

Figure 4.9 shows the position of porous polymers on the same selectivity triangle as for the liquid phases. Porous polymers also tend to exhibit similar nonpolar characteristics, eluting compounds based on molar volume or boiling point. Figure 4.10 adds the $\Sigma\Delta I$ axis and indicates that Porapak T and Chromosorb 104 have the most significant polar character (large $\Sigma\Delta I$) and the most different polar selectivity (proton donor). Therefore, for the porous polymers, the following phases are recommended:

1. Chromosorb 101, 102, or Porapak Q, QS, P
2. Chromosorb 104, 108 or Porapak T

The nonpolar phases listed in number 1 are the most popular and useful. When separation of mixtures of nonpolar and polar solutes is required, or when water is used the solvent, then the polar phases are preferred.

4.6.3 Film Thickness (β)

Stationary phase thickness (or loading) influences capacity and retention/resolution. The phase ratio β relates the film thickness and column

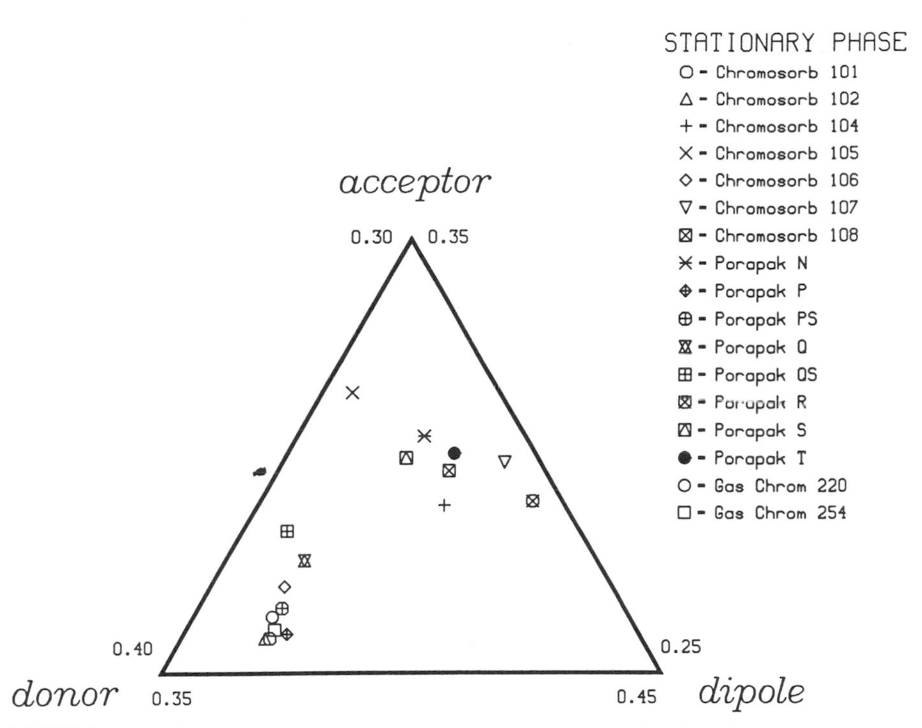

FIGURE 4.9 Expanded selectivity triangle showing relative locations of common porous polymers.

diameter:

$$\beta = \frac{r}{2d_f} \tag{4.2}$$

where r is the column radius, and d_f is the film thickness. Doubling the stationary phase doubles capacity and retention. However, thicker films are not as efficient, so resolution does not improve as much as it does by doubling the column length at the same film thickness. In addition, changes in β can also affect the relative retention of solutes during a temperature program. By selecting columns with identical phase ratio, one can change to a column of different diameter to increase capacity or resolution or analysis speed without affecting overall retention order and selectivity. Most column manufacturers list phase ratio with the specifications for their columns.

The easiest way to select an appropriate film thickness for an initial separation is to determine an appropriate β for the type of sample at hand,

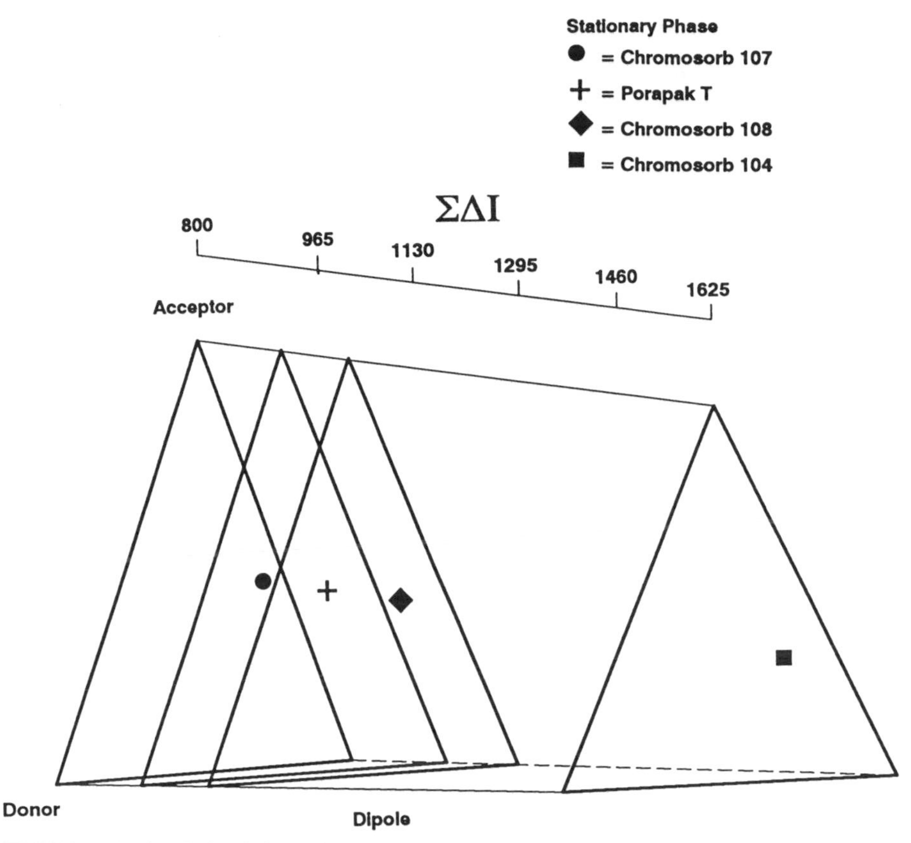

FIGURE 4.10 Selectivity prism showing selectivities and $\Sigma\Delta I$ for the most polar porous polymers.

TABLE 4.8 Approximate Phase Ratio (β) Required for Optimal Separation

Solute Volatility	β
Very volatile (gases)	<100
Medium volatility (liquids)	~250
Low volatility (solids)	>750

and then determine the film thickness that corresponds to that β for the column diameter of interest. Table 4.8 recommends β values based on the volatility of sample components to be separated.

4.7 INFLUENCE OF COLUMN DIMENSIONS AND MATERIAL

The influences of column dimensions on resolution, capacity, and analysis time are summarized in Table 4.2. Additional information can be found in

Chapters 2 and 3. In general, to maximize speed of analysis, reduce column diameter or particle size, stationary phase film thickness, and/or column length. To maximize capacity (sensitivity), increase film thickness, column diameter, and/or length of the column. To maximize resolution, increase column length, decrease film thickness, and decrease column diameter.

4.7.1 Column and Packing Materials

Active column materials can degrade ultimate performance through adsorption (distorted peaks) and sample decomposition. Degradation and adsorption are most pronounced for polar compounds. The more polar the functional group, the higher the potential for adsorption and decomposition (e.g., hydroxy < 3° amine < carboxy < 2° amine < multicarboxy < 1° amine). Therefore, analysis of polar compounds requires inert chromatographic systems. This includes

- Inlets (liners, flow path; see Chapter 8)
- Column (fused silica > glass > Ni > SS > copper)
- Column installation (follow instrument manufacturer's instructions)
- Detectors (fused-silica lined > glass lined > Ni > SS).

4.7.2 Column Dimensions

Column dimensions have a strong influence on capacity and sensitivity, analysis time, and resolution. More details on these aspects can be found in Chapters 2 and 3. The amount of sample that can be injected into a given column before overload (peak shape distortion, shifts in retention time) is proportional to the column capacity. So column capacity is an important variable to maximize when optimizing an analysis for sensitivity.

The capacity of a column is related to the volume of stationary phase in the column. This is in turn a function of the column length, diameter, and the thickness of the stationary phase. For packed columns and capillary columns with a given β larger columns hold a larger amount of stationary phase, and therefore have higher capacity according to the volume of the column:

$$\text{Capacity} \propto \text{Radius}^2 \times \text{Length} \tag{4.3}$$

Doubling the diameter, increases capacity at least fourfold. For capillary columns with constant film thickness, capacity increases proportionally with the diameter (Equation 4.4). Doubling the diameter doubles the capacity:

$$\text{Capacity} \propto \text{Radius} \tag{4.4}$$

For capillary and packed columns, doubling the film thickness or loading doubles the capacity.

$$\text{Capacity} \propto \text{Film thickness} \tag{4.5}$$

Column dimensions also affect ultimate resolution because decreasing film thickness increases efficiency. Separation of two closely eluting peaks is a function of the efficiency of the column. Doubling the column length increases the efficiency (theoretical plate count) twofold and resolution by 1.4 (Equation 4.6). For packed columns, a 2× decrease in particle size results in a 1.7 increase in resolution. A 2× decrease in i.d. of a capillary column doubles resolution:

$$\text{Packed} \qquad \text{Resolution} \propto \left(\frac{\text{Length}}{d_p + d_p^2}\right)^{1/2} \tag{4.6}$$

$$\text{Capillary} \qquad \text{Resolution} \propto \left(\frac{\text{Length}^{1/2}}{\text{Diameter}}\right) \tag{4.7}$$

Of course, a practical limit will be reached as particle size and column diameter decrease, because the pressure necessary to get the required carrier-gas flowrate will be prohibitive.

Analysis time is proportional to the length of the column and the square root of the column diameter. Doubling the column length doubles the analysis time (at constant average linear velocity). The optimum linear velocity of a column is a function of particle size or column diameter: the smaller the particle or the smaller the diameter, the faster the optimum linear velocity. Decreasing particle size two times or halving column diameter increases optimum linear velocity by 1.4 and decreases analysis time also by 1.4 times:

$$\text{Analysis Time} \propto \text{Length} \times \text{Diameter}^{1/2} \tag{4.8}$$

The benefits of using longer columns at faster than optimum flowrates is also illustrated in Figure 4.11. In Figure 4.11a, separation of components 1 and 2 is not possible with the 15-m column at the temperature program used. Lowering temperature program rate may separate peaks 1 and 2, but analysis time would be even longer than the 23 min for the initial separation. By using a 30-m column (Figure 4.11b), components 1 and 2 are separated. By increasing column flowrate using a pressure program, the remaining compounds are separated and eluted in less time than in the original chromatogram. In this example, high efficiency was maintained early in the chromatogram when required, then analysis time was decreased by increasing flowrate during the remainder of the chromatogram where peaks were better separated.

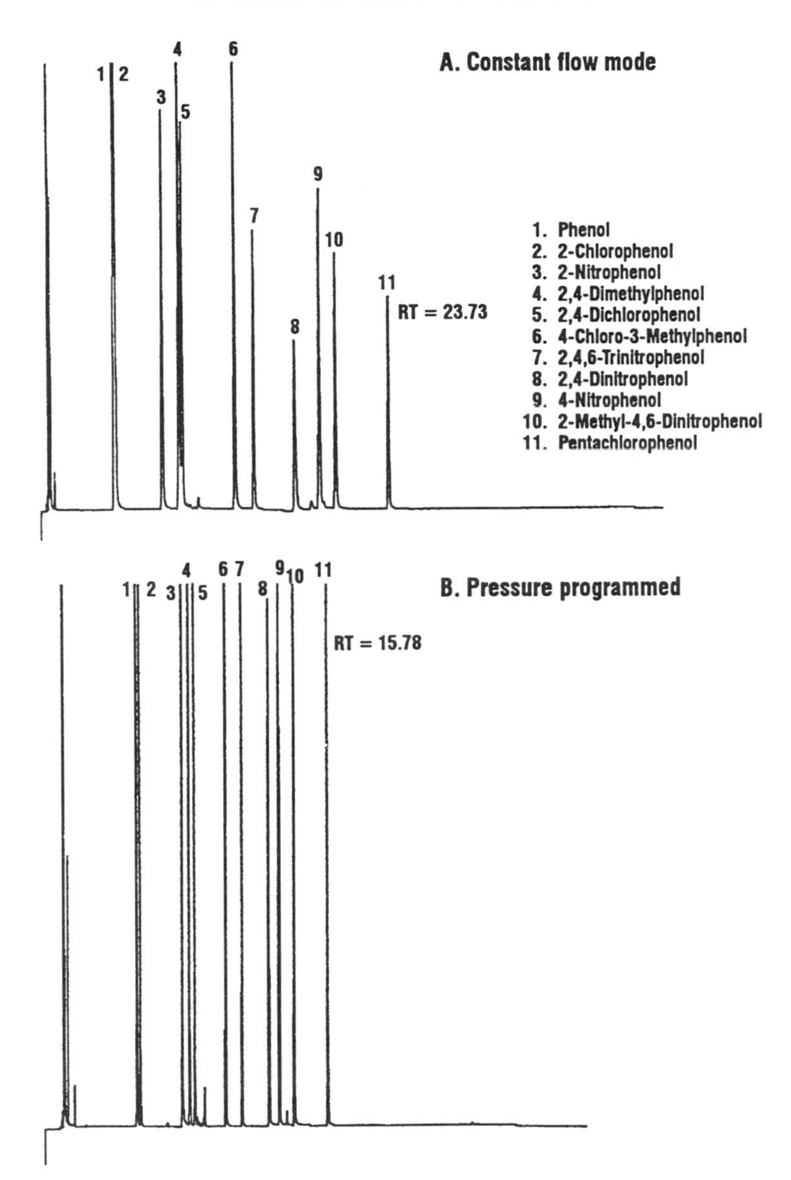

FIGURE 4.11 (a) Separation of phenols on a 15-m, 530-μm-i.d., 0.88-μ HP-1 capillary column. Compounds 1 and 2 are not separated, run time is approximately 24 min. He carrier-gas pressure of 7 psi, constant flow mode (pressure increases as temperature increases to keep outlet volumetric flow constant). Oven temperature program: 35 to 250°C and 5°C/min, hold for 5 min. (b) Separation of phenols on a 30-m, 530-μm-i.d., 0.88-μm HP-1 capillary column. Compounds 1 and 2 are now separated due to the extra column efficiency. Run time is <16 min due to active inlet pressure program. He carrier-gas pressure program from 10 to 25 psi at 1 psi/min, hold for 10 min. Column temperature program: 50 to 250°C and 8°C/min, hold for 5 min. (Reprinted from Reference 28 with permission.)

4.8 SCREENING SAMPLES

4.8.1 Evaluation of Sample and Initial Chromatographic Choices

Once a column, stationary phase, inlet, and detector are chosen and set up appropriately, the process of optimizing a separation can begin. The first step is to screen the sample to assess how close the initial choices came to meeting the analysis goals, and to get the approximate elution temperature range of the sample components. This is done using fast analysis conditions—flowrate at OPGV and a temperature program from low to high at 25–35°C/min. The "low" and "high" temperatures are selected based on knowledge of the sample, solvent boiling points, injection considerations, or from the published temperature limits of the stationary phase being used.

If the peak shapes are poor, first correct any reasons for sample or solvent overload and rerun the screen. If the peaks are still unsatisfactory, switch immediately to a different stationary phase of very different polarity. If peak shapes are good but separation of peaks is poor, adjust the temperature program to bracket the elution temperature range for the components and reduce ramp rate by 1/2. Rerun the sample screen. If separation is still not promising, switch to a different stationary phase before optimization is continued. Once reasonable separation and peak shapes have been achieved, the temperature program can be optimized.

4.8.2 Determine Relative Sensitivity to Temperature

Temperature and temperature program rate are the next most useful parameters second to stationary phase type for optimizing separations. The most common use of temperature programs is to shorten the time of analysis. For this, optimization consists only of adjusting the program rate to yield the fastest analysis while meeting goals of resolution and reproducibility. It should be noted that there is also a trade-off in time savings by temperature programming because of the increased time necessary to cool the oven back down to starting conditions before the next injection.

A little used advantage of temperature programs is the optimization of separations of closely eluting compounds. This is especially useful when the samples contain compounds with different chemical characteristics. Relative retention of homologs will not change significantly as a function of temperature, whereas the relative retention of compounds with different functional groups can change significantly. The largest differences in retention resulting from temperature changers occur for polar solutes, although optimization for mixtures of nonpolar and polar solutes can also be productive. This phenomenon can also be a disadvantage: Trying to reproduce a separation done in one column becomes difficult if column length L and flowrate are not exactly the same.

To evaluate if the separation is sensitive to temperature changes, a new analysis should be done using a different temperature program rate with the same flowrate, or a different flowrate at the same temperature program rate. If relative retention order changes or if more or less peaks show up under the new conditions, then flow/temperature optimization can be very important.

A change of flowrate or flowrate program (if using electronic pressure control) is intimately tied with temperature affects, since both influence the residence time of solutes in the column and the "average" temperature history of each solute before it exits the column.

4.9 FINE TUNING TEMPERATURE PROGRAM AND FLOWRATE

4.9.1 Fundamental Principles of Optimization

Optimization of the temperature (and flowrate) is approached systematically. The affect of temperature and flow changes on relative retention is continuous. In other words, if an increase in flowrate or temperature program rate improves resolution of a poorly resolved pair of compounds, then further increase will improve separation further. This is true only within reasonable operating ranges of temperature program rate and flow, however. For example, a temperature program rate might be chosen that is so fast that solutes effectively go from infinite retention to no retention in a short period of time (less than the void time of the column). Then there would be no chance of partitioning into the stationary phase and no resolution of closely eluting sample components. If an extremely fast flowrate is chosen that exceeds the optimal flow rate for the column by more than 10-fold, then there will be little retention, peaks will be broad, and resolution will suffer.

Optimization of temperature and flow involves a few experiments to determine the direction and magnitude of the dependence of resolution on temperature and flow. Visual comparison of chromatograms may be all that is necessary to fine tune a separation when a few poorly resolved peaks are involved, but plotting of the relationships and use of graphical approaches often provides better insight, shortens optimization time, and facilitates locating the true optimum.

4.9.2 Case in Point: Organochlorine Pesticides

Table 4.9 lists 16 organochlorine pesticides that are typically analyzed by GC with ECD. These pesticides have similar structures and are difficult to resolve using nonpolar stationary phases. One of the stationary phases recommended by the EPA (7, 8) is a 5% phenyl methyl silicone which is very close to methyl silicone in polar strength and selectivity. Two pairs of pesticides show the worst resolution, peaks 9/10 and 15/16. As can be seen

TABLE 4.9 Organochlorine Pesticides

Peak Number	Formula	Name
1	$C_6H_6Cl_6$	α-Hexachlorocyclohexane
2	$C_6H_6Cl_6$	β-Hexachlorocyclohexane
3	$C_6H_6Cl_6$	γ-Hexachlorocyclohexane (Lindane)
4	$C_6H_6Cl_6$	δ-Hexachlorocyclohexane
5	$C_{10}H_2Cl_7$	Heptachlor
6	$C_{12}H_6Cl_6$	Aldrin
7	$C_{10}H_4OCl_7$	Heptachlor Epoxide
8	$C_9H_6O_3Cl_6S$	Endosulfan I
9	$C_{12}H_{10}OCl_6$	Dieldrin
10	$C_{14}H_8Cl_4$	4,4′-DDE
11	$C_{12}H_{10}OCl_6$	Endrin
12	$C_9H_6O_3Cl_6S$	Endosulfan II
13	$C_{14}H_{10}Cl_4$	4,4′-DDD
14	$C_{12}H_8OCl_6$	Endrin Aldehyde
15	$C_9H_6O_4Cl_6S$	Endosufan Sulfate
16	$C_{14}H_9Cl_5$	4,4′-DDT

from Figure 4.12, as temperature program rate is changed, the separation of the first peak pair gets better at higher program rates, whereas the separation of the later peaks is better at lower program rates. This relationship is plotted in Figure 4.13. At the intersection of the two lowest lines, separation of the peak pairs is equal and corresponds to an optimum for this given column and flowrate. However, the optimum temperature program rate is different at different flowrates, with different column lengths and with different phase ratios. If an analysis goal is to mininize analysis time while maintaining a minimum resolution, then the optimization should include flowrate along with temperature program rate. The influence of flow on optimal pressure can be seen by comparing plots in Figure 4.13 for the different head pressures.

Figure 4.14 shows the dependence of the optimum flowrate versus column head pressure. As head pressure (and therefore flowrate) is increased, the optimum temperature program rate also increases. This relationship exists because optimum separation occurs when the solutes experience the same average temperature before eluting that they did at the lower flow rates. In order to do this at higher flow rates, the temperature program rate must also increase. This is fortuitous if the objective is to minimize analysis time, since both higher flowrates and temperature program rates reduce analysis time. The increase in speed is gained at a cost of resolution. The resulting decrease in the resolution corresponding to the optimum temperature program rate is shown in Figure 4.15. To meet a required resolution of 1.2, the analysis time is 20 min, the temperature program rate is 3.6°C/min, and the head pressure is 110 psi.

Several commercial computer programs have recently become available

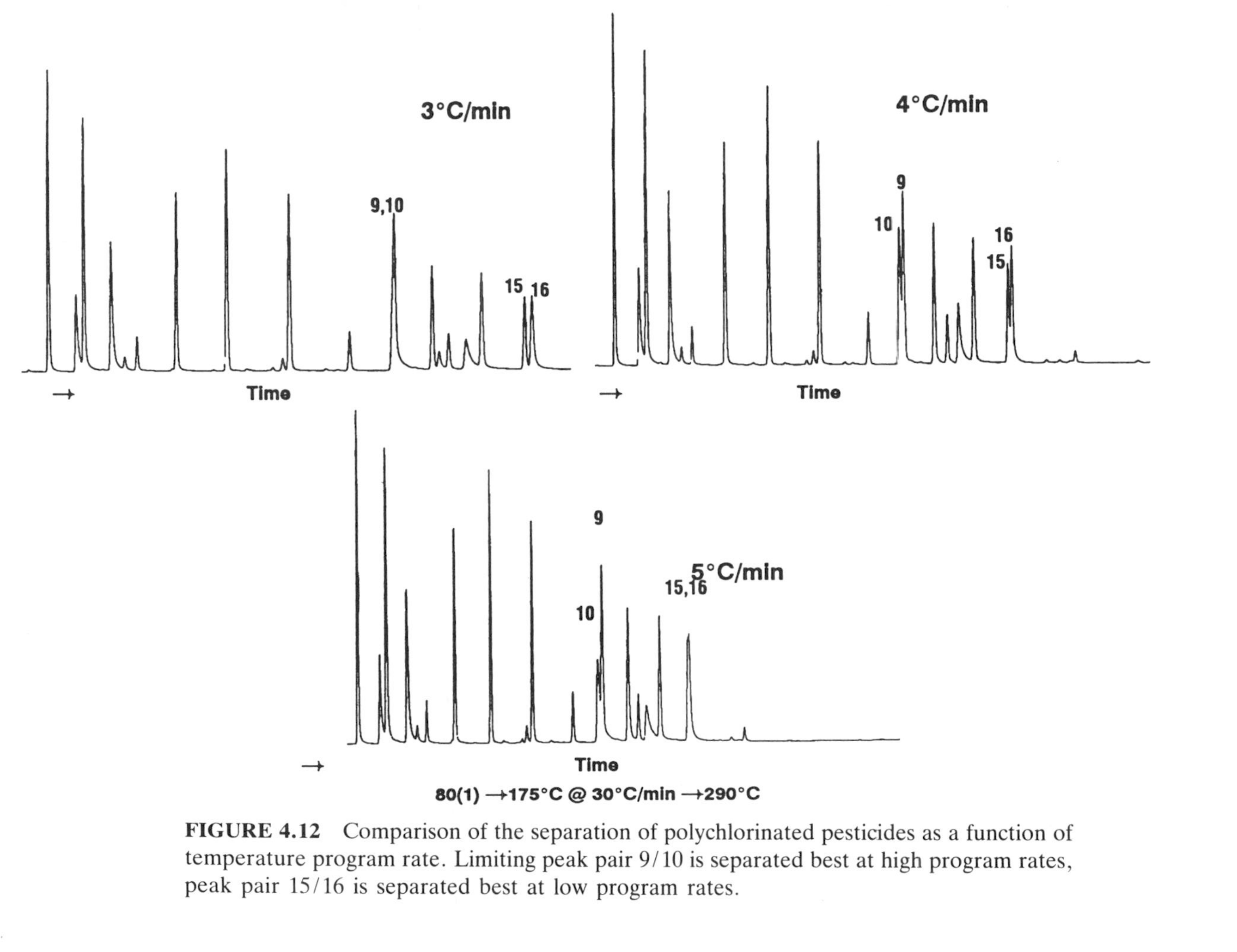

FIGURE 4.12 Comparison of the separation of polychlorinated pesticides as a function of temperature program rate. Limiting peak pair 9/10 is separated best at high program rates, peak pair 15/16 is separated best at low program rates.

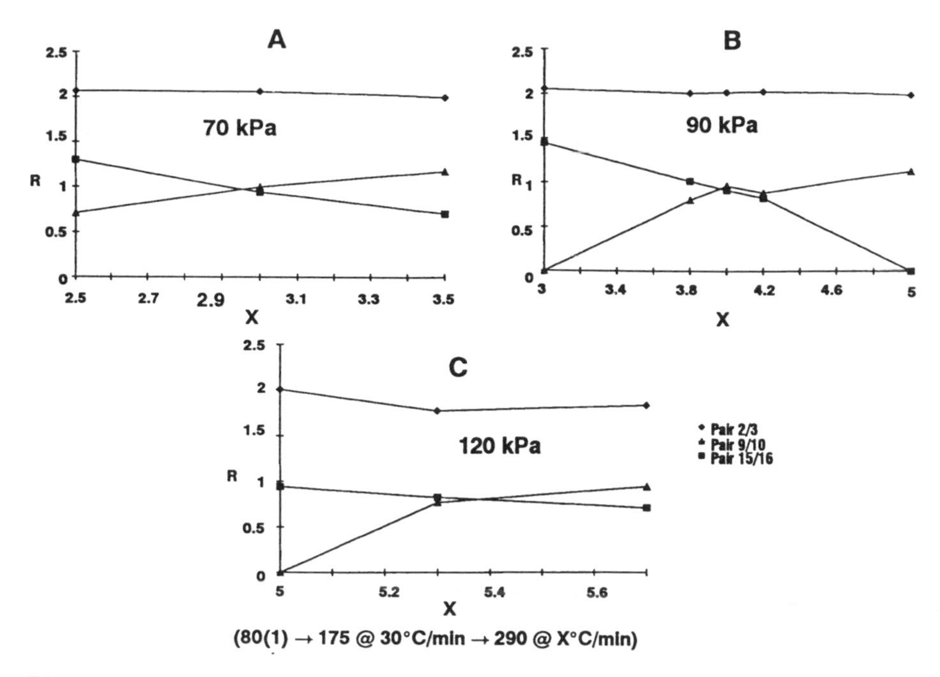

FIGURE 4.13 Relationships of resolution of limiting peak pairs as a function of temperature program rate: (a) at 70 kPa head pressure, (b) at 90 kPa head pressure, (c) at 120 kPa head pressure. Resolution of the limiting peak pairs is the equivalent where the line cross, representing the "optimum" program rate. As pressure (and flowrate) increases, the optimal temperature program rate increases.

to help in optimizing temperature (and sometimes flow) rate. They require that retention data and peak assignment be input for two chromatographic screens at different temperature program rates. The programs then predict the changes in retention and optimal temperature program rate.

4.9.3 Simplex Optimizations

Morgan and Deming introduced the Simplex method (9) for computerized optimization of column temperature in GC (10). They used a chromatographic response function (CRF) to measure the efficiency of separation as temperature was varied in the direction of increasing CRF. The CRF is based on a "response function" proposed by Kaiser (11), and adaptations that include the total number of peaks in the chromatogram and the total analysis time (12, 13) allow computerized sequential optimization of parameters, such as temperature and flowrate. Berridge (13) extended CRF

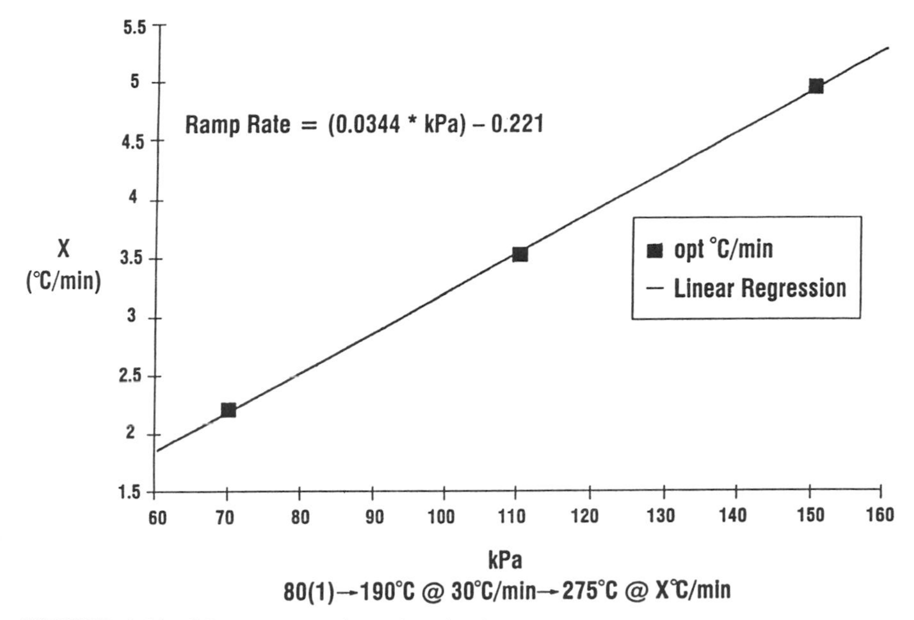

FIGURE 4.14 Linear regression of optimal temperature program rate as a function of column head pressure. The relationship is linear over the range investigated. (Reprinted from Reference 30 with permission.)

by adding weights to time and peak-number factors. He defined CRF as

$$\text{CRF} = \sum_{i=1}^{L} R_i + L^x - a|T_m - T_L| - b|T_0 - T_1| \tag{4.9}$$

where R is the resolution between adjacent pairs (R values under 2 were ignored); L is the total number of peaks detected; T_m is the acceptable analysis time; T_L is the time of the last eluted peak; T_1 is the time of the first peak; T_0 is the minimum acceptable time; and x, a, and b are weights that allow flexibility in weighing the importance of the latter factors.

A major advantage of using a CRF in optimization schemes is that peaks need not be identified, since the CRF responds to the number of peaks and degree of separation. This precludes the use of CRF for certain optimization strategies, since it does not allow prediction of elution order or resolution; it is simply a measure of the overall quality of a separation.

Figure 4.16 illustrates the Simplex principle as presented by Morgan and Deming (10). A separation is done at three temperatures (or temperature program rates or flowrates), denoted as 1, 2, 3 in Figure 4.16. The point with the smallest CRF (point 3) is rejected and reflected through the plane

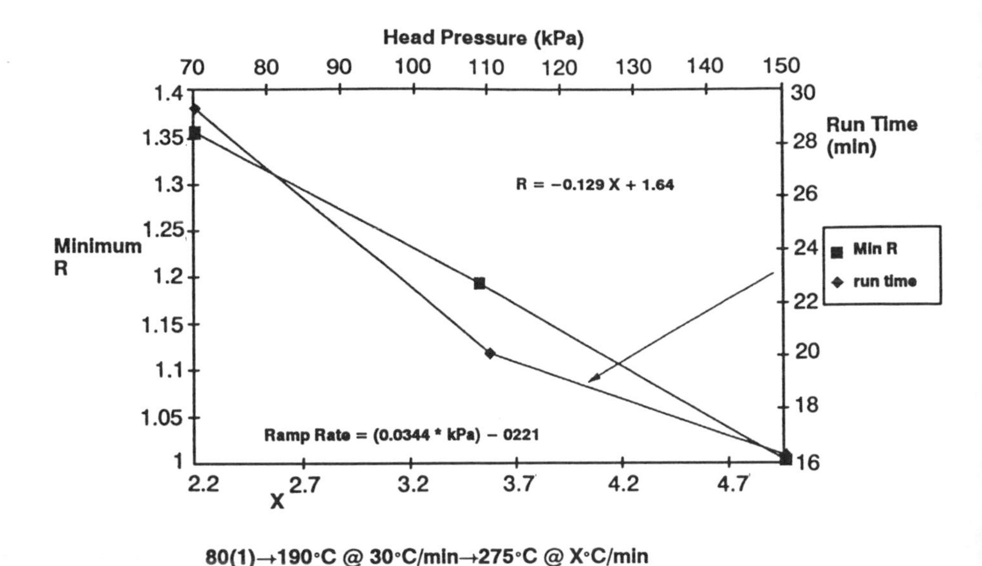

FIGURE 4.15 Relationships of resolution and run time as a function of optimal temperature program rate and head pressure. A reduction of run time by increasing head pressure and optimum temperature program rate leads to a reduction in resolution. A resolution of 1.2 is achieved at a head pressure of 110 kPa and a program rate of 3.6°C/min, and results in a 20-min run time. (Reprinted from Reference 30 with permission.)

of the other two points (1 and 2). A fourth analysis is done with the conditions indicated by the new projected point. If the new response (at point 4) is not significantly larger than the previous two best (1 and 2) or is smaller than the worst of the two, the Simplex size remains the same and is reflected across the best axis (to point 5). If the new response is better than the previous best, the Simplex is expanded in the same direction (point 6). If a boundary condition is exceeded, such as maximum analysis time, the Simplex contracts to an intermediate point and continues until changes in the experimental variables do not cause a significant change in CRF. The Simplex approach is useful as a systematic means of optimizing experimental parameters even with as little as a programmable calculator. The procedure can also be expanded to include multiple variables.

4.9.4 Optimizing Stationary-Phase Mixtures

When the sample is well characterized and standards are available for individual sample components, then a more rigorous optimization of temperature program and flowrate is possible. Difficult separations can also

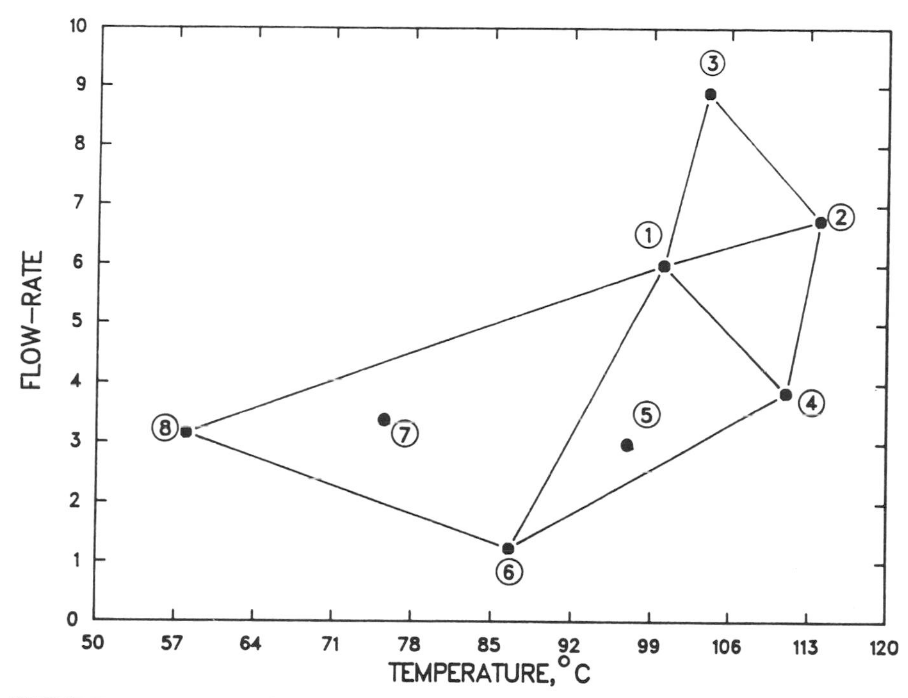

FIGURE 4.16 Progression of Simplex optimization procedure that seeks an increase in chromatographic response function (CRF). See text for details. (Reprinted from Reference 10 with permission.)

be optimized by combining different columns or stationary phases. This has been done by coupling of columns, intimate mixing of stationary phases, or synthesis of new stationary phases with mixed functional groups. It seems clear that the most reproducible and stable approach is the latter, although it requires the most investment of time and resources, and is usually limited to column manufacturers for high-volume analytical methods.

When mixing phases to achieve some intermediate separation, one relies on the additivity of characteristics of the individual phases, and this usually holds true. Practical success of this approach is often limited by the miscibility of the phases, which can be a problem if the phases differ too much in polarity, especially when coating capillary columns. In addition, if two different polar phases are mixed, the result may be quite different than the sum of the individual parts because the phases react with each other or interact with each other more strongly than with the sample components.

The easiest procedure to follow in optimizing stationary phase mixtures is to combine lengths of individual columns (14–17). There is some influence on the order of connecting columns (16, 17) because the linear velocity is slower in the first part of a column than to the last, and because solutes are in different chemical environments during stages of the temperature program. But these can be corrected for through small changes in column lengths once the general retention relationships are known.

4.9.5 Window Diagrams

Optimization of stationary phase mixtures involves determining the retention behavior of the solutes of interest on each phase under consideration. The optimization process than assumes linear additivity in retention according to the following equation:

$$K_D = \Phi_A K^0_{D(A)} + \Phi_S K^0_{D(S)} \tag{4.10}$$

where K_D represents the distribution constant of a solute on a mixture of phases A and S, $K^0_{D(A,S)}$ are the distribution constants of the solute on pure S and A, and $\Phi_{A,S}$ are the molar fractions of phases A and S in a binary mixture (18).

Laub, Purnell, and coworkers introduced one of the most common ways to help in determining the optimum proportion of phases to yield a given separation, the "window diagram" (19). This approach has been reviewed by Laub (20) and is also useful for optimizing temperature, flow, and amount of modifier (salt) added to the stationary phase (20).

To construct a window diagram, capacity factors for all solutes in a mixture are determined with the use of two separate phases. Additivity in retention is assumed and a plot of K_D versus Φ_A is generated (Figure 4.17). Separation factors α are then calculated for all limiting solute pairs. In those cases where elution order changes, Φ is inverted such that all a values are greater than 1. A composite plot of α versus Φ_A is generated. Optimal stationary-phase composition occurs at the highest values (Figure 4.18).

Window diagrams provide sufficient information that logical choices may be made on selecting stationary-phase composition, column length, and analysis time. If more than one acceptable window exists, the Φ_A corresponding to the widest window can be chosen to reduce errors associated with inaccurate mixing of phases or compensation for flowrate gradients. Since α of the poorest resolved pair of peaks for a given Φ_A is provided by the diagram, calculation of the minimum number of plates to affect the separation is also possible (20).

Although early eluting compounds may have acceptable α, they may not be retained enough to be satisfactorily separated. Jennings et al. (17) found it more useful to construct window diagrams based on resolution (R) versus Φ_A than α versus Φ_A. In either case, the procedure is straightforward and effective.

4.10 COMPUTERIZED PEAK DECONVOLUTION

There is growing interest in making use of the power of the computers commonly used for data acquisition, to improve integration and estimation

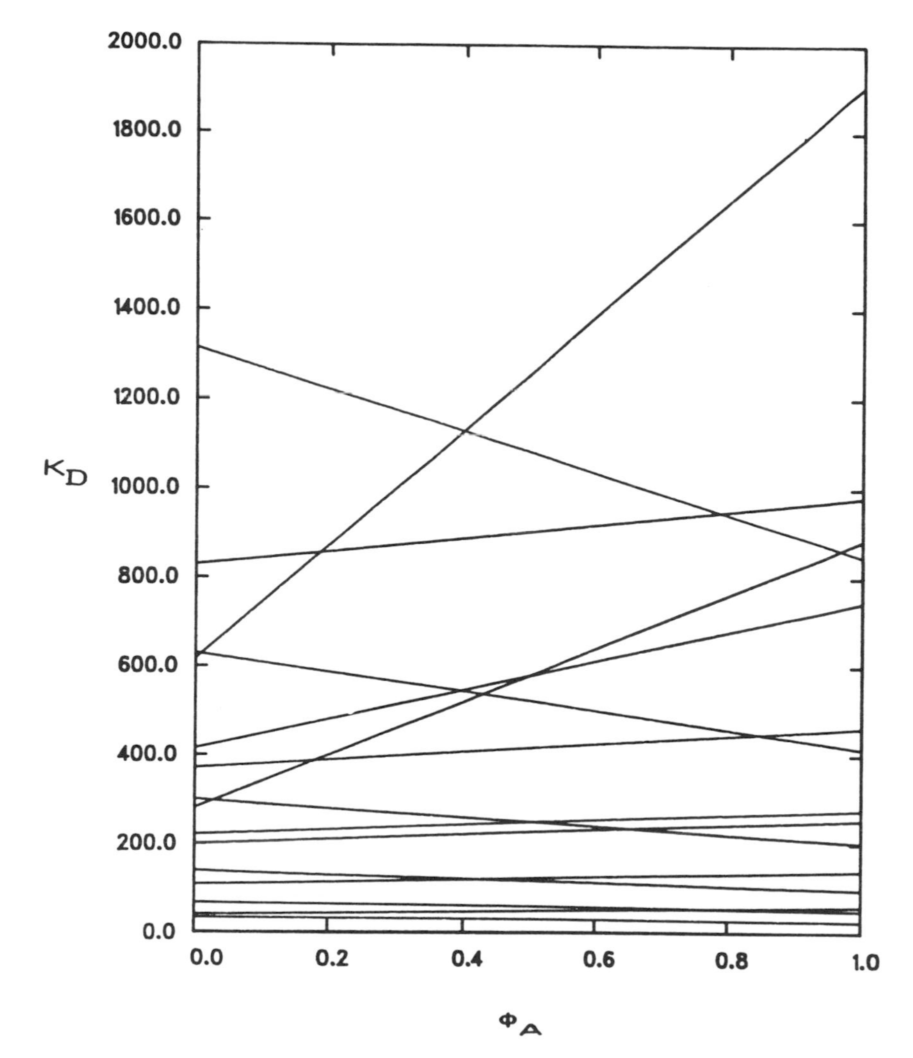

FIGURE 4.17 Plot of the distribution constant K_D of 15 solutes versus mole fraction of stationary phase dinonyl phthalate (Φ_A) in a binary mixture with squalane. The plot was constructed from end-point data (retention on single, unmixed stationary phases). (Reprinted from Reference 19 with permission.)

of peak areas for poorly resolved peaks, through the use of computerized algorithms. This process is most common when detectors with multiple response characteristics, such as mass spectrometers, infrared spectrometers, and atomic emission spectrometers are used. The most flexible algorithms depend on the differences in response of the components making up the unresolved peak. This approach usually involves factor analysis (21–23). If a peak shape model (e.g., Gaussian, Lorentian) fits a particular elution profile, then least-squares analysis can be used to resolve overlapped

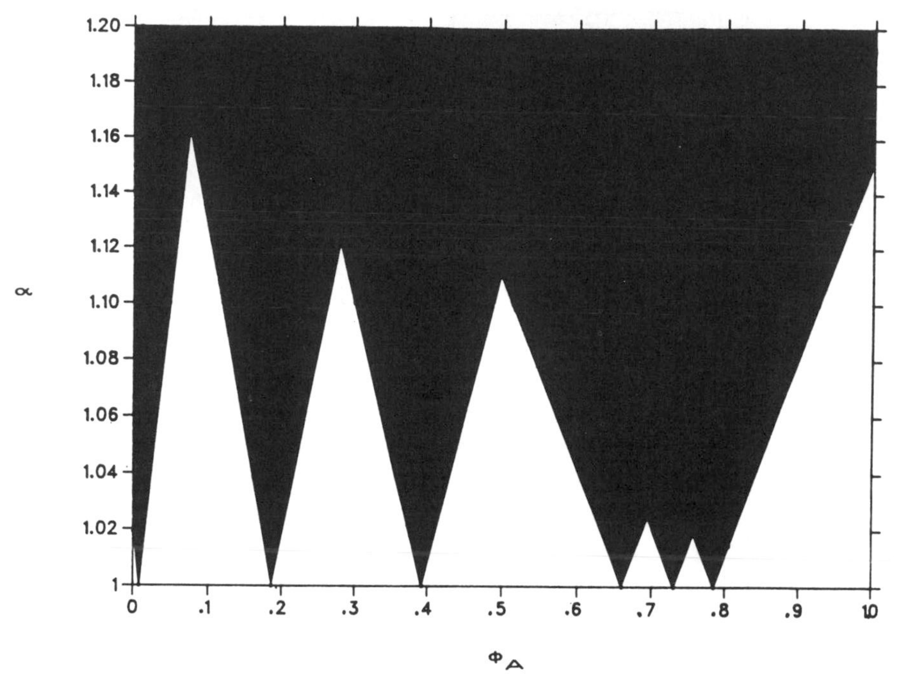

FIGURE 4.18 Window diagram of α versus Φ_A for the same data given in Figure 4.17. The α values were calculated for each solute pair that might overlap; α values <1 (elution order reversed) are inverted so that $\alpha > 1$. Windows of highest α indicate the optimum stationary phase compositions. (Reprinted from Reference 19 with permission.)

peaks (24–27). But this is more dependent on chromatographic reproducibility and stability than is factor analysis and is less apt to find general utility.

REFERENCES

1. R. P. W. Scott and G. S. F. Hazeldean, in *Gas Chromatography*, R. P. W. Scott (Ed.), Butterworths, London, 1960, pp. 144.
2. K. Yabumoto and W. J. A. Vandenheuval, *J. Chromatogr.*, **140**, 197–207 (1977).
3. R. V. Golovnya and T. A. Misharina, *J. High Resolut. Chromatogr., Chromatogr. Commun.*, **3**, 4–10, (1980).
4. W. O. McReynolds, *J. Chromatogr. Sci.*, **8**, 685–691 (1970).
5. M. S. Klee, M. A. Kaiser, and K. B. Laughlin, *J. Chromatogr.*, **279**, 681–688 (1983).
6. M. A. Hepp and M. S. Klee, *J. Chromatogr.*, **404**, 145–154 (1987).

7. "Method 608—Organochlorine Pesticides and PCBs," *Federal Register*, Vol. 49, No. 209, October 1984.
8. Method 505—Analysis of Organochlorine Pesticides and Aroclors in Drinking Water by Microextractin and Gas Chromatography," U.S. Environmental Protection Agency, Cincinnati, OH, September 1986.
9. S. Morgan and S. Deming, *Anal. Chem.*, **46**(9), 1170–1181 (1974).
10. S. L. Morgan and S. N. Deming, *J. Chromatogr.*, **112**, 267–285 (1975).
11. R. Kaiser, *Gas Chromatograpie*, Geest & Portig, Leipzig, 1960, p. 33.
12. M. W. Watson and P. W. Carr, *Anal. Chem.*, **51**(11), 1835–1842 (1979).
13. J. C. Berridge, *J. Chromatogr.*, **244**, 1–14 (1982).
14. J. Krupcik, G. Guiochon, and J. M. Schmitter, *J. Chromatogr.*, **213**, 189–201 (1981).
15. J. Krupcik, J. Mocak, A. Simova, J. Garaj, and G. Guiochon, *J. Chromatogr.*, **238**, 1–12 (1982).
16. G. Takeoka, H. M. Richard, M. Mehran, and W. Jennings, *J. High Resolut. Chromatogr., Chromatogr. Commun.*, **6**, 145–151 (1983).
17. D. F. Ingraham, C. F. Shoemaker, and W. Jennings, *J. Chromatogr.*, **239**, 39–50 (1982).
18. R. J. Laub, J. H. Purnell, D. M. Summers, and P. S. Williams, *J. Chromatogr.*, **155**, 1–8 (1978).
19. R. J. Laub and J. H. Purnell, *J. Chromatogr.*, **112**, 71–79 (1975).
20. R. J. Laub, *Physical Measurements in Modern Chemical Analysis*, Vol. 3, T. Kawana (Ed.), Academic, New York, 1983, pp. 249–341.
21. J. M. Halket, *J. Chromatogr.*, **186**, 443–455 (1979).
22. J. M. Halket, *J. Chromatogr.*, **175**, 229–241 (1979).
23. M. A. Sharaf, and B. R. Kowalski, *Anal. Chem.*, **54**, 1291–1296 (1982).
24. M. Rosenbaum, V. Hancil, and R. Komers, *J. Chromatogr.*, **246**, 1–11 (1982).
25. M. Rosenbaum, V. Hancil, and R. Komers, *J. Chromatogr.*, **191**, 157–166 (1980).
26. J. T. Lundeen and R. S. Juvet, Jr., *Anal. Chem.*, **53**, 1369–1372 (1981).
27. J. Grimalt, H. Iturriaga, and X. Tomas, *Anal. Chem. Acta*, **139**, 155–166 (1982).
28. *Electronic Pressure Control in Gas Chromatography*, Sally Stafford (Ed.), Hewlett Packard #5182–0842, Wilmington, DE, 1993.
29. L. R. Snyder, *J. Chromatogr.*, **92**, 223 (1974).
30. M. S. Klee, "Increasing Lab Throughput With Automated Compound Confirmation Reporting—The Organochlorine Pesticide Analysis System," Hewlett Packard Application Note 228-114 (43-5952-2341), May 1990, 9 pg.
31. *Guide to Stationary Phases for Gas Chromatography*, 12th ed., J. A. Yancey (Ed.), Analabs, CT, 1979, pp. 126–127.

PART II

TECHNIQUES AND INSTRUMENTATION

Books must follow sciences, and not sciences books.

—Proposition touching Amendment of Laws

CHAPTER FIVE

Detectors and Data Handling

LORRAINE HARMANOS HENRICH
ARCO Chemical Company

Modern Practice of Gas Chromatography, Third Edition. Edited by Robert L. Grob.
ISBN 0-471-59700-7

5.1 INTRODUCTION

Part I of this book has covered the theoretical aspects of the gas chromatographic separation process. This chapter is devoted to understanding the "black box" that provides the actual results for our interpretation of that process. Just as there is no ideal liquid phase that meets all separation needs, there is no ideal gas chromatographic detector. A wide variety of physical and chemical properties are used for measuring purposes in gas chromatography. Choosing the right detector for the purpose requires an understanding of the detection mechanisms of each type of detector. Obtaining accurate results with the chosen detector requires an understanding of the dependence of the detector's response on experimental conditions.

Gas chromatographic detectors differ from other analytical instruments in that the input to the detector is a *flow* of separated chemicals. Therefore, it must complete its analysis in a few seconds or less. The sample arrives at the detector completely volatilized and relatively free of matrix effects. Most gas chromatographic detectors have great sensitivity and a huge dynamic range. Most often the detector is used to quantify known compounds. In the cases of the mass spectrometer (MS), the Fourier transform infrared spectrophotometer (FT-IR), and the atomic emission (AE) detectors, compound identification is their primary purpose.

This chapter surveys the most frequently used detectors and their principles of operation for the correct interpretation of the results obtained.

5.2 PARAMETERS OF PERFORMANCE

5.2.1 Signal-to-Noise Ratio

Figure 5.1 shows a portion of a chromatogram illustrating three parameters: the peak height S, the peak width at half its height w_h, and the noise N. The peak height is measured in any convenient units from the base of the peak to its maximum. The peak width w_b is defined in several ways: (1) the base of the triangle that most closely matches the shape of the peak; (2) twice the width of the peak at half its height (illustrated); (3) a multiple of the variance, or second moment, of the peak shape; and (4) the ratio (in consistent units) $2A/S$, where A is the area of the peak. Method 2, although it is not as accurate as 3 or 4, is used here because it is the easiest to measure.

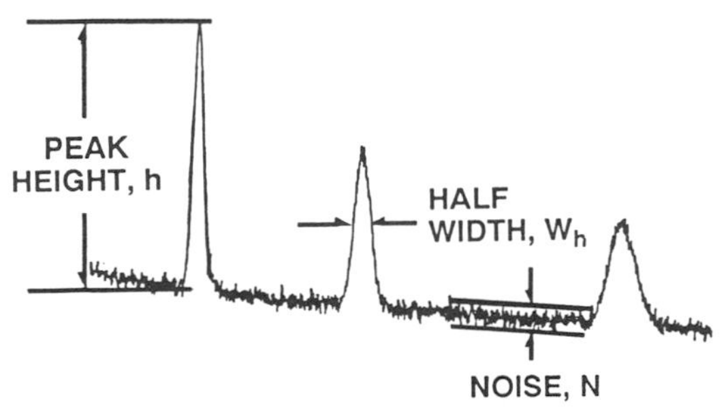

FIGURE 5.1 Portion of a chromatogram illustrating peak height, peak width at half its height, and noise.

Noise is the fluctuation of the detector signal when no sample is present. It can be caused by changes in temperature, gas flowrate, line voltage, stability of electronic circuitry, etc. It is a measurement of the average distance (in the same units as peak height) between the highest excursion of the baseline and the lowest excursion during a period of time. Most modern electronic integrators measure noise automatically.

When a strip chart recorder is used with manual measurements, some precautions must be taken to ensure accurate results. To obtain a statistically significant value for noise, use a section of baseline several peak widths long. Set the gain so that the noise is more than a few percent of full scale. All recorders have some deadbanding, that is, little or no response for signal changes corresponding to a fraction of a percent of full scale. Measuring a noise level of 1% or so of full scale will always result in too low a measurement. At this level the width of the line drawn by the pen will also become significant. Run the chart recording rapidly enough to prevent the noise excursions from overlapping and making a broad line down the paper.

The baseline can have perturbations other than the noise shown in Figure 5.1. Figure 5.2 shows drift and wander. *Drift* is a slow, constant change of the baseline over time. In severe cases, drift can make it difficult to keep the recorder pen on scale until the chromatogram is finished. Integrators can also suffer in performance due to drift. Drift is often due to temperature change or to column and septum bleed and is often controllable. By contrast, noise is usually due to things within the detector and thus is a more fundamental limit in performance. When the baseline varies randomly over times similar to the peak width, it is called *wander* (Figure 5.2). Wander is distinguished from *fast noise* (here taken to be much faster than the peak). Wander interferes with the interpretation of a chart recording and the operation of an electronic integrator. It can come either from the detector or from extraneous sources, such as a small leak at a column fitting. A measurement of noise will include both fast noise and wander, but drift will be ignored.

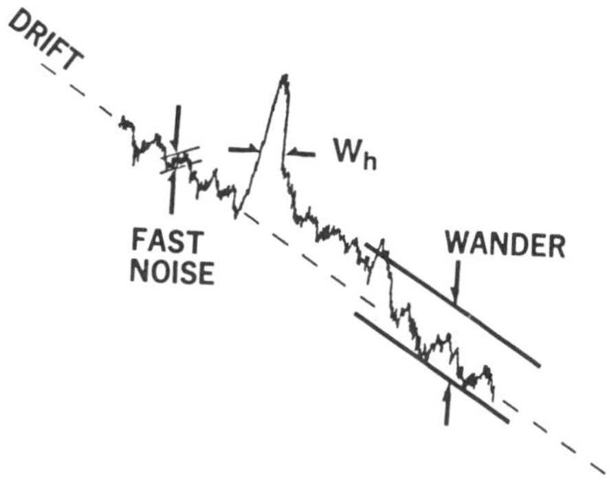

FIGURE 5.2 Portion of a chromatogram showing the distinction between drift, fast noise, and wander relative to the peak width at half its height w_h.

A detector specification frequently encountered is the *signal-to-noise* ratio (S/N). A signal-to-noise ratio of 2 indicates a 95% probability that the signal represents a peak if the noise level was evaluated within a time interval corresponding to ten peak widths. The probability is increased to 99% if the signal is 2.65 times greater than the noise. Monitoring the noise over five peak widths would require a signal-to-noise ratio of 6.72 for 99% probability and 4.08 for 95% probability (1).

5.2.2 Minimum Detectable Level

One of the parameters used to specify detectors independent of column parameters and sample size is the *minimum detectable level* (MDL). It is the quantity or concentration of sample in the detector at the maximum of the peak, when the S/N ratio is 2. This *quantity* is commonly given as g/sec and is a mass flowrate.

The maximum of the peak in Figure 5.1 corresponds to its height S. If it has a width w_b of 20 sec and its area A represents 570 pg of component, then S would equal 2 times 570 pg divided by 20 sec or 57 pg/sec. If the S/N ratio of the peak is 18, the mass flowrate corresponding to an S/N ratio of 2 would be 6.3 pg/sec. This is the minimum detectable level.

Variations of this definition may require other S/N ratios. This parameter may also be defined in terms of *root-mean-square* (rms) noise. Rms noise is the standard deviation of the detector response in the absence of sample. Noise N can be taken as six times rms noise.

There are some detectors (e.g., thermal conductivity) that respond to the *concentration* in the detector, rather than the quantity. With these detectors the MDL is dependent on the column flow. Consider the case of such a detector used with a $\frac{1}{4}$-in. column with a carrier flowrate of 1 mL/sec. If the MDL were measured to be 16 ng/sec, it could be made better by a factor of 2 (8 ng/sec) by going to a $\frac{1}{8}$-in. column with a flowrate of 0.5 mL/sec. The same quantity of sample is going through the detector as before, but there is less carrier gas diluting it, so the detector performs with higher sensitivity.

To make the MDL independent of column flow, it is useful to let it be the *concentration* of the sample in the detector with units of g/mL. In the preceding example, the MDL is 16 ng/mL for both flowrates.

Often the minimum detectable level is expressed in terms of number of molecules, with units of mol/sec or mol/mL. For detectors that respond selectively to certain elements, the MDL is expressed in g/sec or g/mL, but the weight refers to the weight of the atoms that the detector monitors. Thus a nitrogen detector that gives an S/N ratio of 2 for 10 pg/sec of azobenzene has an MDL of 10 pg/sec for the compound and 1.5 pg/sec for nitrogen, since the compound is 15% nitrogen by weight. Elemental MDLs use the notation $g(X)$/sec or $g(X)$/mL, where X is the element being detected, to distinguish them from MDLs of compounds. The MDLs for some common detectors are given in Table 5.1.

TABLE 5.1 Minimum Detectable Levels for Some Common Detectors

Detector	MDL	Sample Compound
Thermal conductivity (TCD)	5×10^{-10} g/mL	Propane
Flame ionization (FID)	10^{-12} g(C)/s	Propane
Electron capture (ECD)	10^{-16} mol/mL	Lindane
Flame photometric (FPD)	10^{-10} g(S)/s	Thiophene
	2×10^{-12} g(P)/s	Tributylphosphate
Alkali flame (AFID)	5×10^{-14} g(N)/s	Azobenzene
	5×10^{-15} g(P)/s	Tributylphosphate

5.2.3 Sensitivity

Sensitivity S is defined as the change in detector response with the change in the amount or concentration of the analyte. It is the slope of the calibration graph, a plot of detector response vs. analyte concentration or analyte quantity.

5.2.4 Response Factor

A response factor is a ratio of *signal-to-sample* size used to characterize a detector. It is independent of carrier flow F and its units reflect the way the detector works, i.e., sensitive to mass or concentration. Depending on whether an integrator is used to determine areas A or whether peak height S and peak width w_b are measured, there are two ways to define response factor. They are equivalent to the extent that $Sw_b/2 = A$. Table 5.2 gives the equations for response factor in terms of mass M of compound injected.

Care must be taken to express peak heights and areas in exact terms. A peak height described as a range setting, an attenuation, and a percentage of full-scale recorder response is ambiguous. To eliminate confusion caused by differences in instrumentation, it is best to describe the peak height in terms of detector output such as nanoamperes.

5.2.5 Selectivity

The choice of one gas chromatographic detector over another for an analysis can depend on a particular characteristic of the detector. In petroleum

TABLE 5.2 Equation for Response Factor in Terms of Weight M of Compound Injected, Peak Width w_b, and Flow F

	Based on Peak Height h	Area A
Mass-Flow-sensitive detectors	$hw_b/2M$	A/M
Concentration-sensitive detectors	$hw_b\bar{F}_c/2M$	$A\bar{F}_c/M$

analysis where detector sensitivity is important, the flame ionization detector is used, because it can measure all the compounds of interest at very low levels. With the analysis of natural gas, where several of the important components, such as N_2 and CO, have little or no response in the FID, sensitivity must be sacrificed for universality. For this analysis the TCD, which is capable of detecting all the components of the natural gas sample, becomes the detector of choice.

When the compounds of interest cannot be resolved from the background, it becomes practical to use a selective detector for the analysis. A *selective* detector responds to a limited number of compounds. Ideally, the detector would have little or no response to the interfering material and very good response to the compounds of interest. Consider chromatogram A in Figure 5.3 (2). This is a mixture of 10 pesticide standards added to milk extract run on an FID. This detector responds to every organic component in the extract, some of the least of which are the pesticides of interest. The FID cannot be used for this analysis without a considerably increased amount of sample treatment. Chromatograms B, C, and D in Figure 5.3

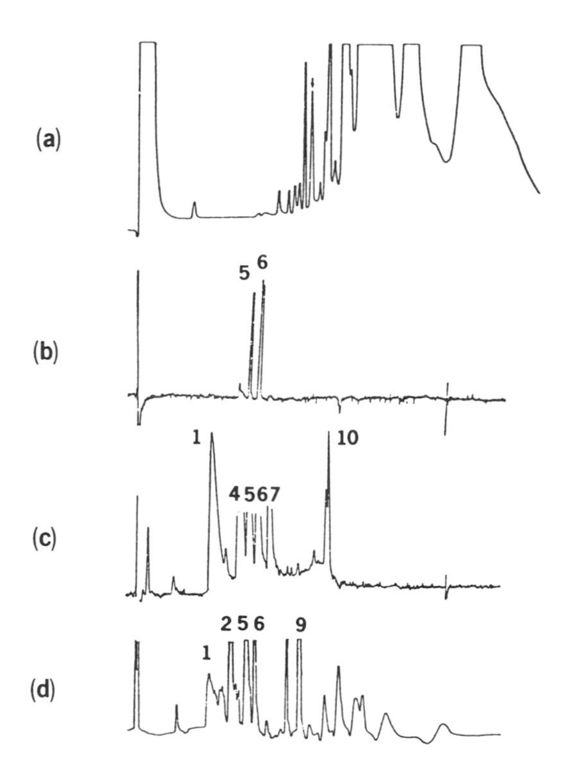

FIGURE 5.3 Chromatograms of a mixture of 10 pesticide standards, extracted from milk, and run on various detectors (Reference 2): (a) flame ionization detector; (b) flame photometric detector, sulfur mode; (c) flame photometric detector; (d) electron-capture detector.

show how other selective detectors allow the pesticides to be analyzed even when they cannot be resolved from the background.

The selectivity of a detector for one compound over another can be measured by the ratio of their sensitivities. Since most applications of specific detectors are similar to that shown in Figure 5.3, that is, small components must be found in a great mass of interfering material, a selectivity less than three orders of magnitude is of marginal value.

5.2.6 Linear and Dynamic Ranges

The *linear range* of a detector is the range of levels of a substance at the detector over which the response of the detector is constant within a specified variation, usually plus or minus 5%. The lower limit of linearity is the MDL of the substance determined separately. The magnitude of this range can be greatly influenced by the choice of test substance. Therefore, the test substance must be specified, along with the variation and the MDL, when presenting the range. Chromatographers responsible for specifying the accuracy of their analyses should keep in mind that the linear range specified by the instrument manufacturer may have been determined under optimal, rather than practical, instrumental conditions and may be much larger than the range attainable under the conditions of their analysis.

Figure 5.4 shows a convenient plot to demonstrate linearity (3). A perfectly linear system (including the column) gives a straight, horizontal line. The arrows indicate the limits of the linear range.

The dynamic range of a detector is greater than the linear range. The *dynamic range* is defined as that range of levels of a substance over which an incremental change in the level produces an incremental change in detector

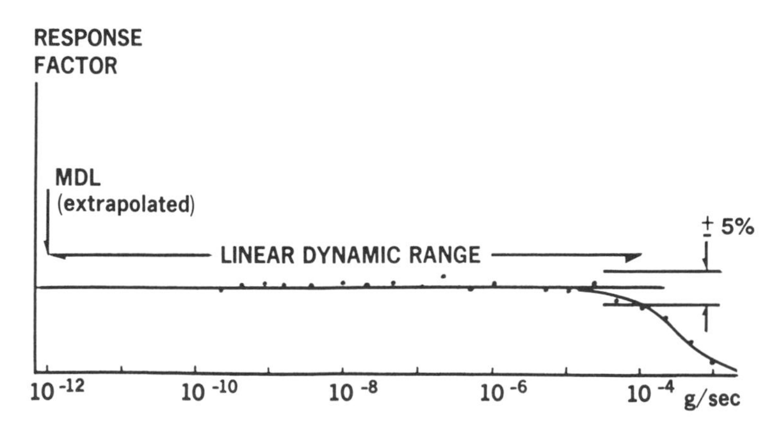

FIGURE 5.4 Method of plotting response factors to demonstrate linearity. Detector is FID; compound is methane (Reference 3).

signal. The MDL of the substance is also the lower limit of the dynamic range. Above the upper limit of the dynamic range, an increase in the amount of the substance gives no observable increase in detector signal.

5.2.7 Quenching and Enhancement

When specific detectors are used to quantify a certain class of compounds in the presence of a background of many nonspecific compounds, it is important that the presence of the nonspecific background does not change the response factors for the compounds of interest. The flame photometric detector, which is sensitive to sulfur and phosphorus, shows diminished response to sulfur compounds if large amounts of hydrocarbons are eluted simultaneously. This occurs even though there is no significant response on the chromatogram from the interfering hydrocarbons. The opposite effect can also occur. In some of the newer types of electron capture detector, low-level pesticide responses are improved by very small increases in the amount of column bleed. This enhancement effect is especially surprising since in the older types of electron capture detector (constant frequency type), the presence of column bleed quenched the pesticide response.

To prevent errors due to quenching and enhancement, it is important to run a standard, a blank, and a spike in addition to the samples. The *standard* is a mixture of the compounds to be determined in known concentrations. The *blank* is the matrix of the samples without the analytes. The *spike* is a blank or a previously analyzed sample to which a known amount of standard is added. The standard is used for calibration; the blank checks that the detector selectivity is adequate; and the spike checks whether the response factor is the same in the presence of background as it is in the standard mixture. This type of assay validation assures meaningful results and is a good practice even when quenching or enhancement are not suspected.

5.2.8 Predictability of Response Factors

There is no widely used detector that has predictable *absolute* response factors. Theoretical or predicted response factors are used only when calibration is not practical, such as with unknowns. The thermal conductivity detector and the flame ionization detector are usually within a factor of 2 of the same response for an unknown *organic* compound as for other peaks in the run. This allows estimates to be made for the amount of unknown peaks. It also ensures that there are no major volatile impurities if a chromatogram does not show them. This is how manufacturers of chemical products specify the level of organic volatile impurities without identifying them in each case.

5.2.9 Practical Operating Hints

The following guidelines apply to all detectors. Refer to sections on specific detectors for information pertinent to particular detectors.

1. To avoid temperature fluctuations in the detector (and in the oven) situate the gas chromatograph away from drafts and heating and air conditioning vents. Locations near poorly insulated exterior walls or in direct sunlight should be avoided.
2. Greater thermal stability is achieved when the instrument covers are left on.
3. Oxygen scrubbers, moisture traps, and the like are worthwhile insurance against a contaminated bottle of gas. However, they must be replaced when exhausted or else they can *add* impurities to the carrier gas instead of removing them.
4. Always check for leaks in the gas lines. While carrier gas is flowing out, air is diffusing *in*.
5. Never condition a column while it is connected to the detector.
6. Check the detector exit tubes for condensation and plugging after running high-concentration or "dirty" samples and clean whenever necessary.
7. Bring the detector up to operating temperature *before* turning on the oven when starting up the gas chromatograph. This will prevent condensation in the detector, since the oven heats up more rapidly than the detector.
8. For a stable baseline in the morning, leave the detector at operating temperature overnight.

5.3 FLAME IONIZATION DETECTOR

5.3.1 Background

The flame ionization detector (FID) was introduced in 1958 by McWilliam and Dewar (4) in Australia and by Harley et al. (5) in South Africa. It has become the most commonly used detector in GC for several reasons:

1. It responds with high sensitivity to virtually all organic compounds.
2. Modest changes in flow, pressure, or temperature have a minimal effect on its response characteristics.
3. In normal operating mode it does not respond to common carrier gas impurities such as water or carbon dioxide.
4. When properly installed, it has a stable baseline.
5. It has a wide linear range, about 10^8.

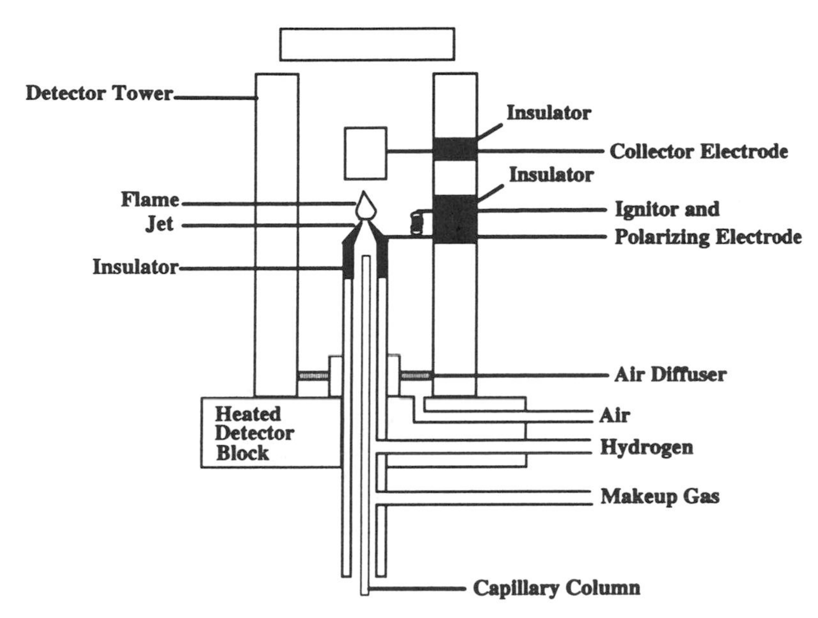

FIGURE 5.5 Schematic diagram of a flame ionization detector.

6. It is easy to adjust.

The detector consists of a small hydrogen–air diffusion flame burning at the end of a jet (Figure 5.5). When organic compounds are introduced into the flame from the column effluent, electrically charged species are formed. These are collected at an electrode and produce an increase in current proportional to the amount of carbon in the flame. The resulting current is amplified by an electrometer. Although this detector has been widely used for over thirty years, the mechanism of its response is still not completely understood and is still under investigation.

5.3.2 Flame Chemistry

The reaction that produces ionization in a flame detector was confirmed in the early 1960s. Most of this work was done by combustion researchers who obtained the direct evidence from mass spectroscopic examination of the interior of flames. Summaries of flame ionization processes have been published by Miller (6) and also Bocek and Janak (7).

The mechanisms of the diffusion flame of the flame ionization detector begin at the tip of the jet (Figure 5.6). The mixture of carrier gas, hydrogen, and makeup gas (if used) flows from the restriction of the jet and expands outward. The air flows at a lower velocity around the outside of the jet gases. As the jet flow approaches the reaction zone (dotted line in Figure

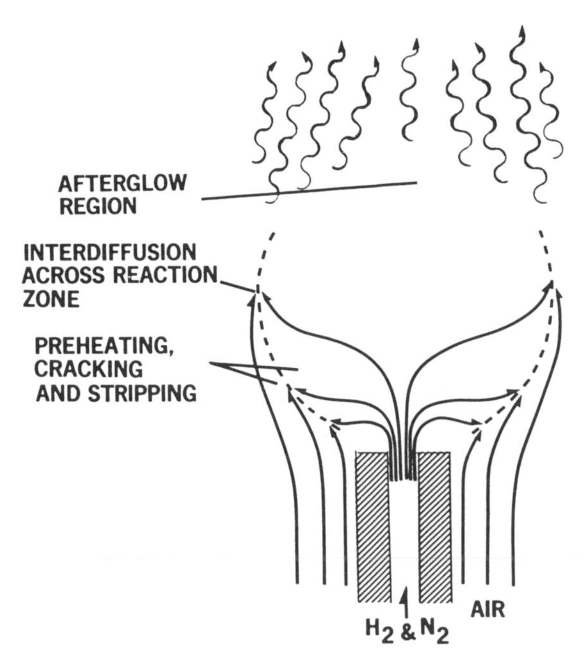

FIGURE 5.6 Schematic diagram of the basic flame processes in the FID.

5.6), it is preheated by back diffusion of the heat energy produced there. Organic materials eluting from the column undergo degradation reactions in this hydrogen-rich region of the flame and form a group of single carbon species. When the two gas flows mix together at the reaction zone and oxygen becomes available, the following chemi-ionization reaction occurs:

$$CH + O \rightarrow CHO^+ + e \quad (5.1)$$

The unstable CHO^+ ions react rapidly with the water produced in the flame to generate hydroxonium ions according to the reaction

$$CHO^+ + H_2O \rightarrow H_3O^+ + CO \quad (5.2)$$

These ions and their polymerized form $(H_2O)_nH^+$ are the primary positive charge carrying species.

This extremely important reaction occurs once for every 100,000 carbon atoms introduced in the flame. It is essentially a quantitative counter of carbon atoms being burned. Therefore, the response of the flame ionization detector is proportional to the number of carbon atoms, rather than the compound weight or moles.

The process by which the organic compounds are broken down into single carbon species is less well understood and several schemes have been proposed (8–11). There must be a good balance among all the flows and diffusions for the flame processes to be efficient. Flame turbulence, whether caused by too high a flow or the wrong diameter jet, must be avoided for proper operation of the detector.

5.3.3 FID Design

Jet. The type of column and the analysis to be performed dictate the size of jet to be used. For packed columns, a jet with a 0.018-in. internal diameter is standard. Capillary columns require maximum detector sensitivity. A jet with an internal diameter of 0.011 in. is used for capillary work, since it produces almost double the sensitivity of the 0.018-in.-i.d. jet (12). If this jet is used with packed columns, however, flame-out and clogging problems may occur. Another packed-column jet with a 0.030-in.-i.d. is used with 0.53-mm capillary columns so that the column can be inserted into the jet to within a few millimeters of the flame. This has been shown to improve peak shape by preventing the loss of peak integrity due to adsorption or catalytic decomposition which can occur when some eluted materials contact hot metal surfaces (13). If the fused silica column is allowed to extend beyond the tip of the jet, the stationary phase will pyrolyze and the polyimide coating of the column will decompose, resulting in noise or spiking.

Since the diffusion of the gases at the tip of the jet is so central, the detailed shape of the jet affects detector operation. The tip should be smooth. It should be kept free from solid support particles from the column, pyrolysis products of samples, SiO_2 deposits from the liquid phase, and other contaminents.

Gas Flowrates. The maximum sensitivity of the flame ionization detector is achieved at a particular ratio of carrier gas (or carrier plus makeup gas) to hydrogen (14). The maximum and the ratio depend on the particular gas. Detector response is much greater with a nitrogen–hydrogen flame than with a helium–hydrogen flame. Therefore, sensitivity can be dramatically increased with capillary columns by using nitrogen rather than helium as the makeup gas. Since variations in detector design may require different gas flowrates, the instrument manual is the best reference for setting these flows. In the absence of this information, 30 mL/min of both hydrogen and carrier gas is a good starting point.

Because not all the air enters the reactive part of the flame in the diffusion flame, it is necessary to use several times the stoichiometric amount of oxygen. Air flows of 300 to 500 mL/min are typical for most detectors. It is important to have a uniform and laminar flow along all sides of the jet. For this reason, many instruments introduce the air through a porous diffuser located well below the tip of the jet. As with the hydrogen and carrier flows, the instrument manual should be consulted when setting the air flow.

Exhaust Flow. The exhaust flow system is usually ignored until problems occur. It removes the gases, heat, soot, and silica from the detector. Soot is formed from the incomplete combustion of chlorinated solvents such as chloroform and methylene chloride in the flame. It can build up in the tower and cause spikes as particles fall back into the flame. Silica, the product of the combustion of silicone-coated column bleed, condenses on the interior of the detector as a fine, white powder. This gives an insulating coating on which an electric charge can build up, changing the electric fields in the detector. Water can also condense if the detector is too cool, causing electrical shorts. Most instruments have some mechanism for isolating the flame from air currents in the vicinity of the instrument. Ambient air should be prevented from backdiffusing into the flame chamber to avoid the detection of solvent vapors or cigarette smoke in the laboratory air.

Thermal Control. There are several considerations in setting the temperature of the flame detector. The block should be heated above column temperature to prevent sample components from condensing in the carrier-gas transfer line. It should also be kept above 100°C to prevent the condensation of water. The temperature of the body of the detector should be kept stable, because it has a slight effect on the detection mechanism. It is more important not to let the detector get too hot. If parts of the detector are heated close to the threshold for significant thermionic ionization, a large sample can increase the temperature and cause a positive detector response, even with samples that normally have negligible response. Under these conditions, water, which reduces the flame temperature, might give a negative peak. Another reason not to overheat the detector is to prevent electrical leakage across the insulators. This can cause instability in the detector current.

Ion Collection. The geometry of the collector electrode and how it affects the distribution of the electrical field and the flow pattern of the gas stream has been extensively investigated. Most collectors are cylindrical and symmetrical and have a relatively large surface area for ion collection. The collector is placed near the top of the flame so that ion collection takes place near the base where the electric field is greatest. When very large samples are burned, the flame can increase drastically in size and change from a diffusion flame to a hydrocarbon-like flame. Ion generation now takes place higher in the collector where the electric field is weaker or completely above it. This results in nonlinearity of response and detector saturation.

5.3.4 Response Factors

The flame ionization detector is mass flow sensitive. Its area response for a compound does not change with small changes in carrier flow like those in temperature-programmed operation. The units for its response factors are coulombs/gram of carbon. It has a high sensitivity to organic carbon containing compounds, most with a response of 0.015 coulombs/g C. The detector is noted for its 10^7 linear range (Figure 5.4). However, its response is adversely affected by the presence of heteroatoms such as O, S, and halogens.

The concept of an effective carbon number, ECN, was introduced to estimate the relative response for any compound. Particular groups of atoms are given a value relative to a reference material, usually an *n*-paraffin, for which the ECN is simply its carbon number. The set of parameters used to calculate effective carbon numbers is given in Table 5.3 (8, 15, 16). The response for a compound is the total of all its atoms and groups. For example, *n*-propanol would have an ECN of 2.4, having three aliphatic carbons (3×1) and one primary alcohol oxygen (-0.6). Likewise, butyric acid would have an ECN of 3.0, having three aliphatic carbons (3×1) and one carboxyl group (0). Several workers have tabulated large numbers of experimental relative response factors (17–20).

One obvious use of the ECN is in determining the relative response factors of compounds that cannot be secured in sufficient purity to de-

TABLE 5.3 Contributions to Effective Carbon Number

Atom	Type	Effective Carbon No. Contribution
C	Aliphatic	1.0
C	Aromatic	1.0
C	Olefinic	0.95
C	Acetylenic	1.30
C	Carbonyl	0.0
C	Nitrile	0.3
O	Ether	−1.0
O	Primary alcohol	−0.6
O	Secondary alcohol	−0.75
O	Tertiary alcohol, esters	−0.25
Cl	Two or more on single aliphatic C	−0.12 each
Cl	On olefinic C	+0.05
N	In amines	Similar to O in corresponding alcohols

termine them experimentally. Discrepancies between experimental and theoretical response factors are a good indication of adsorption or decomposition of analytes in the chromatographic system, especially in the absence of the usual signs of poor peak shape and tailing. Response factor determination, however, can be seriously impaired by improperly set flow parameters, water in the air supply to the detector, and high column bleed. Table 5.4 lists substances that have little or no response in the flame ionization detector.

5.3.5 Modifications of the FID

The FID can be modified to respond to inorganic gases by operating it in a hydrogen-rich mode with oxygen to support combustion. This can be accomplished with most commercial FIDs by introducing the oxygen with the carrier gas and using the normal air inlet to introduce the hydrogen fuel directly into the detector. When modified like this the FID is known as a hydrogen-atmosphere flame-ionization detector (HAFID). Detection limits for some gases are as follows (21):

CH_4	2×10^{-11} g/sec
CO_2	4×10^{-8} g/sec
H_2S	4×10^{-10} g/sec
NO	2×10^{-11} g/sec
O_2	5×10^{-8} g/sec
SO_2	4×10^{-10} g/sec
N_2O	7×10^{-9} g/sec
NO_2	2×10^{-9} g/sec
CO	4×10^{-7} g/sec
He	5×10^{-8} g/sec

Doping the hydrogen atmosphere with hydrides has enabled enhanced detection of organometallic compounds (22) containing iron, tin, lead, molybdenum, and tungsten. Phosphine was more effective than methane, silane, or germane. Silicon-containing compounds have also been found to give enhanced response in HAFID mode, which on doping the flame with ferrocene, produced a silicon-to-carbon selectivity of 10^4 (23).

TABLE 5.4 Substances With Little or No Response in the FID

He	N_2	H_2S	NO	CCl_1
Ar	O_2	CS_2	N_2O	$SiCl_4$
Kr	CO	COS	NO_2	CH_3SiCl_3
Ne	CO_2	SO_2	N_2O_3	SiF_4
Xe	H_2O	HCN	NH_3	$SiHCl_3$

5.3.6 Practical Operating Hints

1. One of the major sources of problems with the FID is an incorrect adjustment of the flame. Both the size of the flame and the ratio of its gases are important. When changing the jet, carrier-gas flow, or sample size, always consider the effect this will have on the flame and make adjustments, if necessary.
2. The second source of problems with the FID is contamination. This can be due to impure gases, water shorting out the detector, or soot or silica deposits in the chimney.
3. For maximum sensitivity with capillary columns, use the proper size jet and use nitrogen as the makeup gas.

5.4 THERMAL CONDUCTIVITY DETECTOR

5.4.1 Background

The thermal conductivity detector (TCD) is one of the most commonly used detectors in gas chromatography. It measures changes in the thermal conductivity of the carrier gas caused by the presence of eluted substances. Chromatographers began using thermal conductivity because devices for measuring this property had been in use since the 1880s. The major uses of these devices included gas analysis and process control. So it was natural to use thermal conductivity in early experiments with gas–solid chromatography.

The TCD was technically mature by the late 1950s. There was always a general dissatisfaction with the problem of sensitivity despite many attempts to improve it. The more sensitive ionization detectors were used in its place for trace analysis and with capillary columns. The advantage of the TCD lies in the detection of gases such as CS_2, COS, H_2S, SO_2, CO, NO, NO_2, and CO_2 in gas–solid chromatographic analyses on packed columns. The introduction of the porous layer open tubular (PLOT) columns, however, brought gas–solid chromatography to capillary columns and established a market for the microflow cell TCD.

5.4.2 Principles of Operation

The property of a material transmitting heat when subjected to a temperature difference is known as thermal conductivity λ. It can be defined in terms of the heat flow Q through a unit thickness of material x of a unit cross section A when the temperatures on each side differ by unity. From Figure 5.7, the heat flow is

$$Q = \frac{A(T_1 - T_2)\lambda}{x} \tag{5.3}$$

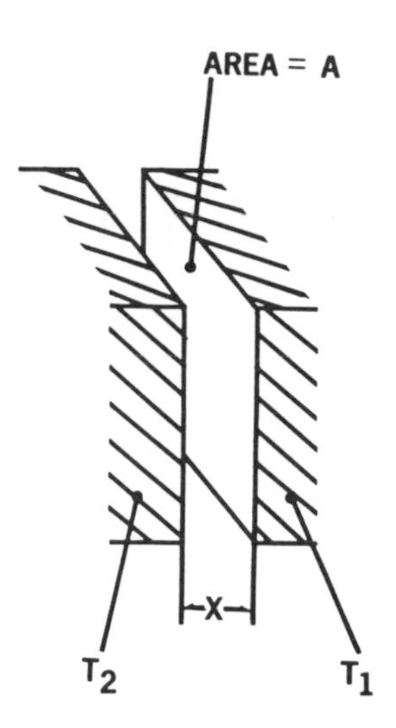

FIGURE 5.7 Thermal conductivity between two planar surfaces of temperatures T_1 and T_2 and cross-sectional area A, separated by a distance x.

If thermal conductivity is measured in a cylindrical geometry, it is much easier to avoid heat losses other than conduction through the gas. This is accomplished by replacing one planar surface with a cylindrical cell of temperature T_c and the other planar surface with a wire filament running through the center of the cell with a temperature of T_f. Replacing the dimensions of the planar surfaces A/x with a geometry factor G in the equation above gives the equation for heat transfer in a cylindrical geometry:

$$Q = G(T_f - T_c)\lambda \tag{5.4}$$

The thermal conductivity can be determined by supplying heat at a known rate Q and measuring the temperatures of the center conductor wire and the outside cell wall. The center conductor can be a wire of known electrical resistance, R_f ohms. The heat flow can then be supplied by an electric current, I amperes:

$$Q = \frac{I^2 R_f}{J} \tag{5.5}$$

where J is Joule's constant (4.183 W/cal-sec). If a wire whose resistance R_f is temperature dependent is used, the temperature of the wire can be measured:

$$R_f = R_f^\circ (1 + \alpha T_f) \tag{5.6}$$

where α is the temperature coefficient of resistance. The wire now doubles as a heating element and as a resistance thermometer. Use the voltage across the wire V to calculate the resistance (at temperature) V/I and substitute in the above equation to find the temperature T_f.

The thermal conductivity can be calculated from the temperatures, heat flow, and geometry. However, in the thermal conductivity detector for GC, the absolute value of λ for the column effluent is not important. The small change in λ when eluents are present in the carrier gas is important because it changes T_f, which can be detected by the change in the resistance:

$$\Delta R_f = \alpha R_f^{\circ} \, \Delta T_f \tag{5.7}$$

and

$$\Delta T_f = \frac{T_f - T_c}{\lambda} \Delta\lambda \tag{5.8}$$

These equations can be combined to give

$$\Delta R_f = \frac{-\alpha R_f^{\circ}(T_f - T_c) \, \Delta\lambda}{\lambda} \tag{5.9}$$

Note that the sensitivity for changes in λ is proportional to the temperature difference across the cell. However, there is a limit to the temperature that the filament can be run. For most commercial TC detectors that use a hot wire, a practical upper limit is 450°C for continuous operation and 500°C for short periods of time. This is a result of oxidation of the filament with trace amounts of oxygen in the carrier gas. A variety of other reactions can occur with the sample. At the maximum filament temperature, sensitivity can be increased by lowering the detector temperature T_c. In practice, the best sensitivity can be obtained by setting the detector block temperature as low as possible without condensing eluted substances in the cell and setting the filament temperature as high as possible without burning it out.

Besides the thermal conductivity of the gas, there are four other processes that contribute to the loss of heat from the filament:

1. Thermal radiation
2. Thermal conductivity of leads and connections (end losses)
3. Forced convection of the gas
4. Free convection and diffusion of the gas

Thermal radiation is dependent on the surface area of the element, on its temperature, and on its material quality. Using the type of detector shown in Figure 5.8 operating under the conditions of Table 5.5, the power loss through thermal radiation can be estimated as 15 mW. At the highest filament temperature, radiation can transfer a few percent of the power.

Although the temperature of the filament is constant over its length, the ends must drop to the cell body temperature. This heat loss is proportional to the difference in temperature between the filament and the body of the cell. For the case cited, this amounts to 45 mW. For the small-diameter wire

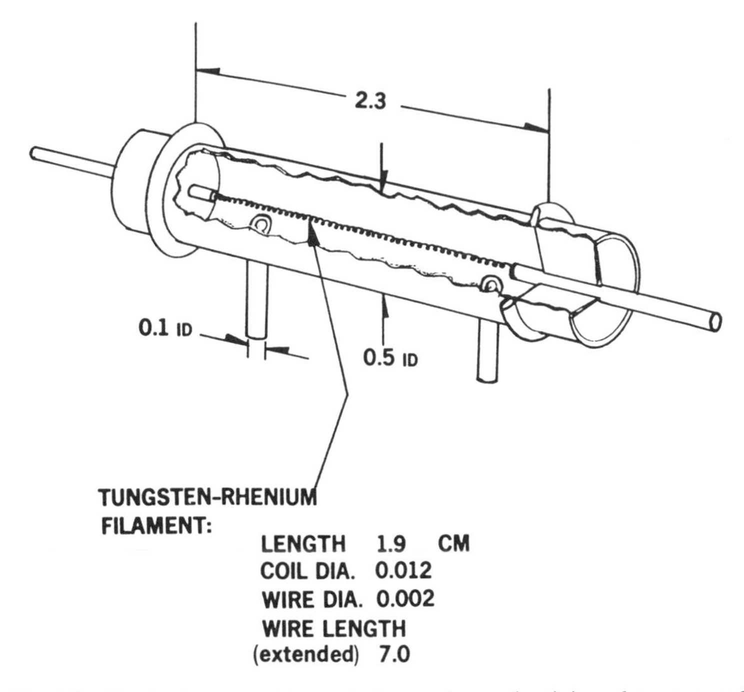

FIGURE 5.8 Typical geometries of thermal conductivity detector cells. The feedthroughs supporting the axial filament are insulated from the stainless-steel body by alumina ceramic. All dimensions are in centimeters.

TABLE 5.5 Typical Operating Conditions for Thermal Conductivity Cells

Temperature	
Body of detector	150°C
Filaments	350°C
Filaments	
Material	Tungsten–rhenium
Temperature coefficient, α	0.0033/°C
Resistance, R_0 (at 0°C)	25 Ω
Resistance (at 350°C)	55 Ω
Electrical (for four-element bridge)	
Current	0.3 A
Voltage	16.5 V
Power (for each filament)	4.95 W
Current	0.15 A
Voltage	8.25 V
Power	1.24 W
Carrier gas	Helium
Flowrate	1.0 mL/sec
Thermal conductivity, λ (at 150°C)	4.4×10^{-4} cal/sec-cm-°C

and high-conductivity carrier gases used today, this heat loss mechanism can usually be neglected.

The heat loss due to the forced convection of the gas is a product of the mass flowrate of the gas, the specific heat of the gas, and the difference in temperature between the gas flowing out of the cell and the detector temperature T_c. For the geometry in the example, the gas temperature exceeds the block temperature by only 2.4% of the 200°C difference across the cell. Assuming uniform axial flow, this gives an approximate heat loss of 7 mW.

The larger the detector cell, the more important the free-convection heat transfer becomes. In modern detector cells this effect is negligible.

Table 5.6 summarizes all the heat-transfer terms considered. All the effects, except for the thermal conductivity of the gas, contribute only a few percent to the heat transfer. However, if nitrogen were used as the carrier gas, the TC term would be six times lower, and the other terms would be comparably more important. The mass flow term is the cause of the flow dependence of the TC cell and thus contributes to the noise and drift of the detector.

5.4.3 TCD Design

Most general-purpose TC detector cells are constructed similar to the one in Figure 5.8. One common variation of this design is to support the filament on posts, both of which are mounted on the same face of the cavity. To obtain higher resistances, coiled filaments are used instead of straight wire. Platinum, platinum alloys, tungsten, tungsten alloys, and nickel can be used for the filament. The choice of filament material depends on mechanical strength and chemical inertness since the resistivity and α of these materials are similar.

The thermal conductivity cell is extremely sensitive to fluctuations of physical variables. Some of these can be canceled out if two cells are used (Figure 5.9), one to detect the sample and the other to serve as a reference. A bridge circuit is used to balance the resistance R_3 of the sample cell

TABLE 5.6 Summary of Heat Transfer Effects With Thermal Conductivity Detectors

Effect	Heat Transfer (mW)
Thermal conductivity ($G = 3.08$)	1130
Radiation	15
End losses	45
Mass flow	7
Free convection	Negligible
Total (measured)	1240

Note. Cell of Figure 5.8 and Table 5.5.

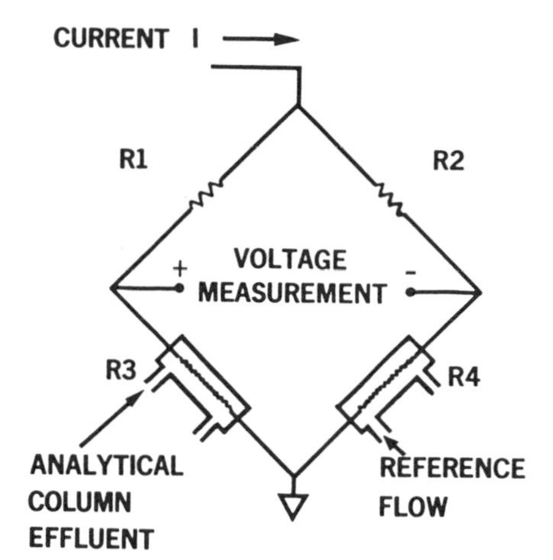

FIGURE 5.9 Bridge circuit used in a two-cell detector. The reference cell R_4 compensates for the drift in the analytical cell R_3 due to flow temperature fluctuations.

against the resistance R_4 of the reference cell. Two conventional resistors, R_1 and R_2, comprise the remainder of the bridge. When no substance is being eluted, all the resistors have the same value and there is no voltage difference to be measured. However, when a sample elutes from the column, the resistance R_3 increases, and a voltage difference is measured. If a change in room temperature affects the resistance of the analytical cell, it will also affect the resistance of the reference cell by nearly the same amount. No significant change in voltage would be measured. This configuration can compensate for column bleed by setting the reference flow equal to the analytical flow and passing the reference flow through an identical column. This is particularly useful with temperature programming.

One serious problem that the designers of this type of detector had to solve was the effect of temperature on the conventional resistors. Either resistors with very low temperature coefficient could be used, or the resistors could be temperature controlled. Figure 5.10 shows another configuration which replaces the conventional resistors with two more filament cells. The effluent from the analytical column is split into two matched cells of resistance R_3. The effluent from the reference column is sent to two matched reference cells of resistance R_4. In some detectors the matched cells are simply two filaments mounted in the same cell cavity. One of the benefits of this approach is that the response factor is doubled since two cells are contributing to the change in signal.

Two important parts of the TCD electronics are not shown in the figures. Variable resistors must be connected in the bridge to null the output voltage before a chromatogram is run. This is necessary because the four cells of the detector rarely match. These resistors are controlled by the "fine" and "coarse" balance adjustments on the front panel of the detector. If a recorder is used instead of an integrator, the attenuator switch must be used to reduce the response to keep the larger peaks on scale.

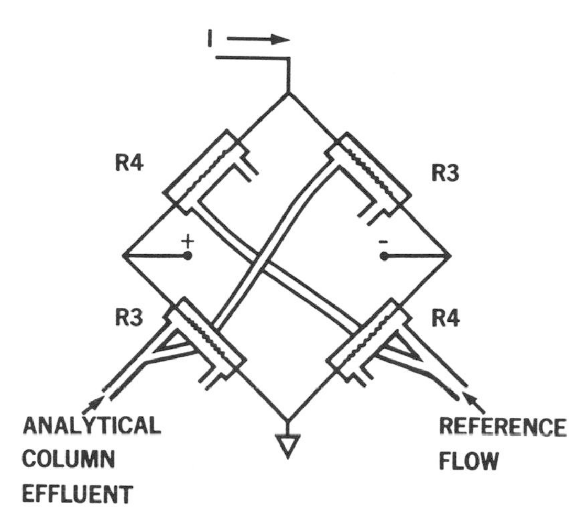

FIGURE 5.10 Bridge circuit used in a four-cell detector. This approach gives twice the response of that shown in Figure 5.9.

The electrical requirements for the TCD are simpler than those for almost any other gas chromatographic detector. The mechanical requirements, particularly thermal control, are, by contrast, unusually demanding. The cells of the TCD are mounted closely together and embedded in metal, with the entire assembly meticulously insulated. It is so important to control the temperature of the detector body that this circuit can often provide better thermal stability than that of the chromatographic oven. The oven should not influence the detector temperature. The detector insulation should prevent heat transfer by thermal conduction. But the carrier gas should also not transfer heat to the detector. If it does, variations in the gas flow can cause noise or drift in the detector.

Of course, the filaments of the detector must match electrically; but they must also match mechanically because of the effect on the geometry factor. Manufacturers commonly select not only matched sets of filaments, but also cells matched for geometry factor. Another stringent requirement is that the cells be gastight, even at high temperatures. Any diffusion of air into the cell will give drift and noise and burn up the filament. Electrical leakage from the filaments to the detector body can cause noise. High-density ceramics have replaced glass as insulators for the filament mounts to prevent this problem.

Another configuration of the TCD is known as a modulated detector. It has only one filament cell. A valve alternates the flow of the analytical column effluent with the reference column effluent. The flow of gases in this cell is illustrated in Figure 5.11. The flows are switched at a rate of ten times per second, faster than any changes due to thermal fluctuations. This design facilitates ease of startup and reduces baseline drift during temperature

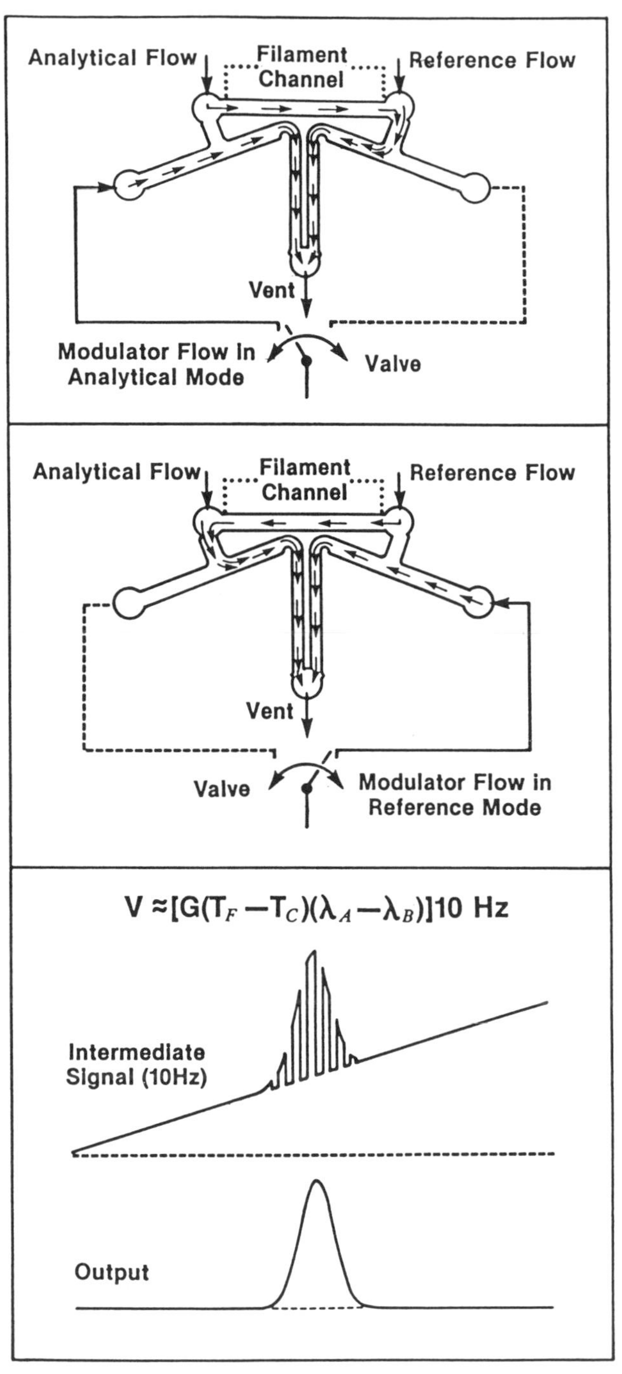

FIGURE 5.11 Diagram of a commercial TCD. Modulated thermal conductivity detector that uses only one filament cell. The analytical and reference gas flows are switched at a rate of 10 times/sec. This design results in a detector with very low drift. This assembly is further insulated on all sides and mounted in a metal enclosure.

programming. The cell volume is on the order of 5 mL, making it suitable for capillary systems.

Another variation in the thermal conductivity detector is the use of thermistors instead of filaments. These metal oxide beads used as temperature-sensitive resistors have been used in detectors since the 1950s. One of the advantages of thermistors is that their temperature coefficient is larger by an order of magnitude, although it is negative. Another advantage is that their resistance R^0 is large, usually thousands of ohms. This makes their response factor much larger than that of a hot wire detector. They are also small—as little as 0.25 mm in diameter—allowing cells to be made with volumes as small as 5×10^{-5} mL. Because the thermistor is a metal oxide glass, it is almost inert to oxidizing conditions. However, thermistors are quite sensitive to reducing conditions. To overcome this problem, thermistors with an inert coating have been developed. The major problem with a thermistor detector is that the sensitivity worsens rapidly as the detector temperature is raised above 50°C. At typical operating temperatures for liquid samples, a hot-wire detector is actually more sensitive. The operating conditions for the thermistor detector are more difficult to set. These detectors are used mostly for capillary columns, which require fast response and low detector volumes, and gaseous samples that can be run at low temperatures.

The thermal conductivity detector is nearly always operated with helium or hydrogen as the carrier gas. Both these gases have a thermal conductivity much higher than virtually everything else. All other carrier gases share several serious problems:

1. Some compounds give positive peaks, while others give negative peaks.
2. Response factors are dependent on temperature in unexpected ways, and linearity is often poor,
3. Minimum detectable levels are usually an order of magnitude lower than with helium or hydrogen.
4. In some cases W-shaped peaks are observed because the thermal conductivity changes with sample concentration.

5.4.4 Response Factors

When a rather large peak enters the detector certain changes take place. Using the detector of Figure 5.8 and the conditions of Table 5.5, these changes are summarized in Table 5.7. The 0.1 mg peak with a width of 10 sec causes a change of voltage on the recorder of 140 mV for the peak height. The response factor is the number of millivolts for the concentration of sample in milligrams per milliliter at the detector. It is sometimes called

TABLE 5.7 Changes in Typical TC Detector (see Table 5.5) Due to Sample

Sample concentration at peak maximum	0.02 mg/mL
Thermal conductivity	4.4–4.14 × 10^{-4} cal/sec-cm-°C
Filament temperature	350–362°C
Filament resistance	55–56 Ω
Voltage across filament	8.25–8.39 V
Recorder response	140 mV
Detector response factor	7000 mV – mL/mg

the *DPS number*, after Dimbat, Porter, and Stross (24), who first proposed it.

One of the best features of thermal conductivity detectors used with helium carrier gas is the ease of quantitative analysis. *Relative* response factors, using sample weight, are independent of the following:

1. Type of detector (filament or thermistor)
2. Cell and sensor temperatures
3. Sample concentration
4. Helium flowrate
5. Detector current

In a series of homologous compounds, the relative response factors change only slightly. The first systematic study of TCD responses in helium was done by Rosie and Grob and is summarized in Reference 25.

A rule of thumb is that all compounds have a weight response close to that of isooctane. The three exceptions to this are that compounds containing heavy metal atoms have unusually low response factors; halogenated compounds tend to have low response factors; and very light compounds, with molecular weights below 35, tend to have high response factors. Of the 171 compounds in Reference 25 (excluding the three types of exceptions) for which there are relative response factors, 88% are within 20% of the response of isooctane and 96% are within 30%. Response factors have also been tabulated for hydrogen (26) and nitrogen (27) carrier gas.

5.4.5 Practical Operating Hints

1. Keep carrier gas flowing whenever the filament current is on. If air is allowed to diffuse into the detector, the filaments will literally burn up. Some instruments are equipped with sensors to turn off the filament current if the carrier flow is interrupted.
2. Turn off the filament current when performing any maintenance that might introduce air into the carrier line, e.g., when changing columns, septa, or gas cylinders.

3. A permanent shift in the baseline following a large peak may be caused by a change in the resistance of the sample filament due to an oxidizing, halogenated, or strongly reducing compound. If this is occurring, reverse the sample and reference sides of the detector periodically to equalize the changes.
4. To extend filament life, turn off the detector current overnight. For the best detector stability in the morning, leave the current on overnight (along with the oven temperature).
5. Set the detector temperature just slightly above the highest column temperature when programming. Higher setpoints sacrifice sensitivity.
6. Use the instrument operating manual for setting the filament current. Higher current settings give better sensitivity; lower current settings prolong filament life.
7. Since response factors are inversely proportional to column flow, set carrier flow as low as is consistent with good column performance and speed of analysis.

5.5 ELECTRON CAPTURE DETECTOR

One of the most popular and powerful gas chromatographic detectors in use today is the electron capture detector (ECD). Based on gas-phase electron capture reactions, this approach to detection can provide response to picogram or even femtogram levels of specific substances in complex matrices. This makes it an ideal detector for environmental and biomedical studies.

5.5.1 Background

The introduction of the electron capture detector in 1960 (28) resulted from a series of developments begun in 1951 with the invention of the β-ray ionization cross-section detector (Section 5.9). In 1958 Lovelock (29) modified this detector to produce the β-ray argon detector (Section 5.8). He substituted argon as the carrier gas and placed a potential of 1000 V across the electrodes. Argon absorbed β-radiation from a radioactive source and formed a metastable species with sufficient energy to ionize most substances. Because anomalies were observed when halogenated compounds were eluted, Lovelock proposed the theory that electronegative species, functionally present in an organic molecule, could capture an electron to form a negatively charged species:

$$\mathrm{CX} + e^- \rightarrow \mathrm{CX}^- + \text{energy} \tag{5.10}$$

These entities would then cause a reduction in the standing or background current. This phenomenon, known as *electron capture*, is observed more readily at lower electrode potentials.

5.5.2 Principles of Operation

Cell Design. The design of the electron capture detector is a simple arrangement of a chamber containing two electrodes with a source of radiation to induce ionization. As expected, the early cells were usually converted from argon ionization detectors. These detectors were of coaxial geometry, where the anode lay along an axis sheathed by the cathode containing the radioactive source. Another design, known as the *plane-parallel electron capture detector*, is illustrated in Figure 5.12. It features a parallel alignment of the anode and the cathode. This design directs the flow of carrier gas in a direction opposite to the motion of the negatively charged species. This is considered to be a more efficient design. The concentric-cylinder ECD has the radioactive foil located in the cathode region and the anode in an isolated region. The pincup detector, illustrated in Figure 5.13, can be considered a modified chamber of the coaxial ECD geometry. All of these design types have been evaluated, but no firm conclusions have been made because of the complex nature of the experimental variables. Figure 5.14 illustrates a special cell designed to promote gas-phase coulometry (30). Coulometry by electron capture occurs when the ratio of electron captured in a second (coulombs) to the number of molecules through the detector in a

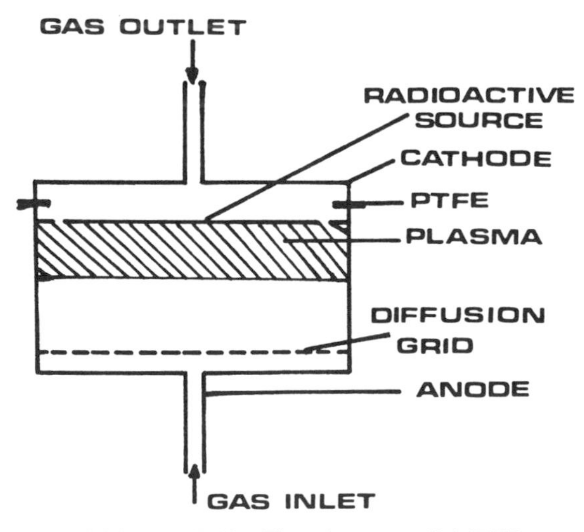

FIGURE 5.12 The plane-parallel ECD.

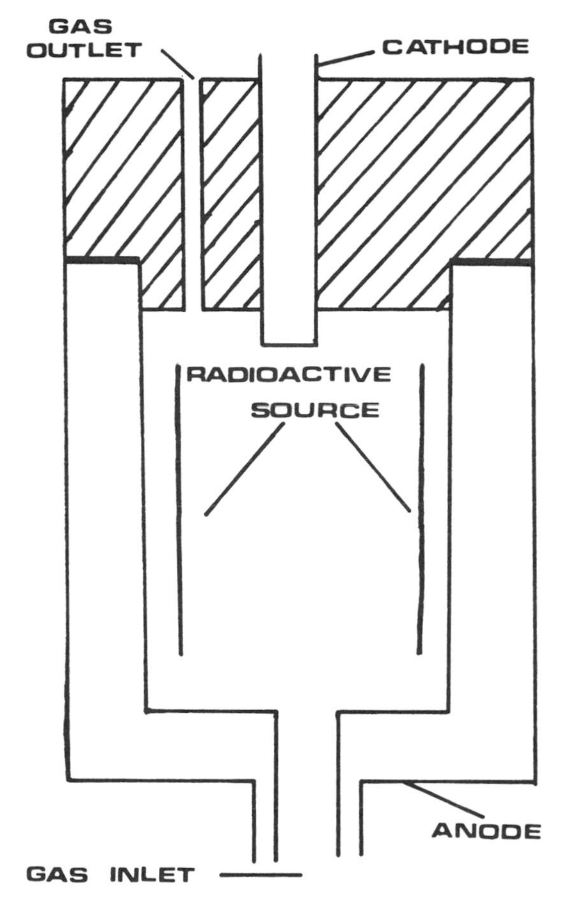

FIGURE 5.13 Pincup ECD.

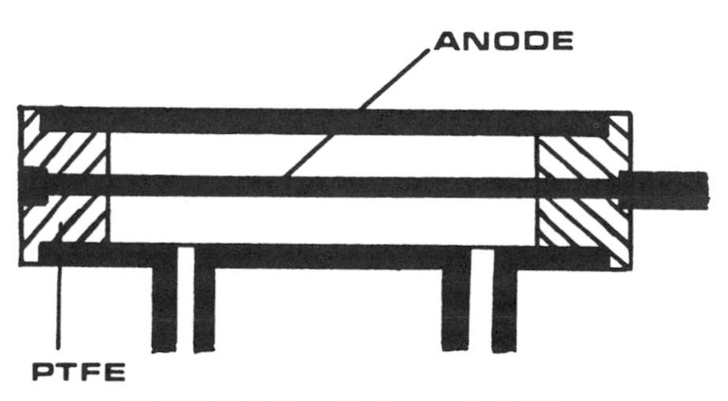

FIGURE 5.14 Coulometric ECD.

second approaches 1. Coulometric detection provides the best sensitivity, but has limited linearity.

Radiation Source. The currently most popular form of the ECD uses ^{63}Ni as the source of radiation. The second most common source is tritium, a weak β emitter. Tritium requires a short cell distance for efficient currents since its β rays have a range of about 2.0 mm. The radiation from ^{63}Ni, however, has a range of 8.0 mm. The maximum operating temperature for tritium is 225°C and for ^{63}Ni the limit is 400°C. The column conditions determine the lower limit. The temperature should be high enough to prevent condensation of high-boiling eluants and stationary-phase bleed. The ionization chamber should be well insulated to keep the detector temperature stable to better than ±0.1°C. This is necessary because the temperature affects the number of electrons emitted from the source, their energies, and the electron capture mechanism. One advantage of tritium is that the flux of radiation is higher, so ionization is more efficient. Although it is more sensitive, the tritium is easily contaminated by adsorbed components on the foil that shields the weak β-rays (18 keV). At high temperatures tritium emanates with the effluent from the detector at a level that constitutes a health hazard according to the Nuclear Regulatory Commission (NRC). Therefore, the effluent should be vented into a fume hood. These problems and inconveniences are minimized with the ^{63}Ni foil. If this source is accidentally overheated, loss of activity occurs by diffusion of the ^{63}Ni into the foil. No radioactive material escapes from the detector. This detector can generally be cleaned by periodically applying high temperature. This source can also be purchased as a "sealed source" for which the vendor holds the necessary radiological license. The user is required to perform only relatively simple radiological tests.

A nonradioactive electron-capture detector has also been described (Figure 5.15) that maintains good detection limits (31). This detector utilizes a thermionic cathode made of platinum coated with barium zirconate and protected by a guard gas to prevent oxidation and solvent overload of the source. Only a few watts of power are necessary to heat the filament to temperatures where electrons are emitted (like a light bulb). The anode, a mesh barrier separator for the carrier gas, sheathes the filament and the protective guard gas. Both gases are maintained at equivalent, nonturbulent flows. The collector electrode confines the carrier gas. This detector works equally well in the direct current (dc) mode and in the pulse mode with a linear dynamic range of six orders of magnitude. The advantages of this rugged, oval chamber detector are that it has a small volume (50 μL), has no temperature limit, is easy to clean, and does not need NRC licensing.

Flow Requirements. The ECD requires a gas that can efficiently attenuate the beta radiation, creating a population of positive ions and electrons in the detector, and rapidly thermalize these secondary electrons. The probability

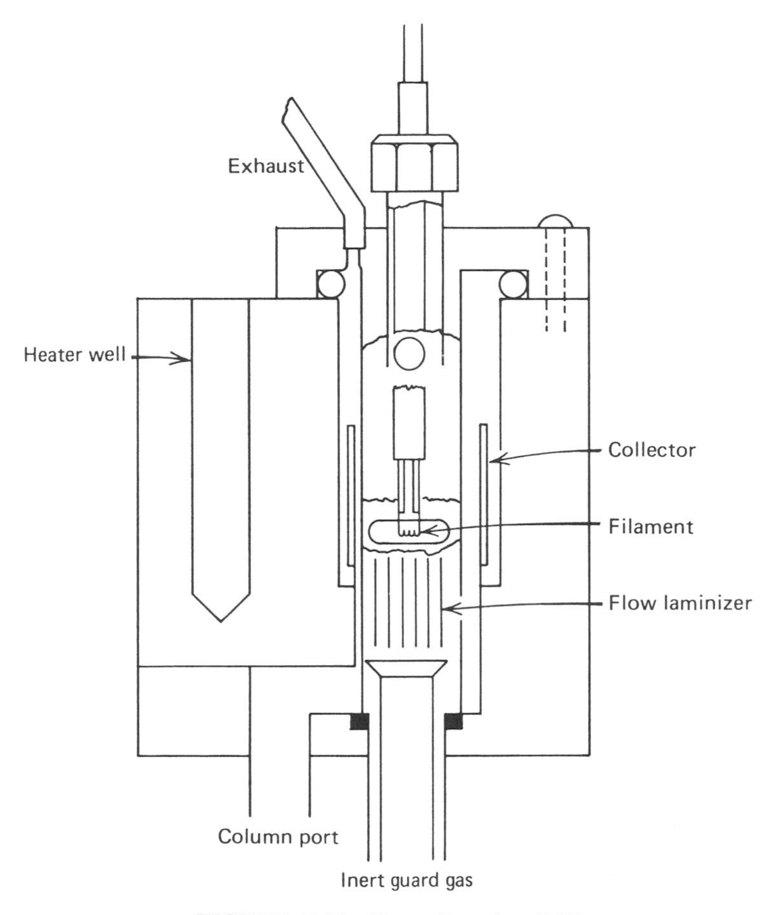

FIGURE 5.15 Nonradioactive ECD.

of electron capture increases with increasing concentration of free electrons. Gases with large ionization cross sections, such as nitrogen or argon containing 5 to 10% methane, give optimum performance of the ECD. Pure hydrogen and helium, with small ionization cross sections, will not work. Operating capillary columns with helium or hydrogen carrier gas at low flowrates does not significantly affect the sensitivity of the ECD as long as nitrogen or methane/argon is used as the makeup gas at a flowrate of 20 mL/min or more. Hydrogen carrier-gas flowrates greater than 5 mL/min significantly affect the sensitivity and linearity of the ECD. Using hydrogen with a tritium source decreases its activity and lifetime. The stability of the baseline and the response of the ECD are affected by fluctuations in the carrier gas. Therefore, constant flow control should be used with capillary column operation. If a pressure regulator is used, the column carrier-gas flow will be reduced as the column temperature is increased in temperature programming. This effect can be minimized by adding makeup gas.

Voltage Requirements. There are several ways to apply the potential to the electrons of the ECD: (1) at constant voltage, (2) under pulsed constant frequency, and (3) under pulsed variable frequency constant current. The magnitude of the applied potential in the dc mode is a critical parameter dependent on the species measured, cell design, carrier-gas composition, and detector contamination. Good operating practice dictates the construction of a voltage response curve. Under dc conditions it is possible to have competing processes (32) so that anomalous behavior results (Figure 5.16). To overcome this, researchers prefer pulsing the voltage. This mode significantly decreases the buildup of charged zones in the detector resulting from differences in the velocity of positively charged ions compared to the mobility of free electrons. The positive ions can form a charged cloud near the cathode and contribute to anomalies when a solute enters the chamber and perturbs the equilibrium. While the concentration of electrons in the dc

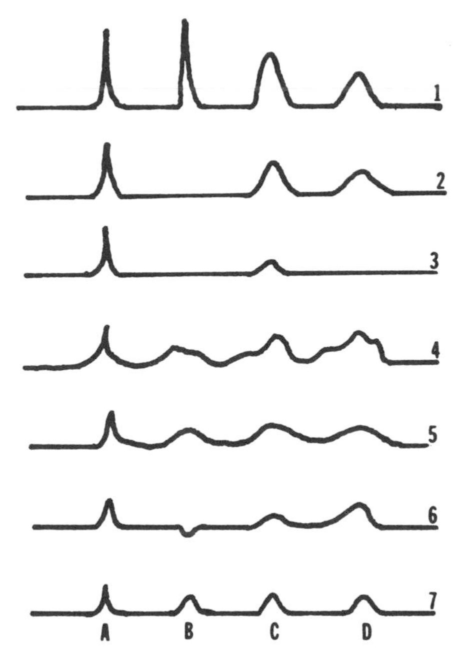

FIGURE 5.16 Anomalous responses that can occur with an ECD operated in the DC mode (from Reference 33, with permission). Chromatograms depicted are illustrative of observed phenomena: (1) chromatogram that is truly representative of a mixture; A and D are peaks of electron absorbers, B is a larger amount of a nonabsorber, and C is an unresolved mixture of absorbing and nonabsorbing compounds; (2) ECD operating correctly; (3) ECD losing peaks as a result of space charge effects; (4) ECD with contact potential (the result of material adsorbed on an electrode) enhancing the applied potential; (5) contact potential opposing the applied voltage (observe increased tailing and false peak at B); (6) ECD acting as if it were an argon detector or a cross-section detector (note inversion of B and reduction of peak C); (7) ECD operating as an electron mobility device and an ECD (note the false peak at B).

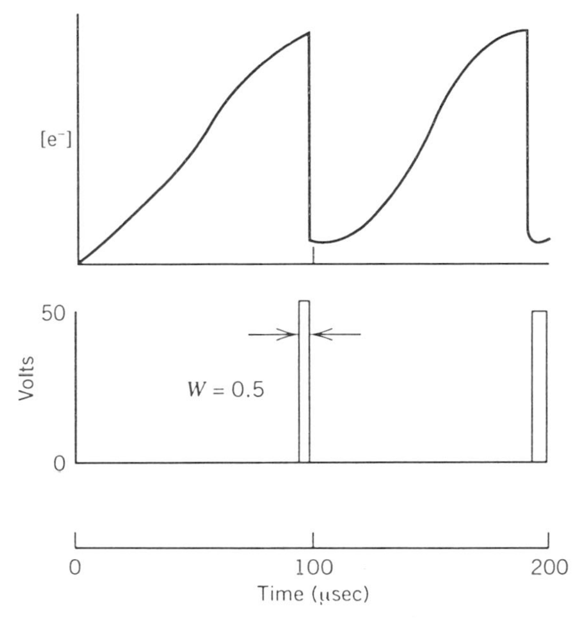

FIGURE 5.17 Pulse mode.

mode is constant, the pulse can be visualized as a process for collecting electrons (Figure 5.17). In general, the pulse reaches 30–50 V over a time interval and is repeated every 100 μsec. If argon/methane is used, the duration of this pulse can be as short as 0.1 μsec; if nitrogen is used, a duration of at least 1.0 μsec is typically required. Another significant difference between dc mode and pulsed mode is that the driving force of the applied potential field is absent and the electrons attain thermal equilibrium. Sensitivities are enhanced using this pulsed-constant frequency or fixed frequency mode.

The mode of signal processing considered to be a superior method of controlling a pulsed ECD holds the current constant and varies the frequency of the pulse. The controlling circuit of the constant current ECD demands a preselected level of current. When an electron-capturing solute enters the detector and decreases the current, the circuit adjusts the frequency of the pulsing to provide that current. The response to sample in this mode is the change in pulse frequency, while the response in the other modes is the decrease in current. The constant current mode has an improved linear dynamic range and increased sensitivity. Some commercial ECDs are available with the fixed frequency mode of operation, but all modern ECDs include the constant current mode.

5.5.3 Response Factors

In the ECD, the reaction of electrons with strong electron-absorbing compounds is a second-order rate mechanism that can be influenced by

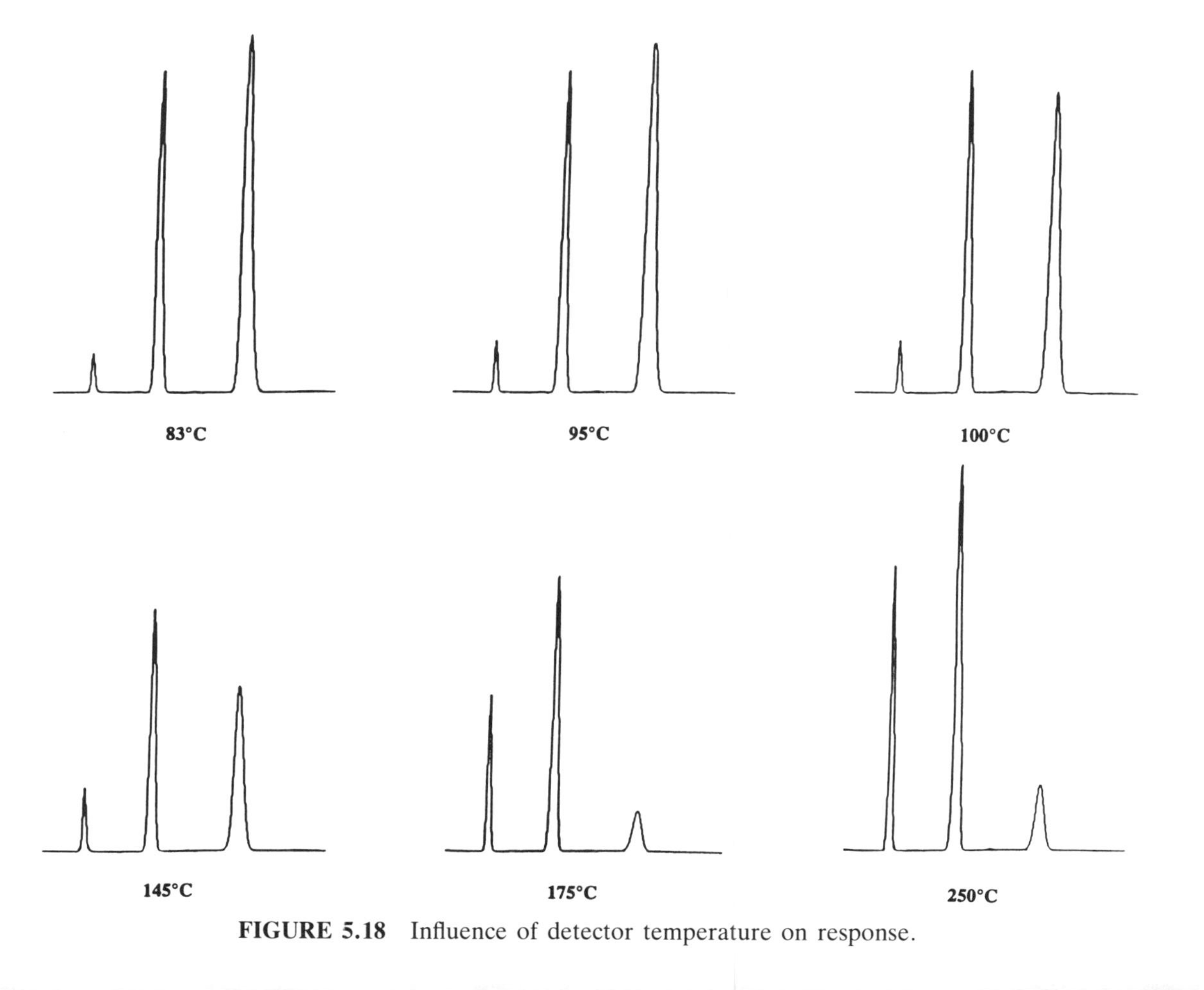

FIGURE 5.18 Influence of detector temperature on response.

temperature (Figure 5.18), the electronegativity of the species itself, the presence of other species, and the energy of the electrons. Variations in comparative detector response among competent laboratories have been observed even with control of these conditions. However, by selecting suitable operating parameters and having a strong electron-absorbing molecule, the forward rate of the reaction would be sufficient for the mechanism to be pseudo-first-order and approach coulometry. If the kinetics of the reaction were known, a response factor could be calculated.

Sullivan (34) proposed a more convenient approach for pulsed electron-capture detection. It was shown that the output frequency F and sample concentration [A] are related by

$$F = \frac{1}{K(k_I[\mathrm{A}] + K_\mathrm{d})} \tag{5.11}$$

where k_I = rate constant for electron capture
K_d = pseudo-first-order rate for reassociation of electrons with positive ions and is generally small
K = constant of proportionality

The output current I can be expressed as

$$I = \frac{I_s}{K(1 - e^{-k})} \tag{5.12}$$

where I_s = current produced by radioactive source
$K = (k_I[\mathrm{A}] + K_\mathrm{d})t_\mathrm{p}$
$t_\mathrm{p} = 1/F$

Substitution of values for I and I_s in the equation and use of reiterative techniques yields an approximate value of K, so that

$$k_I = \frac{KfM \times 10^{12} Z}{Nr} \tag{5.13}$$

where f = flow (mL/sec)
N = Avogadro's number
r = frequency-to-voltage correction factor for analog output (μV/Hz)
M = analyte molecular weight
Z = area response factor (μV-sec/pg)

Wentworth and Chen (35) developed a kinetic model using a parallel-plate tritium source for the electron-capture detector that correlates observed response values with electron affinities derived from reversible half-wave potentials in aprotic solvents (36) and temperature dependence of the kinetics and the thermodynamic properties of the solute.

Since the electron-capture detector is a specific detector, the response factor will vary by the very nature of the solute. However, some generalizations can be made. Table 5.8 lists classes of compounds and their relative responses.

Dopants. Although a scrupulously clean carrier gas is essential for a well-balanced system and coulometric response, a dopant in the carrier gas can enhance the response of certain compounds by in situ reaction. An

TABLE 5.8 Relative Response Values Using Electron Capture Detector

Chemical Classes	K'	Selected Samples
	0.01	
Alkanes, alkenes, alkynes, aliphatic ethers, esters, and dienes		Hexane Benzene Cholesterol Benzyl alcohol Naphthalene
	0.10	
Aliphatic alcohols, ketones, aldehydes, amines, nitriles, and monofluoro and monochloro compounds		Vinyl chloride Ethyl acetoacetate Chlorobenzene
	1.0	
Enols; oxalate esters; and monobromo, dichloro, and hexafluoro compounds		*cis*-Stilbene *trans*-Stilbene Azobenzene Acetophenone
	10.0	
Trichloro compounds, chlorohydrates, acyl chlorides, anhydrides, barbiturates, thalidomide, and alkyl-leads		Allyl chloride Benzaldehyde Tetraethyl-lead Benzyl chloride Azulene
	300	
Monoiodo, dibromo, and trichloro compounds, mononitro compounds, lacrimators, cinnamaldehyde, fungicides, and pesticides		Cinnamaldehyde Nitrobenzene Carbon disulfide 1,4-Androstadiene-3,11,7-trien Chloroform
	1000	
1,2-Diketones; fumarate esters; pyruvate esters; quinones; diiodo, tribromo, polychloro, and dinitro compounds; and organomercurials		Dinitrobenzene Diiodobenzene Dimethyl fumarate Carbon tetrachloride
	10,000	

Note. K' values are based on chlorobenzene = 1.0.

Source. From Ref. 33, with permission.

obvious example is the methane in the argon carrier. Normally the level of oxygen in the carrier gas is reduced to the lowest possible level. Oxygen's weak but positive electron affinity can capture electrons, thus reducing the population available for reaction with the sample. Allowing a sacrifice of electron population, the addition of a few parts per thousand of oxygen added to the carrier gas can significantly increase the response of hydrocarbons and halogenated hydrocarbons by the following scheme:

$$e^- + O_2 \rightleftharpoons O_2^- \tag{5.14}$$

$$O_2^- + AB \rightleftharpoons \text{products} \tag{5.15}$$

Weak electrophores such as benzo[*e*]pyrene exhibit a 400-fold increase in response (37). Other investigators have used N_2O (38):

$$e^- + N_2O \rightarrow O^- + N_2 \tag{5.16}$$

$$O^- + N_2O \rightarrow NO^- + NO \tag{5.17}$$

$$NO^- + AB \rightarrow NO + \text{products} \tag{5.18}$$

$$O^- + AB \rightarrow \text{products} \tag{5.19}$$

and have improved the sensitivity to vinyl chloride by 10^3 and thereby effectively reduced much of the labor-intensive sample workup (39). Nitrous oxide doping was also shown effective in enhancing response to CO_2, H_2, and CH_4(40).

5.5.4 Linearity of Response

The response of the ECD is linear with concentration over only two orders of magnitude in the dc mode. Wentworth and his colleagues (41) proposed a relationship to increase the linear range to about four orders of magnitude by the equation

$$\frac{I_s - I}{I} = kc \tag{5.20}$$

where I_s = standing current
I = current measured when electron-absorbing species is at concentration c
k = constant of the cell and the species present

By using analog converters the linear dynamic range can be extended to five orders of magnitude.

5.5.5 Practical Operating Hints

1. Oxygen, which is an electron absorber, should be scrupulously trapped by molecular sieves.
2. Septa should be baked out in a vacuum oven before installation.
3. All tubing in the system should be cleaned and baked out.
4. Column bleed should be kept to a minimum.
5. Avoid stationary phases with high electron affinity such as trifluoropropylmethyl silicones (OV-202, OV-210, OV-215, etc).
6. Keep the system leaktight to avoid diffusion of gases.
7. Exercise the necessary precautions when dealing with radioactive isotopes.

5.6 FLAME PHOTOMETRIC DETECTOR

5.6.1 Background

In 1966, Brody and Chaney (42) developed a detector based on the principle of flame photometry. Although it is used for halogens and nitrogen-containing compounds, its simultaneous sensitivity and specificity for the determination of sulfur and phosphorus have made it one of the most widely used selective detectors in gas chromatography today. The flame photometric detector (FPD) has also been used for the detection of other elements, such as tin, chromium, selenium, tellurium, and boron (43).

5.6.2 Principles of Operation

Flame photometric detection is based on the formation of chemiluminescent species in a hydrogen-rich flame. These species emit light characteristic of the heteroatoms introduced into the flame. The flame photometric detector measures the intensity of these emission spectra. An optical filter is placed in the radiation path to select the wavelength of light reaching the photomultiplier tube. For sulfur an interference filter with a 394-nm bandpass is used; for phosphorus a 526-nm bandpass is used.

5.6.3 FPD Design

Flame photometric detectors are either single-burner or dual-burner design. Figure 5.19 illustrates the original Brody and Chaney single-burner design. The effluent from the column using nitrogen carrier gas is mixed with oxygen in a proportion similar to air. Excess hydrogen is added to the exterior of the burner tip. The diffusion flame is situated inside the burner tip, which shields the photomultiplier tube (PMT) from a direct view of the flame. With this design, the emission of sulfur and phosphorus species occur

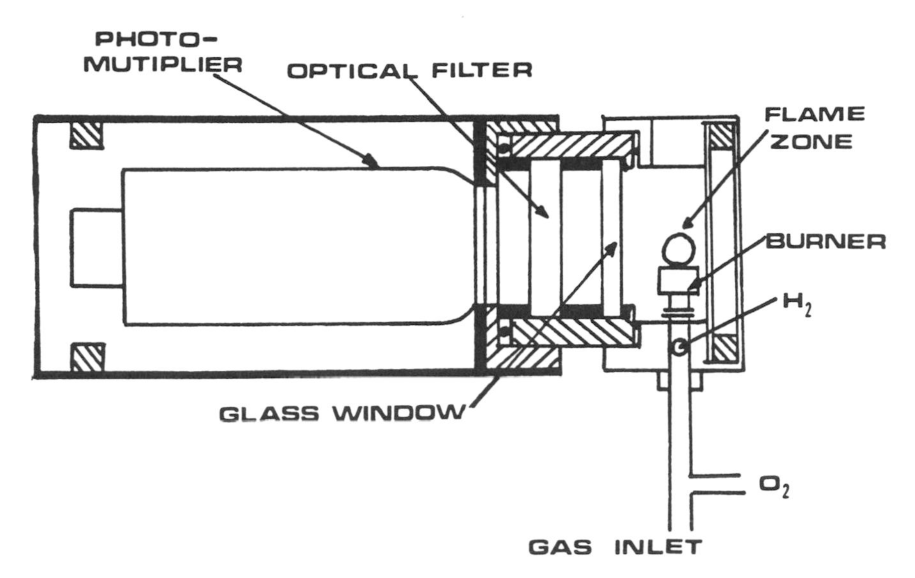

FIGURE 5.19 Flame photometric detector.

above the shielded flame in the direct view of the PMT. Hydrocarbons and other interferences emit their light in the flame itself, inside the burner tip, and are not detected by the PMT.

Both diffusion and premixed flames are employed in flame photometric detectors. Diffusion flames are "cooler" and more suitable for monitoring phosphorus and sulfur. A premixed flame can have forty times the noise of a diffusion flame. However, a large sample size can change the geometry and temperature of the cool single flame, adversely affecting light emission. This is because the flame is used both to decompose the sample and to excite the atoms. Another common problem with this design is known as "solvent flameout." This occurs when the solvent peak elutes from the column, starving the flame of oxygen and extinguishing it. This problem has been eliminated by modifying the detector design so that the hydrogen and oxygen or air inlets to the burner are interchanged. Quenching has also been observed with the single-burner design when water or hydrocarbons are coeluted with the sample.

The dual-flame photometric detector has two optical filters and photomultiplier tubes for the simultaneous monitoring of both phosphorus and sulfur. Figure 5.20 is a schematic diagram of the dual-burner flame photometric detector reported by Patterson and coworkers (44). The column effluent is mixed with air while hydrogen is added at the base of the detector. The lower flame combusts the sample while the upper flame generates the light-emitting excited species. With this design, if the lower flame suffers solvent flameout, it is reignited by a flashback from the upper

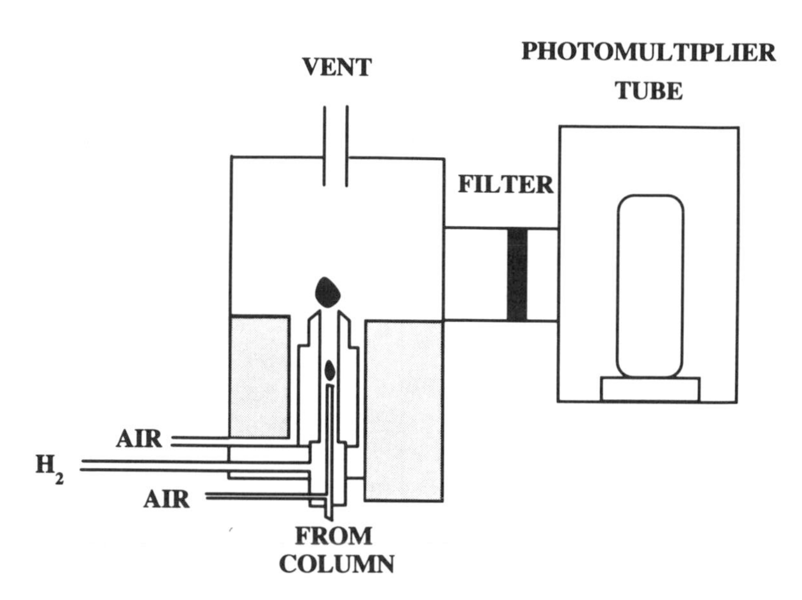

FIGURE 5.20 Dual-burner flame photometric detector.

flame. This design also overcomes some of the quenching problems of the single-burner detector.

5.6.4 Response Factors

The sensitivity of this detector depends on the intensity of the light emitted by the excited species. This intensity increases with decreasing flame temperature. The use of carrier gases with high thermal conductivity, likc helium and hydrogen, instead of nitrogen, will increase the sensitivity by decreasing the flame temperature. The operating range of the FPD is usually 150 to 275°C due to the proximity of the photomultiplier tube to the flame. PMT noise increases with temperature, so the temperature should be set just high enough to prevent condensation of the high molecular weight compounds in the detector. The sensitivity of the FPD increases with excess hydrogen in the diffusion flame; however, excess hydrogen makes the flame unstable and easily extinguished during solvent elution. Given the diversity of detector configurations, individual optimization of the flows is necessary.

The response of phosphorus in the FPD is linear. The response of sulfur, however, varies such that the square root of the response is proportional to the concentration. Sulfur selectivity to hydrocarbons ranges from 10^3:1 at low sulfur content to 10^6:1 at high sulfur content. Better selectivity is obtained if the sulfur is completely separated from other species. Quenching of the sulfur chemiluminescence and other anomalous behavior have been observed when sulfur elutes with other compounds. Self-absorption also occurs at high concentrations, producing deviations from linearity.

Phosphorus selectivity to hydrocarbons is 10^6:1. Phosphorus does not seem to be affected by self-absorption and quenching. What is not favorable is the selectivity of phosphorus to sulfur at 4:1. Occurrence of the crossover phenomena in the phosphorus mode should always be checked against response in the sulfur mode. Fortunately, a distinction can be made since sulfur has a selectivity of 10^3–10^4 for phosphorus. In the event that a compound contains both sulfur and phosphorus, the ratio of the response of phosphorus to the square root of the sulfur response is an excellent diagnostic of their relative amounts (45):

Compound	R_P/R_S
PS	5.2–6.0
PS_2	2.8–3.3
PS_3	1.7–2.3

These ratios are independent of concentration, column temperature, and retention time.

Using an FID in tandem with an FPD in either the phosphorus or sulfur mode can indicate the presence of P or S. If the S/N ratio on the FID is greater than 1000 times the S/N ratio on the FPD, neither P nor S is present. If the S/N ratio on the FPD is greater than that on the FID, they are present.

Response is also a function of the degree of oxidation of the sulfur. The response follows the order: $-S-S- > SO_4^{2-} > S{=}O > SO_3^{2-} > S^{2-}$. Responses for phosphorus containing organics is linear within a homologous series; degree of oxidation and functional group substituents cause variable response. The minimum detectable level for phosphorus is about 0.5 pg/sec, and that of sulfur is on the order of 50 pg/sec. Phosphorus is linear over a 10^4 range, and sulfur, on a log–log scale, is linear over three orders of magnitude.

5.7 NITROGEN–PHOSPHORUS DETECTOR

5.7.1 Background

Today's nitrogen–phosphorus detector (NPD) evolved from an earlier type of gas chromatographic detector known as the alkali flame ionization detector (AFID). Also known as the thermionic detector (TID), this detector was founded on the observation that a metal anode emits positive ions when heated in a gas. A probe with sodium salt deposited on it was inserted above the flame of an FID (Figure 5.21). This detector showed a

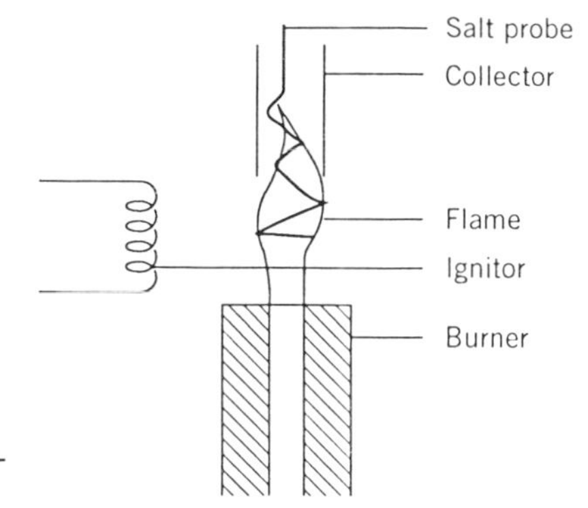

FIGURE 5.21 Alkali flame ionization detector.

specificity for phosphorus- and halogen-containing molecules. However, the selectivity was poor. Karmen (46) obtained a phosphorus:hydrocarbon selectivity of 105:1 by stacking two flames (Figure 5.22). The first flame burned the eluted materials and vaporized sodium deposited on a platinum screen located above the flame. The vapor was then transferred into the second flame where ionization took place. Many design modifications were

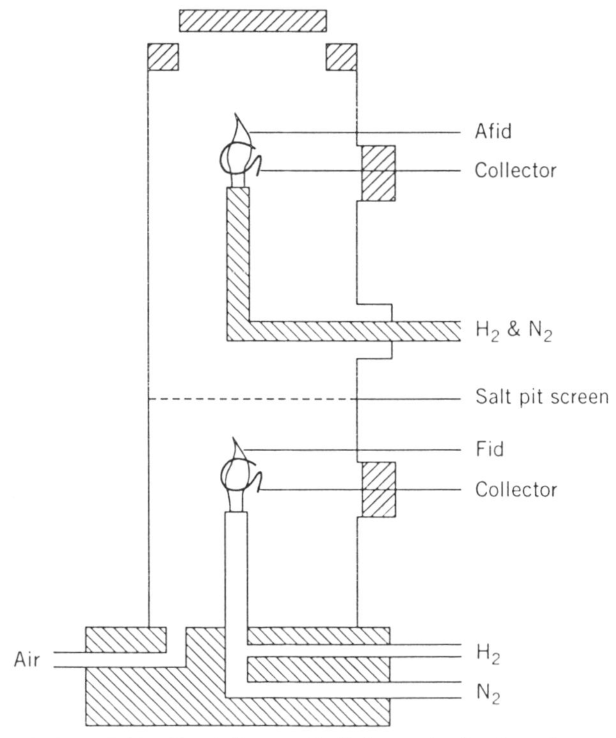

FIGURE 5.22 Dual-flame alkali flame ionization detector.

proposed, but the location of the salt around the burner tip was preferred (Figure 5.23). The salt was not diminished as quickly, equilibrium was established more rapidly, and sensitivity was better than in other designs. However, the AFID was difficult to use because it required frequent adjustments of many interdependent operating parameters. The background signal was unstable and the sample responses were nonreproducible. The alkali metal salt degraded rather rapidly.

The second generation of thermionic detectors arrived in 1974 with Kolb and Bischoff's design for a detector specific for nitrogen and phosphorus compounds (47). This detector had three significant features: (1) a glass bead containing nonvolatile rubidium silicate was used instead of a pellet of

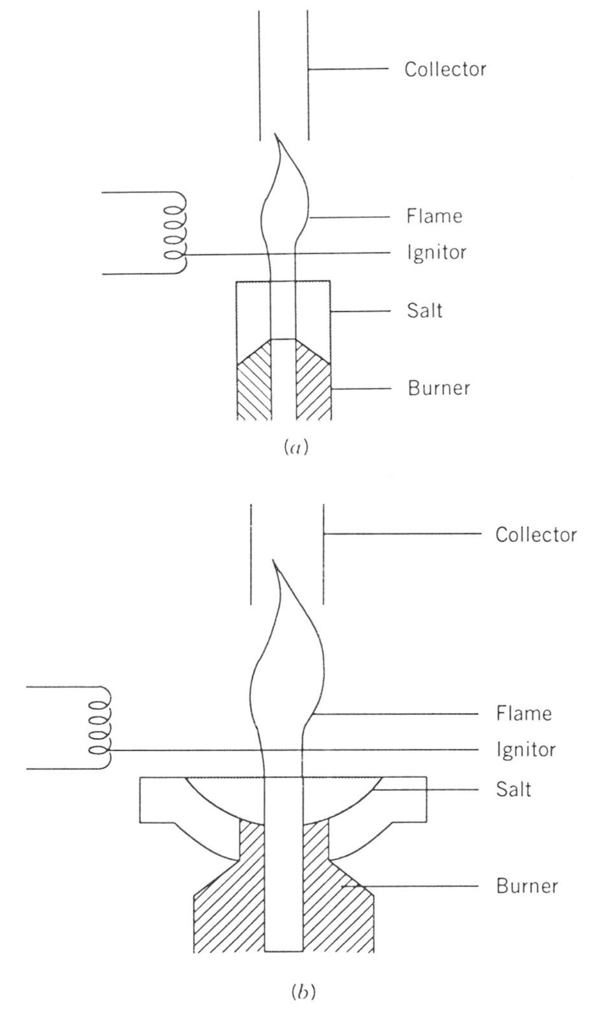

FIGURE 5.23 Modifications of AFID.

volatile alkali metal salt; (2) the bead was fused onto a platinum wire which electrically heated the bead instead of the flame; (3) hydrogen flow was only a few mL per minute for both N and P response, so there was no flame. Figure 5.24 shows a schematic of this nitrogen–phosphorus detector (NPD). Compared with the AFID, the NPD exhibited a longer life for the alkali-impregnated component, better baseline stability, more reproducible response, and better control of key operating parameters. The NPD has replaced the AFID for the determination of N or P compounds.

5.7.2 Principles of Operation

Several theories have been proposed to explain the mechanisms involved in the NPD. The system is complex and does not lend itself to a complete theory because intricate surface phenomena are believed operating. The gas environment of the detector is a dilute mixture of hydrogen in air. The detector does not begin to function until there is enough thermal energy to dissociate H_2 molecules into reactive H atoms, which, through a series of chain reactions, produce a highly reactive chemical environment. This thermal energy is provided by heating the thermionic source. The process is similar to that which occurs in the flame of a conventional FID. However, unlike the FID, there is no self-sustaining flame and the chemistry exists only in a gaseous layer near the hot thermionic source. If the heating current is turned off, the layer ceases to exist and the response disappears. When

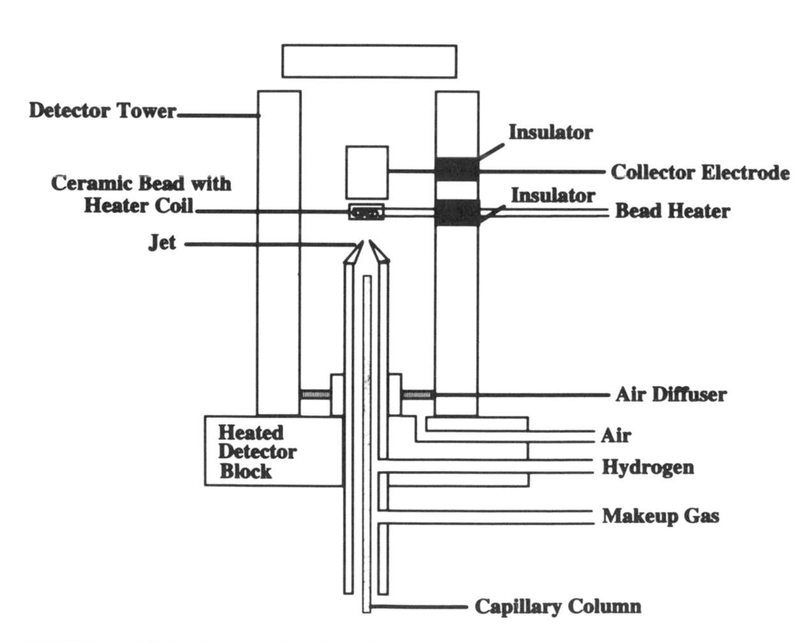

FIGURE 5.24 Schematic drawing of the nitrogen–phosphorus detector.

sample compounds enter the detector, they are decomposed in the hot, chemically active layer and the decomposition products are ionized. The ionization mechanism is not clear. Two different theories have been proposed. The first theory views the NPD as a modification of an AFID and proposes a gas-phase ionization process occurring in the layer immediately adjacent to the hot thermionic source (48). The second theory describes a surface ionization process occurring on the hot surface of the thermionic source (49).

5.7.3 Operating Parameters

The magnitude of the hydrogen flow to the detector and the magnitude of the heating current supplied to the thermionic source are extremely important parameters in the NPD. Both the sensitivity and the specificity of the detector are affected by them. The hydrogen flow affects the concentration of the H atoms in the reactive gaseous layer around the thermionic source, which, in turn, determines the response. The effect that the hydrogen flowrate has on the selectivity of some versions of the detector is illustrated in Figure 5.25.

The sensitivity of most versions of the NPD can be increased by as much as a factor of 10 by increasing the source heating current beyond the base value required to initiate the H_2/air chemistry. The temperature of the source is affected by the heating current, the auxiliary heating of the detector walls, the thermal conductivity of the gas mixture flowing past it, and the volume of the gas flowing past it. Since the source must remain hot enough to produce a reactive chemical environment, changes in any of these parameters are important. The auxiliary heater for the detector walls should be operated high enough to minimize the temperature gradient between the

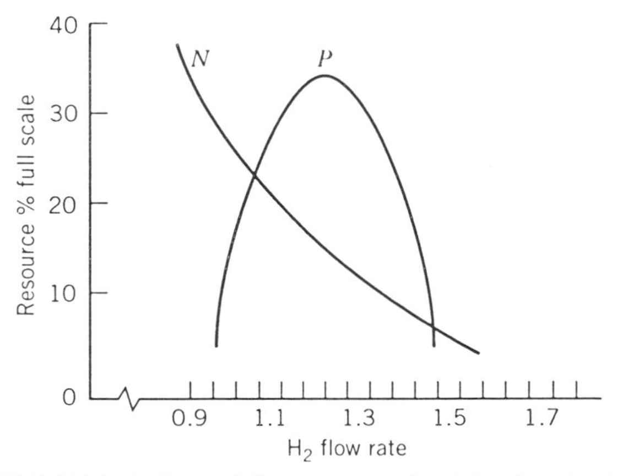

FIGURE 5.25 Effect of flowrate on selectivity for the NPD.

source and the walls. This would also minimize the effect of large concentrations of sample compounds passing through the detector. Helium has a much higher thermal conductivity than nitrogen, so its use as a carrier or makeup gas will require a higher source heating current than when nitrogen is used.

5.7.4 Response Factors

Response factors in the NPD are directly related to experimental conditions and vary for different atoms. Typical selectivities of nitrogen to carbon range from 10^3 to 10^5 g C/g N; for phosphorus to carbon the range is 10^4 to 5×10^5 g C/g P. The linear dynamic range varies from 10^3 to 10^5 and differs for each compound examined.

5.7.5 Practical Operating Hints

1. Avoid column liquid phases that contain nitrogen or phosphorus, such as OV-225, OV-275, FFAP, XE-60, and TCEPE.
2. Avoid halogenated solvents, which decompose and alter the ionization characteristics of the thermionic surface for long periods of time.
3. Avoid silylation reagents, which leave deposits on the thermionic surface and cause a loss of response.
4. If halogenated solvents or silylation reagents must be used, their effects can be minimized by turning off the bead current while the solvent is eluting. The current should not be off longer than 2 min.

5.8 HELIUM AND ARGON IONIZATION DETECTORS

These two detectors are incorporated in the same section because they are identical in principle, differing only in the energy of the metastable species produced. In fact, the first helium ionization detectors were essentially argon ionization detectors used with helium carrier gas. As described in Section 5.5.1, the carrier-gas atoms are excited to a metastable state by β radiation from a radioactive source. The species formed is then capable of ionizing all compounds with a lower ionization potential. The products formed are then subject to an electric field and the change in current is measured. Since the ionization potential of helium (19.8 eV) is significantly higher than that of argon (11.8 eV), helium has the capability of ionizing some species that argon cannot. Because of this extended range of application, helium is now used almost exclusively in metastable detectors.

The most frequently used detector geometries are the symmetrical coaxial configuration and the plane parallel design (Figure 5.12). The electrodes in both these designs are closely spaced, minimizing internal volumes. Typical detector volumes range from 100 to 200 μL, making them suitable for use with conventional capillary columns. Beta-emitting tritium sources in the form of Ti^3H_2 and Sc^3H_3 are usually used as the source of radiation.

The most significant problem associated with these detectors is the purity of the carrier gas. Although helium is commercially available at a minimum purity level of 99.9999%, this may not be its purity at the detector cell. Contaminants due to atmospheric leakage may be introduced at any connection after the cylinder and even through the cell itself. Many detectors are operated in a helium atmosphere to eliminate this problem. A major source of contamination comes from column bleed. For many years these detectors were limited to analyses of gaseous samples performed on packed adsorption columns. The bleed generated from liquid stationary phases, even with capillary columns, was sufficiently great enough to prohibit their use with these detectors. However, a variety of stationary phases bonded or immobilized on fused silica capillary columns are now available commercially. These phases exhibit very low bleed even at high temperatures and can extend the applicability of the detectors.

The parts per billion (ppb) sensitivity for the fixed gases makes the helium detector the most suitable for trace analysis of these gases. This sensitivity also makes it ideal for use with capillary and porous layer open tubular (PLOT) columns.

5.9 CROSS-SECTION DETECTOR

As was mentioned in Section 5.5.1, the cross-section detector was the first of the β-ray ionization detectors. The advantage of this detector is that its response to a substance can be calculated by using published values of the atomic cross sections of the constituent elements. The design of the cross-section detector is similar to that of the argon detector (Section 5.8). Ion pairs are formed in the ionization chamber. The applied potential across the detector is adjusted to minimize losses by recombination and fragmentation at high energies. The current i measured when a component enters the detector is related to the mole fraction X_s of the sample to the carrier gas, the ionization cross section of the sample vapor Q_s, and the ionization cross section of the carrier gas Q_c:

$$i = X_s Q_s + (1 - X_s) Q_c \tag{5.21}$$

The ionization cross section is a constitutive property and is the sum of the atomic cross sections. The response of a gas can be predicted if its elemental composition is known. Hydrogen is used as the carrier gas. This detector is universal and has a wide linear dynamic range of 10^5. It is capable of an MDL of 3×10^{-11} g/sec.

5.10 GAS DENSITY BALANCE

The gas density balance is a nondestructive, universal detector invented by Martin and James and discussed by Martin when he received the Nobel Prize (50). The Gow-Mac Instrument Company (Bethlehem, PA) simplified the design and made it commercially available. The mass chromatograph uses two of these detectors in a dual-column mode for the direct determination of molecular weight. The hydrodynamics of this detector have been studied as they apply to the mass chromatograph (51).

Figure 5.26 is a diagram of this detector. The reference gas, entering the detector at orifice A, is divided and sweeps past both detector filaments. These filaments are the same as those available for thermal conductivity detectors. The column effluent enters the detector through orifice B. Diffusion of the column effluent into the detector filament area is prevented by maintaining the reference gas flow 15 to 20 mL/min faster than the column effluent flow. This feature makes the detector useful for the analysis of corrosive materials. If the effluent gas is of higher density than the reference gas, it will retard the flow of reference gas along the downward path in the detector. This raises the temperature, and the resistance, of the lower detector element, unbalancing a Wheatstone bridge to produce a signal. If the effluent gas is less dense than the reference gas, the flow will be retarded along the upward path with the same effects.

Response for this detector is dependent on the differences in molecular weights of the gas used and the compounds eluted. Therefore, a gas of low molecular weight should be used to determine high molecular weight compounds. Conversely, a high molecular weight gas is best for determining compounds of low molecular weight. Gases that diffuse rapidly, such as hydrogen and helium, are avoided to prevent backdiffusion of the sample stream into the reference stream. Response for this detector is represented

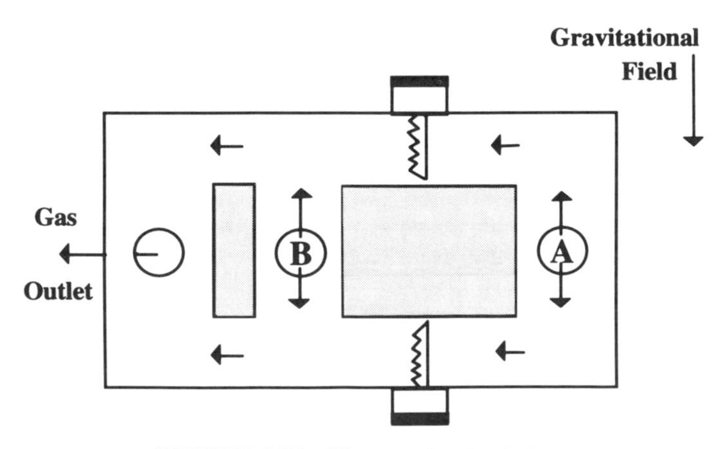

FIGURE 5.26 The gas density balance.

by the following equation:

$$A = WK\frac{M - m}{M}$$

where A = peak area
W = weight of component
K = instrument constant
M = molecular weight of component
m = molecular weight of carrier gas

The mass chromatograph uses two different carrier gases. A sample splitter is used at the injection port to chromatograph a standard of known molecular weight M_s on matched columns. Then the sample is chromatographed. The dual detectors are used to keep K constant under both sets of conditions. The molecular weight of the unknown M_x is determined from the following equation:

$$\frac{A_{x_1}(M_s - m_1)}{A_{s_1}(M_x - m_1)} = \frac{A_{x_2}(M_s - m_2)}{A_{s_2}(M_x - m_2)} \qquad (5.23)$$

Caution should be used when selecting the carrier gas so that comparison of the standard and sample can be done under the best conditions.

5.11 PHOTOIONIZATION DETECTOR

5.11.1 Background

The photoionization detector (PID) is one of the most versatile nonradioactive ionization detectors. Initially proposed by Lovelock (51) in 1960, the detector employed a glow discharge using argon as the source of UV radiation. This early detector did not gain widespread use because it required a vacuum for operation, was unstable, was easily fouled by column bleed, and required a skilled operator. Interest in the detector was revived when researchers (52, 53) found that separating the energy source from the ionization chamber improved the stability. Additional modifications (54) resulting in an improvement in the range from 10^4 to 10^7, lower background, and minimization of the column bleed problem made commercial introduction of the detector possible.

5.11.2 Operating Principles

Photoionization occurs when a molecule absorbs a photon of light energy and dissociates into a parent ion molecule and an electron:

$$R + h\nu \rightarrow R^{+} + e^{-} \tag{5.24}$$

The energy of the photon is dependent on the gas used as the UV emission source; higher-energy lamps using argon or hydrogen are more universal than the more selective lower-energy lamps using krypton or xenon. Separation of the source from the ionization chamber requires an optically transparent window. Glass and quartz are not transparent for high-energy photons, so crystals of alkali and alkaline earth metal fluorides are used as windows. The crystal is chosen with an optical transparency for the emission spectrum of the gas used. The most popular lamp is the 10.2-eV hydrogen source with a magnesium fluoride window. This lamp has the highest photon flux, giving it the greatest sensitivity. Only the argon lamp (11.7 eV) is more universal. Tables of ionization potentials should be consulted to determine whether a particular lamp and window are suitable for the measurement of the target compound.

5.11.3 Response

Freedman has proposed an equation to account for the PID response (55):

$$i = I^{\circ} F \eta \sigma N L [\mathrm{AB}] \tag{5.25}$$

where i = PID response (ion current)
I° = initial photon flux
F = Faraday constant
η = efficiency coefficient of ionization
σ = absorption cross section
N = Avogadro's number
L = pathlength
[AB] = concentration of ionizable substance

This equation states that lamp energy and cell volume are independent variables that can be used to enhance the detector signal.

A study of the sensitivity of a PID with a 10.2-eV lamp showed (56)

aromatics > alkenes > alkanes

polycyclic > monocyclic

branched > nonbranched

and that for substituted benzenes, ring activators increased the sensitivity, while ring deactivators decreased the sensitivity.

5.12 MICROWAVE PLASMA DETECTOR

The principle of the microwave plasma detector (MPD) is atomic emission spectroscopy, giving it the name atomic emission detector (AED). Like the mass spectrometer and the Fourier transform IR detectors, a major function of this detector is qualitative analysis. A microwave-induced plasma is used to cleave the sample molecules into atoms, which are raised to an excited state and emit their characteristic line spectra. A grating or multiple wavelength diode array spectrometer is used for selective element detection of sample components.

The use of a microwave plasma as a gas chromatographic detector was first reported by McCormack et al. (57). Refinements were made by replacing argon at atmospheric pressure with helium at reduced pressure to obtain improved response. A commercially available apparatus based on the investigations of McLean et al. (58) is shown in Figure 5.27. A plasma doped with low levels of oxygen or nitrogen prevents carbon from building up within the discharge tube. The instrument features multielement detection (P, S, Br, Cl, I, C, H, D, N, and O) and improved sensitivity by correcting for background interference. The eschelle grating monochromator provides high optical resolution and enhances sensitivity and

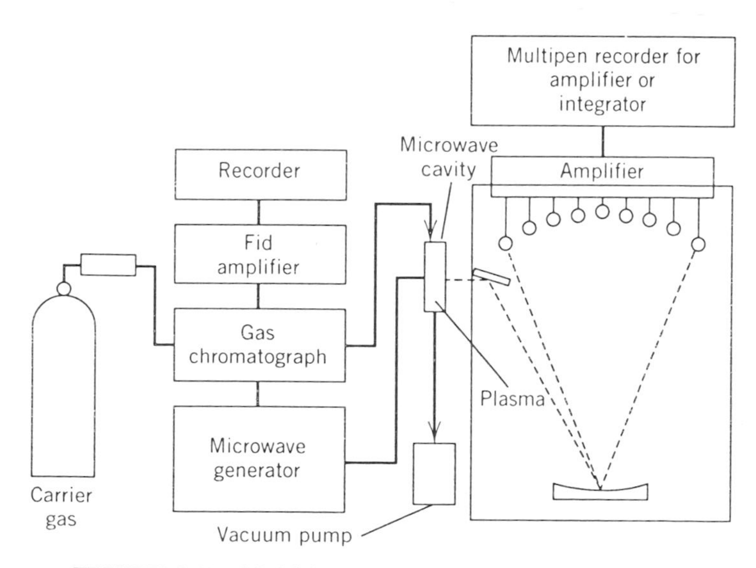

FIGURE 5.27 Multielement microwave plasma detector.

selectivity. Systems such as these are capable of determining empirical formulas as an aid to molecular structural elucidation.

A more recent commercially available instrument (59, 60) incorporates an atmospheric pressure helium microwave-induced plasma with a water-cooled discharge tube to maximize signal to background. A movable photodiode array detector slides along the focal plane from 170 to 780 nm, measuring an array range of approximately 25 nm. Up to four elements can be detected simultaneously and element-specific chromatograms displayed. The placement of the array range determines what combination of elements can be measured in a single chromatogram.

Ultrapure helium and a clean, leak-free chromatographic system are essential since the spectra of impurities can be quite intense. The MPD is capable of very good sensitivity (MDLs of 0.1 pg to 1 ng, depending on the element) and has the added features of aiding structure elucidation and detecting several heteroatoms simultaneously.

5.13 FOURIER TRANSFORM–INFRARED DETECTOR

Infrared spectrometry, with its ability to provide functional group information, has long been considered an ideal qualitative detector for gas chromatographic effluents. However, traditional IR spectrometry is too slow to scan column effluents and too insensitive for trace analysis. Fourier transform IR overcomes these deficiencies. Because the FT-IR sample beam is not dispersed, the detector receives all frequencies simultaneously, providing the speed required for GC. Theoretically, FT-IR should enhance the conventional IR signal-to-noise ratio by 1000. Detector and source performance will act to lower the realized advantage.

The FT-IR spectra of the column effluent is recorded by transferring the gas stream to a reflecting light pipe which is the sample cell. Interferograms are accumulated at the rate of approximately four per second. Typically, 2048 data points/scan are Fourier transformed. The amount of data collected during a GC/FT-IR run greatly exceeds that of other detectors. The computation speed, disk storage capacity, and disk data access time of the computer serve as the limiting factors in data treatment. The chromatographic peaks may be located by the use of a separate detector with effluent splitting or the chromatogram may be reconstructed from the FT-IR measurements alone. The most popular technique is the Gram–Schmidt (G-S) vector orthogonalization (61). A background vector is created when the light pipe is empty and is used to remove background signals from the vector that encodes the analyte information. The resulting chromatogram is analogous to that from total-ion-current detection in GC/MS.

The FT-IR detector can function as both a specific and a universal detector. Its ability to differentiate geometrical isomers is a real advantage.

Computer techniques such as signal averaging, background subtraction, and spectral searches improve its capabilities for many critical analyses.

5.14 ELECTROLYTIC CONDUCTIVITY (HALL) DETECTOR

Coulson (62) made the first commercial gas chromatographic detector based on the electrolytic conductivity of ionic species in water. Material eluted from the column was oxidized or reduced catalytically to form an ionic species that was transferred to a stream of deionized water for detection. The water was kept in a closed circulating system with ion exchange resins to remove the ions after their detection. The oxidative mode has been found to be ineffective because of carbon dioxide dissolving in the liquid stream and producing background interference. Hall developed an improved design to detect picogram quantities of halogen-, sulfur-, and nitrogen-containing compounds (63). Other operating modes have been reported for the selective detection of esters, nitrosoamines, aliphatic halocarbons, hydrogen, and hydrocarbons.

The three principal detection modes of commercial detectors are (1) the halogen mode, detecting HX; (2) the sulfur mode, detecting primarily SO_2, but also SO_3; and (3) the nitrogen mode, detecting NH_3. The detector response depends on the reaction conditions, the solvent and its pH, and the use of a postreaction scrubber. A nickel reaction tube with hydrogen reaction gas at a temperature of 850–1000°C will convert compounds containing chlorine to HCl and other products such as methane and water. The conductivity solvent, *n*-propyl alcohol, will dissolve the HCl and its electrolytic conductivity will change. The nonhalogen products will not dissolve in the propanol or will not change its conductivity to any significant degree. Other halogen compounds will be converted to their corresponding hydrogen halides and be detected likewise. In this halogen mode, sulfur-containing compounds are converted to H_2S, which is poorly ionized in the propanol. Nitrogen-containing compounds are converted to NH_3, which is also poorly ionized in the propanol.

The nickel reaction tube with hydrogen reaction gas at a temperature of 850–1000°C is used in the nitrogen mode of the detector. The conductivity solvent is water containing a small amount of organic solvent to fully ionize the NH_3. Care must be taken to prevent the neutralization of low levels of NH_3 by trace quantities of CO_2 that permeate into the solvent. Some detectors utilize a two-bed ion exchange cartridge to increase the pH of the solvent from bleed from the anion resin. Other systems may allow a small amount of ammonia to permeate into the solvent stream through a short loop of Teflon tubing immersed in a vessel of dilute ammonium hydroxide. Others may simply exclude the CO_2 by using nonpermeable tubing in the solvent delivery system. In addition, the H_2S formed from sulfur-containing compounds and the HX formed from halogen-containing compounds must be removed from the conductivity solvent before it reaches the cell. This is accomplished by placing a postreaction scrubber at the exit of the reactor. For the nitrogen mode this scrubber consists of a length of coiled tubing containing several strands of quartz thread coated with KOH.

In the sulfur mode the nickel reaction tube is used with air as the reaction gas at 850–1000°C to convert sulfur-containing compounds to SO_2. Methyl alcohol containing a small amount of water is used as the conductivity solvent. Nitrogen-containing compounds are converted to N_2 and some nitrogen oxides, which give little or no response. Halogen-containing compounds, however, are converted to HX. These hydrogen halides must be removed with a postreaction scrubber containing several strands of silver wire.

There are two types of cell designs in commercial detectors: (1) the dynamic reservoir cell, which separates the gas and liquid streams before measurement of conductivity, and (2) the mixed phase cell, which allows both the gas and liquid phases to pass through the measurement zone. Because of these differences, the two types of detectors may not give the same performance for a given application.

The detector should be optimized according to reactor temperature, reactant gas flowrate, solvent flowrate, composition and surface area of the nickel catalyst, and cell voltage. The reaction tubes, scrubbers, ion exchange resins, and solvents must be changed from time to time to maintain performance. The detector's high selectivity makes it ideal for the analysis of trace substances such as pesticides and drugs in complex matrices with minimum cleanup.

5.15 DATA HANDLING FROM DETECTOR SIGNAL

There are three main types of data-handling devices for gas chromatography: recorders, integrators, and computers (both personal and mainframe). In some cases, such as the FT-IR detector, only one type of device is suitable—the computer. The signals from most detectors, however, can be handled by any of the three types.

5.15.1 Recorders

The strip chart recorder presents the data generated by the detector in analog form. This device dates back to the early days of chromatography. For quantification, all the peaks of the chromatogram have to be kept on scale by attenuating the detector signal. Therefore, the recorder must be connected to the attenuated signal output of the chromatograph (usually marked REC). Retention times, peak heights, and peak areas must be measured manually. These measurements are tedious, time-consuming, and limited in accuracy and precision. The recorder presents the data in real time and has no memory. With the availability of inexpensive integrators, the recorder has all but disappeared from the lab. It can be used to provide an immediate chromatogram as a supplement to a mainframe-based system that only updates the chromatogram every one or two minutes.

5.15.2 Integrators

The modern integrator is a microprocessor-based device that combines the functions of plotting the chromatogram, peak integration, and final calculation of concentrations. It is a dedicated, highly specialized minicomputer without a monitor. It connects to the unattenuated signal output of the chromatograph (usually marked INT). The integrator contains programs for integration, peak detection, baseline correction, and final calculations. It stores peak identification data, response factors, etc., labels chromatograms, and prints reports. In addition, many integrators have memory enough to store about ten chromatograms. Some integrators feature BASIC programming, instrument control, and communication capabilities to higher-order data systems.

5.15.3 Computers

Computer-based data systems can be as simple as a PC interfaced to one or two instruments. Turnkey programs provide all the capabilities of the integrator. In addition, the computer monitor, mouse, and memory enable the use of expanded integration capabilities such as manual placement of the baseline. Chromatogram storage is greatly increased and limited only by the capacity of the hard drive.

Multichannel computer systems are also available. These systems use a dedicated computer and several terminals. An analog-to-digital converter digitizes the detector signals near the chromatographs and the digital signal is transmitted to the computer where the integration and complete data analysis is done. These systems vary greatly in capabilities and cost from manufacturer to manufacturer.

REFERENCES

1. Jiri Sevcik, *Detectors in Gas Chromatography*, (*Journal of Chromatography* Library, Vol. 4), Elsevier, New York, 1976.
2. H. A. McLeod, A. G. Butterfield, D. Lewis, W. E. J. Phillips, and D. E. Coffin, *Anal. Chem.*, **47**, 674 (1975).
3. H. Oster and F. Opperman, *Chromatographia*, **2**, 251 (1969).
4. I. G. McWilliam and R. A. Dewar, *Nature* (*London*), **181**, 760 (1958).
5. J. Harley, W. Nel and V. Pretorius, *Nature* (*London*), **181**, 177 (1958).

6. W. J. Miller, in *Fourteenth Symposium (International) on Combustion*, The Combustion Institute, Pittsburgh, PA, 1973, p. 307.
7. P. Bocek and J. Janak, *Chromatogr. Rev.*, **15**, 111 (1971).
8. J. C. Sternberg, W. S. Gallaway, and T. L. Jones, *Gas Chromatography*, pp. 231–267, Academic, New York, 1962.
9. J. Peeters, J. F. Lambert, P. Hertoghe, and A. Van Tiggelen, *Symp. (Int.) Combust. [Proc]*, **13**, 321 (1971).
10. A. J. C. Nicholson, *J. Chem. Soc., Faraday Trans. 1*, **79**, 2183 (1982).
11. A. T. Blades, *J. Chromatogr. Sci.*, **22**, 120 (1984).
12. F. J. Yang and S. P. Cram, *J. High Resolut. Chromatogr./Chromatogr. Commun.*, **2**, 487 (1979).
13. K. J. Hyver, in *High Resolution Gas Chromatography*, 3d ed., K. J. Hyver (Ed.), Hewlett–Packard, Avondale, PA, 1989, Chapter 4.
14. I. G. McWilliam, *J. Chromatogr.*, **6**, 110 (1961).
15. G. Perkins, Jr., G. M. Rouayheb, L. D. Lively, and W. C. Hamilton, *Gas Chromatography*, Academic, New York, 1962, pp. 269–285.
16. R. G. Ackman, *J. Gas Chromatogr.*, **2**, 173 (1964).
17. R. Kaiser, *Gas Phase Chromatography*, Vol. 3, P. H. Scott (transl.), Butterworths, Washington, DC, 1963.
18. W. A. Dietz, *J. Gas Chromatogr.*, **5**, 68 (1967).
19. L. S. Ettre, in *Gas Chromatography*, N. Brenner, J. E. Callen, and M. D. Weiss (Eds.), Academic, New York, 1962, p. 307.
20. J. T. Scanlon and D. E. Willis, *J. Chromatogr. Sci.*, **23**, 333 (1985).
21. P. Russev, M. Kunova, and V. Georev, *J. Chromatogr.*, **178**, 364 (1979).
22. M. D. Dupuis and H. H. Hill, Jr., *J. Chromatogr.*, **195**, 211 (1980).
23. M. A. Osman and H. H. Hill, Jr., *Anal. Chem.*, **54**, 1425 (1982).
24. M. Dimbat, P. E. Porter, and F. H. Stross, *Anal. Chem.*, **28**, 290 (1956).
25. D. M. Rosie and R. L. Grob, *Anal. Chem.*, **29**, 1263 (1957); D. M. Rosie and E. F. Barry, *J. Chromatogr. Sci.*, **41**, 237 (1973).
26. D. Jentzsch and E. Otte, *Detektoren in der Gas Chromatographie*, Akademische, Frankfurt, 1970.
27. G. R. Jamieson, *J. Chromatogr.*, **3**, 464 (1960); **3**, 494 (1960); **4**, 420 (1960); **8**, 544 (1962); **15**, 260 (1964).
28. J. E. Lovelock and S. R. Lipsky, *J. Am. Chem. Soc.*, **82**, 431 (1960).
29. J. E. Lovelock, *J. Chromatogr.*, **1**, 35 (1958).
30. W. A. Aue and S. Kapila, *J. Chromatogr. Sci.*, **6**, 255 (1973).
31. A. Neukermans, W. Kruger, and D. McManigill, *J. Chromatogr.*, **235**, 1 (1982).
32. J. E. Lovelock, *Anal. Chem.*, **33**, 162 (1961).
33. E. D. Pellizzari, *J. Chromatogr.*, **98**, 323 (1974).
34. J. J. Sullivan, *J. Chromatogr.*, **87**, 9 (1973).
35. W. E. Wentworth and E. C. M. Chen, *J. Chromatogr.*, **186**, 99 (1979).
36. E. C. M. Chen and W. E. Wentworth, *J. Chromatogr.*, **217**, 151 (1981).

37. E. P. Grimsrud and R. G. Stebbins, *J. Chromatogr.*, **155**, 19 (1978).
38. P. G. Simmonds, *J. Chromatogr.*, **166**, 593 (1978).
39. P. D. Golden, F. C. Fehsenfeld, W. C. Juster, M. P. Phillips, and R. E. Sievers, *Anal. Chem.*, **52**, 1751 (1980).
40. P. D. Golden et al., *Anal. Chem.*, **51**, 1819 (1979).
41. W. E. Wentworth, E. C. M. Chen, and J. E. Lovelock, *J. Phys. Chem.*, **70**, 445 (1966).
42. S. S. Brody and J. E. Chaney, *J. Gas Chromatogr.*, **4**, 42 (1966).
43. W. A. Aue and C. G. Flinn, *J. Chromatogr.*, **158**, 161 (1978).
44. P. L. Patterson, R. L. Howe, and A. Abu-Shumays, *Anal. Chem.*, **50**, 339 (1978).
45. H. W. Grice, M. L. Yates, and D. J. David, *J. Chromatogr. Sci.*, **8**, 90 (1970).
46. A. Karmen, *Anal. Chem.*, **36**, 1416 (1964).
47. B. Kolb and J. Bischoff, *J. Chromatogr. Sci.*, **12**, 625 (1974).
48. B. Kolb, M. Auer, and P. Pospisil, *J. Chromatogr. Sci.*, **15**, 53 (1977).
49. P. L. Patterson, *J. Chromatogr.*, **167**, 381 (1978).
50. A. J. P. Martin, in *Nobel Lectures, Including Presentation Speeches and Laureates' Biographies, in Chemistry, 1942–1962*, Elsevier, New York, 1964, p. 359.
51. J. E. Lovelock, *Nature*, **188**, 401 (1960).
52. J. Sevcik and S. Krysyl, *Chromatographia*, **7**, 375 (1973).
53. J. N. Driscoll and F. F. Spaziani, *Anal. Instrum.*, **13**, 111 (1974).
54. J. N. Driscoll and F. F. Spaziani, *Res./Dev.*, **27**, 50 (1976).
55. A. N. Freedman, *J. Chromatogr.*, **190**, 263 (1980).
56. M. L. Langhorst, *J. Chromatogr. Sci.*, **19**, 98 (1981).
57. A. J. McCormack, S. C. Tong, and W. D. Cook, *Anal. Chem.*, **37**, 1470 (1965).
58. W. R. McLean, D. L. Stanton, and G. E. Penketh, *Analyst*, **98**, 432 (1973).
59. J. J. Sullivan and B. D. Quimby, *J. High Resolut. Chromatogr./Chromatogr. Commun.*, **12**, 282 (1989).
60. R. L. Firor, *Am. Lab.*, **21**, 40 (1989).
61. J. A. de Haseth and T. L. Isenhour, *Anal. Chem.*, **49**, 1977 (1977).
62. D. M. Coulson, *J. Gas Chromatogr.*, **3**, 134 (1965).
63. R. C. Hall, *J. Chromatogr. Sci.*, **12**, 152 (1974).

CHAPTER SIX

Techniques for Gas Chromatography/Mass Spectrometry

JOHN A. MASUCCI and GARY W. CALDWELL

The R. W. Johnson Pharmaceutical Research Institute

Modern Practice of Gas Chromatography, Third Edition. Edited by Robert L. Grob.
ISBN 0-471-59700-7

6.1 INTRODUCTION

Gas chromatography/mass spectrometry (GC/MS) combines high-resolution separation of components with very selective and sensitive detection. Its use in all areas of science has expanded to the point of routine analysis, and thus requires a detailed discussion.

Mass spectrometry was discovered around the turn of the century by the physicist J. J. Thomson in his investigations of *Rays of Positive Electricity* (1). It was further developed by Aston through the early twentieth century where studies focused on using the technique to determine elemental isotopes (2, 3). Although its potential for chemical analysis was recognized very early by Thomson, it was not until the 1950s with the availability of commercial instruments that MS was widely used in support of chemical research, primarily in the oil industry. Gas–liquid chromatography was developed by James and Martin in 1952 (4), with open tubular columns being development by Golay in 1958 (5). In 1957, Holmes and Morrell demonstrated the first coupling of gas chromatography with mass spectrometry (6). The purpose of this chapter is to present a general overview of GC/MS. For more comprehensive treatments, readers are directed to several volumes dedicated exclusively to GC/MS and its applications (7–17).

The schematic diagram of a typical capillary GC/MS system is shown in Figure 6.1. The gaseous effluent from the chromatograph is directed through

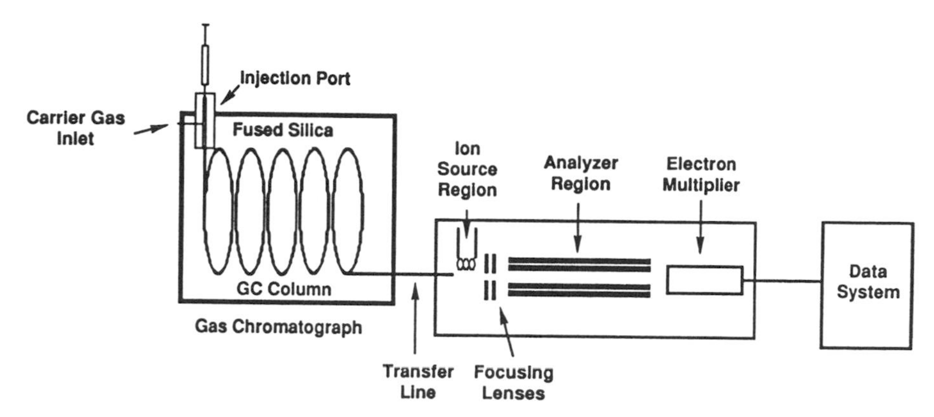

FIGURE 6.1 Schematic diagram of a typical gas chromatograph/mass spectrometer system. Gaseous analytes eluting from the chromatograph are directed into the spectrometer ion source where they are ionized. The ions produced are separated based on their m/z values and detected.

the transfer line into the ion source. The vaporized analytes are then ionized, producing molecular and/or fragment ions which are then mass resolved and detected. The resulting mass spectrum is displayed as a plot of the relative intensity of these ions versus their mass to charge ratio (m/z). Since most ions produced are singly charged, their m/z values are indicative of their masses. Atomic mass units are defined as Daltons (u). A typical mass spectrum of the common analgesic acetaminophen is shown in Figure 6.2. As the gas chromatographic separation proceeds, the mass analyzer is repeatedly scanned. The ion intensities for all m/z values for each scan can then be summed to generate a chromatographic trace commonly called a total ion current chromatogram. This is illustrated in Figure 6.3.

Since MS usually requires analytes that have some volatility and thermal stability, it is well matched with GC, where analytes are vaporized, separated, and sequentially eluted. Volatile derivatives of many analytes have been prepared for analysis, extending the utility of the technique (12).

Mass spectrometers must be operated at low pressures, typically in the range 10^{-5} to 10^{-7} Torr, to minimize ion–molecule collisions. This requirement is the major obstacle for chromatographic coupling, in both GC and liquid chromatography. Since mass spectrometer vacuum systems cannot accommodate the higher carrier-gas volumes required for packed-column GC, various interfaces were developed over the years (18). The advent of capillary GC simplified the coupling of these techniques, since modern mass spectrometric vacuum systems easily accommodate the lower carrier-gas flowrates, eliminating the lower efficiency interfaces required for packed-column GC/MS.

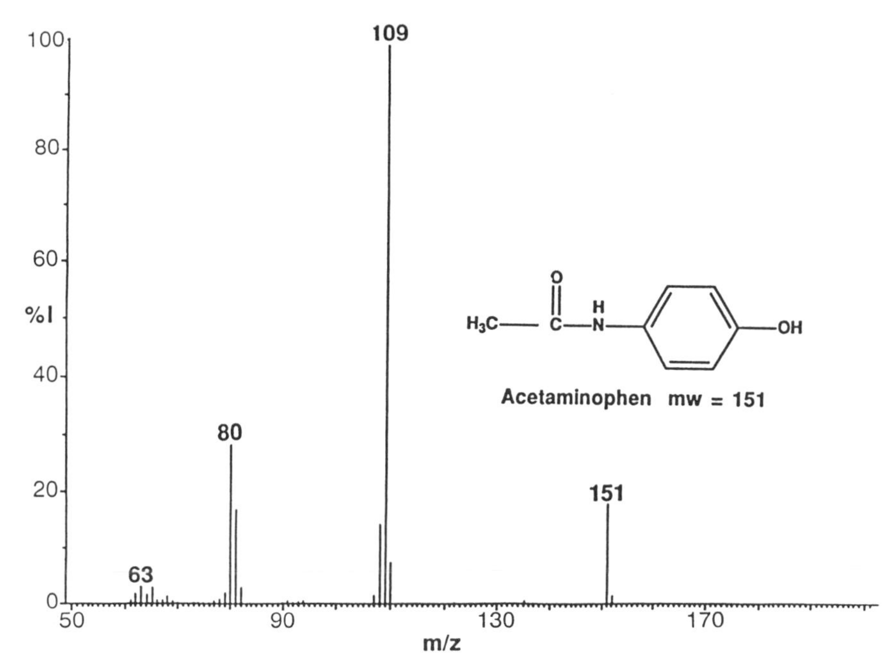

FIGURE 6.2 The mass spectrum of acetaminophen. Spectra are normally plotted with the mass to charge ratio (m/z) on the X axis and the relative intensity ($\%I$) on the Y axis. Since the majority of ions are produced with only one charge ($z = 1$), the m/z is equal to the mass of the ion. Note that the m/z is a dimensionless unit. The intensity of a peak is expressed as a percentage of the base peak. The peak at m/z 151 represents the intact acetaminophen molecule and is referred to as the molecular ion ($C_8H_9NO_2$). The largest peak in a particular spectrum is called the base peak. The peak at m/z 109 is the largest fragment ion produced and is referred to as the base peak of the spectrum. This terminology carries throughout all mass spectrometry.

6.2 GENERAL GAS CHROMATOGRAPHY/MASS SPECTROMETRY CONSIDERATIONS

In this section, the "GC/MS journey" from sample preparation to data presentation will unfold. Practical aspects of each subtopic is covered along the way.

6.2.1 Sample Preparation

In preparing samples for GC/MS, one simple fact must be kept in mind: Everything injected onto the gas chromatographic column will be deposited into the mass spectrometer, with the exception of those sample components

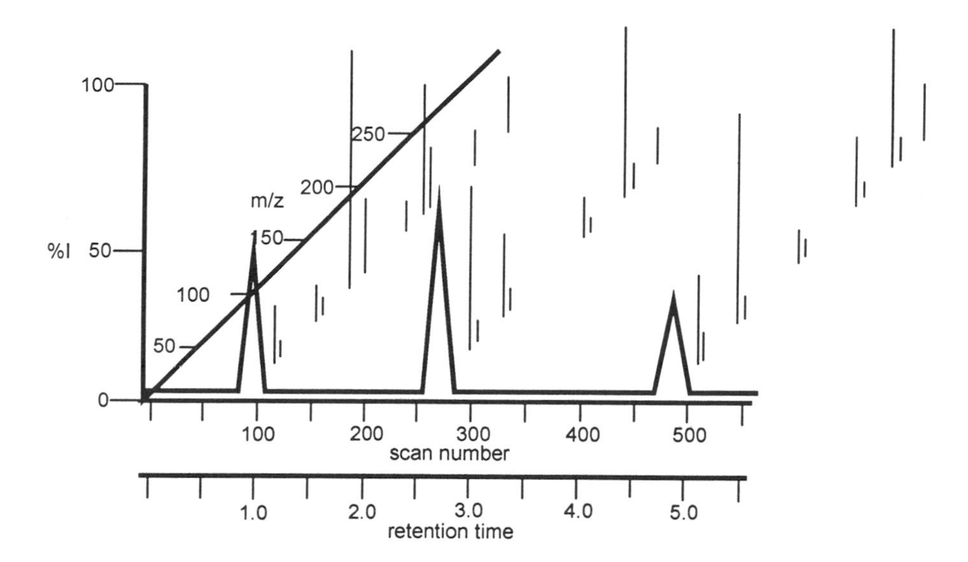

FIGURE 6.3 The data generated from a GC/MS experiment, represented here as a three-dimensional plot of scan number (time) vs. mass/charge (m/z) vs. relative intensity ($\%I$).

that remain in the injection port or on the column. For volatile components that is not a concern, because they are pumped away by the spectrometer vacuum system without consequence. But semivolatile materials may deposit in the ion source of the spectrometer with resultant loss of sensitivity, increased maintenance, etc. It is not uncommon for normal column bleed to eventually degrade system performance.

For particularly valuable samples, such as metabolite extracts, biological samples, or other samples obtained through extensive effort, the contamination threat must be tolerated as the cost of analysis. However, if sample clean up is possible without significant sample alteration, then a reasonable effort should be made to prevent contamination of the spectrometer.

Inorganic nonvolatiles can be removed by such methods as ion exchange or extraction. Polar, organic nonvolatiles can be removed using silica gel or Florisil. These methods require caution to avoid inadvertent loss of analyte.

Solvent selection is also important in GC/MS. Since the mass spectrometer is typically scanned over a wide mass range during data acquisition, it is important to minimize the possibility of interference peaks. This is accomplished by choosing a solvent that does not generate peaks in the mass range of interest. Methanol is a good solvent choice when detecting components with MW < 100 u and detecting peaks close to the solvent front, since it has a low molecular mass of only 32 u. Solvents such as chloroform or methylene chloride, with their higher masses (118 and 84 u, respectively), should be avoided in these cases. Typical spectra of these three solvents are shown in Figure 6.4.

6.2.2 Chromatography

Although considerable work has been done in the past with packed columns, this method is becoming less utilized in GC/MS because of lower chromatographic and transfer efficiencies. Capillary column GC/MS, with its simpler interface design and higher chromatographic efficiency, is typically preferred in all areas of analysis and is exemplified in this discussion.

The limiting factor with respect to column selection is the maximum flowrate that can be accommodated by the spectrometer vacuum system. This is usually no more than 1 mL/min for standard instruments, which limits column diameters to either 0.25 or 0.32 mm i.d. Higher flowrates are possible with certain instruments according to design, pumping capacity, and application, but columns of 0.53 mm i.d. are usually too large in both

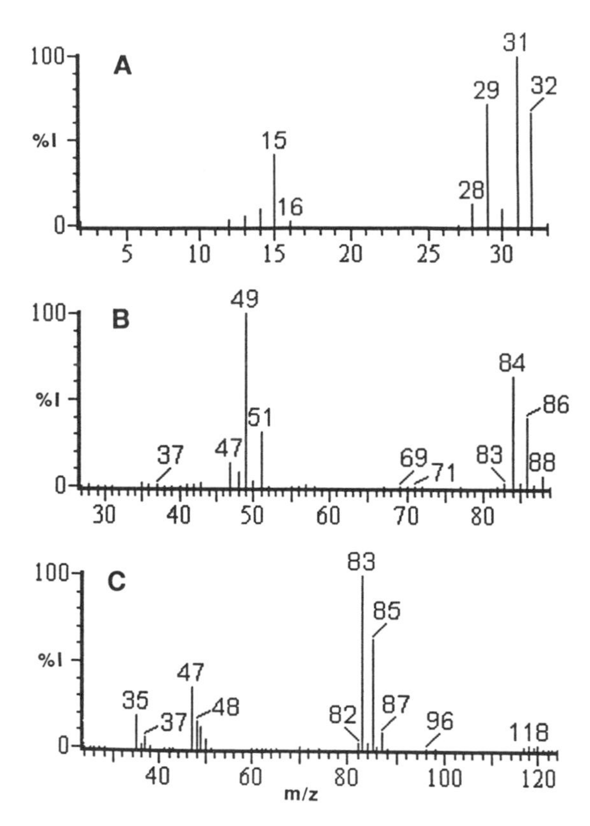

FIGURE 6.4 Electron ionization mass spectra of (A) methanol, (B) methylene chloride, (C) chloroform.

physical size and/or required carrier-gas flow. Either split, splitless, or on-column injections can be used in GC/MS. Split injections are usually avoided in cases where trace-level components are being analyzed. Splitless or on-column injections are preferred for trace component analysis. Often splitless injections (septum purge off) are made with split-injection port liners, which are packed with a small amount of adsorbent to trap nonvolatile, polar materials that could contaminate the column and/or the mass spectrometer. For the same reasons, a retention gap is recommended for on-column injections, because the initial column section can be replaced with little effect on the chromatogram. When using a retention gap, however, care must be taken to insure leak-tight connections, since a small air leak can have greater consequences for the mass spectrometer than for other detectors.

In general, there are no significant limitations on chromatographic operational parameters in GC/MS, with the exception of flowrates, as previously discussed. A few specific minor recommendations include not maintaining columns near their maximum operating temperatures for extended periods, because the increased column bleed may degrade spectrometer performance. In addition, it is usually best to condition columns prior to connecting them to the spectrometer. Split injections should be utilized when possible to avoid solvent contamination of vacuum pumps and prevent premature filament failure. Extended use of corrosive carrier gases, such as ammonia, should be avoided when possible.

6.2.3 Gas Chromatography/Mass Spectrometry Interfaces

Before capillary columns were introduced in GC, it was necessary to eliminate the larger volumes of carrier gas eluting from the chromatograph prior to introduction into the mass spectrometer. Various interfaces were developed for packed-column GC, including the jet, membrane, and effusion separators to name a few. The purpose of all of these devices is to eliminate most of the carrier gas, thereby enriching the analyte concentration. Unfortunately, in many cases, a large percentage of the analyte was removed as well and efficiencies in the 20–50% range were common. Since these devices are no longer in widespread use, the details of their design and operation will not be covered. Comprehensive discussions of these are available to the interested reader (18).

Capillary GC enables direct coupling of the chromatographic column to the mass spectrometer since flowrates are substantially reduced, typically from about 30 to 1 mL/min. The main requirement for these interfaces is that a constant temperature be maintained across the entire length from oven to ion source with no "cold spots" that may cause peak broadening or trapping of high boiling components. As simple as this requirement may seem, it is sometimes difficult to achieve due to ion source geometry, system configurations, etc.

6.2.4 Temperature Problems

As in gas chromatography, thermal degradation of components can occur in GC/MS. This degradation is frequently catalyzed by active sites somewhere in the chromatographic system.

The injection port is normally constructed with a replaceable glass liner. The silanol groups normally present on glass surfaces can cause degradation of sample components. An example of thermal degradation is shown in Figure 6.5 in which a fructose derivative (topiramate) being developed as a new anticonvulsant was analyzed by GC/MS (19). As the injection port temperature was increased, a new peak was detected which was identified as a thermal degradant. It is worthy to note that this degradation was usually detected after repeated injections at elevated temperatures. Typically, injection port liners are deactivated with silanizing reagents, which convert

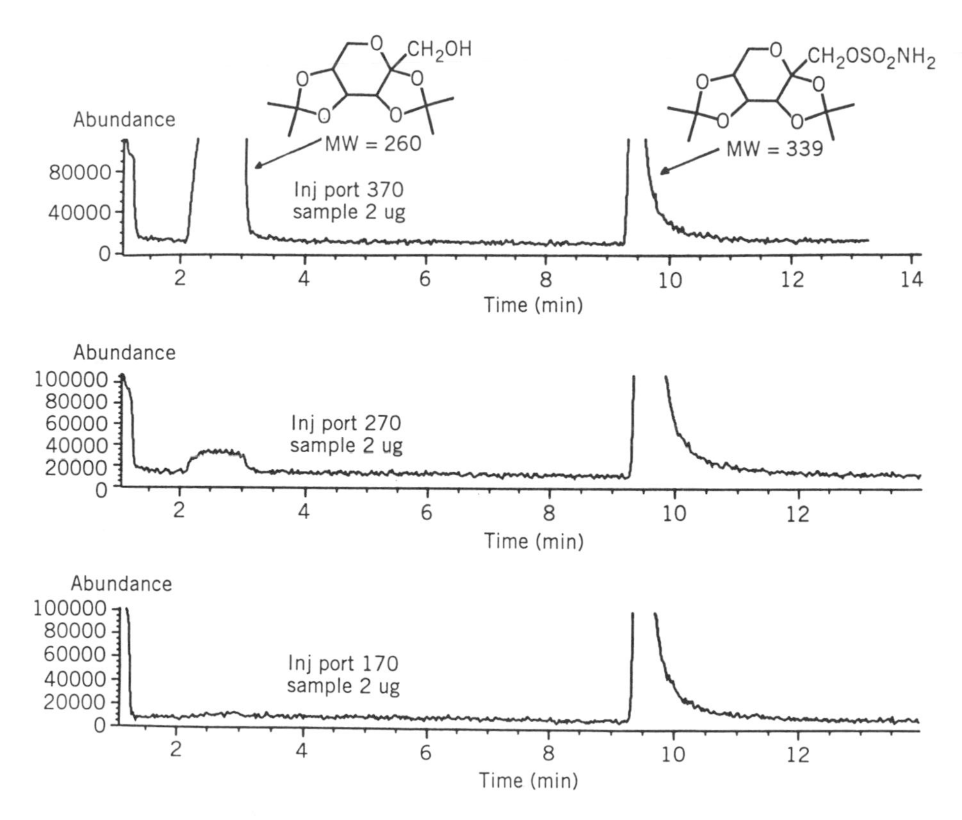

FIGURE 6.5 The total ion current chromatograms (TIC) demonstrate the thermal decomposition observed during gas chromatographic analysis of topiramate (MW = 339 u). As the injection port temperature is increased from 170 to 370°C, decomposition to the alcohol derivative (MW = 260 u) is observed. (Masucci and Caldwell, unpublished data.)

silanol groups to trimethylsilyl ethers. The use of on-column injections is also recommended to prevent thermal degradation, since the fused-silica capillary column is coated with liquid phase and fewer active silanol groups are present and injections are usually performed at lower temperatures.

It is also possible for thermal degradation to occur during separation on the column itself. Deactivation of the column can be performed in a similar manner by injection of a silanizing reagent prior to analysis. Since the interface may be operated at a higher temperature than the column oven, it can also be the location of thermal degradation. This can be minimized by lowering the interface temperature.

Finally, the mass spectrometer ion source itself can cause degradation, since the heated metal surfaces that the vaporized analyte molecules are subjected to can act as a catalyst. Again, decreasing the source temperature, if possible, can minimize this problem.

In GC with other detectors, the identity of the degradant is not generally known without running authentic standards, trapping of peaks, and matching retention times. However, with the additional molecular weight and structural information provided by the mass spectrometer, degradation products can frequently be identified during analysis. The nature of this degradation can often lead to specific chromatographic remedies.

6.2.5 Ion Sources

The purpose of the ion source, as the name implies, is to provide the energy necessary to ionize the analyte molecules, while being maintained at a high enough temperature to prevent analyte condensation. In addition, electrostatic focusing lenses are usually included to accelerate the ions and collimate the ion beam. The two types of ionization normally used in GC/MS are electron ionization (EI) and chemical ionization (CI). The specifics of their operation are covered in Sections 6.3 and 6.4, respectively.

A schematic diagram of a typical EI source is shown in Figure 6.6. It is important that the capillary column extend as close as possible to the ionization region without obstructing the ion or electron beams. This will maximize analyte transport into the ion source and minimize the possibility for thermal degradation.

6.2.6 Mass Analyzers

As ions leave the source, they enter into the mass analyzer where they are separated based on their mass-to-charge ratio. The mass range of interest is scanned causing separation of ions in space or time domains. The two most common mass analyzers, the magnetic sector and the quadrupole, are shown in Figure 6.7. The magnetic sector analyzer utilizes an electromagnet to separate ions in space according to the radius of their trajectories. The

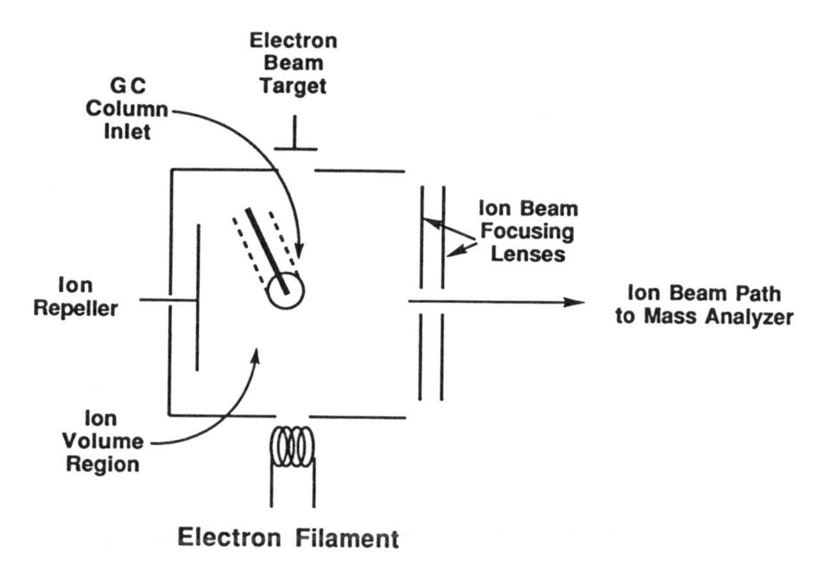

FIGURE 6.6 Diagram of an electron ionization source.

relationship of magnetic field strength to mass/charge is given by

$$\frac{m}{z} = \frac{B^2 r^2 e}{2V} \tag{6.1}$$

where B is the magnetic field strength, r is the radius of trajectory, e is the electron charge, z is the number of charges, and V is the accelerating voltage. In a typical magnetic sector analyzer, the magnetic field strength is varied, directing the ion beam across a narrow slit through which ions of increasing or decreasing m/z are selected. In this way a full-range mass spectrum is obtained. It is important that the magnet be scanned quickly enough to sample a chromatographic peak as it elutes. This was difficult with capillary GC in the past, but today's faster scanning magnets can easily cover the range from 40 to 500 m/z in 0.5 sec. This sampling rate (2 Hz) would yield 20 scans across a 10-sec-wide chromatographic peak.

The other common mass analyzer is the quadrupole, which consists of four cylindrical rods oriented in a square arrangement as shown. Radio frequency (rf) and direct current (dc) potentials are applied to the rods, enabling ions with a specific m/z to have a stable trajectory and pass through to the detector. By simultaneously increasing the rf and dc potentials, ions of increasing m/z will pass through the analyzer and be detected. Equations describing these ion trajectories are discussed elsewhere (20). Two advantages of the quadrupole are its fast scanning rate and lower cost. For these reasons, this is the analyzer most commonly used in GC/MS.

Two other analyzers also interfaced with GC are shown in Figure 6.8.

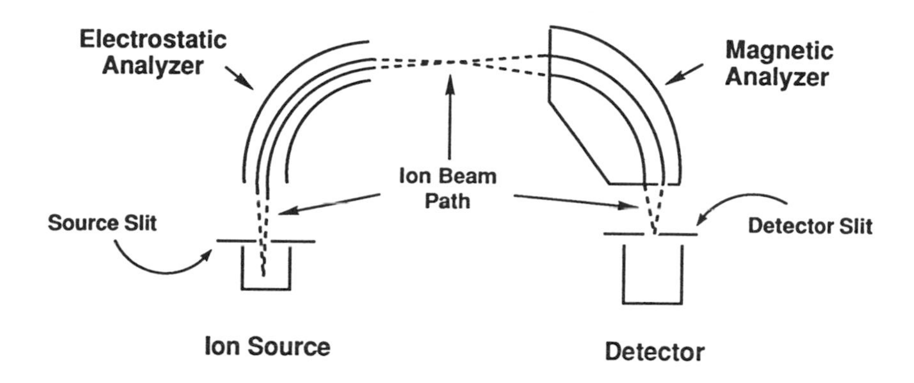

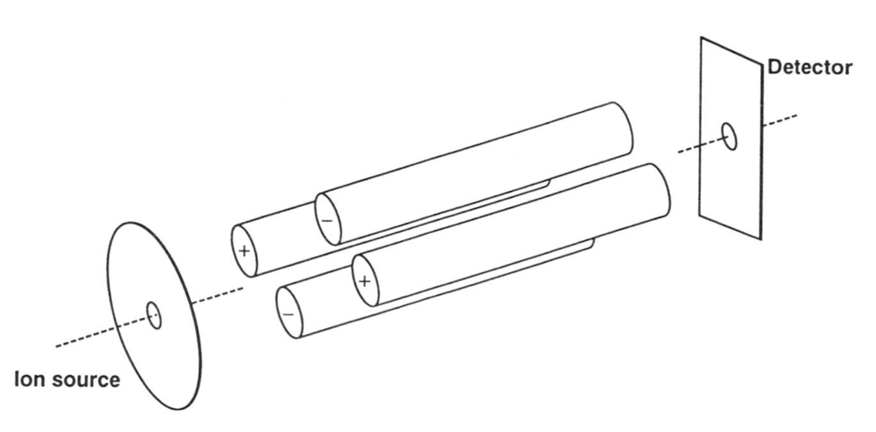

FIGURE 6.7 The two most popular mass analyzers. The magnetic sector instrument is normally configured in tandem with an electrostatic analyzer, which narrows the kinetic energy spread of the ion beam. For this reason, it is called a double-focusing spectrometer.

These include the ion trap analyzer in which ions can be confined by electric and magnetic fields. Ions of a specific m/z value circulate in stable orbits within the analyzer. As the rf is increased, ions of lower m/z values are destabilized and pass into the detector. Ions are typically introduced from the ion source in a pulsed fashion and the rf is quickly scanned to produce a

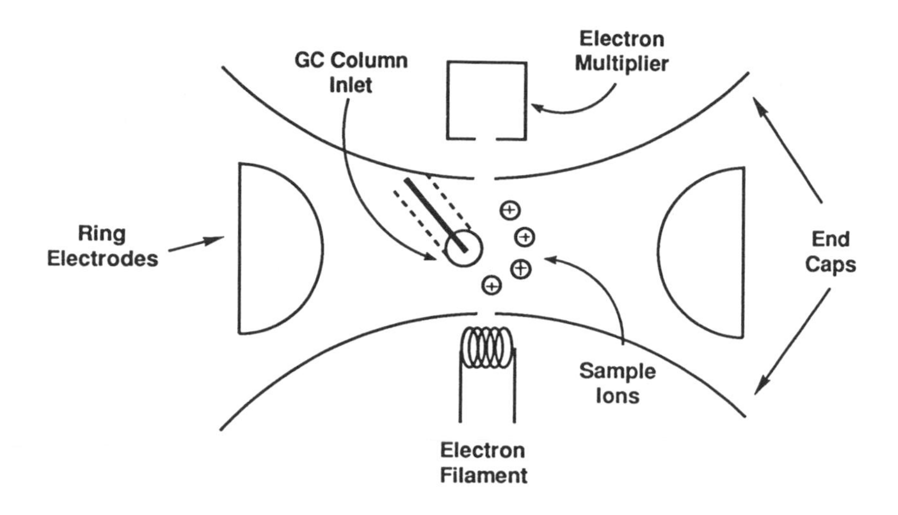

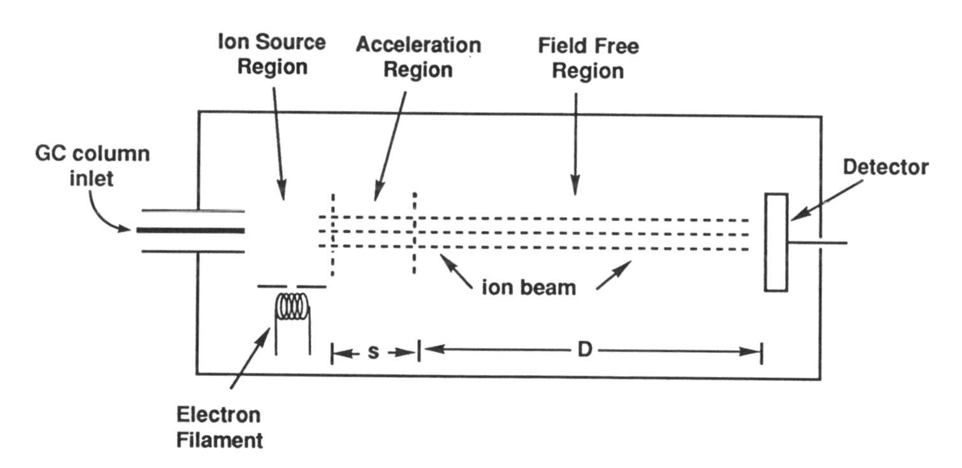

FIGURE 6.8 The ion trap and time-of-flight mass analyzers.

spectrum. These analyzers are the least expensive and most sensitive of all and are continuing to gain popularity as detectors for GC/MS (21).

The time-of-flight analyzers separate ions based on their migration time down a field-free region after acceleration. Since ions with equivalent kinetic energy are produced, those of lower m/z arrive first, followed in sequence by those of higher m/z, according to

$$\frac{m}{z} = \frac{2Ese(t)}{(D)^2} \tag{6.2}$$

In this equation, e is the electron charge, z is the number of charges, E is the magnitude of electrical field over which the ions are accelerated, s is the acceleration distance, t is the ion migration time, and D is the length of the field-free drift region. Although scan rates are fast enough for GC and the cost is low, the mass resolution is generally poorer than with other analyzers.

The degree of mass separation between adjacent ions is referred to as the mass resolution and is defined as $R = M/\Delta M$, where R is the resolution, M is the nominal mass of the ions, and ΔM is the difference in mass. All mass analyzers used in GC/MS are capable of at least resolving unit mass to m/z 1000 for a resolution of 1000. This resolution is sufficient for the majority of applications. However, it is sometimes necessary to accurately determine the mass of an unknown sample component to aid in its identification, by determination of its elemental composition. This is frequently done with a high-resolution analyzer such as the double-focusing magnetic sector instrument shown in Figure 6.7. With an instrument of this type it is possible to differentiate species such as benzene (C_6H_6, MW = 78.0469 u) and dimethyl sulfoxide (C_2H_6OS, MW = 78.0139 u), which require a resolution of 2400 to separate. Resolutions to 40,000 are possible with gas chromatography/Fourier transform mass spectrometry (GC/FTMS); however, the transient nature of chromatographic peaks and the rapid scanning makes the measurement more difficult (22).

6.2.7 Detectors

Mass resolved ions travel from the analyzer to the ion detector. The detectors used in MS are required to have fast response and a large gain to convert the small ion currents generated into recordable signals.

The most popular detector, the electron multiplier, is shown in Figure 6.9. The ions collide with the first of a series of dynodes. The dynodes are operated at between 1 and 3 kV, each one in the series maintained at a higher voltage. The effect is multiplication of the primary ion beam for a current gain of about 10^5. Frequently a higher voltage (5–20 kV) conversion dynode is inserted before the multiplier to increase the ion beam energy prior to detection.

Another detector used in GC/MS is the photomultiplier. These are similar in design to those used in optical spectroscopy. The ion beam collides with a phosphor-coated target, which converts the ions into photons that are subsequently amplified and detected. These detectors are typically operated at lower voltages (400–700 V) and last longer than conventional

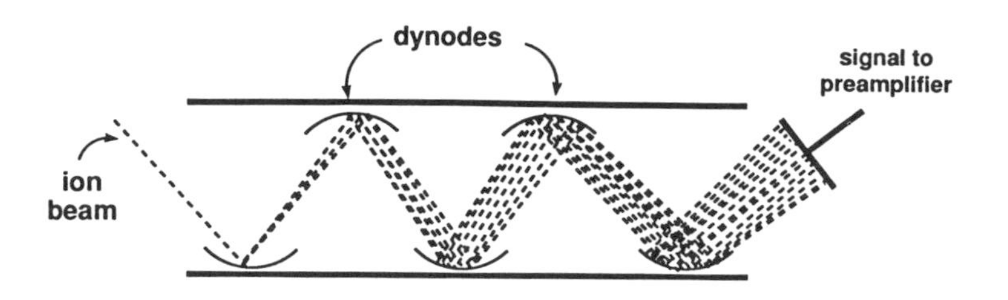

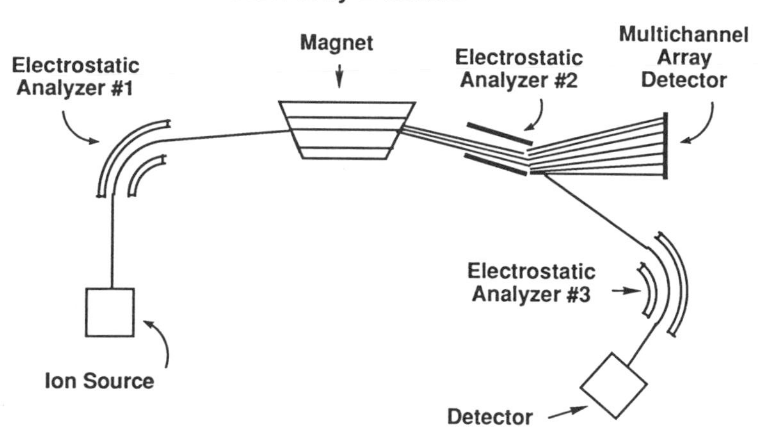

FIGURE 6.9 The electron multiplier and array detectors.

electron multipliers, since they are sealed units that are not subjected to external contamination.

To enable detection of many masses simultaneously, a photographic plate had been used in the past in combination with magnetic sector instruments. Today's substitute, for multiple mass detection, is the multichannel array detector (Figure 6.9). A series of evenly spaced detectors are configured in a linear array. A selected range of ions is directed onto the array, allowing simultaneous recording of many masses. This enhances the overall sensitivity of measurement, extending MS for applications such as trace analysis of unknown components. However, the specialized nature of this detector combined with its substantial cost has limited its use to date.

6.2.8 Scanning Techniques

The mass spectrometer is used to perform many tasks, including analysis of unknowns, trace level analysis, target compound analysis, and accurate mass measurements, to name a few. Each of these analyses has a number of specific instrumental requirements, including a preferred scanning mode.

The most convenient method of mass scanning is full mass range scanning. In this scan mode, the spectrometer is scanned over a mass range covering all predicted molecular and fragment ions produced for a complex multiple-component sample. For most GC/MS of unknowns, this requires scanning the analyzer from about 50 to 600 u. This lower mass (50 u) is chosen to exclude the background ions produced from residual air, carrier gases, and CI reagent gases normally used. The upper limit is based on useful volatility of analytes. Many compounds above this molecular mass (600 u), with the exception of specifically prepared volatile derivatives, have insufficient vapor pressure for analysis. In addition, for many spectrometers this represents the maximum practical mass range for capillary column use, since scanning speeds of 0.5 sec/scan become more difficult beyond this range.

Another scanning mode frequently used is selective ion monitoring (SIM). With this method the mass analyzer can be set to sample a single m/z value over the course of the chromatographic separation. By monitoring only a single m/z, the sensitivity is enhanced up to three orders of magnitude depending on mass range, since the instrument does not spend time sampling undesired masses. This technique is useful for analyses such as quantification of target compounds, where the base peak of the analyte is normally chosen as the monitored m/z. It is also useful for analytes that are only partially resolved, since a unique ion can usually be chosen, which is not produced by the coeluting species. The computerized control possible with modern spectrometers allows the selected ion masses to be changed during the course of a separation, enabling optimization of analyses for all individual mixture components. SIM scanning epitomizes the selectivity and sensitivity possible with GC/MS.

A variation of SIM involves monitoring multiple ions. This scanning method allows selection of several discrete masses. These can be molecular and/or fragment ions of a single analyte or of different analytes. This technique can be used as a compromise between full mass range scanning for best qualitative information and single mass SIM for increased sensitivity, by choosing several characteristic peaks from the desired components. Modern computers have significantly improved this process and specific mass ranges can be selected as the course of the separation proceeds. This allows, for example, shifting the scanned range to higher values as the column oven is heated over the course of a gradient separation, since with many analog series, elution time and/or temperature increases with molecular mass.

6.2.9 Data Presentation

Going hand-in-hand with the scanning techniques described above, specific data presentation formats are used in GC/MS. A summary of some of these is given in Figure 6.10. The most common format used with full mass range scanning is the total ion current (TIC) chromatogram (Figure 6.10a). This signal represents the summed ion current for the peaks detected in each mass spectral scan. Depending on response factors, this chromatogram frequently resembles the flame ionization detector trace. Frequently, the mass spectral scan recorded at the maximum ion current for each chromatographic peak is presented as its characteristic spectrum. This is not usually a problem unless the scan rate is too slow relative to the peak width (less than ten scans across the peak). Alternatively, spectra can be averaged to normalize variations in analyte concentration as each elutes into the ion source.

In cases in which target compound analysis is being performed, chromatograms can be generated in which a single m/z is profiled. This is usually an intense or structurally characteristic peak which will identify the analyte of interest. These traces are often referred to as selected ion or mass chromatograms, and the method of data retrieval is called selected ion extraction. It differs from SIM in that the ion is selected postacquisition from the full mass range data. This technique is also valuable in cases where two or more components are not chromatographically resolved. By selecting ions characteristic of each of these analytes and plotting these mass chromatograms, a broad peak can be deconvoluted into its individual components and peak purity can be determined. Mass chromatograms are shown in Figures 6.10b and c. To complement multiple-ion SIM, multiple-ion extraction can also be performed to differentiate components that exhibit common intense peaks but different minor peaks, such as in the case of a hydrocarbon or other chemical series.

6.2.10 Background Artifacts

There are several sources of contamination in GC/MS. As previously mentioned, many liquid phases used in GC columns have appreciable vapor pressure and thus can bleed into the ion source along with the effluent. Vacuum pump oil also has appreciable vapor pressure and can bleed into the ion source. Another very common contaminant are phthalates such as dioctyl and di-*n*-butyl phthalate, which are used to stabilize plastic or rubber seals. Organic solvents stored in plastic bottles or passed through plastic tubing can be contaminated with these plasticizers. An intense peak at m/z 149, which corresponds to the phthalic anhydride cation, will be observed for phthalates (Figure 6.15). These contaminants and others, such as air and water, make up the background of a GC/MS experiment. Spectra of some of these are shown in Figures 6.11 and 6.12.

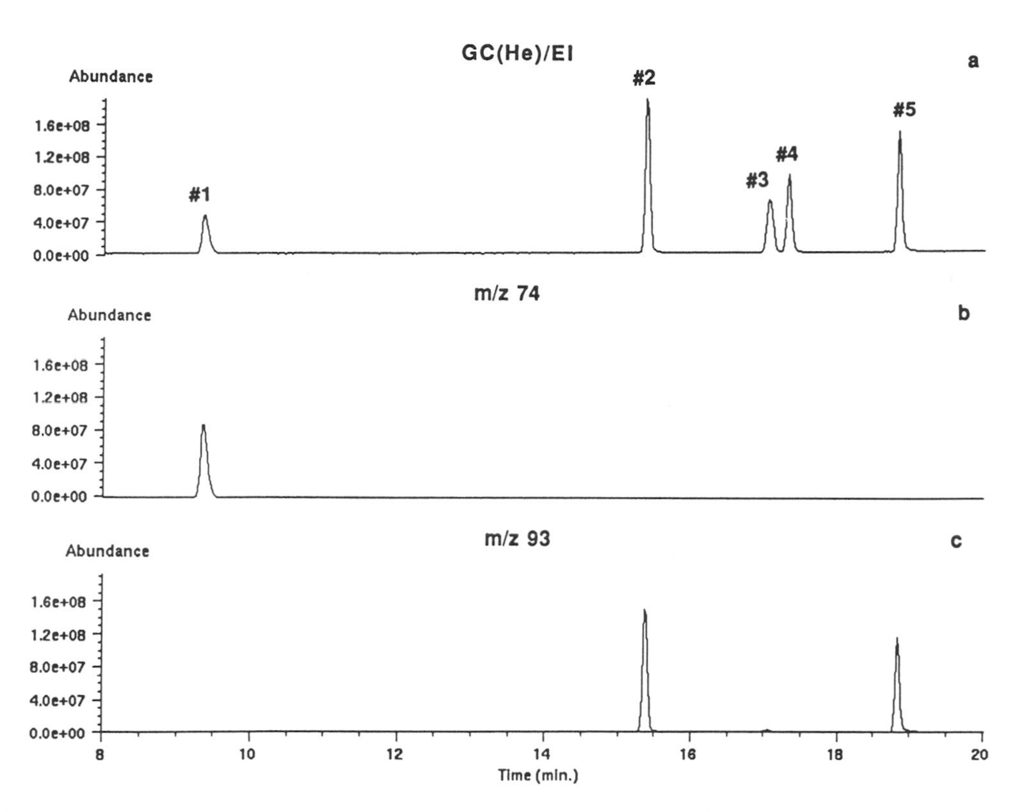

FIGURE 6.10 (a) GC/EIMS total ion current chromatogram (TIC) of a five-component mixture of 1, N-nitrosodimethylamine; 2, bis(2-chloroethyl)ether); 3, bis(2-chloroisopropyl)ether); 4, *N*-nitroso-di-*n*-propylamine); and 5, bis(2-chloroethoxy)methane. GC/EIMS obtained under the following conditions: GC conditions—column, DB-1 (30 m × 0.320 mm); film thickness, 5.00 μm; carrier gas, helium @ 25 cm/sec; oven program, 45°C for 3 min, then 10°C/min to 300°C for 12 min; injection port, 265°C; sample, 1 μL at 2000 μg/μL; solvent, methylene chloride; samples were injected in the splitless mode (0.75-min load). MS conditions—mass range, 50–500 u; electron energy, 70 eV; repeller, 7.0 V; GC/MS interface temperature, 250°C; ion source temperature, 200°C. (b) Mass chromatogram of *m/z* 74. (c) Mass chromatogram of *m/z* 93. (Masucci and Caldwell, unpublished data.)

For a mass spectrum of a particular analyte, the spectrum is a mixture of the background and the analyte. If a representative mass spectrum of the background can be obtained, it is desirable to subtract the background spectrum from the analyte spectrum. The resulting subtracted mass spectrum of the analyte is then of the "pure" substance.

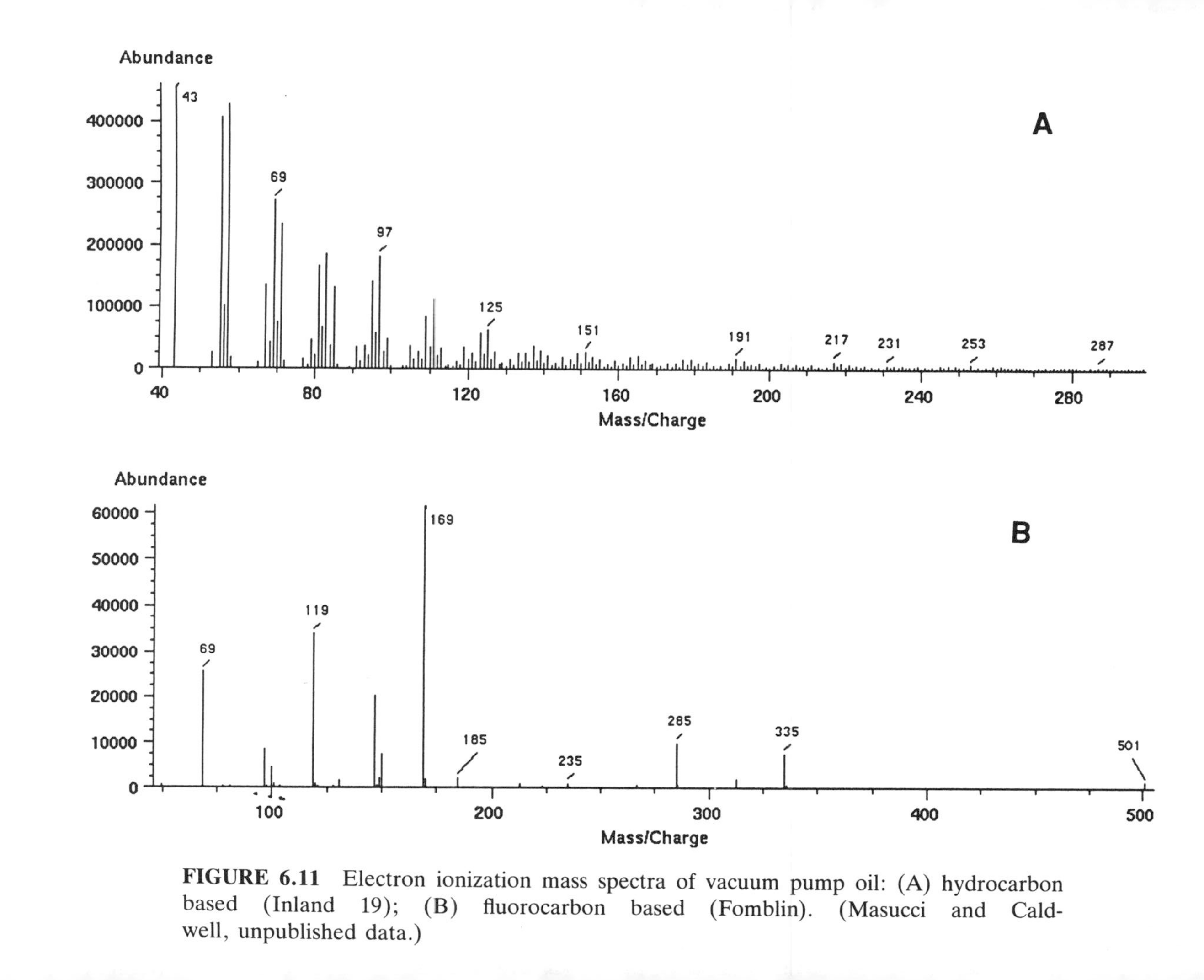

FIGURE 6.11 Electron ionization mass spectra of vacuum pump oil: (A) hydrocarbon based (Inland 19); (B) fluorocarbon based (Fomblin). (Masucci and Caldwell, unpublished data.)

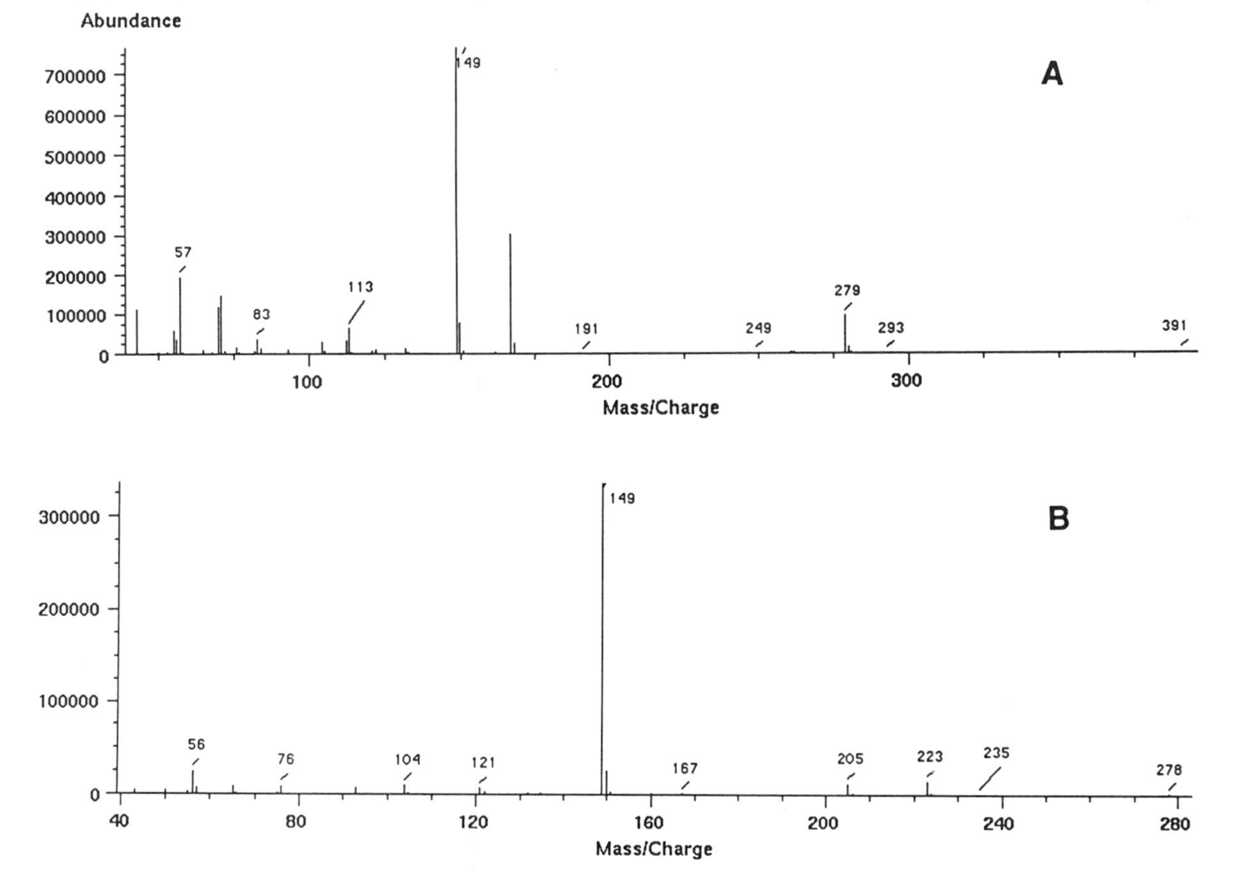

FIGURE 6.12 Electron Ionization mass spectra of phthalates: (A) dioctyl phthalate; (B) di-*n*-butyl phthalate. (Masucci and Caldwell, unpublished data.)

6.2.11 Chemical Derivatizations

In GC/MS, chemical derivatization of the sample molecule often improves peak symmetry, volatility, and thermal stability for gas chromatographic separations and can afford improved selectivity and detection limits for mass spectral analyses. The most common form of derivatization of an active hydrogen atom (–OH, $–CO_2H$, $–NH_2$, –NHR, –SH) in a sample molecule is replacement of this atom with a trimethylsilyl (TMS) group. Many reagents are available for preparation of TMS derivatives such as *N*,*O*-bis(trimethylsilyl)acetamide and bis(trimethylsilyl)trifluoroacetamide. Since each TMS group adds 73 u to the analyte mass, depending on the number of active hydrogens, this could add significantly to the mass of a compound and put it out of range of the spectrometer's upper mass limit. In such case, a fragment may have to be used to characterize the derivative.

The reaction of diazomethane with carboxylic acids to form methyl esters and with phenols to produce aromatic methyl ethers is easy to perform. In addition to providing a more volatile derivative, this method can be used diagnostically to verify the presence of these compounds since the observed mass will increase by 14 for each reactive OH group. This derivatization is frequently used in drug metabolism studies, since polar acids and phenols are often produced by the biotransformation of pharmaceutical agents. For more details on chemical derivatization the reader is referred to Vouros (12) and Drozd (23).

6.3 GAS CHROMATOGRAPHY/ELECTRON IONIZATION MASS SPECTROMETRY

6.3.1 Electron Ionization

There are many general reviews of electron ionization (EI) mass spectrometry (24–28). Here key factors that affect the technique are described.

The neutral molecules [M] that elute from the gas chromatographic column must be charged positively or negatively in order to manipulate them in mass analyzers. As these neutral molecules randomly diffuse throughout the ion source they are bombarded with electrons at 70 electron volts (eV) of energy. If an electron with sufficient energy collides and knocks out of orbit one of the neutral molecule's electrons, a radical cation $[M]^{+\cdot}$ is formed (Reaction 6.3). This $[M]^{+\cdot}$ cation is referred to as the molecular ion. If the molecule captures the electron, a radical anion $[M]^{-\cdot}$ is formed (Reaction 6.4):

$$e^- + [M] \rightarrow [M]^{+\cdot} + 2e^- \tag{6.3}$$

$$e^- + [M] \rightarrow [M]^{-\cdot} \tag{6.4}$$

A few hundredths of a percent of the sample molecules are ionized to molecular ions under these conditions, with the bulk of the sample molecules being removed by the vacuum pumps. Negative ions are 10^4 less abundant than positive ions at 70 eV. One might think, with such a low concentration of ions being formed, that the EI technique would be insensitive. On the contrary, only a few nanomoles of sample are required to be detected.

The electrons that are emitted from the filament in the ion source have a distribution of energies from 0 eV to greater than 20 eV (1 eV = 96.3 kJ mol^{-1}). Thus, some of the radical cations $[M]^{+\cdot}$ and the radical anions $[M]^{-\cdot}$ have considerable excess energy and some have very little. It is this excess energy that makes the radical cations and anions unstable and provides the source of energy that could fragment almost any single bond contained in the molecule. Organic molecules have bond energies approximately in the range 193 to 578 kJ mol^{-1}. The molecular ion with its excess energy undergoes a unimolecular decomposition reaction where one fragment retains the positive charge $[F_1]^+$ and the other is a neutral radical $[N_1]^\cdot$ (Reaction 6.5). The molecular ion can fragment via single bond cleavages to many different fragment types ($[F_1]^+$, $[F_2]^+$, $[F_3]^+$, etc.). In some cases, the molecular ion can rearrange such that two bonds are cleaved and two are formed to produce a radical cation and neutral molecule (Reaction 6.6). Since all chemical bonds are not of equal strength, the fragmentation process is not random and structurally characteristic fragments are produced. The excess energy in the $[M]^{-\cdot}$ ion causes it to convert back to a neutral molecule (Reaction 6.7), which accounts for its low concentration in the ion source. Reactions 6.3–6.6 all occur in the ion source simultaneously.

$$[M]^{+\cdot} \rightarrow [F_1]^+ + [N_1]^\cdot \tag{6.5}$$

$$[M]^{+\cdot} \rightarrow [R]^{+\cdot} + [N] \tag{6.6}$$

$$[M]^{+\cdot} \rightarrow [M] + e^- \tag{6.7}$$

The thermochemical relationship for Reaction 6.5 can be expressed as the ionization energy (IE) of the radical fragment [F·] less the IE of the neutral sample molecule [M] plus the bond dissociation energy [D] of the F–N bond; i.e., ΔH(Reaction 6.5) = IE(F·) − IE(M) + D(F–N). The ionization energy is the amount of energy required to remove an electron from the species under consideration. The bond dissociation energy is the amount of energy required to produce a homolytic bond cleavage. Several sources of thermochemical data can be used to calculate the exothermicity of Reaction 6.5 (29, 30). The importance of positive charge stabilization and bond strength is reflected in ΔH(Reaction 6.5). Thus, a comparison of two competitive fragmentation reactions from the same sample molecule, the difference in the IEs, and the difference in the D's control the abundance of the pathways. In other words, the fragment that can best stabilize the positive charge will predominate the EI spectrum.

The molecular ion is observed in the mass spectrum in varying abundance. When the molecular ion retains a large amount of excess energy, the $[M]^{+\cdot}$ cation may totally fragment and not be observed in the mass spectrum. A lower voltage on the filament, such as 10–15 eV, may increase the abundance of the molecular ion. Fragmentation of the molecular ion decreases since less excess energy is transferred to the $[M]^{+\cdot}$. Thus, lower electron energies drastically alter the overall spectrum as well as maximize molecular ion production. However, the total ion production decreases, causing lower sensitivity. Electron energy variations in the range of 55–85 eV are rather insignificant in the overall appearance of the GC/EI technique. For this reason, EI mass spectrometers are operated at 70 eV electron energy, and spectra from various instruments are comparable.

6.3.2. Qualitative Methods—Structure Elucidation

The EI fragments are pieces of the original molecule and provide a way to determine its structure. McLafferty (27) has created a step-by-step procedure for interpreting an unknown mass spectrum. While it is not the intention to repeat the entire procedure here, useful key steps in this process are reviewed below.

First, the sample's history is probably the most valuable piece of information to have when determining an unknown structure. Time should be taken to obtain as much information as possible. Second, a mass spectrum of the unknown should be obtained free of any artifacts (see Section 6.2.10). Third, the molecular ion and thus, the molecular weight of the sample molecule, should be determined. Since mass spectra frequently do not contain a molecular ion (see Figure 6.4) an alternative ionization technique (see Sections 6.4 and 6.5) or lowering electron energy (see Section 6.3.1) should be attempted to establish the molecular weight. Fourth, the recognition of natural abundance of isotopes in a mass spectrum can provide elemental composition of the peak at an m/z value. Table 6.1 lists the natural abundances of the commonly encountered atoms. Note that fluorine, phosphorus, and iodine (not listed in table) have no stable heavy isotopes. The presence of stable isotopes in the sample molecule results in the mass spectrum containing peaks with multiplicity. Consider the mass spectrum of acetaminophen (Figure 6.2), the peak m/z 151 is the molecular ion that has an elemental composition of $C_8H_9NO_2$. The peak at m/z 152 is due to the naturally occurring ^{13}C isotope of acetaminophen, i.e., $^{12}C_7{}^{13}CH_9NO_2$, and has a relative intensity of 0.011 multiplied by the number of carbon atoms (Table 6.1).

TABLE 6.1 Natural Abundance of Stable Heavy Isotopes

Name	Element	Abundance	Isotope	Abundance	Isotope	Abundance
Hydrogen	^{1}H	99.99	^{2}H (or D)	0.01		
Carbon	^{12}C	98.9	^{13}C	1.1		
Nitrogen	^{14}N	99.6	^{15}N	0.4		
Oxygen	^{16}O	99.76	^{17}O	0.04	^{18}O	0.20
Silicon	^{28}Si	92.9	^{29}Si	4.7	^{30}Si	3.1
Sulfur	^{32}S	95.02	^{33}S	0.76	^{34}S	4.22
Chlorine	^{35}Cl	75.77			^{37}Cl	24.23
Bromine	^{79}Br	50.5			^{81}Br	49.5

Source. Reference 27.

Compounds containing an odd number of nitrogen atoms, such as acetaminophen, will have an odd nominal molecular mass. Those compounds with an even number of nitrogens will have an even nominal mass. This is called the nitrogen rule and is valid because nitrogen has an even atomic mass and an odd valence. The atoms C, H, O, S, Si, and the halogens either have an even atomic mass and an even valence or an odd atomic mass and an odd valence.

Isotopes of Cl and Br are easily recognized in a mass spectrum. In Figure 6.4b is shown the mass spectrum of methylene chloride (CH_2Cl_2). The peak at *m/z* 84 is the molecular ion that has an elemental composition of $CH_2{}^{35}Cl_2$; the peak at *m/z* 86 has an elemental composition of $CH_2{}^{35}Cl^{37}Cl$; the peak at *m/z* 88 has an elemental composition of $CH_2{}^{37}Cl_2$. The base peak at *m/z* 49 is simply the [M–Cl] cation. Note that the ratio of the abundances of the peaks at *m/z* 49 (100%) and 51 (33%) is approximately 3:1. This ratio is consistent with this fragment containing one chlorine atom (Table 6.1). Note that the ratio of the abundances of the peaks at *m/z* 84 (65%), 86 (42%), and 88 (7%) is approximately 9:6:1. This ratio is consistent with this fragment containing two chlorine atoms (27).

As the fifth step, if the molecular ion can be recognized, a strategy for interpretation is to calculate losses from the molecular ion that might account for the fragments. Some common losses are listed in Table 6.2. Again note that the base peak at *m/z* 49 in Figure 6.4 is simply the loss of a Cl atom from the molecular ion at *m/z* 84.

The final strategy for mass spectral interpretation is to recognize mass series of fragments that are unique to a particular functional group. For example, if a series of peaks at *m/z* 43, 57, 71, 85, 99, 113, and 127 (note the 14 u mass difference) were observed, it can be concluded that the unknown sample molecule contains a long-chain, saturated hydrocarbon tail

TABLE 6.2 Common Losses From the Molecular Ion

Molecular Ion	Species	Molecular Ion	Species
M-15	CH_3	M-41	C_3H_5
M-16	O, NH_2	M-42	C_3H_6, CH_2CO
M-17	OH, NH_3	M-43	C_3H_7, CH_3CO
M-18	H_2O	M-44	C_3H_8, CO_2
M-19	F	M-45	CO_2H, OC_2H_5
M-20	HF	M-46	CH_3CH_2OH
M-27	HCN	M-55	C_4H_7
M-28	CO, C_2H_4	M-57	C_4H_9
M-29	CHO, C_2H_5	M-58	C_4H_{10}, $(CH_3)_2CO$
M-30	CH_2O, C_2H_6	M-60	CH_3COOH
M-31	OCH_3	M-73	$(CH_3)_3Si$
M-32	S, $HOCH_3$	M-79	Br
M-34	H_2S	M-89	$(CH_3)_3SiOH$
M-35	Cl	M-127	I

(Figure 6.11). Detailed mechanisms have been published for organic compounds containing a wide variety of functional groups (26–28).

On-line computer comparison of the mass spectrum of an unknown analyte against a reference mass spectral library was developed to aid the interpreter (27). Over the past two decades, there has been a steady growth in the understanding of how to use computers to interpret mass spectral information. It is not surprising that this development of library search routines has been strongly influenced by computer technology. Several successful algorithms for comparing the mass spectrum of an unknown against a library of known compounds have been developed and are available on most commercial mass spectrometer data systems. The most widely used algorithm is the Probability-Based Matching (PBM) software. This software can search 220,000 reference mass spectra in approximately three seconds (27). The PBM uses statistical information gathered from mass spectral databases to assign a "uniqueness" to mass spectral peaks. For example, a fragment at m/z 57 with an abundance of 100% occurs quite frequently in mass spectra. Since this peak has a high probability of occurrence, it is given a low uniqueness value. A fragment at m/z 570 with an abundance of 100% would be given a high uniqueness value since this combination of mass and abundance occurs infrequently in mass spectra. Comparison of the mass spectrum of an unknown sample molecule against a reference mass spectral library is called forward searching. Comparison of the reference library against an unknown is called reverse searching. All mass spectra are a mixture of the analyte and the background. In a forward search, these background peaks are included in the search; however, the reverse searching procedure ignores peaks in the unknown that are not in the reference spectrum. Reverse searching typically gives superior search results.

6.3.3 Quantitative Methods

GC/EI quantitative methods are routinely applied to biological fluids, such as blood plasma and urine (31), environmental (32), energy (33), agricultural (34), law enforcement (35), and biomedical (36) areas. Almost all quantitative applications of GC/EI rely on stable isotope dilution techniques and selective ion monitoring (SIM) recording methods. Stable isotope-labeled analogs of sample molecules serve as ideal internal standards and are added to the sample matrix to correct for losses during sample preparation and GC/EI analyses. Multiply deuterated or ^{13}C-labeled compounds with a mass difference of at least 3 u from the unlabeled compound are preferred since they preclude overlap in simultaneous recording of selected ion pairs. Complex biological matrices, such as urine or plasma, are extracted into organic solvents and purified by chromatographic methods. Due to certain analyte functional groups (e.g., ROH and RCO_2H), these samples may have unfavorable gas chromatographic properties leading to tailing peaks and/or unfavorable mass spectrometric properties leading to insensitive detection. These samples are generally derivatized to improve gas chromatographic properties (peak symmetry, volatility, thermal stability) by replacing any active hydrogen atom (–OH, $-NH_2$, –NHR, –SH) in a molecule with a trimethylsilyl group. A calibration curve is typically constructed by analyzing standard solutions containing varying amounts of the sample of interest and a fixed amount of internal standard. Calibration curves are typically linear with a correlation coefficient >.99

An interesting example of the GC/EI quantification method is the determination of underivatized anabolic steroids utilizing a short (3 to 5-m) gas chromatrographic column and full-scan EIMS (36). The development of an analytical assay for the determination of drugs in biological fluids is a challenge. Because of high polarity, low volatility, and thermal instability, many drugs must be derivatized for GC/EI. Whenever the raw sample is handled either by purification or derivatization procedures, the potential for sample loss is evident. Thus, short-column GC/EI methods are strategies for direct quantification of drugs. It has been determined that GC/MS operating conditions (i.e., injection method; splitless, column inlet pressure, 1–2 psi; column temperature, 150–250°C at 20°C min^{-1}; ion source temperature, 190°C; and electron energy, 40 eV) are critical for maximal chromatographic performance and mass spectral sensitivity. Six anabolic steroids (oxymetholone, testosterone, nortestosterone, stanozolol, methyltestosterone, and dehydrotestosterone) were spiked in urine samples at 2000 ng/mL. The results indicated that testosterone, nortestosterone, methyltestosterone, and dehydrotestosterone had detection limits slightly higher than results for derivatized steroids using a 30-m column and SIM. It was noted that oxymetholone and stanozolol were lost during the sample preparation procedure. While this approach needs further development, it does point out that not all polar compounds require derivatization.

6.3.4 Negative-Electron Ionization

Negative ions generated from electron ionization (EI) generally give poor sensitivity, because negative-ion production requires electrons of much lower energy (ca. 0 eV) to facilitate electron capture and ion-pair production. Halide ions (F^-, Cl^-, Br^-, and I^-) are typically observed as low-level background anions in this mode. However, negative-ion chemical ionization (NICI) can generate negative-ion abundances comparable to EI. A description of NICI is given in Section 6.5.

6.4 GAS CHROMATOGRAPHY/POSITIVE-ION CHEMICAL IONIZATION MASS SPECTROMETRY

6.4.1 Advantages of Positive-Ion Chemical Ionization

A useful alternative to electron ionization mass spectrometry (EI) is positive-ion chemical ionization (PICI) mass spectrometry. The PICI technique was developed by Field and coworkers (37–39) and general reviews of this technique by Munson (40, 41), Harrison (42), and Bartmess (43) have appeared. Briefly, the difference between EI and PICI can be understood simply by considering the amount of energy deposited into the sample molecules during the ionization process. In EI, sample molecules are ionized to radical cations by electrons. The excess energy deposited in these radical cations can range from near zero to the electron energy (typically 70 eV) and can cause extensive fragmentation. In PICI, sample molecules are ionized to cations by other cations and the excess energy deposited in these cations depends on the thermochemistry of the cation/molecule reaction. The energy range of these cation/molecule reactions is much narrower (near zero to ca. 20 eV), which results in much less fragmentation.

As discussed above, sample molecules [M] ionized by electrons (Reaction 6.8) retain some fraction of the 70-eV energy beam internally. Depending on the amount of excess internal energy retained by the $[M]^{+\cdot}$ cations, these cations generally further dissociate to yield fragment cations and neutral radicals (Reaction 6.9). Remember that the amount of energy required to break a bond in a typical organic molecule ranges only from ca. 2 to 6 eV (193 to 578 kJ/mol). In some cases the amount of internal energy retained by the $[M]^{+\cdot}$ cation is so great that the $[M]^{+\cdot}$ cation fragments completely. Thus, the $[M]^{+\cdot}$ cation is sometimes not observed in the EI spectrum. The lack of a molecular cation can greatly complicate the identification of unknowns and is a serious disadvantage of the EI technique:

$$e^- + [\mathrm{M}] \rightarrow [\mathrm{M}]^{+\cdot} + 2e^- \tag{6.8}$$

$$[\mathrm{M}]^{+\cdot} \rightarrow [\mathrm{F}_1]^+ + [\mathrm{F}_2]^+ + [\mathrm{N}]^\cdot \tag{6.9}$$

The PICI technique requires that a gaseous mixture consisting of a reagent gas [R] (e.g., methane, isobutane, or ammonia) and the sample molecule [M] of interest be present in the ion source in a molar ratio of approximately 1000 to 1. Since the reagent gas is in a much larger excess to the sample molecules, virtually all primary cations are produced by direct electron ionization of the reagent gas (Reaction 6.10). The sample molecules are not ionized to any extent by the direct electron beam. These primary cations (C_p^+) will further react with the bulk reagent gas to produce a set of secondary cations $[C]_n^+$ that are unique to the reagent gas and are at a relatively steady-state concentration (Reaction 6.11). The number of secondary cations generated will vary from reagent gas to reagent gas; however, for the common gases there are generally not more than three major cations (42). At some point in time these secondary cations collide with [M] and a cation/molecule reaction occurs. Sample molecules ionized by these secondary cations typically produce protonated molecular cations $[M+H]^+$ (Reaction 6.12), hydride abstraction cations $[M-H]^+$ (Reaction 6.13), charge exchange cations $[M]^{+\cdot}$ (Reaction 6.14), and/or cluster adduct cations $[C+M]^+$ (Reaction 6.15). The abundance of these cations is controlled by cation/molecule reactions and ultimately depends on the specific reagent gas, its pressure, and its source temperature. More details on how these cations are created in PICI are given in Sections 6.4.3–6.4.7.

The $[M+H]^+$, $[M-H]^+$, $[M]^{+\cdot}$, or $[M+C]^+$ cations that are produced via the PICI technique are usually much less energetic than those formed in the EI process, and result in less fragmentation. The fragments produced by PICI are sometimes a different set of fragments than those produced by EI for the same molecule. The fragmentation of $[M+H]^+$, $[M-H]^+$, and $[M+C]^+$ cations usually involves the elimination of an even-electron neutral (a stable molecule) to form an even-electron fragment cation. It should also be noted that $[M+H]^+$, $[M-H]^+$, and $[M+C]^+$ cations are even-electron cations, which, in general, are more stable than the odd-electron cations produced in EI. Therefore, in many cases, the $[M+H]^+$, $[M-H]^+$, or $[M+C]^+$ cation is observed in relatively high abundance with a limited number of fragment cations.

$$e^- + [\mathrm{R}] \rightarrow \mathrm{C}_p^+ + 2e^- \tag{6.10}$$

$$\mathrm{C}_p^+ + [\mathrm{R}] \rightarrow [\mathrm{C}]_1^+ + [\mathrm{C}]_2^+ + \cdots + [\mathrm{C}]_n^+ \tag{6.11}$$

$$\mathrm{C}^+ + [\mathrm{M}] \rightarrow [\mathrm{M}+\mathrm{H}]^+ + [\mathrm{C}-\mathrm{H}] \tag{6.12}$$

$$C^{+} + [M] \rightarrow [M - H]^{+} + [C + H] \tag{6.13}$$

$$C^{+\cdot} + [M] \rightarrow [M]^{+\cdot} + C \tag{6.14}$$

$$C^{+} + [M] \rightarrow [M + C]^{+} \tag{6.15}$$

In Figure 6.13 examples of EI, methane PICI, and an interesting cation/molecule reaction called self-PICI are shown. Figure 6.13a is the 70-eV EI spectrum of an aromatic compound in which the $[M]^{+\cdot}$ cation (m/z 429) is not observed. For this molecule, the amount of internal energy retained by the $[M]^{+\cdot}$ cation after electron ionization was so great that the $[M]^{+\cdot}$ cation totally fragmented (Reactions 6.8 and 6.9).

Figure 6.13b is the methane PICI spectrum of the same compound in which the $[M + H]^{+}$ cation (m/z 430) is the base peak (Reaction 6.12). Note there is less fragmentation in PICI than in EI. Figure 6.13c is an example of a phenomenon called self-PICI, which can show up in spectra if the pressure of the reagent gas is low and the pressure of the sample molecules is high. That is, when the reagent gas is accidentally turned off and a sample molecule at high concentration (ca. 0.1 Torr) is introduced into a CI source, a mixed EI/CI spectrum is obtained. It is easy to see peaks from both the EI and PICI ionization processes in Figure 6.13c. The cation/molecule reaction that produces the self-PICI peak is generalized by Reaction 6.16:

$$[M]^{+\cdot} + [M] \rightarrow [M + H]^{+} + [M - H]^{\cdot} \tag{6.16}$$

$$[M + H]^{+} \rightarrow [M + H - X]^{+} + [X] \tag{6.17}$$

$$[M + H]^{+} + [M] \rightarrow [2M + H]^{+} \tag{6.18}$$

Note that the $[M + H]^{+}$ cation is formed by transfer of a proton from $[M]^{+\cdot}$ to [M] or by transfer of an hydrogen atom from [M] to $[M]^{+\cdot}$. The $[M]^{+\cdot}$ (Reaction 6.9) and $[M + H]^{+}$ (Reaction 6.17) cations can both fragment. The [X] species in Reaction 6.17 is a stable neutral. If the analyte pressure is sufficiently high, dimers ($[2M + H]^{+}$) are also produced (Reaction 6.18). Self-PICI has been noted by several investigators (44–47).

6.4.2 Kinetic and Thermodynamic Considerations

To utilize GC/CI, in general, an understanding of gas-phase kinetics and thermodynamics is necessary. Questions such as What is the best reagent gas? or What parameters effect CI and why? can best be understood from these data. Examination of rate constants for typical PICI reactions precedes the description of cation/molecule thermochemistry. Knowledge of thermochemistry can predict and rationalize many properties of chemical ionization and thereby reduce the effort of trial-and-error searching for optimum analytical conditions.

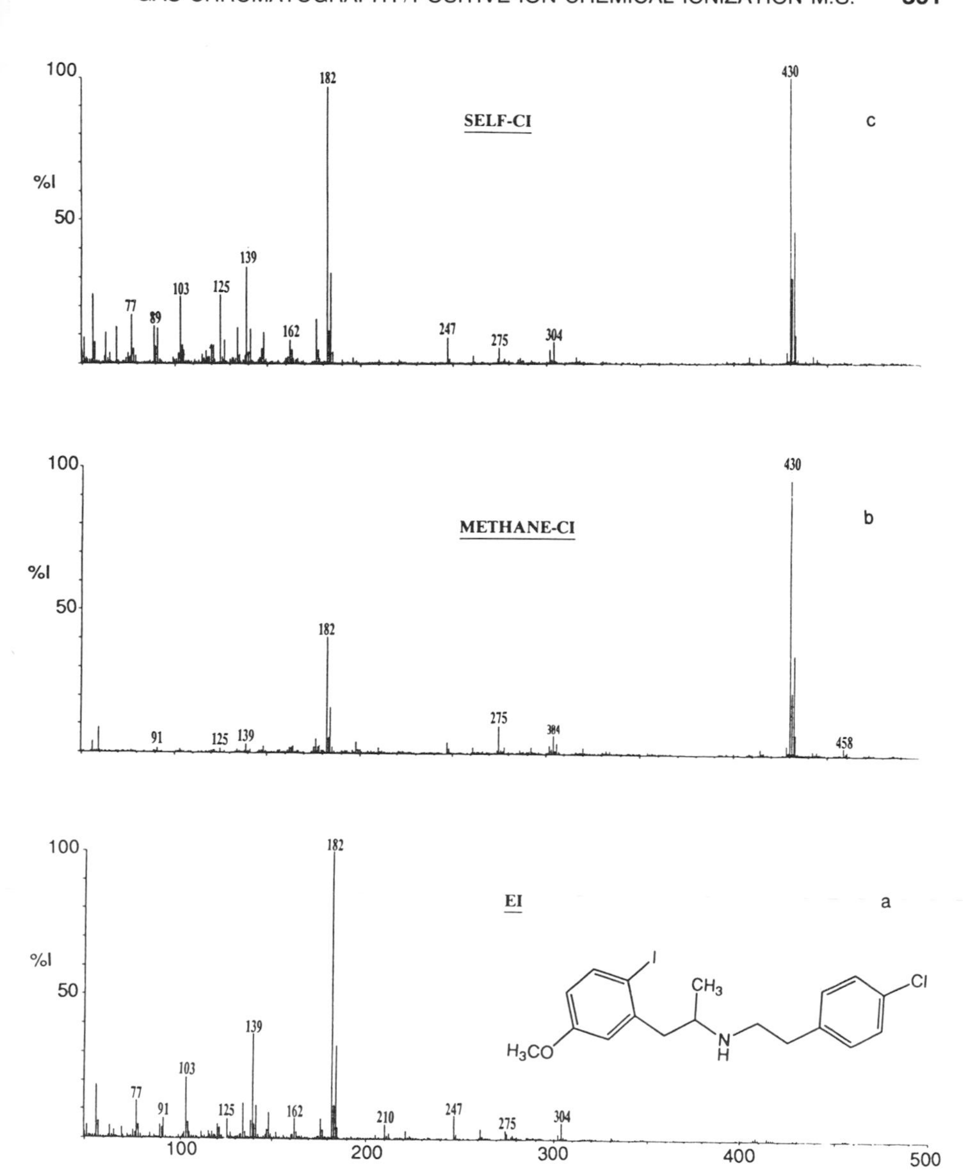

FIGURE 6.13 (a) EI spectrum of [2-(4-chlorophenyl)ethyl]-[2-(2-iodo-5-methoxyphenyl)-1-methylethyl]-amine (**I**) obtained under the following conditions: mass range, 50–500 Da; electron energy, 70 eV; ion source temperature, 200°C; ion exit slit, 1.5 × 7.4 mm. (b) PICI spectrum of (**I**) obtained under the following conditions: mass range, 50–500 Da; electron energy, 200 eV; ion source temperature, 200°C; reagent gas, methane ca. 0.1–0.2 Torr; ion exit slit, 0.1 × 7.4 mm. (c) Self-PICI spectrum of (**I**) obtained under the following conditions: mass range, 50–500 u; electron energy, 70 eV; ion source temperature, 200°C; ion exit slit, 0.7 × 7.4 mm. (Masucci and Caldwell, unpublished data.)

Bimolecular reactions, such as those in Reactions 6.12–6.14, have second-order rate constants (k) typically on the order of 1 to 4×10^{-9} cm^3/molecule sec^{-1} (42). The number of collisions Z (collisions/sec) that occur between cations and sample molecules in the CI source can be estimated by multiplying the rate constant times the density N of molecules in the source (molecule/cm^3). At a pressure of 0.5 Torr and a temperature of 473 K, the density N is approximately 10^{16} molecule/cm^3 and therefore $Z = 1$ to 4×10^7 collisions/sec. The residence time t of most cations in a typical CI source is on the order of 10^{-5} sec. More details on the parameters that effect the residence time of cations are given in Section 6.4.3. A cation undergoes approximately 100 to 400 collisions (Z^*t). This range of collisions permits equilibria to be sufficiently established in order to assume a Boltzman distribution of internal energy of the cations.

The clustering reactions (6.15) have rate constants typically on the order of 10^{-27} cm^6/$molecule^2$ sec^{-1}. Note that this is a third-order rate constant, since it depends on the total pressure of the CI source. When a cation and a neutral molecule collide to form a complex $[M + C]^+$, it is initially in an excited state and must be stabilized by collisions with the reagent gas for observation. In the absence of such stabilization, the complex decomposes. Assuming again that the density N is approximately 10^{16} molecule/cm^3, an effective second-order of 10^{-11} cm^3/molecule sec^{-1} can be defined by multiplying N times the third-order rate constant. While this effective bimolecular rate constant is considerably smaller than the bimolecular rate constant quoted above, Reaction 6.15 is important in PICI for polar compounds and polar reagent gases capable of hydrogen bonding. While the values estimated above will vary somewhat from instrument to instrument, they will be of this order of magnitude. These results suggest that the kinetics are fast enough so that thermodynamic data can be used to examine the energetics that are pertinent to PICI.

The Gibbs free energy ($\Delta G°$) of Reactions 6.12–6.15 is an important way of relating structure and reactivity. That is the net enthalpy ($\Delta H°$) and entropy ($\Delta S°$) changes that occur upon the making of new bonds and the breaking of old ones ($\Delta G° = \Delta H° - T\Delta S°$). For the proton transfer reaction (6.12), for example, a large positive $\Delta G°$ means that the reaction will not take place. If $\Delta G°$ is large and negative, then the proton transfer Reaction 6.12 will occur. The literature contains a great deal of thermochemistry that can be used to calculate the energetics of Reactions 6.12–6.15 (48).

Another way of examining the thermochemistry of Reaction 6.12, for example, is to consider the individual proton affinities (PA) of the reagent gas [C − H] and the sample molecule [M]. The proton affinity (gas phase basicity) is generalized by

$$[M] + [H]^+ \rightarrow [M + H]^+ \tag{6.19}$$

$$C^+ \qquad \leftarrow [H]^+ + [C - H] \tag{6.20}$$

As can be seen, the addition of Reactions 6.19 and 6.20 is simply Reaction 6.12; thus the energetics for the transfer of a proton from the reagent cation to the sample molecule can be calculated by comparing the proton affinities of the reagent gas to the sample molecule (ΔH (Reaction 6.12) = PA (reagent gas) − PA (sample)). If the sample molecule has a greater proton affinity than the reagent gas, then the CI reaction can take place. The fundamental concept of proton affinity (basicity) is well defined within organic chemistry. The variation of proton affinity with structure has been examined by Taft (49) and Aue and Bowers (50). Useful relationships are found that allow unknown proton affinities to be estimated. For example, typical nitrogen-containing species have proton affinities in the range 854 to 1005 kJ/mol, sulfur-containing species are in the range of 829 to 875 kJ/mol, and oxygen-containing species are in the range of 754 to 853 kJ/mol.

Table 6.3 lists the proton affinities of several possible reagent gases, while Table 6.4 lists the proton affinities of several small organic molecules with various functional groups (42, 48). The proton affinities of the reagent gas increase on proceeding down the table. By choosing the proper reagent gas, the PICI technique can selectively protonate molecules. As an example, if the reagent gas is methane and the sample molecule is toluene (ΔH (Reaction 6.12) = 550 − 794 = −244 kJ/mol), the reaction is exothermic and would be observed in the PICI spectra. However, if the reagent gas is ammonia (ΔH (Reaction 6.12) = 854 − 794 = +60 kJ/mol), then the reaction is endothermic and would not be observed. If the difference in PA is large (strongly exothermic), then there is a substantial excess energy in the $[M + H]^+$ cation and fragmentation may also occur. Thus, the degree of

TABLE 6.3 Examples of Proton Affinities (kJ/mol) and Ionization Energies (eV) of Reagent Gases

Species (B)	PA(B)[a]	IE(B)[b]	Cation Formed
He	178	24.6	HeH^+
H_2	423	15.4	H_3^+
CH_4	550	12.5	CH_5^+
C_2H_4	680	10.5	$C_2H_5^+$
H_2O	697	12.6	H_3O^+
H_2S	712	10.5	H_3S^+
CH_3OH	761	10.9	$CH_3OH_2^+$
i-C_4H_{10}	823	10.6	t-$C_4H_9^+$
NH_3	854	10.2	NH_4^+
CH_3NH_2	896	9.0	$CH_3NH_3^+$
Ar		15.8	
N_2		15.6	
O_2		12.1	
NO		9.3	

[a] Reference 48.
[b] Reference 42.

TABLE 6.4 Examples of Proton Affinities (kJ/mol), Hydride Ion Affinities (kJ/mol), and Recombination Energies (eV)

Species (B)	PA(B)[a]	HIA(BH^+)[b]	RE(BH^+)[b]
H_2	423	1255	9.3
HCl	538		
CH_4	550	1130	8.0
C_2H_4	601		
i-C_3H_8	628	1046	7.5
C_2H_4	680		
i-C_4H_{10}	683	967	6.9
H_2O	697	950	6.2
C_6H_6	759		9.3
CH_3OH	761	498	6.0
$C_2H_5NO_2$	773		
CH_3SH	784		
CH_3CN	787		
$C_6H_5CH_3$	794	975	9.2
CH_3COOH	796		
CH_3COCH_3	823		
NH_3	854	816	4.8
$(CH_3)_2NH$	923		

[a] Reference 48.
[b] Reference 42.

fragmentation increases as the PA of the reagent gas decreases (e.g., methane).

It should be understood that the sample molecule can have ca. 4–8 kJ/mol of excess energy not accounted for in the above calculation due to thermal energy that the sample molecule may acquire from the ion source wall or in the course of its chromatography. Therefore slightly endothermic reactions may be observed. It is noted that an entropy change for equilibrium processes such as Reaction 6.12 is usually small and can be neglected for this discussion.

There are no specific reagent gases that involve hydride abstraction as the major or sole ionization reaction. Many PICI spectra of organic compounds show both species. Compounds with a lot of hydrocarbon nature, such as fatty acids and long-chain methyl esters, show abundant peaks for both $[M+H]^+$ and $[M-H]^+$ cations. The thermochemistry of hydride abstraction reactions (Reaction 6.13) can be calculated in a similar manner as outlined above using hydride-ion affinities (HIA) instead of PAs (ΔH (Reaction 6.13) = HIA (reagent cation) − HIA (sample cation)). That is, if the hydride-ion affinity of the reagent cation ($[C]^+$) is higher than the hydride-ion affinity of the cation formed by loss of H^- of the sample molecule ($[M-H]^+$), then the reaction is exothermic and would occur (48). As an example, if the reagent gas was methane and the sample molecule was toluene (ΔH (Reaction 6.13) = −1130 − (−975) = −155 kJ/mol), then the reaction is exothermic and would occur.

The charge exchange reaction (6.14) produces $[M]^{+\cdot}$ cations characteristic of EI. Consequently, the fragmentation observed for charge exchange reactions are the same as those observed in EI. However, the $[M]^{+\cdot}$ cations produced by EI have a distribution of internal energies of 0–70 eV, whereas the $[M]^{+\cdot}$ cations have discrete internal energies defined by the exothermicity of the charge exchange Reaction (6.14). The exothermicity of Reaction 6.14 is determined by the ionization energy (IE; Table 6.3) of the sample molecule [M] less the recombination energy (RE; Table 6.4) of the reagent cation (ΔH (Reaction 6.14) = IE (sample molecule) − RE (reagent cation)). The recombination energy is simply the energy released when an electron recombines with a cation to form a neutral species. The ionization energy is the energy required to remove an electron. Reagent gases for charge exchange include the noble gases, nitrogen, carbon dioxide, carbon monoxide, and hydrogen. As an example, if the reagent gas is hydrogen and the sample molecule is toluene (ΔH(RXN) = 9.2–9.3 = −0.1 eV), the reaction is exothermic and would occur. In contrast, if the reagent gas is methane and the sample molecule is toluene (ΔH(RXN) = 9.2 − 8.0 = +1.2 eV), the reaction is endothermic and would not occur (42). Helium has an RE equal to 24.6 eV and most organic molecules have IEs in the range of 7–12 eV. Correspondingly, when He^+ is used as the reagent cation, complete fragmentation often results.

When reactions such as proton transfer, hydride abstraction, and charge exchange are not thermodynamically favorable, cluster Reactions (6.15) are sometimes observed. For example, in ammonia PICI an intense $[M + H + NH_3]^+$ cation is observed at M + 18 u and sometimes $[M + H + (NH_3)_n]^+$ cations are observed, where $n = 1, 2, \ldots$; in methane PICI the presence of $[M + C_2H_5]^+$ and $[M + C_3H_5]^+$ cations are observed at M + 29 and M + 41, respectively; in isobutane PICI the presence of $[M + C_3H_3]^+$ and $[M + C_4H_9]^+$ cations are observed at M + 39 and M + 57, respectively. When the sample molecule pressure is sufficiently high, dimers ($[2M + H]^+$) are also produced (Reaction 6.18). Dimer cations are particularly prevalent in the PICI spectra of sample molecules such as amines and alcohols, since they are capable of forming hydrogen bonds (51). If the cluster cations that are formed by Reactions 6.15 and 6.18 are unstable, subsequent fragmentation may also be observed (Reaction 6.17).

6.4.3 Instrumentation

The GC/PICIMS technique was developed primarily by Arsenault (52) and Munson (53). The combination of the PICI technique with gas chromatography was a natural outgrowth from the GC/EIMS technique. The general considerations outlined for GC/EIMS (Section 6.3) are the same for GC/PICIMS. However, three major differences between the GC/EIMS and

GC/PICIMS instrumentation deserve attention. First, since PICI utilizes the principle of cation/molecule reactions (Reactions 6.10–6.15) between sample molecules (10^{-3} to 10^{-4} Torr) and a high-pressure (0.2- to 2-Torr) plasma of reagent gas, a specially designed CI source is required. The CI source is usually an EI source that can operate at high pressure. To operate at these high pressures the CI source must be "tight"; thus, the apertures of the electron filament entrance and the ion exit must be kept small (Figure 6.14). Typical examples for the electron filament entrance aperture are EI ($3\,mm^2$) and CI ($0.3\,mm^2$), and for the ion exit aperture are EI ($1.5\,mm^2$) and CI ($0.15\,mm^2$). The ion exit aperture can have a major affect on CI sensitivity and may need to be increased or decreased. Most modern CI sources can be altered to accommodate aperture changes.

Second, since the CI source is at a high pressure, such as 0.5 Torr, electrons with 70-eV energy penetrate a short distance into the source. The high pressure of the CI source requires a much higher electron energy (200–500 eV) to have an overall efficiency—expressed in amperes of ion current per microgram of sample—approaching that of an EI source. Third, MS instruments that operate in the CI mode must be differentially pumped. When the gas chromatographic effluent and the reagent gas (1–5 mL/min) enters the ion source and then exits through apertures in the CI source into the vacuum envelope surrounding it, the CI source housing pressure should be at ca. 10^{-4} to 10^{-5} Torr. The pressure in the MS analyzer region must be maintained at ca. 10^{-5} to 10^{-6} Torr. Diffusion pumps or turbomolecular pumps with pumping speeds in the range of 1000–1500 L/sec are used to maintain these pressures. Diffusion and turbomolecular pumps are backed by rotary pumps. It is important that the MS analyzer remain at low

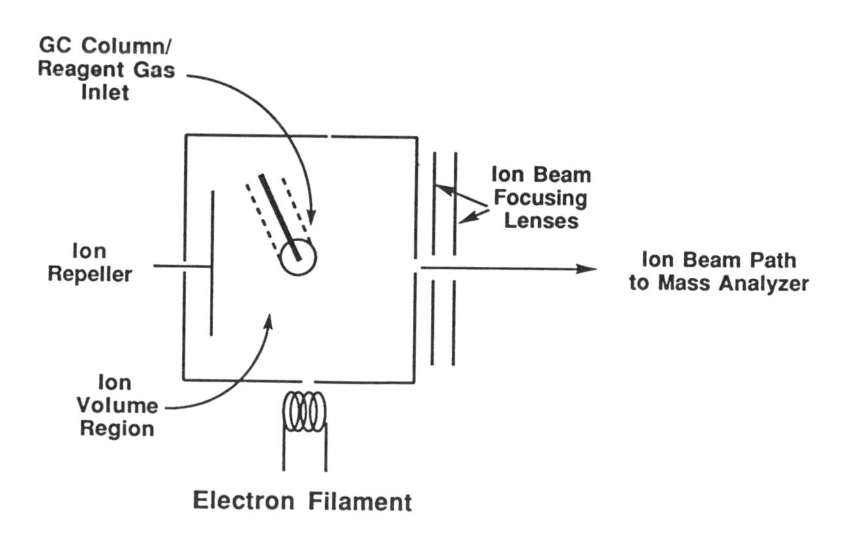

FIGURE 6.14 Diagram of chemical ionization source.

pressure, since the cations must traverse a fairly long path in a highly focused beam. At higher pressures the beam scatters and deteriorates MS performance.

Certain source parameters, such as the repeller voltage, reagent gas pressure, and ion source temperature, can markedly affect the PICI sensitivity and the appearance of the PICI spectra. Each of these parameters affects the residence time of the cations in the ion source, the kinetics, and the thermochemistry of the cations, and thus affects the yields of cations present in Reactions 6.10–6.15. The repeller is used to move cations that are formed in the ion source out of the ion source. A positive voltage (0–10 V) is applied to a repeller plate which is located inside the CI source (Figure 6.14) and is usually tuned to produce the maximum number of reagent gas ionizing species (ca. 7–10 V). The repeller has a major effect on sensitivity. If the repeller is set too low, some low-abundance cations may be absent in the PICI spectrum. The effect of ion source pressure on the PICI spectrum of di-*n*-butyl phthalate is shown in Table 6.5 to illustrate this parameter. At low methane pressure no chemical ionization occurs and the PICI spectra resemble EI spectra (53). There is typically a range of pressures where the change in the overall appearance of the PICI mass spectrum is insignificant. For this example, this is observed in the range of 0.3 to 0.5 Torr. At 1.0 Torr there is an increase in abundance of the high-mass cations and a decrease in abundance of the low-mass cations. This is due to greater collisional stabilization of the $[M+H]^+$ cation, from increased residence time in the ion source, and thus less fragmentation of the $[M+H]^+$ cation. The pressure dependence of PICI spectra is seldom investigated; however, by increasing the reagent gas pressure from 0.3 to 1.0 Torr, a factor of two is gained in the abundance of the $[M+H]^+$ cation at m/z 279. At pressure higher than 1.0 Torr, there is a decrease in the total abundance of cations

TABLE 6.5 Effect of Ion Source Pressure on Methane PICI Spectrum for Di-*n*-Butyl Phthalate

Source Pressure (Torr)	% of Base Peak						
	m/z: 149	177	205	223	$[M+H]^+$ 279	$[M+C_2H_5]^+$ 307	$[M+C_3H_5]^+$ 319
0.1	100	—	7	4	—	—	—
0.3	100	13	97	9	32	3	2
0.5	100	17	99	10	39	4	3
1.0	49	7	100	6	81	14	9

Note. GC/PICIMS spectrum obtained under the following conditions: GC conditions—column, DB-1 (30 m × 0.320 mm); film thickness, 5.00 μm; carrier gas, helium @ 25 cm/sec; oven program, 45°C for 3 min, then 10°C/min to 300°C for 12 min; injector port, 265°C; sample, 1 μL at 2000 μg/μL; solvent, methylene chloride; samples were injected in the splitless mode (0.75-min load). MS conditions—mass range, 50–500 u; electron energy, 200 eV; ion source temperature, 200°C; repeller, 7.0 V; GC/MS interface temperature, 250°C.

TABLE 6.6 Effect of Ion Source Temperature on Methane PICI Spectrum for Di-*n*-Butyl Phthalate

Source Temperature (°C)	% of Base Peak						
					$[M+H]^+$	$[M+C_2H_5]^+$	$[M+C_3H_5]^+$
	m/z: 149	177	205	223	279	307	319
150	78	12	100	9	51	7	4
200	100	17	99	10	39	4	3
250	100	15	85	8	33	1	—
300	100	16	55	5	11	—	—

Note. GC/PICIMS spectrum obtained under the following conditions: GC conditions—column, DB-1 (30 m × 0.320 mm); film thickness, 5.00 μm; carrier gas, helium @ 25 cm/sec; oven program 45°C for 3 min, then 10°C/min to 300°C for 12 min; injector port, 265°C; sample, 1 μL at 2000 μg/μL; solvent, methylene chloride; samples were injected in the splitless mode (0.75-min load). MS conditions—mass range, 50–500 u; electron energy, 200 eV; reagent gas, methane 1.2×10^{-4} Torr (ion source housing); ion source pressure, 0.5 Torr; repeller, 7.0 V; GC/MS interface temperature, 250°C.

formed due to deterioration of the MS performance. The effect of ion source temperature on the PICI spectrum of di-*n*-butyl phthalate is shown in Table 6.6. An increase in the ion source temperature reduces the abundance of the higher-mass cations $[M+H]^+$, $[M+C_2H_5]^+$, and $[M+C_3H_5]^+$ and increases the abundance of the lower-mass cations. If the ion source temperature could be increased to higher temperatures, the PICI spectrum of di-*n*-butyl phthalate would probably mimic the EI spectrum due to excessive cation fragmentation caused by thermal effects (54). Between the ion source temperatures of 150 and 300°C, the abundance of the cations at *m/z* 149 and 205 reverse. The cation/molecule reaction for these cations is summarized in Figure 6.15.

Since *m/z* 149 is produced from *m/z* 205, as the temperature is increased one would expect to observe an increase in the abundance of *m/z* 149 and a decrease in the abundance of *m/z* 205. The cations residence times in the ion source are all decreased at higher temperatures. In addition to the effect of temperature on cation abundance, chemical decomposition of thermally

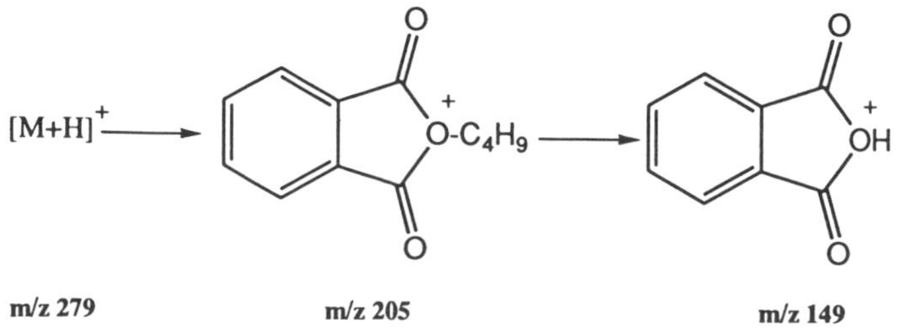

$[M+H]^+$ → $\overset{+}{O}$-C_4H_9 → $\overset{+}{O}H$

m/z 279 m/z 205 m/z 149

FIGURE 6.15 Reaction scheme of di-*n*-butyl phthalate.

labile molecules may also be a problem. There is always a compromise between a lower temperature, which would allow the ion source to contaminate quickly, and a higher range, which would promote higher-energy collisions or decomposition. Only by thorough investigation of temperature and pressure effects can any potential problems be clarified.

6.4.4 Chromatographic Carrier Gas Substituted as the Reagent Gas

The choice of chromatographic carrier gas and CI reagent gas in GC/PICIMS is very important. The correct chromatographic carrier gas should be used to maximize column efficiency, while the correct reagent gas should be used to maximize PICI sensitivity. The chromatographic carrier gas can be substituted as the reagent gas in some cases. Methane, helium, and hydrogen are three examples of chromatographic carrier gases that can also be used as the reagent gas with little effect on chromatographic resolution. In Figure 6.16 are shown three chromatographic traces of a five-component mixture. Figure 6.16a shows the results using methane as the chromatographic carrier gas and the reagent gas. Figure 6.16b shows the same mixture but with helium as the chromatographic carrier gas and methane as the reagent gas. The mass spectra were identical for these two combinations—that is, the cations at $[M+H]^+$, $[M+C_2H_5]^+$ and $[M+C_3H_5]^+$ were present, usually in the ratio of 100:20:5, respectively. With helium as the chromatographic carrier gas and the PICI reagent gas (Figure 6.16c), charge exchange PICI spectra were obtained that resembled EI data. Note that the relative retention times and the resolution were the same for all three combinations. When hydrogen is used (not shown) as both carrier gas and PICI reagent gas, the PICI spectra contain a mixture of proton transfer $[M+H]^+$ and proton abstraction $[M-H]^+$ peaks.

6.4.5 Helium Chromatographic Carrier Gas and Different Reagent Gases

It is more typical in GC/PICIMS to use helium as the chromatographic carrier gas while varying the reagent gas. In Figure 6.17 are shown the total ion current (TIC) traces of a five-component mixture using three different reagent gases. The reagent gas for Figure 6.17a was ammonia, for Figure 6.17b it was isobutane, and for Figure 6.17c it was methane. The mixture consisted of two nitroso-type compounds (Nos. 1 and 4) and three chloro-ether-type compounds (Nos. 2, 3, and 5). The chromatographic carrier gas was helium for all three examples. Note that the absolute retention times and the chromatographic resolution were the same for all three combinations. Also note that components 1 and 4 are much more pronounced than components 2, 3, and 5 when ammonia is used as the reagent gas. This is an example of ammonia selectivity for the components in the mixture containing nitrogen functionalities. The nitroso-type compounds have proton

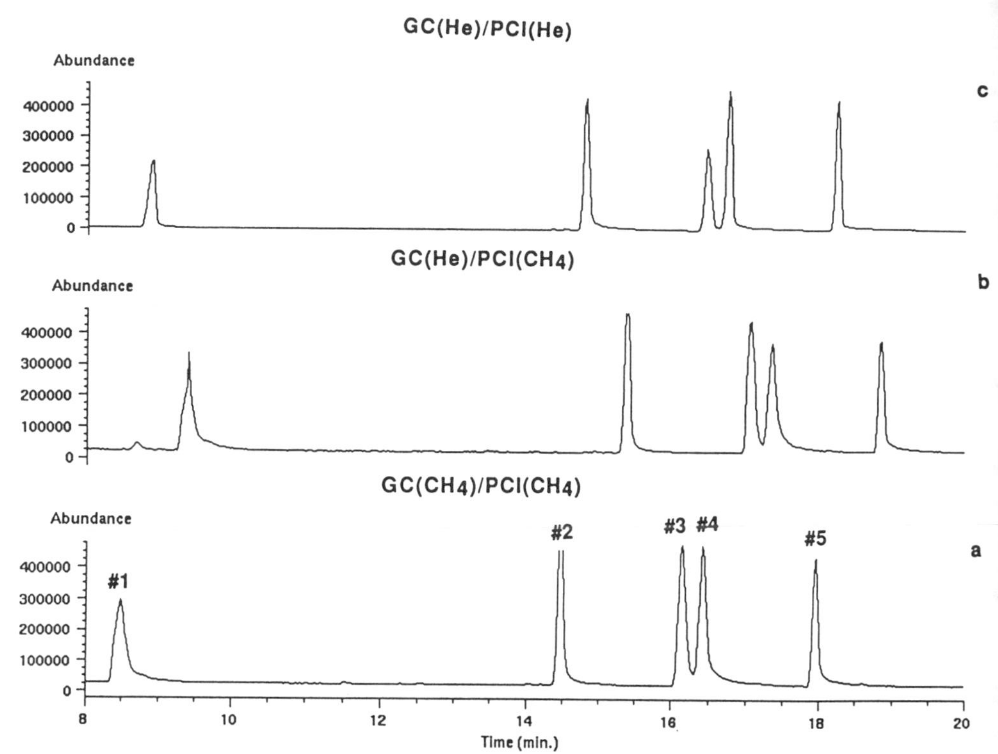

FIGURE 6.16 (a) The total ion current chromatograms of a five-component mixture of 1, *N*-nitrosodimethylamine); 2, bis(2-chloroethyl)ether; 3, bis(2-chloroisopropyl)ether; 4, *N*-nitroso-di-*n*-propylamine; and 5, bis(2-chloroethoxy)methane. GC/PICIMS spectrum obtained under the following conditions: GC conditions—column, DB-1 (30 m × 0.320 mm); film thickness, 5.00 μm; carrier gas, methane @ 25 cm/sec; oven program, 45°C for 3 min, then 10°C/min to 300°C for 12 min; injection port, 265°C; sample, 1 μL at 2000 μg/μL; solvent, methylene chloride; samples were injected in the splitless mode (0.75-min load). MS conditions—mass range, 50–500 u; electron energy, 200 eV; reagent gas, methane, 1.2×10^{-4} Torr (ion source housing); ion source pressure, 0.5 Torr; repeller, 7.0 V; GC/MS interface temperature, 250°C; ion source temperature, 200°C. (b) Same as (a), except the carrier gas was helium @ 25 cm/sec. (c) Same as (a), except the carrier gas was helium @ 25 cm/sec and the reagent gas was helium at 0.5 Torr. (Masucci and Caldwell, unpublished data.)

affinities more similar to ammonia than the chloro-ether compounds (48). The transfer of a proton from the NH_4^+ cation to the chloro-ether compounds is not as exothermic as the proton transfer to the nitroso-type compounds. Thus, the abundance of the $[M+H]^+$ cations for the chloro-ether compounds is less with a decrease in sensitivity.

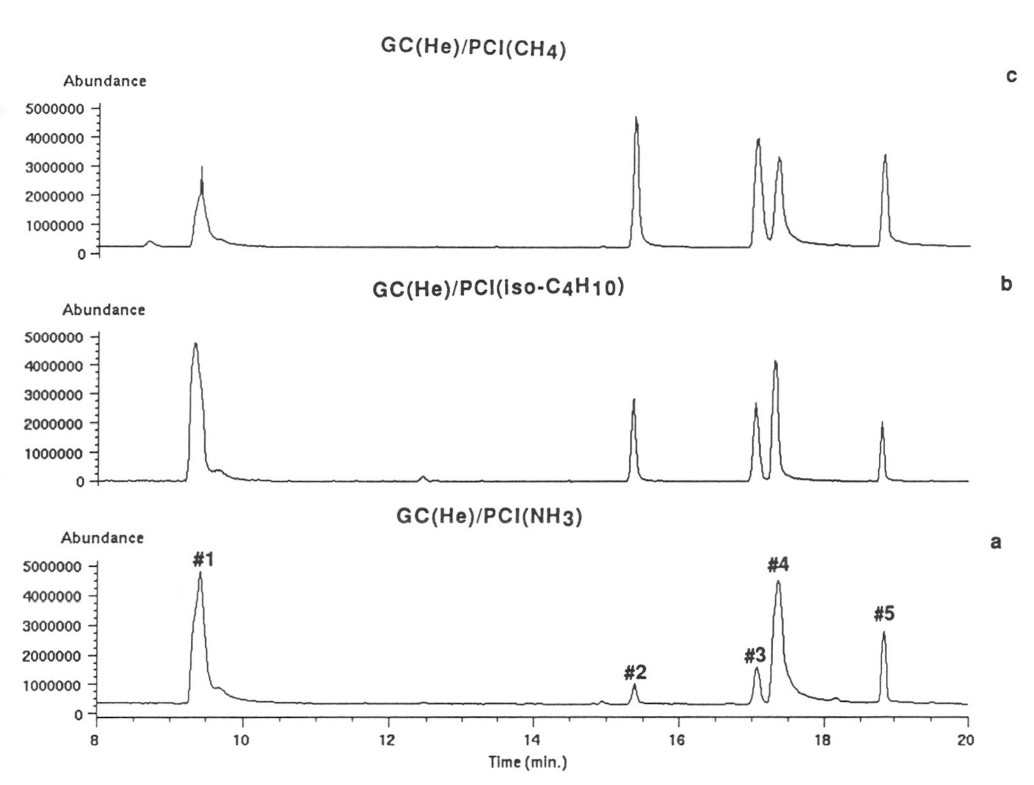

FIGURE 6.17 (a) The total ion current chromatograms of a five-component mixture of 1, *N*-nitrosodimethylamine); 2, bis(2-chloroethyl)ether; 3, bis(2-chloroisopropyl)ether; 4, *N*-nitroso-di-*n*-propylamine; and 5, bis(2-chloroethoxy)methane. GC/PICIMS spectrum obtained under the following conditions: GC conditions—column, DB-1 (30 m × 0.320 mm); film thickness, 5.00 μm; carrier gas, helium @ 25 cm/sec; oven program, 45°C for 3 min, then 10°C/min to 300°C for 12 min; injector port, 265°C; sample, 1 μL at 2000 μg/μL; solvent, methylene chloride; samples were injected in the splitless mode (0.75-min load). MS conditions—mass range, 50–500 u; electron energy, 200 eV; reagent gas, ammonia, 1.2×10^{-4} Torr (ion source housing); ion source pressure, 0.5 Torr; repeller, 7.0 V; GC/MS interface temperature, 250°C; ion source temperature, 200°C. (b) Same as (a), except the carrier gas was isobutane at 0.5 Torr. (c) Same as (a), except the reagent gas was methane at 0.5 Torr. (Masucci and Caldwell, unpublished data.)

It is of interest to examine the mass spectra contained in the TIC peaks (Figure 6.17). Variation of reagent gases can be used for selective ionization, fragmentation, and detection of specific functional groups. For example, the PICI spectra of *N*-nitroso-di-*n*-propylamine (No. 4) and bis(2-chloroethoxy)methane (No. 5), using ammonia, isobutane, and methane as reagent gases, are shown in Figures 6.17 and 6.18, respectively. The reagent NH_4^+ cation transfers a proton to *N*-nitroso-di-*n*-propylamine to produce a $[M+H]^+$ cation at m/z 131 and a clusters $[M+NH_4]^+$ cation at m/z 148 (Figure 6.17a). Note also the presence of the dimer at $[2M+H]^+$. The t-$C_4H_9^+$ cation from isobutane produces a $[M+H]^+$ cation at m/z 131, while the CH_5^+ and the $C_2H_5^+$ cations from methane produce a $[M+H]^+$ cation at m/z 131. The $C_2H_5^+$ and $C_3H_5^+$ cations cluster to produce peaks at m/z 159 and 171, respectively. We can compare these spectra with those in Figure 6.19. Bis(2-chloroethoxy)methane produces only a $[M+NH_4]^+$ cation at m/z 190 (Figure 6.19a) with no significant fragments. When isobutane is used as the reagent gas, a small $[M+H]^+$ cation at m/z 174 and fragments are observed. When methane is used, only fragments are observed in the PICI spectrum.

6.4.6 Hydrocarbon Positive-Ion Chemical Ionization Reagent Systems

Hydrocarbon reagent gases such as methane and isobutane are the most common reagent gases that produce characteristic and abundant cations for determining molecular weights and fragments for structural elucidation. For methane, the following cation/molecule reactions describe the sequence of events in the ion source (38):

$$CH_4 + e^- \rightarrow CH_4^{+\cdot}, CH_3^{+\cdot}, CH_2^{+\cdot}, CH^{+\cdot}, \ldots + 2e^- \qquad (6.21)$$

Methane is ionized by an electron to produce a series of radical cations. The major radical cations are $CH_4^{+\cdot}$ and $CH_3^{+\cdot}$, which comprise approximately 90–95% of the total ionization at ca. 1 Torr. The $CH_2^{+\cdot}$ radial cation is produced at a much lower concentration. These radial cations react with methane to produce the following cations:

$$CH_4^{+\cdot} + CH_4 \rightarrow CH_5^+ + CH_4^{\cdot} \qquad (6.22)$$

$$CH_3^{+\cdot} + CH_4 \rightarrow C_2H_5^+ + H_2 \qquad (6.23)$$

$$CH_2^{+\cdot} + CH_4 \rightarrow C_2H_3^+ + H_2 + H^{\cdot} \qquad (6.24)$$

$$C_2H_3^+ + CH_4 \rightarrow C_2H_5^+ + H_2 \qquad (6.25)$$

Thus, the PICI spectrum of methane contains cations at m/z 17 ($[CH_5]^+$),

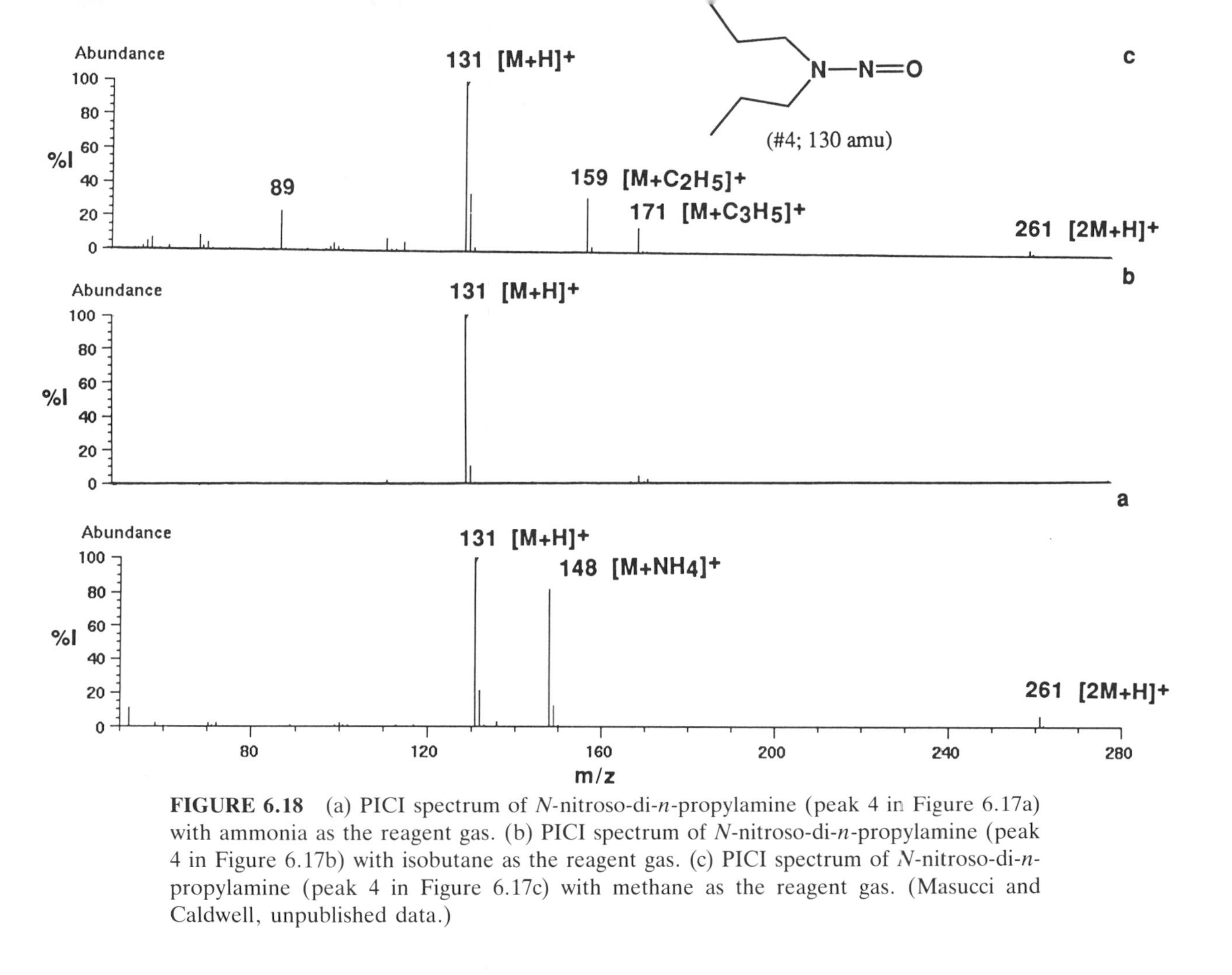

FIGURE 6.18 (a) PICI spectrum of *N*-nitroso-di-*n*-propylamine (peak 4 in Figure 6.17a) with ammonia as the reagent gas. (b) PICI spectrum of *N*-nitroso-di-*n*-propylamine (peak 4 in Figure 6.17b) with isobutane as the reagent gas. (c) PICI spectrum of *N*-nitroso-di-*n*-propylamine (peak 4 in Figure 6.17c) with methane as the reagent gas. (Masucci and Caldwell, unpublished data.)

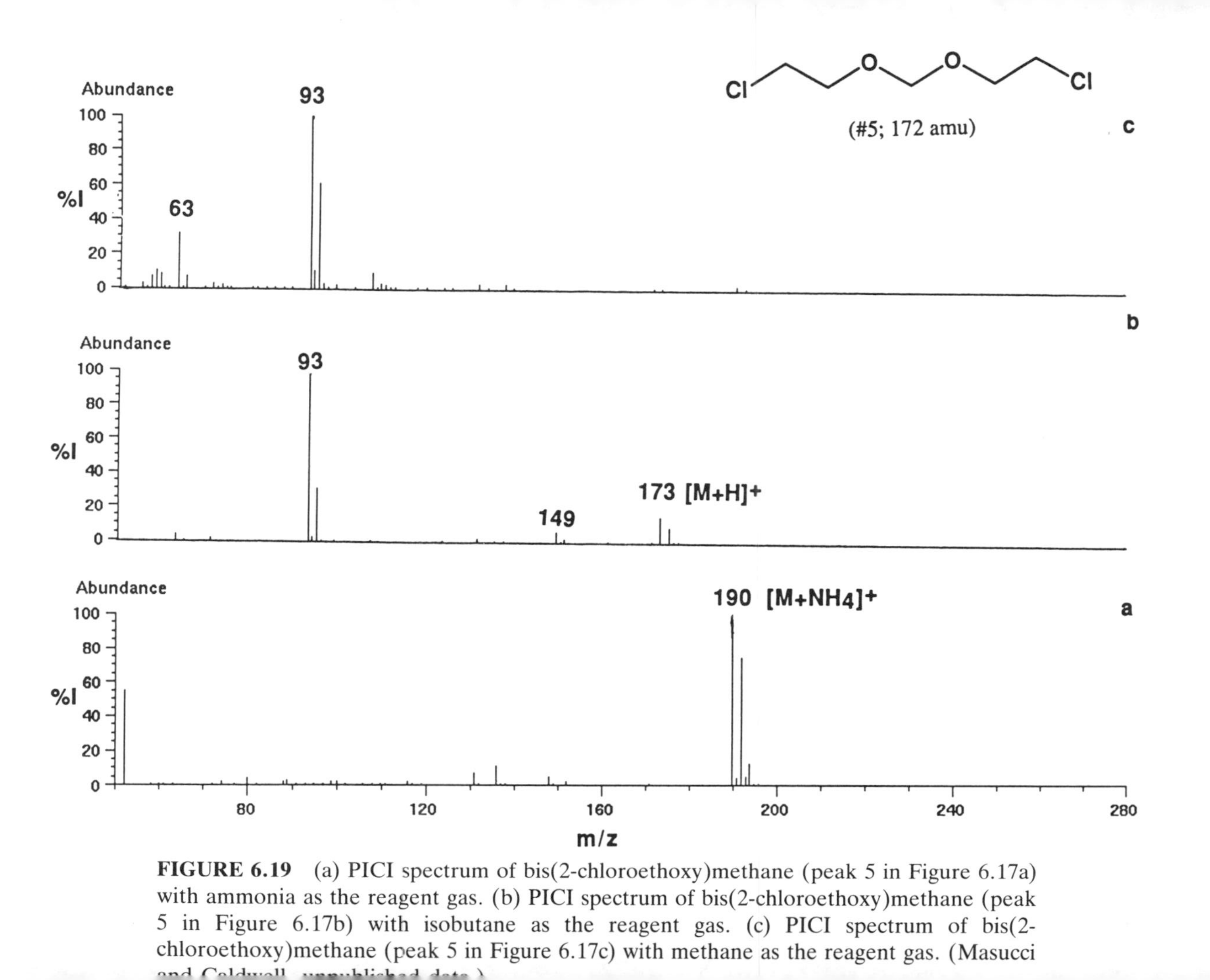

FIGURE 6.19 (a) PICI spectrum of bis(2-chloroethoxy)methane (peak 5 in Figure 6.17a) with ammonia as the reagent gas. (b) PICI spectrum of bis(2-chloroethoxy)methane (peak 5 in Figure 6.17b) with isobutane as the reagent gas. (c) PICI spectrum of bis(2-chloroethoxy)methane (peak 5 in Figure 6.17c) with methane as the reagent gas. (Masucci and Caldwell, unpublished data.)

m/z 29 ($[C_2H_5]^+$), and *m/z* 41 ($[C_3H_5]^+$) with small relative concentrations of $C_2H_3^+$, $C_3H_7^+$, $C_2H_2^+$, $C_3H_3^+$, $C_3H_4^+$, and $C_4H_9^+$. When this series of cations at $[M+H]^+$, $[M+C_2H_5]^+$, and $[M+C_3H_5]^+$ is observed in the methane PICI spectra the molecular weight can be stated with confidence. The CH_5^+ cation can protonate all organic compounds exothermically. When proton transfer occurs with considerable exothermicity, the $[M+H]^+$ cation contains appreciable energy and fragmentation occurs (see Figure 6.18c). Other hydrocarbons, such as methane-d_4 (40), propane (40), *n*-hexane (55), and *n*-octane (55), have also been used as reagent gases.

An interesting artifact that appears in methane PICI spectra involves the following set of reactions (56):

$$[M+H]^+ + [M] \rightarrow [2M+H]^+ \tag{6.26}$$

$$[2M+H]^+ \rightarrow [F+M]^+ + [N] \tag{6.27}$$

where F is a fragment cation and N is neutral. As an example, the methane PICI spectrum of 1,1′-methylenebis[pyrrolidine] contains cations at *m/z* 84 ($[F]^+$; 100%), at *m/z* 153 ($[M-H^+$; 3%), at *m/z* 155 ($[M+H]^+$; 2%), and at *m/z* 238 ($[F+M]^+$; 3%). Note that the $[F+M]^+$ peak is a relatively abundant peak when compared to the $[M-H]^+$ and $[M+H]^+$ cations. The $[F+M]^+$ peak could be mistaken as an impurity or the protonated molecular cation.

6.4.7 Amine Positive-Ion Chemical Ionization Reagent Systems

Amine reagent gases such as ammonia and methylamine are the most common gases that produce characteristic and abundant cations for determining molecular weights. The following cation/molecule reactions describe the sequence of events in the ion source (57) for ammonia:

$$NH_3 + e^- \rightarrow NH_3^{+\cdot}, NH_2^{+\cdot}, \ldots + 2e^- \tag{6.28}$$

Ammonia is ionized by an electron to produce a series of radical cations. The major radical cation is $NH_3^{+\cdot}$, which comprises approximately 90% of the total ionization. The $NH_2^{+\cdot}$ radial cation is produced at a much lower concentration. These radial cations react with ammonia to produce the following cations:

$$NH_3^{+\cdot} + NH_3 \rightarrow NH_4^+ + NH_2 \tag{6.29}$$

$$NH_2^{+\cdot} + NH_3 \rightarrow NH_4^+ + NH^{\cdot} \tag{6.30}$$

$$NH_4^+ + NH_3 \rightarrow N_2H_7^+ \quad (6.31)$$

$$N_2H_7^+ + NH_3 \rightarrow N_3H_{10}^+ \quad (6.32)$$

Thus, the PICI spectrum of ammonia contains primarily a cation at m/z 18 ($[NH_4]^+$) with small relative concentrations of cations at m/z 35 ($N_2H_7^+$) and at m/z 52 ($N_3H_{10}^+$). When this series of cations at $[M+H]^+$ and $[M+NH_4]^+$ are observed in the ammonia PICI spectra one can confidently state the molecular weight (see Figure 6.18a). The ammonia PICI technique is selective for phosphorous and nitrogen bases in the presence of many common solvents. The NH_4^+ cation does not protonate water, acetone, ethyl acetate, methanol, ethanol, halogenated hydrocarbons, tetrahydrofuran, and ethyl ether (48). A more general review of ammonia PICI has been published (57).

6.4.9 Applications—Structure Elucidation and Quantification

Structure elucidation of an unknown sample molecule is probably best accomplished utilizing hydrocarbon-type reagent gases. Hydrocarbons such as methane and isobutane give structural and molecular weight information. Methane PICI produces a series of cations at m/z M + 1, M + 29, and M + 41 and fragments. Therefore, if a series of cations is observed in the methane PICI spectrum that have m/z differences of 28 and 40, then the molecular weight of the unknown can be defined with strong confidence. If uncertainty remains concerning the molecular weight, nitrogen-containing reagent gases can be used. Ammonia PICI spectra will typically contain M + 1 and/or M + 18 cations and fragments; thus a mass difference of 17 u will establish the molecular weight. Combination charge exchange/chemical ionization (Ar/H_2O) reagent gases are useful (58). This type of combination provides soft ionization by H_3O^+ to produce $[M+1]^+$ cations and hard ionization by Ar^+ to produce fragments. Combined GC/accurate mass PICI of the $[M+H]^+$ and $[M+C]^+$ cations has been utilized to determine the elemental composition of sample molecules (59). The methodology of structure eludication by PICI has been published elsewhere (42). The major fragmentation pathway for the $[M+H]^+$ cations is elimination of a stable neutral molecule [X] (Reaction 6.17).

Table 6.7 presents some common neutral loss fragments observed in PICI. Some sample molecules show multiple neutral losses. In some cases combined GC/isotopic exchange PICI can be used in structural studies. Under favorable GC/PICIMS conditions hydrogens bonded to heteroatoms like thiols, amines, amides, carboxylic acids, phenols, and alcohols undergo rapid isotopic exchange for deuterium in the CI source. Reagent gases such as D_2O (60, 61), ND_3 (62, 63), and CD_3OD (64, 65) have been utilized.

TABLE 6.7 Common Neutral Loss Fragments Observed in Reaction 6.17

$[M+H-X]^+$	[X]
$[M+1-128]^+$	HI
$[M+1-80]^+$	HBr
$[M+1-36]^+$	HCl
$[M+1-20]^+$	HF
$[M+1-18]^+$	H_2O
$[M+1-32]^+$	CH_3OH
$[M+1-46]^+$	CH_3CH_2OH
$[M+1-90]^+$	$(CH_3)_3SiOH$
$[M+1-34]^+$	H_2S
$[M+1-48]^+$	CH_3SH
$[M+1-27]^+$	HCN
$[M+1-58]^+$	$(CH_3)_2CO$
$[M+1-17]^+$	NH_3
$[M+1-31]^+$	CH_3NH_2

This approach can be used to differentiate isomers. For example (65), consider the isomers *p*-methylphenol and anisole. When CH_3OH is used as the reagent gas the PICI spectra of *p*-methylphenol and anisole both produce $[M+H]^+$ cations at m/z 109. In contrast, when CD_3OD is used, *p*-methylphenol produces a $[M-H+2D]^+$ cation at m/z 111, while anisole has a $[M+D]^+$ cation at m/z 110.

Quantitative measurement of a known analyte present in a complex matrix is a common application of GC/PICIMS (66). A deuterium-labeled analogue or a homologous sample molecule is usually added as an internal standard to account for sample losses during workup and gas chromatographic separation. To achieve maximum sensitivity, a reagent gas is chosen such that one or more cations that are characteristic of the sample molecule and the internal standard are produced in high yield. These cations may be any of those in Reactions 6.10–6.18 and/or fragments. Quantitative measurement by GC/PICIMS is usually done in the selective-ion monitoring mode or over a limited mass range to achieve the highest sensitivity.

A typical example of a pharmacokinetic investigation is the quantitative determination of linalool and linalyl acetate in plasma (67). The structures of linalool and linalyl acetate are shown in Figure 6.20.

Tiglinic acid benzyl ester was used as the internal standard with ammonia as the reagent gas and helium as the carrier gas. The PICI spectrum of linalool contained major peaks at m/z 81 (100%), 137 (92%; $[M-OH]^+$), and 154 (47%; $[M]^{+\cdot}$), while linalyl acetate had major peaks at m/z 47 (40%), 57 (19%), and 137 (100%; $[M-OCOCH_3]^+$). Note that these cations are probably produced by a charge exchange reaction (6.14), since

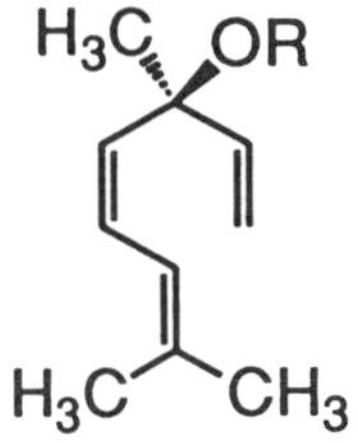

FIGURE 6.20 The structure of linalool (R=H) and linalyl acetate (R=$COCH_3$).

the observed cations are more characteristic of EI. These cations were used for selective ion monitoring and accounted for approximately 90% of the total ion current. The fragrance compounds linalool and linalyl acetate could be detected in the range of 7–9 ng/mL and 1–2 ng/mL, respectively. There are many other examples in the literature of quantitative GC/PICIMS (66–74).

6.5 GAS CHROMATOGRAPHY/NEGATIVE-ION CHEMICAL IONIZATION MASS SPECTROMETRY

6.5.1 Advantages of Negative-Ion Chemical Ionization

The chemical ionization process produces both positive (cations) and negative (anions) ions, thus an alternative to positive ion chemical ionization mass spectrometry (PICI) is negative ion chemical ionization mass spectrometry (NICI). A general review of NICI has been published by Harrison (42) and Budzikiewicz (75). There are two major types of NICI techniques used today: electron-capture and acidity/hydrogen-bonding techniques. While molecular weight information can be obtained from both NICI techniques, there is typically little structural information from either. The NICI electron-capture technique can be 10–100 times more sensitive than the NICI acidity/hydrogen-bonding technique or the PICI technique. Therefore, the NICI electron-capture technique has been utilized extensively for quantification studies. Gas chromatography/negative-ion chemical ionization mass spectrometry (GC/NICIMS) studies are typically performed with helium as the chromatographic carrier gas and a variety of reagent gases. Changing the detection mode has no effect on the chromatographic resolution; Thus, all conclusions drawn from the GC/PICIMS technique (Section 6.4) will apply here.

The NICI electron-capture technique requires that a gaseous mixture consisting of a reagent gas [R_1] (e.g., methane, isobutane, or ammonia) for the production of thermal electrons and the sample molecule [M] of interest be present in the ion source. When the anion mode of detection for the mass spectrometer is setup, only anions are observed. In a typical experiment, the reagent gas $[R_1]$ is ionized by electrons at ca. 200 eV. Cations, anions, and radicals are produced along with thermal electrons (Reaction 6.33). As the name implies, thermal electrons are low-energy electrons that have a very narrow distribution of energies (0 to ca. 5 eV). Thermal electrons react with sample molecules [M] to produce radical (attachment) anions $[M]^{-\cdot}$ (Reaction 6.34), dissociative attachment anions $[M-X]^-$ (Reaction 6.35), and ion-pair anions $[X]^-$ (Reaction 6.36). The sample molecules are not ionized to any extent by the high-energy electrons, since these electrons are converted to low energy quickly. The abundance of $[M]^{-\cdot}$, $[M-X]^-$, and $[X]^-$ anions depends on the reagent gases and the sample molecules. Only certain types of sample molecules (e.g., nitro-aromatic containing compounds, polyhalogenated-containing compounds, and highly conjugated pi systems bearing electron-attracting substituents) will react via Reactions 6.34–6.36:

$$e^- \text{ (200 eV)} + [R_1] \rightarrow C_p^+ + C_p^- + 2e^- \text{ (thermal)} \tag{6.33}$$

$$e^- \text{ (thermal)} + [M] \rightarrow [M]^{-\cdot} \tag{6.34}$$

$$e^- \text{ (thermal)} + [M] \rightarrow [M-X]^- + [X] \tag{6.35}$$

$$e^- \text{ (thermal)} + [M] \rightarrow [M-X]^+ + [X]^- + e^- \tag{6.36}$$

The NICI acidity/hydrogen-bonding technique requires that a gaseous mixture consisting of a reagent gas $[R_1]$ (e.g., methane, isobutane, or ammonia) for the production of thermal electrons, a reagent gas $[R_2]$ for the generation of reactive anions, and the sample molecule [M] of interest be present in the ion source. The second reagent gas $[R_2]$ is used such that the thermal electron reacts with $[R_2]$ to produce reactive anions $[C]^-$ (Reaction 6.37). This reactive anion $[C]^-$ can react with the sample molecules [M] to produce proton abstraction anions $[M-H]^-$ (Reaction 6.38) and/or cluster adduct anions $[M+C]^-$ (Reaction 6.39). The abundance of these anions is controlled by anion/molecule reactions and ultimately depends on the reagent gases and the sample molecules. Sample molecules that contain an acidic proton, such as alcohols, carboxylic acids, and phenols, work well.

$$e^- \text{ (thermal)} + [R_2] \rightarrow [C]^- \tag{6.37}$$

$$[C]^- \quad + [M] \rightarrow [M-H]^- + [C+H] \tag{6.38}$$

$$[C]^- \quad + [M] \rightarrow [M+C]^- \tag{6.39}$$

If the analyte pressure is sufficiently high, dimers are also produced:

$$[M + C]^- + [M] \rightarrow [2M + C]^- \tag{6.40}$$

$$[M - H]^- + [M] \rightarrow [2M - H]^- \tag{6.41}$$

More details on how $[M]^{-\cdot}$ and $[M - H]^-$ anions are generated in the NICI technique are given in Sections 6.5.2–6.5.5. The anions that are produced via both NICI techniques usually produce little or no fragmentation.

In Figure 6.21 is an example of the methane/methyl iodide NICI spectrum of 5,5-dimethyl-1,3-cyclohexanedione. In this example, methane is the $[R_1]$ reagent gas that produces thermal electrons upon electron ionization. Methyl iodide is the $[R_2]$ reagent gas used to generate the reactive iodide anion ($[I]^-$) at m/z 127. This $[I]^-$ anion reacts with 5,5-dimethyl-1,3-cyclohexanedione ($[M] = 140$ u) to produce the cluster $[M + I]^-$ peak at m/z 267 (Reaction 6.14). Note that there is no fragmentation for this NICI technique.

6.5.2 Kinetic and Thermodynamic Considerations

Kinetic and thermochemistry knowledge can predict and rationalize many properties of chemical ionization and reduce the effort of trial-and-error searching for optimum analytical conditions. The electron attachment anions $[M]^{-\cdot}$ (Reaction 6.34), the dissociative electron attachment anions $[M - X]^-$ (Reaction 6.35), and the ion-pair anions $[X]^-$ (Reaction 6.36) are all produced as a result of electron/molecule reactions. These reactions can have rate constants in the range of 10^{-8} to 10^{-7} cm^3/molecule sec^{-1} (42). Since the electron has a higher mobility than a cation (or an anion), these rate constants are considerably higher than those in PICI. As a result of these rate constants, the NICI electron attachment technique can have sensitivity increases between 10 and 100 over those of the NICI acidity/hydrogen-bonding technique or the PICI technique. The proton abstraction anions $[M - H]^-$ (Reaction 6.38) proceed with rate constants on the order of 1 to 4×10^{-9} cm^3/molecule sec^{-1} (42). Note that these rate constants are the same as the PICI rate constants. Thus, there is no gain in sensitivity from changing ionization modes.

At ion source pressures on the order of 0.5 Torr and ion source temperatures of approximately 373 K, the rate constants for electron attachment and proton abstraction suggest there are an adequate number of collisions in the ion source to permit equilibria to be sufficiently established. This is a prerequisite in order to assume a Boltzman distribution of internal

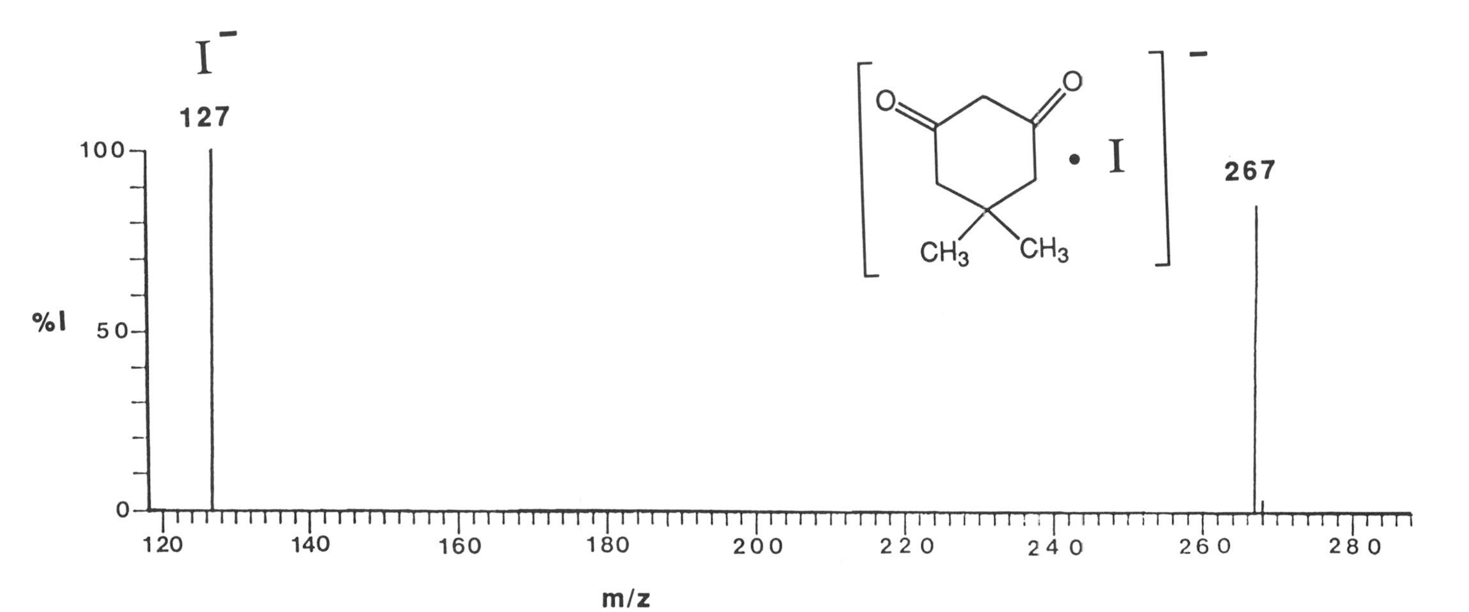

FIGURE 6.21 (a) (CH_4/CH_3I) NICI spectrum of 5,5-dimethyl-1.3-cyclohexanedione. The following conditions were used: reagent gas, methane at 0.4 Torr; reagent gas, methyl iodide at 10^{-6} Torr; electron energy, 500 eV; repeller at −9 V; mass range, 50–500 u; ion source temperature, 115°C; a solid probe was used as the inlet. (Masucci and Caldwell, unpublished data.)

energies of the anions (or cations). Thus thermochemical data, such as electron affinities and the proton affinities of anions, can be used to calculate the energetics of these reactions (48). The cluster adduct anion $[M+C]^-$ (Reactions 6.39–6.41) have third-order rate constants. Where comparisons can be made, the magnitude of positive and negative-mode, third-order rate constants are similar (42). The clustering reactions are important in NICI spectra for polar compounds in the presence of polar molecules such as water and alcohol.

The electron affinity (EA) of an anion is defined as the lowest energy required to remove an electron. An equivalent definition is the affinity of the sample molecule [M] for an electron:

$$[M] + e^- \rightarrow [M]^{-\cdot} \tag{6.42}$$

The EA of many compounds have been measured (48, 76–78) and selected examples are presented in Table 6.8. If a sample molecule has a negative EA, such as benzene, then the electron attachment (Reaction 6.42) is not energetically favorable. However, if a sample molecule has a positive EA, such as perfluorotoluene, then the electron attachment reaction is energetically favorable. From the EA data in the literature (48, 76–78), structural features that must be contained in a molecule for that molecule to

TABLE 6.8 Examples of Electron Affinities (kJ mol^{-1}) and Anion Proton Affinities (kj mol^{-1})

Species (B)	EA (B)[a]	Species (BH)	PA (B^-)[b]
H^-	75	H_2	1675
$O^{-\cdot}$	142	$\cdot OH$	1599
CH_3^-	754	CH_4	1744
CI^-	349	HCl	1395
Br^-	327	HBr	1354
HO^-	176	H_2O	1635
$C_6H_6^{-\cdot}$	<0		
CH_3O^-	156	CH_3OH	1592
c-$C_5H_5^-$	161	c-C_5H_5	1481
CH_3S^-	183	CH_3SH	1493
$mCF_3^-C_6H_4NO_2^{-\cdot}$	136		
CH_2CH^-	141	CH_3CN	1560
$C_6H_5CH_2^-$	87	$C_6H_5CH_3$	1593
CH_3COO^-	296	CH_3COOH	1459
$C_6H_5NO_2^{-\cdot}$	97		
$CH_3COCH_2^-$	180	CH_3COCH_3	1544
NH_2^-	72	NH_3	1689
$C_6F_5CF_3^{-\cdot}$	164		

[a] References 48 and 78.

[b] Reference 48.

have a positive EA may be extrapolated. For example, compounds with functional groups containing highly conjugated pi systems, such as large polycyclic aromatic rings or polycarbonyl-type species, should have positive EAs. Aromatic rings bearing electron-attracting substituents, such as nitro, carbonyl, cyano, and trifluoromethyl groups, should have positive EAs. Two or more halogens on an aromatic ring or polyhalogenated hydrocarbons typically have positive EAs (48).

While a positive EA is a necessary criterion for production of the radical anion $[M]^{-\cdot}$, the observation of $[M]^{-\cdot}$ anions in NICI spectra also depends on the lifetime of these radical anions in the NICI source. In other words, molecules with positive electron affinities may not be observed because they lose an electron (autodetachment) due to collisions with the reagent gas. Autodetachment of the radical anion is important for small molecules and organic molecules with small positive (<50 kJ/mol) EAs (48). Also, once a sample molecule has captured an electron, it may follow a dissociative electron attachment reaction pathway to produce $[M-X]^-$ anions (Reaction 6.35) or may produce ion-pair $[X]^-$ anions (Reaction 6.36) (42).

Once the radical anion is formed by Reaction 6.42 it may not be observed due to a charge exchange:

$$[M]^{-\cdot} + [N] \rightarrow [N]^{-\cdot} + [M] \qquad (6.43)$$

The charge exchange (Reaction 6.43) will be exothermic provided the EA of [N] is greater than the EA of [M]. For example, if one is interested in detecting nitrobenzene ($[M] = C_6H_5NO_2$) in the presence of perfluorotoluene ($[N] = C_6F_5CF_3$), then ΔH (Reaction 6.43) = EA[M] − EA[N] = $97 - 164 = -67$ kJ mol^{-1}. The reaction is strongly exothermic and the $[M]^{-\cdot}$ anion would not be observed. Charge exchange reactions are important when oxygen-containing impurities are present in the reagent gas or perfluoro-containing calibration gas is present as background in GC/MS.

The anion proton affinity (gas-phase acidity) is generalized by Reactions 6.44 and 6.45. The fundamental concept of anion proton affinity (acidity) is well defined in the gas phase (48):

$$[M] \rightarrow [M-H]^- + [H]^+ \qquad (6.44)$$

$$[C]^- + [H]^+ \leftarrow [C+H] \qquad (6.45)$$

The addition of Reactions 6.44 and 6.45 is simply Reaction 6.38, and thus the energetics for the abstraction of a proton from the sample molecule [M] by the reactive anion $[C]^-$ can be calculated by comparing the gas-phase acidities of the sample molecule to the gas-phase acidities of [C + H]. That is, ΔH (Reaction 6.38) = PA (sample) − PA (C + H). If the reactive anion has a greater anion proton affinity (gas-phase acidity) than the sample molecule, then the NICI reaction can take place. For example, if the reactive anion is Cl^- ([C + H] = HCl) and the sample molecule is toluene ([M] = $C_6H_5CH_3$), then ΔH (Reaction 6.38) = 1593 − 1395 = +198 kJ/mol; The reaction is endothermic and would not occur. However, if the reactive anion is NH_2^- ([C + H] = NH_3) and the sample molecule is toluene (ΔH (Reaction 6.38) = 1593 − 1689 = −96 kJ/mol), the reaction is exothermic and would occur. If the reaction is strongly exothermic, there is substantial excess energy in the [M − H] anion and fragmentation could occur or the electron could detach.

6.5.3 Instrumentation

Instrumentation considerations for NICI are the same as those outlined for PICI (Section 6.4.3). A tight CI source, an electron energy that can operate at ca. 200 eV, and good pumping speeds are required. To operate in the NICI mode, the polarity of the voltages applied to the repeller, the lenses of the ion source, and the detection system are reversed. A negative voltage is applied to the repeller and positive voltages are applied to the lenses. Under these conditions anions are expelled from the ion source. The ion exit aperture has a major affect on NICI sensitivity and may need to be increased or decreased. Most modern NICI sources can be altered to accommodate aperture changes. The detection system varies from instrument to instrument and has different configurations. The change over from positive to negative mode is usually computer controlled and selected through the data system software.

An interesting artifact observed in quadrupole mass spectrometers is the appearance of a $[M + CH_5]^-$ anion in the NICI spectra when methane is used as a reagent gas (79). In some quadrupole mass spectrometers set up for negative mode, cations can be formed outside the ion source and are recorded along with anions that were formed in the ion source. These spectra contain peaks that are produced by cations and anions.

The repeller voltage, the sample molecule to reagent gas ratio, the ion source pressure, and temperature can influence the sensitivity of the NICI technique and the appearance of the NICI spectra. These parameters more strongly influence the NICI technique than the PICI technique. The repeller has the same effect on sensitivity as described for the PICI technique; if the repeller is set too low, the overall sensitivity significantly decreases and/or low abundant anions may be absent in the NICI spectrum. The NICI spectrum of α-chlordane was investigated with different concentrations of methylene chloride and varying amounts of α-chlordane (80). The results indicated that the $[M + Cl]^-/[M]^{-\cdot}$ ratio varied with the α-chlordane/methylene chloride ratio. The effect of ion source pressure on the sensitivity and NICI abundances for α-chlordane has been studied over the range 0.1–0.3 Torr (80). There is usually an increase in sensitivity as the ion source pressure is increased. This increase in sensitivity has been attributed to an increase in collisional stabilization of the anion. At higher reagent gas pressure (>1 Torr), the sensitivity of NICI generally decreases, which is attributable to deterioration of the mass spectrometer performance. In general, the ion source pressure has a large effect on sensitivity; however, the appearance of the NICI spectrum changes slightly with pressure. Most NICI experiments are conducted with ion source temperatures in the range of 100–250°C.

In Table 6.9 is shown the effect of ion source temperature on the NICI spectrum of *p*-toluic acid (81). As the temperature is increased, the overall sensitivity of the NICI technique decreases. The ion counts for the radical anion decreased by approximately a factor of two. By increasing ion source temperature, the abundances of low mass fragments increase relative to the $[M + H]^-$ anion at m/z 135. Thus, the ion source temperature has a significant effect on the appearance of NICI spectra and the sensitivity. The abundance of the cluster anion (e.g., $[2M - H]^-$) decreases with increasing temperature and increases with increasing ion source pressure.

6.5.4 Electron-Capture Techniques

Gas chromatography/electron-capture negative-ion chemical-ionization mass spectrometry (GC/ECNICIMS) has become widely used for quantification of compounds at low concentration levels in complex biological matrices (82–94). The electron-capture technique shows very large variations in sensitivity among different compound types, thus permitting specific compound detection in complex matrices. As noted (Section 6.5.3), small changes in the concentrations, the ion source pressure, and temperature from one analysis to another can have significant effects on the sensitivity and fragmentation in ECNICI spectra.

GC/ECNICI analytical quantification procedures usually contain many of the following steps. An internal standard is added to the complex matrix to correct for losses during sample preparation, cleanup steps, and GC/MS analyses. Complex biological matrices, such as urine or plasma, are extracted into organic solvents and purified by chromatographic methods. Due to certain functional groups (e.g., ROH and RCO_2H), these isolated

TABLE 6.9 Effect of Ion Source Temperature on the NICI Spectrum of *p*-Toluic Acid

Source Temperature (°C)	% of Base Peak						
	m/z: 91	118	$[M - H]^-$ 135	150	156	171	$[2M - H]^-$ 271
190	3	5	100	7	8	12	50
220	16	5	100	5	3	1	3
250	36	6	100	10	2	—	—

Source. Reference 81.

samples may have unfavorable gas chromatographic properties, leading to tailing peaks, and/or unfavorable mass spectrometric properties, leading to insensitive detection. The sample is generally derivatized to improve gas chromatographic properties (peak symmetry, volatility, thermal stability) by replacing any active hydrogen atom (–OH, $-NH_2$, –NHR, –SH) in a molecule with a trimethylsilyl group. Examples of derivatization to improve electron-capture properties include the use of pentafluorobenzoyl chloride for reaction with phenols and amines, tetrafluorophthalic anhydride for reaction with amines, and pentafluorobenzaldehyde for reaction with aromatic amines (82).

The combination of limited fragments, high sensitivity, and selectivity makes GC/ECNICI ideal for quantification studies of metabolites or drug substances from physiological fluids (82–94). A typical example of pharmacokinetic investigation is the quantitative determination of the antibiotic chloramphenicol present in cow's urine, muscle, and eggs (Figure 6.22). The isolated samples are converted to their corresponding trimethylsilyl derivatives. ($^{37}Cl_2$)Chloramphenicol was used as the internal standard with methane as the reagent gas and helium as the carrier gas. Note that the chloramphenicol molecule contains a nitroaromatic group—thus, it has a positive EA—and two chlorine atoms—thus, there is a ^{35}Cl and a ^{37}Cl atom. The NICI spectrum contained peaks at *m/z* 466 ($[M]^{-\cdot}$), at *m/z* 376 ($[M - HOSi(CH_3)_3]^{\cdot}$, and at *m/z* 342 ($[M - OSi(CH_3)3 - Cl]^{-\cdot}$. The mass pairs *m/z* 466(^{35}Cl)/468(^{37}Cl) and 376(^{35}Cl)/378(^{37}Cl) were used for selective ion monitoring. These four anions accounted for approximately 60% of the total ion current. The internal reference was monitored at *m/z* 470 ($^{37}Cl_2$). These conditions lead to a limit of detection of 2 pg at a 3:1 signal/noise ratio. The method was suitable for the determination of chloramphenicol in urine, muscle, and eggs at 3 μg/L. The NICI technique was at least ten times more sensitive than EIMS (83).

The assay of the 3-hydroxy-3-methylglutaryl-coenzyme A reductase inhibitors lovastatin, pravastatin, and simvastatin in plasma is an example of a pharmacokinetic investigation in which the sample is derivatized to improve gas chromatographic properties and electron-capture properties (85). These molecules were isolated, converted to their acid forms, and subsequently derivatized by trimethylsilylation of the hydroxyl groups and by pentafluoro-benzylation of the carboxyl group. Lovastatin is shown in

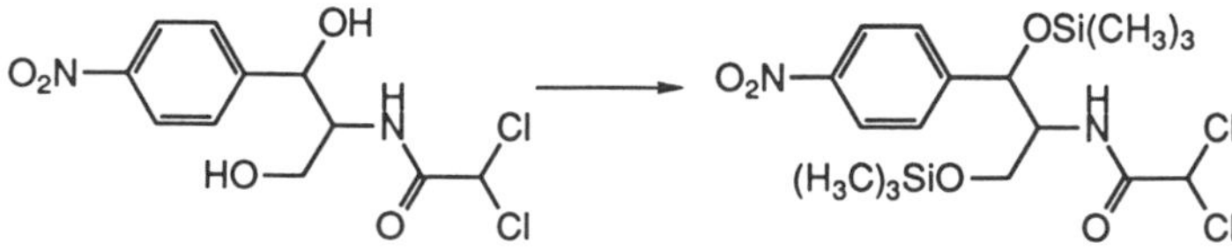

FIGURE 6.22 The antibiotic chloramphenicol, which is converted to its corresponding trimethylsilyl derivative.

Figure 6.23. Derivatized 6′β-hydroxymethylsimvastatin acid was used as the internal standard with ammonia as the reagent gas and helium as the carrier gas. The NICI spectrum of lovastatin contained a major peak at m/z 565 ($[M - CH_2C_6F_5]^-$), while the internal standard had a major peak at m/z 639 ($[M - CH_2C_6F_5]^-$). These anions were used for selective ion monitoring and accounted for approximately 90% of the total ion current. The presence of the radical anion $[M]^{-\cdot}$ was not observed in the NICI spectrum. The base peak anion $[M - CH_2C_6F_5]^-$ results from electron transfer and cleavage of the pentafluorobenzyl functionality with charge retention on the carboxyl group. Note that this reaction is simply a dissociative electron attachment (Reaction 6.35). The method was suitable for the determination of lovastatin in plasma at 0.2 μg/L.

6.5.5 Acidity and Hydrogen-Bonding Techniques

The acidity/hydrogen-bonding NICI technique gives primarily proton abstraction anions $[M - H]^-$ (Reaction 6.38) and/or cluster adduct anions

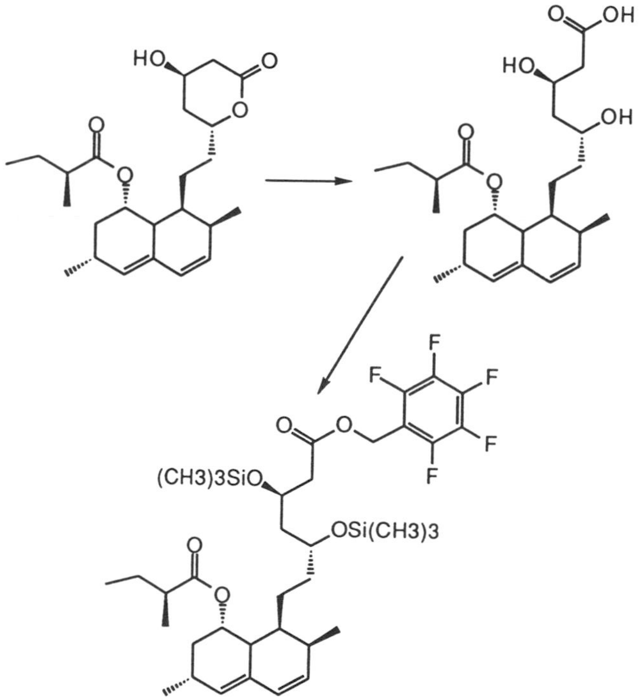

FIGURE 6.23 Lovastatin is isolated and converted from the lactone to its acid form. This molecule is subsequently derivatized by trimethylsilylation of the hydroxyl groups and by pentafluorobenzylation of the carboxy group.

$[M+C]^-$ (Reaction 6.39). Exactly what anion is observed in the NICI spectrum depends on the reactive anion $[C]^-$ and the sample molecule. The NICI spectra can be rationalized or predicted utilizing thermochemical data (96). Many different reagent gases can be used to generate reactive anions $[C]^-$. A variety of gases or gas combinations are listed in Table 6.10 that when reacted with thermal electrons produce reactive anions $[C]^-$ (Reaction 6.37). A qualitative prediction of the concentration of these anions in the CI source is also given.

The analytical potential of the NICI acidity/hydrogen-bonding technique can best be understood by considering an example where methane is used for the production of thermal electrons and one of the halogen gases listed in Table 6.10 is used for the generation of a reactive halide anion ($[C]^- = [X]^-$ = F, CI, Br, and I). The sample molecule in this example contains an acidic proton ([M] = [MH]). Common experimental conditions include an ion source temperature of ca. 100°C, an ion source pressure of ca. 1 Torr, and small superimposed repeller electric fields (7–9 V). Under these conditions, an approach toward equilibrium can be obtained. The following reactions can take place in the CI source (Figure 6.24).

Each of the reaction pathways is denoted and a short-lived excited intermediate complex is further denoted by an asterisk. The NICI spectra obtained from this experiment contain either only the $[M-H]^-$ anion or the $[M+C]^-$ anion or a mixture of both. Nucleophilic displacement (S_{N^2}) reactions are not considered here.

TABLE 6.10 Reagent Gases and Their Dissociative Attachment Anions

Species	Reactive Anions $[C]^-$	Yield of Anions
NH_3	H^-, NH_2^-	Medium
H_2O	H^-, HO^-	Medium
N_2O	$O^{-\cdot}$, NO^-	Medium
O_2	$O^{-\cdot}$, $O_2^{-\cdot}$	Low
RONO	RO^-, NO_2^-	High
$RONO_2$	NO_2^-	High
H_2S	H^-, HS^-	Medium
RSSR	RS^-	Low
HCN	CN^-	Medium
NF_3	F^-	High
SO_2F_2	F^-	High
SF_6	F^-	High
CCl_4	Cl^-	High
CF_2Cl_2	Cl^-	High
CF_3Br	Br^-	High
CH_3I	I^-	High
CH_3CN/H_2O	CN^-, $^-CH_2CN$	Medium
N_2O/H_2	HO^-	Medium

Source. References 42, 75, and 95.

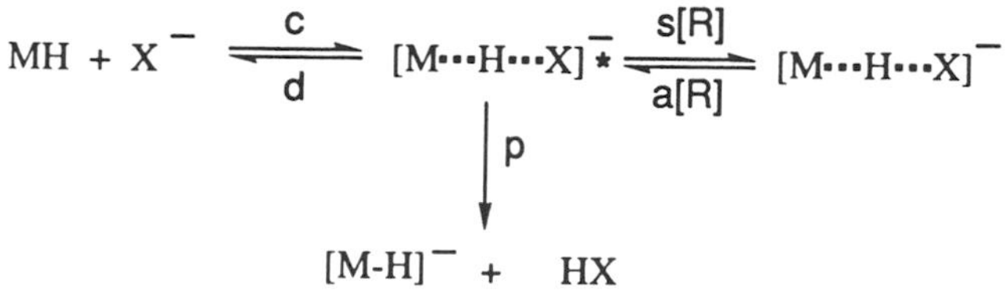

$$MH + X^- \underset{d}{\overset{c}{\rightleftharpoons}} [M \cdots H \cdots X]^{-*} \underset{a[R]}{\overset{s[R]}{\rightleftharpoons}} [M \cdots H \cdots X]^-$$

$$[M \cdots H \cdots X]^{-*} \xrightarrow{p} [M\text{-}H]^- + HX$$

FIGURE 6.24 NICI reaction scheme, where MH = sample molecule, R = methane, X = F, CI, Br, I, and the reaction path notation is c = condensation, d = dissociation, a[R] = collisional activation, s[R] = collisional stabilization, and p = proton transfer.

The gas-phase thermochemistry of thc halide anions X^- (F, CI, Br, and I) with a given MH sample molecule is highly dependent on the halide anion radius. For example, the gas-phase basicity of the halide anions decreases with increase of anion radius (97). Thus, proton transfer from MH to F^- (Figure 6.24; reaction paths c and p) is exothermic for many organic acids ($\Delta H_{(acid)} > 1555$ kJ/mol), while proton transfer from MH to I^- is endothermic for many organic acids (Table 6.11). Since the extent of fragmenta-

TABLE 6.11 Halide Binding Energies[a] (kJ/mol) and Anion Proton Affinities[b] (kJ/mol) of a Variety of Functional Groups

Species [MH]	F^{-c}	Cl^{-d}	Br^{-e}	I^{-e}	$\Delta H°$ (acid)
CH_4					1744
NH_3		44		31	1689
C_6H_6		42		38	1677
H_2O	98	62	53	42	1635
$C_6H_5CH_3$					1593
CH_3OH	124	59		47	1592
CH_3CN	67	56	54	50	1560
HF	161	91	71	63	1554
$(CH_3)_2CO$		57		50	1544
$C_6H_5NH_2$	131			54	1533
H_2S	145			37	1469
C_6H_5OH	173	109			1461
CH_3CO_2H	185	90		71	1459
$(CN)_2CH_2$				79	1405
HCl	251	99	82	60	1395
$(CF_3)_3COH$	242			97	1388
HBr	272	123	87	67	1354
HI	301	128	106	71	1315

[a] Values obtained for reaction $[MH + X]^- \rightarrow X^- + HM$, i.e., D(X − HM).
[b] Values obtained for reaction $[MH] \rightarrow [M - H]^- + H^+$, i.e., $\Delta H°$ (acid).
[c] Reference 101.
[d] Reference 102.
[e] Reference 96.

tion depends primarily on the proton transfer exothermicity, it is reasonable to assume that fragmentation in halide NICI spectra should decrease with increase of anion radius for a given MH (see Figure 6.21). The binding energy of the hydrogen-bonded $[M\cdots H\cdots X]^-$ anion decreases with an increase of ion radius (98–100). Generally, when comparing binding energies to a given X^- anion, binding energy increases with the gas-phase acidity of MH. This is shown in Table 6.11.

While there are exceptions to these generalizations (99, 100), several conclusions can be drawn. When F^- is used as the reactive anion, the NICI spectra of organic acids ($\Delta H_{(acid)} > 1555$ kJ/mol) will contain large yields of the proton transfer $[M-H]^-$ anion and some fragmentation. When Cl^- or Br^- are used as the reactive anions, the NICI spectra will contain a mixture of the proton transfer $[M-H]^-$ anion, fewer fragment anions, and the hydrogen-bonded $[M\cdots H\cdots X]^-$ anion. When I^- is used as the reactive anion, the NICI spectra will contain large yields of the hydrogen-bonded $[M\cdots H\cdots X]^-$ anion and no fragmentation or $[M-H]^-$ anion (see Figure 6.21).

6.6 MULTIDIMENSIONAL (GAS CHROMATOGRAPHY)^m/(MASS SPECTROMETRY)^n

6.6.1 Advantages of Multidimensional (Gas Chromatography)^m/(Mass Spectrometry)^n

Multidimensional techniques such as (gas chromatography)^m/(mass spectrometry)^n (GC^m/MS^n), where m or n equals 1, 2, etc., can improve the sensitivity, separation, and selectivity of trace-level compounds in complex biological and environmental matrices. In analyses where the lowest detectable amount is limited by interference from endogenous components in the matrix, performance can be improved by using different ionization techniques, by increasing the separation at the chromatographic stage, or by increasing the selectivity at the mass separation stage. The advantages of different ionization modes have been pointed out in previous Sections (6.3–6.5). One important advantage of the GC^2 technique is the ability to inject large quantities of samples on the first column and then divert a fraction of the effluent to the second column. Thus, the chromatographic resolution is maintained. The analyte exiting the second column, which reaches the mass spectrometer, is much purer than with single-gas chromatographic separations and ion source contamination is reduced. The specificity of the analysis in GC^2 is also improved greatly by establishing two retention times per component. One important advantage of the MS^2 is its versatile platform for a broad range of experiments to provide sensitive and selective analysis of complex mixtures. Multidimensional experiments where the ionization modes are changed between electron ionization (EI), positive and negative chemical ionization (CI), and tandem chromatography and mass spectrometry techniques are discussed below.

6.6.2 Gas Chromatography/Mass Spectrometry With Simultaneous or Alternating Ionization Modes

Simultaneous recording of different ionization modes give complementary information of structural knowledge. These techniques are advantageous where a limited amount of sample is available and speed of analysis is important. Gas chromatography/electron ionization/positive-ion chemical ionization/mass spectrometry (GC/EI/PICI/MS) (103) produces simultaneous or alternate EI-CI spectra. The mass spectrometer is designed with a single ion source with two ionization chambers—one CI and one EI source. The CI and EI sources are in series, with the chromatographic effluent entering the CI source then exiting to the EI source. Methane is used as the carrier and reagent gas. In the alternating mode, the ion source is changed between CI and EI repeatedly, with the number of alternating scans depending on the width of the chromatographic peak. In the simultaneous mode, a mixed CI/EI spectrum is obtained.

Pulsed positive-ion–negative-ion chemical ionization/mass spectrometry (PPINICI/MS) (104) has been demonstrated for the simultaneous recording, on separate channels, of positive and negative chemical ionization mass spectra. The CI ion source is pulsed at 10 kHz by alternating the source and lens potentials such that packets of cations and anions are ejected from the CI source. A quadrupole mass filter is used to mass analyze these packets, since it transmits cations and anions with equal facility. After mass analysis the cations and anions are detected using a conversion dynode system (105). Several different reagent gases can be utilized. For example, when methane is used as the reagent gas, the cations CH_5^+ and $C_2H_5^+$ are generated to produce $[M+H]^+$ cations (Section 6.4.6) along with thermal electrons, which can be used to produce $[M]^{-\cdot}$ radical anions (Section 6.5.4). If methylene chloride is added to the methane reagent gas, the Cl^- anion is produced, which generates $[M-H]^-$ and $[M+Cl]^-$ anions without changing the cation reagents (Section 6.5.6).

6.6.3 Gas Chromatography^2/Mass Spectrometry

Gas chromatography/gas chromatography/mass spectrometry (GC^2/MS) techniques are particularly useful for the analysis of complex mixtures with severely overlapping chromatographic peaks (106–117). The technique allows the selectivity of two sequentially linked columns with different liquid phases to be combined in a single analysis for improved chromatographic separation. The second column is terminated directly into the source of the MS. There are two main GC^2 configurations: mechanical valve and pneumatic switching devices. For example, two columns (a 30-m × 0.25-mm

DB-1 with 1.0-μm film thickness and a 30-m $\times$ 0.25-mm DB-WAX with 0.5-μm film thickness) are connected through a mechanical valve-based system and both columns are located in a single-gas chromatographic oven. Carrier gas flows into the first column through the valve and exits the second column. Peak "heart cutting" from the first column onto the second column can be accomplished by simply turning the valve (106). The major problem with this mechanical valve-based system is peak tailing due to solvent/analyte interaction within the valve assembly. A pneumatic switching system is based on the pressure balance of the column system. That is, the carrier-gas flowrate through any two columns is determined by the pressure differential at the ends of the columns. This pressure difference is controlled by solenoid valves that are not in the solvent/analyte pathway, and thus this method does not have the above-mentioned limitations. The pneumatic switching system can be configured for heart-cutting, backflushing, and cold-trapping processes (107).

A typical example of GC^2/MS is the analysis of polychlorodibenzodioxins and polychlorodibenzofurans in complex matrices such as fly ash (107, 108). These compounds can have many isomers, depending on the degree of chlorination. For example, chlorination of dibenzofuran can yield 135 unique structures. Due to the different toxicity of different polychlorodibenzodioxins and polychlorodibenzofurans isomers, especially the seventeen "2,3,7,8" isomers, all components in the sample, must be separated with good resolution for quantification studies. A single-gas chromatographic separation is not possible to fulfill this requirement. In a two-dimensional experiment, specific components eluting from the first column are diverted into a cold trap and then transferred to the second column where the components are further separated. Compounds exiting the second column are analyzed by electron ionization in the selective ion monitoring mode. ^{13}C-labeled internal standards of the species of interest are added to the sample to allow quantification by isotope dilution and to cnsurc that thc correct region was collected from the first column. The GC^2/MS techniques led to detectable concentrations for polychlorodibenzodioxins and polychlorodibenzofurans at 400 fg/μL when 25-μL solutions were injected onto the first column.

6.6.4 Gas Chromatography/Mass Spectrometry^2

Gas chromatography/mass spectrometry/mass spectrometry (GC/MS^2) techniques are useful for the analysis of complex mixtures when selectivity of detection is required (118–123). An MS^2, or tandem mass spectrometer, consists of an ion source, a mass analyzer, a collision cell, a second mass analyzer, and an ion detector (124, 125). In a typical experiment, the components of a mixture are separated by GC prior to mass analysis. A component that elutes from the column is ionized in the ion source to form characteristic ions. The separation of these ions is achieved by selecting a particular m/z (parent) ion with the first mass analyzer. This parent ion is accelerated, with energies in the range of 10–20 eV, into the collision cell containing ca. 10^{-3} Torr of argon. It then undergoes collisionally activated dissociation (CAD) through collisions with the argon gas molecules to yield various fragment (product) ions. The separation of the product ions is achieved by the second mass analyzer. A product mass spectrum is recorded, which can be used for identification of the component. It is interesting to note that MS^7 has been achieved in sequence using a Fourier transform mass spectrometer (126).

The four scan modes that are widely used for MS^2 include product, neutral loss (gain), parent, and selected reaction monitoring (127, 128). The specific scan mode is chosen based on the type of information desired. The scan mode described above is called a product scan and is useful for structural determination of unknown compounds. The neutral loss scan is useful for rapidly screening mixtures for classes of compounds that have a common neutral loss fragmentation pathway in the ion source and the collision cell. For example, polyhalogenated (X = F, Cl, Br, and I) organic molecules sometimes lose X or HX during collisionally activated dissociation. Thus a neutral loss scan requires that both mass analyzers are scanned with a constant mass difference of 19 or 20 u for fluoro-compounds, 35 or 36 u for chloro-compounds, 79 or 80 u for bromo-compounds, and 127 or 128 u for iodo-compounds. Note that the isotopes ^{37}Cl and ^{81}Br of the chloro- and bromo-compounds could also be used. A parent scan is useful for rapidly screening mixtures for compounds that produce common fragments in the ion source and/or collision cell. The second analyzer is set to transmit the common fragment while the first analyzer is scanned. As discussed in Section 6.4.3, under positive-ion chemical ionization (PICI) conditions di-n-butyl phthalate produces cations at m/z 279 $[M + H]^+$, 205 $[M + H - C_4H_9OH]^+$, and 149 $[M + H - C_4H_9OH - C_4H_8]^+$. When the cations at m/z 279 and 205 are CAD in the collision cell, both fragment to the cation at m/z 149. Therefore, by scanning the first mass analyzer over the mass range of 1000–190 u and selecting only the m/z 149 cation with the second mass analyzer, a method would be developed for screening for phthalates. This technique is also useful in detection of drug metabolites, since structurally related compounds frequently produce common fragment ions. The selected reaction monitoring scan mode is useful for trace analyses. Here a limited number of parent ion and product ion pairs are monitored for each component. The selected reaction monitoring (SRM) scan mode is analogous to selective ion monitoring (SIM) in GC/MS experiments.

An example of a pharmacokinetic investigation using these techniques is the assay of indomethacin(1-(4-chlorobenzoyl)-5-methoxy-2-methyl-1H-indole-3-acetic acid (Figure 6.25) in plasma and synovial fluid (122). This analyte was isolated and subsequently derivatized by pentafluorobenzylation of the carboxy group. Deuterium labeled d_2-indomethacin PFB was used as

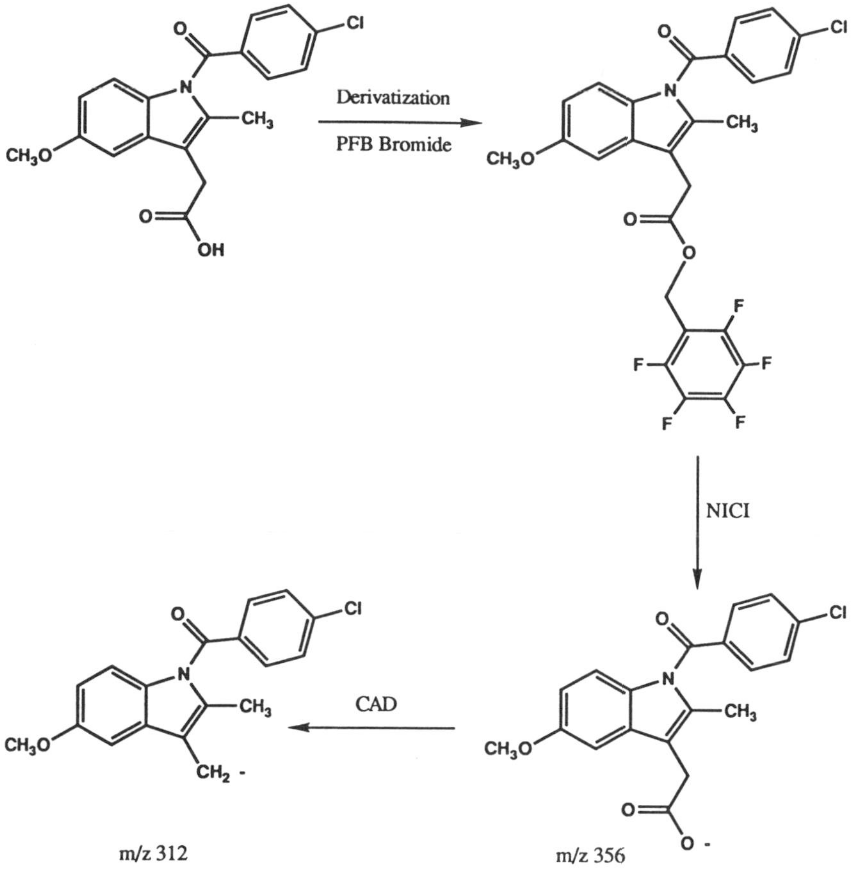

Derivatization

PFB Bromide

NICI

CAD

m/z 312

m/z 356

FIGURE 6.25 Indomethacin(1-(4-chlorobenzoyl)-5-methoxy-2-methyl-1H-indole-3-acetic acid) is derivatized to the pentafluorobenzyl (PFB) ester. The negative-ion chemical ionization (NICI) mass spectrum contains a peak at *m/z* 356. The ions at *m/z* 356 are accelerated with an energy of 10 eV into the collision cell at 1.5×10^{-3} Torr, where they undergo collisionally activated dissociation (CAD) through collisions with the argon gas molecules to yield a peak at *m/z* 312. Note that the anion at *m/z* 312 probably rearranges to delocalize the negative charge.

the internal standard with methane as the reagent gas and helium as the carrier gas. The NICI spectrum of indomethacin PFB contained major peaks at *m/z* 356 ($[M - CH_2C_6F_5]^-$) and 312 ($[M - CH_2C_6F_5 - CO_2]^-$), while the internal standard had major peaks at *m/z* 360 ($[M(d_2) - CH_2C_6F_5]^-$) and *m/z* 316 ($[M(d_2) - CH_2C_6F_5 - CO_2]^-$). These anions accounted for approximately 50% of the total ion current. The presence of the radical anion $[M]^{-\cdot}$ was not observed in the NICI spectrum. The base peak anion $[M - CH_2C_6F_5]^-$ results from electron transfer and cleavage of the pentafluorobenzyl functional group with charge retention on the carboxy group. Note that this reaction is simply the dissociative electron attachment (Reaction 6.35). The anion at m/z 356 was fragmented in the collision cell to produce m/z 312; this pair along with the deuterium-labeled d_2-indomethacin PFB peaks at m/z 360 and 316 were used for selective reaction monitoring. The method was suitable for the determination of indomethacin in plasma and synovial fluid at 0.1 ng/mL.

The effect of ion source temperature on the NICI spectrum of indomethacin PFB was investigated over the range of 100–180°C. As the ion source temperature was increased from 100 to 160°C, the abundance of m/z 356 increased slightly. At temperatures above 160°C, decarboxylation occurred, the abundance of m/z 356 decreased, and the peak at m/z 312 increased. The effect of collision energy and collision pressure was also investigated on the CAD spectrum of m/z 312. Maximum sensitivity was obtained at a collision energy of 10 eV and a collision pressure of 1.5×10^{-3} Torr.

An interesting experiment has been designed where the gas chromatograph was interfaced to the collision cell instead of the ion source of a MS^2 instrument (129). In this experiment, the components of a mixture are separated by gas chromatography prior to mass analysis. The components that elute from the column enter the collision cell. A reagent ion is selected with the first mass analyzer and is accelerated into the collision cell. The reagent ion reacts with the effluent in the collision cell and the resulting ion/molecule products are separated by the second mass analyzer. The chromatographic integrity was less than that obtained when the effluent directly entered into the ion source; however, the chromatographic peak shape was expected to improve with a separately heated collision cell. Charge exchange and positive ion chemical ionization were investigated (Section 6.4.2). The technique offers many advantages over normal GC/MS experiments. One of the more important advantages is the ability to enhance selectivity by choosing a reagent ion that will ionize only compounds of interest.

6.6.5 Gas Chromatography^2/Mass Spectrometry^2

Recently the ability to analyze complex mixtures with severely overlapping chromatographic peaks and the ability to have selective detection has been combined in a gas chromatography/gas chromatography/mass spectrometry/mass spectrometry (GC^2/MS^2) instrument (130). The chromatographic system was designed such that separations could be carried out on two columns of identical or different diameters. The two columns were placed in two chromatographs separated by an interface. The design of the interface between the gas chromatographs allowed for direct transfer, backflushing, heart cutting, intermediate cold trapping, trace enrichment, and selective sample introduction. The MS^2 consisted of an ion source, a

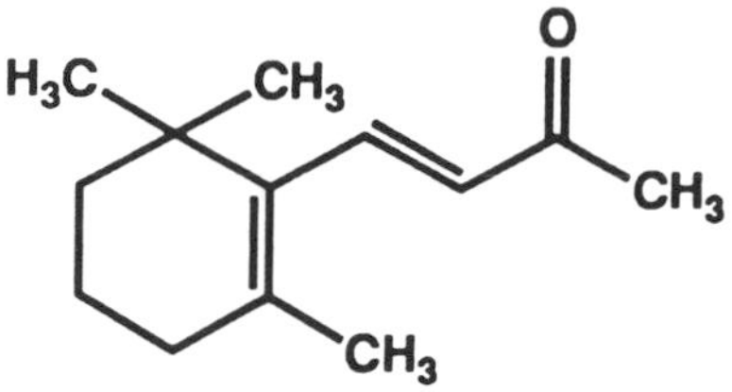

FIGURE 6.26 Structure of β-ionone.

mass analyzer, a collision cell, a second mass analyzer, and an ion detector (124, 125). The detection of β-ionone (Figure 6.26) in the essential oil of *Buddleia salvifolia* was demonstrated utilizing this technique. The essential oil was injected onto the first column and a portion of the effluent of the first column was transferred to the second column after enrichment. The existence of β-ionone in this fraction was established with a series of MS^2 experiments. This multidimensional instrument could provide solutions for solving problems that are encountered in highly complex mixtures.

REFERENCES

1. J. J. Thomson, *Rays of Positive Electricity*, Longmans, Green, London, 1913.
2. F. W. Aston, *Isotopes*, Edward Arnold, London, 1922, 1924.
3. F. W. Aston, *Mass Spectra and Isotopes*, Edward Arnold, London, 1942.
4. A. T. James and A. J. P. Martin, *Biochem. J.*, **50**, 679 (1952).
5. M. J. E. Golay, in *Gas Chromatography*, V. J. Coates, H. J. Noebels, and I. S. Fagerson (Eds.), Academic, New York, 1958.
6. J. C. Holmes and F. A. Morrell, *Appl. Spectrosc.* **11**, 86 (1957).
7. W. H. McFadden, *Techniques of Combined Gas Chromatography/Mass Spectrometry*, Wiley-Interscience, New York, 1973.
8. R. S. Melville and V. F. Sobson, *Selected Approaches to Gas Chromatography–Mass Spectrometry in Laboratory Medicine*, U.S. Dept. of Health, Education and Welfare, Natl. Inst. Health, Bethesda, MD, 1975.
9. B. J. Gudzinowicz, M. J. Gudzinowicz, and H. F. Martin, *Fundamentals of Integrated GC-MS, Parts I–III*, Dekker, New York, 1976.
10. Y. Masada, *Analysis of Essential Oils by Gas Chromatography and Mass Spectrometry*, Wiley, New York, 1976.
11. B. J. Gudzinowicz and M. Gudzinowicz, *Analysis of Drugs and Metabolites by Gas Chromatography–Mass Spectrometry, Vols. 1–7*, Dekker, New York, 1980.
12. P. Vouros, *Mass Spectrometry, Part B*, C. Merritt, Jr., and C. N. McEwen (Eds.), Dekker, New York, 1980, p. 129.
13. S. I. Goodman and S. P. Markey, *Diagnosis of Organic Acidemias by Gas Chromatography–Mass Spectrometry*, Liss, New York, 1981.
14. G. M. Message, *Practical Aspects of Gas Chromatography/Mass Spectrometry*, Wiley, New York, 1984.

15. G. Odham, L. Larson, and P. A. Mardh, *Gas Chromatography/Mass Spectrometry Applications in Mibrobiology*, Plenum, New York, 1984.
16. H. Jaeger, *Capillary Gas Chromatography–Mass Spectrometry in Medicine and Pharmacology*, Alfred Huethig, Heidelberg, 1987.
17. C. J. Bierman and G. D. McGinnis, *Analysis of Carbohydrates by GLC and MS*, CRC Press, Boca Raton, FL, 1989.
18. Reference 9, p. 58.
19. G. W. Caldwell et al., *Org. Mass Spectrum.*, **24**, 1051 (1989).
20. P. H. Dawson (Ed.), *Quadrupole Mass Spectrometry and Its Applications*, Elsevier Scientific, New York, 1976.
21. R. J. Strife, J. R. Simms, and M. P. Lacey, *J. Am. Soc. Mass Spectrom.*, **1**(3), 265 (1990).
22. R. L. White and C. L. Wilkins, *Anal. Chem.*, **54**, 2443 (1982).
23. J. Drozd, *J. Chromatogr.*, **113**, 303 (1975).
24. R. I. Reed, *Ion Production by Electron Impact*, Academic, New York (1962).
25. K. Biemann, *Mass Spectrometry: Organic Chemical Applications*, McGraw-Hill, New York, 1962.
26. H. Budzikiewicz, C. Djerassi, and D. H. Williams, *Mass Spectrometry of Organic Compounds*, Holden Day, San Francisco, 1967.
27. F. W. McLafferty, *Interpretation of Mass Spectra*, University Science Books, Mill Valley, CA, 1980.
28. J. T. Watson, *Introduction to Mass Spectrometry*, Raven, New York, 1985.
29. H. M. Rosenstock, K. Draxl, B. W. Steiner, and J. T. Herron, *J. Phys. Chem. Ref. Data*, **6** (Suppl. 1), 783 (1977).
30. R. D. Levin and S. G. Lias, *Natl. Stand. Ref. Data Ser.*, Natl. Bur. Std. (U.S.), 71 (1982).
31. J. Deutsch, L. Hegedus, N. H. Greig, S. I. Rapoport, and T. T. Soncrant, *J. Chromatogr.*, **579**(1), 93 (1992).
32. G. Durand and D. Barcelo, *Anal. Chim. Acta*, **243**(2), 259 (1991).
33. E. Matisova, S. Vodny, S. Skrabakova, and M. Onderova, *J. Chromatogr.*, **629**(2), 309 (1993).
34. A. Golan-Goldhirsh, A. M. Hogg, and F. W. Wolfe, *J. Agric. Food Chem.*, **30**, 320 (1982).
35. G. D. Reed, *J. High Resolut. Chromatogr.*, **15**(1), 46, (1992).
36. S. A. Rossi, J. V. Johnson, and R. A. Yost, *Biol. Mass Spectrom.*, **21**(9), 420 (1992).
37. F. H. Field, J. L. Franklin, and F. W. Lampe, *J. Am. Chem. Soc.*, **79**, 2419 (1957).
38. M. S. B. Munson and F. H. Field, *J. Am. Chem. Soc.*, **88**, 2621 (1966).
39. F. H. Field, *Accounts Chem. Res.*, **1**, 42 (1968).
40. B. Munson, *Anal. Chem.*, **43**, 28A (1971).
41. B. Munson, *Anal. Chem.*, **49**, 772A (1977).

42. A. G. Harrison, *Chemical Ionization Mass Spectrometry*, CRC Press, Boca Raton, FL, 1983.
43. J. E. Bartmess, *Mass Spectrom. Rev.*, **8**, 297 (1989).
44. C. F. Kuhlman, T. L. Chang, and G. L. Nelson, *J. Pharm. Sci.*, **64**, 1581 (1975).
45. F. Hatch and B. Munson, *Anal. Chem.*, **49**, 169 (1977).
46. S. Ghaderi, P. S. Kulkarni, E. B. Ledford, Jr., C. L. Wilkins, and M. L. Gross, *Anal. Chem.*, **53**, 428 (1981).
47. G. W. Caldwell, J. A. Masucci, and R. R. Inners, *Org. Mass Spectrom.*, **23**, 86 (1988).
48. S. G. Lias, J. E. Bartmess, J. F. Liebman, J. L. Holmes, R. D. Levin, and W. G. Mallard, *J. Phys. Chem. Ref. Data*, **17** (Suppl. 1) (1988).
49. R. W. Taft, Gas-phase proton-transfer equilibria, in *Proton Transfer Reactions*, E. F. Caldin and V. Gold (Eds.), Chapman & Hall, London, 1975.
50. D. H. Aue and M. T. Bowers, *Gas Phase Ion Chemistry*, M. T. Bowers (Eds.), Academic, New York, 1979.
51. P. Kebarle, *Annu. Rev. Phys. Chem.*, **28**, 445 (1977).
52. G. P. Arsenault, J. J. Dolhun, and K. Biemann, *Chem. Commun.*, 1542 (1970).
53. D. M. Schoengold and B. Munson, *Anal. Chem.*, **42**, 1811 (1970).
54. H. M. Fales, G. W. A. Milne, and R. S. Nicholson, *Anal. Chem.*, **43**, 1785 (1971).
55. T. Y. Yu and F. Field, *Org. Mass Spectrosc.*, **8**, 267 (1974).
56. C. A. Maryanoff, G. W. Caldwell, and S. Y. Chang, *Org. Mass Spectrosc.*, **23**, 129 (1988).
57. J. B. Westmore and M. M. Alauddin, *Mass Spectrom. Rev.*, **5**, 381 (1986).
58. D. F. Hunt and J. F. Ryan, *Anal. Chem.*, **44**, 1306 (1972).
59. D. L. Lawrence, *Rapid Commun. Mass Spectrom.*, **4**(12), 546 (1990).
60. D. F. Hunt, C. N. McEwen, and R. A. Upham, *Anal. Chem.*, **44**, 1292 (1972).
61. M. Svoboda and S. Musil, *J. Chromatogr.*, **520**, 209 (1990).
62. D. F. Hunt, C. N. McEwen, and R. A. Upham, *Tetrahedron Lett.*, 4539 (1971).
63. P. A. D'Agostino and L. R. Provost, *J. Chromatogr.*, **600**(2), 267 (1992).
64. W. Blum, E. Schlumpf, J. G. Liehr, and W. Richter, *Tetrahedron Lett.*, 565 (1976).
65. M. V. Buchanan, *Anal. Chem.*, **56**(3), 546 (1984).
66. B. J. Millard, *Quantitative Mass Spectrometry*, Heyden & Son, London, 1978.
67. L. Jirovetz, W. Jaeger, G. Buchbauer, A. Nikiforov, and V. Raverdino, *Biol. Mass Spectrom.*, **20**(12), 801 (1991).
68. B. D. Dulery, J. Schoun, and K. D. Haegele, *J. Chromatogr.*, **571**(1–2), 241 (1991).
69. H. K. Lim, S. Zeng, D. M. Chei, and R. L. Foltz, *J. Pharm. Biomed. Anal.*, **10**(9), 657 (1992).

70. J. Deutsch, L. Hegedus, N. H. Greig, S. I. Rapoport, and T. T. Soncrant, *J. Chromatogr.*, **579**(1), 93 (1992).
71. Z. H. Huang, D. A. Gage, L. L. Bieber, and C. C. Sweeley, *Anal. Biochem.*, **199**(1), 98 (1991).
72. M. R. Harkey, G. L. Henderson, and C. Zhou, *J. Anal. Toxicol.*, **15**(5), 260 (1991).
73. B. Geypens, Y. Ghoos, M. Hiele, P. Rutgeerts, G. Vantrappen, E. Joosten, and W. Pelemans, *Anal. Chim. Acta*, **247**(2), 243 (1991).
74. M. Ohtani, F. Shibuya, H. Kotaki, K. Uchino, Y. Saitoh, F. Nakagawa, and K. Nishitateno, *J. Chromatogr.*, **487**(2), 469 (1989).
75. H. Budzikiewicz, *Mass Spectrom. Rev.*, **5**, 345 (1986).
76. G. Caldwell and P. Kebarle, *J. Chem. Phys.*, **80**(1), 577 (1984).
77. E. P. Grimsrud, G. Caldwell, S. Chowdhury, and P. Kebarle, *J. Am. Chem. Soc.*, **107**, 4627 (1985).
78. P. Kebarle and S. Chowdhury, *Chem. Rev.*, **87**, 513 (1987).
79. A. Poppe, E. Schroder, and H. Budzikiewicz, *Org. Mass Spectrom.*, **21**, 59 (1986).
80. E. A. Stemmler and R. A. Hites, *Anal. Chem.*, **57**, 684 (1985).
81. D. Stockl and H. Budzikiewicz, *Org. Mass Spectrom.*, **17**, 470 (1982).
82. A. J. Lewy and S. P. Markey, *Science*, **201**, 741 (1978).
83. E. van der Heeft, A. P. J. M. de Jong, L. A. van Ginkel, H. J. van Rossum, and G. Zomer, *Biol. Mass Spectrom.*, **20**, 763 (1991).
84. A. Changchit, J. Gal, and J. A. Zirrolli, *Biol. Mass Spectrom.*, **20**, 751 (1991).
85. M. J. Morris, J. D. Gilbert, J. Y.-K. Hsieh, B. K. Matuszewski, H. G. Ramjit, and W. F. Bayne, *Biol. Mass Spectrom.*, **22**, 1 (1993).
86. B. W. Christman, J. C. Gay, J. W. Christman, C. Prakash, and I. A. Blair, *Biol. Mass Spectrom.*, **20**, 545 (1991).
87. C. Prakash, A. Adedoyin, G. R. Wilkinson, and I. A. Blair, *Biol. Mass Spectrom.*, **20**, 559 (1991).
88. A. J. R. Teixeira, G. van de Werken, J. F. C. Stavenuiter, and A. P. J. M. de Jong, *Biol. Mass Spectrom.*, **21**, 441 (1992).
89. D. G. Watson, J. M. Midgley, and C. N. J. McGhee, *J. Chromatogr.*, **571**, 101 (1991).
90. M. Nagasawa, H. Sasabe, T. Shimizu, and H. Mori, *J. Chromatogr.*, **577**, 275 (1992).
91. N. Goto, T. Kamata, and K. Ikegami, *J. Chromatogr.*, **578**, 195 (1992).
92. M. J. Bartels and P. E. Kastl, *J. Chromatogr.*, **575**, 69 (1992).
93. H. J. Leis, P. T. Ozand, A. Al Odaib, and H. Gleispach, *J. Chromatogr.*, **578**, 116 (1992).
94. J. Turk, A. Bohrer, W. T. Stump, S. Ramanadham, and M. J. Mangino, *J. Chromatogr.*, **575**, 183 (1992).
95. A. T. Lehman and M. M. Bursey, *Ion Cyclotron Resonance Spectrometry*, Wiley Interscience, New York, 1976.

96. G. W. Caldwell, J. A. Masucci, and M. G. Ikonomou, *Org. Mass Spectrom.*, **24**, 8 (1989).
97. J. B. Cumming and P. Kebarle, *Can. J. Chem.*, **1**, 78 (1978).
98. R. Yamdagni and P. Kebarle, *J. Am. Chem. Soc.*, **93**, 7139 (1971).
99. G. Caldwell and P. Kebarle, *J. Am. Chem. Soc.*, **106**, 967 (1984).
100. G. Caldwell and P. Kebarle, *Can. J. Chem.*, **63**, 1399 (1985).
101. J. W. Larson and T. B. McMahon, *J. Am. Chem. Soc.*, **109**, 6230 (1987).
102. J. W. Larson and T. B. McMahon, *J. Am. Chem., Soc.*, **109**, 517 (1984).
103. G. P. Arsenault, J. J. Dolhun, and K. Biemann, *Anal. Chem.*, **43**, 1720 (1971).
104. D. F. Hunt, G. C. Stafford, Jr., F. W. Crow, and J. W. Russell, *Anal. Chem.*, **48**, 2098 (1976).
105. G. C. Stafford, *Environ. Health Perspect.*, **36**, 85 1980.
106. E. R. Kennedy, P. F. O'Connor, and A. A. Grote, *J. Chromatogr.*, **522**, 303 (1990).
107. F. David, P. Sandra, A. Hoffmann, and J. Gertel, *Chromatographia*, **34**, 259 (1992).
108. W. V. Ligon, Jr., and R. J. May, *J. Chromatogr.*, **294**, 77 (1984).
109. K. K. Himberg and E. Sippola, *Chemosphere*, **27**(1–3), 17 (1993).
110. M. Woerner and P. Schreier, *Phytochem. Anal.*, **2**(6), 260 (1991).
111. A. Bernreuther and P. Schreier, *Phytochem. Anal.*, **2**(4), 167 (1991).
112. E. Sippola and K. Himberg, *Fresenius, J. Anal. Chem.*, **339**(7), 510 (1991).
113. L. I. Lamparski, T. J. Nestrick, D. Janson, and G. Wilson, *Chemosphere*, **20**(6), 635 (1990).
114. E. L. White, M. S. Uhrig, T. J. Johnson, B. M. Gordon, R. D. Hicks, M. F. Borgerding, W. M. Coleman III, and J. F. Elder, Jr., *J. Chromatorg. Sci.*, **28**, 393 (1990).
115. J. F. Elder, Jr., B. M. Gordon, and M. S. Uhrig, *J. Chromtorg. Sci.*, **24**, 26 (1986).
116. W. V. Ligon, Jr., and R. J. May, *J. Chroamtogr.*, **294**, 87 (1984).
117. W. V. Ligon, Jr., and R. J. May, *Anal. Chem.*, **52**, 901 (1980).
118. A. D. Jones, C. K. Winter, M. H. Buonarati, and H. J. Segall, *Biomed. Mass Spectrom.*, **22**, 68 (1993).
119. P. A. D'Agostino, L. R. Provost, and P. W. Brooks, *J. Chromatogr.*, **541**, 121 (1991).
120. R. L. M. Dobson, D. M. Neal, B. R. DeMark, and S. R. Ward, *Anal. Chem.*, **62**, 1819 (1990).
121. H. Schweer, C. O. Meese, and H. W. Seyberth, *Anal. Biochem.*, **189**, 54 (1990).
122. M. Dawson, M. D. Smith, and C. M. McGee, *Biomed. Mass Spectrom.*, **19**, 453 (1990).
123. M. Dawson, C. M. McGee, P. M. Brooks, J. H. Vine, and T. R. Watson, *Biomed. Mass Spectrom.*, **17**, 205 (1988).
124. S. J. Gaskell, *Biomed. Mass Spectrom.*, **21**, 413 (1992).
125. F. W. McLafferty (Ed.), *Tandem Mass Spectrometry*, Wiley, New York, 1983.
126. D. B. Jacobson and B. Freiser, *J. Am. Chem. Soc.*, **106**, 4623 (1984).

127. J. C. Schwartz, A. P. Wade, C. G. Enke, and R. G. Cooks, *Anal. Chem.*, **62**, 1809 (1990).

128. J. Johnson and R. A. Yost, *Anal. Chem.*, **57**, 758A (1985).

129. M. E. Hail, D. W. Berberich, and R. A. Yost, *Anal. Chem.*, **61**, 1874 (1989).

130. S. G. Claude and R. Tabacchi, *J. High Resolut. Chromatogr./Chromatogr. Commun.*, **11**(2), 187 (1988).

CHAPTER SEVEN

Qualitative and Quantitative Analysis by Gas Chromatography: Sample Preparation and Trace Analysis

FREDERICK J. DEBBRECHT

Deceased

Part 1 Qualitative Analysis

Modern Practice of Gas Chromatography, Third Edition. Edited by Robert L. Grob.
ISBN 0-471-59700-7

Part 1 Qualitative Analysis

7.1 DISCUSSION OF CHROMATOGRAPHIC DATA

Inherently, two important pieces of data can be obtained from a gas chromatograph. The output of the detector is processed either electronically or placed on a simple strip chart recorder. The first piece of data obtained is simply the time it took for a given component to travel through the column. This is the time from the point of injection to the maximum of the peak as it passes through the detector. This time is referred to as the retention time t_R. It is this retention time information that is used in qualitative analysis. The second important piece of information that is obtained is simply the size of the peak. Size data is discussed in Part 2 of this chapter, "Quantitative Analysis." A third piece of information that can be obtained from the chromatograph is the shape of the peak. This is available only if the chromatogram is displayed for the individual. This information is lost in many of the electronic integrators that are in use today. The shape of the peak may give some information for both qualitative and quantitative analysis. This peak shape is discussed where it is important in both of these parts. The chromatogram also provides information about the chromatographic operation of the system and any degradation that may have occurred with time or with a particular sample.

7.2 IDENTIFICATION FROM GAS CHROMATOGRAPHIC DATA ONLY

7.2.1 Retention Data

Qualitative analysis by gas chromatography (GC) in the classical sense involves the comparison of retention data of; an unknown sample with that of a known sample. The alternative approach involves a combination and comparison of gas chromatographic data with data from other instrumental and chemical methods of analysis. The simplest qualitative tool is simply the comparison of retention data from known and unknown samples. A chromatogram illustration the commonly used retention nomenclature is given in Figure 7.1. The retention time t_R is the time elapsed from injection of the sample component to the recording of the peak maximum. The retention volume V_R is the product of the retention time and the flowrate of the carrier gas. Generally, the adjusted retention time t'_R or adjusted retention volume V'_R and the relative retention $r_{a/b}$ are used in qualitative analysis. Adjusted retention time (volume) is the difference between retention time (volume) of the sample and an inert component (usually air). The relative retention is the ratio of the adjusted retention time (or volume)

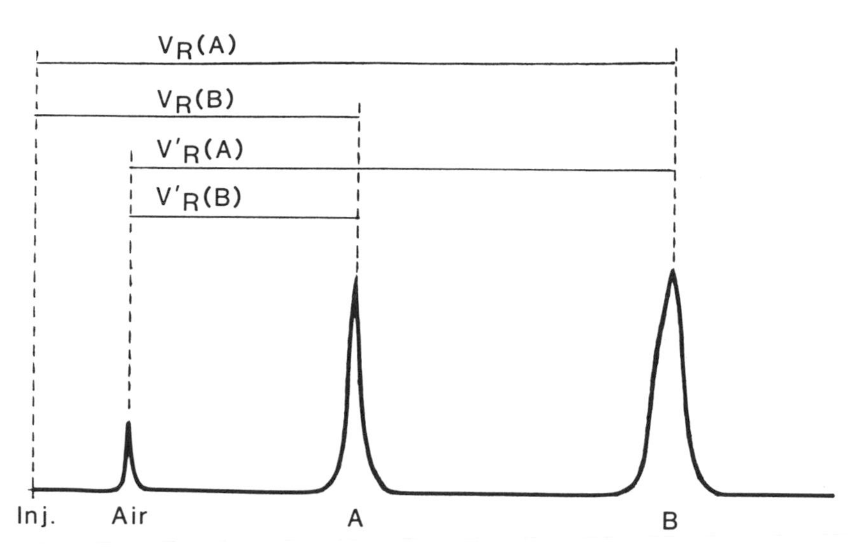

FIGURE 7.1 Chromatogram illustrating retention nomenclature: V_R = retention volume, V'_R = adjusted retention volume, $V_{A/B}$ = relative retention = $V'_{R(A)}/V'_{R(B)}$.

of a standard to the adjusted retention time (or volume) of the unknown (see Chapter 1).

There are three fundamentals concerning retention times obtained on a given instrument with a given column operating under fixed operating conditions. These fundamentals must be known, understood, and believed before useful qualitative data can be obtained from gas chromatographic information only. The first and most important principle is simply that if the retention time of component A is equal to the retention time of an unknown component, this does not prove that the unknown component is component A. This is the biggest pitfall of qualitative analysis and is the statement that prevents gas chromatography from being an exceptional qualitative tool. The rest of the first part of this chapter is devoted to ways and means of supplementing retention data to obtain qualitative analysis. The second fundamental is simply that if the retention time of component A does not equal the retention time of an unknown component, then indeed with absolute certainty we can say that the unknown component is not component A. The third important fundamental is that if we have no discernible peak at the retention time of component A, we can say with certainly that no component A is present in the sample to our limits of detection.

Many factors must be considered in comparison of any retention measurements. The precision of the data generally depends on the ability of the instrument to control the temperature of the column and the flowrate of the carrier gas. A change in the temperature of approximately 30°C changes the retention time by a factor of 2. Thus to maintain a 1% repeatability in retention measurements, one must hold the column temperature to within 0.3°C. A 1% change in the carrier-gas flowrate affects the retention time by approximately 1%.

Sample size also plays an important role (see Figure 7.2). If too much sample is introduced onto the column for its diameter, "leading peaks" will appear. These leading peaks are distorted, giving a slow rise to the peak and a fast drop. As shown in Figure 7.2, the actual time of the peak maximum shifts to longer times, causing the retention time to actually increase for more of a particular component. This phenomenon is caused by column overload. It can be most apparent in gas–solid chromatography, where the action is simply a surface action. In gas–liquid chromatography it is more important at very low loadings on the column packing. Higher loadings will not cause column overload to occur as rapidly as the component amount in the sample is increased. There are cases where the column temperature is operating above the boiling point of the component. Instead of seeing

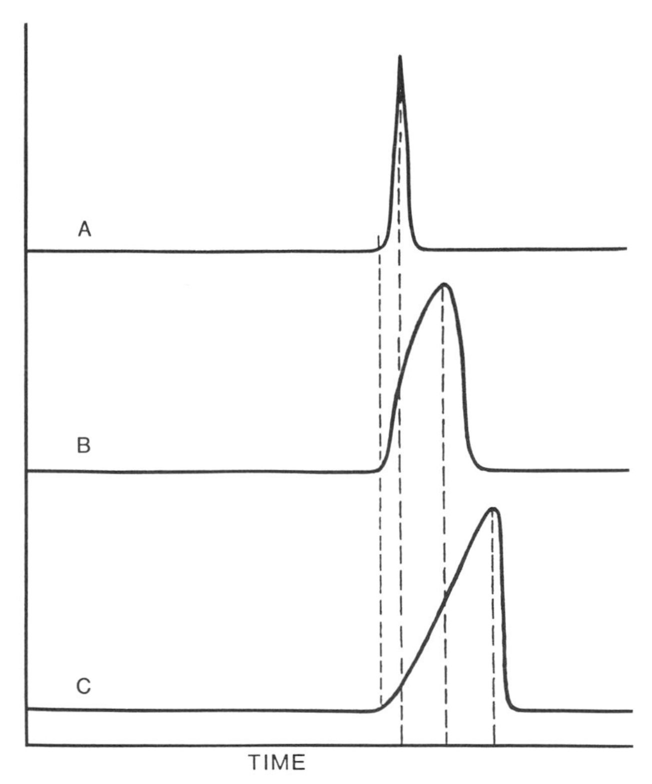

FIGURE 7.2 Effect of sample size on retention time: (a) column not overloaded; (b) column slightly overloaded; (c) column severly overloaded.

leading peaks, we actually see tailing peaks where the front edge of the peak is very sharp and the back edge of the peak slopes. In this case the retention time moves to shorter times under a column overload condition. In comparing retention times for qualitative analysis, one should be alert to this overload condition and test it simply by cutting the sample size in half and injecting the sample again. If retention times stay constant, both conditions could be said to be under a nonoverload situation. If the retention time changes for the reduced sample size, however, the sample size must be reduced once again to ensure that the system is operating in a nonoverload or ideal condition.

Attempts to compare retention times on two different columns of the same type can be difficult at best. Differences in packing density, liquid loading, activity of the support, age and previous use of the packing, and variations in the comparison of the column wall can lead to large differences in retention time measurement between the two columns. Thus tabulations of absolute retention times are of not much value in qualitative analysis. However, there are a number of solutions, to this dilemma. The first and simplest solution is the use of relative retention times. The relative retention of a component is simply its adjusted retention time divided by the adjusted retention time of a reference material. This is indicated in Figure 7.1, where the reference material is assumed to be peak B. Relative retention data are much less subject to variation from column to column and for slight changes in temperature and flow changes. It is also quite simple to obtain relative retention data.

7.2.2 Plot of Log Retention Time Versus Carbon Number

A linear dependence exists between the logarithm of the retention times for compounds in homologous series and the number of carbon atoms in the molecule. This relationship has been shown to hold for many classes of compounds such as alkanes, olefins, aldehydes, ketones, alcohols, acetates, acetals, esters, sulfoxides, nitro derivatives, aliphatic amines, pyridine homologs, aromatic hydrocarbons, dialkyl ethers, thiols, alkyl nitrates, substituted tetrahydrofurans, and furan. A typical series of plots of the logarithm of the retention time versus the carbon number is given in Figure 7.3. It must be reemphasized that this method of identification is valid only for members of a homologous or pseudohomologous series. However, if plots such as that shown in Figure 7.3 are known for a given column under a given set of operating conditions, this method can be extremely useful in helping to identify unknown components. In many cases the first member, and even in some cases the second member, of the series may deviate slightly from this strictly linear relationship. In general, however, one does not have a column so well defined at a fixed set of operating conditions that a large number of these curves are available. It is reasonably easy to obtain

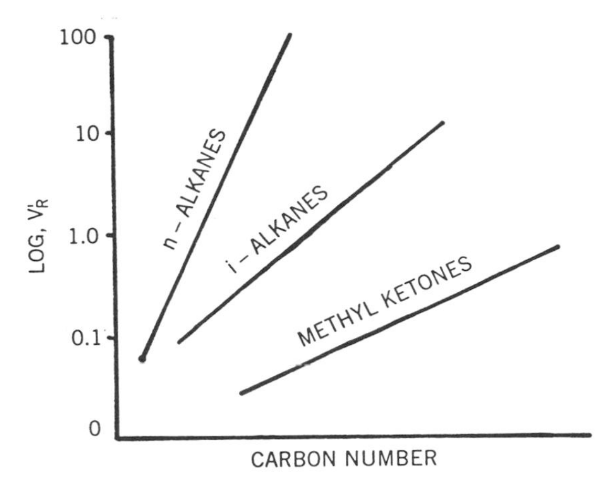

FIGURE 7.3 Logarithm of adjusted retention time versus carbon number.

these curves because strictly speaking, only two compounds in the series are needed to define the curve. These curves can be quite useful in at least eliminating certain classes of compounds relative to known peaks in a chromatogram. For instance, if the retention time of an unknown peak falls between the seven- and eight-carbon straight-chained alkanes, it is impossible for the unknown to be a straight-chained alkane since fractional carbon atoms are not allowed in the molecule. This technique can eliminate a number of potential materials.

7.2.3 Kovats Index

Wehrli and Kovats (1) introduced the concept of the retention index to help confirm the structure of the organic molecules. This method utilizes a series of normal alkanes as a reference base instead of one compound as in the relative retention method. Identification can be assisted with the use of the retention index I:

$$I = 100N + 100\left[\frac{\log V'_{\mathrm{R}}(A) - \log V'_{\mathrm{R}}(N)}{\log V'_{\mathrm{R}}(n) - \log V'_{\mathrm{R}}(N)}\right] \qquad (7.1)$$

where N and n are the smaller and larger n-paraffin, respectively, that bracket substance A, and V'_{R} is the adjusted retention volume. The retention indices for n-alkanes are defined as 100 times the number of carbon atoms in the molecule for every temperature and for every liquid phase (e.g., octane = 800, decane = 1000).

In practice, the retention index is simply derived from a plot of the logarithm of the adjusted retention time versus carbon number times 100 (Figure 7.4). To obtain a retention index, the compound of interest and at

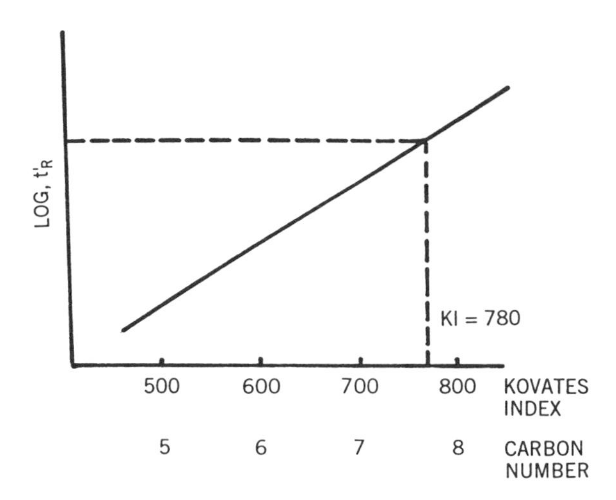

FIGURE 7.4 Plot of logarithm of adjusted retention time versus Kovats index.

least three hydrocarbon standards are injected onto the column. At least one of the hydrocarbons must elute before the compound of interest and at least one must elute after it. A plot of the logarithm of the adjusted retention time versus the Kovats index is constructed from the hydrocarbon data. The logarithm of the adjusted retention time of the unknown is calculated and the Kovats index is determined from the curve (Figure 7.4).

Many factors can influence the Kovats index, which make it unreliable at times for characterization of gas chromatographic behavior, although it generally varies less than relative retention with temperature, flow, and column variation. For many, however, the Kovats index is the preferred method of reporting retention data.

A number of attempts have been made to correlate retention index and molecular structure (2). Success here can greatly enhance the use of the retention index in qualitative analysis.

7.2.4 Multiple Columns

The use of two or more columns improves the probability that the identity of an unknown compound is the same as that of a compound with identical retention times. However, these data alone are not conclusive proof. The reliability of the identification depends on the efficiency and polarities of the column used. With efficient columns the probability of having two or more components under one peak diminishes and the peaks are generally well resolved. Care must be taken in selecting columns to be certain that columns have different selectivities and not just different names. The McReynolds constants (see Chapter 3) must be compared and should be quite different

TABLE 7.1 Pesticide Relative Retention Times

	Columns					
Pesticide	1	2	3	4	5	6
Lindane	0.44	0.46	0.47	0.44	0.74	0.81
Heptachlor	0.78	0.79	0.79	0.78	0.85	0.87
Aldrin	1.00	1.00	1.00	1.00	1.00	1.00
Dieldrin	1.88	1.84	1.83	1.93	2.70	3.00
Endrin	2.12	2.06	2.05	2.18	3.19	3.56
P,P'-DDT	3.19	3.10	3.03	3.50	3.63	4.07

Column Packing	Column Size	Temperature	Reference
1 3.8% UCW-09	7 ft × 2.2 mm i.d.	195°C	3
2 3% SE-30	6 ft × 2.2 mm i.d.	180°C	3
3 10% DC-200	6 ft × 4.0 mm i.d.	200°C	4
4 3% OV-1	5.9 ft × 4.0 mm i.d.	180°C	5
5 5% OV-210	6 ft × 2.2 mm i.d.	180°C	3
6 5% OV-210	6 ft × 4.0 mm i.d.	180°C	5

for each column. Table 7.1 shows the relative retention times for a number of chlorinated pesticides on six different columns. From the relative retention data shown it would certainly appear that the first four columns are handling the pesticides in basically the same fashion. If two of these columns were selected to help confirm the identity of an unknown by using two different columns, therefore, we would expect these not to show differences and thus give a confirmation. In fact, four of these columns could be used and we could be quite convinced that the unknown is the same as the component whose retention time it matches on these four different columns. However, an examination of McReynolds constants (see Chapter 3) for these four columns certainly indicates that regardless of the name attached, which are trade names, the materials are essentially all the same. Indeed, they are methyl silicone polymers. Table 7.1 certainly indicates that OV-210 is a different column from the other four. However, this same piece of information could be determined very readily by published McReynolds numbers.

One problem in using retention time to identity unknown components occurs in a multicomponent mixture where more than one component in the mixture may have the same retention time on even two or three different columns. Laub and Purnell (6, 7) have described a systematic technique of using multicomponent solvents in the gas chromatographic column to optimize separation of mixtures. This technique should not be overlooked in qualitative analysis since it can be fairly useful in spotting two or more components contributing to the same peak (see Chapter 4).

It should also be noted that in addition to retention time measurements obtained on two or more column systems, if reasonable care has been exercised, quantitative measures of the suspect compound should also correspond, thus providing additional secondary identification. In other words, regardless of what the known compound is, it cannot be a mixture of two components on one column and a single component on the second column without quantitative measure detecting this fact. The value of this particular observation is commonly ignored. Information on the structure of an unknown peak can be obtained from the difference in the retention indices on polar and nonpolar stationary phases:

$$\Delta I = I_{\text{polar}} - I_{\text{nonpolar}} \tag{7.2}$$

For a particular homologous series ΔI is a specific value that is determined by the character of the functional group(s) of the molecule. Takacs and coworkers (8, 9) calculated the Kovats index for paraffins, olefins, cyclic hydrocarbons, and homologs of benzene on the basis of molecular structures. The index was divided into three additive portions: atomic index, bond index, and sample stationary phase index components. (See also Reference 2).

7.2.5 Relative Detector Response

Selective Detectors. Comparison of the relative detector response from two or more detectors can aid in the identification or classification of an unknown component. Generally the component is chromatographed on one column and the effluent split and fed to two or more detectors. Commonly used pairs of detectors are the phosphorus and electron capture, flame ionization and radioactivity, and flame-ionization and phosphorous detectors. The electron capture detector allows the identification of substances containing atoms of phosphorous, oxygen, nitrogen, and halogens in a complex mixture while remaining quite insensitive to other substances. Flame photometric detectors are useful with phosphorus- or sulfur-containing compounds. The flame-ionization detector (FID) is especially sensitive to virtually all organic materials, but especially hydrocarbons. (For a complete discussion of specific and nonspecific detectors, see Chapter 5.)

Molecular Weight Chromatography. The molecular weight of a component can be obtained through mass chromatography. This relies on two gas density detectors, two columns, and two carrier gases. A diagram of a typical mass chromatographic system is given in Figure 7.5. The sample is introduced into the injection chamber by syringe, gas sampling valve, pyrolysis unit, or reaction chamber and trapped on two separate trapping columns. After the sample has been trapped it is displaced from the traps by backflushing and heating and swept onto two matched chromatographic columns using two different carrier gases. The carrier gases are chosen on the basis of significant difference in molecular weight; for example, CO_2

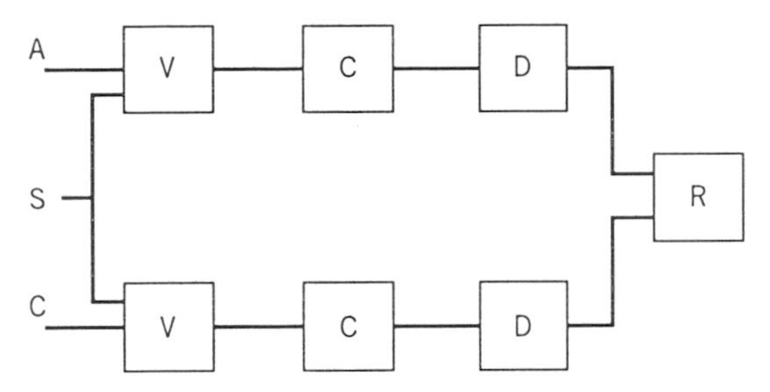

FIGURE 7.5 Mass chromatograph: A, carrier A inlet; S, sample inlet; B, carrier B inlet; V, valve–trap system; C, chromatographic column; D, gas density detector; R, recorder.

(44 g/mol) and SF_6 (146 g/mol). The sample is then separated on the column and the eluate is passed through each gas density detector (Figure 5.26). Thus two peaks are recorded for each component (Figure 7.6). The molecular weight of a component is obtained from the ratio of the two peak heights or areas by use of the following equation:

$$\mathrm{MW} = \frac{K(A_1/A_2)(\mathrm{MW}_{\mathrm{CG2}} - \mathrm{MW}_{\mathrm{CG1}})}{K(A_1/A_2) - 1} \tag{7.3}$$

where K is an instrumental constant; A_1 and A_2 are the area responses of the unknown component from detectors 1 and 2, respectively; and $\mathrm{MW}_{\mathrm{CG1}}$ and $\mathrm{MW}_{\mathrm{CG2}}$ are the molecular weights of carrier gas 1 and carrier gas 2, respectively. In practice, A_1 and A_2 are measured for known compounds

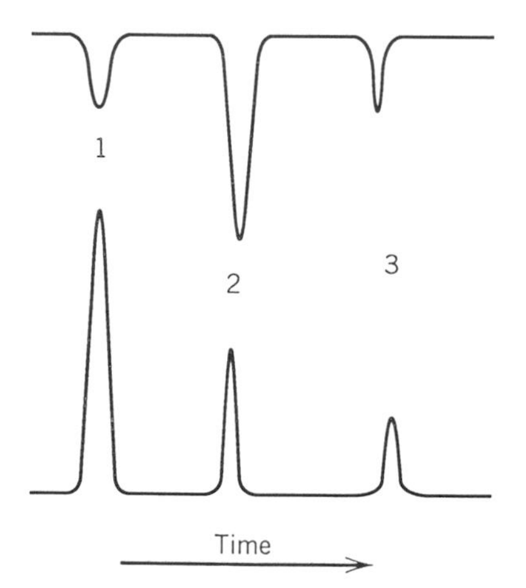

FIGURE 7.6 Mass chromatogram.

and K is determined for the experimental conditions. Then the molecular weight of the unknown is determined by obtaining its area ratio and using the K previously obtained for known compounds. The molecular weight and the Kovats retention index can then be combined to aid in the identification of the component. A linear relationship exists between the molecular weight and retention index for a homologous series of compounds. The relationship varies for each class of compound; thus a clue can be obtained regarding the type of compound present, which can be verified by some other technique.

7.2.6 Simple Pretreatment

A few minutes devoted to simple pretreatment of the sample can save many hours of interpretation of the complex data. Procedures such as filtration, extraction, or distillation can be readily accomplished and will simplify the identification of separated components.

Extractions. Simple partition phases can add another valuable piece of information about the sample. A gas chromatographic analysis before and after extraction indicates the character of the components present. For example, carboxylic acids are readily separated from the phenolic compounds by extracting a nonaqueous solution of the sample with dilute aqueous sodium bicarbonate. The carboxylic acids are almost completely transferred to the aqueous phase, whereas phenolic constituents remain in the organic layer. Additional information on extractants for specific classes can be found in most organic analysis textbooks or inferred from solubility tables.

Beroza's p Value. An additional mechanism of identification has been reported by Beroza and Bowman (10–13). The technique involves a measurement of the distribution of the unknown component between two nonmiscible liquids. This work was directed primarily toward pesticide residue analysis. With reference to pesticide residue analysis the sample processing invariably provides the pesticides in an extraction solvent of hexane or isooctane. Occasionally this may be another solvent. In any case, the prepared sample extract is chromatographed by the use of the appropriate instrumental sensitivity settings to generate a properly measurable chromatogram. A portion of the remainder of the prepared extract is then equilibrated with the same volume of a nonmiscible solvent, such as acetonitrile or an acetone–water mixture. Again a portion of the hexane or isooctane phase after equilibrium is chromatographed under precisely the same conditions. The peak height or area of the compound of concern is determined by both chromatograms. The ratio of signal following the equilibrium divided by the signal before equilibrium has been defined as the

p value:

$$p \text{ value} = \frac{\text{Peak height (or area) after partition}}{\text{Peak height (or area) before partition}} \tag{7.4}$$

The p value for standards that have the same retention time under the same chromatographic conditions are then determined. The p value for the unknown component and one of the standards should be the same if the unknown and standard are the same compound. If two or more of the standards have closely similar p values and corresponding retention time, the experiment should be repeated using a different solvent pair.

For convenience, the p value was selected to designate the component distribution in solvent systems of equal volumes. If different volumes of the two solvent phases are used, appropriate corrections must be made (13). Ideally, the solvent systems should be chosen so that the p values for components of interest range between 0.25 and 0.75 to provide the greatest precision and assurance of identity or nonequivalence with the standard.

The p values for over 100 pesticides and related substances were established by studying the extraction behavior of those compounds in a wide range of binary solvent systems. As a result of these studies, Beroza and Bowman concluded the following:

1. Each pesticide exhibits a characteristic distribution ratio (p value).
2. Distribution ratios are practically independent of pesticide concentration over any range of concern in trace analysis.
3. Other components extracted from the original sample do not appreciably affect this ratio.
4. Compounds other than pesticides can be identified by the use of this p value and the gas chromatographic technique.

The use of distribution coefficients or their simplified equivalents as p values is not new and is based on sound chemical principles. Its particular value, however, is that it can be applied as a confirming means of identification where the component of concern is not available in sufficient quantity for the more common analytical identification techniques, such as mass spectrometry, IR spectroscopy, elemental analysis, or physical property measurements (14). Its elegance rests in its simplicity.

Water–Air Equilibrium. McAuliffe (15) introduced a multiple-phase equilibrium procedure for the qualitative separation of hydrocarbons from water-soluble organic compounds. For n-alkanes, more than 99% was found to partition in the gas phase after two equilibriums with equal volumes of gas and aqueous solution. Cycloalkanes require three equilibriums to be essentially completely removed, and oxygen-containing organic compounds (e.g., alcohols, aldehydes, ketones, and acids) remain in the aqueous layer.

After equilibrium with equal volumes of gas, an immediate clue is given regarding the identification of the compound. More details of this technique can be found in Chapter 14. This technique also provides two additional pieces of information: the distribution coefficient (D_s or D_g) and the initial concentration of the unknown component.

7.2.7 Tandem Gas Chromatographic Operation

Two Columns in Series. Some of the problems associated with obtaining a retention time or index on two different columns individually can be overcome by running two columns in series. A good example of this is the analysis of benzene in gasoline. On a methyl silicone column benzene will elute between *n*-hexane and *n*-heptane. Some major components of the gasoline also have retention times in this area such that the benzene would be completely swamped by these components. Similarly, a very polar column such as 1,2,3-Tris(2-cyanoethoxy) propane (TCEP) will have benzene eluting between undecane and dodecane. There are sufficient hydrocarbons in gasoline in this range also that would obscure or make the confirmation of the presence of benzene extremely difficult. Thus, even though two columns were used, the presence of benzene certainly is not proven. If these same two columns were worked in series, however, the sample could be introduced into the silicone column and the effluent from that column directed into the TCEP column. Following the retention time of heptane on the silicone column the higher hydrocarbons still remaining on the silicone column after heptane could be either backflushed or eluted forward through the column but not allowed to enter the TCEP column. A second carrier-gas flow would then elute the components from the TCEP column. In this case all the hydrocarbons up through heptane would emerge well before the benzene, and in this case the benzene peak would be isolated and completely identifiable. This technique of two columns in series should be considered when one is attempting to confirm the presence or absence of a specific component in a very complex mixture.

Subtractive Precolumns. For many applications the mixture to be analyzed is so complex that the only reasonable method of analysis requires the removal of certain classes of compound. This process can be easily implemented by the use of a reactive precolumn. For example, a precolumn of potassium hydroxide can be used to remove acid vapors. The mixture could then be chromatographed with and without the precolumn to identify peaks of acid character. A discussion of precolumn reagents is given by Littlewood (16). Potential packing materials for precolumns may also be found in the trace analysis literature.

Carbon Skeleton. The technique of precolumn catalytic hydrogenation can be applied to reduce certain unsaturated compounds to their parent hydrocarbons. Other classes of compounds also analyzed by this technique include esters, ketones, aldehydes, amines, epoxides, nitriles, halides, sulfides, and fatty acids. Fatty acids usually give a hydrocarbon that is the next-lower homolog than the parent acid. For most systems utilizing hydrogenation, hydrogen is also used as the carrier gas. Usually 1% palladium or platinum on a nonadsorptive porous support such as Chromosorb P-AW is used as the catalytic packing material. This operation can be performed with two columns in series such that only a single component or a selected range of retention times of components from the first column is directed to the hydrogenation catalyst, which is then followed by a second column to observe the hydrogenated products of that particular segment. If one has a relatively pure material and is attempting identification, the injection can be made directly into the catalytic column followed by a column to identify the reduced hydrocarbon species.

Controlled Pyrolysis. The principle of controlled pyrolysis or controlled thermolytic dissociation for the identification of chromatographic effluents lies with the examination of the pattern ("fingerprint") produced. The peak selected for identification from the first column is transferred with continuous flow from the gas chromatograph through a gold coil reactor helically wound on a heated stainless-steel core, and then through a second gas chromatograph for the identification of the pyrolysis products. The products are identified by comparing the Kovats retention indices to those of standard compounds and by enhancing the peaks with selected standard compounds. The fingerprint can also be obtained with increased certainly by coupling a mass spectrometer to the second chromatograph. The controlled pyrolysis technique can be especially useful in forensic and toxicological applications when direct comparison is necessary. Information concerning functional groups absent or present in the molecules can be obtained by determining the concentration ratios of the small molecules produced on pyrolysis (CO, CO_2, CH_4, C_2H_4, C_2H_6, C_3H_6, NH_3, H_2S, and H_2O). "Large-molecule" pyrograms (C_4H_8 and larger) in combination with "small-molecule" pyrograms can give additional information regarding the functional groups present. Neither technique can yield a priori identification of molecules by pyrolysis at this time.

7.3 IDENTIFICATION BY GAS CHROMATOGRAPHIC AND OTHER DATA

An unequivocable identification of an unknown component is unlikely by chromatographic process alone. Not the least of the reasons for this is the need for the comparisons of standards, thereby assuming reasonable prior assurance of the possible identity of the unknown. Certainly the more discrete pieces of information obtainable concerning an unknown compound, the easier it will be to obtain confident identification. Microchemical tests such as functional group classification, boiling point, elemental analysis, and derivative information, as well as infrared spectroscopy, coulometry, flame photometry, and ultraviolet (UV)-visible spectroscopy are also useful aids when used in conjunction with gas chromatographic data.

7.3.1 Elemental and Functional Group Analysis

The major reason why GC is not generally used for qualitative analysis is that it cannot differentiate or identify indisputably the structure of the molecule. Therefore, it is necessary to perform additional tests on the separated peak to ascertain its functionality and elemental composition. Many books and articles are available regarding microanalysis, so this method is not extensively reviewed here. Usually it is necessary to trap the peak, then perform whatever specific microanalysis techniques are necessary to confirm the identity of the peak. Several commercial instruments are available for elemental analysis (usually carbon, hydrogen, sulfur, and halogens). These instruments usually require 0.1–3 mg of sample and often employ trapping systems for quantitative analysis.

Hoff and Feit (17) reacted samples in a 2-mL hypodermic syringe before injection onto the gas chromatographic column. Reagents were selected either to remove certain functional groups or to alter them to obtain different peaks. Reagents used included metallic sodium, ozone, hydrogen, sulfuric acid, hydroxylamine, sodium hydroxide (20%), sodium borohydride (15%), and potassium permanganate (concentrated).

A stream splitter attached to the exit tube of a thermal conductivity detector can be used to identify the functional groups of gas chromatographic effluents. Table 7.2 lists functional groups tests and the limits of detection. A review of elemental analysis is given by Rczl and Janak (18).

Crippen's excellent book (19) gives an extensive compilation of the techniques of organic compound identification with the assistance of GC. It includes a stepwise account of the preliminary examination, physical property measurements, and functional group classification tests. There are numerous graphs relating retention times with various physical properties such as melting point, boiling point, refractive index, and density. This book is a must for any extensive gas chromatographic laboratory dealing with unknown samples. Gas chromatographic methods for qualitative analysis of complex systems such as biological materials and bacteria proteins, steroids, and triglycerides also have been developed.

7.3.2 Coupling Gas Chromatography and Other Instrumental Techniques

A technique that has become more common for the identification of compounds is the combination of gas chromatography and mass spec-

TABLE 7.2 Functional Group Classification Tests

Compound Type	Reagent	Type of Positive Test	Minimum Detectable Amount (μg)
Alcohols	$K_2Cr_2O_7$-HNO_3	Blue color	20
	Ceric nitrate	Amber color	100
Aldehydes	2,4-DNP	Yellow ppt.	20
	Schiff's	Pink color	50
Ketones	2,4-DNP	Yellow ppt.	20
Esters	Ferric hydroxamate	Red color	40
Mercaptans	Sodium nitroprusside	Red color	50
	Isatin	Green color	100
	$Pb(OAc)_2$	Yellow ppt.	100
Sulfides	Sodium nitroprusside	Red color	50
Disulfides	Sodium nitroprusside	Red color	50
	Isatin	Green color	100
Amines	Hinsberg	Orange color	100
	Sodium nitroprusside	Red color, 1^{c} Blue color, 2^{c}	50
Nitriles	Ferric hydroxamate-propylene glycol	Red color	40
Aromatics	HCHO-H_2SO_4	Red-wine color	20
Aliphatic unsaturation	HCHO-H_2SO_4	Red-wine color	40
Alkyl halide	Alc, $AgNO_3$	White ppt.	20

Source. Reprinted with permission from *Anal. Chem.*, **32**, 1379 (1960).

trometry. This is due in part to the decreasing cost, increasing sensitivity, and decreasing scan time of mass spectrometry equipment. Read Chapter 6 for a complete discussion of this most important technique. Now it is possible to obtain not only a complete mass spectral scan of a gas chromatographic peak "on the fly," but also the mass spectra at various portions of the peak such as the front edge, the heart of the peak, and the tailing edge. This is especially useful in helping to ascertain whether a given peak is a single- or multiple-component peak, in addition to determining what those components are. This technique in general does not require a prior knowledge, or reasonable suspicion even, of the identity of the component to be identified. The most conclusive identification will be the recreation of the same mass spectrum from a known standard. The spectrum obtained from an unknown, if not immediately decipherable, will provide a significant number of clues to the probable identity, thus limiting the need either for searching reference spectra or for the generation of a reference spectra.

There is really a question of semantics regarding the mass spectrometer coupled to a gas chromatograph. Most gas chromatographers would consider the mass spectrometer as simply another of several selective detectors available for use in helping to identify compounds. It would then be considered as entirely a gas chromatographic technique as opposed to GC coupled with other analytical techniques. The other side of the story is simply that mass spectroscopists would certainly consider the gas chromatograph as just another one of many inlet systems that that have available for their mass spectrometer.

The second most used instrumental technique is infrared spectroscopy. In general the first instrumental method to consider is the one most readily available. In a few cases, notably mass spectrometry, the technique may be used in tandem with the gas chromatograph, but in general most techniques require trapping of the peaks as discussed in Section 7.3.3.

Consideration must be given to the quantity of the sample needed for the minimum detection limits of the instrumental technique used. A number of techniques have been ranked in order of increasing amounts of material needed as follows: mass spectrometry (1–10 μg), chemical spot tests (1–100 μg), infrared and ultraviolet spectroscopy (10–200 μg), melting point (0.1–1 mg), elemental analysis (0.5–5 mg), boiling point (1–10 mg), functional group analysis (1–20 mg), and nuclear magnetic resonance spectroscopy (1–25 mg).

7.3.3 Trapping of Peaks

Trapping a sample as it elutes from the column followed by some other identification or classification technique is often utilized with gas chromatographic analysis. The most common trapping devices are the cold trap, the gas scrubber (gas-washing bottle), the evacuated bulb, and the absorbent postcolumn. A simple cold trap can be constructed from the small-diameter glass tubing, such as melting-point capillary tubing, and connected with some flexible inert tubing to the outlet port of the chromatograph (Figure 7.7a). Part of the coil should be immersed in a liquid coolant such as liquid nitrogen (−196°C), dry ice–acetone (−86°C), sodium chloride ice (1:2) (−21°C), or ice-water slush (0°C). One should not use liquid nitrogen when air or oxygen is being used as the carrier gas because of the explosion hazard as liquid oxygen accumulates. The upper part of the coil should be above the coolant liquid so that loss of sample due to too rapid cooling (fogging) can be avoided.

A gas-washing bottle (Figure 7.7b) may also be used for trapping. This technique is especially useful in conjunction with infrared analysis. The sample is simply bubbled through the anhydrous solvent as it exits the chromatographic column. The solution is then placed in a liquid sample infrared cell. A matching cell containing only the solvent is placed in the reference beam. An infrared spectrum of the sample may then be recorded.

Evacuated bulbs (Figure 7.7c) are generally used for trapping volatile components. Since this technique does not concentrate the sample, additional sample preparation may be required. For substances with high infrared

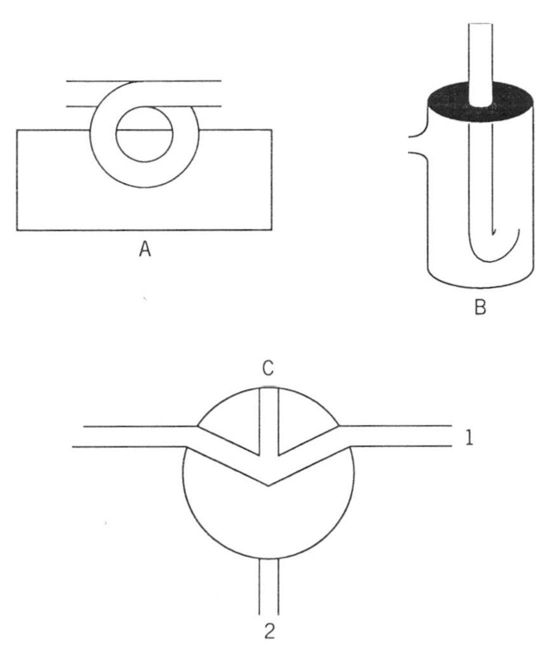

FIGURE 7.7 Traps: A, simple coil cold trap; B, gas scrubbing trap; C, evacuated bulb trap; 1, to outlet of gas chromatography; 2, to evacuated gas sampling bulb.

absorptivity, the sample may be trapped directly in an evacuated infrared gas cell and analyzed directly. For nonvolatile samples that may condense on the outside walls, the cells must be heated before analysis.

An adsorbent postcolumn can also be used to trap eluting peaks. Packing materials such as Tenax-GC (Enka N.V., The Netherlands), Porapak N and Porapak R (Waters Associates), Carbosieve B, and 20% DC-200 have been tested as sampling tubes for concentrating organic compounds in air. Tenax-GC and Porapak N seem to have the widest general applicability. Tenax-GC was more suitable for higher-boiling compounds, and Porapak N was more suitable for lower-boiling organics (20–100°C). In many cases, in order to trap sufficient amounts of materials for subsequent analysis, many repetitive injections must be made into the chromatograph and more sophisticated trapping techniques may be required.

7.4 QUALITATIVE ANALYSIS WITHOUT PEAK IDENTIFICATION

7.4.1 General Fingerprints

There are many cases, especially in the analysis of very complex mixtures of materials, where identification of the source of material can be determined without the individual identification of any single chemical entity in the mixture. In these cases the chromatograms of the mixtures are simply used as somewhat of a fingerprint analysis. The general appearance of the peak, as far as retention times versus response is concerned, is the first piece of information, and in some cases ratios of peaks at given retention times can be used to facilitate the identification procedure. The power to resolve a complex mixture into its components and thus provide this fingerprint is an important technique of GC that should not be overlooked. There are many instances of the use of this technique. For instance, the origin of a particular spice can be determined by comparing its fingerprint chromatogram against the fingerprint of a sample known to be genuine. Paint chips can be compared by using a pyrolysis technique ahead of the gas chromatograph. In this particular instance the pyrolysis products of the paint chip are used as a fingerprint and in many cases can be used to identify the source of a given paint chip. Two other techniques where this fingerprinting of the gas chromatograph has been used to advantage are given in a little more depth.

7.4.2 Arson Accelerants

In investigations for arson, GC has become a very powerful tool (also see discussion in Chapter 13). It can be first used to indicate the presence of solvents that were used for accelerants, which would definitely indicate the possibility of arson. The fingerprint obtained can be used to identify which accelerant was used. This can help in narrowing the search for the source of the accelerant and thus the possible suspect. Even following a fire the residue from the fire will contain sufficient unburned solvent for this to be recovered and detected on a gas chromatograph. The solvent is released from the debris either by heating in a closed container and then sampling the headspace for gas chromatographic analysis, or by addition of a solvent that will remove the accelerant from the debris, followed by gas chromatographic analysis of the solvent. Typical fingerprint patterns can be obtained that can identify the presence of gasoline, kerosene, paint solvents, and the like. These can be distinguished from each other by having known patterns for these various commercial mixtures. Examination of debris from various parts of the fire can indicate where the fire was started. In some cases it is possible to identify a particular chemical entity that was used as the accelerant, but in many cases the mixtures are very complex and only a fingerprint is needed to determine the type of accelerant used. This technique is being used by many state crime laboratories for the proof of arson and the identification of the accelerant.

7.4.3 Oil Spills

In many cases of oil spills it is possible that one of several sources could be the cause of the spill. The use of GC as a fingerprinting technique can help to identify the source of the oil spill. In this case the sample of the spill itself is chromatographed and samples from any suspected source are chromatographed under the same conditions. The resulting fingerprints can be extremely useful in identifying the source of the spill. In many cases these fingerprint chromatograms for oil spills are generated on a dual-detector system. In this system flame ionization is used for the detection of all hydrocarbon components and a flame photometric detector working in the sulfur mode is employed to detect the sulfur-containing species. In effect, then, each sample has two different fingerprints that can be compared. Proof of the source of the oil spill has been admitted as court evidence and is sufficient proof of the source of the spill. This is important to determine the liability of the expense of cleanup of the spill. In this particular instance there is no component-by-component identification of the peaks. Simply the fingerprint of the chromatogram itself is used to determine the source.

7.5 SPECIAL TECHNIQUES FOR TRACE ANALYSIS

To emphasize the need for simplicity in trace analysis methods, it is important to recognize the similarities in steps that apply, regardless of the specific analytical problem at hand. In every case, additional thought and concern must be addressed to the problem of sampling and transfer and storage of the sample because of the concern for stability and interactions with the components sought. Nearly all samples intended for trace analysis will subsequently require partial isolation of the components sought and a significant concentration beyond that natively present in the sample. Because the components sought represent such a minor proportion of the sample, it is likely that other portions may constitute an interference in the planned method of analysis. Therefore, the processes that may be applied to eliminate or modify the effects of interfering materials must be considered. Finally with trace analytical methods, it is desirable or necessary to convert the components sought to another compound that is more readily determined by gas chromatographic means. The compounds sought may be derivatized to form compounds that are responsive to one or another of the selective detectors. Derivatization may also be required to provide for the separation from an interfering component or to improve the chromatographic behavior of the component sought for more accurate analysis.

7.5.1 Sampling and Sampling Devices

Bulk Products, Tissue (Animal and Vegetable), Crops, Other Solids, and Body Fluids. With samples of this nature, the sampling devices are either dictated by the nature of the sample source such as body fluids or can otherwise be the simplest types of containers. It is still necessary to ensure that the sample is either sufficiently homogeneous to remove a representative small portion comfortably, or that a sufficiently larger sample is taken randomly and this larger sample reduced in particle size by one of the number of means available (grinding, blending, etc.); and finally a representative portion of blended material is taken for the analytical procedure. The primary concern in these circumstances is that the container and other utensils used to obtain the sample be scrupulously clean and that the sample, once taken, is stored until processing so that no deterioration or chemical change occurs. Freeze-drying is widely used as a preliminary treatment for tissue and crop samples where this process is available and applicable.

Water Sampling. For pressurized or public water systems, the only real concern in obtaining the sample is to avoid the portion that has been standing quiescent in the system prior to sampling. In other words, water that may have been standing for an unknown period of time in contact with valve packing, check valves, pump fittings, and other lubricants may have dissolved materials from this exposure in proportions much greater than the bulk flowing water when in use. Therefore, the system should be run for some short period of time before the sample is taken. The size of the sample may be dictated by the known sensitivity limit if the analytical level of contamination or the specified allowable upper limit of contamination. Again the sampling device is seldom critical other than the clear necessity for cleanliness. For this type of analysis, glass containers are superior to any other alternative. The cap or closure should be lined with metal foil or Teflon (Du Pont trademark for polytetrafluoroethylene) to avoid any concern over leaching of organic material from the closure.

As before, the time of holding before processing and the conditions of storage are important. If processing is not to be carried out immediately, the sample should be refrigerated in the interim. Preservatives such as mineral acids, toluene, or boric acid, which are sometimes used for clinical analysis, should be avoided, since they generally complicate the subsequent analysis. Because of the ever-present potential for bacterial action to change the sample composition, analysis should follow soon after the sample has been obtained, ideally one day or less. Specialized containers are used for the determination of volatile components in water (see Chapter 14).

Sampling of waste or natural streams is generally a more complex problem. For waste streams from manufacturing plants, the composition of components of concern can change dramatically over short time periods. In addition, in many systems different waste lines enter the same stream at different points and in varying volume and velocity. This requires that the chemist have some knowledge of these entries to the stream and their locations. The possibility of incomplete mixing of these various streams at the point chosen or available for sampling must be considered. For the materials sought or suspected, the relative solubilities and densities should be noted. If the suspected or possible concentration exceeds the solubility limit, the excess will either float to the surface or sink to the bottom, and a sample taken solely of flowing solution below the surface will not necessarily represent the true level of contamination. Similarly, if insufficient mixing has occurred, laminar portions of the flowing stream may vary significantly from other portions.

If single or "grab" samples are taken, the sampling containers again are relatively simple, although in this case some effort must be exerted in an attempt to obtain proportionate volumes from different parts of the cross section of flow if this appears necessary or desirable. This type of sampling can also be accomplished automatically with commercially available instrumentation. In general, the automatic systems will not providc for as largc a sample as may sometimes be desired. Usually, an automatic system will sample from one single point in the stream and it will be necessary to assure that this sampling point is representative. The samples thus obtained can be analyzed singly and the variation of concentration with time subsequently determined.

This type of measurement can be extremely useful even in cases where the compounds detected are not fully identified. For example, if there is a marked increase of one or several components at the same time daily, this may be correlated to the operation of a leaky plant valve and the economic loss that this represents corrected before it becomes serious. Conversely, if each single sample is too small for adequate analysis, the automatically collected group of samples can later be combined in one container (a composite sample), and all or a portion of this composite sample can be subsequently analyzed. This, of course, eliminates the ability to recognize variations with time. Some automatic systems are designed to provide a composite sample directly; that is, each portion that the system removes from the stream is transferred to a single container that accumulates all samples taken over some fixed period of time.

For natural waters, a flowing stream can usually be treated in much the same fashion as a plant stream, again with the need to account for the entry of subsidiary streams. The most difficult water-sampling problem involves quiescent ponds, lakes, marshes, and other bodies of water. In this case the exchange of composition from one portion to another can be extremely slow; thus a sample taken at one point will yield an analysis markedly different from a sample taken at another location in the same body of water. This is even more true in circumstances where the contaminant sought is incompletely soluble. If it floats on the surface, wind action will tend to concentrate the material against the downwind shore. If the contaminant tends to sink, the concentration of this component nearest its point of entry to the body of water will generally be higher than at any other point. These last circumstances present the greatest difficulty in assuring that a sample is properly representative. This also is a circumstance in which subjective judgment enters before any analytical processing can occur. In such a circumstance, multiple sampling at a variety of locations is highly desirable, if not imperative.

Air or Gas Sampling. For this type of sample, the problems, as well as sampling devices, are generally more complex. The primary reason is the fact that the sample itself is invisible and is quite likely inhomogeneous. Two fundamentally different circumstances are encountered. One involves sampling of plant stack output, which is usually done from some point within the stack or immediately over the top. Under these circumstances, the sample itself is essentially driven past the sampling point and some opportunity exists to assure mixing ahead of the sampling point or to design mathematically a variation in locating the sampling point.

The second circumstance is in quiescent or slowly moving air, such as that encountered in testing pollution in the region of a highway or on plant grounds. Here, as in the case of a pond, a single sample taken at one point in time is unlikely to represent a correct picture of the total circumstance. The sampling point should be varied and samples should also be taken over a significant period of time before the analytical results can be expected to have significant meaning. It is obvious, for example, that the pollution level generated by motor traffic will be highly variant throughout a normal day.

For both types of air or gas sampling, a wide variety of sampling devices is available or can be devised from available items. The simplest must be the system used for "grab" sampling. This may be nothing more than a sealed, evacuated-glass container (Figure 7.8) that can be opened at the site chosen for sampling and reclosed after the sample has entered the container. A manual or electric pump can also be used to draw sample into the container after opening on site. This type of device can be obtained with or without stopcocks at both ends. Although the stopcocks are convenient, they do not provide certain insurance against leaks or alteration of the sample by

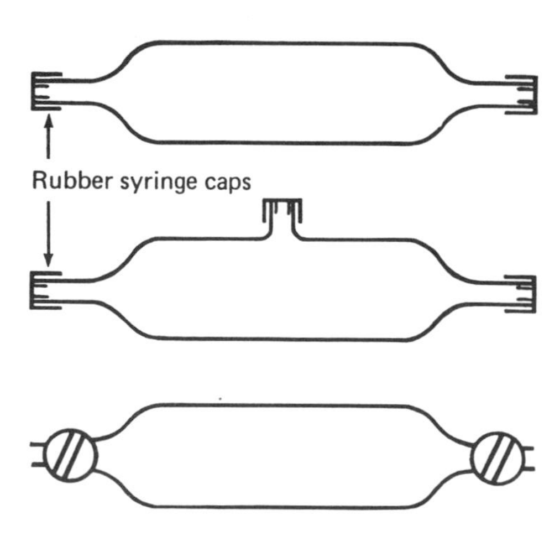

FIGURE 7.8 Gas sampling containers.

diffusion. Teflon stopcocks are strongly recommended for this type of closure. The lubricant required for glass stopcocks must always be suspected of preferentially absorbing or dissolving some one or another of the sample components that may be of critical concern. Rubber septum seals tend to be better for this purpose, although less convenient. This same possibility exists with rubber closures, although not to the same degree. If the compounds sought in the sample are clearly not oil soluble, lubricated stopcocks can be used. Olefins and aromatic compounds are generally more readily soluble and more easily lost under these circumstances than are saturated hydrocarbons or oxygenated compounds.

Plastic bags are also used to obtain gas samples. The most reliable of these are Teflon, although Mylar (Du Pont trademarks for polyester film and polyvinyl fluoride film) are also used. The simplest means of obtaining the sample is to enclose the empty bag in a cardboard or other container that is reasonably well sealed. The entrance to the bag is brought through the box at some point and sampling tube attached if a tube is used. The bag is filled by creating a partial vacuum within the box. This forces the bag to expand either to its capacity limit or to the interior volume of the box. This operation is illustrated in Figure 7.9.

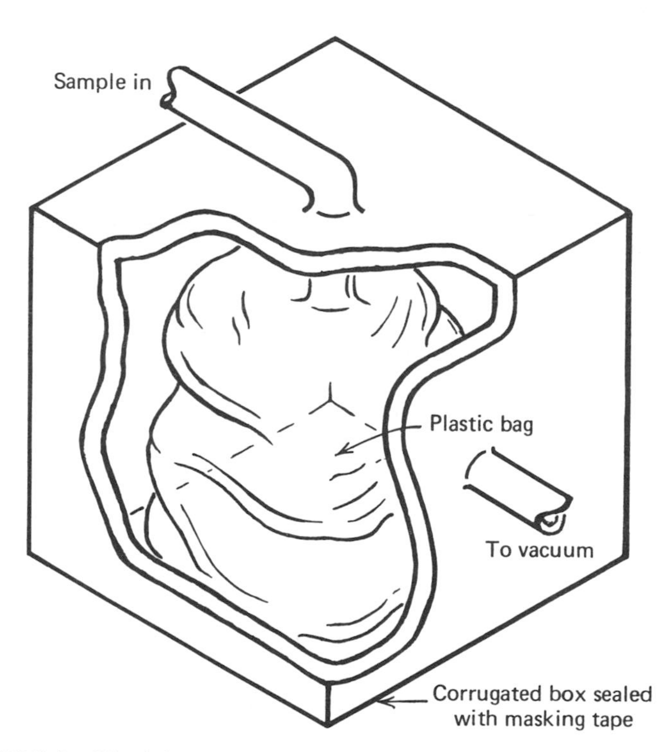

FIGURE 7.9 Obtaining a gas sample by expansion of a plastic bag.

It should be noted in all cases that the driving mechanism causing the air or gas to enter the sample container is the creation of a vacuum or partial vacuum. This contrasts with pressuring the sample and driving it into a container. In other words, the driving mechanism for transferring the sample from its source to a container should be downstream from the sample container in the flow pattern. The purpose is to avoid inadvertent contamination of the sample with lubricants or other components. A device that is designed contrary to this general rule is a battery-powered, programmable device for taking multiple air samples in plastic bags at predetermined times. The pumps used are of special construction in which all gas-contacting surfaces are of Teflon.

A large syringe in many cases can be adequate device for obtaining and holding a sample. It can also serve as the means for drawing a sample into a container as long as the container volume is not significantly larger than the volume of the syringe. At least three container volumes should be displaced to obtain a proper sample.

Each sampling method described to this point is primarily aimed at gas samples in which the components sought are directly detectable without concentration. However, there will be circumstances where the components sought may be present in a concentration that cannot be directly detected or measured. Thus sampling must also include a concentration process. For this purpose a variety of approaches have been used.

7.5.2 Isolation and Concentration Techniques

In a very few cases a sample intended for trace analysis can be measured directly as received. Wherever this is possible it should be the method of choice, since the problems inherent in sampling processing are automatically eliminated. As indicated previously, however, most trace analysis problems will require higher concentration of a component than that in its original sample matrix. All the various techniques that are used to accomplish this step in the general literature and as discussed in this chapter are highly similar processes. It is important to recognize the high degree of similarity that exists so that the combination of techniques is most readily applied to any given analytical problem. For example, it is common to consider headspace analysis techniques as entirely different from column adsorption techniques for the isolation and concentration of a given component. The similarity in concept and physical processes involved should be recognized. In all cases an attempt is being made to transfer the sought components from one matrix to another with a higher concentration in the second matrix, either by weight or by volume, while leaving behind major parts of the original matrix. Since most of these transfers depend on the physicochemical properties of the compound in question and the major composition of the matrix phases, each process can be considered an extraction and values corresponding to distribution coefficients can be developed in many cases. These will provide an extra dimension of qualitative information.

In an effort to emphasize the common relationships that should be recognized among nearly all trace analysis procedures, the concentration step should be regarded as an extraction step. Each process listed in Table 7.3 can be properly considered an extraction step from one matrix to another. For purposes of trace analysis, each process (listed in Table 7.3) of transfer of the sought component from the original to the new matrix should also function as a concentration mechanism. The processes can be either static or dynamic. Static processes imply fixed volumes held in contact for a sufficient time to achieve an equilibrium distribution of the components sought between the two phases. These processes, irrespective of the nature of the two phases, will thereby allow for the determination of a distribution coefficient or distribution ratio. This value can serve as an additional item of qualitative information.

Dynamic processes will not yield this kind of information. A dynamic process implies either that the sample phase or the extracting phase is passed continually through the second phase so that the total of the second phase is never in continuous contact with the other phase and equilibrium is never attained. Dynamic extraction from a liquid or solid sample matrix into a gas phase is usually accomplished by purging the sample with an inert gas. This gas phase, in turn, is subsequently passed through an adsorbing solid. In such cases the dynamic extraction process may be considered the sequential transfer of sample components from a liquid or solid phase to a gaseous phase and then to a solid phase.

Clearly, all transfers of sample components that require movement in a gas phase will also require that the components of concern have significant vapor pressure at room temperature or at the operating temperature of the extraction process. In other words, transfer to, or extraction in, a gas phase is applicable only to compounds customarily considered volatile.

Where the extraction is into a gas phase that is subsequently to be analyzed directly, the concentration ratio is more difficult to calculate since concentration units in condensed phases are not directly comparable to concentration units commonly used in gas phases. Nevertheless, an

TABLE 7.3 Extraction Processes

Original Matrix	New Matrix	Description
Gas	Liquid	Dissolution (or reaction)
Gas	Solid	Adsorption
Liquid	Gas	Vaporation
Liquid	Liquid	Selective solution
Liquid	Solid	Adsorption
Solid	Gas	Vaporization
Solid	Liquid	Dissolution

adequate transfer from a liquid or solid phase to a gas phase accomplishes two things that relate to the ultimate sensitivity of a gas chromatographic analysis. The new gas phase customarily is inert and not detected; hence there is no solvent peak and no solvent impurity peaks. Therefore, no interference with the chromatographic analysis is expected from the new matrix. In addition, even though the phase volumes may be the same, the proportionate volume of a gaseous phase that may be taken as the analytical aliquot is larger than the volume of a liquid phase aliquot by an approximate factor of 1000. If the component transfer to the gas phase is near quantitative, therefore, the amount presented to the detector can be as much as 1000 times greater.

Solid adsorbents packed in a small tube have been widely used to obtain and concentrate components from air or gas samples. Adsorbents that have been used are charcoal, porous polymers such as Tenax and Porapak, alumina, silica gel, and stationary phase-coated solids. One example of this type of sampling device is used in conjunction with a small, battery-powered pump for monitoring worker exposure to solvent vapors or vinyl chloride monomer (Figure 7.10). Alternative devices that do not require a pump are diffusion-controlled monitoring badges (20). These accumulate organic vapors through a diffuser unit with adsorption on an activated charcoal element. Transfer to the adsorbent is controlled by the design of the diffuser unit. Devices of this type are marketed by E.I. Du Pont de Nemours and Company (Inc.) and are illustrated in Figure 7.11.

Alternately, single or series trapping systems that cause the sample to bubble through a liquid are also used (Figure 7.12). These are referred to as impingers or midget impingers. The liquid can be an effective solvent for the component of concern. It could also contain a reactant to convert the component to a new compound. The liquid solution that results when sampling is completed is then analyzed. Trapping systems are also used for aqueous samples. Activated charcoal has been used in the past, but the most common in current use have been the porous styrene-divinyl benzene polymers. These are marketed as Chromosorb Century Series (Johns Manville), the XAD series (Rohm and Haas), and the Porapak materials (Water Associates). A large sample volume of up to 150 L can be passed through a small column of these porous polymers, with organic compounds very effectively held. The adsorbed organic compounds are then sub-

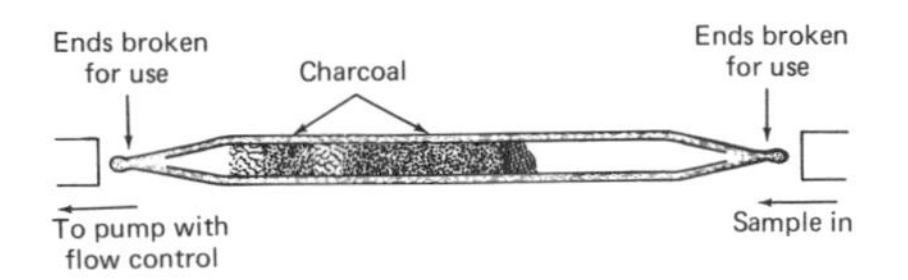

FIGURE 7.10 Approved charcoal adsorber tube for industrial hygiene sampling.

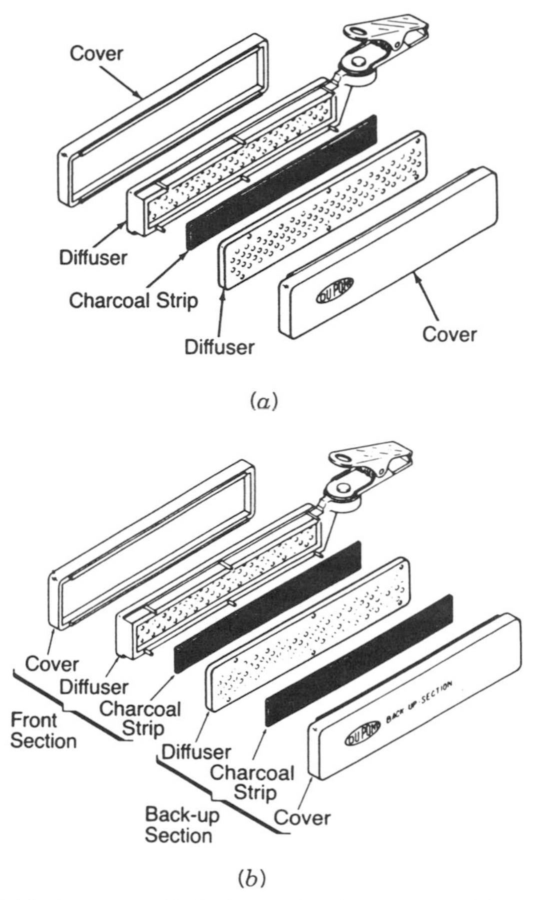

FIGURE 7.11 Diffusion-controlled air monitoring badges: (a) low concentrations; (b) moderate to high concentrations. (Reproduced by permission of E. I. du Pont De Nemours & Co., Inc.)

sequently eluted from the small column with as little as 20 mL of an organic solvent such as diethyl ether or methanol.

In nearly all cases of trace analysis the sample or sample concentrate should be analyzed as soon as possible. In water and soil samples the possibility of bacterial action altering the composition must be considered. In air or gas samples the potential loss by either adsorption on the container walls or reaction with oxygen or other components of the sample will severely limit the allowable storage time. In samples of tissue and body fluids, these same concerns must be recognized, although some type of stabilizing additive is used in a number of clinical analyses. Because of the

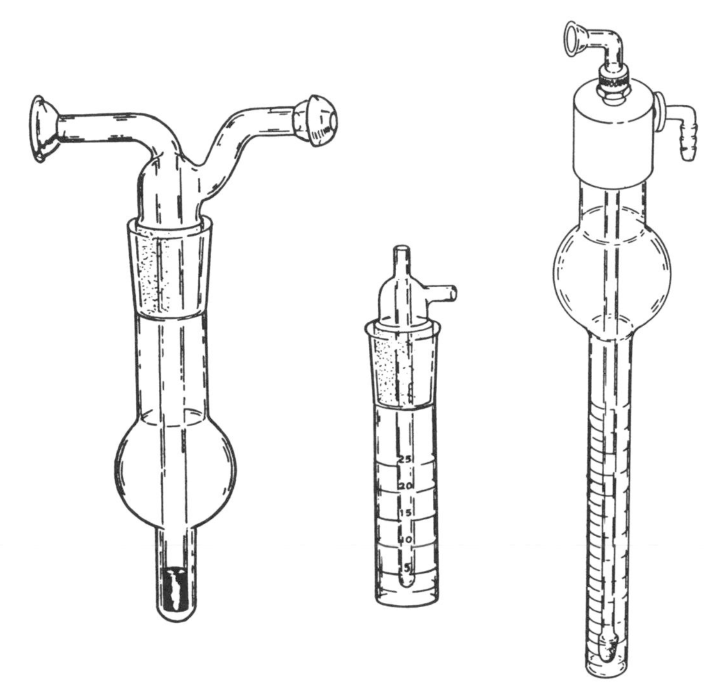

FIGURE 7.12 Impingers for concentrating gas samples in a liquid absorber.

very low concentration of the components sought, mass action alone could be considered as an interfering mechanism, since other reacting compounds must be assumed to be present in great excess over the component of concern. Thus reactions that should normally be considered slow may consume the component more rapidly than anticipated.

7.5.3 Elimination of Interferences

After the partial isolation and concentration steps have been completed, many published methods include what is commonly referred to as a "cleanup" step. This step is presumed to remove extraneous and interfering components from the concentrated sample without serious loss of the components of concern. This usually involves passing the sample through some type of adsorption column such as alumina, silica gel, or Florisil. The components of concern may be intended to pass through while the other components are retained, or they may be retained while the interfering components are presumed to pass through. Either alternative may require the successive passage of different volumes of solvent of varying composition and collection of an appropriate fraction. The reliability of these methods is relatively limited because the physical properties of the adsorbents used or recommended are rarely uniform. Furthermore, these properties will vary significantly, depending on the type of treatment and storage afforded the adsorber material (21). To the extent possible, therefore, this type of processing step should be avoided.

Proper attention to and use of the fullest efficiency of the gas chromatographic system can normally provide a means of avoiding analytical interferences. After all, chromatography itself is one of the most effective separating mechanisms. This is too often overlooked. Proper initial processing in isolating and concentrating the components of concern can clearly be shown to be effective in minimizing interferences without the so-called cleanup step.

It is still necessary to consider the type of column and detector to be used to accomplish the necessary measurement. The detector chosen will ideally be a selective detector that is highly responsive to the component of concern and has very limited response to other compounds. These various detectors are described in Chapter 5. One resorts to the more universally responsive flame ionization detector (FID) only when one of the selective detectors is not appropriate for the analysis at hand. With this detector (FID), the selection of column and other operating conditions is increasingly important in order to obtain both the sensitivity and selectivity required. In such cases increasing use is made of capillary columns, normally in the splitless or on-column injection modes. Many problems of trace analysis will require the more standard packed chromatographic column. This is because there are still components in the processed extract present in high concentration, although these are not of direct analytical concern. Some of the components will be carried through the chromatographic column and may likely represent a severe overload of a capillary column and even possibly of a packed column with a low stationary phase loading. The result of such overloading of the column is a peak for that component that is excessively broad and may interfere with the analysis of a lesser component.

Also, with many trace analysis problems the prepared sample usually is not only fairly complex but includes materials with a wide range of boiling points even though not all the compounds, not necessarily even the full boiling point range, is of concern. Since the chromatograph is to be operated normally at or near its sensitivity limit, significant components from the first sample injected may show up in the chromatograms of the second, third, or fourth sample. It will become increasingly difficult to trace the origins of stray component peaks when a long series of trace analyses is to be made in sequence. Therefore, the use of temperature programming, where possible, or elevating the column temperature after the components of concern have been eluted is fairly common. Normally, this will mean that the chromatographic column should be chosen to give no significant bleed problems. This is another area where the newer capillary column systems and bonded phases have merit if the sample components do not severely overload the column.

7.5.4 Derivatization Techniques

There are a number of reasons for converting a compound sought at trace concentration levels to another compound. Any attempt at chemical conversion of the sample must be justified on the basis of providing a faster, more convenient, or more accurate final analysis. These goals may be served by forming a new compound that is more readily extractable, more readily chromatographed, more accurately or sensitively measured, or more easily separated from interfering components, either in the steps of sample processing or in the final chromatography of the prepared sample. As a standard rule, extractability, as well as achievement of most nearly ideal chromatography, is accomplished by conversion of more polar compounds to less polar compounds. One of the most dramatic examples would be the conversion of carboxylic acids to the corresponding ester derivatives. The acids tend to have a moderate water solubility and present generally severe problems in achieving good quantitative accuracy with GC. This is particularly true at trace levels.

Other polar compounds that present similar problems are alcohols, amines, and aldehydes. Fortunately, these compounds are also the ones most readily derivatized to less polar compounds. Silylether derivatives can generally be prepared from organic compounds that contain the functions –OH, –SH, or –NH. The –OH and –NH compounds can be converted to acetates or other acyl derivatives with some ease. Carboxylic acids are commonly converted to esters, usually methyl esters, by a wide variety of methods, each of which may have merit in a specific circumstance. Carbonyl compounds can be converted to oximes by reaction with hydroxylamine or derivatives. They can also be reacted with primary amines to form Schiff bases, or with alcohols in certain cases to form acetals or ketals. A number of references that review a variety of derivatization methods are available (22–25).

For trace analysis, it is common to form a derivative that not only aids the desired chromatographic or separation factors, but also provides sensitivity to one or another of the specific detectors available. It is common to use a halogen-substituted derivatizing reagent to provide simultaneously decreased polarity or reactivity, better chromatographic separation, and more selective, sensitive detection by use of electron capture. Thus, for example, trichloracetic anhydride is used to form acetate derivatives instead of using anhydride.

7.6 LOGIC OF QUALITATIVE ANALYSIS

The most important factor in qualitative gas chromatographic analysis is the collection of as much information as possible about the sample before beginning any laboratory work. This information is first gathered by the people involved in the collection of the sample. The sampling location, the person taking the sample, the method of sampling, and sample handling should be known. The sample matrix (solvent, etc.) should be investigated to determine the source of chromatographic peaks. A pure sample should be utilized to compare with the unknown sample. The technique of running blanks on solvents should certainly not be overlooked since the solvent used to work up a sample may be the contributing factor to unknown peaks. Furthermore, the chemist should always be alert to unknown peaks originating from simple decomposition in storage or decomposition or isomerization under chromatographic conditions. All of the above are important considerations, especially in the area of trace analysis. Many times impurities can be in excess of the amount of trace components being analyzed. One should keep in mind that the identification of an unknown by GC can easily turn into a major research project.

Part 2 Quantitative Analysis

7.7 GENERAL DISCUSSION

The gas chromatographic technique is at best a mediocre tool for qualitative analysis. As has been previously shown, it is best used with other techniques to answer the question of what is present in the sample. The rapid growth of GC over the last four decades cannot be explained by ease of operation, the simplicity of the technique, the relative low cost of the instrument, or the wide range of the types of samples being handled. That growth comes from the fact that GC has all these attributes and provides an answer to the question How much? Its reason for existence is that it is an excellent quantitative analytical tool regardless of whether one is quantifying micrograms of heptachlor in a liter of water or one volume carbon monoxide in a million volumes of air.

Sometimes we get carried away with the latest advancement in instrumentation or with the perfectly symmetrical peaks obtained with a certain system. These are only means to an end, perhaps very necessary means, but they are not the end. The end is a number that tells us how much of a component is in a sample. Without the ability of GC to supply that number with reasonable accuracy, this entire book would not be written. The tremendous advances in instrumentation, theory, columns, applications, and technique are all justified because they provide more accurate and precise analyses, analyses for materials not previously possible, or much more rapid analysis.

The remainder of this chapter deals with the techniques used to obtain the answer to the question How much? from the information given by the chromatograph. The quantitative principle of GC depends on the fact that the size of the chromatographic peak is proportional to the amount of material. The first aspect to be considered is the technique of determining peak size. Next, the problem of relating peak size to quantity of material is discussed. Finally, factors that influence peak size and thus introduce errors are considered.

7.8 PEAK SIZE MEASUREMENT

The size of a chromatographic peak is proportional to the amount of material contributing to that peak, and the size of this peak can be measured by a number of ways. Each of these is considered individually. Two basic concepts can be used for peak size. The first is simply the measurement of the height of the peak. The second involves the measurement of area with a wide variety of methods available.

7.8.1 Peak Height

Peak height is the simplest and easiest of the measurement techniques. As shown in Figure 7.13, the baseline is drawn in connecting the baseline segments both before and after the peak (line *AB* in the diagram). This line would be the best estimate of the detector output if there had been no

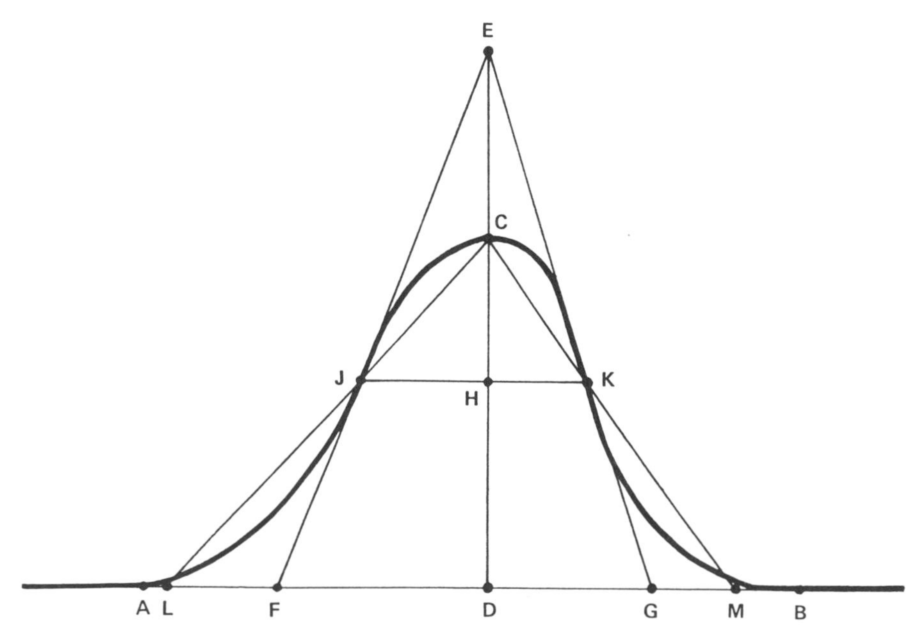

FIGURE 7.13 Constructions for peak size measurements.

detectable amount of material present that contributed to that peak. The height of the peak is then measured from this baseline vertically to the peak maximum (line CD). This height is proportional to amount of material contributing to the peak if nothing in the system changes that could cause a change in the width of the peak between sample and standard.

Factors that can influence the peak width are generally instrumental or technical in nature. The temperature of the column changes the retention time of the material, thus changing the width of the peak. To a first approximation the ratio of retention time to peak width will stay constant for a given component on a given column. Temperature can influence retention time by approximately 3%/°C. A 1°C change in column temperature between the standard and the unknown chromatograms can cause a 3% change in peak width. This change in width will be accompanied by a compensating change in peak height such that height times width remains constant. The height then will change by 3%/°C. This means that to maintain analysis at 1% accuracy by use of peak height measurement, the temperature of the column must be controlled within ±0.3°C and preferably to better than ±0.1°C, assuming the temperature change of the column to be the only factor producing error. Detector temperature may also affect the peak height measurement since the detector's response may be temperature sensitive. This problem will affect any measurement of peak size generally in an indetermined way. Thus excellent control of detector temperature is also important in any quantitative use of a gas chromatograph.

The carrier-gas flow also produces a change of retention time and thus peak width and peak height. To a first approximation, a 1% change in flow will change retention time 1%; thus peak height and peak width are changed by 1%. Use of peak height measurement thus requires that the flowrate between the standard and the sample chromatograms be constant within 1% to maintain an accuracy of 1%. The previously mentioned consideration regarding the effect of flow on the error of peak height measurement is independent of the major error consideration regarding constant flow (see Chapter 5). Several detectors, notably thermal conductivity and photoionization, are flow sensitive; that is, the sensitivity or electrical output for a given amount of material varies with flow. This flow effect affects any method of peak measurement and is really not an error of size measurement. It simply says that flow control is needed regardless of the method of peak measurement.

Reproducibility of peak height is also quite dependent on the reproducibility of the sample injection. This is especially important on early, and thus normally quite sharp, narrow peaks. On such early peaks, the width of the peak is controlled more by the injection time rather than the chromatographic process. A fraction-of-a-second increase in injection time can double the width of these peaks and reduce peak height by 50%. The peaks most subject to error in peak height measurement from injection problems are those with retention volumes 1–2 times the holdup volume V_m of the column. Peaks beyond 5–10 times the holdup volume are negligibly affected by injection technique.

When there is column adsorption of a particular component in the system, the peak will show some tailing. This may not be evident at high concentrations of a component, but with low concentrations a significant portion of the component may be in the tail. This means that at low concentrations the relationship of peak height to amount of materials may not be linear because of the amount of the material in the tail. For quantitative analysis, it is best to avoid adsorption by a better choice of column regardless of the technique of peak measurement. However, with adsorption, peak height may give a significant error with low amounts of material.

The final consideration of peak height measurement is the phenomenon of column overload. When a large amount of a component is injected onto a chromatographic column, the liquid or adsorptive phase becomes saturated with the material, causing a broadening of the peak. This causes reduction in the height, contributing to a nonlinear relationship between peak height and amount of material with high amounts of material. This is independent of any detector nonlinearity at high concentrations. Overloading can be observed by careful observation of the peak shape. There is sloping front edge with a sharp tail, or in some cases, a sharp front with a sloping tail. The peak maximum also moves with this distortion to longer times with the sloping front and to shorter times with the sloping tail (see Chapter 2 and Section 7.2.1). This overload distortion is a function of the amount of liquid phase per unit length of column. It occurs more readily than on small-diameter columns and on packings with low percentage of liquid phase.

7.8.2 Height and Width at Half-Height

Contrary to peak height measurement, a number of techniques are used for peak area measurement. Some of these are manual techniques, and others make use of instrumental accessories to provide an area measurement. The discussion that follows considers all of these techniques from the manual through the instrumental, in that order.

In height–width measurement the area is determined by multiplying the height of the peak by the width of the peak at one-half the height. This technique requires the construction of the baseline (line AB in Figure 7.13) and the measurement of the height of the peak CD as in the peak height technique. Point H is then determined as being halfway between points C and D such that DH is one-half the height CD. Line JK is then drawn parallel to AB and through H. The distance JK is thus the peak width at half-height. The product of CD and JK is the exact area of the triangle CLM. It is a close approximation of the true area of the chromatographic peak. It includes an area below the line JL not a part of the peak, but excludes some peak area above the line JC that is a part of the peak. To the extent that these areas compensate each other, the area of triangle *CLM* is equal to the area of the chromatographic peak.

If the baseline is sloping for any reason, the measurement becomes a bit more complicated. Figure 7.14 is constructed as Figure 7.13, with the same parts of the construction labeled with the same letters. The baseline *AB* is constructed as the best extension of the baseline before and after the peak. The peak height *CD* is constructed vertically from the peak maximum to the baseline. The midpoint *H* is located as before. Line *JK* is then drawn through point *H* and parallel to the baseline *AB*. The desired peak height is the distance *CD*. However, the width at half-height is then the distance *NP*. This would be the width measured with no slope in the baseline. Note that the distance *JK* could be used if it corrected by the cosine of the angle *JHN* or the angle of the baseline to a truc horizontal. The important point here is that points *J* and *K* are the true points on the peak one-half the height of the peak up from the baseline. What is wanted for the width measurement then is the real-time separation of points *J* and *K*, which is given by the horizontal component only of the distance between them.

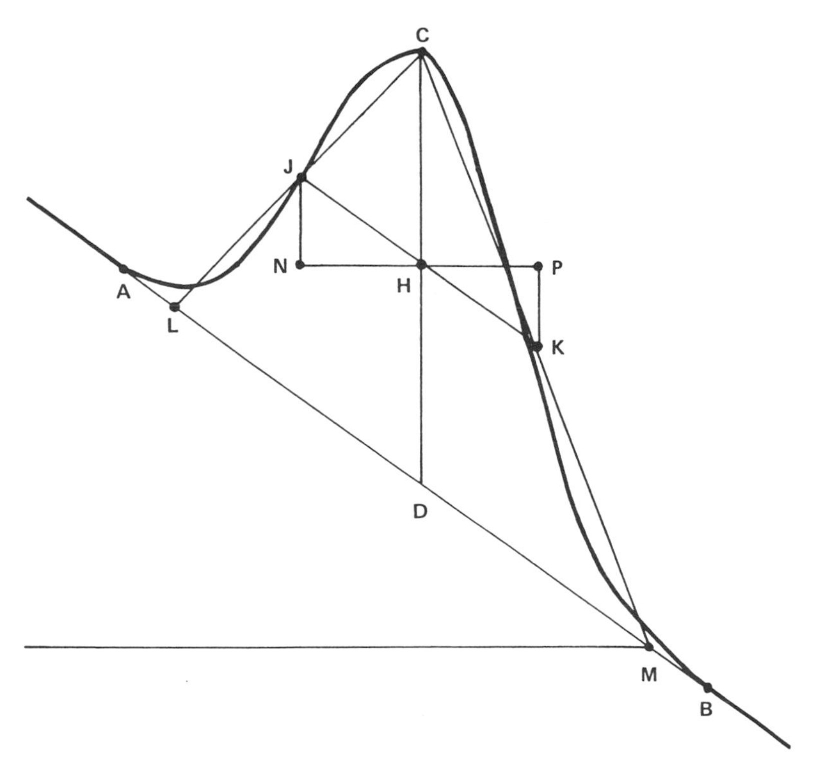

FIGURE 7.14 Constructions for peak size measurements on sloping baseline.

The various errors of this technique are summarized at the end of this section for all area techniques.

7.8.3 Triangulation

Triangulation always involves construction of the baseline *AB*. Tangents to the peak are then drawn at the inflection points of the peak. These tangents are lines *EF* and *EG* in Figure 7.13 and, along with the baseline, form the triangle *EFG*. The area of this triangle is the height *ED* times one-half the base *FG*. This area closely approximates the area of the peak. Comparison of the various area techniques, presented later, includes a discussion of the problem with this technique. For this reason a more detailed discussion here regarding sloping baselines is not warranted.

7.8.4 Cut and Weigh

This technique, sometimes referred to as "paper dolls," involves drawing of the baseline *AB* as before. Then the peak is carefully cut out of the chart paper and weighed on an analytical balance. This weight is then converted to an area by weighing a known area cut from the same chart near the peak. The major advantage of the technique is that it can accommodate distorted and tailing peaks, giving a true measure of the area. The major problem is the inhomogeneity of the paper and the destruction of the chromatogram. Certainly copies can be made prior to cutting, or the copy itself could be cut. Ball et al. (26) indicate that the chart paper itself was much better than bond used for copies as far as homogeneity was concerned.

The general errors associated with this technique are reserved for comparison at the end of this section.

7.8.5 Planimeter

Like the cut-and-weigh method, the planimeter method is a perimeter method that makes use of a surveying or drafting instrument called a planimeter. In this technique the baseline is drawn as usual. The perimeter of the peak is then traced using the eyepiece containing cross hairs of the planimeter. When the starting point is reached, the dial reads a number proportional to the area. On some planimeters the number is proportional to area with application of a settable scale factor. On other instruments the factor must be determined by measuring a known area. The major advantage of the technique is the ability to handle distorted and tailing peaks to produce a true area. The major problems are the painstaking nature of tracing the peak and the use of a tool not normally found in a laboratory.

7.8.6 Disk Integrator

The simple ball–disk integrator (briefly described here for historical reasons) attaches to the recorder as illustrated in Figure 7.15. The integrator pen draws a trace on about 10% of the chart, leaving 90% for chromatogram as drawn by the recorder pen. The integrator pen is linked mechanically to the ball through the cam and roller, and the ball rides on the disk that rotates at a fixed speed. When the recorder pen deflects the ball (which is linked to the recorder pen drive), it moves away from the center of the disk and begins to rotate at a speed proportional to its distance from the center. The roller begins rotating at the speed of rotation of the ball. The cam then causes the integrator pen to oscillate on the chart at rate proportional to the ball rotation. Thus the number of integrator pen excursions between the beginning and the end of the chromatographic peak is directly proportional to the peak area. A single excursion is assigned a value of 100 counts. A partial excursion generally can be estimated to ±1 count.

7.8.7 Electronic Integrators and Computers

In general, electronic integrators are fed the detector signal directly without attenuation. Following amplification, this voltage signal may be converted to frequency such that the output pulse rate is proportional to voltage and the pulse sum is proportional to area. Generally, however, with microprocessor-base integrators the amplified voltage signal is simply sampled several times (2–10) per second, and the voltages are then summed to produce a number proportional to the area. A slope detector in either case detects when the peak begins and ends. A more extensive discussion of electronic integration is presented in Chapter 5.

The major advantages of the electronic integrator are the speed and accuracy with which the area is obtained. These devices operate on the detector signal only and thus are limited only by the detector. Their wide dynamic range permits the integration of both trace and major components without attenuation. The high count rate and sensitive voltage detection ensure accuracy well beyond any other mode of peak measurement. The only drawback of electronic integrators, if indeed it can be called a drawback, is their cost. The minimum cost is around $2000.00. In general, the cost is proportional to the features and capability of the instrument. However, integrators are a minimum cost when compared to instrument investment. They are becoming more versatile and less expensive.

Many integrators today have the ability to store calibration data and permit a report that gives areas or peak heights, retention times, and the final concentration of one or more components in the sample, using any of the techniques discussed in Section 7.9. Many gas chromatographs today have built-in electronic integration and data-handling capabilities. Finally,

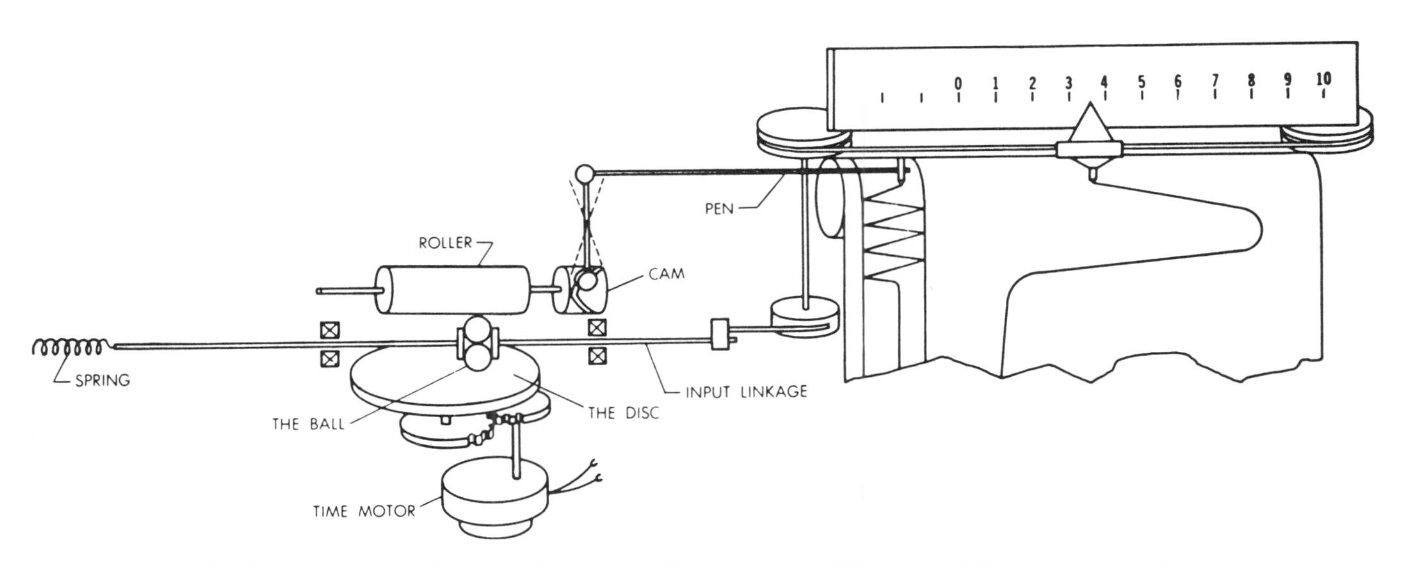

FIGURE 7.15 Mechanical schematic of disk integrator. (Courtesy of Disc Instruments.)

there are network systems whereby a number of gas chromatographic detector signals are fed into a central computer for area measurement and data reduction. Again, the discussion presented in Chapter 5 should be read for a more thorough presentation. For further detailed information, the manuals, application notes, and product bulletins provided by the instrument manufacturers should be consulted.

7.8.8 Comparison of Peak Size Measurements

In Fall 1966 a survey of over 1600 practicing gas chromatographers in the United States was made and reported by Gill and Habgood (27) on measurement techniques then in use. These are reported in Table 7.4. Even though almost 30 years have passed since this survey, the gas chromatographic technique had been fairly well established by then. The large number of respondents would also tend to support the validity of the data. In this survey the technique of triangulation would have to be assumed to include the technique of height times width at half-height. Speculation as to how this has changed over the years since the survey would lead to the conclusion that use of the direct integration techniques of electronic integrators, computers, and tape systems has certainly increased. This is probably at the expense of planimeter, ball–disk integrator, cut and weight, and triangulation, in that order. Personal surveys of today's chromatographers would indicate that more than 90% use electronic methods in quantitative analysis. The ball and disk integrators are virtually gone as are planimeters. Manual methods use peak height, height and width at half-height, and "paper dolls."

Ball et al. (26, 28, 29) in a series of papers have considered the various manual techniques for the peak size measurements. The treatment was both theoretical and experimental in that standard peaks were given to a sizable group for manual measurement. Five techniques were studied: peak height, height and width at half-height, triangulation, planimetry, and cut and weight. The actual measurement errors are summarized in Table 7.5. Peak

TABLE 7.4 Peak Size Technique in Use in 1966

Technique	Relative Usage (%)
Peak height	28.0
Triangulation	16.9
Planimeter	15.5
Cut and weigh	6.4
Disk integrators	20.8
Digital electronic integrators	8.5
Computers	2.4
Tape systems	1.4

Source. Data of Gill and Habgood (27).

TABLE 7.5 Conditions for Least Error in Peak Size Measurements

	Relative Least Error (%)		Peak Shape for Least
Measurement Technique	1.5-cm^2 Area	15-cm^2 Area	Error[a]
Peak Height	<1	<0.5	<1
Height × width at half-height	2	0.5	5
Triangulation	3.5	1.5	1
Planimeter	4	0.6	1–10
Cut and weigh	3.2	2	1–10

[a] Peak shape is defined as peak height/width at half-weight.

shape is defined as the peak height divided by the peak width at half-height. In manual methods of peak measurement, accuracy and precision of measurement degrade considerably as the peak shape becomes extreme. This occurs with very sharp peaks (peak shape < 10) or very flat peaks (peak shape < 0.5). Thus there is an optimum peak shape that the chromatographer should strive to achieve. The attenuator of the chromatograph can be used to adjust peak height, and the chart speed to adjust peak width. In all cases of measurement the bigger the area, the better is the precision of measurement. In the case of peak height, the narrower the peak for a given area, the better is the precision. In all cases it is best to record the peak at the minimum attenuation (maximum sensitivity) and still maintain the peak on scale. Since most chromatographic attenuators work in steps of two, this places the peak height between 50 and 100% of the chart width. For manual area measurements the chart speed should then be selected to give an optimum peak shape between 1.0 and 10. Note that an increase in the chart speed for peak height measurements accomplishes nothing.

It should be stressed here that the errors mentioned in Table 7.5 are for the measurement technique only. They do not represent the precision expected of the analysis.

Condal-Busch (30) has pointed out that triangulation gives 97% of the true area of a Gaussian peak, whereas height and width at half-height give only 90% of the true area. Since chromatographic peaks are not truly Gaussian, however, the error is less. Condal-Busch also concludes that since standards and unknowns are measured in the same way, this error becomes insignificant compared to the actual measurement error itself.

McNair and Bonelli (31) report a study made comparing a number of techniques for area measurement wherein the entire chromatographic system was analyzed. An eight-component sample was used. The relative standard deviation of 10 replicate analyses by the different techniques is recorded in Table 7.6. In general, the data in Tables 7.5 and 7.6 are consistent if one remembers that Table 7.5 contains data on measurement technique precision only and Table 7.6 has data on the entire system precision.

TABLE 7.6 Precision of Area Measurement

Measurement Technique	Relative Standard Deviation (%)
Planimeter	4.06
Triangulation	4.06
Height × width at half-height	2.58
Cut and weigh	1.74
Ball and disk	1.29
Electronic	0.44

Source. Data of McNair and Bonelli (31).

The problem of peak measurement on a sloping baseline must be considered. On peak height, planimeter, and cut–weigh techniques, this is only a problem of drawing the proper baseline under the peak. As discussed previously, it is more complex for height–width and ball–disk techniques. In the case of electronic integration, severe error may be introduced or accurate correction may be made simply depending on the features and capability of the particular integrator. Errors of this type and their solutions have been discussed in detail (32–36).

A final point of consideration is the measurement time required. Certainly electronic integration is by far the fastest, and the ball–disk integrator would also be considered fast. The manual methods in increasing slowness would be peak height, height and width, triangulation, planimeter, and finally the cut–weigh technique. In evaluating all of these observations, (References 26–36, especially References 26 and 29), and the current practices, the following conclusions regarding peak size measurements seem evident:

1. Where the time saved and accuracy obtained can justify the expense, electronic or computer integration is the preferred approach. In general, an integrator capable of handling drifting baseline and fused peaks accurately, although more expensive, is much preferred to the less-expensive integrators that lack these capabilities.
2. The ball–disk integrator is capable of excellent results on all but the most exacting analyses. The recorder used with it should be of top quality and in excellent working order to obtain full capability of the integrator.
3. Peak height, because of its simplicity, speed, and inherent measurement precision, is the preferred manual method. Chromatographic conditions are much more critical here than in any other measurement technique. Current instrumentation helps in this regard. However, more frequent standardization is the real solution.
4. The time required and the difficulty of accurate tangent construction

makes triangulation a method that cannot be recommended under any circumstance.

5. Height–width is the preferred manual area technique assuming reasonable peak shapes.
6. Perimeter methods should be used on irregularly shaped peaks.
7. The cut–weigh method is quite time-consuming. However, with adequate control of variable paper density, it has real value for irregularly shaped peaks.

The subject of peak size measurement cannot be left without the mentioning of required reading for anyone attempting to do quantitative analysis by chromatography. This is a book by Norman Dyson called *Chromatographic Integration Methods* (37). The first paragraph of the Preface says, "This book is about the measurement of chromatographic peaks. In particular, it describes and discusses the manual and electronic techniques used to make these measurements, and how to use integrators. The aim of the book is simply to help analysts extract more data from their chromatograms, and to help them to understand how integrators work so that results are never accepted unquestioningly." The quality of the book and the logic contained therein are best given in the final conclusion in the book: "It should be clear now that integrators are like any tool—an excellent thing in the right hands. What they do best is measure peaks which are suitable for measuring, rapidly and without tedium. If these measurements are worth making then all subsequent calculations are worth noting and perhaps acting upon. As long as integrators use perpendiculars and tangents and draw straight baselines beneath peaks, they are of use only in controlled circumstances, when the chromatography is good. They cannot improve bad chromatography: only the analyst can do that—and at the end of the day that's what he is paid for."

7.9 STANDARDIZATION

7.9.1 General

With techniques of peak measurement in hand, the next important step in quantitative analysis is to convert the size of the peak into some measure of the quantity of the particular material of interest. In some fashion this involves chromatographing known amounts of the materials to be analyzed and measuring their peak sizes. Then, depending on the technique to be used, the composition of the unknown is determined by relating the unknown peaks to be known amounts through peak size.

There is always the question of standards (known amounts of material generally in a matrix) regarding their preparation in the laboratory versus the purchase of ready-made standards. In general, standards should be as close to the unknown samples as possible not only in the amounts of the materials to be analyzed, but also in the matrix of the sample itself. In all cases this requirement would dictate the preparation of standards in the laboratory. There is also the question of stability of the standards. With elapsed time, loss of either the matrix (e.g., hexane evaporation from a solution of pesticides in hexane) or the components of interest (e.g., adsorption of xylene on container walls of 50-ppm standard of xylene in air) cause the standard to be unreliable. In general, in the absence of prior knowledge this dictates that standards be prepared, used, and then discarded all within a short period of time.

Generally, it is much easier to purchase gas standards already prepared and analyzed. Experience here would indicate that these standards be viewed skeptically until credibility has been established for a given source. Certainly rather specialized equipment is needed to prepare a gas mixture with known concentrations of components, but in some cases this is the only reliable way to obtain standards.

The question of purity arises regarding materials used to prepare standards. Two problems occur here: the purity of the component of interest and the purity of the matrix. Fortunately, GC can be used to check purity of chemicals in a reasonable fashion, If a small ($1 - \mu$L) sample of a "pure" liquid is injected into a chromatograph and the detector system is operated at reasonably high sensitivity, impurities will be observed. Without even identifying these impurities, it is generally possible to make some comment on the purity of the chemical relative to its use in a standard. This does not require the use of a general-type detector (e.g., thermal conductivity) rather than any of the specific detectors. If no impurities are observed where one might be expected to see approximately 0.05% of most materials, it is not unreasonable to assume that the purity is better than 99.5%. This could certainly be used to prepare a standard well within $\pm 1\%$ accuracy, assuming no other problems. There are a number of loopholes in this approach. Certainly the column system is overloaded for 1 μL of essentially a pure component. This causes the major peak to broaden and possibly to obscure an impurity very close to the major peak. In general, suspected impurities will be close to the major constituent as a result of similarities in chemical properties and boiling points. The advantage of checking compounds to be used as standards by chromatography is that contamination can be detected. This contamination may have been introduced by previous users of the chemical not using good analytical technique, the inadvertent use of unclean containers, and possibly by mislabeling.

In general, one cannot be too critical regarding standard purity. However, a realistic approach must be taken. If an analysis is required at the 10-ppm level to ± 1 ppm (10% relative), it is not reasonable to spend time and money obtaining standards with reliability to better than 1% when perhaps even 5% would be sufficient.

A reasonable approach to any standard preparation is to obtain the best accuracy in the standard that one can obtain quickly, and then see whether this accuracy will be the limiting factor in the final analysis. If it is limiting, however, further work is needed to improve the accuracy of the standard. In this light the separating power of GC should not be overlooked. The trapping techniques discussed for qualitative analysis by other techniques can be used to isolate small amounts of pure material for standard preparation.

With this introduction to standardization, three techniques are now discussed and the use of standards is covered in each technique.

7.9.2 External Standardization

The technique of external standardization involves the preparation of standards at the same levels of concentration as the unknowns in the same matrix as the unknowns. These standards are then run chromatographically under ideal conditions as the sample. A direct relationship between peak size and composition of one or more components can then be established, and the unknowns can be compared graphically or mathematically to the standards for analysis.

This technique allows the analysis of only one component or several in the same sample. Standards can be prepared with all components of interest in each standard, and the range of composition of the standards should cover the entire range expected in the unknowns. The peak size is then plotted against either absolute amounts of each component or its concentration in the matrix, generally the latter.

Figure 7.16 shows a typical calibration curve for four methyl ketones in an air matrix in which peak heights were used as the size measurement. Note that at some of the higher concentrations, the actual chromatograms were obtained at sensitivity (attenuation) settings different from the lower concentrations. These are all related to a fixed attenuator setting by multiplying all sizes by the attenuator setting used for that peak. The peak sizes are then the values (either height or area) that would be obtained if the chromatogram were run at an attenuator setting of 1, the maximum sensitivity, and the recorder chart large enough to keep the peak on scale. This is why chromatographers refer to attenuator settings as "times 32" or "times 512" and why attenuators are marked with increasing numbers for decreasing sensitivity. The peak size multiplied by the attenuator setting gives the peak sizes at a constant sensitivity (attenuator set at 1). In some case it is more convenient to relate the peak size to an attenuation other than 1. This can be done by dividing the size at attenuation 1 by the desired attenuation, as was done in Figure 7.16 for an attenuation of 16.

Five separate standards were used to prepare Figure 7.16, and all four components were present in each standard. Two different calibration scales were used to separate the curves for ease of identification. Two very important items can be learned from the calibration curve. In general (and

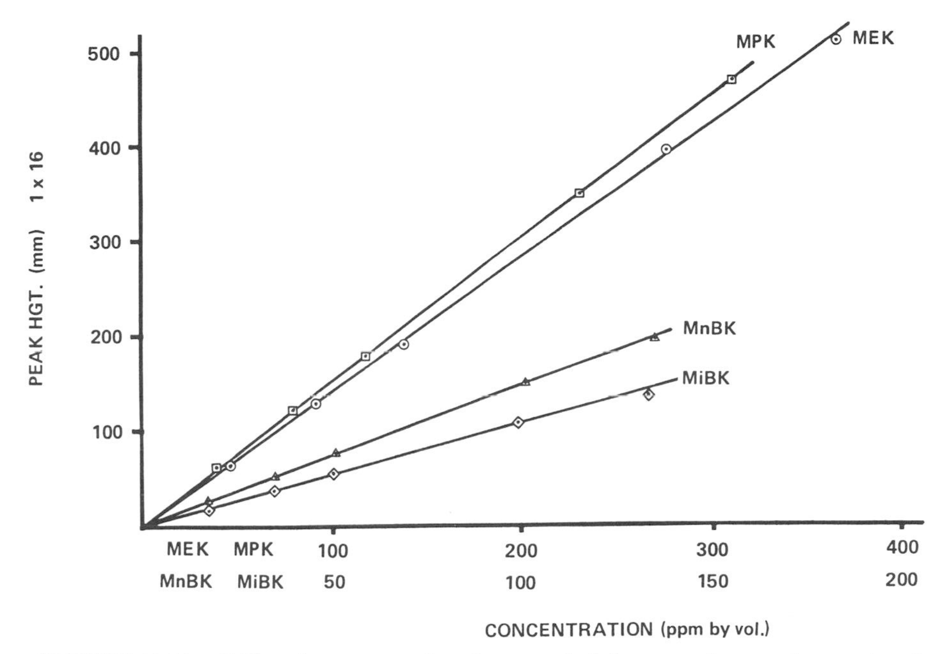

FIGURE 7.16 Calibration curves for four methyl ketones in an air matrix: 2-butanone (MEK), 2-petanone (MPK), 2-hexanone (MnBK), 2-methyl,4-pentanone (MiBK).

in Figure 7.16), the curves are straight lines, and they pass through the origin. These two requirements are most important, for they determine that under the conditions of analysis and over the concentration range covered (1) the column has not been overloaded, (2) the detector has not been overloaded, (3) the electronics are responding linearly, and (4) there is no apparent component adsorption in the injection port, the column, the detector, and associated plumbing.

At some point in any system, as the amount of component doubles, the peak size will not quite double. The column may overload, distorting peak shape; the detector capacity may be exceeded; or some other phenomena may occur. Where possible, one should operate below this point by using a smaller sample size or by diluting the sample. Although it is possible to perform quantitative analysis in a region where the system is nonlinear; this requires that the calibration curve be very well defined in the nonlinear region, meaning a large number of standards. It also means that the calibration curve must be redefined each time unknowns are analyzed. This obviously is quite time-consuming and should be avoided if at all possible.

Adsorption problems and/or sample degradation are generally the cause of failure of the calibration curve to extrapolate through the origin. These complications can often be avoided by proper sampling handling and by proper choice of columns, both materials of construction and packing. It is possible to work with a calibration curve that does not pass through the origin, but this also requires that the calibration curves be generated quite frequently.

It is generally possible to obtain calibration curves as in Figure 7.16, where the concentration region of interest is linear and where the plot extrapolates through the origin. When one is satisfied that these two conditions are indeed met in a given analytical system, it is not necessary to regenerate these curves frequently by running various concentrations of standards. Slight changes of flowrates and temperatures of the detector and column may change the sensitivity of the system and perhaps even the response relationship between various components in the sample, but they will not change the linearity and the origin situation. For day-to-day calibration of the same system, one need to run only one standard and simply ratio concentrations and peak size for each component of interest between the standard and the unknown. What is involved, in effect, in this approach is recalculation of the slope of the calibration curve with a new standard. With any new system or any new analysis, however, the two basic requirements should again be verified by running several concentrations and plotting the calibration curve.

The major problem with external standardization is that the sample size of the standards and unknowns be known accurately. One should attempt to make them equal so that the size of the standard and the size of the unknown divides out of the calculation. If the sample size varies slightly, the peak size must be corrected to unit sample size for standards before the calibration curve is plotted and for unknowns before the calculation is made. Sample size obviously enters into the calculations. As stated earlier, reproduction and measurement of sample size constitute the biggest single error in quantitative analysis by GC. Considerable attention is given to this technique of sample injection later in the discussion of general errors. It should be noted that, in the generation of calibration curves, it is absolutely unacceptable to vary the amount of the component injected by varying the amount of a single standard rather than by using the same amount of different standards having different concentrations. There is no doubt that doubling of the sample size results in doubling of the absolute amounts of each component injected into the chromatograph. But there is no guarantee that the chromatographic system will double the response obtained in the presence of double the amount of matrix. Sample sizes for all standards and unknowns should be kept the same within the errors of size measurement.

Because of the ease of reproducing injection volumes of gas with a gas-sampling valve and the difficulty of applying the technique of internal standardization discussed in the next section for gas samples, external standardization is the preferred approach to the analysis of gas samples. For these reasons considerable attention is given to the preparation of gas standards and the problems associated with gas analysis. In many cases this touches also on the area of trace analysis, since much of the gas analysis done today is the analysis of trace components in an air matrix.

Static Gas Standards. All static methods involve mixing of known amounts of gases or vapors together in some form of a container. These amounts may be measured by volumes or pressures, depending on the types of equipment available. Mixes of the permanent gases in the percentage range are generally reliable. However, they should not be used as primary standards without verification and prior experience (38). These mixtures are generally analyzed and thus become secondary standards.

Difficulties are encountered with these mixtures as the concentration of some of the components approaches the 1- to 100-ppm (v/v) range. Reaction and adsorption become problems even for gases normally considered fairly nonreactive. One report (39) of two CO standards certified at 26 and 41 ppm by the same supplier gave 51 ppm for the second one (25% error) with the instrument calibrated using the first one. Two conclusions arise from this: (1) at least one "certified" standard is wrong, and (2) even "certified" standards should not be trusted implicitly without verification.

Recently, introduction of pretreated cylinders with proprietary coatings or treatments have shown some promise of overcoming reaction and adsorption of even some reaction gases (40). Even assuming that a mixture stays constant in such cylinders, the true concentration must still be known. If the mixture does not remain constant, the situation is impossible.

With the use of compressed gas standards, extreme care is needed in the hardware used between the cylinder and the injection of the standard into the chromatograph. In many cases the cylinder supplier can recommend the proper valving and regulators to use. The question here is not merely What is safe? but also What will not add to or subtract from the standard gas passing through it? In some cases valves rather than regulators should be chosen. For the sake of safety, one should not rely on the cylinder valve for control.

Standards are available today in small pressurized cans that are extremely convenient to use with a gas-sampling valve for injection. Again, supplier reliability and verification are a must.

Laboratory preparation of standard mixtures can be made. In general, the static methods are used only for low concentrations in a matrix gas. Fixed-volume containers made from inert materials, capable of being sealed, and having a resealable septum system can be used. One-gallon glass jugs with lids modified for a septum are very common. On small containers the volume can be determined by weighing the container before and after filling with water and then converting the weight of water to a volume. In some cases quite large containers are used, and here the volume is generally calculated from measured dimensions. In either case some means must be provided to facilitate the mixing of the mixture to provide homogeneity. Diffusion is not sufficient. In small containers this can be a piece of heavy-gauge aluminum foil that can be shaken in the container. In large containers it is generally a fan blade or a blower. The container is thoroughly flushed with the matrix gas until it is reasonable to assume that the container has matrix gas only. It is then sealed and a small volume is withdrawn through the septum for analysis by GC. This is to ensure that the matrix in the container is free of the component to be added to within the error of the needed standard. Failure to perform this simple check can result in many problems and wasted effort.

For gases, a gastight syringe is flushed thoroughly with the component to be added, filled with the needed amount of pure component, and then emptied into the container through the septum. The concentration is simply a ratio of the volumes:

$$\%\text{A} = \frac{\text{volume A added}}{\text{container volume}} \times 100 \tag{7.5}$$

or

$$\text{ppm A} = \frac{\text{volume A added}}{\text{container volume}} \times 10^6 \tag{7.6}$$

The concentrations are volume or mole percent or parts per million (v/v). This is the usual method of presenting gas concentrations, as opposed to those on a weight basis. The container must be thoroughly mixed to ensure homogeneity. The two major sources of error of this technique result from inadequate mixing and lack of assurance regarding whether the syringe volume used contained 100% of the desired component. One never knows when both conditions have been satisfied; therefore overcaution is the word.

Known concentrations of vapors can be prepared in the same way by injecting a known volume of a volatile liquid into the container, using a microliter syringe normally used for liquid sample injection into a gas chromatograph. The density and molecular weight of the component are needed for the calculation:

$$\text{ppm A} = \frac{\text{volume A} \times \text{density} \times 24.25 \times 10^6}{\text{MW}_\text{A} \times \text{container volume}} \tag{7.7}$$

where volume A = μL of A added, as a liquid
density A = density of A (g/mL or mg/μL)
24.45 = molar volume at 25°C and 760 Torr (L/mol or mL/mmol)
MW = molecular weight of A (g/mole or A mg/mmol)
container volume = volume of container (mL)

The liquid syringe must be touched against the side of the container or the foil to obtain the final amount of injected liquid off the needle prior to its withdrawal from the container. The container must again be thoroughly mixed to ensure both complete evaporation of the liquid and a homogeneous mixture. If the temperature and absolute pressure of the matrix gas in the container are different than 25°C and 760 Torr (the conditions of molar volume used), either the container volume must be corrected to these conditions or the molar volume must be corrected to the conditions of the matrix gas. Differences of 3°C or 7 Torr cause a 1% error. Generally, larger differences should be corrected. The important point to remember is that the volume of the vapor (calculated in Equation 7.7 by applying data for liquid volume, density, molecular weight, and molar volume) must be at the same temperature and pressure as the matrix gas for calculation of a volume ratio such as volume percent or volumc part pcr million.

Several gases or vapors may be added to the container by either technique to provide standards for a number of components. A disadvantage of a fixed-volume container is that the sample is depleted as withdrawals are made. Generally, about 10 mL would be withdrawn to adequately flush a 1-mL volume of gas-sampling valve. Two such withdrawals will deplete a 2-L container by 1%. This depletion will cause a dilution of the standard by air, either from small leaks in the container or as the syringe is withdrawn from the sample under reduced pressure. Adsorption with time can be a serious problem, especially with vapors. Generally, the best practice is to prepare the standard using intermittent mixing over a period of 15–30 min. Then the chemist should use the standard, perhaps in duplicate or triplicate, and discard the standard. Unless experience has indicated a longer period of stability for a given system, these static standards should be trusted no longer than 1 h.

Plastic bags have been used to overcome the problem of fixed volume (41–43). However, other problems are introduced. The volume of the matrix gas must now be measured accurately each time a standard is made. This is usually done by filling the bag with a constant flowrate for a fixed period of time. Components can be added to the matrix as it is flowing into the bag. Mixing may be done by gently kneading the bag. Calculations are the same as those for the fixed-volume container. Adsorption problems can be considerable, depending on the components and the bag material. Bags are also susceptible to small leaks, which can cause serious error, especially in the volume of gas matrix added. It would be reasonable to apply the same time frame of standard preparation and use in fixed-volume containers to bags without other experience (i.e., 1 h).

Dynamic Gas Standards. Dynamic methods are basically flow dilution systems providing a continuous flowing calibration gas. In this approach two or more pure gases flow at a constant, known flowrate into the mixing junction. Dynamic standards have two major advantages that make the technique desirable and worth the effort to set them up. The first is that adsorption problems are virtually eliminated in the generation and sampling systems because of the constant flowing system. However, adsorption is not eliminated. It is still present, but very soon the amount absorbed is in equilibrium with the concentration in the flowing stream. Thus a standard, known concentration is exiting the system. This is extremely important in the preparation of trace standards of reasonably polar and adsorptive materials. The second advantage is that the flowrate of one or more of the components can generally be easily changed, thus providing various concentrations of standards for calibration curves. This becomes important in the initial evaluation of a system for analysis.

For mixtures in the percentage range, the dynamic mixing technique is reasonably straightforward. Flows can be accurately controlled and, with the use of a technique such as a soap-film flowmeter, can be measured reasonably accurately. In general, however, continuous in-line flow meters are used, the most common of which are rotameters. It is a very unusual rotameter than can be read and set to within 1% accuracy over even 50% of its scale. Too often the rotameter is read and the value for flowrate is assumed accurate without full appreciation of the reading error involved. In general, the greatest error in dynamic standards is the lack of accuracy in one or more of the flowrates. Again, one should be concerned about pressure and temperature of the gases and that these are the same either actually or by calculation, and also how these two variables affect the means used to measure the volumetric flowrate.

As mentioned earlier, the simple form of dynamic dilution works well in the percent range. However, attempts to produce a 5-ppm methane in air standard, by mixing a 1-mL/min methane flow with a 200-L/min airflow, fail simply because of the problems of measuring the high and low flows accurately and conveniently. Generally, a double-dilution technique works here. First, a dynamic standard of 2000 ppm is generated by a flow of 15 mL/min of methane and 7.5 L/min of air. Then 20 mL/min of the 2000-ppm standard is mixed with 8-L/min air to produce 5 ppm. Properly, the airflow to produce the 2000-ppm standard should be 7.485 L/min (7.5–0.015). The total flow of the 2000 ppm is then 7.5 L/min. The same applies, of course, for the second dilution. The equation used to keep these concentrations straight in successive dilutions is as follows:

$$F_1 \times C_1 = F_2 \times C_2 \tag{7.8}$$

where F_1 (mL/min) is the flowrate of concentration C_1 and F_2 (mL/min) is the flowrate of concentration C_2. Thus the second dilution given above is

$$20\ \text{mL/min} \times 2000\ \text{ppm} = 8000\ \text{mL} \times 5\ \text{ppm}$$

The accuracy of multiple dilutions fades as an increasing number of dilutions are made because of the added errors of additional flow measurements. In the double dilution given above, four flow measurements are needed, two for each dilution. Fortunately, however, multiple dilutions are used to produce low concentrations where an analysis accuracy of perhaps ±10% would be acceptable.

Low flowrates of gases can be delivered to a larger volume flowrate of a diluent gas by the use of small motor-driven syringes (44). This is one way of accurately delivering low volumetric flowrates. Generally, periods no longer than an hour are used since the syringe must be refilled. Backdiffusion of the diluent gas into the syringe volume at low delivery rates is a problem here. Also, the downstream pressure of the standard thus prepared cannot change since this can cause a pumping action in and out of the syringe volume.

A technique of making known vapor concentrations of reasonably volatile liquids in a diluent gas involves the use of the vapor pressure of the liquid (45). The diluent gas is passed through successive thermostatted bubblers to obtain a mixture determined by the saturation vapor pressure (SVP). Thus for ethanol, if the bubblers were maintained at 20°C (ethanol vapor pressure at 20°C is 43.9 Torr) and the diluent gas flow were maintained low enough to ensure saturation, a dynamic standard is generated with the following concentration:

$$C = \frac{\text{SVP}}{\text{Total P}} \times 100 = \frac{43.9}{760} \times 100 = 5.78\% \tag{7.9}$$

At this temperature the vapor pressure changes by about 5%/°C, requiring bubbler thermostatting to better than ±0.2°C for a 1% standard accuracy. It is also important to know accurately the total pressure at the final bubbler, since this is also used in the calculation. This was assumed to be 760 Torr, to illustrate the preceding calculation, but must be measured in practice.

If the vapor pressure technique is used, two methods can be used to change the concentration:

1. The vapor pressure can be changed by changing the temperature. This can be quite time-consuming in that true thermal equilibrium is required for each concentration. Also, the temperature must be kept lower than any subsequent temperature the developed standard will see to prevent condensation and thus a loss of the standard.
2. The total pressure under which the bubbler system is working can be changed. Since this can be done only for pressures greater than any subsequent pressure to which the standard will be exposed, it can require a sophisticated experimental setup.

In general, for multiple concentrations one standard is prepared in this fashion and is then diluted by a second diluent gas stream. This requires that both the original bubbler flow and the diluent flow can be accurately measured, whereas with single concentration provided by the vapor pressure system, the bubbler flow need not be known accurately because it does not enter into the calculations. One only has to be assured that the gas is saturated.

Another approach to vapor standards is to use the diffusion of vapor through a capillary to add small amounts of vapor to a flowing gas stream (46–48). The theory and practice are reasonably well defined. The concentration is determined by knowing the rate of diffusion and using the following equation:

$$C = \frac{R \times K}{F} \tag{7.10}$$

where C = concentration (ppm v/v)
R = diffusion rate (ng/min)
F = diluent gas flowrate (mL/min)
K = 24.45/MW (nL/ng) at 25°C and 760 Torr

Once again the diluent flowrate must be at the same conditions as the K factor used or vice versa. To ratio gas volumes, they must be measured at the same temperature and pressure. The K factor simply converts the diffusion rate in weight per unit time to vapor volume per unit time.

Theory predicts the diffusion rate by the following equation (46):

$$r = 2.303 \frac{DMPA}{RTL} \log \frac{P}{P - p} \tag{7.11}$$

where r = diffusion rate (g/s)
D = diffusion coefficient (cm^2/s)
M = molecular weight (g/mol)
P = total air pressure (atm)
A = diffusion cross-sectional area (cm^2)
p = partial pressure of sample at T° (atm)
R = gas constant (mL − atm/mol − K)
T = temperature (K)
L = length of diffusion path (cm)

By incorporating R into the constant, converting both pressures (P and p) into Torr from atmospheres, and converting the rate into ng/min from g/s this equation is obtained:

$$R = 2.216 \times 10^6 \frac{DMPA}{TL} \log \left(\frac{P}{P - p} \right) \tag{7.12}$$

where R = diffusion rate (ng/min)
P = total pressure (Torr)
p = partial pressure of sample (Torr)

All other terms are as above.

By use of this equation with vapor pressures and diffusion coefficients from data in the literature and very accurate measurements of area and length, the diffusion rate generally can be calculated to within about 5%. Thus the only way to build dynamic standards using the diffusion technique is to determine the rate in a given system. One such system is to use a diffusion tube, as shown in Figure 7.17. The bulb of the tube is loaded with liquid to about 80% of its capacity (perhaps 5 mL). The capillary length is variable up to about 7 cm and the capillary diameter perhaps 0.2 cm. This tube is placed in a thermostatted chamber permitting a dilution gas flow across the tube. The diffusion rate is then determined by weight loss over several days using a good analytical balance. Only the gas flowrate need be measured then to generate a primary standard. Diffusion rates can be measured during the life of one filling while the diffusion tube is in use. Different materials can be filled in the same tube, or the tube can be refilled with the same material. Only pure materials can be used, not mixtures. Several tubes, however, can be put in the same gas stream to generate a multiple standard. The concentration of the standard may be varied over a wide range by variation of the dilution gas flowrate. This is preferred to a temperature change of the diffusion tube. Again, the temperature control of the diffusion tube is critical. General practice is to maintain the temperature constant to within ±0.1°C.

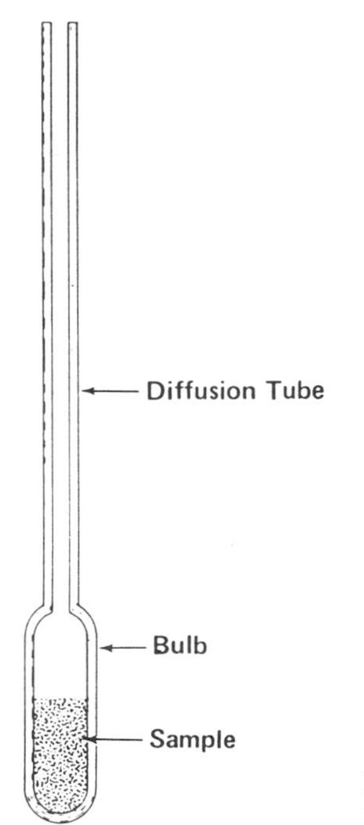

FIGURE 7.17 Cross-sectional diagram of a diffusion tube (Courtesy of Analytical Instrument Development.)

TABLE 7.7 Diffusion Tube Data

Chemical	Diffusion Rate[a] (ng/min) Measured	Diffusion Rate[a] (ng/min) Calculated	Concentration at 150 mL/min ppm (v/v)
Benzene	13,500	14,300	28.2
Toluene	4,550	4,490	8.1
n-Octane	2,160	1,917	3.1
Methanol	15,100	13,700	76.9
Ethanol	6,950	6,840	24.6
Ethyl acetate	15,100	14,200	28.0
Chloroform	45,600	52,200	62.3
Carbon tetrachloride	29,200	33,000	30.9
Acetone	34,500	32,100	97.1

[a] All diffusion tubes were 7 cm × 2 mm i.d., operating at 30°C.

Table 7.7 illustrates the types of standards that can be generated by diffusion tubes. The measured and calculated rates give some indication of the errors in attempting to use Equation 7.12 to calculate the diffusion rate. These are primarily the accuracy to which the diffusion coefficient and vapor pressure are known at the operating temperature and the accuracy to which the diameter of the diffusion path and its consistency are known. The measured rate by weight loss is as accurate as the balance used and the ability to hold the temperature constant between weighings. It can easily be within 1% accuracy. The concentrations shown can, of course, be changed by changing the flowrate. Certainly higher concentrations can be developed by using a wider bore tube, a shorter tube, or higher temperatures. However, the temperature should be held at least 20°C below the boiling point of the material.

A clever means of dynamic generation of standards at the ppm level involves permeation through a polymer. In 1966 O'Keefe and Ortman (49) described this technique primarily for air-pollution standards. A condensable gas or vapor is sealed as a liquid in a Teflon tube under its saturation vapor pressure, as shown in Figure 7.18. After an initial equilibrium period, the vapor permeates through the tube wall at a constant rate. This rate is determined by weight loss over a period of time. Temperature must be controlled to within ±0.1°C to maintain 1% accuracy. In use, the tube is thermostatted in a chamber that permits a diluent gas to fully flush the chamber. The concentration is then determined by the same equation used for diffusion tubes:

$$C = \frac{R \times K}{F} \tag{7.13}$$

where C = concentration (ppm, v/v)

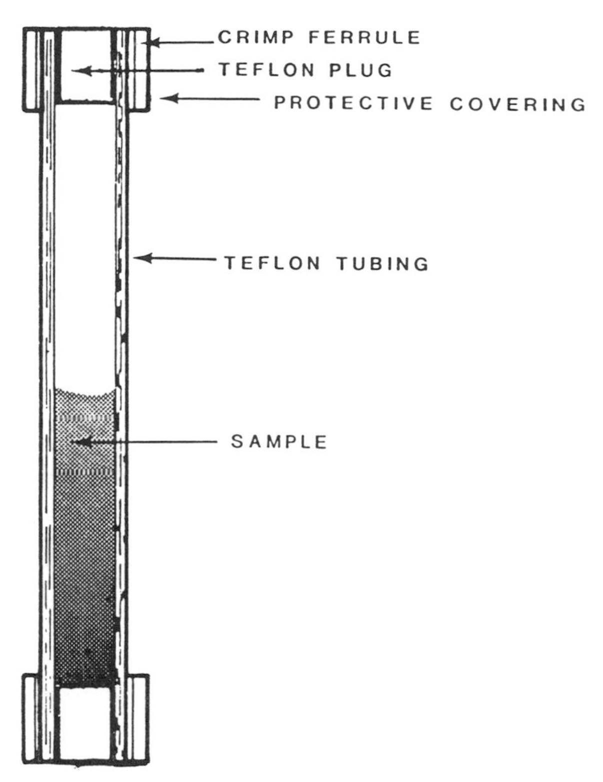

FIGURE 7.18 Cross-sectional diagram of a permeation tube (Courtesy of Analytical Instrument Development).

R = permeation rate (ng/min)
F = diluent gas flowrate (mL/min)
K = 24.45/MW (nL/ng) at 25°C and 760 Torr

Again, to accurately ratio gas volumes they must be at the same temperature and pressure. Either F is corrected to 25°C and 760 Torr, or the K factor is adjusted to the conditions under which F was measured.

Typical materials available in permeation tubes for operation at 30°C are listed in Table 7.8 along with average rates per centimeter length of tube and the K factor. As the length of the tube increases, the permeation rate increases in reasonable proportion. The data in Table 7.8 are for tubes of 0.25 in. o.d. The wall thicknesses are shown in the table. The final column gives the concentration available from a 5-cm tube using 1-L/min flow. For instance the SO_2 rate is 240 ng/min cm^{-1}. For a 5-cm tube, the rate would be 1220 ng/min. At 1L/min the concentration is

$$C = \frac{1200 \times 0.382}{1000} = 0.458 \text{ ppm} \qquad (7.14)$$

TABLE 7.8 Permeation Rates for Chemicals in Permeation Tubes at 30°C

Chemical	K	Thick Wall Thickness (In.)	Permeation Rate (ng/min cm^{-1})	Concentration at 1 L/min ppm (5-cm Tube)
Sulfur dioxide	0.382	0.062	240	0.46
Nitrogen dioxide	0.532	0.062	1000	2.66
Hydrogen sulfide	0.719	0.062	240	0.86
Chlorine	0.345	0.062	1250	2.15
Ammonia	1.439	0.062	210	1.51
Propane	0.556	0.062	100	0.28
Butane	0.422	0.030	24	0.05
Methyl mercaptan	0.509	0.030	65	0.16
Ethyl chloride	0.379	0.030	56	0.11
Vinyl chloride	0.391	0.030	400	0.78

If the flow is increased to 2.0 L/min, the concentration is cut in half to 0.229 ppm. Longer tube lengths, thinner tube walls, and higher temperatures all increase the permeation. Generally, to prevent the diluent gas from cooling the tube, in practice, a low flow is passed across the tube and is then diluted with a higher flow downstream from the tube. The sum of both flows must be used in the calculation.

Tubes for higher temperature operation containing some common industrial solvents have been introduced. Some of these are listed in Table 7.9. These permit low concentration standards to be prepared for some industrial hygiene-type analyses. Some of these tubes at 70°C begin to

TABLE 7.9 Permeation Rates for Chemicals in Permeation Tubes at 70°C

Chemical	K	Permeation Rate (ng/min cm^{-1})	Concentration, ppm at 200 mL/min (10-cm tube)
1,1,1-Trichloroethane	0.183	112	1.03
Trichloroethylene	0.186	1060	9.86
Chloroform	0.205	713	7.31
Carbon tetrachloride	0.159	220	1.75
Acetone	0.422	270	5.70
Methyl ethyl ketone	0.340	100	1.70
Benzene	0.313	260	4.07
Toluene	0.266	120	1.60
o-Xylene	0.231	40	0.46
Cyclohexane	0.291	20	0.29
n-Hexane	0.284	160	2.27
Methanol	0.764	216	8.25
Vinyl acetate	0.284	700	9.94

overlap the type of standards developed with diffusion tubes as shown in Table 7.7. Thus the combination of permeation tubes and diffusion tubes provides means of preparing standards of common solvents from below one to several hundred ppm.

Permeation tubes are not refilled, have a limited life, and cannot be turned off. However, their life can be prolonged during periods of nonuse by storing them in a refrigerator to reduce the permeation rate. Not many materials are practical for use in permeation tubes. When the technique can be used, however, it is generally preferred as a means of standard preparation.

Liquid Standards. Significant space has been devoted to gas standards because of the difficulty in preparing known standards. The fact that such a wide variety of techniques are in use attests to the problem. On the other hand, liquid standards are quite straightforward, and reasonable analytical technique can ensure reliable standards.

In general, liquid standards are prepared in a solvent matrix, which should be the same as the matrix of the unknown. In many cases the liquid may be an extraction solvent or simply a dilution solvent, depending on the type of analysis (as opposed to being prescribed by a procedure). The solvent should be chosen such that it does not interfere with any of the potential sample components. For trace analysis, it is important that the solvent be checked for impurities and that these impurities will not be confused with sample components. Chromatographing the solvent at the maximum sensitivity to be used in the analysis is referred to as "blanking the solvent." It is very important to blank the solvent each time it is used to ensure that it has not been inadvertently contaminated. Also, in trace analysis it is preferred to have a solvent elute from the column following the sample components of interest rather than ahead of the sample.

Standards are prepared by adding known weights of materials to a volumetric flask and then diluting to volume with the solvent or matrix. The approach is best illustrated with an example of the analysis of benzene in toluene at the 0.01% (weight) level. A standard is prepared by weighing 100 mg of benzene into a 10-mL volumetric flask. This is diluted to the mark with benzene-free toluene. This can be used as a master standard. Each milliliter of solution contains 10 mg of benzene. The master standard is then used to prepare several additional standards as follows:

Standard	Master Standard (μL)	Benzene (mg)	Wt/vol (%)	Wt/wt (%)
1	50	0.50	0.0050	0.00577
2	75	0.75	0.0075	0.00866
3	100	1.00	0.0100	0.01155
4	150	1.50	0.0150	0.01732

Each standard is prepared in a 10-mL volumetric flask, and the proper amount of master standard is then diluted to the mark with benzene-free toluene, giving the concentrations shown in the preceding table. The weight/weight percent simply assumes a toluene density of 0.866 g/mL. The calibration curve can now be run by applying the four standards. Peak size can be plotted against absolute weight of benzene in the injected sample or against weight percent, depending on the final form needed for the unknown. Assuming a 1-μL injection, standard No. 3 would provide a benzene peak for 0.1 μg of benzene. This is the convenience of preparing the dilution to volume and calculating weight percent by density. If it is assumed that the density of the solution is the same as the pure toluene, an error of no more than 0.01% relative can be introduced at this level. In this case it is assumed that the density of the unknown is the same as that of the toluene. These assumptions should not be made for solutions in the percent range. At this level the standards should be prepared by weight of each component.

There are several advantages of the double-standard preparation as used above. Significant amounts of toluene are conserved, and the standards are prepared by volume measurements (except for the one weight measurement). Also, other components can be added to the second set of standards at the time of preparation from other master standards. These components can then be varied independently of each other.

Obviously, solid samples can be made up by weight in a solvent as above. This is generally the technique used for such materials as pesticides.

Reliability of liquid standards over a period of time is generally quite good if they are kept in sealed containers. They should not be stored for any length of time in volumetric flasks, but small vials are quite convenient. However, a word of caution about the vial caps is in order. Plastic or plastic-coated cardboard liners in vial caps pose serious problems in most cases. Solvents dissolve or leach a number of materials from these caps, generally causing gross interference with the standards. In general, foil-lined caps should be used unless these are known to produce problems.

Tightness of the seal is important to prevent selective evaporation of components or solvent from the vial. Homemade inert cap liners may be inert but seldom adequately seal the vial. Evaporation is generally the major reason why liquid standards become nonstandards. Chemical knowledge of the components should also be considered as far as reactivity and adsorption are concerned in terms of the useful life of standards.

Many small container designs are available today that form a tight, inert seal and allow sample to be withdrawn by a syringe through a septum. These containers have been known to maintain standards up to a year without change. However, preparation of standards more frequently than once a year is certainly recommended.

7.9.3 Internal Normalization

In the internal normalization technique a sample is injected into the chromatograph and peaks are obtained for all the sample components. Generally, area measurements are used for all peaks, although peak heights can be used. The basic calculations are shown in Table 7.10 for an assumed sample containing four components, A, B, C, and D. If the peak areas are simply added, one can calculate the area percent. This, however, does not account for the fact that different materials will have different responses in the detector for a given weight. These different responses may be determined either absolutely as concentration per unit area or relative to each other for a given analysis. If one used the area percent as the weight percent, the assumption is that all the components would respond in the detector with exactly the same sensitivity; that is, a given weight of any of the components will give exactly the same area. This might be justified in some cases when one is attempting to check purity of a substance, such as mentioned for standards in Section 7.9. If the major component is 99 + %, the error introduced is small.

The data in Table 7.10 are typical of the data for internal normalization. The standard is prepared by adding known weights of the pure components to each other and calculating the weight percent as shown. The standard is then chromatographed and the areas of the four peaks are measured. The area percents are listed to show their relationships to weight percents for this mixture. The weight percents are then divided by areas to give the concentration per unit area. Component A was chosen as a reference and assigned a response factor of 1.000. The other response factors are determined by dividing the concentration per unit area by 0.005138.

TABLE 7.10 Internal Normalization

	Standard					
Component	Taken (g)	Weight (%)	Peak Area	Area (%)	Weight % / Area	Response Factor F
A	0.3786	21.74	4231	22.41	0.005138	1.000
B	0.4692	26.94	5087	26.94	0.005296	1.031
C	0.5291	30.38	5691	30.14	0.005338	1.039
D	0.3648	20.94	3872	20.51	0.005408	1.053
Total	1.7417	100.00	18881	100.00		

	Unknown				
Component	Peak Area	Weight %	Normalized Weight %	Area × F	Weight %
A	3862	19.84	19.66	3862	19.66
B	5841	30.93	30.66	6022	30.66
C	4926	26.29	26.06	5118	26.06
D	4406	23.83	23.62	4640	23.62
Total	19035	100.89	100.00	19642	100.00

Generally these response factors should be constant as long as the operating conditions of the detector remains constant. The FID is relatively insensitive to flow and temperature changes, making it almost ideal for internal normalization. One should be able to reproduce these response factors over long periods of time with this detector. As the response time increases, the detector is less sensitive for that component.

The unknown sample can now be chromatographed and the areas measured. Both approaches mentioned earlier for calculation will be shown. In the first case, each area is multiplied by the weight percent per unit area to obtain the raw weight percents in the unknown. It should be stressed that the sample size injected into the chromatograph was the same in both standard and unknown. When the weight percents are added, however, it is found that the total is greater than 100%. Why? The answer is that the sample sizes were not identical. Sample size is about the biggest error in gas chromatographic analysis next to some of the manual area measurement techniques. The technique of internal normalization corrects for this sample size error. Each weight percent value is divided by the total percent (100.89% in this case) and multiplied by 100 to provide the normalized weight percent. The second approach multiplies each area by the response factor, thus correcting each area for the individual component response factor in the detector. The weight percent is then simply the response-corrected area percent.

Even though this technique can correct the variation in sample size, one should still make the attempt to keep sample size the same. The same sample size then requires a uniform effort on the part of the chromatographic system regarding injection, vaporization, sample loading on the column, and response in the detector. For improved accuracy, component levels in the unknown are bracketed in the standards. Results obtained with the use of this technique on round-robin samples were reported by Emery (50). Emery's paper also provides some excellent data on various methods of peak measurement.

The major disadvantage of this technique is that the entire mixture must be separated and detected in the chromatographic system. All peaks must be standardized by response factors regardless of whether their analysis is needed. Internal normalization also requires that a detector be used that responds somewhat uniformly to all components. This technique cannot be used with electron capture and flame photometric detectors, for instance.

With care, internal normalization can be used where peak size is measured by height instead of area, although this is rare. The response factor is now subject to slight variations in column temperature, injection technique, carrier flow, and the like, all mentioned in the discussion of peak measurement previously. This approach requires that the standard mix for response factors be run as close in time to the unknown as possible. Response factors determined from area measurement are in no way the same as those determined from peak height.

The preceding sample has four components in approximately the same concentration. This is certainly not necessary and in practice is seldom attained. However, a major concern with the use of this technique is that the chromatographic system can handle the absolute amounts injected of all components in a linear fashion. This means that the detector systems must still be responding linearly to the absolute amount of each component even if one represents 99% of the sample and is not the component of interest. Certainly smaller sizes can be used, but here again practicality enters in.

One way to avoid the nonlinear problem is to dilute the standard and unknown with a compatible solvent that is fully resolved chromatographically from all the sample components. These dilutions need not be accurately made or be identical for the standard and the unknown. Good practice dictates that they be approximately the same for each. This is merely a technique for injecting a smaller amount of the standard and sample into the chromatograph. Since calculations do not involve sample size, this dilution is not a factor; the solvent and any solvent impurity peaks are not measured and are not to be considered in the calculation.

In theory, the internal normalization technique may appear ideal. But in analysis of real-life samples that may contain many components, some of which may be unresolved chromatographically and of no interest to the analyst, one of the other two techniques offers more advantages and is generally employed. One analysis using this technique and performed hundreds of times each day is the component-by-component analysis of natural gas. A complete analysis is needed since the analysis is used to calculate the heating value of the sample. Thus it is natural to normalize the results.

7.9.4 Internal Standardization

The technique of internal standardization may best be understood by referring to Table 7.10, which outlines the method of internal normalization. It is assumed in this instance that only component C is of interest for analysis and that the unknown contains no component A. If a standard containing known weights of both A and C is prepared and chromatographed, the response factor F can still be determined. This is shown in Table 7.11, assuming the same weights and areas as before. In practice, several standards should be made, with a plot of area as abscissa and weight ratio as ordinate. This plot must be linear for the particular system. Once the linearity is established for a given sample type and system, only one standard mix need be used to define the slope of that plot. Note that the response ratio R is the slope of that line. Therefore, the standard is actually used to determine the ratio R. Note that the response factor F for C in Table 7.10 is the reciprocal of the R ratio in Table 7.11.

TABLE 7.11 Internal Standardization

Component	Weight	Weight Ratio C/A	Area	Area Ratio C/A	$R = \frac{\text{Area Ratio}}{\text{Weight Ratio}}$
A	0.3786		4231		
		1.398		1.345	0.962
C	0.5291		5691		

The unknown is now ready to be run. Since no A is present in the unknown, a known weight of component A is added to a known weight of the sample. This mixture is then chromatographed and the area ratio of components C:A is measured. Knowing R, the ratio of the area ratio and weight ratio, and the area ratio in the unknown, one can calculate the weight ratio of the unknown:

$$\frac{W_C}{W_A} = \frac{A_C}{A_A} \times \frac{1}{R} \tag{7.15}$$

where W_C and W_A = weights of C and A, respectively, and A_C and A_A are the areas of C and A, respectively; R is the response ratio. Since the weight of A added to the sample is known, the weight of C in the sample can be calculated:

$$W_C = \frac{A_C}{A_A} \times \frac{1}{R} \times W_A \tag{7.16}$$

And since the weight of the sample is known,

$$\%C = \frac{W_C}{\text{Sample weight}} \times 100 \tag{7.17}$$

In practice, a master standard of component A and one of component C are prepared on a weight/volume basis in a solvent. Mixture of known volumes of each of these two standards can provide a variety of weight ratios of the two materials for the initial linearity check. The standard of component A can also be used to add a known amount of A to a known weight of sample.

In the preceding example, area was used to measure peak size since that was the technique used in the example for the internal normalization. Peak height can be used as the size measure just as well as peak area. The same advantages of peak height measurement are present in this method of standardization as in any other. Likewise, the same requirement for frequent standardization is present.

In this instance component A is referred to as the internal standard. All the advantages of the internal normalization technique, such as lack of knowledge regarding the exact sample size and the noncritical aspects of dilution, carry over to this technique. The major disadvantage of internal normalization, namely, the necessity of measuring all the components of the sample, does not carry over into this technique. The cautions under internal normalization regarding system overload apply, but only to the components of interest and the internal standard, not to the entire sample.

Again as with internal normalization, even though the sample size is theoretically not critical, attempts should be made to use the same sample size for both standards and unknowns. This constant load on the chromatographic system gives one the best shot at the high accuracy the technique of internal standardization is capable of producing.

With attention to the purity of the standards and to the lack of interference of any solvent impurities, the precision of the internal standard method is controlled by the ability to quantify peak size. That certainly qualifies this technique as the most precise method of quantitative analysis by GC, and where precision is paramount, the internal standard technique should be applied. Its advantages far outweigh the slight increase in effort required for standard and sample preparation. An excellent, detailed, "how to" approach for the internal standardization technique as applied to a practical problem is given by Barbato et al. (51).

The preceding discussion of sample storage of external liquid standards certainly applies to the standards prepared for the internal standardization technique. There is one further consideration in this regard and that is in the proper selection of the internal standard for a given analysis. The first step is to chromatograph a typical sample and identify the component or components to be analyzed. The internal standard is then chosen such that it must

1. Elute from the column adequately separated from all sample components.
2. Elute as near as possible to the desired component(s) and ideally, before the last sample peak so that analysis time is not increased.
3. Be similar in functional group type to the component(s) of interest. If such a compound is not readily available, an appropriate hydrocarbon should be substituted.
4. Be stable under the required analytical conditions and nonreactive with sample components.
5. Be sufficiently nonvolatile to allow for storage of standard solutions for significant periods of time.

Several attempts may be necessary to find the best internal standard for a given analysis, but the effort is worthwhile if highest precision is needed.

7.9.5 Standardization Summary

In all three methods of standardization, standards and samples are chromatographed and the standards are known but the samples are unknown. Peak sizes can then be determined for both. The difference in the three methods is in the second piece of information needed to relate the standard to the sample:

1. In internal normalization this relationship is that in both the standard and the unknown, the analyzed peaks total 100%.
2. In external standardization this relationship is the accurately known amounts of standard and unknown actually injected into the chromatograph.
3. In internal standardization this relationship is the accurately known amount of different material added to an accurately known amount of the standard and unknown.

The errors associated with standardization have been discussed throughout the above section, but should be summarized:

1. Standard purity and known standards must be checked and not assumed.
2. Linearity of response versus absolute amount injected must be confirmed for each different sample type and for each different set of chromatographic operating conditions. This linearity cannot be assumed. Nonlinearity may result from column overload, detector overload, or adsorption problems.
3. Proper attention to good analytical practices is important but most especially as it regards proper "blanking" of solvents, syringes, and all sample handling equipment. The high sensitivity for small amounts of material in most detector systems increases the importance of cleanliness.

7.10 QUANTITATIVE ERROR

7.10.1 General Discussion

Attention has already been given to the errors associated with peak size measurement and standardization. There are many other places in the chromatographic process where errors enter into quantitative analytical GC. Detailed analysis of most of these error sources is not possible, especially in the confines of this chapter, but they should be and are mentioned and briefly discussed. Most of the error sources are generally obvious; it may indeed seem even ridiculous that some have to be mentioned. However, the mere fact that they are obvious tends to slowly place them in the overlooked category. One has to be constantly reminded of these errors until the consideration of them becomes habitual with each problem. These general errors can be grouped into two categories: (1) the general area of sampling, involving problems of getting the sample from where it is into the gas chromatograph, and (2) the gas chromatographic system itself.

7.10.2 Sampling Techniques

The methods used to obtain samples and physically transport them to the gas chromatograph are really no different for GC than for any analytical technique. However, since GC has the inherent capability to do trace analysis, it becomes even more critical to observe the best analytical sampling techniques. Some major areas of concern are obvious.

The sample taken must be the sample that one wants to analyze. Since very little sample is required for gas chromatographic analysis, it is very easy to take a small sample that stands a good chance of not being representative of the environment to be analyzed. Small differences in homogeneity, or lack thereof, become quite apparent on two small samples supposedly taken from the same bulk sample.

Problems of adsorption, evaporation, and reaction of samples following the sampling procedure, prior to analysis, must be considered. The discussion regarding storage and handling of gas and liquid standards under external normalization certainly applies even more to the unknown samples. Time between sampling and analysis must be kept to a minimum. In addition, this time element should be checked with standards to ensure that samples do not change with time, or to at least define the extent of the error if no other solution is possible.

Containers for sampling, and indeed all sampling equipment, must be checked to determine the contribution to error. This becomes especially important if the sample must undergo some processing prior to the analysis. This processing may be extraction, preliminary cleanup by column chromatography or, even chemical reaction such as esterification. All of these steps must be proved in a given system or known to ensure either quantitative sample handling or the reproducibility of the processing. It is not sufficient to assume that "since Joe Blow at Podunk obtained 82.3% efficiency in the methyl esterification of adipic acid 3 years ago," the same efficiency is valid for a procedure that attempts to duplicate Joe Blow's procedure today. Reaction or extraction efficiencies must be reestablished.

7.10.3 Sample Introduction

As mentioned previously, when a known sample size is required, as in the external standardization technique, the measurement of that sample size will generally be the limiting factor in the analysis. However, improper sample injection can introduce into the analysis errors other than those pertaining to sample size. Thus it will be beneficial to examine the various methods of sample injection and both types of error associated with them.

Syringe Injection. The use of a syringe is by far the most common mode of sample introduction into the chromatograph. Today there are a number of excellent syringes on the market designed for GC. The most common syringe in use today for liquids has 10 μL of total volume. With the current greater use of smaller-diameter and lower-loaded columns, coupled with better and more sensitive detectors, sample sizes continue to decrease. Generally, liquid samples of about 1-μL are used. In a sample of this size, a component of interest should be less than 1% of the injected sample. For concentrated samples, this means sample dilution with a compatible solvent. An error can be introduced here if the solvent contains impurities that have the same retention time as any component of interest or if it contains even some of the same material. As with any of solvents in GC, the solvent has to be "blanked" before it is used.

The use of a 10-μL syringe to deliver a 1-μL volume has a certain error associated with the accuracy to which the syringe markings can be read and the plunger set. This uncertainty alone can contribute a 2–5% error in a 1-μL volume. Many users of gas chromatography are acquainted with the problem of injecting a volatile liquid into a hot injection port of a gas chromatograph. The error associated with this phenomenon outweighs the reading error without use of the proper technique. The basic problem is this: With the syringe properly loaded to the 1-μL mark, the amount of liquid contained in the syringe is that of 1 μL in the barrel plus the amount in the needle. When the liquid is injected, the 1 μL enters into the chromatograph, but any of the liquid remaining in the needle after injection and prior to withdrawal also evaporates. This may be the entire volume in the needle, which will be approximately 0.8 μL. The actual volume in the needle can be determined by loading a syringe with a liquid, running the plunger to zero, wiping the droplet off the needle, slowly drawing the plunger back until the liquid–air interface can be seen in the barrel, and then measuring the liquid slug in volume on the syringe. Knowledge of this total holdup on a given syringe can permit one to measure the amount actually injected. If the needle volume is 0.8 μL and the plunger is set at 1.0 μL, the total liquid in the syringe is 1.8 μL. Following the injection the plunger is withdrawn and the amount of liquid remaining in the needle measured. If this now is 0.3 μL, an amount of 1.5 μL was injected, a 1.0 μL by actual injection and 0.5 μL by evaporation from the needle.

There are two problems here. First, four syringe readings are needed (plunger and liquid–air interface, each on initial and final syringe loading), thus giving rise to two reading errors. The second error is worse in that its magnitude cannot be known with certainty. That is, the amount that is evaporated from the needle may not (and generally is not) representative of the true sample concentration due to selective evaporation of the more volatile components of the sample. A technique used to overcome this selective evaporation is to draw some pure solvent into the syringe (say, 1.5 μL), then about 1 μL of air, about 1 μL of sample, and finally about 1.5 μL of air. The sample slug is then measured in the barrel between the two liquid–air interfaces (two syringe readings). When this material is injected, only pure solvent is left in the needle and the amount that evaporates is not important. All the measured sample volume will be injected.

Another solution to liquid injection is the use of a 1-μL total-volume syringe. This syringe uses the internal volume of the needle for the sample volume. The plunger is a fine wire extending the full length of the needle. The volume readout is actually accomplished on a glass barrel with an indicator inside the barrel much the same as any other syringe. However, the actual liquid held in the syringe is in the needle only. Initially these syringes were not leaktight. For some time now, however, with improvements in design, they have been performing satisfactorily. The accuracy of a 1-μL injection is generally within 1% with the use of these syringes, but these syringes are more expensive.

Finally, proper handling technique is very important, especially wiping the outside of the needle and the droplet at the tip of the needle prior to injection. Any residual liquid on the outside of the needle will be caught in the septum puncture and will slowly enter the column. This produces broad tailing, especially of the solvent, making separations difficult as well as introducing an unknown amount of sample. On the other hand, liquid in the needle can be removed by the capillary action of the wiping towel.

All the preceding points regarding liquid injection should be considered even with the use of a standard technique that does not require an accurate volume to be known. Selective evaporation cannot be tolerated even with the internal standard method. The size measurement errors obvious from the preceding discussion certainly point to the substantial advantage of the internal standard technique for accurate analysis.

There are reasonably good syringes available today for injection of gas samples. Generally, gas samples are in the range of 1 mL in size. These syringes have a very snug-fitting Teflon plunger, allowing a gastight seal between the plunger and the barrel. A gas sample of known volume can be injected into the gas chromatograph with this technique. Even though there is a small (<1%) reading error with these syringes, there is a different form of sample injection error. Common pressures at the head of columns of average length at optimum flows will be approximately 30 psig or 3 atm absolute. When the syringe is inserted into the chromatograph through the septum, initially the carrier gas rushes through the needle into the syringe volume until the pressure in the syringe is 3 atm. Even the smallest leak between barrel and plunger will cause a significant sample loss as a result of this increased pressure, and only erratic results are any indication of this.

Next, the plunger is depressed, injecting a mixture of carrier gas and sample. The residual volume of the syringe, which may be 0.1–0.2 mL in a 1-mL syringe, was at atmospheric pressure when the sample was measured but is now at 3 atm. If the carrier had completely and homogeneously mixed with the sample at 3 atm of pressure prior to sample injection, no error would be introduced. However, this actually never happens, and the extent of mixing and the error introduced by the lack of mixing are not known. Assuming no mixing and a 0.2-mL residual syringe volume, the sample remaining would be 0.2 mL at 3 atm, or 0.6 mL at atmospheric pressure. Since the syringe contained 1.2 mL originally, only 0.6 mL was introduced instead of 1 mL. This is a 40% error on sample size!! This is certainly the worst case, but the magnitude indicates that the level of error is significant even with a perfectly functioning syringe. Gas work with syringes should be carried out with a syringe and needle that have a minimum residual volume.

The leak around the plunger and barrel in gastight syringes was mentioned earlier. A tight, stiff-acting plunger is necessary but not sufficient for a gastight seal. If 99% of the seal is tight, the entire sample can still be lost out of the 1% of the seal that does leak. A tight plunger can give rise to another error. If the needle plugs as a result of particulates in the gas sample, septum coring, or whatever, no sample will enter the syringe and the gas chromatograph. Many "detector malfunctions" have been corrected by syringe needle replacement. A stiff plunger makes a plugged needle difficult to notice. One final comment on gastight syringes. A large number of these have replaceable needles using a standard Luer fitting. This is very convenient for economical needle replacement if the needle becomes burred, bent, broke, or plugged. However, most glass-to-metal Luer fittings will leak gas at 3 atm of pressure. This is a leak source not normally considered but should be the first check placed on a new gastight syringe. One solution is the availability of plastic (Kel-F or equivalent) Luer fittings on syringes and plastic hubs on needles. This combination can greatly reduce leakage problems.

Before concluding our discussion on syringes, we should mention septum problems. The septum is the necessary evil through which a syringe injects samples. Practically every chromatographer has been plagued at one time or another by septum problems. However, it is surprising that these problems are as few as they are considering the function of a septum. A septum must seal, gastight, pressures up to several atmospheres. It must often seal against helium (the worst case next to hydrogen). It must do this at high temperatures (≤300°C). And finally it must, under these conditions, maintain a seal during and reseal following repeated piercing in virtually the same place.

The first problem with septa is leakage. Leaks in the system completely destroy the ability to quantify. Some instruments have "head pressure" gauges that can indicate a leak in the system. In others, simply a change in the chromatogram must be used to detect a leak. In general, retention times will increase. Septa should be changed on a routine basis (daily) to avoid loss of valuable time and samples. Septum bleed will cause noise and sometimes drift in isothermal operations. In addition to these, in programmed temperature, septum bleed will produce extraneous peaks not dissimilar from those in an impure solvent. Both sensitivity and quantification may suffer. When bleed is a problem, high-temperature, low-bleed, septa should be used even though they are considerably more expensive.

If the septum container on the chromatograph is tightened too much (in an effort to ensure leak-free operation), the septum may extrude. This makes it more difficult to pierce with the needle (resulting in bent needles) and invariably results in coring the septum with the core inside the needle. This means a new needle or a new syringe if the needle is permanent. Extruded septa are subject to pieces of the septum breaking off on the chromatographic side, causing increased restriction and possible plugging of carrier flow. It can also cause severe adsorption of sample components on the septum material, making quantification impossible. Tender, loving care of septa with attention to the problems associated with them will provide the chromatographer with peace of mind and much more reliable and expedient quantitative results.

Gas Sampling Valve. With all the problems associated with syringe injection of gas samples, it is not surprising that a more accurate way of injecting gas samples has been found. This system makes use of a gas sampling valve. There are a number of these valves on the market using either rotary or push–pull actuation. Interchangeable volumes are standard. A schematic for a rotary valve is shown in Figure 7.19. In the load position, the volume of the valve is connected to the "in" and "out" load ports. In use, the sample is pulled through the valve by a pump, squeeze bulb, or even a syringe used in the suction mode. If the gas is under pressure, it is allowed to flow through the valve. Sufficient volume of gas is needed to ensure that the "loop" or valve volume contains the sample to be analyzed. For a 1-mL sample loop volume, generally 10 mL of gas is a sufficient flush. The valve is then rotated to the inject position. This action places whatever is in the valve loop into the carrier-gas flow, where it is carried directly to the column for separation. The biggest error, namely, that of volume, is now fixed. If standards are run with the same fixed volume as the sample, the actual volume need not be known with a high degree of accuracy since it is the same for both standard and unknown and will cancel out of the calculation. However, two other parameters must also be held constant in the use of the gas-sampling valve to ensure that the same amount of sample is injected of standard and unknown: temperature and pressure. Either a 3°C difference or a 7-Torr difference will cause a 1% change in the amount of gas sample. The practical solution of these variables is simply to run standards and unknowns as close in time as possible such that these parameters do not vary significantly. If the samples are hot, such as stack gases, it may be necessary to maintain all sample lines and the gas sampling

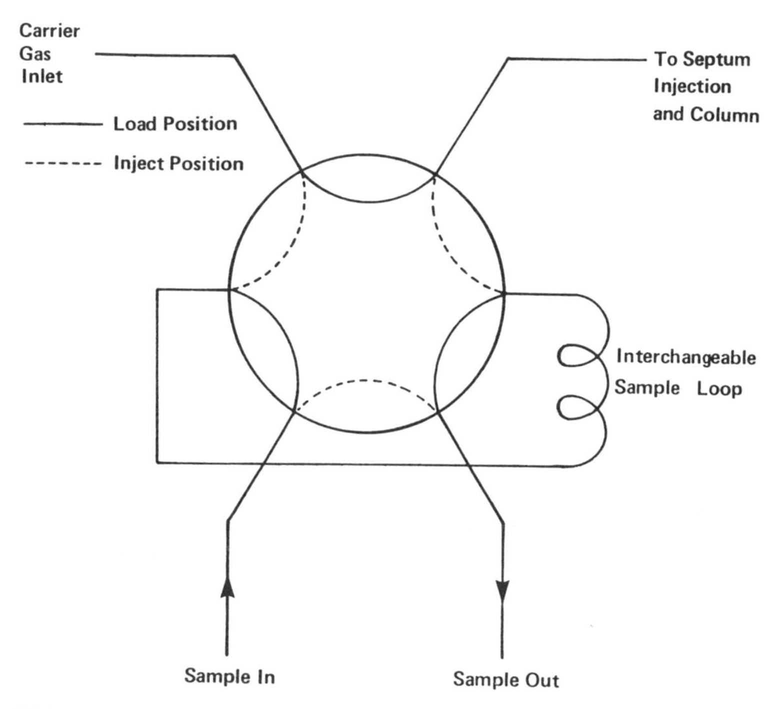

FIGURE 7.19 Flow schematic of a typical rotary gas sampling valve.

valve at an elevated temperature. Obviously, standards must be sampled at the same elevated temperature. If the sample is under reduced pressure, the pressure is then usually measured by use of a manometer to provide proper correction, or, preferably, to permit the standard to bc handled at the same pressure.

It may be tempting to increase the loop volume to increase the amount of sample for trace analysis. Before this is done, the system should be examined. If $\frac{1}{8}$-in.-diameter columns are being used, a reasonable flowrate is 30 mL/min at atmospheric pressure. But if the pressure at the head of the column is at 3 atm (which is not unreasonable), the volumetric flow at the head and through the sample loop is only 10 mL/min, or 6 s/mL. If the loop volume is increased to 5 mL, it will take 30 s to sweep the sample onto the column. Thus no peak can be any narrower than 30 s. A large-volume loop can completely destroy the separating efficiency of the chromatographic process. Again, as with any analytical problem, a common-sense, logical examination of the whole picture will pinpoint problem areas.

Miscellaneous Sampling Devices. There are a number of specialized ways of introducing the sample into the chromatograph and these are generally designed for a specific problem. A liquid-sampling valve is valuable for high-pressure liquid samples and preserves the sample integrity. These valves work the same in principle as the gas-sampling valve, except that the much smaller volume is held internally in the valve as opposed to an external loop.

To increase the precision of the amount of sample injected, a known weight of sample, solid or liquid, can be introduced through the use of a sealed indium tube. When the tube is placed in the hot injection port, the tube melts, releasing the total sample into the chromatograph. The injection port is generally modified in such a way as to allow insertion of a tube through a gaslock system. The same approach has also been used with sealed glass tubes that are crushed in the injection port.

All of these different techniques have some problems of their own but can be quite precise in terms of amounts of sample actually injected. Since their use is not general, but for rather specific sampling problems, no further discussion is warranted in this chapter.

7.10.4 Gas Chromatographic System Errors

Most of the problems associated with the processing of the sample through the column and then its detection are basically covered in specific chapters of this book. However, some areas deserve special mention as they relate to quantitative analysis.

The major concern is that the character of the sample is not changed in the injection port, the column, or the detector prior to its actual detection. Thermal decomposition, catalyzed or thermal reaction, and adsorption of part or all of the sample will contribute to error in the analysis. Problems such as these may be determined by using the chromatograph itself first to detect possible problems by unexpected results and then confirmation of the problem by variation of the actual operating parameters of the chromatograph.

Adsorption problems are generally indicated by failure of the calibration curve to pass through the origin, and in some cases by nonlinearity of the curve. A change of the column may be the answer. Perhaps increased temperature will reduce the problem to a workable level. Even though it is not desirable, some adsorption can be tolerated and still give quantitative results, but frequent recalibration is critical.

Sometimes thermal decomposition and reaction can be shown by variation of injection port temperature, and possibly column temperature. The only real solution is to operate at as low a temperature as possible and perhaps use "on-column" injection. Low-loaded columns sometimes help. Use of glass columns and glass injection port liners often relieve the problem of unwanted thermal degradation and may help in some cases. However, in all cases the precision and accuracy of the quantitative analysis will be affected until a solution is found or a decision is made to "live with it."

Detector errors are basically concerned with the time constant of the detector and its linearity. The time constant certainly can affect the peak height on narrow, sharp peaks, and this may or may not show up as nonlinearity. Assuming a good detector system, the basic linearity concern is with overload. This points to the necessity of initially establishing a calibration curve and assuring its linearity over the entire range of the samples. Extrapolation is dangerous.

All manual methods of quantifying peak size make use of the recorder tracing. Some consideration has already been given to the peak height and width as determined by recorder chart width and chart speed. Both should be maximized for the size measurement technique used. In addition, the recorder may have a limiting time constant as far as response to rapid peaks are concerned. This possibly should be considered with the detector when time constant problems are suspected.

In summary, GC is an excellent analytical tool for quantitative analysis. However, common sense must be used in handling problems, and the entire system should be understood. The best technique should be used to standardize and for sample handling. The "weakest link" concept is no more pronounced that it is in quantitative GC.

REFERENCES

1. A. Wehrli and E. Kovats, *Helv. Chim. Acta*, **42**, 2709 (1959).
2. T. Gornerova, *Petrochemia 1983*, **23**, 172 (1983).
3. *Chlorinated Pesticides in Water*, AN 135 Analytical Instrument Development, Avondale, PA, 1977.
4. J. Burke and L. Giuffrida, *J. Assoc. Off. Agr. Chemists*, **47**, 326 (1964).
5. *Methods for Organochlorine Pesticides in Industrial Effluents*, MDQARL, Environmental Protection Agency, Cincinnati, OH, 28 November 1973.
6. R. J. Laub and J. H. Purnell, *Anal. Chem.*, **48**, 799 (1976).
7. R. J. Laub and J. H. Purnell, *Anal. Chem.*, **48**, 1720, (1976).
8. J. Takacs, C. Szita, and G. Tarjan, *J. Chromatogr.*, **56**, 1 (1971).
9. J. Takacs, Z. Talas, I. Bemath, G. Czako, and A. Fisher, *J. Chromatogr.*, **67**, 203 (1972).
10. M. Beroza and M. C. Bowman, *Anal. Chem.*, **37**, 291 (1965).
11. M. Beroza and M. C. Bowman, *J. Assoc. Off. Agr. Chemists*, **48**, 358 (1965).
12. M. C. Bowman and M. Beroza, *J. Assoc. Off. Agr. Chemists*, **48**, 943 (1965).
13. M. Beroza and M. C. Bowman, *Anal. Chem.*, **38**, 837–841 (1966).
14. G. R. Umbreit, in *Theory and Application of Gas Chromatography in Industry and Medicine*, Grune & Stratton, New York, 1968, pp. 54–67.
15. C. McAuliffe, *Chem. Tech.*, **1**, 46 (January 1971).
16. A. B. Littlewood, *Chromatographia*, **1**(3/4), 133 (1968).
17. J. E. Hoff and E. D. Feit, *Anal. Chem.*, **35**, 1298 (1963).

18. V. Rezl and J. Janak, *Chromatogr.*, **81**, 233 (1973).
19. R. C. Crippen, *Identification of Organic Compounds With the Aid of Gas Chromatography*, McGraw-Hill, New York, 1973.
20. Operating Manual, Pro-Tek, E-40430, G-AA/G-BB "Organic Vapor Air Monitoring Badges," E. I. du Pont de Nemours & Co., Wilmington, DE 19898.
21. H. P. Burchfield and E. E. Storrs, *J. Chromatogr. Sci.*, **13**, 205 (1975).
22. M. Beroza and R. A. Goad, in L. S. Ettre and A. Zlatkis (Eds.), *The Practice of Gas Chromatography*, Interscience, New York, 1967, pp. 488–492.
23. J. C. Cavagnoland and W. R. Betker, in L. S. Ettre and A. Zlatkis (Eds.), *The Practice of Gas Chromatography*, Interscience, New York, 1967, 72–106.
24. K. Hammarstrand and E. J. Bonelli, *Derivative Formation in Gas Chromatography*, Varian Aerograph 6/68: A-1006, Varian Aerograph, Walnut Creek, CA 94598, 1968.
25. Pierce Chemical Company, *Handbook of Silylation*, Handbook GPA-3 (1970), Rockford, IL 61105.
26. D. L. Ball, W. E. Harris, and H. W. Habgood, *J. Gas Chromatogr.*, **5**, 613 (1967).
27. J. M. Gill and H. W. Habgood, *J. Gas Chromatogr.*, **5**, 595 (1967).
28. D. L. Ball, W. E. Harris, and H. W. Habgood, *Anal. Chem.*, **40**, 1113 (1968).
29. D. L. Ball, W. E. Harris, and H. W. Habgood, *Anal. Chem.*, **40**, 129 (1968).
30. L. Condal-Busch, *J. Chem. Ed.*, **41**, A235 (1964).
31. H. M. McNair and E. J. Bonelli, *Basic Gas Chromatography*, Varian Aerograph, Walnut Creek, CA, 1968, p. 156.
32. L. Mikkelsen and I. Davidson, Paper 260, Pittsburgh Conference on Analytical Chemistry and Applied Spectroscopy, 4 March, 1971.
33. L. Mikkelsen, *J. Gas Chromatogr.*, **5**, 601 (1967).
34. F. Baumann and F. Tao, *J. Chromatogr.*, **5**, 621 (1967).
35. H. A. Hancock, L. A. Dahm, and J. F. Muldoon, *J. Chromatogr. Sci.*, **8**, 57 (1970).
36. J. D. Hettinger, J. R. Hubbard, J. M. Gill, and L. A. Miller, *J. Chromatogr. Sci.*, **9**, 710 (1971).
37. N. Dyson, *Chromatographic Integration Methods*, The Royal Society of Chemistry, Cambridge, UK, 1990.
38. B. E. Saltzman, *The Industrial Environment: Its Evaluation and Control*, U. S. Department of Health, Education and Welfare, National Institute for Occupational Safety and Health, Washington, DC, 1973, Chapter 12.
39. D. T. Mage, *J. Air Pollut. Control Assoc.*, **23**, 970 (1973).
40. H. A. Grieco and W. M. Hans, *Industr. Res.*, **39** (March 1974).
41. F. J. Schuette, *Atmos. Environ.*, **1**, 515 (1967).
42. R. A. Baker and C. R. Doerr, *Int. J. Air Pollut.* **2**, 142 (1959).
43. W. K. Wilson and J. Bluchberg, *Ind. Eng. Chem.*, **50**, 1705 (1958).
44. G. O. Nelson and K. S. Griggs, *Rev. Sci. Instrum.*, **39**, 927 (1968).
45. R. M. Nash and J. R. Lynch, *J. Am. Ind. Hygiene Assoc.*, **32**, 802 (1971).
46. A. P. Altschuller and I. R. Cohen, *Anal. Chem.*, **32**, 802 (1960).

47. J. M. H. Fortuin, *Anal. Chim. Acta*, **15**, 521 (1956).
48. J. M. McKelvey and J. E. Hollscher, *Anal. Chem.*, **29**, 123 (1957).
49. A. E. O'Keefe and G. C. Ortman, *Anal. Chem.*, **38**, 760 (1966).
50. E. M. Emery, *J. Gas Chromatogr.*, **5**, 596 (1967).
51. P. C. Barbato, G. R. Umbreit, and R. J. Leibrand, *Internal Standard Technique for Quantitative Gas Chromatographic Analysis*, Applications Lab Report 1005, August 1966, Hewlett–Packard, Avondale, PA 19311.

CHAPTER EIGHT

Inlet Systems In Gas Chromatography

MATTHEW S. KLEE

Hewlett-Packard Company

Modern Practice of Gas Chromatography, Third Edition. Edited by Robert L. Grob.
ISBN 0-471-59700-7

8.1 OVERVIEW

Inlets provide an interface through which samples are introduced into the chromatographic column. Proper choice of inlet is essential to acquiring the highest quality data. There are several choices in inlets, each of which is designed for a specific purpose and, therefore, each of which has certain limitations. Improper choice or use of an inlet reduces system performance and quality of data, and may even prevent important sample components from reaching the column.

Selection of an inlet is most often an exercise in defining the best set of compromises for achieving analysis goals. When using packed columns for analyses, the decision is easier because packed columns have much higher capacities and carrier-gas flowrates than capillary columns, which minimizes inlet requirements and complexity. As a result, manageable sample volumes can be injected into a simple inlet, from which sample vapors are cleared very quickly, yielding narrow initial peak widths. Capillary columns, on the other hand, must rely on special inlet designs and/or focusing phenomena to reduce the sample load on the column and to ensure narrow solute bandwidths.

There are two general classes of inlets: (1) hot, or vaporizing inlets, and (2) cold, or cool inlets. These terms refer to the temperature of the inlet when sample is injected. Hot inlets evaporate the sample upon injection. Cold inlets accommodate condensed sample, then heat up to completely evaporate the sample during the run.

8.1.1 Vaporizing Inlets

In general, hot or vaporizing inlets are held at some elevated temperature during injection so that the solvent and sample components evaporate quickly as they migrate through the inlet to the column entrance. Hot inlets typically

- Are used at constant elevated temperature
- Have relatively large internal volumes
- Help protect the column
- Can cause sample discrimination
- Can cause sample degradation
- Can overload (cause flashback) and lose sample

When using a vaporizing inlet, its temperature is usually set hot enough to "instantaneously" evaporate or "vaporize" the sample upon injection. However, the actual time to complete evaporation depends on the difference between the boiling point of the solvent and the inlet temperature, the volume injected, and the inlet configuration. Fast evaporation requires temperatures 50 or more degrees above the boiling point of the solvent and major components; the higher the temperature, the faster the evaporation. However, if the inlet temperature is set too high, then sample degradation, discrimination, and flashback increase.

When a liquid evaporates, its volume increases two to three orders of magnitude, depending on its composition and the temperature and pressure of the inlet. Vaporizing inlets usually have internal volumes between 250 and 100 μL to contain sample vapors, but this can easily be exceeded by injecting too much sample. When the volume of the vaporized sample exceeds that of the inlet, called "flashback," excess vapors can reach the top of the inlet where they condense on the cool septum, backdiffuse through carrier-gas lines and condense on cool surfaces, and/or exit through the septum purge line. Flashback leads to loss of sample, ghosting on subsequent injections, sample discrimination, and poorer precision of data.

Table 8.1 lists the theoretical gas volumes of several common solvents as a function of temperature and pressure of the inlet. Referring to this chart will help in selecting appropriate inlet conditions and injection volumes. Even if the theoretical volume of the liner is not exceeded, the most volatile sample components may still diffuse to the top of the inlet if residence time in the inlet is too long (flowrate and/or pressure are too low), or if temperature is too high. Vaporizing inlets help protect the column by retaining nonvolatile sample components and by reducing the potential of stationary-phase stripping resulting from contact of condensed solvents with the stationary phase.

8.1.2 Discrimination and Degradation

Discrimination is the selective loss of some sample components during injection. Discrimination can happen in two places during the injection process: (1) in the syringe needle and (2) within the inlet. *Needle discrimination* occurs when heavy sample components remain in the needle as lighter components evaporate out, resulting in increasing amounts of early eluting components relative to late eluting components. This is illustrated in Figure 8.1. The higher the inlet temperature relative to the boiling point of the solvent, the wider the boiling point range of solutes in the sample; and the longer the syringe sits in the injection port, the more pronounced is needle discrimination. Recently, commercial autoinjectors have become available that inject very rapidly and have very short residence times in the inlet ($<$500 ms). These significantly reduce needle discrimination by completing the injection process before significant evaporation occurs from the syringe needle (Figure 8.2).

TABLE 8.1 Approximate Gas Volumes of Common Solvents (μL per 1 μL) Injected at Several Inlet Temperatures and Pressures

Solvent	Head Pressure (kPa)	Inlet Temperature 100°C	200°C	300°C
Ethyl acetate	69 (10 psig)	186	236	286
	138 (20 psig)	133	168	203
	207 (30 psig)	103	131	158
Hexane	69	139	177	214
	138	100	126	152
	207	77	98	119
Isooctane	69	110	140	170
	138	79	99	121
	207	61	78	94
Methanol	69	460	584	708
	138	329	415	503
	207	255	324	392
Methylene chloride	69	284	360	437
	138	203	256	311
	207	158	200	242
Methyl t-butyl ether	69	153	194	235
	138	109	138	167
	207	85	108	130
Water	69	1010	1282	1554
	138	722	910	1105
	207	561	710	860

Inlet discrimination is more complex than needle discrimination in that it can be caused by several factors and is manifested in several ways: inlet discrimination can reduce the relative amounts of low boilers, high boilers, or both. High residence times of vapors in the inlet, high inlet temperatures, low inlet pressures, and overloading of an inlet by excessive sample cause the most volatile sample vapors to selectively migrate to the top of the inlet where they can be condensed on cool inlet surfaces (e.g., septum, carrier-gas lines) or be vented with the septum purge flow. This reduces the amount of low-boiling components relative to the higher-boiling components, as illustrated in Figure 8.3.

High-boiling sample components can selectively adsorb on inlet surfaces due to low inlet temperatures or interaction with active surfaces. This reduces the amount of high-boiling components relative to lower-boiling components (high-end discrimination in Figure 8.3). Under some injection conditions, discrimination of both low-boiling and high-boiling components is possible.

Sample degradation can be caused by pyrolysis into smaller molecules,

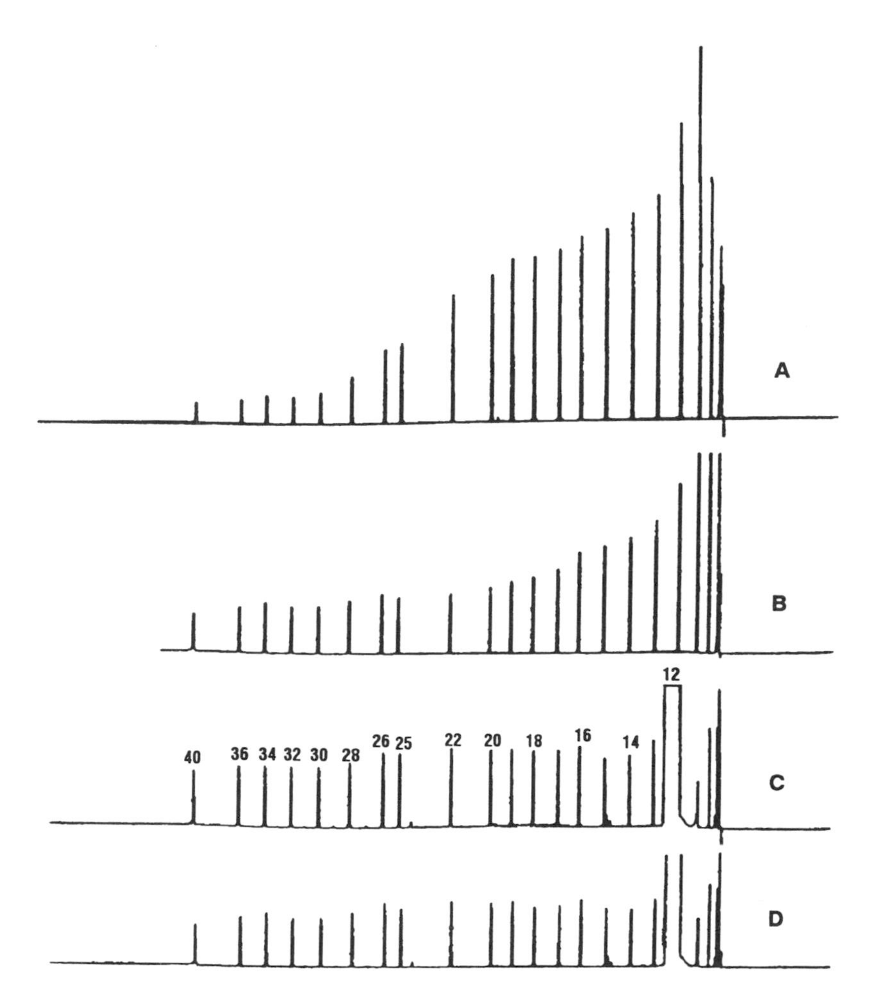

FIGURE 8.1 Needle discrimination as a function of solvent and inlet temperature: (A) *n*-octane solvent, 310°C inlet; (B) *n*-octane solvent, 210°C inlet; (C) *n*-dodecane solvent, 310°C inlet; (D) *n*-dodecane solvent, 210°C inlet temperature. The higher-boiling solvents and lower inlet temperatures yield less needle discrimination. The boiling point of dodecane is 216°C and that of octane is 126°C. (Reproduced with permission from Reference 1.)

molecular rearrangement, and/or by chemical changes in functional groups. Generally, the more polar the molecule, the less stable it is. High inlet temperatures, long residence times, and active inlet surfaces aggravate sample decomposition. Figure 8.4 is a chromatogram showing endrin and DDT degradation. In active inlets and with retention gaps that have not been deactivated, endrin rearranges to endrin aldehyde and endrin ketone, and DDT rearranges to DDE and DDD.

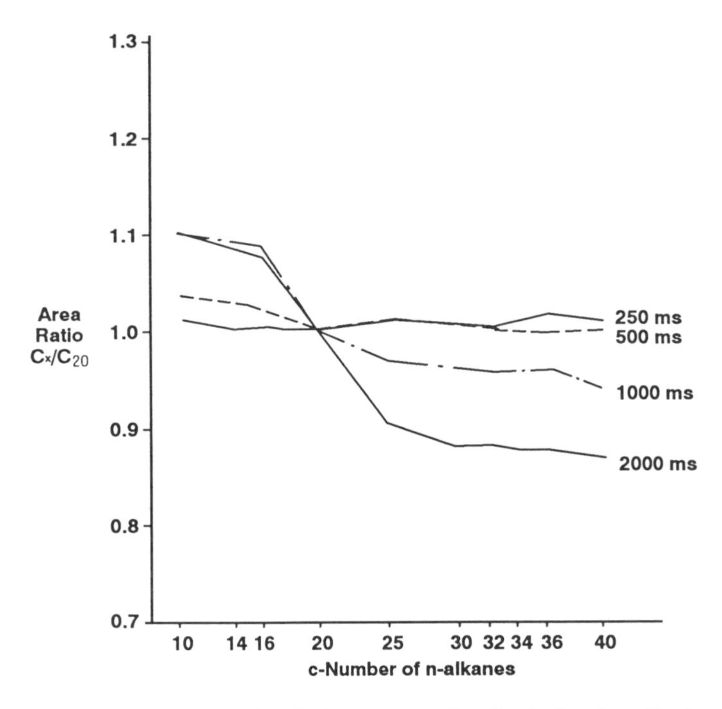

FIGURE 8.2 Effect of needle dwell time on needle discrimination. Peak areas are normalized to that of C_{20}. The longer the syringe needle stays in a heated inlet, the more pronounced is the discrimination. Some commercial autosamplers have a total dwell time in the inlet of less than 500 ms. Instrument conditions: 10 m × 530 μm HP-1, 60°C for 2 min, 15°C/min to 310°C for 5 min, packed-column inlet, 1 μL injection of *n*-alkanes in hexane. (Reproduced with permission from Reference 2.)

To minimize degradation and discrimination with vaporizing inlets, it is critical to set the temperature appropriately for the sample at hand. The temperature must be set high enough to quickly evaporate the sample and minimize discrimination against high boilers, but low enough to minimize decomposition, overload, and discrimination of low boilers. To optimize temperature of vaporizing inlets, screen the sample at an intermediate inlet temperature (e.g., 250°C), then repeat at significantly higher and lower temperatures (e.g., ±100°C) to see if there is any change in relative peak areas or peak shapes. If there is no change, then inlet temperature is not an important consideration, and lower temperatures should be chosen to prolong septum lifetime and minimize stationary-phase degradation at the head of the column. If the chromatograms change significantly, then further experiments should be performed to determine if the differences are due to positive influences (decreased discrimination or degradation) or negative influences (increased degradation or discrimination).

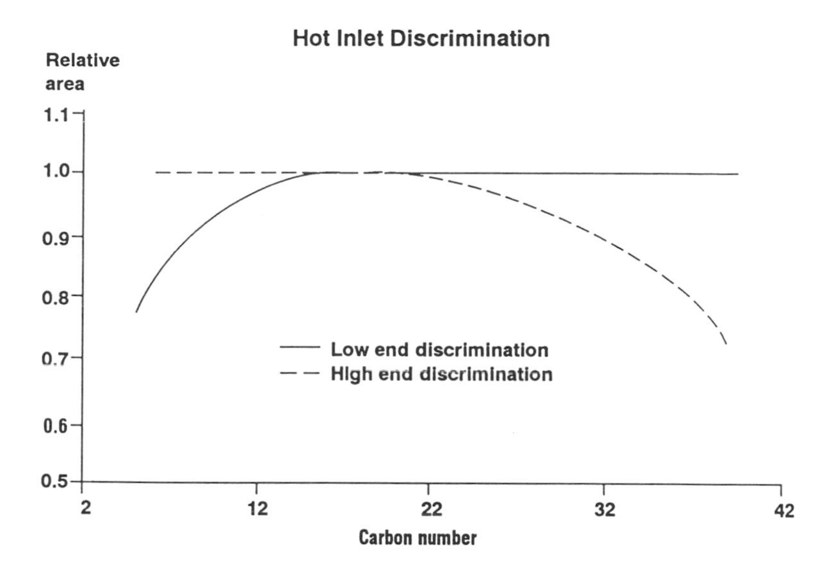

FIGURE 8.3 The relationship of two forms of inlet discrimination. *Low-end discrimination*, or discrimination against the most volatile sample components, is usually caused by overload of the inlet by sample vapors, allowing the most volatile components to vent with the septum purge. *High-end discrimination*, or discrimination against the least volatile compounds, is caused by many variables, including adsorption on cool or active surfaces in the inlet, insufficient inlet temperature, and unmixed flow paths in the inlet.

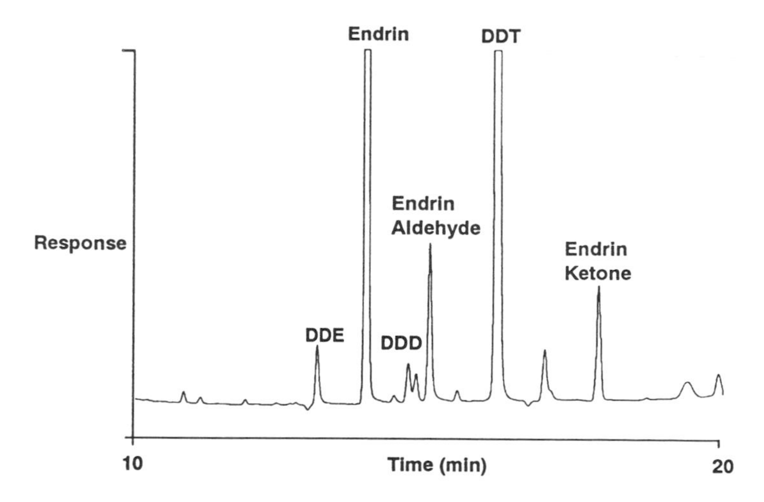

FIGURE 8.4 Standard of 50 ppb endrin and DDT in isooctane. Activity in the inlet causes decomposition of endrin into endrin aldehyde and ketone, and DDT into DDD and DDE. Peaks resulting from inlet-induced degradation have normal peak widths. (Reproduced with permission from Reference 3.)

8.1.3 Cool Inlets

Inlets that do not evaporate the sample upon injection are called cool, cold, or nonvaporizing inlets. In this case the inlet, and sometimes column temperature, is held below the boiling point of the solvent during injection, after which the temperature is raised to evaporate the sample and initiate chromatography. Examples of cool inlets include cool on-column and programmed temperature vaporizing (PTV) inlets. In general, cool inlets

- Minimize sample degradation
- Minimize sample discrimination
- Yield high sensitivity
- Have high reproducibility
- Increase chance of column overload

Since all of the sample enters the inlet at low temperature, thermal degradation is minimized, needle and inlet discrimination are minimized, and sensitivity is maximized. Table 8.2 lists the most common inlets and their general characteristics.

8.1.4 Focusing Phenomena

Sample focusing happens whenever the leading edge of a solute band is slowed relative to the trailing edge, causing the band to narrow. This

TABLE 8.2 Comparison of Inlets for Gas Chromatography

Inlet type	Merits	Debits
Packed-column direct	Vaporizing inlet, usually used for packed columns, easiest, least expensive	Not compatible with most capillary columns, possible sample degradation and needle discrimination
Split/splitless split mode	Universal capillary column inlet, most rugged, ideal for auxiliary samplers, column protecting, simple	Possible sample discrimination and sample degradation
Split/splitless splitless mode	High-sensitivity capillary column inlet, column protecting	Highest sample decomposition and flashback, possible discrimination, optimization necessary, requires most maintenance
Cool on-column	Deposits sample capillary directly into column, most inert, least discrimination, high sensitivity	Higher column degradation and overload possible, difficult to use with columns of $<250\ \mu m$ i.d., most complicated automation
PTV	Cool injection, most flexibility, low sample degradation, high sensitivity	Complex, most expensive, many parameters to optimize, possible loss of volatiles

happens when there is a sharp gradient in either the phase ratio or temperature along the column. Common names for sample focusing are thermal focusing, stationary-phase focusing, β focusing, and solvent focusing. During any temperature-programmed analysis, some focusing occurs as late eluting solutes (high k') trap at the head of the column because of their high affinity for the stationary phase at the initial temperature. Even if the bandwidth of the sample vapors exiting the inlet is large, solutes that elute at temperatures at least 50°C higher than the starting column temperature will focus at the head of the column. As the temperature is programmed during the run, these solutes will start to migrate and will elute as narrow peaks, often with peak widths less than the early eluting peaks (Figure 8.5).

When using vaporizing inlets, better focusing occurs with

- A larger difference between the initial oven temperature and the boiling point of the solvent
- A larger difference between the oven temperature and the elution temperature of the solute

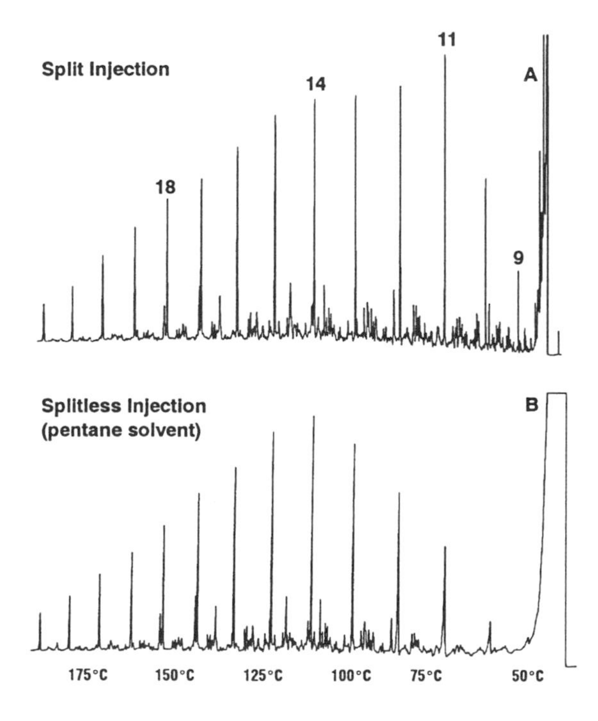

FIGURE 8.5 (A) Split injection yields narrow initial peaks and is being used as a reference. (B) Splitless injection yields wide initial peaks and also loss of peaks for minor components in the solvent tail. However, solutes that elute above 100°C were thermally focused and have peaks as narrow as with split injection. (Reproduced with permission from Reference 4.)

- A lower β (thicker stationary phase)

In other words, there is a combination of thermal and stationary-phase or β focusing. Stationary-phase or β focusing can happen even more dramatically when using retention gaps (see section 8.1.5).

Solvent focusing can narrow peaks in two ways, of which one is almost exclusively related to the splitless injection technique. This first form of solvent focusing happens when a vaporized sample migrates from a hot inlet to a cool column and is condensed. This requires that the column temperature be significantly lower than the boiling point of the solvent. As the solvent vapors enter the column, they cool and condense on the inner surface of the column, reducing the bandwidth several hundredfold (due to the phase change from gas to liquid) and form a "flooded zone."

The second form of solvent focusing happens as the temperature is programmed during the run. Solvent starts to evaporate from the inlet end of the flooded zone. Early eluting solutes remain dissolved in the condensed solvent, which is continuously decreasing in volume as it evaporates, concentrating them and narrowing their initial bandwidths even further. The solutes more volatile than the solvent elute before all of the solvent has evaporated, so they will not benefit fully from solvent focusing and will be broad relative to peaks after the solvent. As the last bit of solvent evaporates, the remaining solutes elute on the solvent tail as very narrow bands, as illustrated in Figure 8.6. The least volatile solutes remain behind,

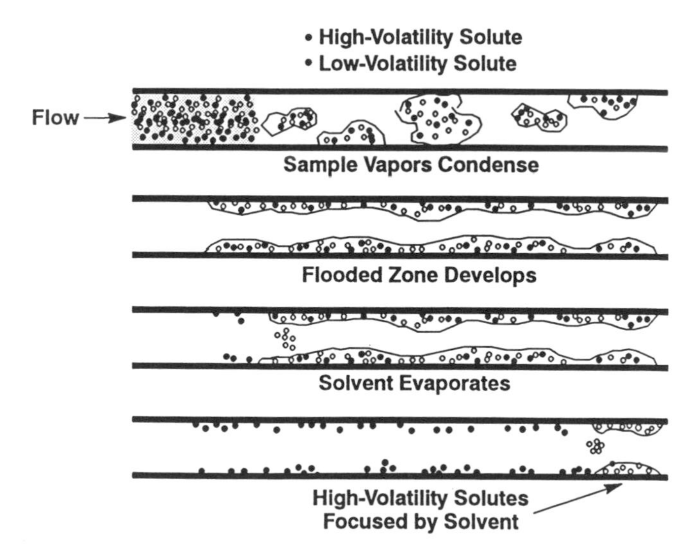

FIGURE 8.6 Solvent focusing occurs in two stages: (1) as the vaporized solvent recondenses from a gas to a liquid in a column held significantly below the boiling point of the solvent, and (2) as the solvent evaporates slowly after onset of a temperature ramp.

focused by stationary phase, and elute later during the temperature program.

Solvent focusing can be problematic if the solvent volume is too small or too large, or if the solvent is of different polarity than the stationary phase. Optimal solvent focusing occurs when the condensed solvent forms a narrow uniform film inside the column; however, this will happen only if sufficient solvent is injected and the oven temperature is cool enough to condense the solvent in the head of the column. The solvent must also have sufficient solubility in the stationary phase ("like unto like") to wet it properly and to minimize the length of the "flooded zone." If the volume is too large, or if the solvent and stationary phase do not have similar polarities, then the solvent will form droplets in the column that lead to distorted and sometimes split peaks. This can be effectively overcome by using retention gaps.

8.1.5 Retention Gaps

Retention gaps are uncoated but deactivated lengths of capillary tubing that are attached to the inlet of the analytical column. Retention gaps provide a volume to hold condensed sample directly after injection. Retention gaps solve several problems inherent in injection techniques that deposit condensed sample into the column. They reduce or eliminate the problems associated with injecting large sample volumes and solvents that are not compatible with the stationary phase. Condensed liquid spreads on the inner surface of the retention gap from approximately 25 cm/μL miscible solvent (similar polarity to the surface) to 2 m/μL immiscible solvent (often as isolated droplets or pools). As the oven temperature increases, the solvent evaporates from the inlet side of the flooded zone and travels at carrier-gas velocity to the analytical column. Volatile solutes focus by the solvent effect and eventually travel with the solvent vapors until they encounter the inlet of the analytical column, where they are focused due to the very large drop in β going from the uncoated retention gap to the analytical column. The less volatile solutes remain behind in the retention gap until they have sufficient vapor pressure during the temperature program, then they too travel at carrier-gas velocity to the analytical column and are focused. This process is illustrated in Figure 8.7.

Retention gaps can help protect the column from contaminants and reduce the impact that nonvolatile sample components have on subsequent analyses. Since the solutes are not retained in the retention gap with contaminants as long as they would be in a coated column, the sample is in contact with the contaminants for much less time and there is less chance of degradation and peak shape distortion.

Retention gaps are usually attached to the analytical column with a glass "butt" connector, although some chromatographers prefer metal connectors with vespel/graphite ferrules, especially if using temperature programs that

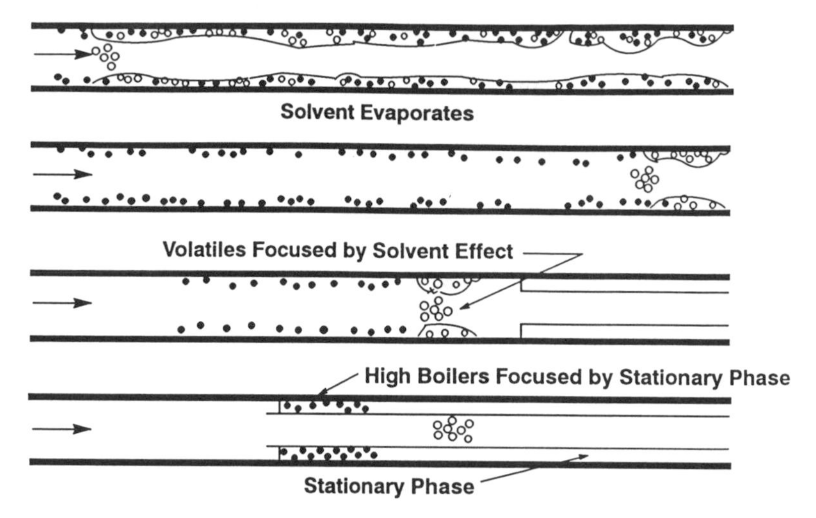

FIGURE 8.7 Retention gaps help improve bandwidths through both solvent focusing and β focusing as solutes pass from the retention gap, which has an insignificant film thickness, to the analytical column, which has significant film thickness.

end above 300°C. Glass connectors are more inert and can couple tubing of very different diameters; however, some people find it more difficult to make a long-lasting leak–free connection than with the metal unions. The most important factors in establishing a leak-free union with glass butt connectors are to ensure that the ends of the tubing are cut square, and then to force the ends into the union hard enough so that the polyimide coating on the outside of the capillary tubing forms a seal with the connector, effectively gluing the tubing in place. Forcing the tubing in too hard can fracture the end of the tubing, after which the union is best discarded and a new one used with freshly trimmed tubing ends.

8.1.6 Septa

Most inlets are sealed with a septum that is punctured by a syringe needle during injection. An ideal septum would have no volatile impurities to bleed into the inlet, would allow an infinite number of injections without leaking, and would seal at very high inlet pressures and at all inlet temperatures. No septum material has all these attributes, and the objective in choosing a septum is to find the best compromise for the inlet and experimental conditions to be used.

Septa are usually mixtures of thermally stable polymers, mostly silicones. Septa are often categorized by the expected temperature of the inlet on which they will be used. The higher the inlet temperature, the higher molecular weight and the less flexible the polymer used in the septum. Some septa have a Teflon coating to minimize adsorption of sample components, but this lowers their upper usable temperature limit.

Pieces of a septum that fall into the inlet can increase sample degradation, discrimination, peak tailing, and ghosting. Eventually, all septa will fail, allowing carrier gas to leak out. This increases sample loss and discrimination and reduces reproducibility of data. It is always a good idea to establish a routine of regular inlet maintenance that includes replacing the septum before it starts to leak.

Several alternatives to septa have recently become available. These replace the seal provided by a polymeric septum by a device that allows nondestructive needle penetration. Two of these "septumless heads" are illustrated in Figure 8.8.

8.2 PACKED-COLUMN DIRECT INLETS

8.2.1 Description

Packed-column direct inlets are very simple in design, as illustrated in Figure 8.9. Injection is usually made through a septum at one end of the inlet and sample vapors flow with carrier gas to the other end where the column is connected. A heater and heating block hold the inlet at constant elevated temperature. Carrier-gas flow is controlled by a mass-flow controller. Some improvements in this basic design include a removable glass liner for easier inlet maintenance and lower activity, more uniform heating from top to bottom of the inlet, a septum purge line to decrease ghosting by compounds bleeding off the septum, and electronic mass-flow control.

Mass-flow controllers are used because the resistance of packed columns changes considerably as temperature increases, decreasing flowrate significantly at constant head pressure. Mass-flow controllers increase their outlet pressure in response to increasing resistance, thereby maintaining constant mass flow. These mechanical devices are reliable from 10 to 90% of their nominal upper limit, which depends on the pressure supplied to them (and the resistance of the internal flow restrictor, if used). The response of mass-flow controllers has a significant temperature coefficient, so the most reproducible inlet designs enclose the flow controllers in thermostatted or insulated compartments.

Most packed column direct inlets allow packed glass columns to be inserted through the inlet, usually in place of a liner or insert (Figure 8.10). This minimizes dead volumes in connections and minimizes decomposition of labile compounds and tailing of polar sample components. Typical internal volumes of packed column inlets are only a few hundred microliters. Since packed-column direct inlets are designed for use with packed columns with flowrates typically about 30 mL/min, the linear velocity toward the column is relatively fast and there is little concern of overloading

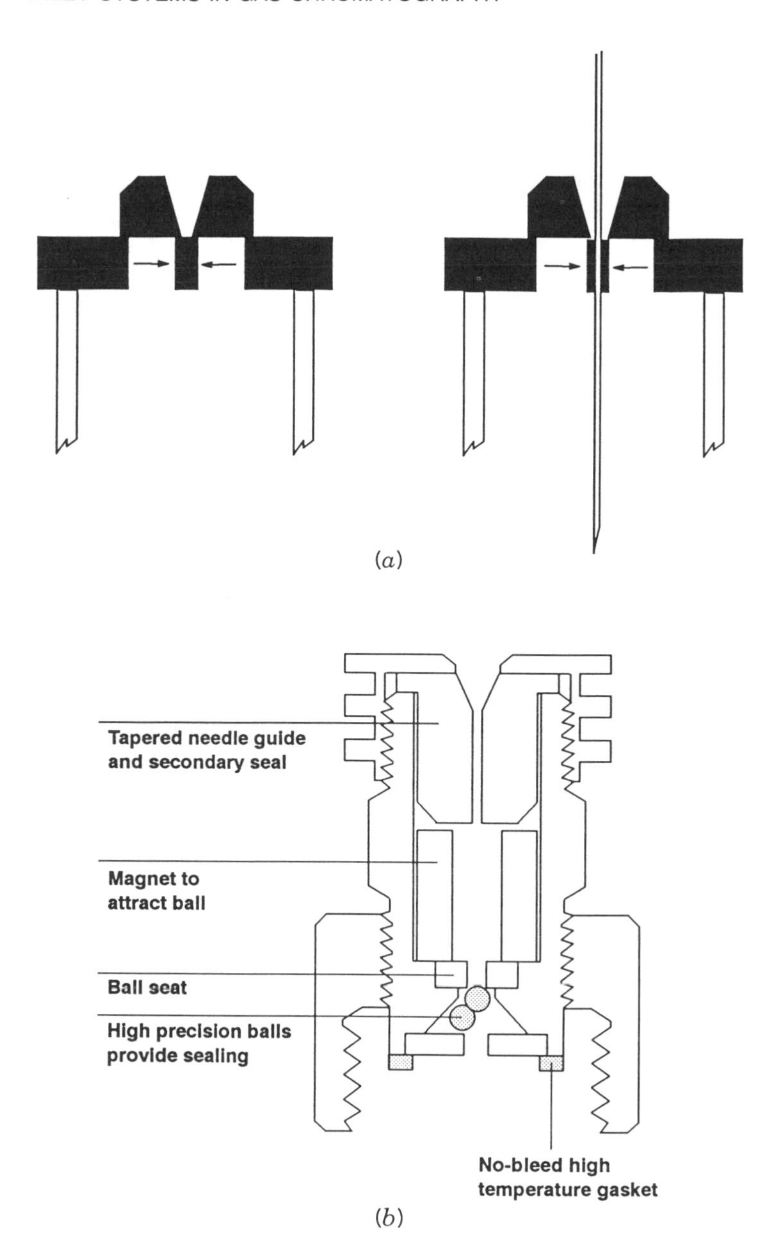

FIGURE 8.8 Alternatives to polymeric septa. (a) A "duckbill valve" uses an elastomeric flap that is held closed by carrier-gas head pressure, but that allows easy, nondestructive penetration by a syringe. (b) A "Jade valve" uses a magnet and two steel balls to seal pressure in the inlet. The balls are displaced during syringe introduction and the tapered needle guide provides the backup seal with the syringe needle. To prevent excessive leakage, the needle guide must be matched closely with the syringe needle.

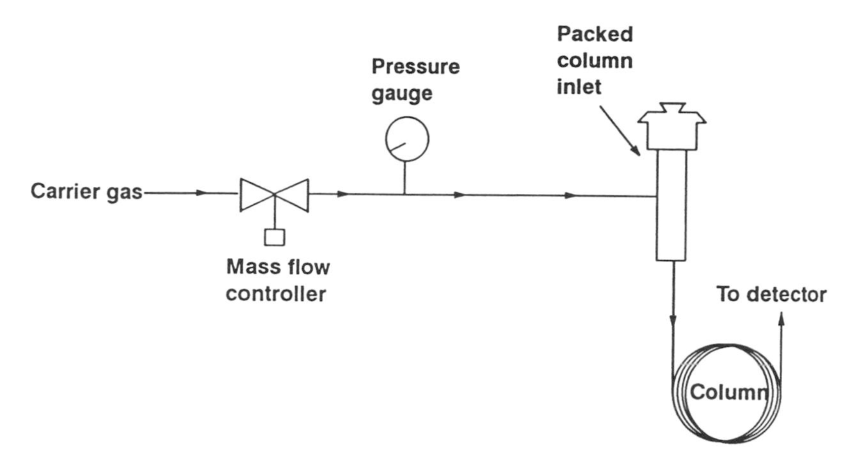

FIGURE 8.9 Flow schematic for a packed-column direct inlet.

the inlet with sample vapors or of overly broad initial peak widths; most inlet volumes are cleared in less than one second.

Even though packed-column inlets are designed for packed stainless steel or glass columns, it is possible to use them with large-bore capillary columns ($>500\ \mu m$ i.d.) as an inexpensive way to get some of the benefits of capillary columns without having to invest in new hardware. Since these columns can be used at carrier-gas flowrates approaching those of packed columns, reasonably good results can be achieved—a 10 m large-bore capillary at 20 mL/min He flow can yield higher plate count than a 2 m packed column. However, the results will not be as good as if a true capillary inlet were used with a flowrate closer to optimum (3 mL/min).

8.2.2 Typical Use

For direct injection, the temperature of the inlet is usually set >50°C above the boiling point of the sample solvent or equal to the elution temperature of the latest eluting component. If the temperature is set too high, then the probabilities of sample decomposition, needle discrimination, and flashback of sample vapors are increased. To determine if an appropriate temperature has been selected, repeat the analysis at a higher and lower inlet temperatures and compare the results to the original analysis.

The most inert configuration to use is with glass columns that have been inserted into the inlet. This is important when analyzing labile or polar solutes. When using packed columns in this manner it is important that the

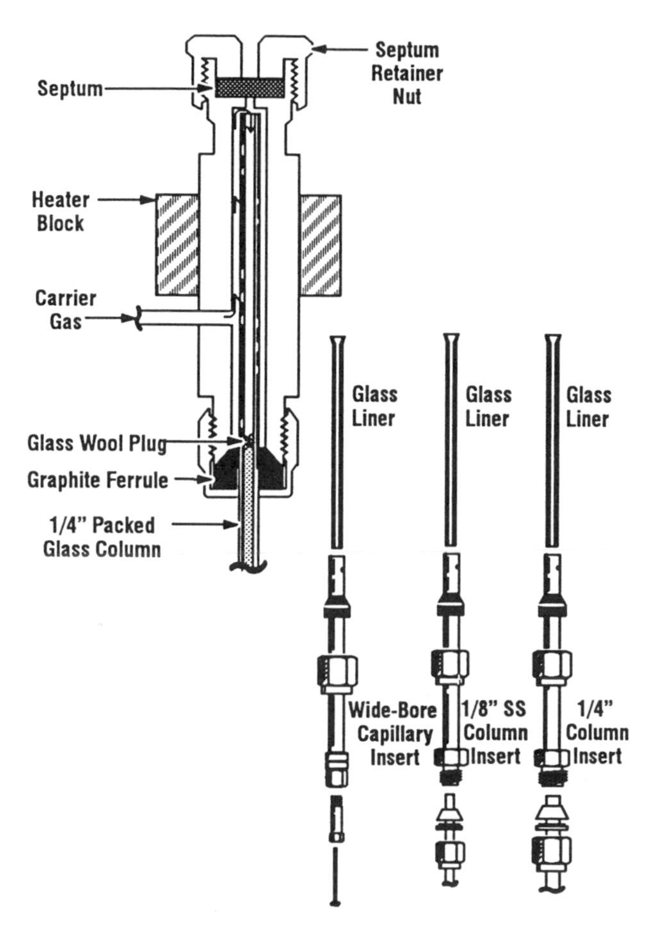

FIGURE 8.10 Packed-column direct inlet with replaceable inserts. Glass columns can be inserted directly into the inlet for maximum inertness. Inserts are used for other types and sizes of columns, from packed to wide-bore capillary columns.

head of the column be empty so that there is sufficient space for the expanding sample vapors after injection.

8.2.3 Disadvantages

When glass columns are inserted into the inlet, nonvolatile sample components and septum fragments can be more problematic to remove than if a removable glass liner were used. Removal of foreign objects from the head of the column often disturbs the packing material and requires replacing the amount that is removed with new packing. Packed-column inlets are hot, so they can degrade labile sample components, especially if the sample comes in contact with metal surfaces. Trying to use capillary columns of $<500\ \mu m$

i.d. with packed-column inlets is not usually successful, especially if the sample contains early eluting components, because band broadening becomes significant. In addition, mechanical mass-flow controllers do not work well at the low flowrates typical of most capillary columns, so reproducibility decreases drastically.

8.3 SPLIT/SPLITLESS INLET; SPLIT MODE

The most popular capillary-column inlet is the split/splitless inlet. It can be used in a split mode to reduce the amount of sample reaching the column and to produce very narrow initial bandwidths. It can also be used in a "splitless" mode to maximize sensitivity.

8.3.1 Description

The split injection mode, or "sample splitting," is the most popular and forgiving injection technique for capillary column analyses. Split inlets are vaporizing inlets; the sample is evaporated in the inlet, flows down the liner, and is split between the column and a split vent.

Figure 8.11 shows a schematic of a popular split/splitless inlet design. A

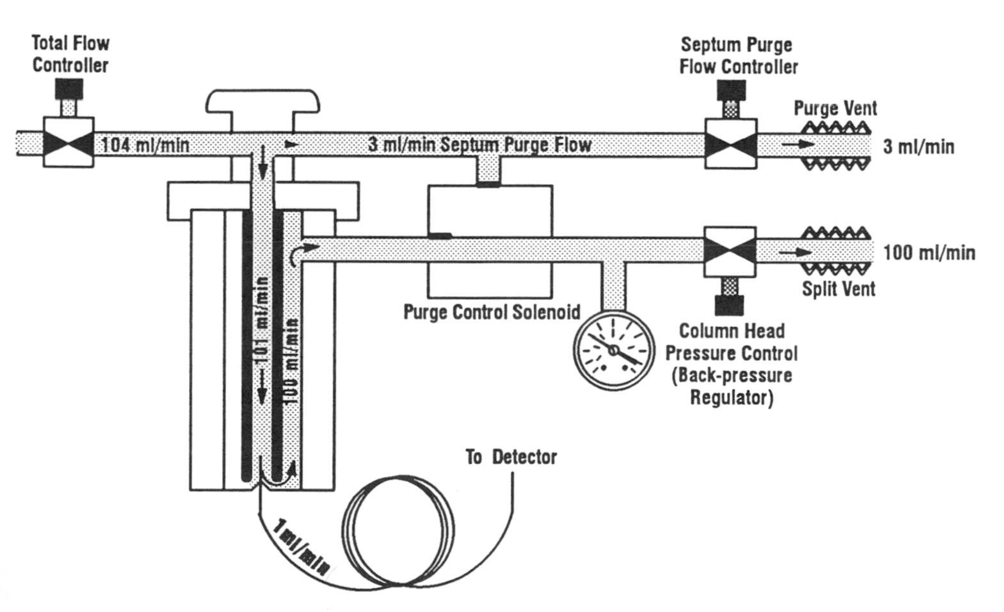

FIGURE 8.11 Schematic of a typical split inlet. Total flow (controlled by a mass-flow controller) is divided between septum purge, split vent, and column streams. A majority of the sample vapors flowing down the inlet are vented, and a minority enters the column in proportion to the relative flowrates of the two paths.

mass-flow controller provides a constant flow to the inlet and acts as a flow limiter should a leak develop in the flow path. This flow is divided into three flow paths: the septum purge, the column, and the split vent. The septum purge helps minimize ghosting from previous injections and septum bleed. A backpressure regulator in the split vent line controls column head pressure. In recent designs, the backpressure regulator has been replaced by an electronic version, providing the user with the capability of electronic control and programming of column head pressure.

The liner is an important part of the split inlet and can limit its ultimate performance if not selected properly. For proper splitting of the sample, the sample must be completely vaporized and quantitatively transported past the entrance of the column, or sample discrimination can occur. The internal volume of split inlet liners is typically 1 mL to contain the rapidly expanding sample vapors. To ensure optimal performance, liners are usually packed with some silanized glass wool or have an integrated glass component so that the stream of injected liquid hits a surface in the inlet to help sample evaporation and minimize discrimination.

8.3.2 Typical Use

An important parameter to know and record when using a split inlet is the "split ratio." In Figure 8.11 the total flow is 104 mL/min, the septum purge flow is 3 mL/min, the column flow is 1 mL/min, and the split vent flow is 100 mL/min. The split ratio is the ratio of the split vent flow to the column flow, and in this example is 100:1. This says that for every sample injected, 100 parts is vented and 1 part enters the column. The column flow is measured at the detector and the split vent flow is measured at the split vent port, which is usually accessible on the outside of the gas chromatograph.

Split inlets are used for all types of capillary columns, including columns of very small inner diameters because they generate very narrow initial peak widths (the flowrate through the inlet is very high) and sample vapors are cleared quickly. They are also excellent choices for general analysis and for screening new samples, because the split ratio can be changed very quickly in response to changes in samples or analysis goals. By adjusting the total mass flow, shown in Figure 8.11, flowrate to the inlet changes. The septum purge flow is controlled by a flow controller and therefore does not change as the total flow to the inlet is changed. Head pressure does not change because it is controlled by the backpressure regulator, so flow to the column is unchanged. Therefore, as the total flow to the inlet is changed, only the flow through the split vent changes, changing the split ratio directly.

Split ratio can be changed, measured, and documented in a matter of minutes, and is considerably quicker and simpler to do than making up new samples. For analyses of major components and when using small-bore capillary columns, a high split ratio is appropriate. For analyses of dilute solutions and when using large-bore capillary columns, a low split ratio is required. Split inlets are indispensable for analyses involving very small diameter columns (≤100 μm i.d.), since the capacity of the columns is very low and they require very narrow initial peak widths. For capillary columns of 100 μm i.d., split ratios of 500:1 are typical.

Split inlets are ideal for interfacing auxiliary sampling devices like headspace, purge and trap samplers, and gas-sampling valves to the gas chromatograph. These devices require high flowrates to quickly transfer the gas samples from the samplers to the columns while maintaining narrow initial bandwidths. This is not an issue when using packed columns with 30 mL/min carrier-gas flowrates, because the transfer lines and connections are cleared quickly with little band broadening. But for capillary column flowrates of closer to 1 mL/min, significant band broadening would result and would be catastrophic to analysis of early eluting compounds that do not focus. By interfacing through a split inlet that has a higher total flowrate, samples can be transferred from the auxiliary sampling device to the column while maintaining narrow initial peak widths.

Split inlets can successfully be used for analysis of some thermally labile compounds because of the short residence time of vapors in the inlet. With higher split vent flowrates (>100 mL/min) the solutes are not exposed to the high inlet temperature for more than a few seconds. However, there is still a higher chance of degradation than if a cool injection technique were used.

8.3.3 Disadvantages

Split inlets have a high probability of suffering from sample discrimination. All forms of discrimination can be found: needle, high-end, and low-end discrimination. It is, therefore, important to select the appropriate liner (including packing material and location in the liner), temperature, injection technique (fast autoinjection is recommended), sample solvent, column installation, and split ratio. An initial starting point for these choices is usually given by the manufacturer of the instrument, but a complete method development should include investigating the influences of some or all of these variables on the results. Clearly, screening runs and qualitative analyses do not require as much optimization of inlet parameters as do validated methods, which require full documentation and tests of ruggedness.

To minimize degradation and discrimination, the inlet temperature should also be optimized. High temperatures aggravate discrimination, decomposition, and overload of the liner with sample vapors. Low temperatures reduce the evaporation rate of late eluters and increase discrimination against them. To investigate the influence of inlet temperature on a given sample and analysis, vary the temperature by at least 75°C above and below the original run, and repeat the analysis. Compare the results and adjust the inlet temperature accordingly.

8.4 SPLIT/SPLITLESS INLET; SPLITLESS MODE

8.4.1 Description

In the splitless mode of injection, the split vent of the split/splitless inlet is turned off by a solenoid valve during sampling. Splitless injection is used with capillary columns to maximize sensitivity of an analysis and is very similar to direct injection in that all, or nearly all, of the sample vapors enter the column. Due to the low flowrates of capillary columns, however, the transfer of sample vapors from the inlet to the column causes very broad initial bandwidths. This is compensated for in splitless injection by solvent focusing, stationary-phase focusing, and a delayed venting or "purge" of the inlet.

Figure 8.12a shows a typical split/splitless inlet during a sample injection. Liquid sample is injected into the liner, where it is vaporized and travels to the capillary column. The speed at which the sample vapors move from the inlet to the column depends on the column flowrate and the volume of the liner being used. The inlet purge, which is actually the split vent, is turned on after a delay to vent vapors that remain in the inlet (Figure 8.12b). During sample transfer, the column temperature is held significantly below the boiling point of the sample solvent, to take advantage of solvent focusing (Section 8.1.4). The oven temperature is then increased, evaporating the solvent in the column and initiating separation.

8.4.2 Typical Use

The original motivation for doing splitless injections was to maximize sensitivity and lower detection limits of capillary analyses. This results directly from transferring most of the injected sample to the column. The most prevalent use of splitless injection is for environmental analysis. Environmental analyses deal with trace solute concentrations and often involve dirty samples which might contaminate the column. Since splitless injection is a vaporizing technique, nonvolatile components are retained in the inlet liner.

With splitless injection, trace sample components that elute on the solvent tail can be much sharper than theory predicts and can be very sensitively analyzed by splitless injection. This is a direct result of solvent focusing as described in Section 8.1.4, and is illustrated in Figure 8.13.

All inlet and injection variables influence performance of splitless injection and the optimal purge delay time. These variables include

- Inlet temperature
- Column temperature
- Column head pressure
- Sample solvent

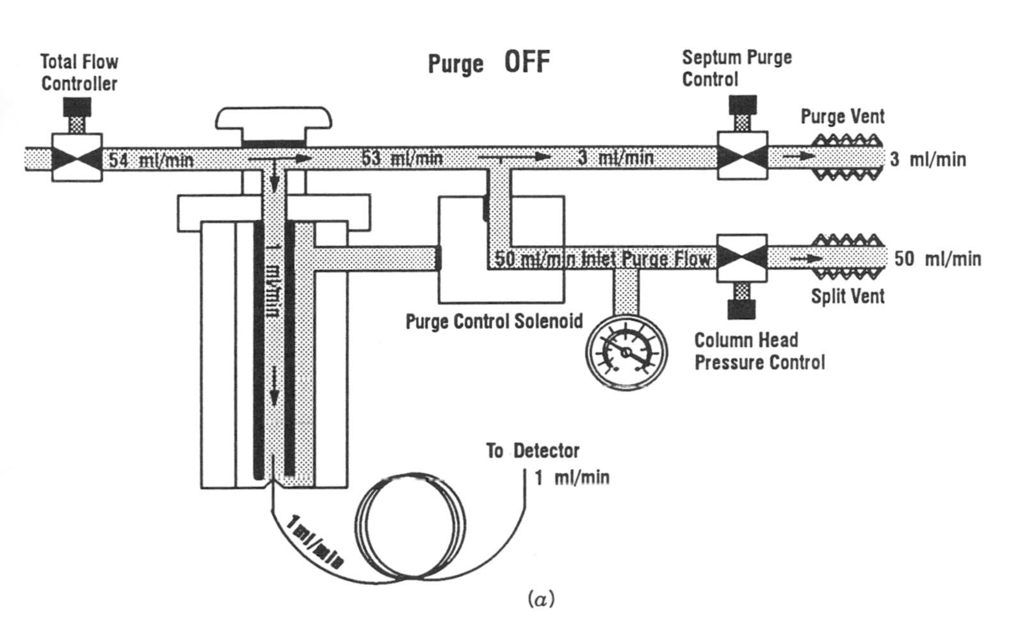

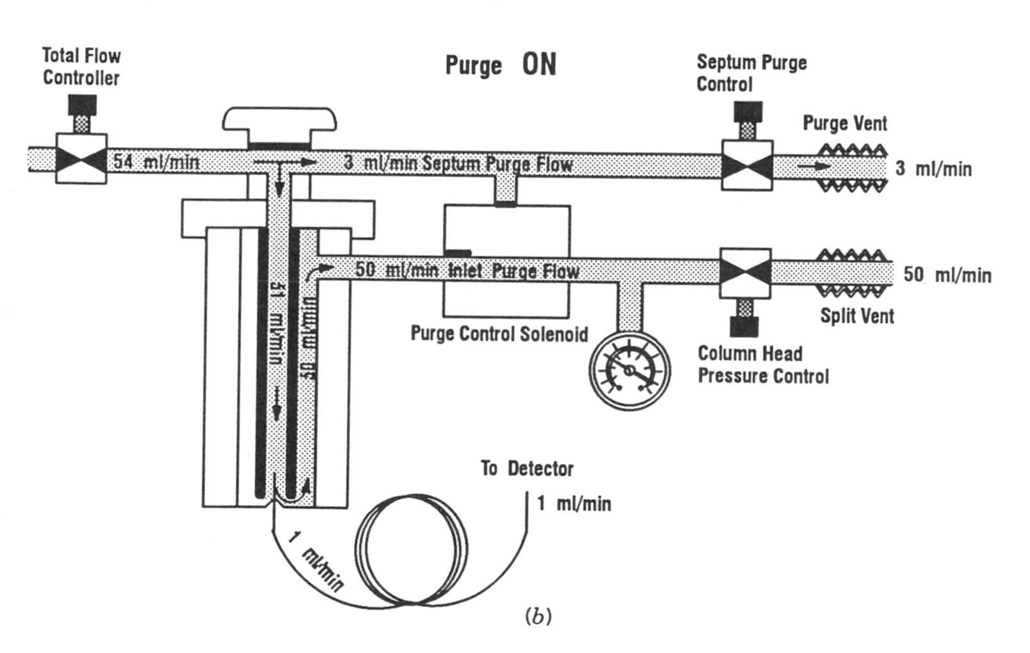

FIGURE 8.12 (a) Splitless inlet during injection. The inlet purge is turned off so that all of the sample vapors flowing down the inlet enter the column. (b) After a predetermined time, the split vent is opened to purge the last bit of sample vapors from the inlet, sharpening the tail of the solvent peak.

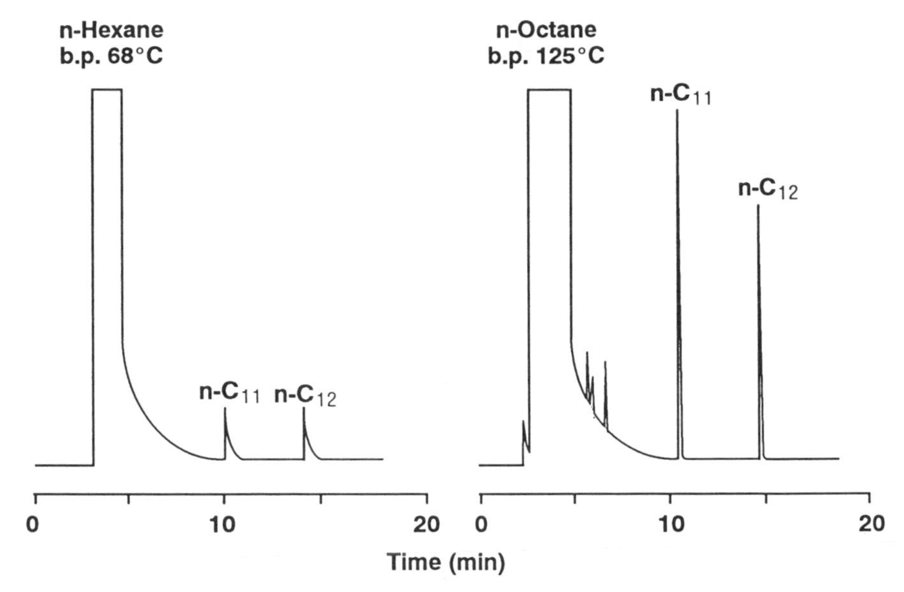

FIGURE 8.13 The solvent effect focuses peaks eluting immediately after the solvent. For proper focusing, the oven temperature must be below the boiling point of the solvent. With a starting oven temperature of 115°C, hexane does not recondense and there is no solvent focusing.

- Injection volume
- Injection speed
- Liner volume

To maximize sensitivity and minimize solvent tail, all of these variables should be considered. To accurately determine the "purge delay time", all of the other variables must be established first. If, for example, a different sample solvent were used, it would expand to a different volume in the liner and would take a different amount of time to be transferred to the column. If the head pressure is changed, both the volume of expanded gas and the column flowrate change, resulting in a shorter residence time of vapors in the inlet and a different optimal purge delay time.

In choosing an appropriate liner, one must first consider liner volume. Small liners take less time to clear and may yield narrower initial peak widths than large volume liners, but they cannot accommodate the sample volume that larger liners can, so less sample can be injected, analyses are less sensitive, and flashback is more probable. For most splitless analyses, a liner volume of approximately 1 mL is recommended (7).

Once a liner is chosen and initial choices have been made for all other inlet variables, the purge delay time can be optimized. The solvent peak is large and tailed with splitless injection because of column overload and slow

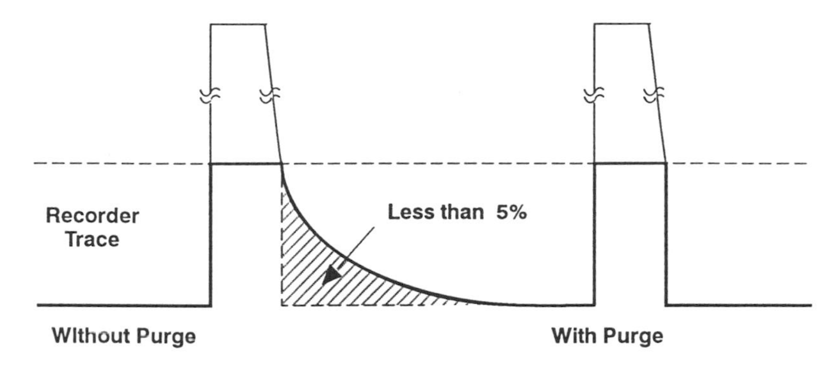

FIGURE 8.14 By turning on the inlet purge after 95% of the sample has entered the column, the solvent tail is sharpened considerably.

removal of vapors from the inlet. By venting the last 5% or so from the liner, the solvent tail is reduced considerably, facilitating detection and integration of compounds eluting just after the solvent (Figure 8.14). For maximum sensitivity, the inlet purge is delayed until at least 95% of the sample vapors have entered the column. Figure 8.12b shows the inlet being purged with the split vent solenoid valve open.

8.4.3 Optimizing Purge Delay Time

Optimization of split vent time requires repeating a few experiments, with all variables except purge delay time held constant. It is important to use the same solvent and injection volume for all standards and samples. First, a solution is made of one or more solutes that are highly retained relative to the solvent. These compounds can be the same as those expected to be in samples or can be surrogates, and they should bracket the elution times of all important sample components. Highly retained solutes are not influenced as much by the solvent because they are focused by thermal focusing, making optimizations easier. They also elute far away from the solvent tail, simplifying peak integration. In addition, it can take longer for the least volatile solutes to transfer from the inlet to the column than for more volatile solutes, so peak areas of late-eluting peaks indicate the maximum required purge delay for maximum sensitivity.

The test solution is injected with the purge vent off during the whole analysis. The areas for the late-eluting peaks are recorded and represent 100% of the sample reaching the column. The solvent peak will be broad and tailed due to overload and exponential dilution in the inlet. The experiment is then repeated with the purge delay time set between 60 and 90 s. The areas for the late peaks are compared to the original areas. If they are the same, then the experiment is repeated with the purge delay time set to 1/2 of the previous one, and so on until a smaller area is noticed. The vent time is fine tuned up or down until 95–99% of the original areas are obtained. At this point, 95–99% of the sample reaches the column, and the solvent tail should drop sharply to baseline. The progress of this process is illustrated in Figure 8.15.

If the solutes of interest are only late eluters, then optimization of the purge vent delay time is less critical, just as with the optimization test solution. In fact, it makes sense to have an excessively long purge delay in this case to maximize sensitivity and reproducibility by ensuring that all of the sample vapors reach the column.

8.4.4 Disadvantages

Because of all the negatives associated with splitless injections, selection of the splitless injection technique over others should probably be viewed as a last resort, when all other choices have been excluded (e.g., lack of alternative hardware or lack of choice because its use is dictated in a regulated method). Splitless injection has the most negative aspects of any injection technique. In addition to the need for optimizing purge delay time, splitless injection suffers from needle discrimination, inlet discrimination,

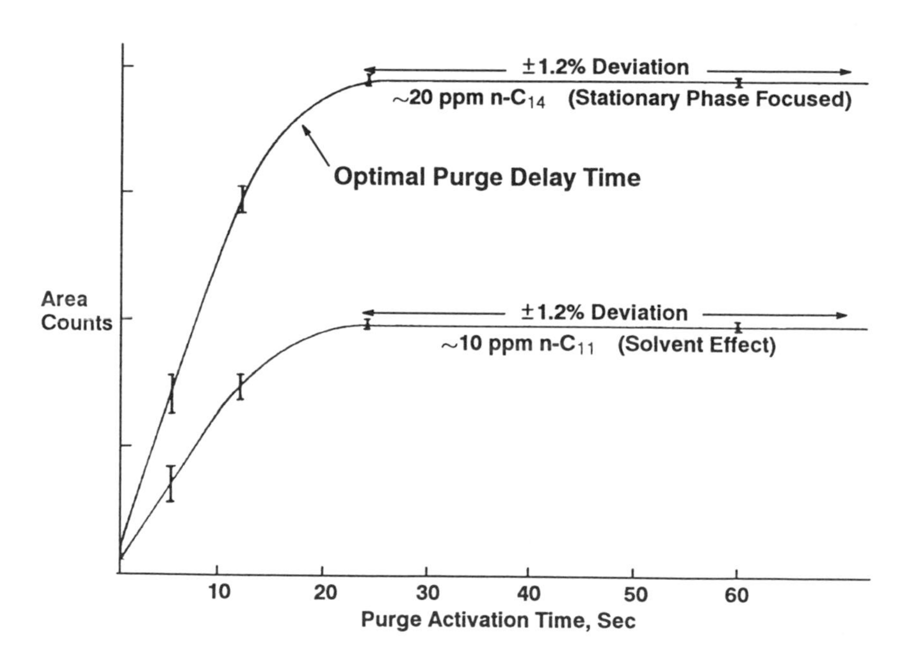

FIGURE 8.15 Optimizing splitless purge delay time requires altering the purge time on successive injections of a standard of late-eluting compounds. Optimal purge delay time corresponds to a transfer of from 95 to 99% of sample vapors. Instrument conditions: isooctane solvent, 1.3 mL manual injection, 16.5 m × 250 mm SE-54 capillary column, 80°C for 0.5 min to 170°C at 15°C/min.

broad initial bandwidths, solvent overload of the column, and the highest potential for sample degradation and flashback. Many of these inherent problems exist with splitless injection because the sample vapors stay in the hot liner for a long time.

To minimize the potential problems with splitless injection, the user should

- Use a deactivated liner packed with a small amount of deactivated glass wool.
- Use a solvent with relatively high boiling point.
- Inject as small a sample volume as necessary, for sensitivity and focusing, making sure it is consistent with the volume of the liner and injection speed of the sample.
- Adjust inlet temperature only as high as necessary to transfer the late eluting sample components to the column.
- Use as high an inlet pressure as practical for the column chosen.
- Use a deactivated retention gap.

Because of the very broad width of the solvent peak, it is difficult, if not impossible, to develop a good assay for compounds that elute before the solvent with splitless injections. These components do not benefit from solvent and stationary phase focusing mechanisms and remain broad and misshapen.

8.5 COOL ON-COLUMN INLET

8.5.1 Description

The cool on-column inlet is a capillary column inlet that allows direct deposition of liquid sample into the column. Both the inlet and the column are held substantially below the boiling point of the solvent during the injection, after which both temperatures are increased to vaporize sample and initiate chromatography. A typical design is shown in Figure 8.16. An elastomeric seal ("duckbill") is shown that can be used with fused-silica needles. Column head pressure helps force the flaps against the needle to form a seal during injection. A septum and nut are also shown and can be used only with stainless steel needles.

Cool on-column inlets typically have low thermal mass to facilitate fast heating at the start of the run and cooling before the next run. The temperature of the inlet should be controlled independently from the column temperature for highest reproducibility and flexibility.

Since the sample is deposited directly into the column without being evaporated first, cool on-column inlets can have the highest reproducibility

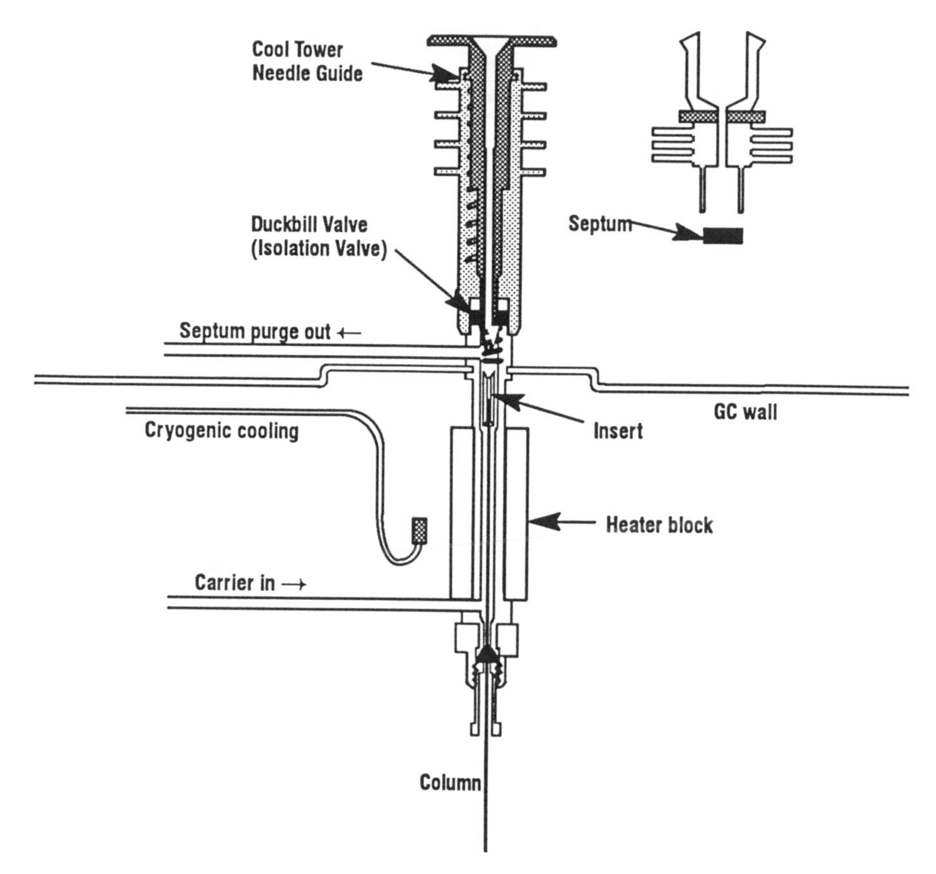

FIGURE 8.16 Schematic of a typical cool on-column inlet. The analytical column or retention gap extends through the cool inlet to the top. Sample is deposited directly into the column, after which the inlet is heated to initiate chromatography. A "duckbill valve" is used with fused-silica needles for manual injection. A thin septum is used for syringes with stainless steel needles and autosamplers.

and lowest discrimination and decomposition of any inlet. However, it is tricky to inject samples directly into small diameter capillary columns. Table 8.3 lists the sizes of syringe needles that fit into common capillary columns. Stainless steel needles are easier to use because they can puncture sample vial and inlet septa, and can be fed into the column through simple needle guides in the inlet. Autosamplers are available for injections using stainless steel needles into columns with internal diameters down to 0.25 mm, but the lifetime and reliability of such small-diameter needles is not as high as with larger-diameter needles. Clearly, the easiest way to do cool on-column injection is when using large-diameter columns or retention gaps (>500 μm i.d.), because more rugged needles, septa, and autosamplers can be used.

Stainless steel needles smaller than 33 gauge are not practical, so for

TABLE 8.3 Needle Sizes Required for Injection Into Common Capillary Column

Column i.d. (μm)	Needle Size (μm)	Needle Size (aug)	Needle Type	Autosampling Possible
530	360	28 aug	SS	Yes
320	240	32 aug	SS	Yes/limited
250	230	33 aug	SS, FS	Yes/most limited
200	170		FS	No
100	—		NA	No

Note. aug, gauge size; SS, stainless steel; FS, fused silica, NA, not commonly available (must use larger-bore retention gap).

columns of smaller than 250 μm i.d., needles made from small-diameter fused silica are used. These flexible needles cannot easily puncture common inlet septa, so novel methods must be used to allow access to the head of the column while still providing a pressure seal.

8.5.2 Typical Use

Cool on-column injection is useful for all types of analyses, but excels for analysis of labile compounds and samples with a large boiling-point range. Simulated distillation analyses, where the boiling points of sample components can range from <50 to over 500°C, is a classical example of when cool on-column inlet is preferred. Because it is so inert and virtually discrimination free, the cool on-column inlet is often a reference inlet by which one can compare results from other inlets when there is a question of discrimination or decomposition.

Since the entire sample is deposited into the column with cool on-column injection, analytical sensitivity is very high and detection limits are at least as good if not better than with splitless injection. Flashback is unusual with cool on-column injection because of the low initial temperatures. Loss of volatile sample components may occur with cool on-column inlets if they have a septum purge and if the initial inlet and column temperatures are not significantly below the boiling points of the most volatile sample components, so some inlet designs include options for cryogenic cooling. Cryogenic cooling also helps reduce the time required for the inlet to cool down before the next run.

8.5.3 Disadvantages

The main disadvantage of cool on-column injection is the sampling process itself, especially if manual techniques are employed. Handling of the small-diameter needles required for small-diameter capillary columns can be frustrating because of bending and the delicate manipulations required. If the inner diameter of capillary columns is not closely controlled by the manufacturer, one can occasionally get a column that is so small that the syringe needle cannot be inserted into it.

Since the condensed sample is injected into the column, cool on-column injection can suffer from solvent overload, peak splitting, premature degradation of the stationary phase, and contamination of the column from nonvolatile sample components. All of these limitations are effectively avoided by using a deactivated retention gap ahead of the column. Retention gaps of slightly larger inner diameter than the analytical column can be used to simplify injection as well. However, if the retention gap is too large relative to the column, the much slower linear velocity in the retention gap may prevent proper refocusing of the solute bands on the analytical column, and peak widths may be wider than expected.

As with splitless injection, sample components that are retained less than the solvent are difficult, if not impossible, to elute with good peak shape. Sample components eluting on the tail of the solvent will benefit from solvent focusing, but the solvent peak may be more tailed than with splitless injection due to the lack of an analog to the splitless inlet purge.

8.6 PROGRAMMABLE TEMPERATURE VAPORIZER INLET

8.6.1 Description

Programmable temperature vaporizing (PTV) inlets are a combination of split/splitless inlet and cool on-column inlet. They are cool injection inlets, temperature programmable, and have a split/purge vent and a timer/controller. PTV inlets are column protecting because nonvolatile sample components are deposited in the inlet and do not migrate to the column. Also, dirty samples tend to cause less problems with PTV injcctions than with hot injection techniques, so time between inlet maintenance is increased.

One design of a PTV is shown in Figure 8.17. During sample injection, the sample is deposited into a packed liner, which is held below the boiling point of the solvent. After a preset time, the inlet temperature is raised, evaporating the solvent and sample components. While sample evaporates, the split vent is electronically controlled to be on or off, or some sequential combination of the two.

If the split vent is left on after injection, then sample is split between the column and vent flow paths as would happen with standard split sampling, and the sample load to the column is reduced in proportion to the split ratio. Evaporation is gradual, however, so the sample is exposed to lower average temperature than with the classical split injection technique. If the split vent is off while the inlet temperature is raised to evaporate sample, then a splitless injection is made.

In a hybrid technique called "solvent vent mode," the split vent is held open to vent solvent vapors at an initial temperature just above the solvent

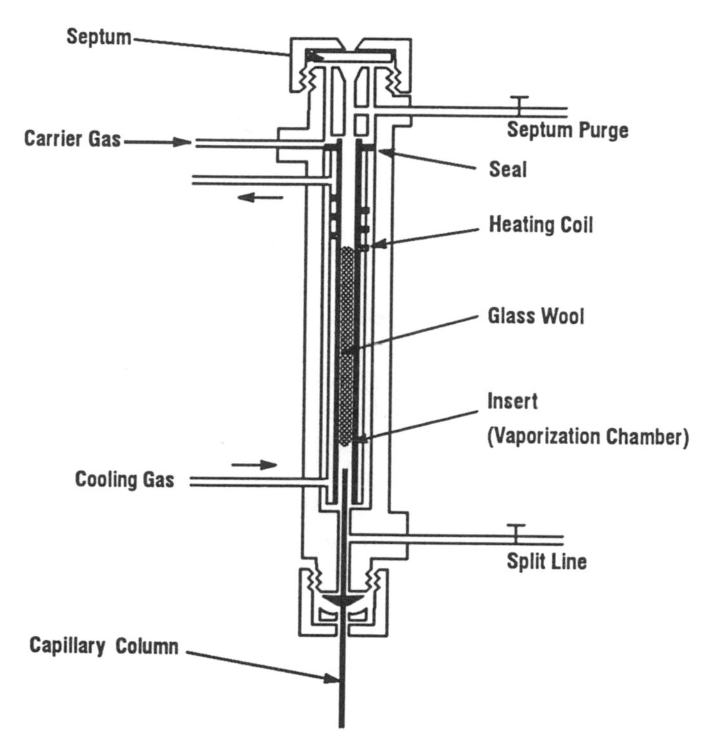

FIGURE 8.17 Schematic of a typical programmable temperature vaporizer (PTV) inlet. The inlet liner is packed with deactivated glass wool to hold liquid sample during the programmed vaporization. The inlet parameters (split vent, temperature) can be programmed to give split, splitless, and solvent vent modes of sample introduction. (Figure courtesy of Gerstel, Gmbh.)

boiling point, concentrating less volatile sample components that remain. The vent is then turned off, the inlet temperature is increased, and the higher-boiling sample components transfer to the column. After transfer is complete, the vent is turned on again to reduce ghosting during the run.

8.6.2 Typical Use

Since PTV is a cool injection technique, there is no needle discrimination. Therefore, PTV inlets are ideal for samples with a wide boiling-point range. Even though the ultimate temperature of the PTV at the end of sample transfer may be very high, the heating process is gradual, and sample components transfer out of the inlet as a function of their boiling points. There is no ballistic pressure pulse at the time of sample injection as there is with hot vaporizing inlets. This reduces sample exposure to high inlet temperatures and, therefore, the potential of sample degradation, and allows much larger injection volumes than other inlets do.

The choice of PTV liner and packing is important. The liner must have enough surface area inside it to hold the liquid sample until it has evaporated and transferred to the column. Yet excessive packing or packing activity might increase adsorption of polar solutes and degradation of labile sample components. Liner volume is another important variable to consider. Too little volume limits sample size and increases the possibility of discrimination and sample loss. Too large of a liner volume yields wider initial bandwidths and poorer heat transfer to the sample during evaporation.

Solvent vent injection can benefit trace analysis if the components of interest have significantly higher boiling points than the solvent. Since the majority of the solvent is vented before it reaches the column, overload of the column by solvent is minimized and much larger injection volumes can be accommodated, decreasing detection limits proportionally. The higher boiling solutes are quantitatively transferred to the column as their boiling points are reached and are focused by the stationary phase, yielding high sensitivities. Care must be taken to verify that there is no discrimination against solutes of interest, or reproducibility and sensitivity of the analysis may be reduced.

PTV split sampling has advantages over classical split sampling because it lacks needle discrimination, has a lower probability of flashback, and has lower inlet discrimination due to pressure transients. However, the initial peak widths are not as narrow as with classical vaporizing split injection, so early eluting components may have broader peaks. PTV splitless sampling has the advantages of no needle discrimination, lower probability of flashback, and less thermal degradation of samples relative to conventional splitless techniques, so it offers a solution to many of the problems associated with classical splitless injection.

8.6.3 Disadvantages

PTV inlets require many hardware components and an electronic or computerized controller. This increases its cost and reduces reliability significantly over classical inlets. However, the inlet capabilities are very powerful and the potential benefits are significant if the inlet is used to full advantage.

Because of their flexibility, programmable temperature vaporizers require considerable time to set up and optimize. To get the most out of a PTV, the user must optimize conditions for each injection mode used. The easiest injection mode to optimize is split injection. The most time-consuming to optimize is solvent vent injection. One must also remember that, as with splitless injection, whenever any sampling variable is changed after optimal conditions had been determined, the inlet parameters must be reoptimized.

REFERENCES

1. G. Schomburg, R.Dielmann, H. Borwitsky, and H. Husmann, in *12th Int. Symposium of Chromatography, Baden-Baden, 1976*, G. Schomburg and L. Rohrschneider (Eds.), Elsevier, Amsterdam, 1978, p. 166.
2. M. S. Klee, "GC Inlets—An Introduction," Hewlett–Packard, Wilmington, DE, No. 5958-9468, ISBN 1-880313-00-6.
3. S. Stafford (Ed.), "Electronic Pressure Control in Gas Chromatography," Hewlett–Packard, Wilmington, DE, No. 5182-0842.
4. K. Grob, Jr., *Classical Split and Splitless Injection in Capillary GC*, 2d ed., Heuthig, Heidelberg, 1988.
5. K. Grob, *On-Column Injection in Capillary Gas Chromatography*, Heuthig, Heidelberg, 1987.
6. J. V. Hinshaw, Jr., *Sample Introduction in Capillary Gas Chromatography*, Vol. 1, P. Sandra (Ed.), Heuthig, Heidelberg, 1985.
7. K. Grob and M. Biedermann, *J. High Resolut. Chromatogr. Chromotogr. Commun.*, **12**: 89–95 (1989).

PART III

APPLICATIONS

Science is nothing but trained and organized common sense, differing from the latter only as a veteran may differ from a raw recruit. And its methods differ from those of common sense only as far as the guardman's cut and thrust differ from the manner in which a savage wields his club.

—Thomas Henry Huxley (1825–1895)
Collected Essays, iv, The Method of Zadig

CHAPTER NINE

Physicochemical Measurements by Gas Chromatography

MARY A. KAISER
Du Pont Company

and

CECIL DYBOWSKI
University of Delaware

Modern Practice of Gas Chromatography, Third Edition. Edited by Robert L. Grob.
ISBN 0-471-59700-7

Gas chromatography (GC) is usually regarded in the context of analysis of complex mixtures. The subtle differences in the manner of interaction of the component of a mixture with the stationary phase in the gas chromatographic column are the primary features that makes this such a generally useful technique. Most often, the goal of a gas chromatographic experiment is to separate, identify, and quantify the components with only passing regard to the nature of this process; however, the fundamental physical and chemical properties of the chromatographic system and the materials that comprise it are amenable to investigation by gas chromatographic experiments. The sensitivity to subtle differences in molecular structure and detail that has been exploited in its use in separation is also exhibited in the detailed information that GC can give on the physicochemical properties of the stationary phase and the materials injected onto the column. Although the measurements of physical parameters such as specific surface area and heat of adsorption are often available from other experiments, gas chromatographic techniques have advantages of speed, reliability, and versatility, in that one apparatus may be configured to measure physicochemical properties over a wide range of conditions. Recent advances in gas chromatographic hardware, especially columns and detection systems, have made rapid measurements possible on very small quantities of sample, making GC an even more powerful tool for physicochemical measurements.

9.1 SPECIFIC SURFACE AREA AND ADSORPTION

The specific surface area of a solid is the quantity of surface available for a particular application, per unit weight, usually given as m^2/g or, if volumetric, m^2/mL. The specific surface area is one of the quantities that must be known if any physicochemical interpretation of the behavior of the material as an adsorbent is to be made. Typical specific surface areas of some gas chromatographic packings and supports are given in Table 9.1(1).

The best known method for determining the specific surface area of powders was developed by Brunnauer, Emmett, and Teller (BET) (2). The BET method involves the determination of the quantity of gas taken up through adsorption by a solid adsorbent and in equilibrium with the gas phase at a pressure *P*. This classical method involves determination of the amount of gas adsorbed in equilibrium with a pressure of gas at a temperature near the condensation point. After a known quantity of gas has been admitted to a chamber containing the sample, adsorption occurs, resulting in a pressure decrease until equilibrium between the adsorbed and gas phases is reached. The quantity of gas adsorbed is determined by the difference between the amount of gas originally admitted and the amount remaining in the gas phase at equilibrium. The quantity of gas originally admitted is calculable from the initial pressure because the volume above the adsorbent was calculated previously.

TABLE 9.1 Specific Surface Areas of Some Common Gas Chromatographic Supports and Packings

Packing	Specific Surface Area (m^2/g)	Packing	Specific Surface Area (m^2/g)
Alumina F-1	223	Chromosorb 750	0.8
Carbopack A	13.2	Molecular Sieve A	230
Carbopack B	81.9	Molecular Sieve 13X	91.2
Carbosieve B	615	Porapak N	518
Chromosorb T	6.94	Porapak P	119
Chromosorb W	1.0	Porapak Q	515
Chromosorb 101	35.4	Porapak R	544
Chromosorb 102	33.8	Porapak S	411
Chromosorb 103	20.3	Porapak T	266
Chromosorb 104	122	Porapak PS	108
Chromosorb 105	452	Porapak QS	453
Chromosorb 106	764	Silica gel	673
Chromosorb 107	416		
Chromosorb 108	162		

Source. Reference 1.

The amount of gas adsorbed depends on the pressure of the gas with which it is in equilibrium. Thus determination of the equilibrium for various amounts of gas originally admitted will give different values of the equilibrium pressure and the amount adsorbed. These two parameters are related by the BET equation:

$$\frac{P}{V_a(P_0 - P)} = \frac{(C-1)P}{V_m C P_0} + \frac{1}{V_m C} \tag{9.1}$$

where V_a = volume of gas adsorbed (reduced to STP) per gram of adsorbent
P = equilibrium pressure
P_0 = saturation vapor pressure of adsorbate at temperature of adsorption
C = a constant determined by energy of adsorption
V_m = volume (at STP) of gas (per gram of adsorbent) that fills one monolayer

A plot of $P/[V_a(P_0 - P)]$ versus P/P_0 allows one to determine V_m from the slope and intercept. Once known, this value may be converted to specific surface area S, in m^2/g, if one knows the average area occupied by one molecule by the following equation:

$$S = \frac{V_m \sigma N}{V_0} \tag{9.2}$$

where σ = average area occupied by one molecule (in m^2/molecule)
N = Avagadro's number (6.02×10^{23} molecules/mol)
V_0 = STP volume of one mole of gas (22.410 L/mol)

Gas chromatographic determination of specific surface area has several advantages over the traditional BET adsorption method. It is more accurate when the adsorption involves low surface coverage and can be readily adapted to reflect conditions other than those near the condensation point; for example, contact time and temperature may easily be varied. The gas chromatographic measurement is usually more convenient to set up and use.

The first demonstration of the use of GC to determine specific surface area was made by Nelson and Eggertsen (N/E) (3). In the N/E method, a quantity of the adsorbent is outgassed at high temperature under helium flow to remove volatile components. This procedure may do more than remove volatile constituents; if the conditions are sufficiently severe, the structure of the surface may actually be altered by outgassing. After cooling to room temperature, the sample, usually in a U tube, is connected to the chromatograph. A flow of helium and nitrogen is sent over the sample. After the gas composition is stable, indicated by no change of the gas chromatographic detector response, the sample is immersed in a liquid nitrogen bath. Adsorption of nitrogen out of the gas stream onto the adsorbent occurs, causing the thermal conductivity detector to show what is known as an *adsorption peak* (Figure 9.1). The sample comes to equilibrium

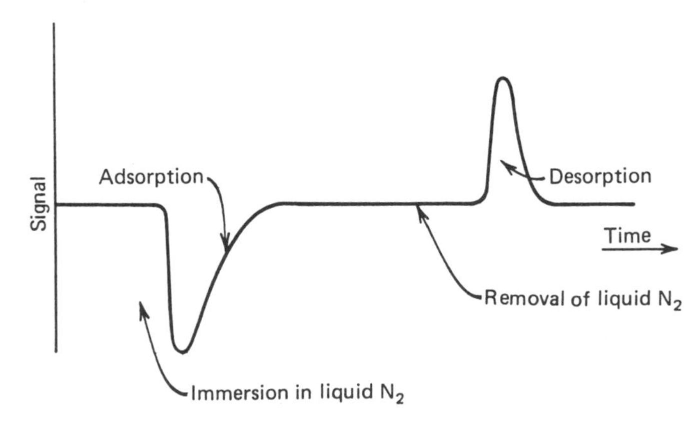

FIGURE 9.1 Schematic detector response in the N/E determination of specific surface area.

at this temperature, indicated by a settling of the detector response. Removal of the liquid nitrogen bath results in desorption of the previously adsorbed material. The gas chromatographic detector indicates this with a *desorption peak*. The area of the desorption peak is measured to determine the amount desorbed. This can be done readily if the detector had been previously calibrated with known amounts of nitrogen. Although the adsorption peak could, in principle, be used to measure the same effect, tailing of the adsorption peak makes calculation of the amount adsorbed from the desorption peak preferable. This adsorption corresponds to a nitrogen pressure with which the adsorbed phase is in equilibrium. This partial pressure is calculable from the total flowrate F_c, the nitrogen flowrate F_N, and the barometric pressure P_B, through Equation (9.3):

$$P = \frac{F_N}{F_c} P_B \tag{9.3}$$

The measurement is repeated for different values of the nitrogen flowrate to establish the isotherm, which may be plotted and analyzed by applying Equations (9.1) and (9.2) to obtain the specific surface area. The N/E method is used for a variety of adsorbents. Improvements to the method extend the range to the measurement of low specific surface areas (down to $0.07\ m^2/g$), such as by stopped-flow techniques. Table 9.2 gives comparisons of specific surface areas determined by the N/E method and by the classical BET method, which indicate that the agreement between the two is quite good.

The shape of the front may be related to the adsorption process. For example, Kuge and Yoshikawa (4) related peak shape to the beginning of multilayer adsorption (Figure 9.2). At injections of very low volume, the peak was symmetrical; however, injections of larger volumes produced a peak with a sharp front, a diffuse tail, and a defect at the front of the top of the peak. For extremely large injections, the peak has a rather diffuse front and a sharp tail. By using repeated injections, those authors were able to determine the injection volume for which the transition from one behavior to another occurs. This corresponds to point *B* on a BET type II isotherm,

TABLE 9.2 Comparison of Surface Area Using BET and N/E Methods

Sample	BET (m^2/g)	N/E (m^2/g)
Furnace Black	24	25.7
Silica–alumina cracking catalyst, used	103	101
Silica–alumina cracking catalyst, new	438	455
Alumina	237	231
Firebrick	3.1	3.4

Source. Reprinted from Reference 3 with permission. Copyright by the American Chemical Society.

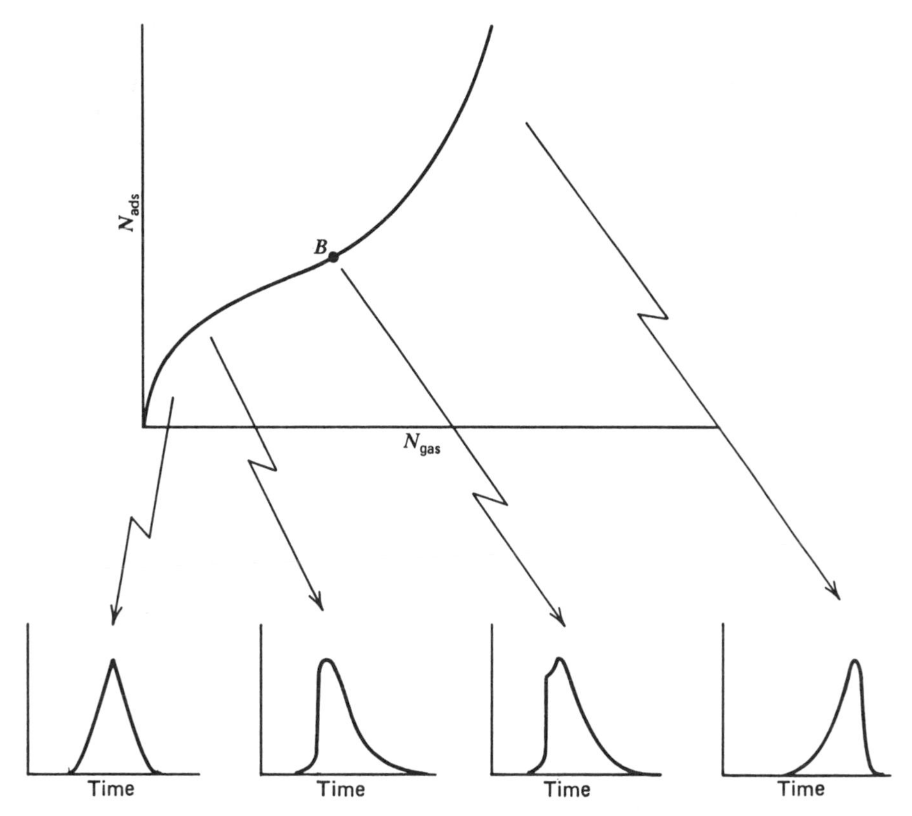

FIGURE 9.2 Peak shape in a multilayer adsorption isotherms. (From Reference 4.)

from which the authors were able to calculate the specific surface area. A complete analysis of the front can enable one to determine the adsorption isotherms, and hence specific surface area, particularly if aided by the use of a computer (5). The description by Kiselev and Yashin (6) is particularly elegant, and we briefly recapitulate it here.

Under certain conditions, the dominant source of broadening of peaks in a gas chromatographic experiment is the adsorption–desorption process. Consider a small volume of gas moving through a packed column. In this volume is a certain concentration of the substance to be sampled by the detector. As it moves through the column, the concentration is changed in one of two ways: (1) by net flow of the substance into or out of this small volume from the nearby gas phase or (2) by the adsorption–desorption processes, which transfer material from the adsorbed phase to the gas phase. The net result of these two processes is that, at equilibrium, the gas-phase concentration c and the adsorbed-phase concentration c_a are related by

Equation 9.4:

$$\left[\frac{\delta_{c_a}}{\delta_c}\right]_X = \left[\frac{F_c - F}{F}\right]\frac{V}{V_a} \tag{9.4}$$

where F_c = flowrate of carrier gas
F = flowrate of a volume of concentration c
V = gas-phase volume
V_a = adsorbed-phase volume

Uptake is usually expressed not as a volumetric concentration, but as a quantity taken up per gram of adsorbent a. The result of conversion is Equation 9.5:

$$\left[\frac{\delta_a}{\delta_c}\right]_X = \frac{V'_R}{m} \tag{9.5}$$

$$a(c) = \frac{1}{m}\int_0^c V'_R(c')\,dc' \tag{9.6}$$

where V'_R is the adjusted retention volume, and m is the mass of the adsorbent. To find the uptake a from a gas phase of concentration c, therefore, one integrates Equation 9.5. Equation 9.6 is the relationship needed to express uptake in terms of the gas chromatographic observables. If the detector is concentration sensitive, the integral of Equation 9.6 can be evaluated directly from the chromatogram, since the height y of the chromatographic detector response corresponds to a given gas-phase concentration. A few algebraic manipulations lead to Equation 9.7:

$$a(c) = \frac{M_a}{m_a m A}\int_0^y \{t_c(y') - t_0\}\,dy' \tag{9.7}$$

where M_a is the molecular weight of the adsorbate, m_a is the mass of adsorbate injected, and A is defined in Equation 9.8:

$$A = \int_0^\infty y(t')dt' \tag{9.8}$$

These integrals are shown in Figure 9.3. The pressure to which this uptake corresponds is calculable, assuming an ideal gas phase, as

$$P = \frac{m_a y R T}{M_a F_c A} \tag{9.9}$$

where R is the gas constant, and T is the absolute temperature. One may, therefore, construct the adsorption isotherm from such an analysis.

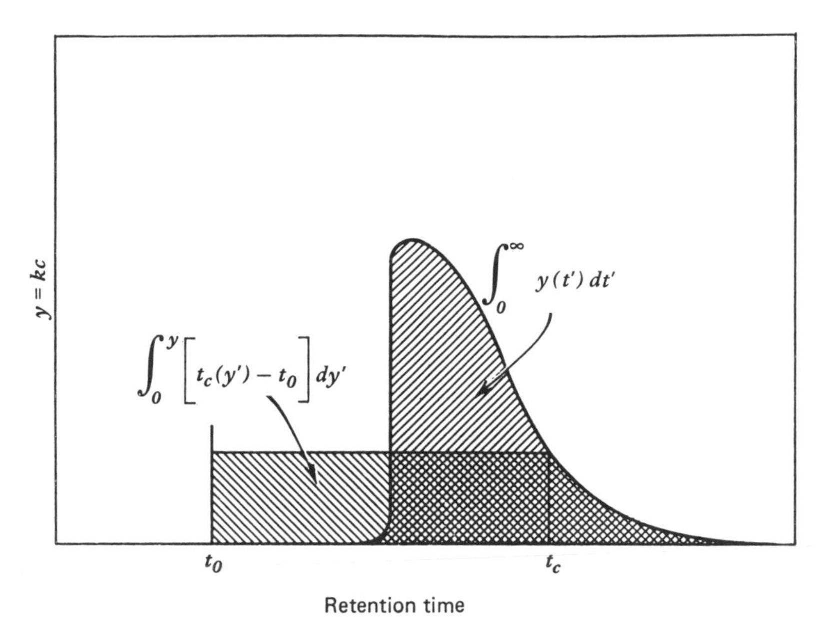

FIGURE 9.3 Schematic of a chromatographic response, showing the areas corresponding to the integrals in Equation 9.7.

Both permanent gases and adsorbates that are liquids above room temperature could be investigated by this method. With the use of specific detectors, gas chromatographic analysis can provide unique adsorption data on interesting systems, including the study of adsorption of hydrocarbons with the use of a flame-ionization detector (FID).

A variety of types of surface area may be determined by GC. For example, the adjusted retention time of a probe molecule, such as n-octane, is fairly linear in the weight of polymer adsorbent in a coated column, so a plot of retention time versus weight in such a column can give the specific surface area from the slope of the plot. In catalyst surface area determinations, the emphasis is usually on measurement of catalytically active surface area rather than on the total surface area. Commonly, chemisorption of gases such as hydrogen, oxygen, or carbon monoxide is used to measure the "active surface area" in a Langmuir adsorption experiment. Karnaukov and Buyanova (7) developed a gas chromatographic method for determining the active surface area of a complex catalyst, by repeated small injections until breakthrough is observed. The total volume chemisorbed is the total volume injected up to breakthrough. Masukawa and Kobayashi (8) used a packed column to determine the "size" of an adsorbate in the adsorbed state. The method is time-consuming, but estimates of the "size" of the molecule according to a theoretical model can provide information on the structure of the adsorbent.

There are many sources of error in surface-area determinations. Lobenstein (9) has delineated the problems associated with lack of additivity, as applied to coadsorption of mixtures, and has derived equations for interactions among coadsorbates to give a more realistic value for surface areas under the conditions of coadsorption. Dollimore and Heal (10) have shown that it is possible to calculate an apparent specific surface area that is less than the actual area because the adsorbate fits snugly in pores of the adsorbent. Thus changes in the "size" of the adsorbate molecule change the surface area determined, with the "true" surface tending to be that found for larger adsorbates. In using flow techniques care must be taken to ensure that repeated injections are performed with all parameters controlled. The flowrate is an exceptionally critical parameter. Since specific surface area depends on the probes used to measure it, the analyst should always measure the area under conditions that mimic as closely as possible the conditions under which the adsorbent will be used.

9.2 SURFACE THERMODYNAMICS

Adsorption isotherms constructed from gas chromatographic experiments of the type discussed in Section 9.1 contain substantially more information on the adsorption–desorption behavior than the specific surface area. The use of temperature as a parameter that can be varied yields information on the thermodynamics of the adsorption–desorption process. A growing body of data on adsorption and thermodynamics (11) provides the basis for predicting sample separations under a variety of conditions, a goal of separation science.

One of the most important of the thermodynamic parameters measurable by gas chromatographic techniques is the isosteric heat of adsorption ΔH^0_{st}. Measurements of heats of adsorption provide a quantification of the interactions occurring between adsorbate and the adsorbent. Although there exists a spectrum of behaviors, the process of adsorption can be roughly divided into three categories: physical adsorption due to weak van der Waals interactions, reversible chemical adsorption, and irreversible chemical adsorption. The last two types occur because of a much stronger specific interaction, such as that of unsaturated hydrocarbons with transition metals or the adsorption of water on activated alumina. It is usual to distinguish physical and chemical adsorption on the basis of the heat of adsorption with a typical value of 15 kcal/mol (62.8 kJ/mol) serving as the arbitrary dividing line between physical and chemical adsorption. Measurements of heats of adsorption from a flow system, as is done in GC, result from a condensation of the equilibrium that exists between the gas phase and the adsorbed phase.

The process of adsorption–desorption occurring in the gas chromatograph is written as

$$A(\text{gas}; P,T) \leftrightharpoons A(\text{adsorbed}; a,T)$$

The equilibrium constant K_D for this process is given by Equation 9.10:

$$\ln K_\mathrm{D} = -\frac{\overline{\Delta G^\circ}}{RT} \tag{9.10}$$

From the relationship between the free energy and enthalpy of adsorption, one obtains a very useful result:

$$\left[\frac{\partial \ln P}{\partial(1/T)}\right]_a = \frac{\overline{\Delta H^\circ_\mathrm{st}}}{R} \tag{9.11}$$

A plot of $\ln P$ versus $1/T$ at constant coverage a gives a straight line, the slope of which is the heat of adsorption divided by the gas constant. This heat is the *isosteric* heat of adsorption, in that it is evaluated at constant coverage. The prescription for a determination is as follows:

1. Determine the isotherms by the prescription in Section 9.1 for several temperatures.
2. For a given isostere (cross section of constant coverage), plot $\ln P$ versus $1/T$.
3. Determine $\Delta H^\circ_\mathrm{st}$ from the slope of this plot.

The heat determined in this way depends on the value of a, the coverage; hence the coverage dependence of $\Delta H^\circ_\mathrm{st}$ can be examined by calculation of $\Delta H^\circ_\mathrm{st}$ for different values of a. This is a useful indicator of the state of the adsorbed system since the surface heterogeneity and the magnitude of the direct interaction of adsorbates produce this coverage dependence. In principle, gas chromatographic determination of this dependence on coverage gives detailed information on the subtle features of surface interactions.

Further thermodynamic parameters may be calculated from the temperature and coverage dependence of $\Delta H^\circ(a, T)$. From thermodynamics, one obtains Equation 9.12, which relates $\Delta G^\circ(a, T)$ to $\Delta H^\circ(a, T)$ (12):

$$\left[\frac{\partial \overline{\Delta G^\circ}/T}{\partial T}\right]_{P,a} = -\frac{\overline{\Delta H^\circ}(a, T)}{T^2} \tag{9.12}$$

The standard free energy for adsorption is then given (at constant coverage) as

$$\overline{\Delta G^\circ}(a, T) = \overline{\Delta G^\circ}(a, T)\left(\frac{T}{T_0}\right) - T\int_{T_0}^{T} \frac{\overline{\Delta H^\circ}(a, T)}{T^2}\, dT \tag{9.13}$$

Once $\Delta H^\circ(a, T)$ and $\Delta G^\circ(a, T)$ are known, the entropy of adsorption can be directly calculated by the thermodynamic relationship:

$$\overline{\Delta S^\circ}(a, T) = \frac{\overline{\Delta H^\circ}(a, T) - \overline{\Delta G^\circ}(a, T)}{T} \tag{9.14}$$

The value T_0 is a temperature for which the standard free energy is defined. The use of GC to yield these thermodynamic parameters has been exploited (10) and should lead to an understanding of the processes underlying the separation process.

Frequently, the stationary phases used in gas chromatographic separations are much more uniform than the general theory given in Section 9.1 implies. Additionally, the amount of material injected is extremely small, corresponding to very low surface coverage. The resultant chromatogram shows little band broadening from the adsorption–desorption effects, and the retention of the injected material corresponds to the case of an infinitely dilute adsorbate in the stationary phase. The distribution constant K_D may be related to the retention volume, the mobile phase volume, and the stationary phase volume. The total number of moles injected onto the column n is divided between the stationary and mobile phases:

$$n = n_s + n_M \tag{9.15}$$

where $n_s = ma =$ the total number of moles in the stationary phase, and $n_M =$ the number of moles in the mobile phase. The distribution constant K_D is defined as

$$K_D = \frac{a_s}{a_M} = \frac{\gamma_s}{\gamma_M} \frac{n_s / V_s}{n_M / V_M} \tag{9.16}$$

where a_s is the activity in the stationary phase, a_M is the activity in the mobile phase, V_s is the volume of the stationary phase, V_M is the volume of the mobile phase between the point of injection and the point of detection, γ_s is the activity coefficient in the stationary phase, and γ_M is the activity coefficient of the adsorbate in the mobile phase. The velocity of the substance u_s is related to the velocity of an unretained substance u_u by the following equation:

$$u_s = u_u f \tag{9.17}$$

where f represents the fraction of time spent by the substance in the mobile phase. On average, f is equivalent to n_M / n and

$$u_s = \frac{n_m}{n} u_u \tag{9.18}$$

By substitution

$$\frac{n_{\mathrm{m}}}{n} = \frac{n_{\mathrm{M}}}{n_{\mathrm{M}} + n_{\mathrm{s}}} \tag{9.19}$$

$$\frac{n_{\mathrm{s}}}{n} = n_{\mathrm{M}} K_{\mathrm{D}} \left(\frac{V_{\mathrm{s}}}{V_{\mathrm{M}}}\right)\left(\frac{\gamma_{\mathrm{M}}}{\gamma_{\mathrm{s}}}\right) \tag{9.20}$$

$$\frac{n_{\mathrm{M}}}{n} = \frac{1}{1 + (K_{\mathrm{D}} V_{\mathrm{s}} \gamma_{\mathrm{M}} / V_{\mathrm{M}} \gamma_{\mathrm{s}})} \tag{9.21}$$

The average velocity of an unretained component u_{u} and the average velocity of the substance u_s are related to the flowrate of the carrier gas F_c and the column length L by the following expressions:

$$u_{\mathrm{u}} = \frac{F_c L}{V_{\mathrm{M}}} \tag{9.22}$$

$$u_{\mathrm{s}} = \left(\frac{F_c L}{V_{\mathrm{M}}}\right)\left[\frac{V_{\mathrm{M}}}{V_{\mathrm{M}} + K_{\mathrm{D}} V_{\mathrm{s}} (\gamma_{\mathrm{M}} / \gamma_{\mathrm{s}})}\right] = \frac{F_c L}{V_{\mathrm{M}} + K_{\mathrm{D}} V_{\mathrm{s}} (\gamma_{\mathrm{M}} / \gamma_{\mathrm{s}})} \tag{9.23}$$

The average velocity of the substance is also defined by the following expression:

$$u_{\mathrm{s}} = \frac{F_c L}{V_{\mathrm{R}}} \tag{9.24}$$

where V_{R} is the retention volume of the substance. Therefore,

$$u_{\mathrm{s}} = \frac{F_c L}{V_{\mathrm{M}} + K_{\mathrm{D}} V_{\mathrm{s}} (\gamma_{\mathrm{M}} / \gamma_{\mathrm{s}})} \tag{9.25}$$

and from Equation 9.16,

$$K_{\mathrm{D}} + \frac{(V_{\mathrm{R}} - V_{\mathrm{M}}) \gamma_{\mathrm{s}}}{V_{\mathrm{s}} \gamma_{\mathrm{M}}} = \frac{V'_{\mathrm{R}} \gamma_{\mathrm{s}}}{V_{\mathrm{s}} \gamma_{\mathrm{M}}} \tag{9.26}$$

where V'_{R} is the adjusted retention volume. This equation is similar in form to Equation 9.7, assuming that all molecules have exactly the same retention volume.

The volume of the stationary phase remains constant; thus variations of K_{D} with temperature are reflective of variations of V'_{R}, γ_{s}, and γ_{M} with temperature. Equation 9.10 gives such a relationship:

$$\ln V'_{\mathrm{R}} - \ln V_{\mathrm{s}} = -\frac{\overline{\Delta G^{\circ}}}{RT} - \ln \frac{\gamma_{\mathrm{s}}}{\gamma_{\mathrm{M}}} \tag{9.27}$$

Differentiation of Equation 9.27 with respect to $1/T$ and comparison with Equation 9.12 gives the important result:

$$\left.\frac{\partial \ln V'_{\mathrm{R}}}{\partial(1/T)}\right|_{P} = \frac{\overline{\Delta H^{\circ}_{\mathrm{st}}}}{R}\,(a=0) \tag{9.28}$$

The heat of adsorption corresponds to infinite dilution on a homogeneous stationary phase. Thus, for sufficiently homogeneous stationary phases (whether they be solid or nonvolatile liquid is not germane to the treatment), the slope of a plot of the adjusted retention volume (V'_{R}) versus $1/T$ can be used to calculate ΔH° at infinite dilution. Because the value of $\ln V_{\mathrm{s}}$ is found from the intercept of such a plot, ΔG° can be calculated by using Equation 9.27 and ΔS° is then calculable from Equation 9.14.

To account for hydrodynamic factors due to a finite pressure drop over the length of the column, the adjusted retention volume is corrected by the pressure gradient correction factor j to give the net retention volume V_{N}:

$$V_{\mathrm{N}} = jV'_{\mathrm{R}} \tag{9.29}$$

and

$$j = \left[\frac{3}{2}\,\frac{(P_{\mathrm{i}}/P_{\mathrm{o}})^2 - 1}{(P_{\mathrm{i}}/P_{\mathrm{o}})^3 - 1}\right] \tag{9.30}$$

where P_{i} is the pressure of the inlet and P_{o} is the pressure at the outlet. As long as the pressures are adjusted to be constant at different column temperatures, the plots of $\ln V_{\mathrm{N}}$ and $\ln V'_{\mathrm{R}}$ versus $1/T$ should produce the same results for $\Delta H^{\circ}_{\mathrm{st}}$. This variation of the hydrodynamic energy is a potential source of error in calculating thermodynamic quantities.

The material injected onto the column may not be ideal; but in cases where it is

$$\gamma_{\mathrm{s}} = \gamma_{\mathrm{M}} = 1 \tag{9.31}$$

if it is ideal in both phases. If it is nonideal in one phase but ideal in the other, one of the activity coefficients must be unity.

Roles and Guiochon (13) developed a numerical method for determining the adsorption energy distribution function from adsorption isotherm data using gas–solid chromatography. They studied a variety of surface hetergeneous solids, such as aluminum oxide ceramic powders, which they coated on the walls of open tubular columns. Using the classical method of elution but with larger volumes of organic vapors, they were able to calculate the distribution of adsorption energy of the probes on the surface.

9.3 SOLUTION THERMODYNAMICS

Although the chromatographic systems used for solution thermodynamic investigations generally have little analytical separations utility, the results do provide interesting evaluations of the interactions between solutes and solvents. In gas–liquid chromatography (GLC), one generally has the interaction of a "solvent" that is a liquid phase coated on an "inert" support with a "solute" probe injected onto the column. Since only small quantities are injected onto the column, one approximates the conditions of infinite dilution. Retention results from the interaction occurring during dissolution of the "solute" in the "solvent." In this case, distribution of the solute between the mobile phase and the stationary phase results in equilibrium described by Raoult's law:

$$a_{\mathrm{s}}^{\mathrm{u}} = \frac{f}{f_0} = a_{\mathrm{M}}^{\mathrm{u}} \tag{9.32}$$

where f is the fugacity of the gas phase, and f_0 is the standard-state fugacity. Replacing the fugacity by the activity coefficient multiplied by the activity coefficient γ_{s}, times the mole fraction X_{s}, one obtains

$$P = \frac{\gamma_{\mathrm{s}} X_{\mathrm{s}}}{\gamma_{\mathrm{M}}} \mathrm{P}^{\circ} \tag{9.33}$$

Assuming that the gas phase is ideal ($\gamma_{\mathrm{M}} = 1$) and remembering that this situation corresponds to infinite dilution ($\gamma_{\mathrm{s}} = \gamma^{\infty}$), one obtains

$$P = \gamma^{\infty} X_{\mathrm{s}} P^{\circ} \tag{9.34}$$

The activity coefficient at infinite dilution accounts for deviations from the ideal Raoult's law. If the deviation is positive, $\gamma > 1$, and the retention time of the solute is decrease. If the deviation is negative, $\gamma < 1$, and the retention time of the solute is increased. The adjusted retention volume and the pressure are related by the ideal gas law:

$$PV'_{\mathrm{R}} = n_{\mathrm{s}} RT \tag{9.35}$$

Thus we find the expression for the activity coefficient at infinite dilution ($n_{\mathrm{s}} \ll N$, the number of solvent molecules):

$$\gamma^{\infty} = \frac{(wt)RT}{V'_{\mathrm{R}} M P^{\circ\prime}} = \frac{RT}{V_{\mathrm{g}} M P^{\circ}} \tag{9.36}$$

An example calculation is shown in Table 9.3.

The substituent constant of the Hammett equation has been related successfully to the logarithm of the activity coefficient ratio at infinite dilution for a series of *m*- and *p*-phenyl isomers. Hammett stated that a

TABLE 9.3 Sample Activity Coefficient Calculation

Data

$V_R = 600$ mL	wt = 2 g	$M = 114$ g/mol
$V_M = 100$ mL	$T = 29°C$	$P° = 140$ Torr

Preliminary calculations

$$V_g = (V_R - V_M)/\text{wt} = (600\text{ mL} - 100\text{ mL})/2\text{ g} = 250\text{ mL/g}$$

$$T(\text{K}) = T(°\text{C}) + 273.15 = 29 + 273.15 = 302.15\text{ K}$$

$$P°(\text{atm}) = P°(\text{Torr})/760\text{ Torr} = 140/760 = 0.184\text{ atm}$$

$$R = 82.0573\text{ mL/atm K mol}$$

$$\frac{RT}{V_g MP°} = \frac{(82.0573\text{ mL/atm K mol})(302.15\text{ K})}{(250\text{ mL/g})(114\text{ g/mol})(0.184\text{ atm})} = 4.73$$

where R = gas constant
T = absolute temperature
V'_R = adjusted retention volume
M = molecular weight of solvent
wt = weight of solvent on column
$P°$ = vapor pressure of the pure solute

free-energy relationship should exist between the equilibrium or rate of behavior of a benzene derivative and a series of corresponding meta- and para-monosubstituted benzene derivatives. The Hammett equation may be written as

$$\log \frac{K'_x}{K'_0} = \sigma\rho \tag{9.37}$$

where K'_x, K'_0 = rate (or equilibrium) constants for a reaction of substituted and unsubstituted benzene derivatives respectively
σ = Hammett substituent constant
ρ = Hammett reaction constant

The constant ρ may be used to establish a selectivity scale for stationary liquid phases in GLC through the following relationship:

$$\log \frac{\gamma_0^\infty}{\gamma_x^\infty} = \sigma_c\rho + b \tag{9.38}$$

where b is constant. The chromatographic substituent constant σ_c is obtained from the following equation:

$$\sigma_c = 0.09 + 0.621 \log \frac{\gamma_0^\infty}{\gamma_x^\infty} \tag{9.39}$$

These relationships were shown to be capable of predicting the types of liquid phases necessary for optimum resolution.

Other important parameters measured by GLC are enthalpy, free energy, and entropy of solution. Although by no means all reactions, many chemical reactions are produced at constant pressure and temperature in solution. The free-energy change of such a reaction determines whether it is spontaneous. Reactions with negative free energy changes are spontaneous; those with positive free energy changes are not. The dissolution process is important in this mechanism, and GLC can measure thermodynamic properties of this type.

The enthalpy (or heat) of solution is defined as the quantity of heat accompanying the process:

$$\mathrm{A(l)} \rightarrow \mathrm{A\ (dissolved\ in\ B)}$$

If the solution consists of two liquids, the resultant heat is sometimes called the *heat of mixing*. The free energy and entropy of solution are analogously defined with respect to this process. Gas–liquid chromatography can be used to measure these thermodynamic parameters if a system is constituted in which the stationary phase is the "solvent" B and the "solute" A is injected onto the column. Under these conditions, the gas–liquid chromatographic retention measures this equilibrium between the solute and dissolved solute. In general, one injects rather small quantities of A at very low dilution in B, that is, at infinite dilution. The free-energy change may be obtained directly from the partition coefficient K_D by Equation 9.10.

From Equation 9.26, we see that each of the quantities may be considered to be divided into an "ideal" ($\gamma_s = \gamma_M = 1$) and a "nonideal" correction:

$$\overline{\Delta G^\circ} = \overline{\Delta G_i^\circ} + \overline{\Delta G_e^\circ} \tag{9.40}$$

$$\overline{\Delta H^\circ} = \overline{\Delta H_i^\circ} + \overline{\Delta H_e^\circ} \tag{9.41}$$

If we assume that the gas (or mobile) phase is ideal, we have the results from GLC:

$$\overline{\Delta G_e^\circ} = -RT \ln \gamma^\infty \tag{9.42}$$

$$\overline{\Delta S_e^\circ} = \frac{\overline{\Delta H_e^\circ} - \overline{\Delta G_e^\circ}}{T} \tag{9.43}$$

Another instance in which GC is useful is in measuring the relative strengths of interaction of two solutes with a single stationary phase. Determination of the differences in the free energies of the two components can be made readily from the chromatogram:

$$\overline{\Delta G_2^\circ} - \overline{\Delta G_1^\circ} = \Delta(\overline{\Delta G^\circ}) = -RT \ln \frac{t'_{R2}}{t'_{R1}} \tag{9.44}$$

where t'_{R1} is the adjusted retention time of substance 1 and $\Delta(\Delta G^\circ)$ is unaffected by flowrate, percent loading, or solvent molecular weight and may be used to study differences in interaction of isomers with a liquid phase or the comparison of isotopically substituted molecules.

Headspace GC takes advantage of the closed-vessel equilibrium between either a liquid or a solid and a gas. In the headspace analysis an aliquot of the equilibrated gas phase is removed from the vessel and the aliquot analyzed by GC. This technique is very convenient when direct sampling (solid or liquid) is difficult using traditional gas chromatographic techniques. In addition to the analysis application, headspace GC can also be used for determining physicochemical data (14). Labows (15) used automatic headspace GC to determine the solubilization behavior of volatile organic compounds in detergent surfactants. Since the sensory intensity and character of the flavor or fragrance depends on the solubility interactions in the system, it is important to quantify the interactions between the components and the surfactants used in the product. Using headspace GC, Labows determined the solubilization site within the micelle for a solute, the effect of the solute on the critical micelle concentration, the solute partition coefficient, and the effect of cosolvents on the critical micelle concentration.

Headspace GC can also be used to determine activity coefficients. Hussam and Carr (16) carried out a meticulous study that produced a device for automated measurements of both solute activity coefficients and vapor pressure (see also Section 9.4). The device allowed for rapid sample analysis, better than 0.01°C temperature control, the ability to work at low sample concentrations (mole fractions < 0.01), minimal equilibrium perturbation by the sampling process, and the ability to vary solvent composition automatically.

9.4 VAPOR PRESSURE AND THERMODYNAMICS OF VAPORIZATION

The equilibrium between a liquid and its vapor may be described by the *vapor pressure*, that is the amount of substance in the gas phase. The first accurate gas chromatographic thermodynamic measurements were of vapor pressure and heats of vaporization made by Mackle et al. (17) using a gas chromatograph with a bypass sampling system. In this system the sample is placed in a sample tube and cooled with dry ice, after which it is warmed to a specific temperature. After equilibrium is attained, the liquid sample is isolated from its vapor by a valve. The vapor is swept into a gas chromatograph by a carrier gas. The gas chromatographic detector is, in effect, measuring the vapor pressure at the temperature. Measurements for different temperatures give differing amounts, that is, different vapor pressures. Plots of the logarithm of thc peak area (or peak height for symmetrical peaks) versus $1/T$ can be analyzed to obtain ΔH_v°, the heat of vaporization from the slope. A rapid but approximate means of determining ΔH_v° is to extract it directly from a plot of the logarithm of the net retention volume versus the reciprocal of the column temperature. Table 9.4 gives some classes of compounds whose vapor pressures have been determined by GC.

The measurements made by Hussam and Carr were made via headspace GC and were limited by the lower limit of the detector's lower limit. Their technique is particularly valuable for low-volatility materials where conventional techniques are difficult due to the requirement that all gases and isomers of similar volatility be removed. Headspace GC circumvents those

TABLE 9.4 Vapor Pressures of Compounds Determined by GC

Compound	Reference
Inorganic chlorides and oxychlorides	a
Propanol, butanol, 2-butanone, 3-butanone, *n*-heptane, *p*-dioxane	b
Benzene, toluene, butyl acetate	c
Metal carbonyls	d
Fatty acids, fatty esters, fatty alcohols, chloroalkanes	e
Alcohols	f
Perfumes	g
Hydrogen saturated with methanol	h,i
Pentane, hexane, heptane, octane, benzene, toluene, nitromethane, acetonitrile, methylethylketone, dioxane, ethanol	j

[a] S. Se, J. Bleumer, and G. Rijnders, *Sep. Sci.*, **1**, 41 (1966).
[b] G. Geisler and R. Janash, *Z. Phys. Chem.*, **233**, 42 (1966).
[c] A. Franck, H. Orth, D. Bilinmaier, and R. Nussbaum, *Chem. Z. Chem. Appl.*, **95**, 219 (1971).
[d] C. Pommier and G. Guiochon, *J. Chromatogr. Sci.*, **8**, 486 (1970).
[e] A. Rose and V. Schrodt, *J. Chem. Eng. Data*, **8**, 9 (1963).
[f] F. Ratkovics, *Acta Chim. Acad. Sci. Hung.*, **49**, 57 (1966); through *G. C. Abstr.*, 442 (1967).
[g] S. A. Voitkevich, M. M. Schedrina, and N. P. Soloveva, *Rudol'fi Maslo-Zh., Prom.*, **37** 27 (1971); through *Chem. Abstr.* **76**, 251 (1972).
[h] F. Ratkovics, *Acta Chim. Acad. Sci. Hung.*, **48**, 71 (1966); through *G. C. Abstr.* 443 (1967).
[i] F. Ratkovics, *Magyar Kem. Foluoirat*, **72**, 186 (1966); through *Chem. Abstr.* **65**, 1450 (1966).
[j] A. Hussam and P. W. Carr, *Anal. Chem.*, **57**, 793 (1966).

potential problems by removing interferences by the separation power of the chromatographic column.

A method for the rapid determination of relative vapor pressure by capillary GC was reported by Westcott and Biddleman (18), which is based on comparison of the adjusted retention volume of the sample to that of a substance (V'_{R2}) whose vapor pressure at the column temperature is known to be P_2° (19). The adjusted retention volumes are related to the known vapor pressures:

$$\ln \frac{V'_{R_1}}{V'_{R_2}} = A \ln P_2^\circ - C \tag{9.45}$$

where A and C are constants, if the heats of vaporization of the two components are assumed constant. Thus A and C are determined from the slope and intercept of the straight-line plot of $\ln(V'_{R1}/V'_{R2})$ versus $\ln P_2^\circ$. Knowledge of these constants allows one to calculate the vapor pressure of the test material P_1°:

$$\ln P_1^\circ = (1 - A) \ln P_2^\circ + C \tag{9.46}$$

This method is useful for narrow temperature ranges (<70°C) and when the reference and test materials are similar, with the vapor pressures of the reference material obtained from other techniques (20).

9.5 COMPLEXATION

Metal ions can act as electron–pair acceptors, reacting with electron donors (ligands) to form coordination compounds or complexes. The ligand must have at least one pair of unshared electrons with which to form the bond. Chelates are a special class of coordination compound, resulting from the reaction of a metal ion with a ligand having two or more donor groups.

Complexes often form in steps, with one ligand added in each step:

$$M + L \overset{k_1}{\leftrightharpoons} ML$$

$$ML + L \overset{k_2}{\leftrightharpoons} ML_2$$

$$ML_n + L \overset{k_{n+1}}{\leftrightharpoons} ML_{n+1}$$

$$ML_{n-1} + L \overset{k_n}{\leftrightharpoons} ML_n$$

The stepwise constants for the equilibria specified by this sequence of reactions are called the formation or stability constants:

$$k_1 = \frac{[ML]}{[M][L]} \tag{9.47}$$

$$k_2 = \frac{[ML_2]}{[ML][L]} \tag{9.48}$$

The more stable a complex, the larger is its formation constant. The reciprocal of the formation constant is the instability constant. The determination of these formation constants is relevant to understanding the chemistry of such systems. Like many other equilibrium constants, such stability constants may be determined by gas chromatography.

Purnell (21) surveyed numerous gas chromatographic approaches to the study of complex equilibria and developed generalized retention theories for each kind. His classification system, which is summarized in Table 9.5, greatly simplifies the approach to complexation reactions.

The important point is that, if the gas chromatograph is set up in such a way that the retention is dominated by the interaction of the ligand with the electron acceptor, the gas chromatographic retention data can be used to measure the stability constant of the complex. As an example, consider the determination of formation constants of complexes of aromatic electron donors and di(n-propyl)tetrachlorophthalate in an inert solvent [Class A(II)]. If a one-to-one complex is assumed, the reactions occurring on the column are

$$\mathrm{A(gas)} \underset{K_R^0}{\leftrightarrows} \mathrm{A} \tag{9.49}$$

$$\mathrm{A(S)} + \mathrm{D(S)} \overset{K_1}{\rightleftarrows} \mathrm{AD(S)} \tag{9.50}$$

where A = volatile di(n-propyl)tetrachlorophthalate
D = nonvolatile electron donor in the stationary phase
S = solvent phase (D + I)
I = nonvolatile inert solvent in which D is dissolved

TABLE 9.5 Purnell Classification System for Complexes

Class	Type I	Type II	Type III
A	AX_n	XA_m	A_mX_n
B	SX_n	XS_p	S_pX_n
C	X polymerizes in solution	X depolymerizes in solution	
D	$SA_{m,m+1\ldots}$	$AS_{p,p+1\ldots}$	

Note. A, additive; X, solute; S, solvent; $n \geq 1$, $m \geq 1$, $p \geq 1$.

K_R° = the distribution constant (partition coefficient) of uncomplexed A between S and the gas phase
K_1 = formation constant of AD in solution

If AD(S) and A(S) are assumed to approach infinite dilution (a reasonable assumption in a gas chromatographic system), the activity coefficients approach 1, and

$$K_1 = \frac{(C_{AD})}{(C_A a_D)} \tag{9.51}$$

where a_D is the activity of the donor in the stationary phase. The apparent gas chromatographic distribution constant, assuming ideal solution, is

$$(K_R)_S = K_R^\circ(1 + K_1 C_D) \tag{9.52}$$

where C_D is the concentration of electron donor in the stationary phase. This constant is related to the corrected net retention volume V_N by

$$V_N = (K_R)_S V_S \tag{9.53}$$

where V_S is the total solvent volume. Thus a determination of the gas chromatographic retention volume of the acceptor on columns containing varying amounts of the donor allows determination of the formation constant. From Equation 9.52 it is clear that a plot of $(K_R)_S$ versus concentration C_D will be linear with an intercept K_R° and a slope of $K_1 K_R^\circ$.

Martire and coworkers (22, 23) developed a method of determining complexation constants that is much less time-consuming than the Cadogan–Purnell method (24) but makes additional assumptions. They have demonstrated that the specific retention volume of A is related to the complex formation constant:

$$(V_g^\circ)_D = \frac{273R(1 + K_1 a_D)}{\gamma_A^D P_A^\circ M_D} \tag{9.54}$$

where $(V_g^\circ)_D$ = specific retention volume of A in a system of pure D
R_D = gas constant
γ_A = activity coefficient of uncomplexed A in D at infinite dilution
P_A° = vapor pressure of A
M_D = molecular weight of D

In the Martire–Riedl method (22), only two columns are used, one containing D and the other containing a reference liquid phase R of approximately the same molecular size, shape and polarizability as D [Class

TABLE 9.6 Comparison of Cadogan–Purnell and Martire–Riedl Methods of Complex formation Constant Determination[a]

Electron Donor	Electron Acceptor	Complex Formation Constant at 40°C		
		C-P Method	M-R	Method
Di-*n*-octylamine	$CHCl_3$	0.405 ± 0.19[b]	0.403[a]	0.392[c]
Di-*n*-octylamine	CH_2Cl_2	0.179 ± 0.014	0.187	0.181
Di-*n*-octylamine	CH_2Br_2	0.222 ± 0.004	0.219	0.224

[a] From (24). Used with permission from *Anal. Chem.*, **45**, 2087 (1973). Copyright American Chemical Society.
[b] From Reference 23.
[c] From J. P. Sheridan, D. E. Martire, and F. P. Banka, *J. Am. Chem. Soc.*, **95**, 4788 (1973).

B(II)]. Utilizing a noncomplexing solute N on R and D, one may measure all the various specific retention times.

Assuming

$$\frac{\gamma_N^R}{\gamma_N^D} \approx \frac{\gamma_A^R}{\gamma_A^D} \tag{9.55}$$

(which occurs because of the assumed simularity of R and D), one obtains an equation of the form

$$\frac{(V_g^\circ)_D^A (V_g^\circ)_R^N}{(V_g^\circ)_R^A (V_g^\circ)_D^N} = 1 + K_1 a_D \tag{9.56}$$

where, generally, $(V_g^\circ)_j^i$ is the specific retention volume of i on a column of j. Liao et al. (25) used the Martire–Riedl (M/R) and the Cadogan–Purnell (C-P) method for three-electron donor systems to make a comparison. As Table 9.6 shows, the two methods are in excellent agreement.

A theoretical study to compare the methods of GC and mass spectrometry for measuring weak complexation constants has been performed by Eon and Guiochon (26), who showed that both methods lead to the same results if properly used. Discrepancies can usually be attributed to misinterpretation of the chromatographic measurement.

9.6 VIRIAL COEFFICIENTS

The mathematical relationship between pressure, volume, temperature, and the number of moles of gas at equilibrium is given by its equation of state. The best-known equation of state is the ideal gas law, $P\bar{V} = RT$, where P is the pressure of the gas, $\bar{V}$ is its molar volume (V/n), n is the number of moles of gas, R is the ideal gas constant, and T is the temperature of the gas. Many modifications of the ideal-gas equation of state have been proposed so that the equation can fit P–V–T data of real gases. One of these equations, the virial equation of state, accounts for nonideality by utilizing a power series in the density:

$$\frac{P\bar{V}}{RT} = 1 + B\rho + C\rho^2 = D\rho^3 + \cdots \qquad (9.57)$$

where $\rho = 1/\bar{V}_m$, and B, C, D, and so on are called the second, third, fourth, and so on virial coefficients, respectively. The values of the virial coefficients are functions of temperature and the particular gas under consideration but are independent of density and pressure.

9.6.1 Second Virial Coefficients

The virial coefficient of state is especially important since its coefficients can be modeled in terms of nonideality resulting from interactions between two molecules. Thus a link is formed between the bulk properties of the gas (P, V, T) and the individual forces between molecules. For a multicomponent mixture, a virial coefficient is needed to account for each possible interaction. The second virial coefficient for a two-component mixture are B_{11}, B_{12}, and B_{22}, where B_{11} represents the interaction between two molecules of component 1, B_{12} represents the interaction between a molecule of 1 and a molecule of 2, and B_{22} represents interaction between two molecules of 2. A tabulation of some compounds whose virial coefficients have been measured by GC is given in Table 9.7.

TABLE 9.7 Second Virial Coefficients Measurements by GC

Compounds	Reference
H_2, N_2, O_2, CO_2, hydrocarbons	a
Hydrocarbons	b,c,d
Hydrocarbons, permanent gases	e
Benzene–gas mixtures	f
Benzene–CO_2, benzene–N_2	g
Higher hydrocarbons and their derivatives	h
Benzene–N_2, cyclohexane, *n*-hexane, di-isodecylphthalate	i

[a] D. H. Desty, A. Goldup, G. Luckhurst, and W. Swanton, in *Gas Chromatography*, M. von Sway (Ed.), Butterworths, London, 1962.
[b] L. Che Kalov and K. Porter, *Chem. Eng. Sci.*, **22**, 897 (1962).
[c] E. M. Dentzler, C. Krobles, and M. L. Windsor, *J. Chromatogr.*, **32**, 433 (1968).
[d] R. L. Pecsok and M. L. Windsor, *Anal. Chem.*, **40**, 1238 (1968).
[e] P. Chavin, *Bull. Soc. Chim. Fr.*, 1964, 1800; through *G. C. Abstr.*, 965 (1966).
[f] C. R. Coan and A. D. King, *J. Chromatogr.*, **44**, 429 (1969).
[g] A. J. B. Cruickshank, B. W. Gainey, C. P. Hicks, T. M. Letcher, R. W. Moody, and C. L. Young, *Trans. Farad. Soc.*, **65**, 105 (1969).
[h] M. Vidergauz and V. Semkin, *Zh. Fiz. Khim.*, **45**, 931 (1971).
[i] B. K. Raul, A. P. Kudchadker, and D. Devaprabhakova, *Trans. Farad. Soc.*, **69**, 1821 (1969).

The method of determining virial coefficients by GC consists of measuring the retention volumes at various carrier-gas pressures and extrapolating to zero pressure. Three procedures of extrapolation have been suggested, although they do not give the same results. The method due to Cruickshank et al. (27), which takes into account carrier-gas flowrate and local pressure, is most promising.

The least complicated equation for determination of B_{12} is given by Cruickshank et al. (28):

$$\ln V_N = \ln V_N^\circ + \beta P_o J_3^4 \tag{9.58}$$

where V_N is the net retention volume; P_o is the outlet pressure; $\beta = (2B_{12} - V_1)/RT$; V_1 is the partial molar volume of the sample at infinite dilution in the stationary (liquid) phase; 1 refers to the sample, 2 to the carrier gas, and 3 to the stationary liquid; and J_3^4 is a function of the column inlet and outlet pressures P_i and P_o. Thus:

$$J_n^m = \frac{n}{m}\left[\frac{(P_i/P_o)^{m-1}}{(P_i/P_o)^{n-1}}\right] \tag{9.59}$$

Thus B_{12} may be obtained from the slope of the plot $\ln V_N$ versus $P_o J_3^4$.

9.6.2 Gas–Solid Virial Coefficients

The determination of gas–solid virial coefficients can be useful in explaining the interaction between an adsorbed gas and solid surface. The terms are defined so that the number of adsorbate molecules interacting can be readily ascertained. For example, the second-order gas–solid interaction involves one adsorbate molecule and the solid surface; the third-order gas–solid interaction involves two adsorbate molecules and the surface, and so on. The number of adsorbed molecules under consideration is expanded in a power series with respect to the density of the adsorbed phase, as is done for a bulk gas.

Few determinations of gas–solid virial coefficients have been made. Halsey and coworkers (29, 30) used the temperature dependence of the first gas–solid virial coefficients to calculate the potential energy curve for a single molecule in the presence of a solid. Hanlan and Freeman (31) showed that this coefficient may be obtained from frontal analysis of gas chromatographic data. Rudzinski et al. (32, 33) first used the second and third gas–solid virial coefficients obtained from gas chromatographic data to estimate surface areas. The surface area of silica gel determined by use of virial expansion data was greater than that obtained by application of the BET method. The discrepancy was explained by noting the BET method does not take the lateral interactions into account. These interactions have an effect of decreasing the effective area of the adsorbent, thus making the calculated BET area less than it should be.

9.7 KINETICS

Chemical kinetics describes the progress of a chemical reaction. The most common description of the progress is given by the term "rate of reaction"—a positive quantity that expresses how the concentration of a reactant or product changes with time. For the reaction A→P, the rate may be expressed as proportional to the appearance of product per unit time or the disappearance of reactant per unit time:

$$\left[\frac{\Delta(\mathrm{P})}{\Delta t}\right] \alpha \quad \text{rate} \quad \alpha - \left[\frac{\Delta(\mathrm{A})}{\Delta t}\right] \tag{9.60}$$

Gas chromatography can be a versatile tool in studying many reactions, especially in multicomponent systems, process reaction studies, or catalytic reactions. Samples can be taken from a reaction mixture at different time intervals and chromatographed, and the rate calculated from changes in concentrations. Alternatively, the chromatograph can be interfaced directly to the reactor where kinetic studies are performed directly (34).

The chromatographic column has been used as a reactor to study kinetics of dissociation. The reactant is introduced as a pulse at the head of the column and is continuously converted to product and separated as it travels through the column. The apparent rate constant (k_{app}) is a function of the rate of the liquid (stationary) phase reaction (k_1), the rate of the gas (mobile) phase reaction (k_g), the residence time in the gas phase (t_g), and the residence time in the liquid phase (t_1):

$$k_{\mathrm{app}} = k_1 + \left(\frac{t_g}{t_1}\right) k_g \tag{9.61}$$

A mathematical statement of the dependence of the rate on the concentrations of reactants is called the rate equation; for example, rate = $k(\mathrm{A})(\mathrm{B})$, where k = the rate constant and (A) and (B) represent the concentrations of reactants. From the rate equation one can frequently extract information on the mechanism (i.e., the path followed to convert reactants to products). Improvements in the gas chromatographic measurement of kinetic data have followed improvements in microchemical techniques and improvements in gas chromatographic instrumentation. Bertsch et al. (35) showed how microscale techniques can be applied to on-line reaction GC. They developed techniques at the nanogram scale for both on-line and postcolumn reactions.

Economopoulos et al. (36) used reversed-flow GC to study the kinetics of alcohol fermentation. In reversed-flow GC extra chromatographic sample peaks are created by reversing the carrier-gas flow direction for short time intervals during the course of the experiment. These narrow extra peaks are superimposed on normal elution curves to elicit information that is used in conjunction with measurements of suspended particles in the fermenting medium. The reversed-flow technique reduces two potential sources of error: the impact from lengthy high-temperature interactions on the column, and the minimization of interaction with the gas chromatographic packing material. Both of these potential sources of error could alter the composition of the sample in this complex biosystem.

Activation energies may be derived from gas chromatographic data. Activation energy is an emperical constant with units of energy that can be visualized as the quantity of energy needed before a reaction may begin. A plot of log k versus $1/T$ is linear and the slope equals $-E_a/2.303R$, where E_a is the activation energy, and R is the gas constant (1.99 cal/K mol). A useful expression relating the rate constants k_2 and k_1 at two different temperatures T_2 and T_1 is

$$\log \frac{k_2}{k_1} = -\frac{E_a(T_2 - T_1)}{2.303RT_1T_2} \tag{9.62}$$

Therefore, if E_a and k_1 are known, a rate constant for any temperature may be calculated.

9.8 PYROLYSIS

Pyrolysis gas chromatography (PGC) was one of the first combination gas chromatographic techniques, yet it is still plagued by problems of accuracy and repeatability of pyrolysis conditions and laboratory-to-laboratory reproducibility. There are three major devices for PGC: (1) heated wire or ribbon, (2) tube furnace, and (3) Curie-point filament. The heated-wire or ribbon apparatus uses resistive heating to provide flash pyrolysis from ambient temperature to 1400°C. It can be controlled to reach the maximum temperature in milliseconds or at some fixed rate and the can hold the top temperature for a settable fixed time. These devices can be placed directly in the injection port (for vertical injection ports). These are high-precision devices. The temperature reading should be checked from time to time to ensure accuracy.

The classical tube furnace is the oldest and simplest. A sample is placed in a boat (e.g., quartz or platinum) and the boat is placed in a quartz tube. The furnace is either moved over the sample, or the sample boat is pushed into the furnace. Carrier gas is swept through the tube and the pyrolysis products are swept into the chromatograph through a sampling valve. This method suffers from reproducibility of sample introduction and temperature lag. The chromatographic peak shape is generally broader than in other PGC methods. Newer microfurnace pyrolyzers give products that can be directly (on-line) analyzed via high-resolution GC (37).

Curie-point pyrolysis involves coating of the sample on a ferromagnetic conductor (wire or capillary tube). The conductor is inductively heated to a specific temperature when exposed to a radio-frequency field. The composition of the conductor determines the Curie temperature (300–1000°C). The major advantage of the Curie-point PGC is the ability to heat samples reproducibly to accurately defined temperatures in milliseconds. The major disadvantage is the inability to vary temperature since a different rod is needed for each point.

By far the largest, most successful applications of PGC have been to the characterization of synthetic polymer microstructure. PGC of these types of compounds yield much information, such as monomer identity and content, purity, and presence of additives. PGC is even more powerful for solving these types of problems when coupled with spectroscopic detectors. For example, Sahota et al. (38) showed that single-step PGC coupled with mass spectrometry could be used to measure the DNA content of cultured mamilian cells.

9.9 OTHER APPLICATIONS OF GC TO PHYSICOCHEMICAL MEASUREMENTS

9.9.1 Catalysis

Gas chromatography has been used to measure catalyst diffusivities, surface area, active surface area, kinetics, thermodynamics of adsorption, and pore size distribution and to study mechanisms and follow catalyst performance. Some examples are given in Table 9.8. The detection limit of about 100 ng DNA suggested that this technique might be useful for determining other cellular constituents in these complex biological systems.

9.9.2 Photochemistry

Gas chromatography may be used to separate photochemically derived species either on-line or off-line. If one uses a glass or quartz column or vaporizer, the study might be done on the column itself. Table 9.9 gives some examples of GC used in photochemistry.

9.9.3 Inverse Gas Chromatography

Inverse gas chromatography (IGC) is different from conventional gas chromatography in that the stationary phase is the analyte. The mobile phase is used to convey probes of known characteristics. The output from

TABLE 9.8 Applications of GC to Catalysis

Application	Reference
Acidity and catalytic selectivity in the Na–H–mordenite system	a
Displacement of H_2 from Rh surface by CO	b
Catalyst activity evaluation during waste-gas cleaning	c
Hydrogenation of olefins on a Pt/Ir GC/MS interface	d
Destruction of polychlorodibenzene-*p*-dioxins	e
Solid–liquid phase transfer catalysis-carboxylic acid alkylation	f
Thiophene poisoning of copper chromite	g
Thermoprogrammed reduction of cobalt oxide catalysts	h
Hydrodenitrogenation catalysis by reversed-flow GC	i

[a] P. Ratnasamy, S. Sivashkar, and S. Vishmoi, *J. Catal.*, **69**, 428 (1981).
[b] W. K. Jozwiak and T. Parjczak, *React. Kinet. Catal. Lett.*, **18**, 163 (1981).
[c] Y. Mezhiritski, A. Yu, G. A. Fakina, and I. S. Kolbasin, *Khim. Tekhnol. Topl. Masel.*, **12**, 34 (1982), through *Chem. Abstr.* **98**, 77456 (1983).
[d] G. C. Jamieson, J. *High Resolut. Chromatogr., Chromatogr. Commun.*, **5**, 632 (1982).
[e] D. C. Ayres, *Nature*, **290**, 323 (1981).
[f] A. Arbin, H. Brink, and J. Vesiman, *J. Chromatogr.*, **196**, 255 (1980).
[g] V. R. Choudhary and S. D. Sansare, *J. Chromatogr.*, **192**, 420 (1980).
[h] T. Paryjczak, J. Rynkowski, and S. Karski, *J. Chromatogr.*, **188**, 254 (1980).
[i] A. Niotis and N. A. Katsanos, *Chrommatographia*, **34**, 398 (1992).

TABLE 9.9 Applications of Gas Chromatography in Photochemistry

Application	Reference
Photochemistry within a glass gas chromatographic column	a
Photodecomposition of sulfonamides and tetracyclines	b
Photoreduction of methylviologen adsorbed on cellulose	c
Radiolysis of D,L-tryptophan	d
Photolysis of dichlorofluanid	e

[a] W. G. Laster, J. B. Pawliszyn, and J. B. Phillips, *J. Chromatogr. Sci.*, **20**, 278 (1982).
[b] W. H. K. Sanniez and N. P. Ipel, *J. Pharm. Sci.*, **69**, 5 (1980).
[c] M. Kaneko, J. Motoyoshi, and Y. Yamada, *Nature*, **285**, 1468 (1980).
[d] W. A. Bonner, N. E. Balir, and J. J. Flores, *Nature*, **281**, 150 (1979).
[e] T. Clark and D. A. M. Watkins, *Pest. Sci.*, **9**, 225 (1978).

the experiment (retention time, peak shape, and so on) is monitored to glean information about the stationary phase. The stationary phase may be composed of fibers, polymer pellets, minerals, or a substance coated on an inert chromatographic support of the wall of the chromatographic column (39). IGC has been especially useful for characterizing polymeric species (40). Measurements such as degree of crystalinity, glass and melting transition temperatures, solubility parameters, diffusion properties, interaction parameters with polymer blends, and interfacial and surface properties have been carried out on a variety of systems. Table 9.10 gives a variety of examples.

9.9.4 Simulated Distillation

Gas chromatographic retention data can be used to simulate the results of fractional distillation. The separation takes place on a chromatographic column whose interactions (between sample components and the stationary phase) give a sample elution profile that can be correlated with its boiling point distribution. Boiling range distribution profiles are especially important in the petroleum industry where such information may be used to control refining operations and specifications testing, to determine the commercial value of crude oil to a refiner, to calculate vapor pressure of gasoline or gasoline fractions (used to describe automobile performance parameters), or as a "fingerprint" (41) to help identify the source of a spill

TABLE 9.10 Applications of Inverse Gas Chromatography

Application	Reference
Polyester fiber structure	a
Interactions between solvents and linear or branched polystyrene	b
Column preparation	c
Organic solute solubility in polymer films	d
Surface properties of active carbons	e
Interaction of polyetherpolyurethane with solvents and solubility parameter	f
Polystyrene-hydrocarbon interaction parameters and solubility parameter	g
Polycarbonate surface energies and interaction characteristics	h
Surface heterogeneity of alumina oxide ceramic powders	i,j
Surface energy distribution	k

[a] A. Seves and G. Prati, *Tinctoria*, **79**, 376 (1982).
[b] M. Galin, *Polymer*, **19**, 596 (1978).
[c] T. Inui, Y. Marakami, T. Suzuki, and Y. Takegami, *Polym. J. (Tokyo)*, **14**, 261 (1982); through *Chem. Abstr.*, **97**, 39599e (1982).
[d] J. E. G. Lipson and J. E. Guillet, *J. Coating Technol.*, **54**, 89 (1982).
[e] F. J. Lopez-Garzon, M. Pyda, and M. Domingo-Garcia, *Langmuir*, **9**, 531 (1992).
[f] A. M. Faarooque and D. D. Deshpande, *Eur. Polym. J.*, **28**, 1547 (1992).
[g] E. Ozdemir, A. Acikses, and M. Coskun, *Macromol. Rep.*, **A29**, 63 (1993).
[h] U. Panzer and H. P. Schreiber, *Macromolecules*, **25**, 3633 (1992).
[i] J. L. Roles and G. Guiochon, *J. Chromatogr.*, **233**, 591 (1992).
[j] J. L. Roles and G. Guiochon, *Anal. Chem.*, **64**, 25 (1992).
[k] J. L. Roles and G. Guiochon, *J. Phys. Chem.*, **95**, 40098 (1991).

or leaking underground storage tank. The chromatographic procedure is often called SIMDIS for "simulated distillation" analysis.

In SIMDIS the sample is introduced onto a chromatographic column that separates components in boiling-point order. The column temperature is programmed up and the area under the chromatogram recorded. When a suitable calibration mixture is used, the retention time axis may be replaced with boiling temperature. From the boiling temperature and the chromatographic areas, the boiling range distribution of the sample is obtained.

Two gas chromatographic methods are designated by the American Society for Testing and Materials, ASTM. ASTM Method D 2887 (42) is used for determining the boiling range distribution of petroleum fractions with a final boiling point of 538°C or lower. ASTM Method D 3710 (43) is used for determining the boiling range distribution of gasoline and gasoline fractions with a final boiling point of 260°C or lower. Method D 3710 is sometimes referred to as gas chromatographic distillation (GCD). Both methods also recommend how to prepare calibration standards for SIMDIS.

Since neither ASTM method makes specific recommendations regarding the gas chromatographic column, any column can be used that meets the method's specifications for separation in order of boiling point and certain column performance requirements regarding resolution, system noise and sensitivity, drift, and so on. Consequently, high-resolution GC has been increasingly applied to SIMDIS analyses (44, 45, 46), since most commercial high-resolution columns can surpass the recommended column performance characteristics.

9.10 ACCURACY, PRECISION, AND CALIBRATION

The accuracy and precision of physicochemical measurements by GC rely on the ability of the gas chromatograph to control and measure all parameters relating to the required chromatographic data. Temperature, in general, should be controlled and known to at least ±0.005°C. In some cases the stationary phase mass must be known within ±0.2%.

Most chromatograph manufacturers claim that temperature can be controlled within 0.25°C on modern chromatographic instruments. That does not mean that the temperature is accurate within the range, however. Furthermore, we have seen that temperature gradients exist both within the oven and even the column itself. For most physicochemical measurements, therefore, the instrumentation must be tested to determine the precision and accuracy of the essential settings.

It is not easy to determine which factors play the greatest role in obtaining good accuracy and precision. One must consider the assumptions inherent in the theory as well as the chemical, mechanical, and instrumental parameters. In general, gas chromatographic methods agree within 1–5% with other physicochemical methods. For example, Hussam and Carr (16) showed that in the measurement of vapor/liquid equilibria via headspace GC, complex thermodynamic and analytical correction factors were needed. These often came from other experimental measurements that were not necessarily accurately known. Another source of significant error can be in the determination of the mass of stationary phase contained within the column (47). Other sources of error include measurement of holdup time (48), flowrate, sample mass, response factors, peak area or baseline fidelity, and so on.

The general issues around chromatographic calibration have been addressed in Chapter 7. In the measurement of physicochemical properties by GC, one must also address how the data were obtained on the "known" species. This usually entails some careful literature inspection, including determining the mechanism and underlying principles used in the method. Another important factor in making relative measurements is how closely the model compound resembles the "unknown." In any event, the chromatographer needs to consider all possible types of interactions that might occur from sample preparation through detection. With all these caveats, however, the speed and simplicity of the gas chromatographic method should continue to make it very attractive for making physicochemical measurements.

REFERENCES

1. R. L. Grob, M. A. Kaiser, and M. J. O'Brien, *Am. Lab.*, **7**, 40 (1975).
2. S. Brunnauer, P. Emmett, and E. Teller, *J. Am. Chem. Soc.*, **60**, 309 (1938).
3. F. M. Nelsen and F. T. Eggertsen, *Anal. Chem.*, **30**, 1387 (1958).
4. Y. Kuge and Y. Yoshikawa, *Bull. Chem. Soc. Jpn.*, **38**, 948 (1965).
5. M. F. Burke and D. Ackerman, *Anal. Chem.*, **43**, 573 (1971).
6. A. V. Kiselev and Y. Yashin, *Gas-Adsorption Chromatography*, Plenum, New York, 1969.
7. A. P. Karnaukov and N. E. Buyanova, in *Surface Area Determinations*, Butterworths, London, 1970, p. 165.
8. S. Masukawa and R. Kobayashi, *J. Gas Chromatogr.*, **6**, 257 (1968).
9. W. V. Lobenstein, *J. Biomed. Mater. Res.*, **9**, 35 (1975).
10. D. Dollimore and G. Heal, *Nature*, **208**, 1092 (1965).
11. T. Paryjczak, *Gas Chromatography in Adsorption and Catalysis*, Ellis Harwood Halsted, West Sussex, New York, 1987.
12. J. H. Noggle, *Physical Chemistry*, 2nd ed., Scott-Foresman, Chicago, 1990.
13. J. L. Roles and G. Guiochon, *J. Phys. Chem.*, **95**, 4098 (1991).
14. B. Kolb, in *Applied Headspace Gas Chromatography*, B. Kolb (Ed.), Heyden, London, 1980, p. 1.
15. J. N. Labows, *JAOCS*, **69**, 34 (1992).
16. A. Hussam and P. W. Carr, *Anal. Chem.*, **57**, 793 (1985).

17. H. Mackle, R. Mayrick, and J. Rooney, *Trans. Farad. Soc.*, **56**, 115 (1960).
18. J. W. Westcott and T. F. Biddleman, *J. Chromatogr.*, **210**, 331 (1981).
19. D. J. Hamilton, *J. Chromatogr.*, **195**, 75 (1980).
20. T. E. Jordan, *Vapor Pressure of Organic Compounds*, Interscience, New York, 1954.
21. J. H. Purnell, in *Gas Chromatography, 1966*, A. B. Littlewood (Ed.), Institute of Petroleum, London, 1967, p. 3.
22. D. E. Martire and P. Riedl, *J. Phys. Chem.*, **72**, 3478 (1968).
23. J. P. Sheridan, D. E. Martire, and Y. B. Tewari, *J. Am. Chem. Soc.*, **94**, 3294 (1972).
24. D. F. Cadogan and J. H. Purnell, *J. Phys. Chem.*, **73**, 3489 (1969).
25. H. Liao, D. E. Martire, and J. P. Sheridan, *Anal. Chem.*, **45**, 2087 (1973).
26. C. Eon and G. Guiochon, *Anal. Chem.*, **46**, 1393 (1974).
27. A. J. B. Cruickshank, M. L. Windsor, and C. L. Young, *Proc. R. Soc. (London), Ser. A*, **295**, 271 (1966).
28. A. J. Cruickshank, M. L. Windsor, and C. L. Young, *Proc. R. Soc. (London) Ser. A*, **295**, 259 (1966).
29. J. R. Sams, G. Constabaris, and G. D. Halsey, *J. Chem. Phys.*, **36**, 1334 (1962).
30. W. A. Steele and G. D. Halsey, *J. Phys. Chem.*, **59**, 57 (1955).
31. J. F. Hanlan and M. P. Freeman, *Can. J. Chem.*, **37**, 1575 (1959).
32. W. Rudzinski, Z. Suprynowica, and J. Rayss, *J. Chromatogr.*, **66**, 1 (1972).
33. W. Rudzinski, A. Waksmundzki, Z. Suprynowica, and J. Rayss, *J. Chromatogr.*, **72**, 221 (1972).
34. R. L. Grob and J. B. Leasure, *J. Chromatogr.*, **197**, 129 (1980).
35. W. Bertsch, J. Correa, and G. Holzer, abstract 402P, 1993 Pittsburgh Conference and Exposition on Analytical Chemistry and Applied Spectroscopy, Atlanta, 1983.
36. N. Economopoulos, N. Athanassopoulos, N. A. Katsanos, G. Karaiskakis, P. Agathonos, and Ch. Vassilakos, *Separation Sci. Technol.*, **27**, 2055 (1992).
37. T. Usami, F. Keitoku, H. Ohtani, and S. Tsuge, *Polymer*, **33**, 3024 (1992).
38. R. S. Sahota, S. L. Morgan, and K. E. Creek, *J. Anal. Appl. Pyrolysis*, **24**, 107 (1992).
39. A. Voelkel, *CRC Crit. Rev. Anal. Chem.*, **22**, 411 (1991).
40. D. R. Lloyd, T. C. Ward, H. P. Schreiber, and C. C. Pizana, *Inverse Gas Chromatography*, ACS Symposium Series 391, Americam Chemical Society, Washington, DC, 1989.
41. J. J. Kosman and R. G. Luko, *J. Chromatogr. Sci.*, **31**, 193 (1993).
42. *Annual Book of ASTM Standards*, **5.02**, 192 (1993).
43. *Annual Book of ASTM Standards*, **5.02**, 558 (1993).
44. S. Trestianu, G. Zilioli, A. Sironi, C. Saravalle, and F. Munar, *High Resolut. Chromatogr. Chromatogr. Commun.* **8**, 771 (1985).
45. L. A. Luke and J. E. Ray, *High Resolut. Chromatogr. Chromatogr. Commun.* **8**, 193 (1985).
46. D. S. Workman, F. Noel, and M. R. Watt, *J. Chromatogr. Sci.*, **31**, 95 (1993).
47. J. L. Roles, *J. Chromatogr.*, **591**, 245 (1992).
48. M. A. Klemp and R. D. Sacks, *J. Chromatogr. Sci.*, **29**, 507 (1991).

CHAPTER TEN

Petroleum and Petrochemical Analysis by Gas Chromatography

E. F. SMITH
Exxon Chemical Company

Modern Practice of Gas Chromatography, Third Edition. Edited by Robert L. Grob.
ISBN 0-471-59700-7

10.1 INTRODUCTION

Gas chromatography has been developed into a key analytical tool for the petroleum and petrochemical industry. This chapter serves as an introduction to the application of gas chromatography in this field. For petroleum, it covers the exploration, production, and refining of crude oil. Applications for the major derivatives of petroleum that are basic to the chemical industry are discussed in Section 10.4. Process gas chromatography is also discussed. In addition to the routine analyses, an attempt has been made to indicate the potential for the further development of gas chromatography in the petroleum industry. For a more detailed review of chromatography in the petroleum industry, the reader is referred to the book edited by K. H. Altgelt and T. H. Gouw (1). Also, a general review of the chemical and technology of petroleum and petrochemicals has been provided by Speight (2).

10.1.1 Historical Perspective

The analysis of hydrocarbons in petroleum and its products began in the mid 1800s. The original methods were based on physical properties such as boiling point and specific gravity. In 1928, the American Petroleum Institute (API) initiated Project 6 to separate, identify, and determine the chemical constituents of commercial petroleum fractions. From this program, column adsorption chromatography was developed for separating components by hydrocarbon type. Because of the military needs of World War II in the 1940s, petroleum laboratories quickly developed spectroscopic methods. Mass spectrometry was introduced in 1943 for gas analysis. It then began to replace the low-temperature fractional distillation method for light hydrocarbons that was developed by Podbielniak (3). Then in the early 1950s, gas chromatography (GC) was developed.

The development of GC and the analysis of petroleum and petrochemicals have enjoyed a mutually beneficial relationship. Indeed, the first international symposium on vapor-phase (gas) chromatography was sponsored by the British Institute of Petroleum in 1956 (4). Papers describing the analysis of refinery gas, solvents, aromatics in coal-tar naphthas, and samples from the internal combustion engine were presented. Most of the work included in these presentations was done on "homebuilt" chromatographs. The first commercial gas chromatograph or "vapor-phase fractometer" was also described, along with the first ionization detector.

The rapid development of GC continued to parallel the refinement of petroleum applications. Eggertsen and his coworkers (5) in 1956 described a 50-ft column packed with carbon black containing 1.5% squalene to separate 10 major C_5–C_6 saturates in 2 h. By 1958 the same workers used other column packings and extended the analysis of the saturates to C_7. All but one of the 25 saturates were resolved in 12–16 h.

The second international symposium was held in 1958 and was again sponsored by the British Institute of Petroleum (6). Improved techniques allowed Scott (6) to separate the C_7 and C_8 paraffin isomers, using a column with 30,000 theoretical plates. Golay (6) also described the potential of open tubular capillary columns. These highly efficient columns were readily adopted by the petroleum industry. In 1961, Desty et al. (7) reported on the use of a 900-ft glass open tubular column coated with squalene to resolve 122 peaks from C_3 to C_9 in 20 h. This work was part of the API Project 6. In 1968, Sanders and Maynard (8) published a method for C_3 through C_{12} hydrocarbons in gasoline. They performed this analysis in less than 2 h on a 200-ft squalene column by using both flow and temperature programming.

The recent trend has been toward the increased use of GC in the petroleum industry. This is due to the relatively low cost of a gas chromatographic system that can provide more detailed analyses as well as the detection of trace components. Gas chromatography has also provided a means for on-line process monitoring. Future utilization of GC will involve the establishment of basic relationships between composition and performance parameters. This will allow the substitution of this rapid and reliable chromatographic analysis for the empirical methods from the past.

10.1.2 Standardization of Analyses

The concept of standardized testing is important in the petroleum industry, as it is in many others. Its usefulness for commodity-type products that are widely bought, sold, and exchanged is apparent. The American Society for Testing and Materials (ASTM) has provided these standardized procedures that are required for product specification testing.

The ASTM is an international, nonprofit, technical, scientific, and educational society that was formed in 1898. Its purpose is the development of standards on the characteristics and performance of materials, products, systems, and services and the promotion of related knowledge. Because of the support of the petroleum industry, the ASTM represents a source of voluntary consensus standards for hydrocarbon analyses.

Although the first committee was formed in 1904, most of the work of the ASTM for petroleum products has been done since 1940. The D-2 technical committee on petroleum products and lubricants is responsible for almost all petroleum products. The D-16 committee is responsible for aromatic hydrocarbons and related chemicals. The E-19 committee for gas chromatography was established in 1961. More recently, it has been expanded to include all types of chromatographic analyses. As such, this committee works closely with the individual product type committees.

The ASTM methods for gas chromatographic analyses usually describe a generalized procedure. They allow for a choice of instrumentation and columns, but set standards for sample preparation, column resolution, and analytical quantification. They therefore provide the analyst with some flexibility to adjust for specific laboratory, company, or personal preferences. At the same time, they maintain standards of practices that can be used to "referee" analytical results between the supplier and customer.

The origins of tests published by the ASTM vary significantly. Many are developed within organizations such as the American Petroleum Institute (API), U.S. Bureau of Mines, and the National Institute of Standards and Technology (NIST, formerly the National Bureau of Standards (NBS)). Regardless of their origin, they represent voluntary consensus standards.

A manual on hydrocarbon analysis is published by the ASTM (9). It is a compilation of all the ASTM standards relating to hydrocarbons and includes the applicable gas chromatographic techniques. A manual on gas chromatographic methods has also been published (10). Volumes 5.01, 5.02, and 5.03 include the updates for petroleum products and lubricants; aromatic hydrocarbons are covered in Volume 6.03; and Volume 14.01 contains the methods for chromatography. For further details on the test procedures discussed in this chapter, the most recent annual book should be consulted.

The petroleum industry, like many other industries, has developed its own terminology. The terms used in this chapter reflect those commonly used in the petroleum industry. They are consistent with those adopted by the Division of Petroleum Chemistry of the American Chemical Society. These terms include the following:

- *Aromatics:* All hydrocarbons containing one or more rings of the benzenoid type
- *Naphthenes:* Saturated cyclic hydrocarbons or cycloalkanes
- *Olefins:* All alkenes
- *Paraffins:* All noncyclic saturated hydrocarbons, including both normal and branched alkanes
- *Saturates:* All naphthenes and paraffins

10.2 EXPLORATION AND PRODUCTION

Petroleum-derived fossil fuels consist of a wide variety of components ranging from methane to high molecular weight multifunctional altered biochemical molecules. Some of these molecules contain the heteroatoms oxygen, nitrogen, and sulfur, which are able to complex trace metals such as vanadium, nickel, and iron. The composition, source, and quality of such oils are determined by a complex set of variables. These include the type of organic material originally deposited, the depositional environment, and the time and depth of burial—which, in turn, determines the amount of thermal alteration. Furthermore, the fossil fuels are recoverable only by migrating from their original source to a porous and permeable trap or reservoir structure from which they can be economically produced.

The quantitative perspective involved in the earth's carbon chemistry is

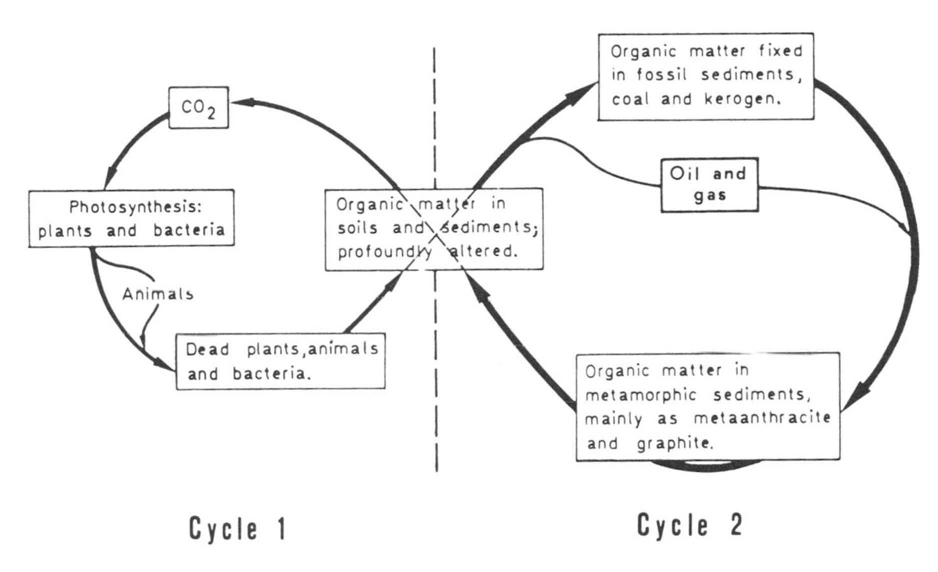

FIGURE 10.1 The two major cycles of organic carbon on earth. Most organic carbon is recycled within Cycle I. The crossover from Cycle I to Cycle II is only a tiny leak that amounts to only 0.01–0.1% of the total primary productivity. (Reprinted with permission from Reference 12, Springer-Verlag, Copyright 1978.)

interesting. There are two parts to the earth's carbon cycle, as illustrated in Figure 10.1. Cycle I is the organic carbon cycle and involves approximately 3×10^{12} tons (T) of fixed organic matter. It has a half-life of several million years and contains a relatively small fraction of recoverable hydrocarbon. Current estimates of 2.1×10^{11} T of recoverable oil and gas means that only 0.003% of the fixed carbon is available for exploitation.

To better understand the complex composition of petroleum and its analysis, it is useful to briefly survey the processes that convert deposited biological detritus to recoverable fossil fuels. The fact that petroleum is derived from biological material is clear from its composition. Carbon isotope data and the presence of optical isomers strongly suggest a biogenic origin as does the presence of "biomarkers" or geochemical fossils. For example, there are a whole series of terpene-derived molecules in petroleum, such as pristine, phytane, and cholestane, which originate from five-carbon isoprene units. Pristine and phytane are altered forms of the phytol chain of chlorophyll. Cholestane is an altered form of cholesterol. Other biomarkers have been identified as derivatives of fatty acids, proteins, carbohydrates, and biological pigments.

The variables that determine how much organic matter is transferred from the organic carbon cycle to the fixed carbon cycle include the quantity of organic matter originally deposited, the type of this organic matter (terrestrial, marine, etc.), and the depositional environment. The last term includes a number of additional variables such as the rate of deposition, whether the depositional zone was aerobic or anaerobic (with or without oxygen), and the degree and type of biological activity during deposition. The biological activity that alters the original matter is the beginning of the formation of petroleum.

The conversion of deposited organic matter to petroleum is clearly a complex process, which is summarized in Figure 10.2. Diagenesis, the first step, is the process by which biological molecules are first altered through microbial activity into smaller molecules. These then polymerize into successively naturally occurring organic functional groups. Kerogen is not soluble in organic solvents and can be isolated by dissolving away the mineral matrix with HCl and HF. Thus isolated, kerogens have been studied and classified. Some general characteristics of kerogen are summarized in Table 10.1. Note that the origin of the organic matter from which the kerogen is formed strongly impacts on the composition of the kerogen. This is important because petroleum is formed from kerogen by a process called *catagenesis*.

Catagenesis is the thermal degradation of kerogen to produce oil and some natural gas. Burial depths necessary to provide temperatures of 60–150°C are needed to optimize oil generation, depending on the type of kerogen present and the time of burial. During catagenesis, the hydrogen/carbon ratio for the kerogen decreases as hydrocarbons are released to the surrounding matrix. The organic matter that can be extracted with solvents make up the bitumen, which also consists of a very small amount of unaltered hydrocarbons from the originally deposited organic matter. As catagenesis progresses, the amount of bitumen increases as hydrocarbons are generated from the thermal degradation of kerogen. Hydrocarbons generated early in catagenesis are higher molecular weight than those generated later in the process. Methane is the predominate hydrocarbon generated in the very latest stages. The point at which kerogen is being cracked to methane and carbon is the beginning of *metagenesis*. Metagenesis is counterproductive toward the formation of oil but is the source of much of the natural gas produced. Petroleum geologists refer to sediments that have reached this stage as overmature. Sediments exposed only to catagenesis are referred to as "mature" and represent the optimum oil sources.

The time scale for these processes is on the order of millions of years; however, another important process is also occurring. During and after the various states of oil generation, hydrocarbons are constantly migrating. Although the mechanisms of oil migration still require better understanding, two types of migration have been defined. *Primary migration* refers to the movement of dispersed hydrocarbons out of the sedimentary matrix in which they originated, that is, the source rock. *Secondary migration* refers to the movement of still highly dispersed hydrocarbons through rock layers other than source rock. These "other" rocks generally have a higher porosity than found in the source rock or have a network of microfractures

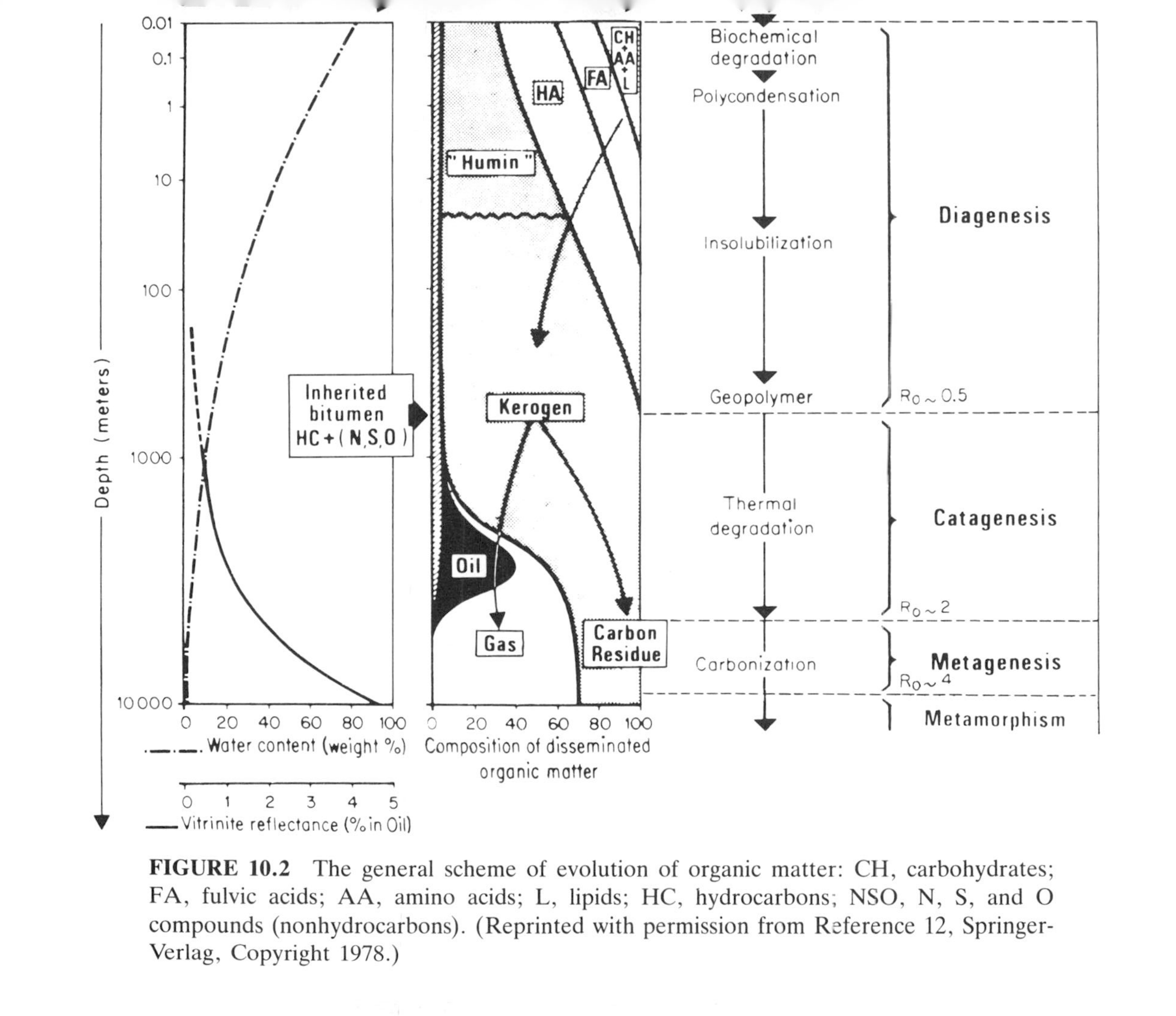

FIGURE 10.2 The general scheme of evolution of organic matter: CH, carbohydrates; FA, fulvic acids; AA, amino acids; L, lipids; HC, hydrocarbons; NSO, N, S, and O compounds (nonhydrocarbons). (Reprinted with permission from Reference 12, Springer-Verlag, Copyright 1978.)

TABLE 10.1 Characteristics of Kerogen Based on Origin

	Marine	Terrestial
Hydrogen:carbon ratio	>1.5	<1.3
Biological markers		
Odd–even preference		
C_{15}–C_{21} *n*-alkanes	Odd	None
C_{27}–C_{35} *n*-alkanes	Even or none	Odd
Pristane:phytane	Low (<1)	High (>3)
Pristane:*n*-C_{17}	Low	High
Terpenoids	Low	High
Fatty acids		
C_{12}–C_{18}	High	Low
C_{24}–C_{36}	Low	High

Source. Reference 13

and cracks. If this secondary migration is impeded by an impermeable layer of geologic structure, the petroleum or natural gas will accumulate in a reservoir. In the absence of a trapping mechanism, migration continues both horizontally and vertically upward toward the earth's surface at exceedingly slow rates. Because of the enhanced concentration of hydrocarbons in reservoirs, these reservoirs are the object of investigation for petroleum and structural geologists. But it is with the help of petroleum geochemists that we better understand the source of the petroleum, are able to correlate an oil to its source, and can better predict other potential reservoirs.

10.2.1 Geochemical Studies

The modern petroleum geochemist is interested primarily in identifying oil sources or potential sources as well as correlating a source with reservoired oil. By combining the information about source with other geologic data, the explorationist can better predict the location of promising petroleum reservoirs. In searching for the source rock, the geochemist is faced with a unique sampling problem. A relatively small percentage of the samples available are oils and an even smaller percentage are cores. The latter are expensive to obtain and require special drilling equipment and procedures. They do, however, provide an intact sample of sedimentary layers from easily identifiable strata and accurately determined depth. Most of the samples available to the petroleum geochemist are rock chips known as *drill cuttings*. These cuttings are brought to the surface suspended in the drilling fluid and are separated from the fluid by wet sieving in a shale shaker. Samples are periodically scooped at intervals that are correlated to the drilling depth. These are frequently contaminated with cave-ins and particles that are recirculated within the well before reaching the surface. Drilling fluid additives are also potential sources of contamination. Clearly, a thorough understanding of such a nonideal sample source is necessary for the reasonable interpretation of the analytical results.

Analysis of light hydrocarbons adsorbed to cuttings or cores is frequently used to screen samples and can provide useful geochemical information. For such an analysis, the samples (most often cuttings) must be preserved wet in tightly sealed containers that contain a biocide to prevent bacteriological alteration of the sample. The C_4 and lighter hydrocarbons are then sampled either from the container headspace or by one of a combination of thermal, mechanical, or chemical extraction methods. Subsequent gas chromatographic analysis frequently includes backflushing the less volatile C_4+ hydrocarbons in order to minimize analysis time while providing the desired analysis of C_1–C_4 hydrocarbons. Samples with more than 50% methane are called "dry," whereas samples relatively high in C_2–C_4 hydrocarbons are referred to as "wet." The latter are most often associated with the presence of oil, whereas the former are associated with natural gas from either immature or overmature sources.

A study of source rocks of western Canada by Evans and Staplin (14) used cuttings gas analyses to help map areas of immature, mature, and overmature petroleum sources. The progression from immature to mature shown in Figure 10.3 occurs abruptly at a narrow interval beginning at a depth corresponding to a temperature of 90°F (32°C). The temperatures shown are the paleotemperatures or the maximum temperatures to which

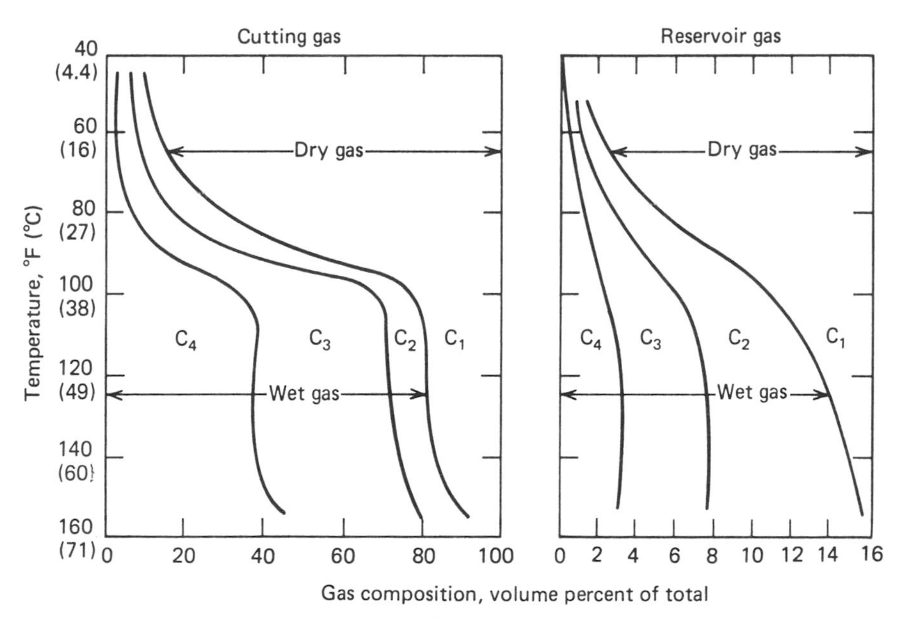

FIGURE 10.3 The composition of Cretaceous gas in source and reservoir rocks versus subsurface temperature. (Reprinted with permission from Reference 14.)

the rock was exposed. These are 70–100°F higher than current temperatures because of the uplifting that was occurred.

Another study of the same general area by Bailey et al. (15) offers an interesting comparison of samples representing the progression from immature to overmature sources. Figure 10.4 shows the analysis of C_1–C_4 hydrocarbons in typical well log form, that is, as a function of well depth. Well 1 shows methane and very little C_2–C_4 hydrocarbons. This, along with a very light-colored kerogen, indicates an immature source. Wells 2 and 3 show increasing amounts of wet gas and have an intermediate kerogen color, indicating optimum maturity. Well 4, like Well 1, shows primarily methane in the cuttings gas, but a very dark kerogen color indicates that this dry gas is the result of an overmature source. Samples of the same locations were also analyzed for C_4–C_7 hydrocarbons, that is, the gasoline range (Figure 10.5). These results closely parallel the cuttings gas results.

The analysis of the C_4–C_7 hydrocarbons of cuttings or cores generally involves solvent extraction with a heavy solvent that can be backflushed with the C_7 + hydrocarbon fraction. A typical instrument and chromatogram are

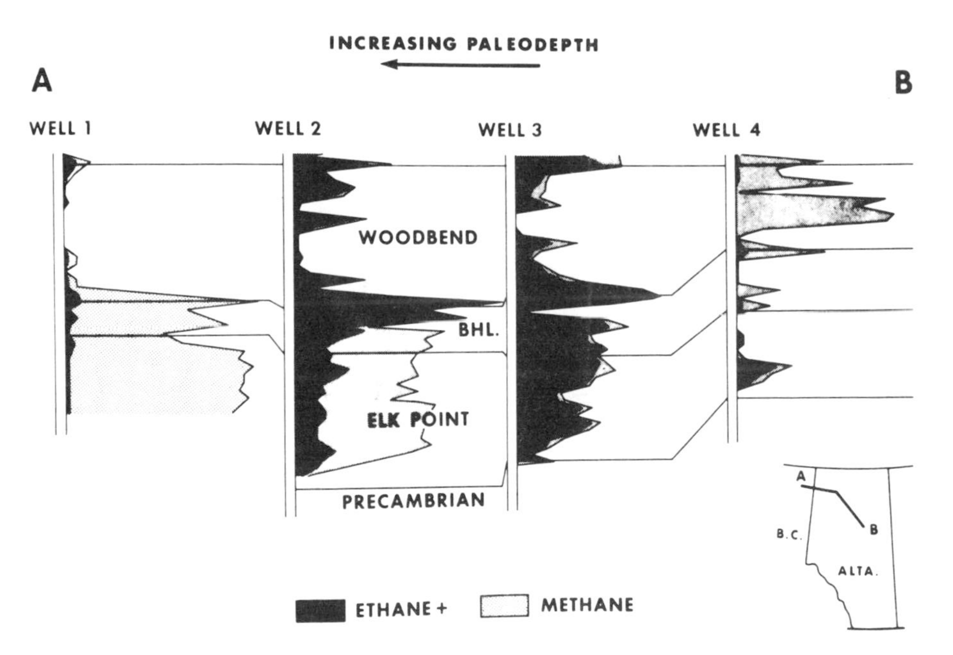

FIGURE 10.4 Cuttings-gas (C_1–C_4) composition (in log cross-section form) of Upper and Middle Devonian strata. The maximum paleodepth increases from east to west. The initial decrease and subsequent decrease in wet gas in this direction illustrate the transition from immaturity to maturity to metamorphism with increasing temperature. (Reprinted with permission from Reference 15.)

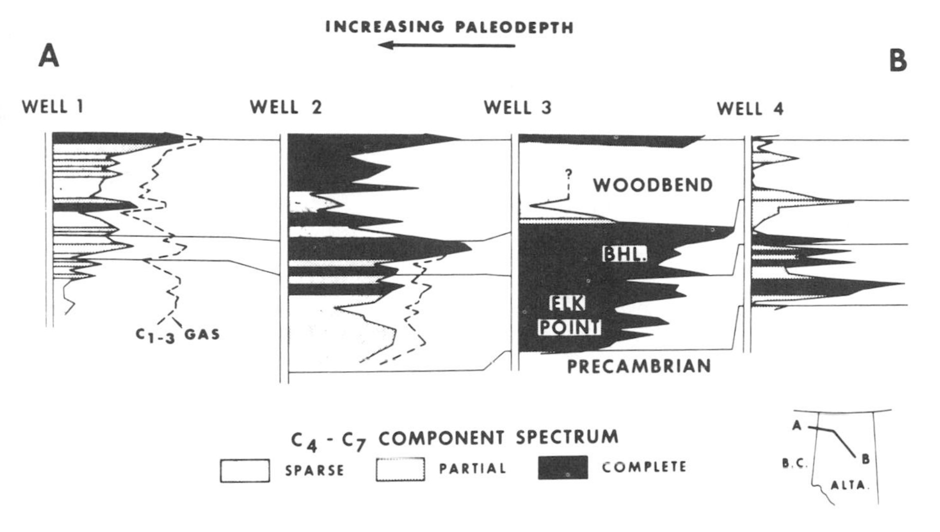

FIGURE 10.5 Gasoline-range hydrocarbon (C_4–C_7) composition (in log cross-section form) for the same wells shown in Figure 10.4. An increase and subsequent decrease both in richness and completeness of the array of components in this fraction illustrates increasing maturation culminating in metamorphism. (Reprinted with permission from Reference 15.)

shown in Figure 10.6. Generally, the choice of solvent is what limits the upper range of the hydrocarbons that are analyzed and so represents a limitation of the method.

An interesting alternative to solvent extraction of samples is direct stripping of the hydrocarbons by the carrier gas as developed by Schaefer et al. (16, 17). In addition to extending the range of hydrocarbons that can be analyzed, this approach reduces sample preparation time. In this analysis, a small sample (<1 g) of rock chips is stripped of adsorbed hydrocarbons by placing the sample directly in the carrier-gas stream (Figure 10.7). The hydrocarbons are concentrated in a cryogenic trap and then vaporized onto a capillary column. A Deans-type flow controller (18) allows backflushing of the C_{10} + hydrocarbon fraction while the analytical separation proceeds on the downstream segment of the column. This method has been shown to compare favorably to the more conventional methods and provides the advantages of extended range, faster analysis, and reduced sample requirements. A typical chromatogram is shown in Figure 10.8. This procedure has been used by Thompson (19) to develop maturation correlations for a variety of samples of similar source characteristics. A paraffin index was defined to simplify and summarize the data obtained in the gas chromatographic analysis. The index is a ratio of normal to isoalkanes for different

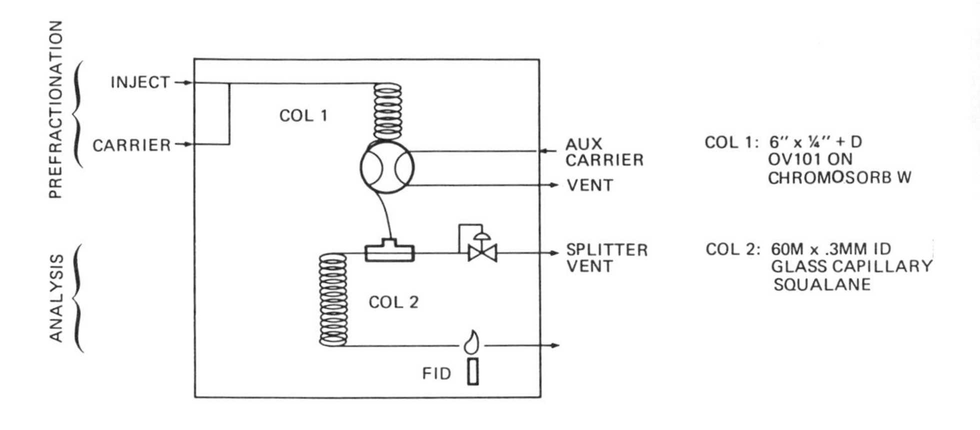

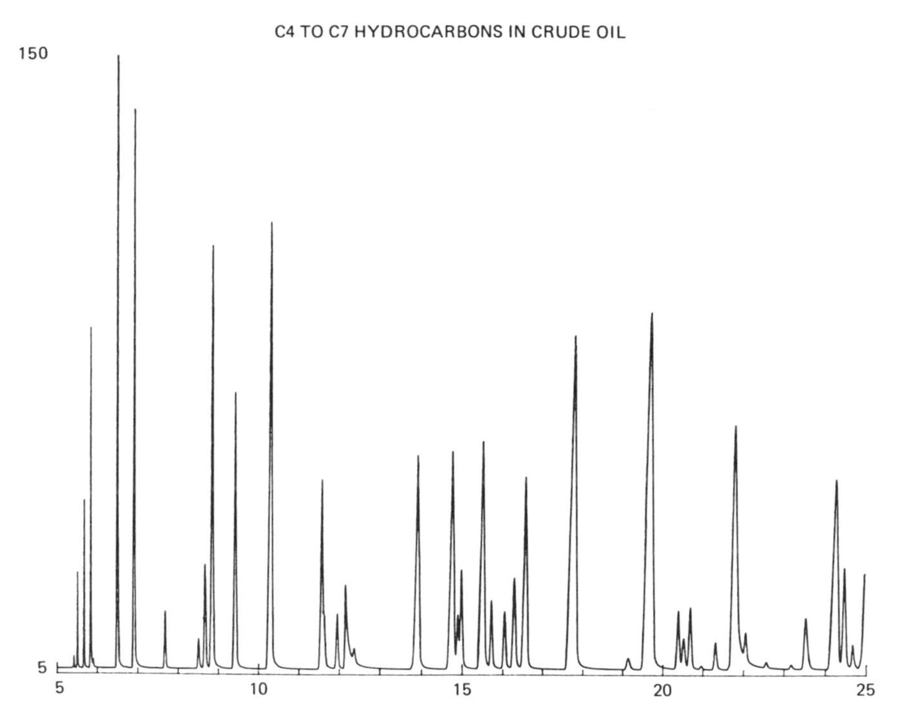

FIGURE 10.6 Chromatographic system and typical chromatogram for the analysis of C_4–C_7 hydrocarbons in crude oil extracts. The solvent is backflushed with the C_{8+} fraction.

carbon number ranges (C_6 and C_7). They correlate with the maturity of samples from different sources and suggest a degree of universality of catagenetic reaction mechanisms for similar sources.

Except for "fingerprinting," the chromatograms of rock extracts or oils

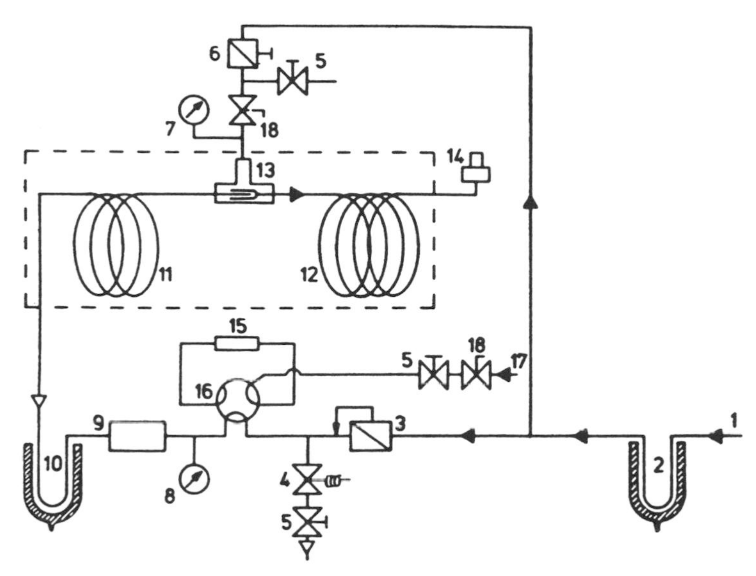

FIGURE 10.7 Modified capillary gas chromatographic system for light hydrocarbon analysis of rock samples by hydrogen stripping: 1, hydrogen inlet; 2, hydrogen purification trap; 3, flow controller; 4, solenoid valve; 5, needle valve; 6, pressure regulator; 7, 8, pressure gauges; 9, sample tube; 10, cold trap; 11, 12, capillary column; 13, T-union; 14, FID; 15, gas loop; 16, six-port valve; 17, inlet for external standard; 18, valve: -▶-direction of gas flow, -▷-reversed flow during backflush. (Reprinted with permission from Reference 17, *Analytical Chemistry*, Copyright 1978, American Chemical Society.)

become too complex to obtain much useful information for the $C_{10}+$ hydrocarbon fraction without some form of sample pretreatment. For the small samples available from rock extracts, the most common way to simplify the gas chromatographic analysis is to fractionate the same sample by hydrocarbon type by liquid chromatography. Asphaltenes are separated from the bulk sample by precipitation from pentane or hexane. The pentane soluble fraction is then chromatographed on an alumina–silica column from which the saturate, aromatic, and NSO (nitrogen-, sulfur-, and oxygen-containing heterocompounds) fractions are eluted by using progressively more polar solvents. The four fractions are then available for subsequent treatment (e.g., urea adduction to further separate the normal alkanes from the other saturates) or gas chromatographic analysis.

The high resolution provided by open tubular glass (and silica) capillary GC has provided valuable geochemical information. A good example of the application of GC to petroleum geochemistry is provided by Wehner and Teschner (20). Techniques of GC, gas chromatography–mass spectrometry (GC/MS), and high-performance liquid chromatography (HPLC) were used to establish oil–oil and oil–source correlations in the Molasse Basin of southern Germany. Typically the saturate hydrocarbon fraction is chromato-

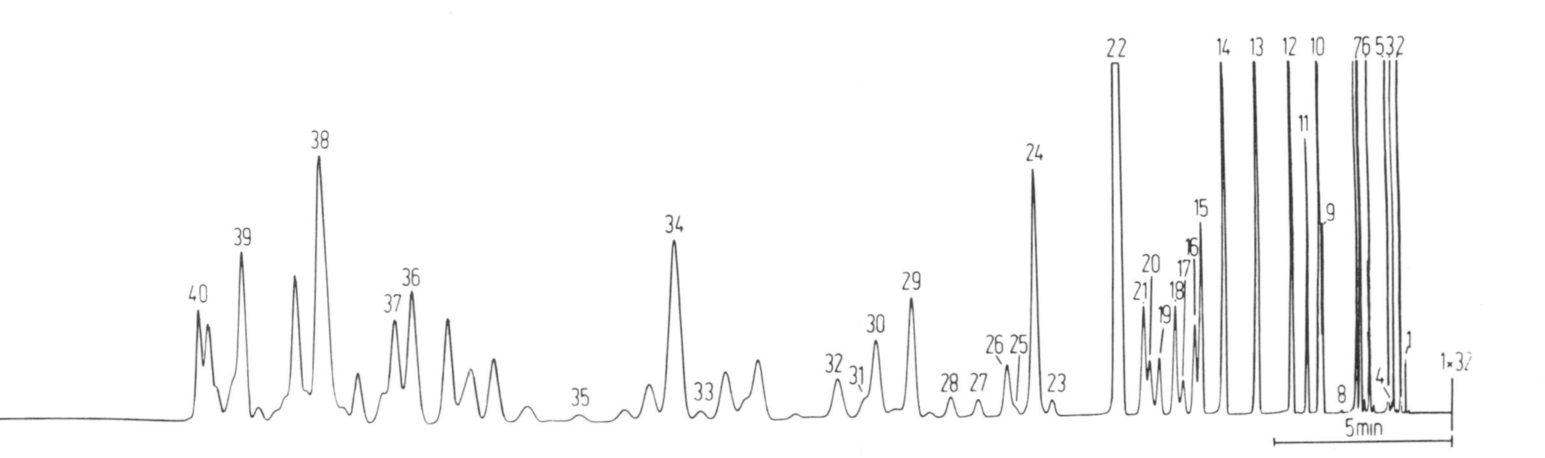

FIGURE 10.8 Typical light hydrocarbon distribution obtained by the hydrogen stripping of a sandy shale core sample (1190-m depth, Lower Cretaceous, northwestern Germany). Peaks: 12, hexane; 14, benzene; 22, heptane; 29, toluene; 34, octane. Complete identification can be found in original reference. (Reprinted with permission from Reference 17, *Analytical Chemistry*, Copyright 1978, American Chemical Society.)

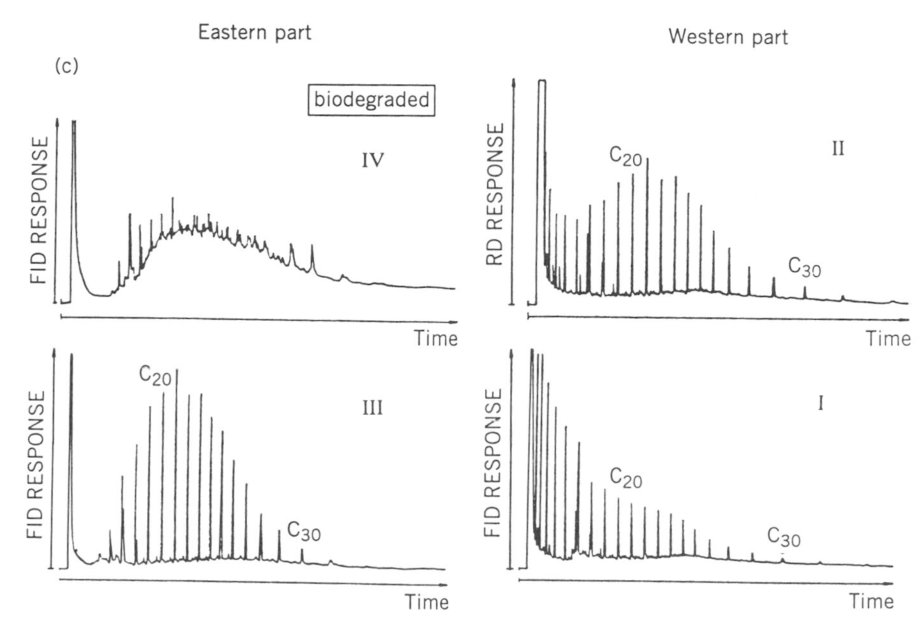

FIGURE 10.9 Gas chromatographic analysis of the saturate hydrocarbon fraction of crude oils from regions in the southern Germany Molosse Basin. Oils labeled II and III have similar patterns of *n*-alkanes. Oil from area IV shows almost no *n*-alkanes, characteristic of biodegradation. (Reprinted with permission from Reference 20, *Journal Chromatography*, Copyright 1981.)

graphed on a 30-m SP-2100 capillary column with temperature programming from 100 to 270°C. The capillary column analysis of the saturate fraction showed evidence of biodegradation (Figure 10.9). Thus a conclusive determination could not be made from comparison of the biodegraded sample with the others. Liquid chromatographic analysis of the aromatic fractions and gas chromatographic–mass spectrometric analysis of the sterane biomarkers in the saturate fractions provided the additional data needed to complete the correlations.

The effects of biodegradation were studied in more detail in a separate study by Jobson et al. (21). Bacteriological degradation proceeded rapidly under controlled conditions. After 21 days, essentially all of the *n*-alkanes were metabolized, leaving behind unaltered iso- and cyclic-alkanes and aromatics in the extractable fraction (Figure 10.10).

Albrecht et al. (22) used capillary columns (45 m Apiezon L or SE-30) to study the diagenetic effects of source rocks in Cameroon. Oil generation occurred in a narrow band between 1500 and 2500 m as indicated by the total organic content of sediments in this interval. Gas chromatographic analyses of the C_{15}^{+} hydrocarbon fraction (Figure 10.11) show a parallel increase in the yield of hydrocarbon as well as a shift in the saturate type

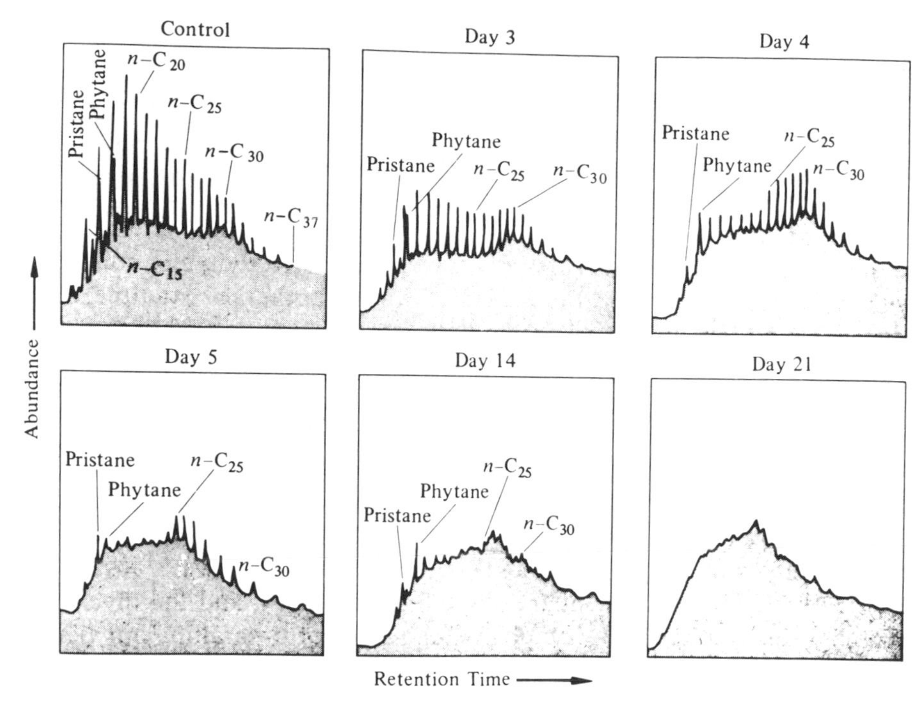

FIGURE 10.10 Gas chromatographic analysis of whole oil shows the disappearance of *n*-alkane peaks, first in the C_{15}–C_{25} range and later in the entire range during incubation with a mixed microbe population at 30°C. (Reprinted with permission from Reference 21.)

and *n*-alkane molecular weight distribution. Saturates at shallower depths are mostly iso- and cyclic-alkanes (represented by the unresolved background), indicative of immature sources, whereas the normal alkanes are the predominant type of alkane at more mature intervals. The carbon number distribution also shifts to higher values for the more mature sources. Analyses of the aromatic fractions of these same samples also show a shift in molecular weight with increasing maturity. Figure 10.12 shows a decrease in the five- and four-ring aromatic compounds as a function of depth. Artificial maturing of samples from shallower intervals by heat treating them in the laboratory produced hydrocarbons that showed the same trends as the samples taken from greater depths.

With the availability of GC/MS, steranes and terpanes have been used for biomarker fingerprinting of crudes (23). Prior to this, most of the attention in geochemistry was directed toward the saturated fractions, particularly the *n*-alkanes below C_{40}. By using either urea adduction or molecular sieving, a fraction dominated by branched and cyclic hydrocarbons can be isolated. From this fraction, the ratio C_{27}/C_{29} steranes have been used to indicate the relative amount of marine versus higher plant

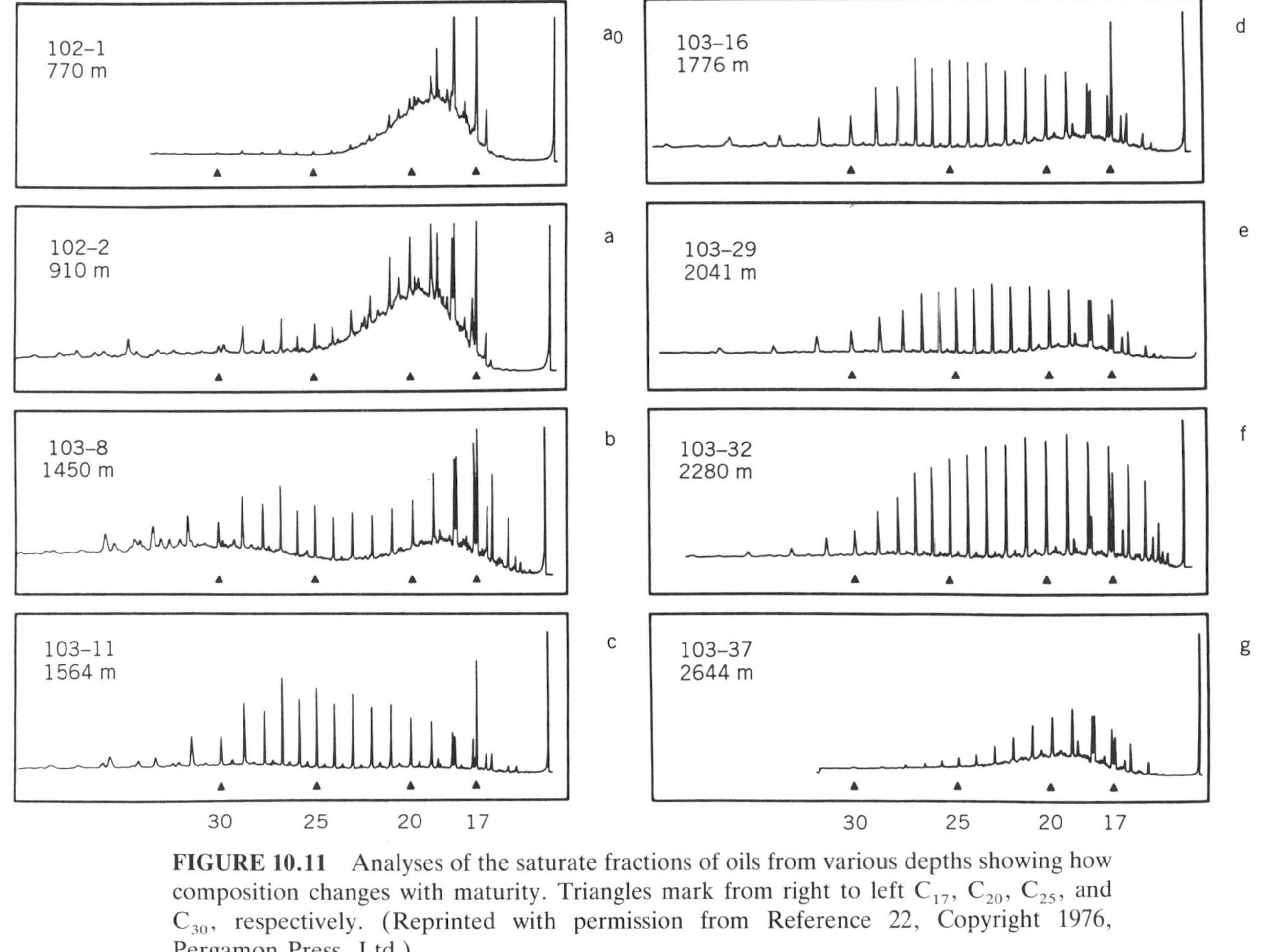

FIGURE 10.11 Analyses of the saturate fractions of oils from various depths showing how composition changes with maturity. Triangles mark from right to left C_{17}, C_{20}, C_{25}, and C_{30}, respectively. (Reprinted with permission from Reference 22, Copyright 1976, Pergamon Press, Ltd.)

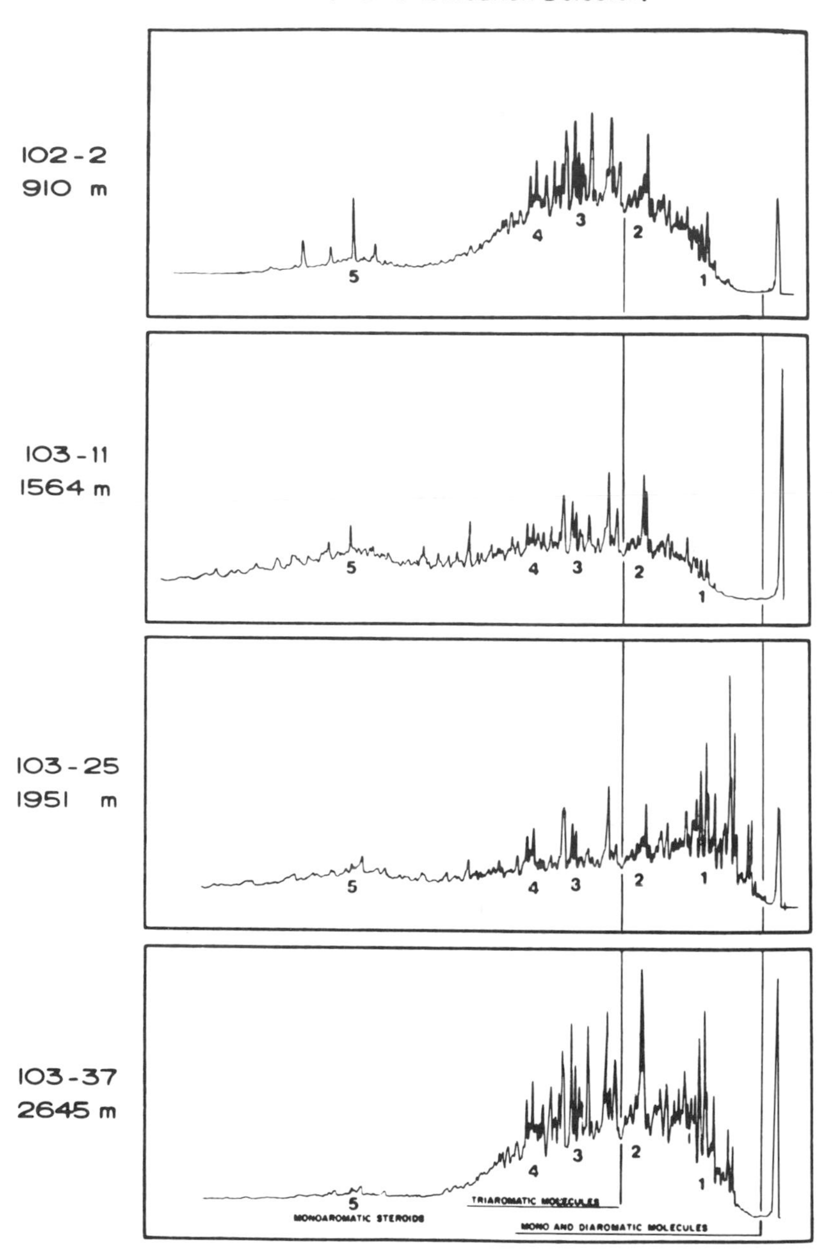

FIGURE 10.12 Analyses of the aromatic fraction of some of the same oil shown in Figure 10.11 showing the change in composition and distribution as a function of maturity. (Reprinted with permission from Reference 22, Copyright 1976, Pergamon Press, Ltd.)

source material. The ratios of sterane isomers have also been used to determine the maturity of samples. The tricyclic and pentacyclic terpanes provide complimentary information to that provided by the steranes. The main use of biomarker fingerprints is to characterize migration patterns of oil reservoirs. Similar biomarker distributions will be obtained for samples of similar maturity and burial history if they had similar source materials (24). In doing these characterizations, quantification is difficult because of differences in response factors with different biomarker families and the variety of GC/MS systems. Hwang (25) has addressed this issue in showing the long-term reproducibility of a benchtop gas chromatograph with a mass selective detector.

10.2.2 Synthetic Crude Oil

Concern regarding the diminishing crude oil supplies has created a great deal of interest in alternative fuel and petrochemical sources. The more similar to crude petroleum the alternative is, the more easily existing processing equipment can be adapted to handle these new feedstocks. The gap between refinery operations and process compatibility is closing from both sides. Refineries are adding new processes to maximize the useful yields from increasingly heavy and low grade crude oils. Research on "synthetic" crude oil from shale or coal is concentrating on processes that will produce a product as close to natural crude oil as possible.

Shale oil probably has the greatest potential for becoming an alternative to natural crude oil because of its similarity to natural petroleum. The processes of converting a kerogen-rich shale to an oil source parallels the natural maturation process that occurs during diagenesis. Thermolysis or heatsoaking of immature shales and isolated kerogen has been used by geochemists to characterize petroleum source rocks and to understand the maturation process. Generation of oil from shale represents a massive scale-up of this same process. Clearly, the application of gas chromatographic analyses to such products would be the same as those described in the previous section. Burnham et al. (26) determined that the laboratory thermolysis of shales could be used to accurately predict the yields and distribution of hydrocarbons in shale oil production. Chromatograms like those in Figure 10.13 are more complex than those of crude oils because of the presence of alkenes resulting from the more severe conditions used in producing shale oil.

Gas chromatography is seldom applied directly to the analysis of tar sands. The bitumen from tar sands represents the heaviest components remaining after the loss of lighter hydrocarbons. Likewise, gas chromatographic analysis is applicable only to the light fractions of coal liquefaction and gasification. Analysis of the gaseous streams of the Solvent Refined Coal (SRCII) process is monitored with an automated gas chromatograph (Figure 10.14.). This instrument is a commercially available gas chromato-

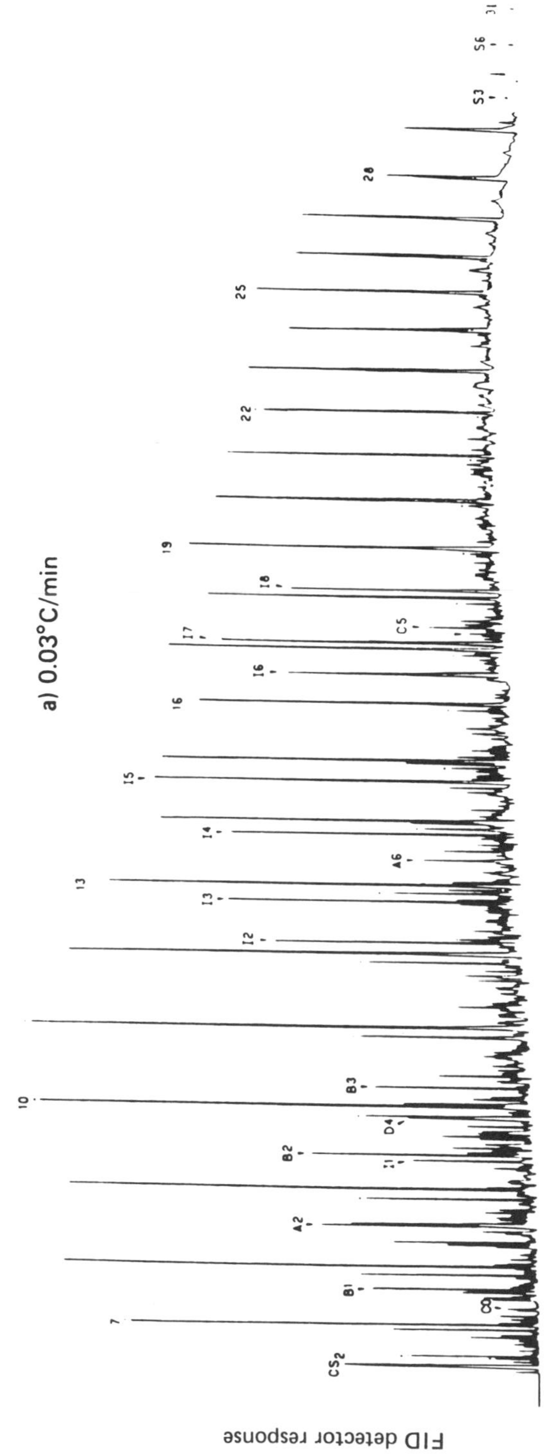
a) 0.03°C/min
FID detector response
CS2
7
B1
A2
B2
10
B3
D4
13
I2
I3
A6
I4
I5
16
I6
C5
I7
I8
19
22
25
28
S3
S6

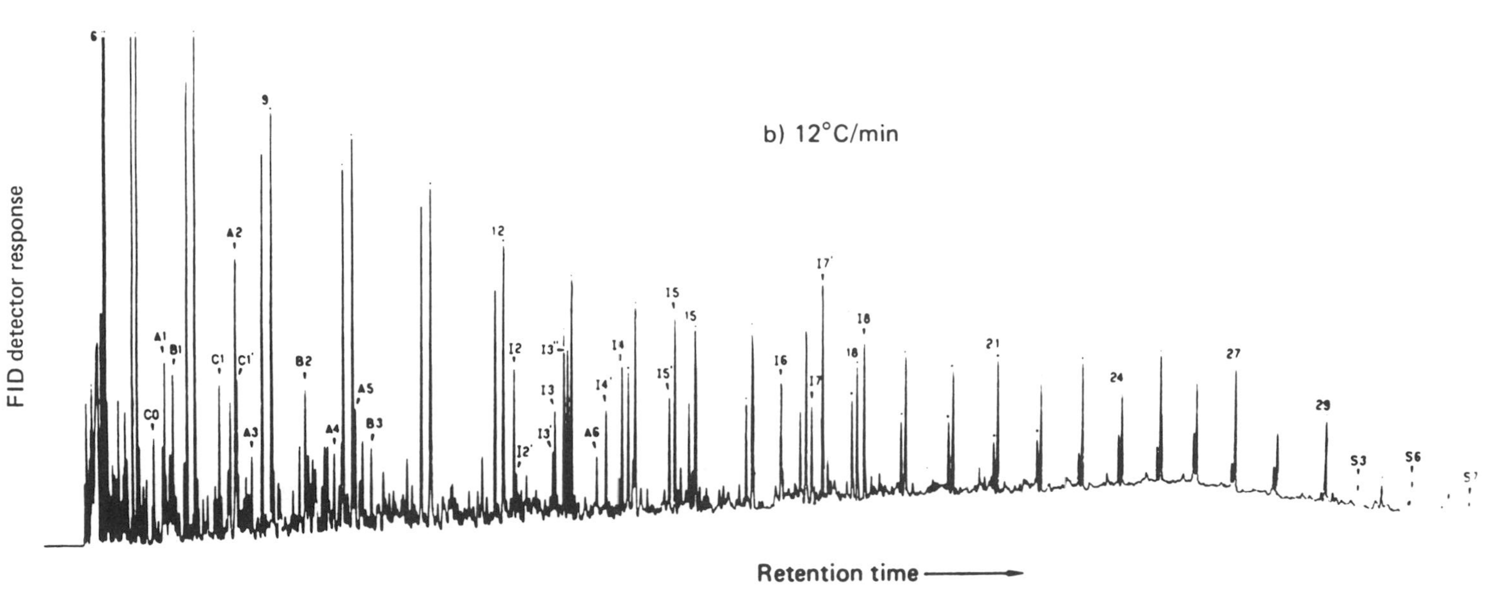

FIGURE 10.13 Chromatograms of shale oil produced at 0.03 and 12°C/min. The shale sample came from Anvil Points, Colorado, and contained 92 mL oil/kg shale. Complete identifications available in the original reference. Normal 1-alkenes and alkanes are indicated by a dot, with the 1-alkenes preceding the corresponding *n*-alkane. (Reprinted with permission from Reference 26, Copyright 1982, Pergamon Press, Ltd.)

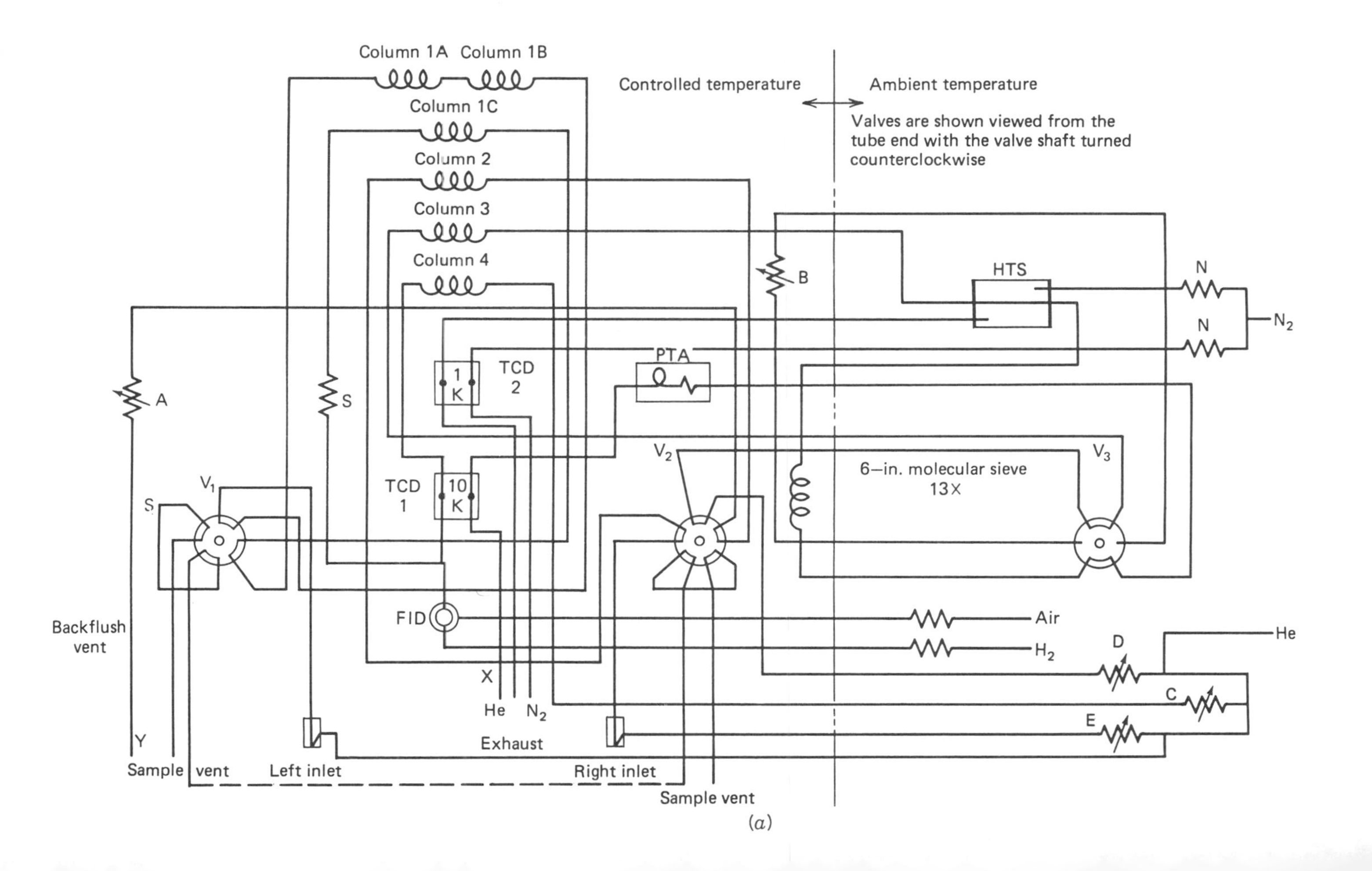
Column 1A
Column 1B
Column 1C
Column 2
Column 3
Column 4
Controlled temperature
Ambient temperature
Valves are shown viewed from the tube end with the valve shaft turned counterclockwise
B
HTS
N
N
N_2
A
S
1 K
TCD 2
PTA
V_2
V_3
6-in. molecular sieve 13X
V_1
S
TCD 1
10 K
FID
Air
H_2
He
D
C
E
Backflush vent
X
He
N_2
Exhaust
Y
Sample
vent
Left inlet
Right inlet
Sample vent

(*a*)

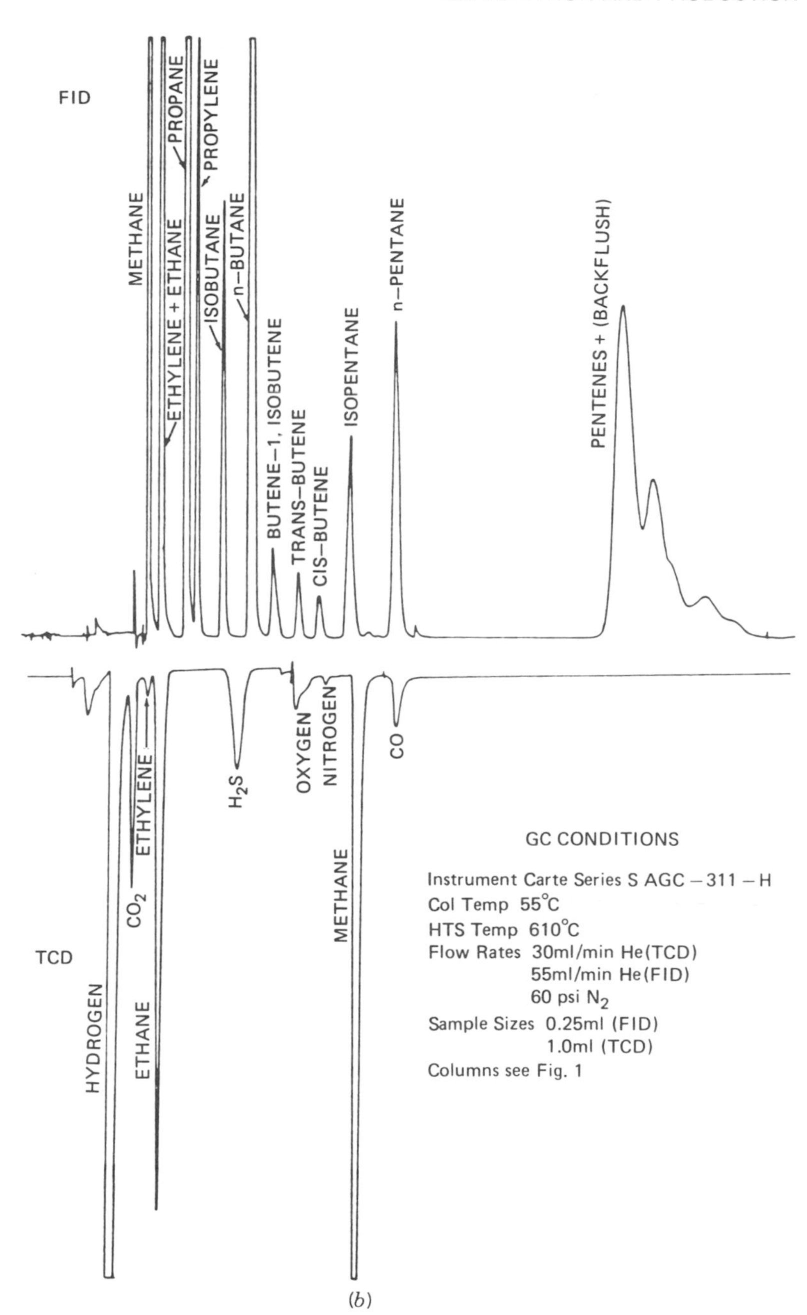

(*b*)

IGURE 10.14 Typical chromatogram from the automated gas chromatographic analysis f coal liquefaction gases. Column 1A: 18 ft—27% bis (EE) A + 3.5% Carbowax 1540 + .5% DC 200/500 mesh. Column 1C: 12 in. on Chromosorb P-AW 60/80 mesh. Column B: 3 ft—OPN/Porasil C 80/100 mesh. Column 2: 6 ft—80% Porapak N + 20% ORAPAK Q 50/50 mesh. Column 3: 7 ft—Molecular sieve 13 × 45/60 mesh. Column 4: ft—8% OV-101 on Chromosorb W-AWDMCS, 80/100 mesh. (Reprinted with permission om Reference 27, *Journal of Chromatographic Science*, Copyright 1982, Preston ublications, A Division of Preston Industries, Inc.)

graph that was originally designed for refinery gas analysis. This type of analysis is discussed in more detail in Section 10.3.1.

Lee and his coworkers (28) combined adsorption chromatography and capillary column GC to characterize the liquid fraction from the SRCII process. A fused-silica column (20 m × 0.3 mm coated with SE-52) was used along with a flame ionization detector (FID) and either a nitrogen–phosphorous detector or a flame photometric detector. Four hydrocarbon fractions were isolated and characterized. They were found to contain the following functionalities:

- Aliphatic hydrocarbons
- Neutral polycyclic aromatic compounds
- Nitrogen containing polycyclic aromatic compounds
- Hydroxy polycyclic aromatic compounds

Novotny et al. (29) demonstrated the application of glass capillary columns to the detailed analysis of coal tar samples. An extensive liquid–liquid partition scheme was developed to separate the crude coal-tar sample into basic, acidic, and "neutral" fractions. High-resolution gas chromatography of each fraction yielded a detailed analysis of the original sample. Gas chromatography–mass spectrometry was used to identify fraction components. Identifications were confirmed with authentic compounds where possible. For coal tars, the aromatic fraction represents more than 50% of the original sample. Figure 10.15 shows the chromatogram of this fraction. Figure 10.16 is the chromatogram of the saturate fraction, which according to the accompanying identification contains a number of substituted naphthalenes.

10.3 REFINING

Petroleum refining is the process that converts complex crude oils into usable fractions. The process consists of initial separation of the crude into gases, narrow-boiling-range distillates, and bottoms. Some of the fractions are then converted into more desirable components that must be subsequently separated by fractionation. The final refinery products, such as gasoline, kerosene, solvents, lubricating oils, and others, are formed by blending of the various fractions.

A simplified flow diagram of the refinery process is shown in Figure 10.17. The initial crude separation is accomplished by two stages of fractionation. An atmospheric pressure tower (commonly referred to as a *pipe still*) separates preheated crude into the following fractions:

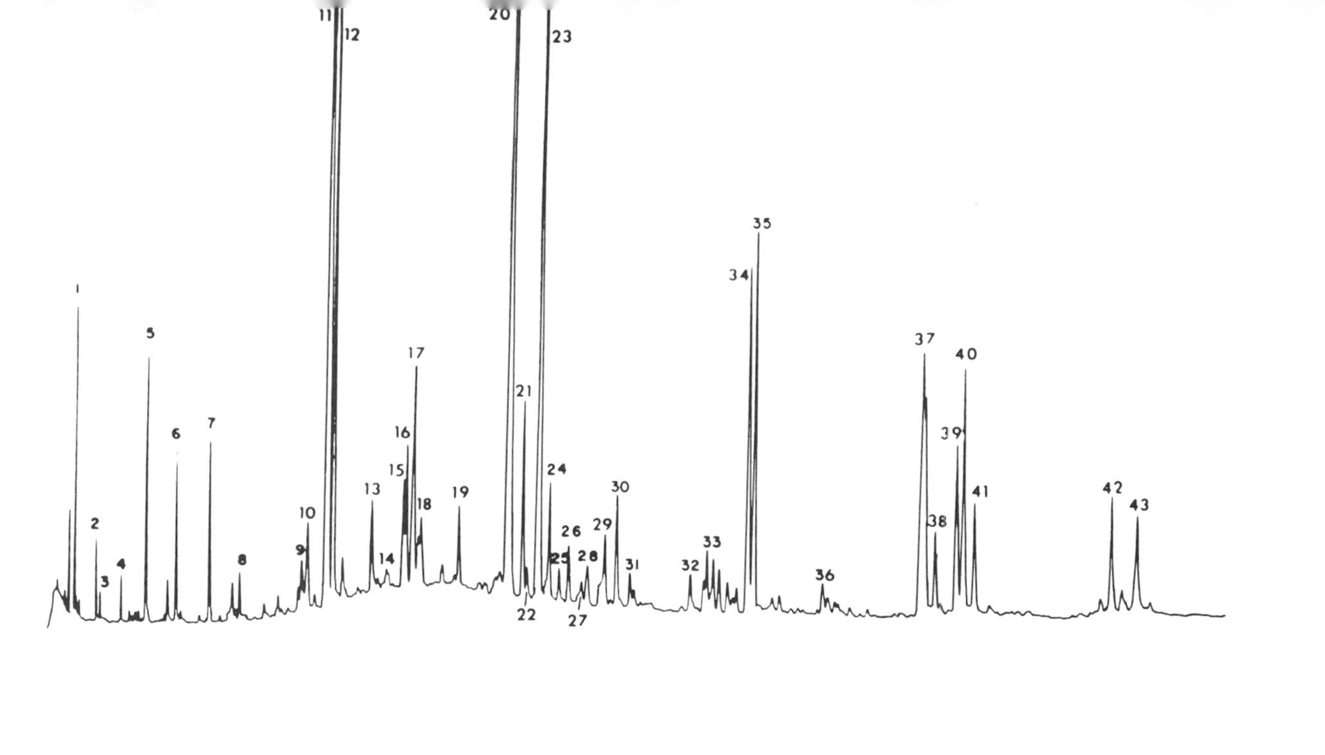

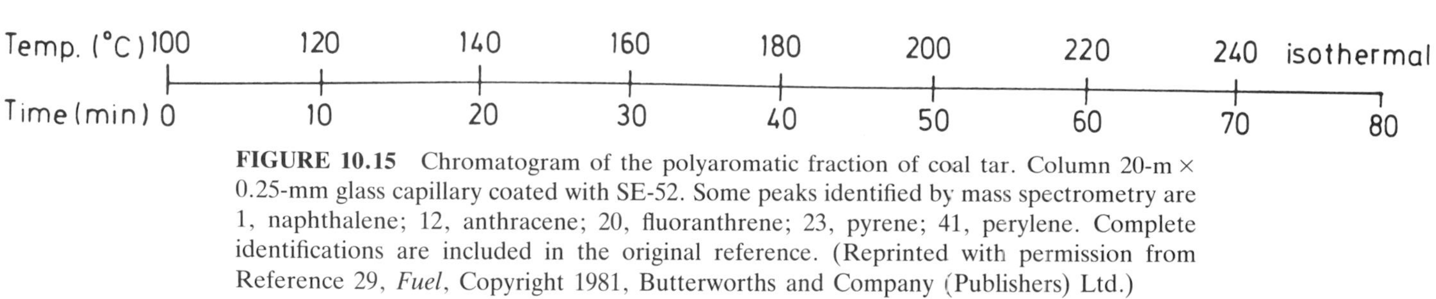

FIGURE 10.15 Chromatogram of the polyaromatic fraction of coal tar. Column 20-m × 0.25-mm glass capillary coated with SE-52. Some peaks identified by mass spectrometry are 1, naphthalene; 12, anthracene; 20, fluoranthrene; 23, pyrene; 41, perylene. Complete identifications are included in the original reference. (Reprinted with permission from Reference 29, *Fuel*, Copyright 1981, Butterworths and Company (Publishers) Ltd.)

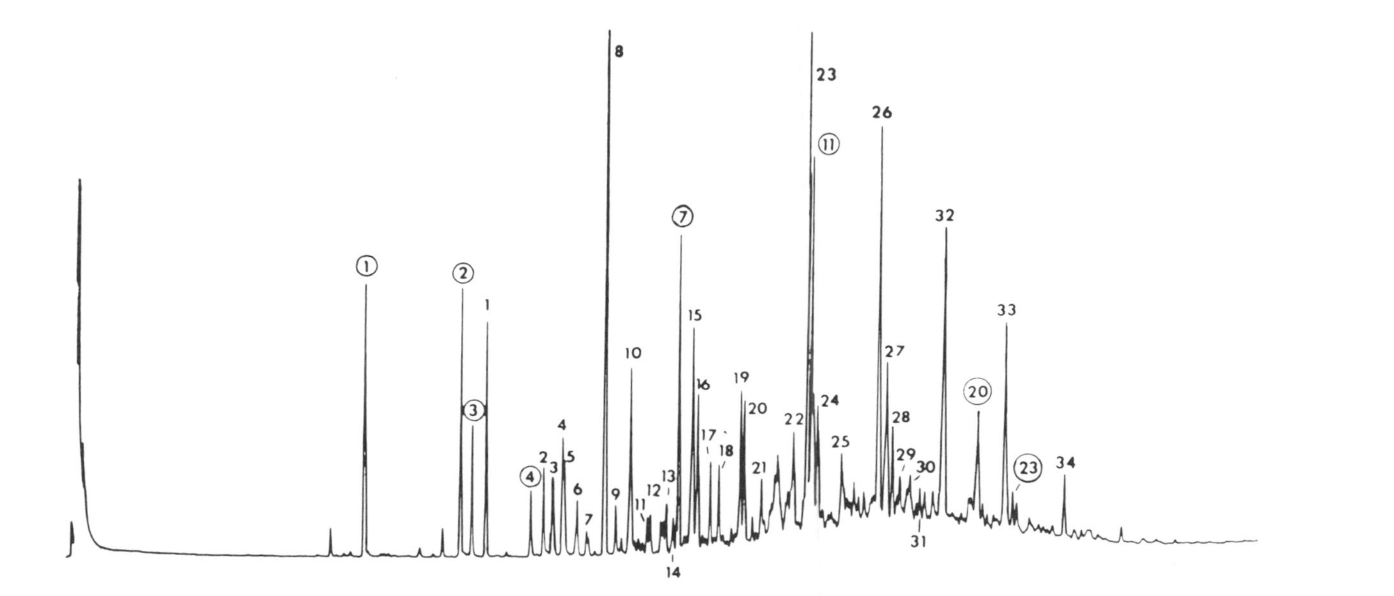

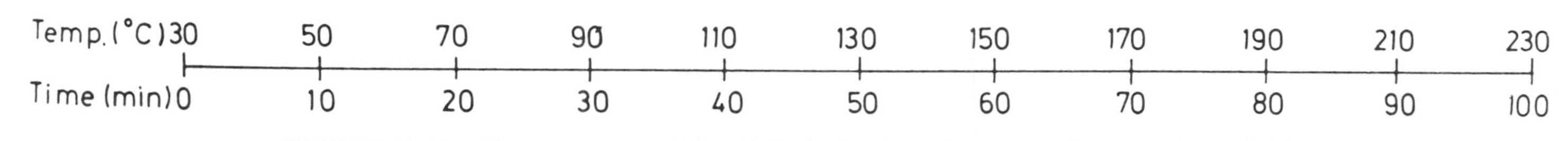

FIGURE 10.16 Chromatogram of the aliphatic fraction of coal tar. Column 20-m × 0.25-mm glass capillary coated with OV-101. Many identified compounds are alkyl-substituted naphthalenes. Peaks: 24, pristane; 27, plythane; 23, n-C_{17}; 26, n-C_{18}; 32, n-C_{19}; 34, nC_{20}. Complete identifications are included in the original reference. (Reprinted with permission from Reference 29, *Fuel*, Copyright 1981, Butterworths and Company (Publishers) Ltd.)

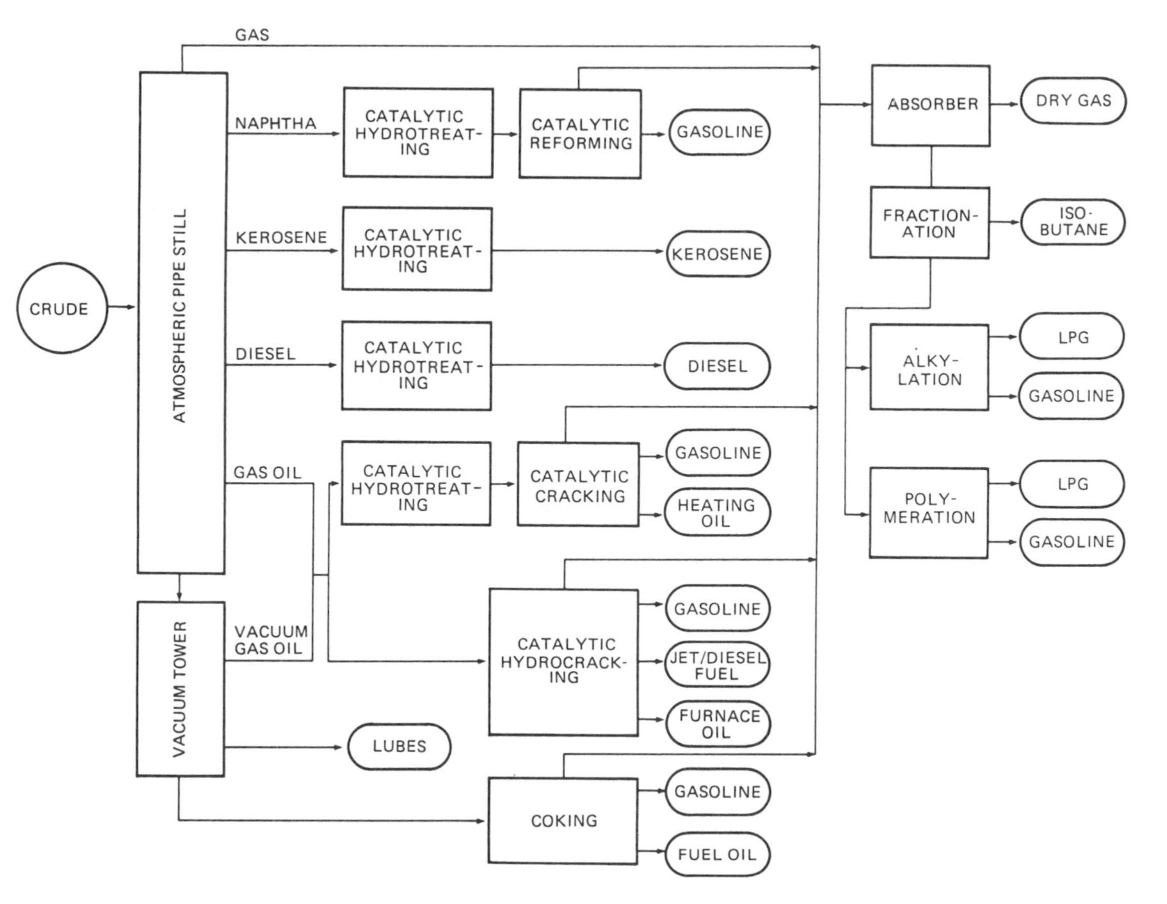

FIGURE 10.17 Simplified flow diagram of a petroleum refining process.

Fraction	Boiling Point
Refinery gas–liquefied petroleum gas (C_1–C_4)	90°F (25°C)
Naphtha-gasoline (C_5–C_{12})	90–400° (25–204°C)
Kerosene-diesel fuel (C_{10}–C_{19})	300–525°F (149–274°C)
Light gas oil (C_{12}–C_{21})	400–650°F (204–343°C)

The bottoms from the pipe still are reheated and further separated in a vacuum tower. Heavy gas oil and lubricating oil cuts are obtained from this tower. The bottoms from the vacuum tower are referred to as *reduced crude* or *residuum* and are used for asphalt and coking. Other common terms used for the distillation fractions include *middle distillates* for light gas oil and *heavy distillates* for heavy gas oil.

Following the initial separation, the various fractions are either sent to blending, separated further, or chemically modified. For gasoline production, which is one of the main purposes of a refinery, the crude fractions are chemically converted to the proper boiling range for gasoline blending. At the same time, this conversion is directed toward the production of higher octane compounds. Octane is a measure of the impact of a compound on engine performance. In general, octane rating follows the progression: aromatics > naphthene > isoparaffins > *n*-paraffins. Thus the paraffins in the initial distillation cuts are chemically upgraded through a decomposition process known as *cracking*.

Because of their effects on cracking catalysts as well as the undesirable effects on the final product, sulfur compounds are removed from the distillation cuts before further processing. This is accomplished by catalytic hydrogen treating (hydrotreating). In this step, the sulfur is removed as H_2S.

Cracking is basically a process that reduces the size of the molecules and in turn produces a high yield of unsaturates. This is accomplished through high temperatures. High pressures are also required to maintain a totally liquid phase. Catalytic cracking utilizes a solid supported catalyst in a fluidized bed. This accelerates the thermal cracking process by three orders of magnitude. Hydrocracking is a type of catalytic cracking in which hydrogen is added to produce isoparaffins and aromatics. Reforming is a milder operation for lighter fractions that also utilizes a catalyst. It dehydrogenates naphthenes and isomerizes naphthenes and paraffins. Aromatics are the predominate product from this step. Coking is the final type of cracking that thermally decomposes the heaviest fractions. This is accomplished through extensive recycling of the heavy components.

Additional fractionation is required to separate the cracking by-products into their appropriate boiling ranges for product blending. All the C_4 and lighter hydrocarbons are compressed and sent to absorbers. The butanes are removed by absorption into gasoline blend cuts. The lighter components are then fractionated. Propane and butane are mainly sold as liquefied petroleum gas (LPG). The lightest components are burned as fuel in place of natural gas.

The olefinic hydrocarbon gases, propylene and the butylenes, are sent to polymerization and alkylation. Polymerization forms dimers and trimers of these olefins for use in gasoline blending or for petrochemicals. Alkylation is a more common process in which the olefins are reacted with isobutane. This step produces the more desirable isoparaffins for gasoline blending.

From this brief overview of the refining process, it becomes apparent that the products as well as the process streams for petroleum refining are very complex. Most of these streams are well suited for gas chromatographic analysis. Some of the main applications of GC to these streams and products are discussed in the following sections.

10.3.1 Refinery Gases

Refinery gas analysis involves the determination of permanent gases, hydrogen, all the individual C_1–C_5 hydrocarbons, and the hexane and heavier content. Streams with varying levels of these components must be analyzed for process control. Natural gas and other streams used for furnace firing must also be analyzed for heat content. Compliance with environmental regulations requires analysis of flue gases and other emission sources. Besides the diversity in composition and origin of samples, sampling is an additional problem. Sampling of multiphase streams over a wide range of temperatures and pressures is often required. Reviews of these sampling and analytical problems have been published by Harvey (30) and Cowper and DeRose (31).

For samples near atmospheric pressure, a glass sample cylinder or rubber bladder may be used. The preferred method, however, is the use of Teflon- or Mylar-coated bags. These bags are easy to handle and are inert to sulfur compounds. Most streams, however, require sampling with stainless-steel cylinders. These are also available with Teflon linings for reactive components. The metal cylinders may be installed on an in-line basis with a slip stream of the process stream to ensure representative sampling. Care must be taken to allow for a vapor space in the cylinder when sampling liquids or high-pressure liquefied gases. This prevents overpressuring due to liquid expansion with temperature changes. Safety valves are often installed for this purpose. With proper safety precautions, metal sample cylinders may be heated in an oven to revaporize samples that may condense at ambient conditions.

Analysis of specific components or classes of components in refinery gases can be accomplished with single-column analyses. However, combinations of columns and valving are required for more complete analyses. The various aspects of hydrocarbon gas analysis have been discussed by Thompson (32). Applicable columns for these applications can also be found in column supplier catalogs and the reviews by Mindrup (33) and Liebrand (34).

Analyses for the fixed gases and light hydrocarbons are required for monitoring of stack or flue combustion gases. The concentration of hydrogen in samples is important for control of cracking and hydrotreating. Oxygen and carbon monoxide must be determined to avoid combustion and other side reactions. A 5A or 13X molecular sieve column with argon or helium as the carrier gas and a thermal conductivity detector is commonly used for the analysis of H_2, O_2, and CO in hydrocarbon streams. When a molecular sieve column is used, the hydrocarbons heavier than methane are normally backflushed from the column. Care must be taken to avoid the deactivation of the molecular sieves with water and large amounts of CO_2. Also, isobutane can interfere with the determination of oxygen. Several alternative column packings include Porapak Q, which is also capable of determining water, carbon dioxide, and the other hydrocarbons. Chromosorb 102 and Carbosieve S columns are also suitable.

For determination of the fuel value of refinery gas streams or natural gas, the inert gases must be determined along with the hydrocarbon components. By determining the mole percent concentration of each component, the calorific (heating) value and specific gravity of a gas can be calculated. This information is used to determine the sales value of natural gas. Stufkens and Bogaard (35) used a Porapak R column for the analysis of methane-rich natural gas. A thermal conductivity and FID were used in series to determine the nonmethane components. The response of the two detectors was normalized based on the ethane concentration. The ASTM methods for fuel value determinations use two columns. Analysis of natural gas by Method D-1945 (11) specifies a molecular sieve adsorption column for O_2, N_2, and methane. The C_2–C_5 hydrocarbons and CO_2 are then determined with a partition column such as BMEE (bis-(2(2-methoxyethoxy)-ethyl)ether), silicone 200/500, or diisodecylphthalate dimethylsufolane. For reformed gas containing only C_2 and lighter components, Method D-1946 (11) uses a Porapak Q column for the C_2 hydrocarbons.

Because of the harmful effccts of sulfur compounds on cracking catalysts, refinery distillation cuts are hydrotreated to convert the sulfur compounds primarily into H_2S. The sulfur content of stack gases must also be monitored for compliance with air pollution standards. Because of their high polarity and reactivity, inert sampling and column materials must be used to avoid losses and peak distortion. For higher levels, a thermal conductivity detector can be used with a silica gel, Porapak Q, or Carbosieve B column. Levels below 50 ppm require the use of a flame photometric detector (FPD). Pearson and Hines (36) used an FID in series with an FPD for determining trace levels of H_2S, COS, CS_2, and SO_2. They used the FID to verify that the hydrocarbons in the sample were completely separated from the sulfur compounds. This is necessary because hydrocarbons reduce the signal of the FPD. For streams containing C_1–C_4 hydrocarbons, several columns were used to achieve resolution of the sulfur compounds. These columns included polyphenyl ether-H_3PO_4 on Chromosorb G, silica gel, and QF-1 on Porapak QS.

Complete systems have been developed for the total analysis of refinery and natural gases. These all utilize automatic column switching and multiple detectors. The Universal Oil Products (UOP) method (37) was one of the first of these systems. It utilized three columns (diethylene glycol adipate plus diethylene glycol sebacate, Porapak Q, and 13X molecular sieve) with valves for backflushing and a single thermal conductivity detector (TCD). Hewlett–Packard (38) has developed an automated system that also has postrun calculation capabilities. Simultaneous dual injection is used with a single 13X molecular sieve column and TCD dedicated to hydrogen analysis. Four other columns and another TCD are utilized for analysis of the other components. These columns are packed with Porapak Q, 13X molecular sieve, and sebaconitrile on Chromosorb P. A schematic of this system is shown in Figure 10.18, and a chromatogram is shown in Figure 10.19.

An excellent example of a more recently developed system for refinery gas analyses is available from Wasson–ECE Instrumentation (39). A Hewlett–Packard gas chromatograph equipped with three independent systems that operate simultaneously provides a total analysis in 25 min. One subsystem uses a gas sampling/backflush valve, packed columns, and an FID for the analysis of olefins with an initial C_5=/C_6+ composite backflush. The second subsystem uses a gas sampling/switching valves, packed columns, and a TCD for CO_2, ethylene, ethane, acetylene H_2S, O_2, N_2, methane, and CO. The third subsystem uses a gas sampling/switching valve, packed columns, and a second TCD for analysis of hydrogen down to 100 ppm. A typical chromatogram is shown in Figure 10.20. Finally, an extended refinery gas analysis is available, which utilizes a capillary column in the first subsystem to resolve the C_1 through C_6 paraffins and olefins.

10.3.2 Simulated Distillation

Because refining is primarily a distillation process, laboratory distillations are commonly used to characterize crude oils and process streams. Gas chromatographic data are now being used with increasing frequency to provide the same type but better quality data than that provided by more time-consuming manual distillations. Although not without some disadvantages, simulated distillation continues to gain general acceptance as the method is improved and correlations with manual distillations are developed. The ASTM Method D-2887 was established in 1973 to standardize the use of simulated distillation for distillate fractions. This method was intended to supplement the manual distillation procedures long used in the petroleum refinery: ASTM D-86 (Engler distillation), ASTM D-1160 (low-pressure version of D-86 for heavier products), and ASTM D-2892 (a complete "true-boiling-point" distillation). ASTM method D-3710 has also been established for determining the boiling-range distribution of gasoline and its fractions by GC (11).

Differences between the simulated and physical distillation procedures lie

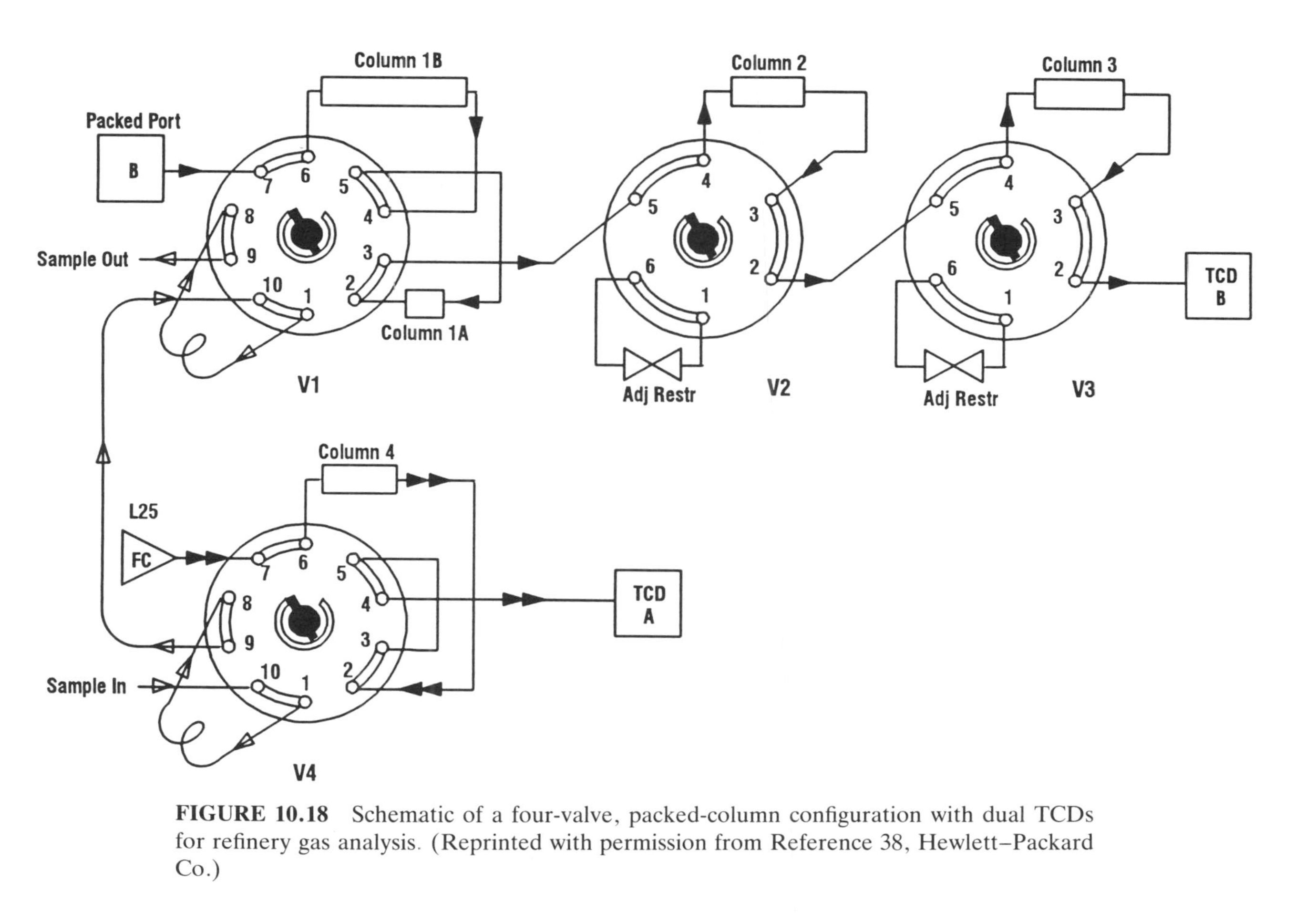

FIGURE 10.18 Schematic of a four-valve, packed-column configuration with dual TCDs for refinery gas analysis. (Reprinted with permission from Reference 38, Hewlett–Packard Co.)

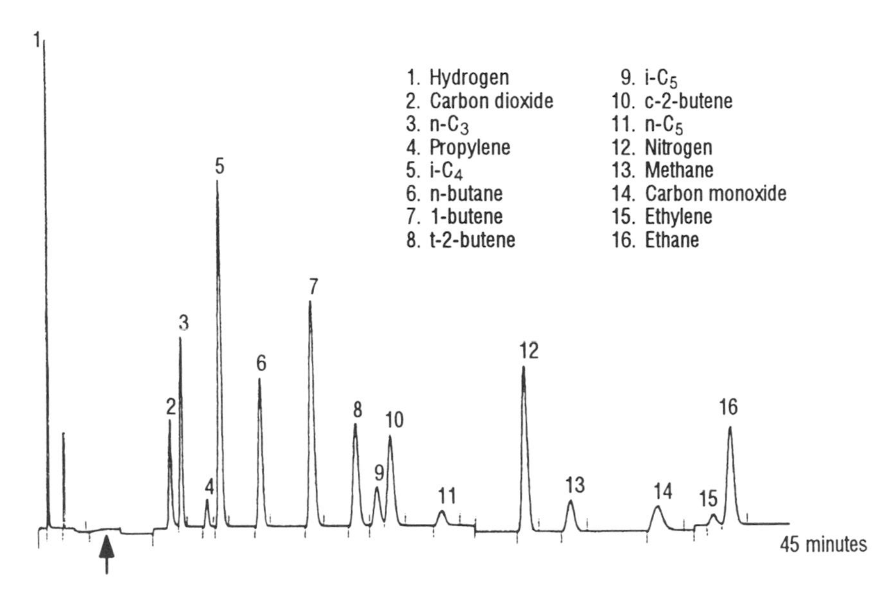

FIGURE 10.19 Refinery gas checkout chromatogram showing fixed gases and light hydrocarbons. The arrow indicates where C_6 plus the C_5 olefins would elute as backflush. Signal switch from TCD A to TCD B at 1.5 min. (Reprinted with permission from Reference 38, Hewlett–Packard Co.)

in the imperfect nature of each. The most frequently used D-86 and D-1160 are relatively fast single-plate distillations that closely approximate the refinery processes. The large scale of the refinery processes are of necessity imprecise. Thus the compromise between analysis time and the need for information of adequate precision for process control was met by the single-plate laboratory distillations. The introduction and use of simulated distillations from chromatographic data paralleled the need for more precise and detailed data to optimize the refinery process. Refinery operations have become more costly with the rise in energy costs and the increased value of refinery products.

Eggerston et al. (40) in 1960 first reported that low-resolution, temperature-programmed, gas chromatographic data could be used to simulate the more time-consuming true-boiling-point distillation. Retention times were correlated to boiling point and detector response was correlated to the amount of material "distilled." This was confirmed by Green et al. (41), who first used the phrase "simulated distillation by gas chromatography."

Separations by boiling point are typically obtained on columns with silicone gum liquid phases. These liquid phases include SE-30, OV-1, OV-101, UC-W98, and Supelco 2100. Relatively high liquid-phase loadings (10%) are often used to increase the sample capacity of the column. This minimizes column overloading, which results in a dependence of retention

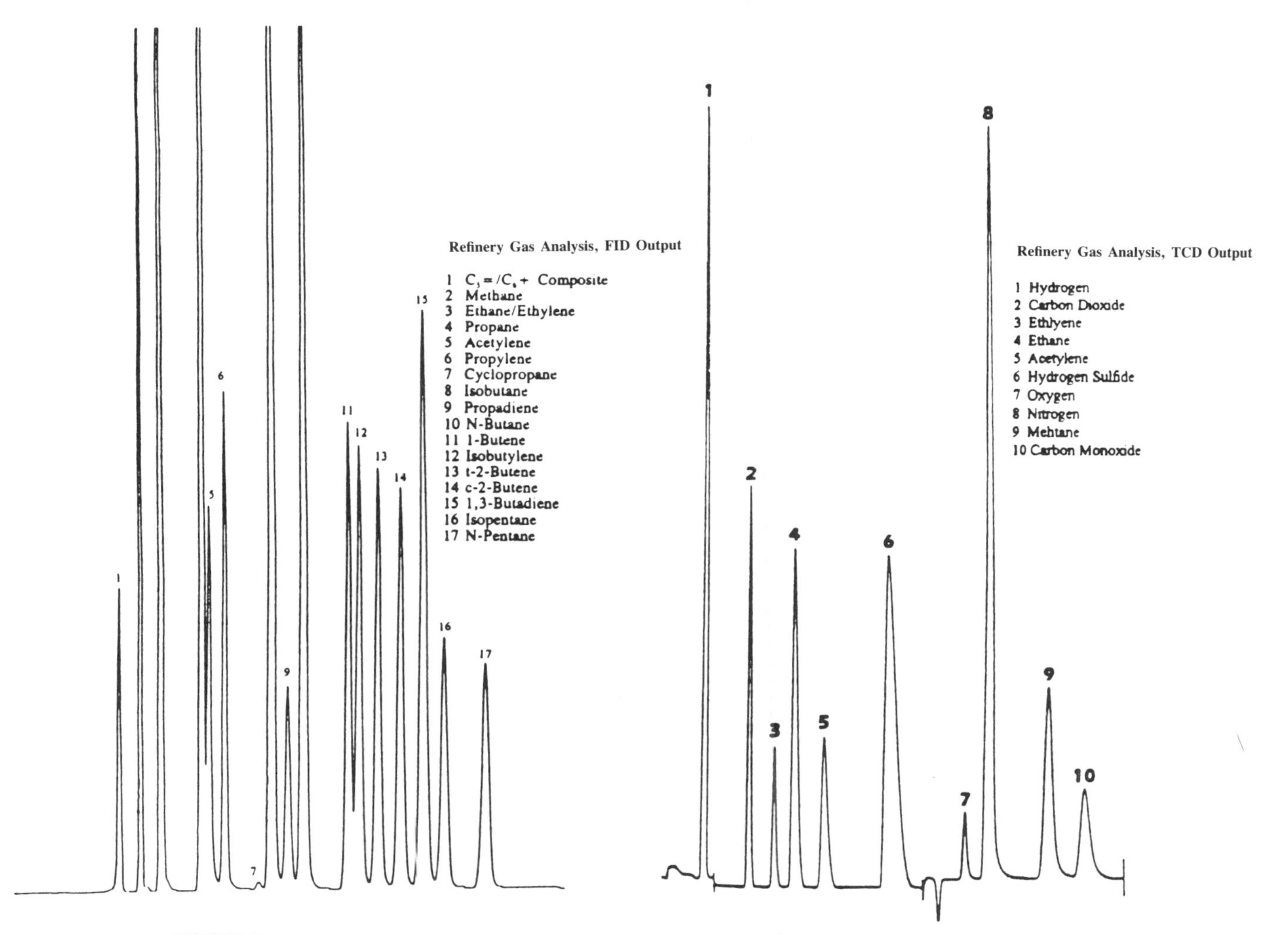

FIGURE 10.20 Typical chromatogram from a standard refinery gas analyzer with

time with the concentration of the sample components. Through the use of temperature programming, a calibration is established on the basis of retention times for a series of *n*-paraffins versus their boiling point, as shown in Figure 10.21. For determining lighter components (C_3–C_5 hydrocarbons), programming from an initial temperature of −30°C is required. The calibration blend can also be used to establish relative response factors. Column resolution must then be monitored along with baseline drift. To compensate for column bleed, a blank run is determined, stored, and then subtracted from subsequent sample runs. Aromatic compounds are used to verify the low polarity of the column.

Sample analyses do not require complete resolution of individual components. In fact, the column length and packing are chosen to obtain a limited resolution that will give a good comparison with the boiling-point distribution. A minimum first peak retention time is required to allow the SIMDIS software to establish a proper baseline to insure an accurate initial boiling point (IBP). An inert packing is used to elute components according to their boiling point and to avoid skewing. Finally, a stable baseline is

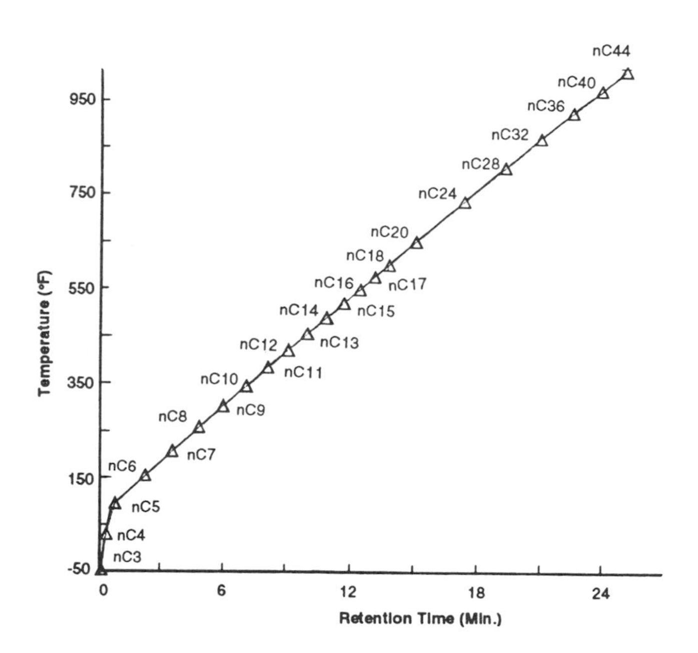

FIGURE 10.21 Simulated distillation calibration curve using a Petrocol C column, 20-ft × $\frac{1}{8}$-in. SS, column temperature: −25 to 350°C at 15°C/min. (Reprinted with permission from Reference 42, Supelco, Inc, Bellefonte, PA 16823.)

required to determine when the sample has completely eluted and to determine the final boiling point (FBP).

A typical chromatogram for gasoline by ASTM Method D-3710 is shown in Figure 10.22. The boiling-point distribution is determined by integration of the chromatogram by sequential time slices. A distillation curve can then be constructed from the calculated percent off versus boiling-point data. Correlations can also be made with ASTM Method D-86 (manual distillation) and Reid vapor pressure. The volume percent concentrations of the C_3–C_5 hydrocarbons are calculated to determine the vapor pressure of the gasoline sample. Many of the chromatographic data systems are capable of handling these calculations automatically.

For heavy distillates, the boiling-point distribution is determined by ASTM D-2887, as shown in Figure 10.23. This analysis is based on a C_5–C_{44} hydrocarbon calibration. A low liquid phase loading (3% versus 10%) is required to allow for elution of samples with final boiling points as high as 1000°F. With this high-temperature range, a stable baseline is critical. A

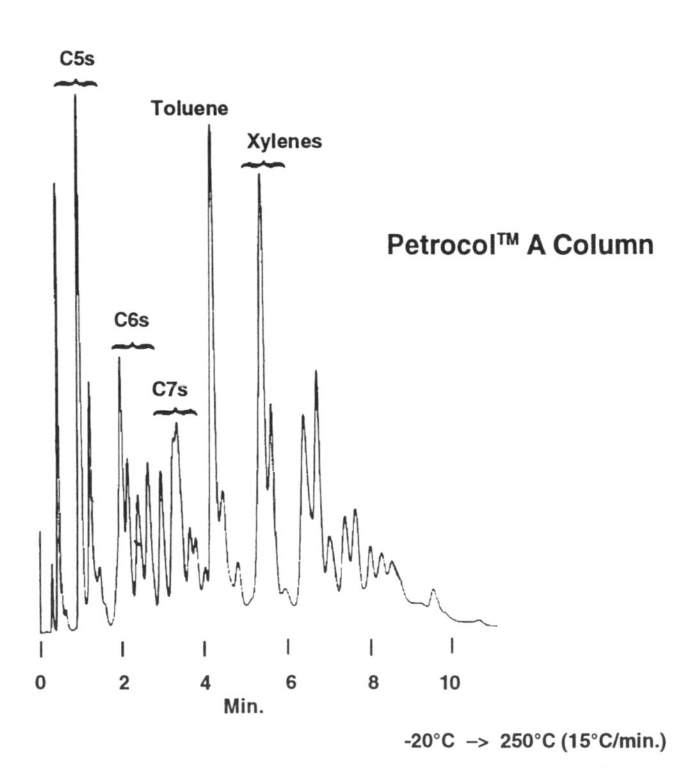

FIGURE 10.22 Simulated distillation chromatogram of a commercial gasoline according to ASTM D3710 using a 20-in. × $\frac{1}{8}$-in. SS column packed with 10% SP-2100 (methylsilicone) on 80/100 Supelcoport. (Reprinted with permission from Reference 42, Supelco, Inc, Bellefonte, PA 16823.)

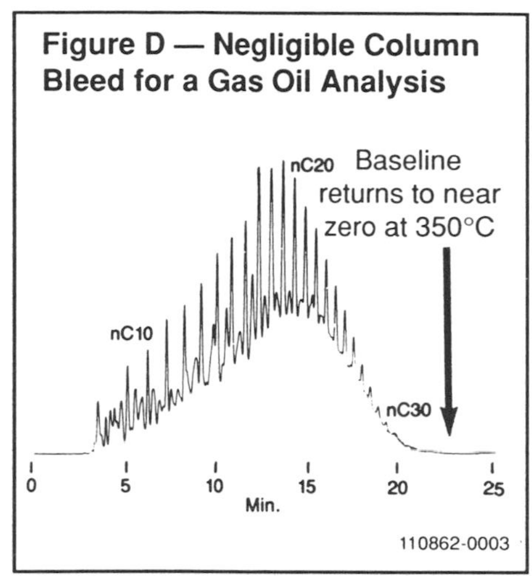

Petrocol B column, Col. Temp.: -25°C to 350°C at 15°C/min., Flow Rate: 30ml/min., He, Det.: FID (256 x 10^{-10} AFS), Sample: 0.1µl reference gas oil.

FIGURE 10.23 Simulated distillation chromatogram of a reference gas oil sample according to ASTM D2887 using a Petrocol B column (20-in. × $\frac{1}{8}$-in. SS column with 3% SP2100 (methylsilicone) on 80/100 Supelcoport). (Reprinted with permission from Reference 42, Supelco, Inc, Bellefonte, PA 16823.)

correlation to ASTM D-1160 (manual vacuum distillation) rather than ASTM D-86 is used for this sample because of its high boiling range.

The ASTM Methods are being updated to include capillary column technology. A SIMDIS chromatogram of a gasoline sample is shown in Figure 10.24. Capillary columns with bonded stationary phases offer advantages including inertness and a more stable baseline due to low column bleed. Longer column life is obtained because large volume sample injections cannot wash the stationary phase away. The higher resolution of capillary columns presents some problems with low-resolution correlations. Column conditions and SIMDIS software can be used to minimize differences. Otherwise, the higher resolution can be used to provide compositional information on the sample.

For the heavier petroleum fractions as well as crude oil, analysis problems are encountered as a result of the presence of solids and materials that will not elute from the column. To account for the total sample, an internal standard mixture is used. For heavy refinery cuts, an internal standard mixture, such as C_{10}, C_{11}, C_{12} *n*-paraffins, can be used such that they elute before the sample. This allows analysis of the sample in a single run. Because of the complexity of crude oils, there are no sections in the sample chromatogram that are void of sample components. Thus an analysis of the sample with and without internal standard (C_{14}–C_{17} normal hydrocarbons) is required. Typically, results are reported for these samples up to 538°C (1000°F). Use of a Dexsil 300 column allows for determination of final boiling points up to 600°C. This extends the limit of the analysis from C_{40}–C_{60} hydrocarbons.

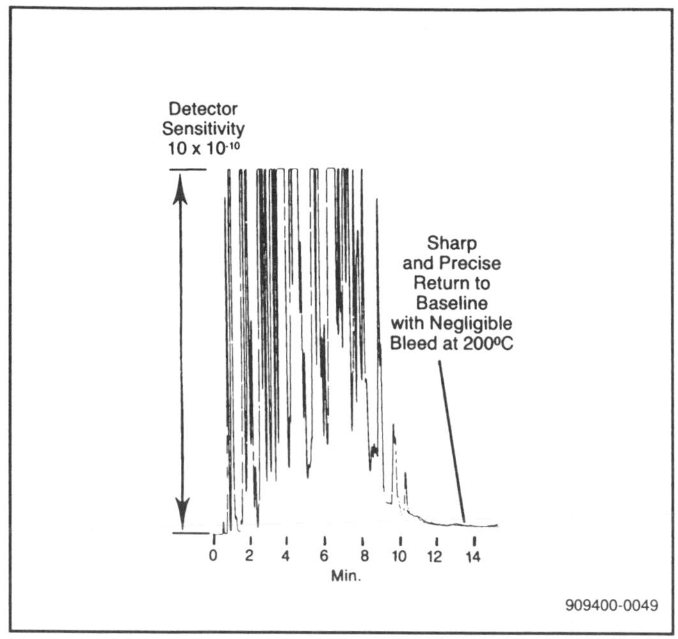

Petrocol 3710, 10m x 0.75mm ID glass, 5.0μm film, Col. Temp.: -20°C to 200°C at 20°C/min., Flow Rate: 15ml/min., N_2, Sample: 0.1μl of a commercial gasoline.

FIGURE 10.24 Simulated distillation chromatogram of a commercial gasoline using a Petrocol 3710 capillary column (10-m × 0.75-mm-i.d. glass, 5.0-μm film). (Reprinted with permission from Supelco, Inc, Bellefonte, PA 16823.)

Relatively minor differences do exist, however, between simulated and "true" distillations. Because of the historical dependence on D-86 and D-1160 for process control and product specifications, efforts have been made to establish more accurate correlations (43). For example, ASTM D-86 has inherent sources of imprecision, including column holdup and incomplete condensation. Gas chromatographic distillation cannot measure the initial boiling point in quite the same way as physical distillation. The initial boiling point is the temperature at which the vapor pressure of the bulk sample is equal to the barometric pressure. Also, even very nonpolar column liquid phases exhibit a slight discrimination between alkanes and other hydrocarbon types. This slightly distorts the boiling point curve. However, this becomes significant only with highly naphthenic or aromatic samples.

Stuckey (44) developed calculation response factors that take into account the chemical nature of the sample. The UOP characterization factor is a measure of chemical character and is used along with boiling point to determine the appropriate response factor (45).

Particular care must be taken when performing a simulated distillation on gasoline samples. Although either a TCD or an FID can be used with appropriate response factors, high relative concentrations of some light components such as butane (which can be as high as 10%) could exceed the saturation limit for a FID.

Not withstanding these relatively minor difficulties, simulated gas chromatographic distillation has expanded greatly. The basic precision and ease of automation represent significant cost savings. Furthermore, gas chromatographic distillation data are being used to replace other manual tests (46). Models have been developed to estimate Reid vapor pressure for gasoline with good reliability. Such data are used to predict engine performance and are included in the product specification for gasoline. Engine performance parameters such as starting index, vapor-lock index, and warmup index can also be calculated from the boiling-range distribution.

10.3.3 Hydrocarbon Type Analysis

Along with boiling-point distribution, refinery streams are normally characterized by a hydrocarbon type analysis. This analysis is used to determine the saturate, olefin, and aromatic (SOA) content of a sample. The original approach to this analysis was to separate a sample into three distinct fractions. These fractions were then analyzed for individual components. Another more recent approach has been to utilize high-resolution GC to provide a total analysis. Individual components are then totaled to provide a type analysis. This analysis allows for the determination of paraffins, olefins, naphthenes, and aromatics (PONA). Regardless of the methodology, hydrocarbon type analyses are necessary for valuing feedstocks as well as optimizing reforming and cracking conditions for naphthas and gas oils. They are also important for the characterization of the quality of product fuels, including gasoline. For instance, aromatic content is important for the octane rating of unleaded gasolines and for smoke elimination in jet fuels.

The standard procedure for hydrocarbon type analysis is ASTM D-1319, the fluorescent indicator adsorption (FIA) method (11). This method covers determination of petroleum fractions that distill below 600°F (315°C). The sample is separated into three fractions by use of a silica column. A mixture of fluorescent dyes is also added with the sample. The dyes are then separated selectively with the hydrocarbon types. The volume percentage of each hydrocarbon type is calculated from the length of each zone as indicated by the dyes. Further analysis of each of the three fractions has traditionally been performed by mass spectrometry. In addition to the disadvantage of being time-consuming, this column chromatographic analysis cannot determine the C_5 and lighter hydrocarbons because of their volatility.

Both liquid and gas chromatographic techniques have been developed to improve this analysis. Suatoni and coworkers (47, 48) have performed most of the work by utilizing liquid chromatography. Soulages and Brieva (49) developed a gas chromatographic analysis that relies on selective adsorption of components. The sample is split onto three parallel columns. The saturates content is determined from one column that has a mercuric perchlorate–perchloric acid absorber for olefins and aromatics. The second column is merely a delayer for a portion of the total sample. The third column has a mercuric sulfate–sulfuric acid absorber to retain the olefins. Figure 10.25 is a typical chromatogram from this system.

Since the individual types are determined by difference, the error in the analysis must be considered when one fraction is small relative to the others. A somewhat simpler system that does not rely on trapping of components has been developed by Ury (50). Trapping can cause errors when only small amounts of an individual component are present. An *N*,*N*-bis(2-cyano-ethyl)formamide (CEF) column was used to delay the aromatics. The olefins were then separated from saturates with a column of cupric sulfate on silica gel. A typical chromatogram for a gasoline sample is shown in Figure 10.26.

Systems have also been developed to give more detailed analyses of the individual compound classes in addition to the basic analysis by type. Stavinoha (51) used a CEF column for preseparation as well as total analysis of the aromatic components. The perchlorate and sulfate absorbers mentioned previously were again used for determination of the saturates and olefins. Mathews et al. (52) utilized dual porous-layer open tubular (PLOT) capillary columns to analyze the aromatics in gasoline and light oils. The

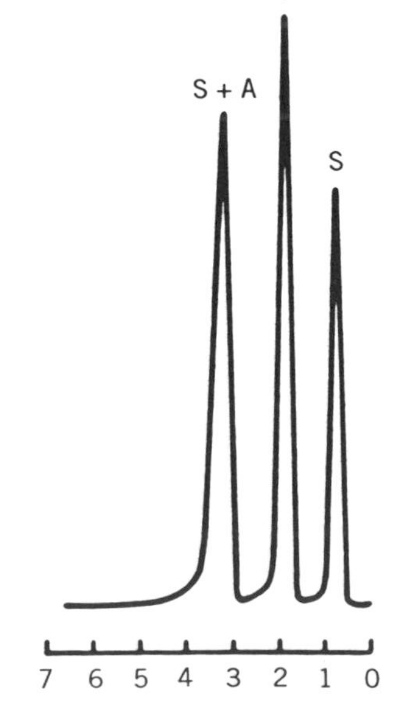

FIGURE 10.25 Chromatogram from hydrocarbon type analysis of gasoline. (Reprinted with permission from Reference 49, *Journal of Chromatographic Science*, Copyright 1982, Preston Publications, A Division of Preston Industries, Inc.)

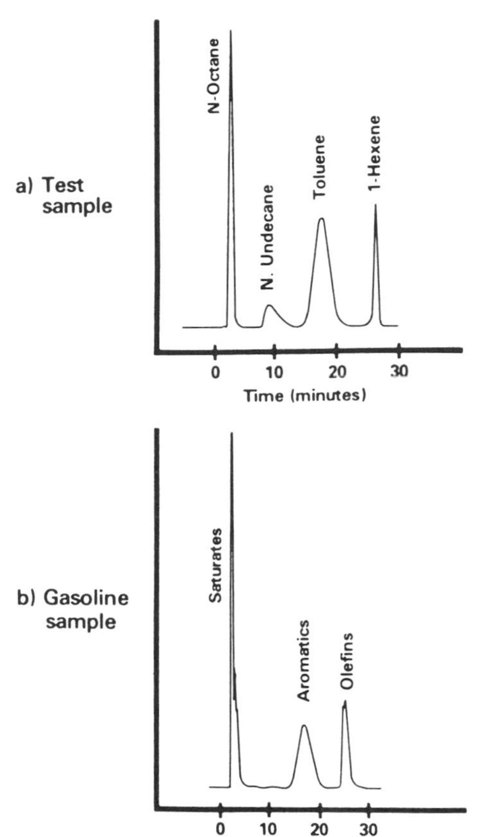

FIGURE 10.26 Chromatogram from hydrocarbon type analysis of a test sample and a typical gasoline (Reprinted with permission from Reference 50, *Analytical Chemistry*, Copyright 1981, American Chemical Society.)

first column is loaded with CEF to retain the aromatics. It can also be used alone to give a total saturate–aromatics split. A second column with a mixed liquid phase of di-*n*-propyl tetrachlorophthalate (DPTCP), Carbowax 400, and methyl abietate separates the individual aromatics. Figure 10.27 contains chromatograms comparing the aromatics in three commercial gasolines.

The ASTM procedures for aromatics in gasolines use either a single- or a dual-column approach. Method D-2267 is used for determination of benzene, toluene, and C_8 aromatics in aviation gasoline as well as reformate feed and product samples (11). A polar column is used to hold up the aromatics and an internal standard (*n*-undecane or *n*-dodecane) is added. A typical chromatogram with a tetracyanoethylated pentaerythritiol column is shown in Figure 10.28. Other suggested columns are diethyl glycol succinate, polyethylene glycol, and 1,2,3-tris(2-cyanoethoxy)propane. For mineral spirits, a similar procedure is suggested through the use of a polar column such as an *N*,*N*-bis(2-cyanoethyl)formamide column and cyclohexanone as the internal standard. This analysis is important to verify the aromatic content of solvents due to air pollution regulations. Finally, benzene and

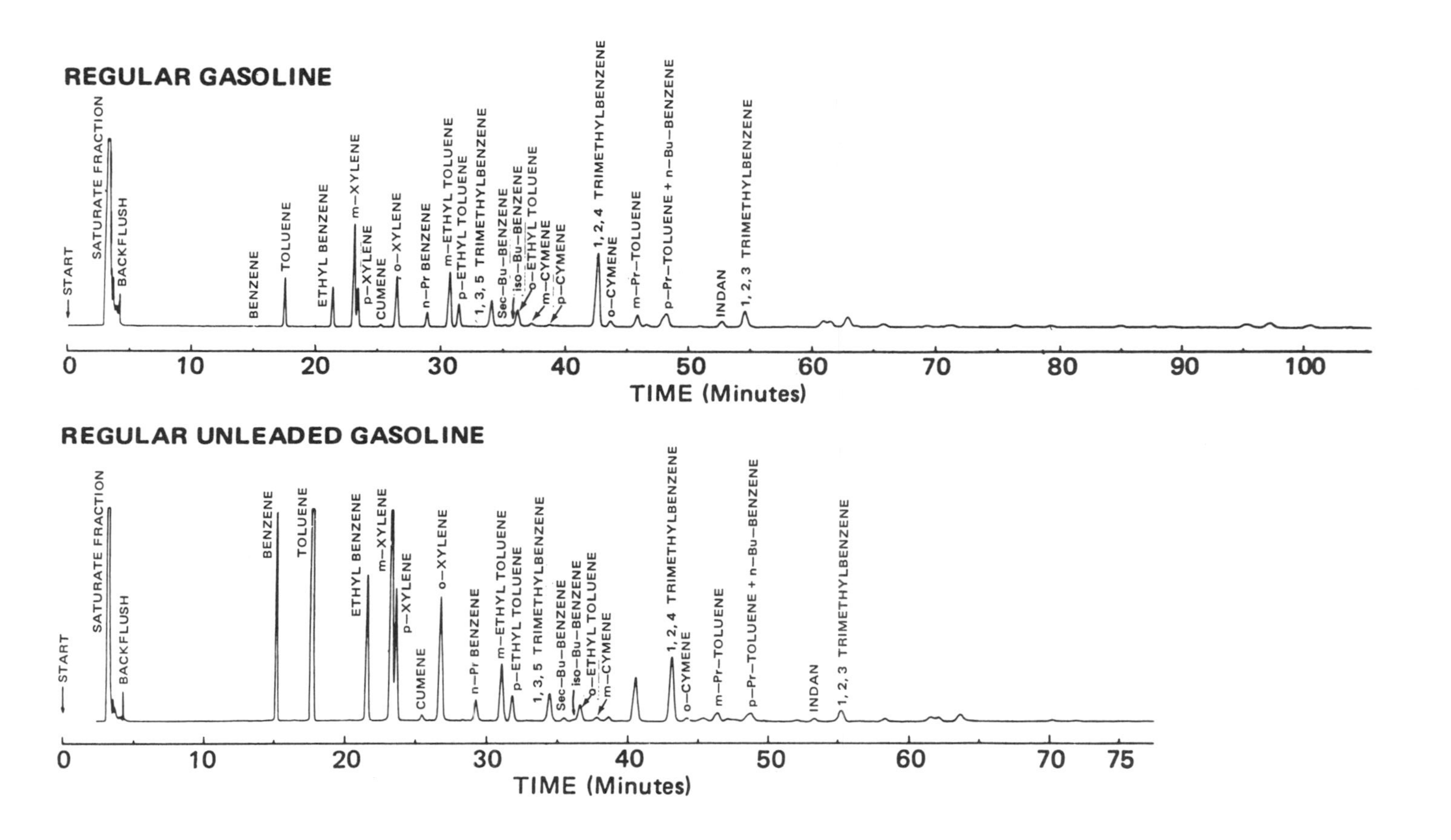

REGULAR GASOLINE
START
SATURATE FRACTION
BACKFLUSH
BENZENE
TOLUENE
ETHYL BENZENE
m–XYLENE
p–XYLENE
CUMENE
o–XYLENE
n–Pr BENZENE
m–ETHYL TOLUENE
p–ETHYL TOLUENE
1, 3, 5 TRIMETHYLBENZENE
Sec–Bu–BENZENE
iso–Bu–BENZENE
o–ETHYL TOLUENE
m–CYMENE
p–CYMENE
1, 2, 4 TRIMETHYLBENZENE
o–CYMENE
m–Pr–TOLUENE
p–Pr–TOLUENE + n–Bu–BENZENE
INDAN
1, 2, 3 TRIMETHYLBENZENE
0
10
20
30
40
50
60
70
80
90
100
TIME (Minutes)
REGULAR UNLEADED GASOLINE
START
SATURATE FRACTION
BACKFLUSH
BENZENE
TOLUENE
ETHYL BENZENE
m–XYLENE
p–XYLENE
CUMENE
o–XYLENE
n–Pr BENZENE
m–ETHYL TOLUENE
p–ETHYL TOLUENE
1, 3, 5 TRIMETHYLBENZENE
Sec–Bu–BENZENE
iso–Bu–BENZENE
o–ETHYL TOLUENE
m–CYMENE
1, 2, 4 TRIMETHYLBENZENE
o–CYMENE
m–Pr–TOLUENE
p–Pr–TOLUENE + n–Bu–BENZENE
INDAN
1, 2, 3 TRIMETHYLBENZENE
0
10
20
30
40
50
60
70
75
TIME (Minutes)

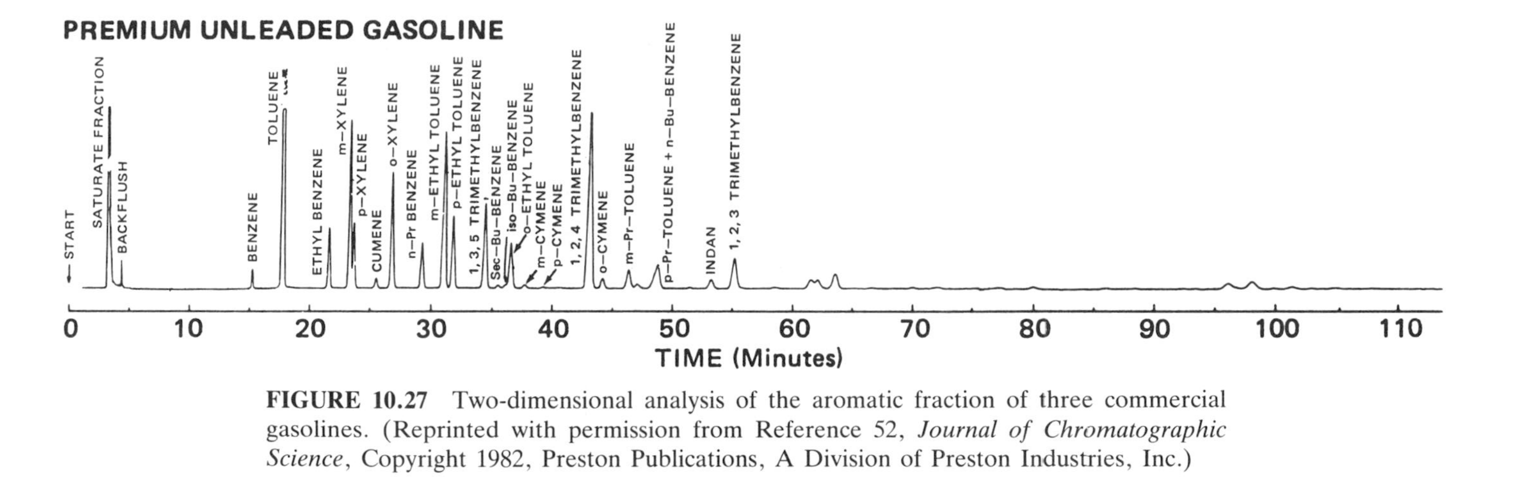

FIGURE 10.27 Two-dimensional analysis of the aromatic fraction of three commercial gasolines. (Reprinted with permission from Reference 52, *Journal of Chromatographic Science*, Copyright 1982, Preston Publications, A Division of Preston Industries, Inc.)

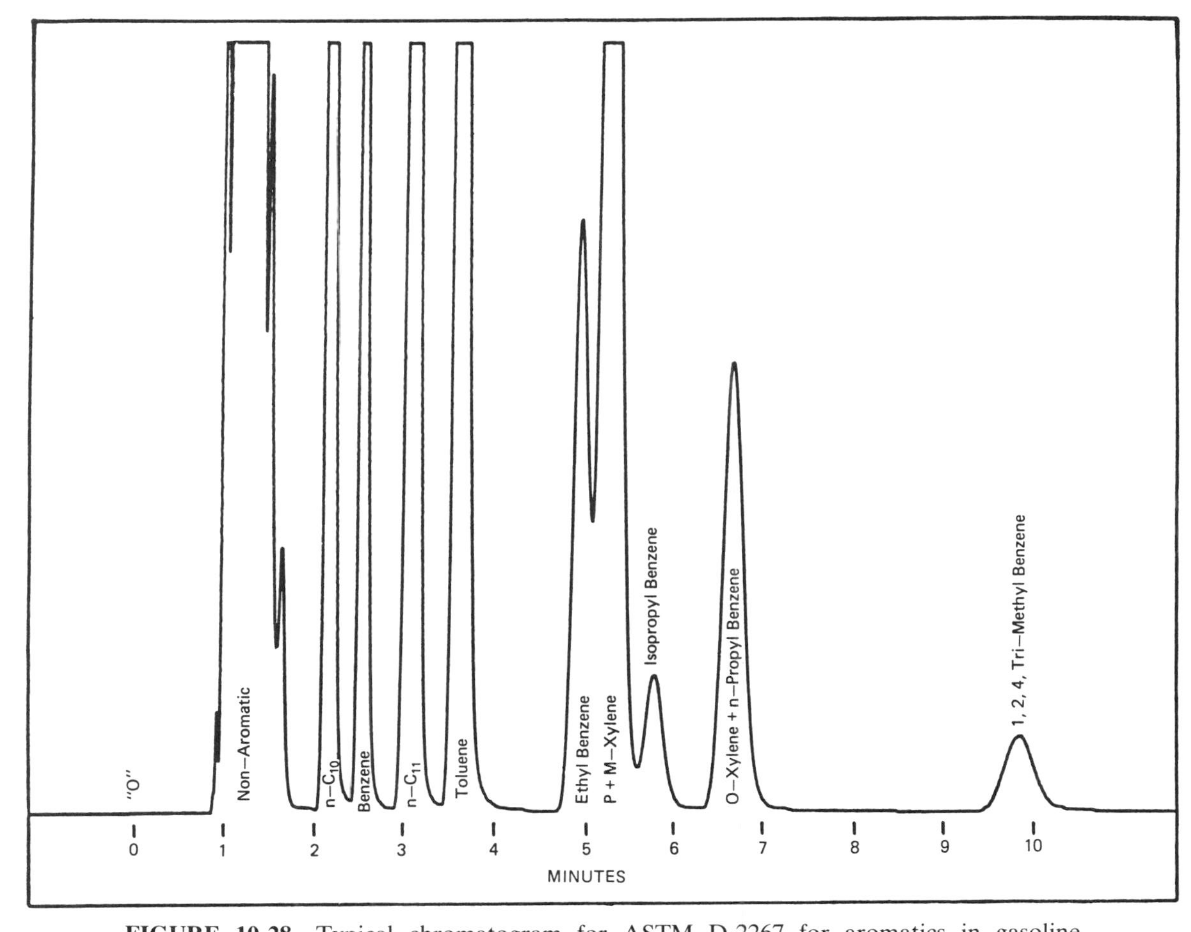

FIGURE 10.28 Typical chromatogram for ASTM D-2267 for aromatics in gasoline. (Reprinted with permission from Reference 11, Copyright 1990, ASTM.)

toluene in motor and aviation gasoline are determined with a two-column system. Method D-3606 suggests the use of a methyl silicone column to backflush the heavier components (11). The octane and lighter components are passed on to a polar column such as a 1,2,3,-tris(2-cyanoethoxy)propane column to separate the benzene, toluene, and methyl ethyl ketone internal standard. The importance of this test is to monitor benzene levels because of its toxicity.

Molecular sieve columns have been used to characterize samples by selectively removing one class of compounds. Folmer (53) and Stuckey (54) utilized this for subtractive chromatography of *n*-paraffins by using parallel columns. In one column, a section of 5A molecular sieve removed the *n*-paraffins, and a SE-52 packed column separated the other components. The other column contained a section of Celite in place of the 5A sieve. With the use of a thermal conductivity detector, a differential signal is obtained for the *n*-paraffins, as shown in Figure 10.29. Mortimer and Luke (55) accomplished the same analysis by first using molecular sieves to remove the *n*-paraffins from vaporized samples. They then destroyed the sieve with hydrofluoric acid, extracted with isooctane, and analyzed the paraffins with a short silicone gum column. A 13X molecular sieve column

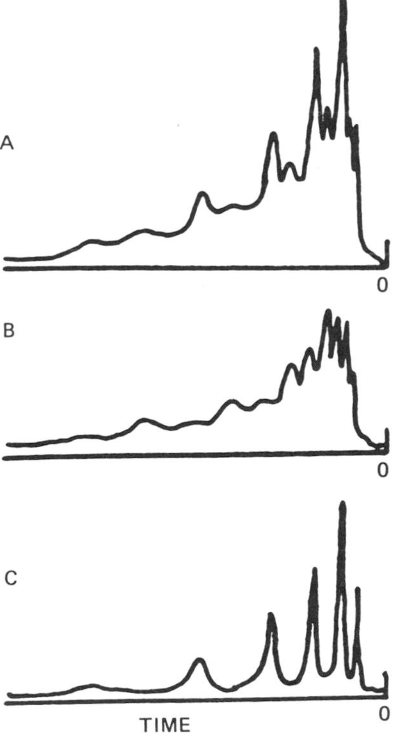

FIGURE 10.29 Substractive chromatographic analysis of a kerosene sample: (A) total sample, (B) nonnormal paraffins, (C) normal paraffins—difference between (A) and (B). (Reprinted with permission from Reference 53, *Analytica Chimica Acta*, Copyright 1972.)

was used by Garilli et al. (56) for fractions with boiling points up to 185°C. This column separated components by carbon number but partially and irreversibly adsorbed aromatics. As shown in Figure 10.30, they used this analysis to monitor the conversion of a naphtha stream into higher octane paraffins in a platforming (reforming) unit.

A method for a more complete analysis of all of the components was developed by Boer and van Arkel (57). A three-column system with automatic valve switching and cold traps was utilized to determine paraffins, naphthenes, and aromatics. Thus it is commonly referred to as a PNA analysis. A polar column of tris(cyanoethyl)nitromethane was used to retard the aromatics. A 13X molecular sieve column then separated the paraffins and naphthenes. A nonpolar column of UCCW-982 provided further separation of the aromatics and naphthene components. Figure 10.31 shows a typical chromatogram along with an indication of the column switching involved.

Boer et al. (58) modified their original PNA analyzer to allow analysis of samples with final boiling points above 200°C (up to 275°C). These modifications have minimized problems with cold-trapping fractions by requiring only medium- and high-boiling components to be trapped along with no flow reversal through the trap. The system is basically two chromatographic systems coupled by a switchable cold trap. One system contains the polar OV-275 and the 13X molecular sieve columns in series. The other contains the nonpolar OV-101 column. A Pt/Al_2O_3 hydrogenator has also been added to saturate olefins.

The overall analytical cycle is basically the same as in the original analyzer. The sample is injected on the polar column. The paraffins and mononaphthenes are trapped as they elute onto a 13X molecular sieve column. The low-boiling aromatics and higher-boiling components are eluted from the polar column into a trap. These trapped components are then separated on the nonpolar column while the high-boiling compounds (b.p. > 200°C) are backflushed as a single peak. The higher-boiling aromatics are then eluted from the polar column, trapped, and analyzed on the nonpolar column. The highest-boiling aromatics are backflushed from the polar column and then analyzed on the nonpolar columns. The nonpolar column is backflushed following both of the previous cycles, and the final step involves temperature programming.

Forty-seven peaks representing single or multiple components are obtained from this analysis. A single rather than dual FID is used to increase the accuracy and repeatability of this analysis at the expense of doubling the analysis time. The inherent repeatability, accuracy, and reproducibility of this multicolumn technique was evaluated by van Arkel et al. (59).

In improving this PNA method, Curvers and van der Sluys (60) reduced the analysis time from 2 h to 70 min. They accomplished this through optimization of operating conditions while also improving resolution. By modifying the configuration around the nonpolar OV-101 column, the

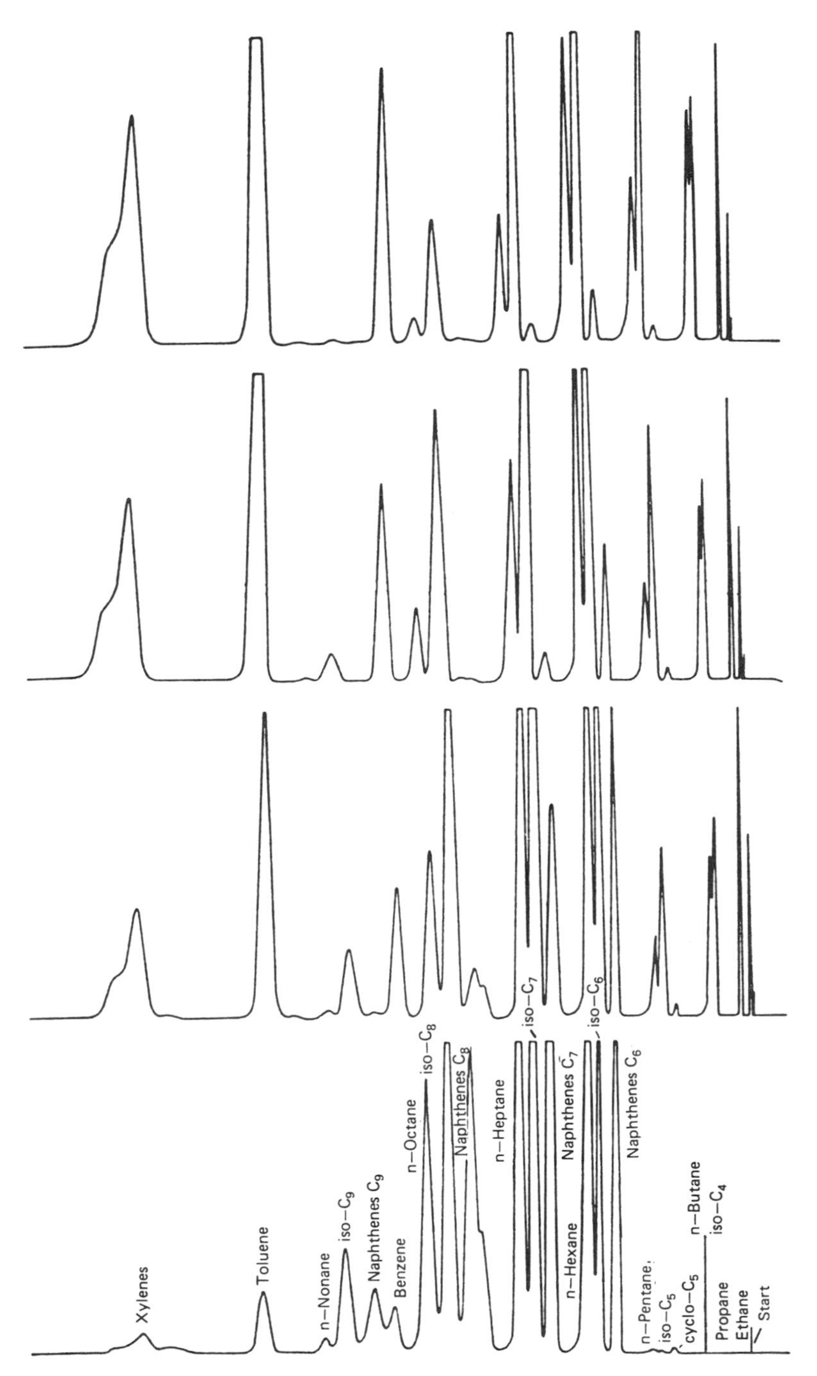

FIGURE 10.30 Four chromatograms, from bottom to top, the first platforming reactor inlet and the outlets from the first, second, and third reactors. (Reprinted with permission from Reference 56, *Journal of Chromatography*, Copyright 1973.)

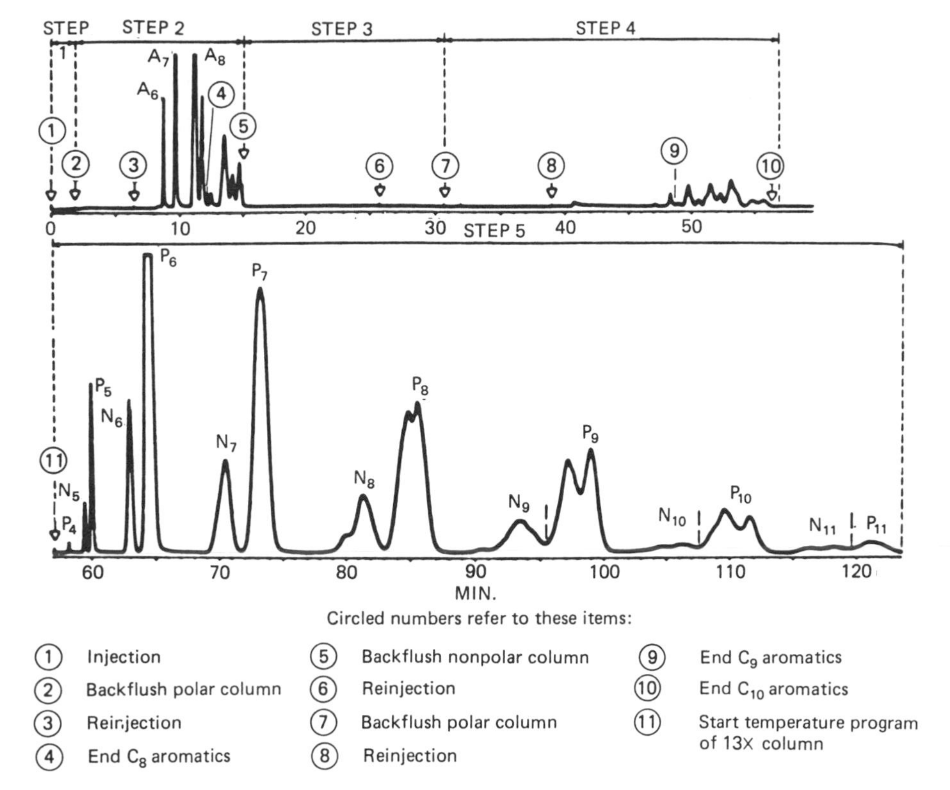

FIGURE 10.31 Typical chromatogram for a heavy naphtha by a PNA analyzer. (Reproduced with permission from Reference 57, *Hydrocarbon Processing*, February 1972.)

desorption of aromatics from the Tenax column trap was changed to backflushing rather than foreflushing through the OV-101 column. Also, the carrier gas flow rate was optimized for the OV-101 column. Finally, nonlinear temperature programming was used for the 13X column.

With this PNA analyzer, olefins can be determined only by dual injections. One analysis is made with hydrogenation of the unsaturates. The other is made after adsorption of the unsaturates. The aromatics are used as the internal standard, but this technique suffers from inaccuracy. A solution to this problem is the reversible olefins trap developed by Boeren et al. (61). Curvers and van den Engel (62) have utilized this reversible olefins trap with their optimized conditions to provide a PONA analysis in 1.5 h, as shown in Figure 10.32. In this system, the nonaromatics eluting from the OV-275 column are flushed through the olefins trap. The unsaturated components are retained, while the saturates are separated by the 13X column. After the elution of dodecane, the olefins are released with rapid heating of the trap.

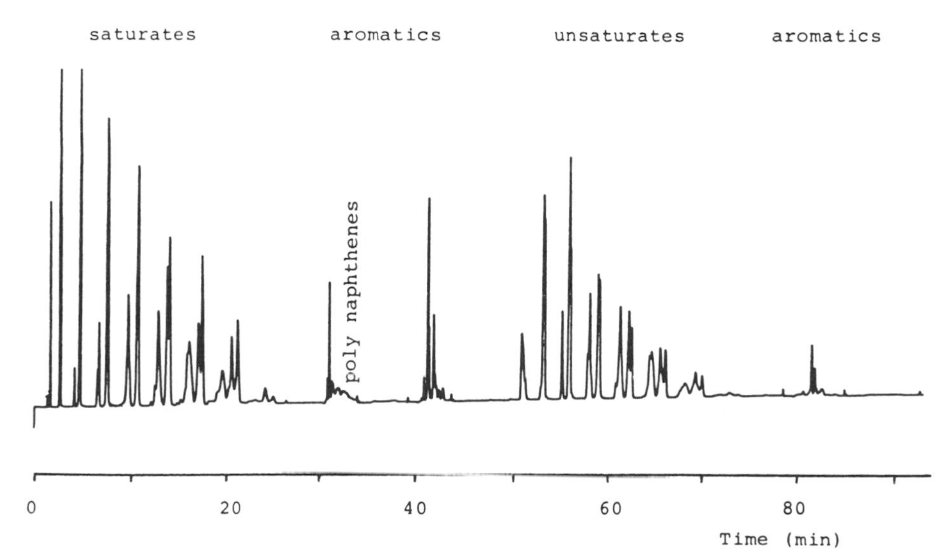

FIGURE 10.32 Typical PONA chromatogram of a visbreaker naphtha. Saturates and unsaturates are separated into naphthalenes and paraffins according to carbon number. (Reproduced with permission from Reference 62, *Journal of Chromatographic Science*, Copyright 1988, Preston Publication, A Division of Preston Industries, Inc.)

With the advances in high-resolution GC, it is possible to obtain a hydrocarbon type analysis for gasolines and naphtha using a single column. A 50-m fused-silica capillary column coated with a cross-linked dimethylsiloxane phase has been developed for this analysis (63). This PONA analysis is capable of resolving individual paraffins, olefins, naphthenes, and aromatics, as shown in Figure 10.33. A data system then identifies each of the peaks and combines them into their respective groups. The system is calibrated with a known mixture of 103 compounds. It is optimized for C_3–C_{11} components, and only the lighter olefins are completely resolved. For routine analyses, care must be taken since a slight shift in relative retention times will dramatically affect the results. Even without full calibration, this method is somewhat easier and more informative than a typical hydrocarbon type analysis. This is due mainly to its capability to determine relative differences in plant process samples.

With the complexity of the more than 230 components in samples such as gasoline, positive identification by retention time is often difficult. Even with a mass spectroscopy detector, identification of overlapping peaks is not always possible, especially mono-olefins and naphthenes, which have the same molecular weight. This problem has been alleviated by Shimoni et al. (64) who utilized precolumn sulfonation to trap olefins and aromatic compounds. The precolumn is a glass injection port packed with a small

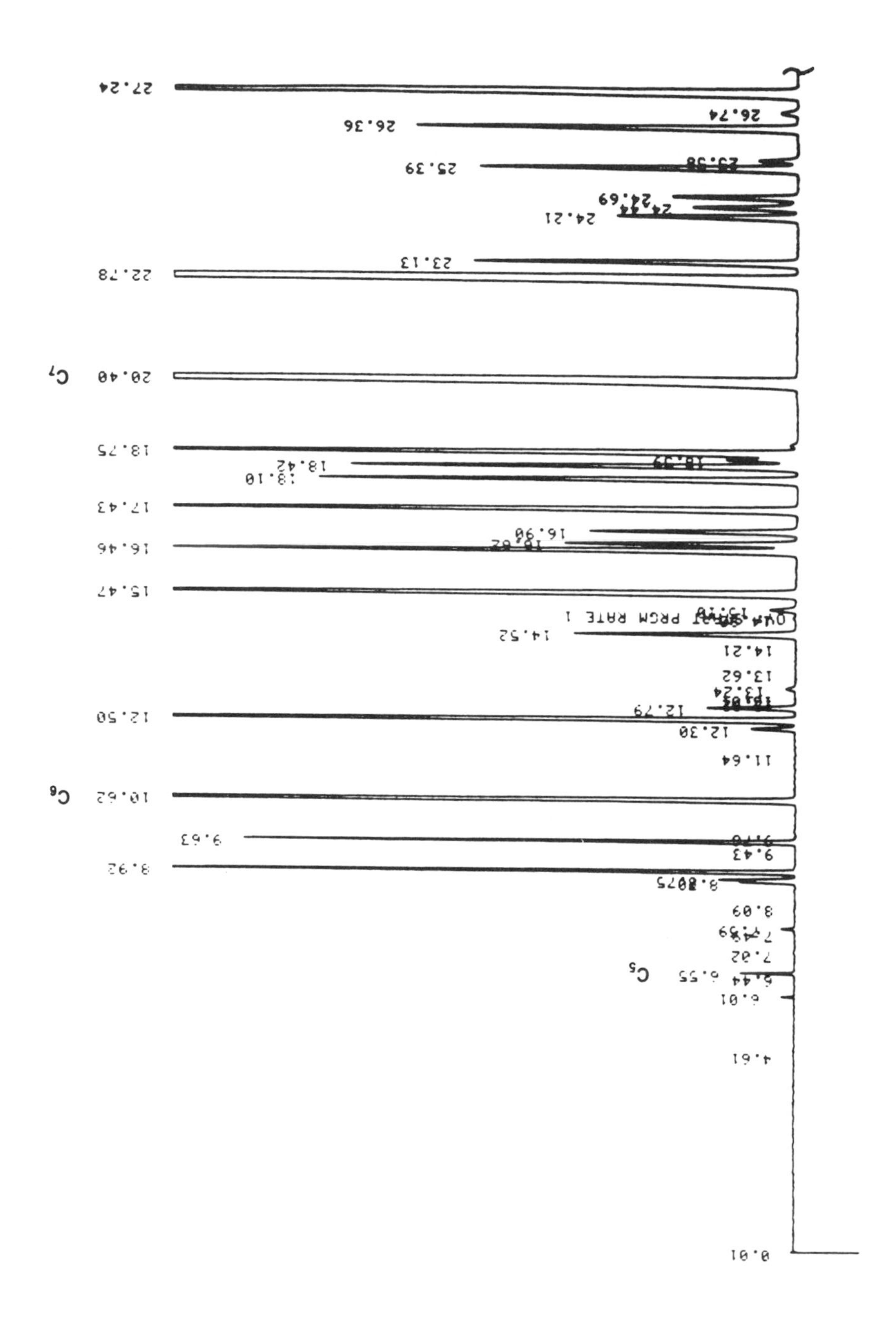
27.24
26.74
26.36
25.39
24.69
24.21
23.13
22.78
20.40
C7
18.75
18.42
18.10
17.43
16.90
16.46
15.47
START PRGM RATE 1
14.52
14.21
13.62
13.24
12.79
12.50
12.30
11.64
10.62
C6
9.63
9.43
8.92
8.09
7.02
6.55
C5
6.44
6.01
4.61
0.01

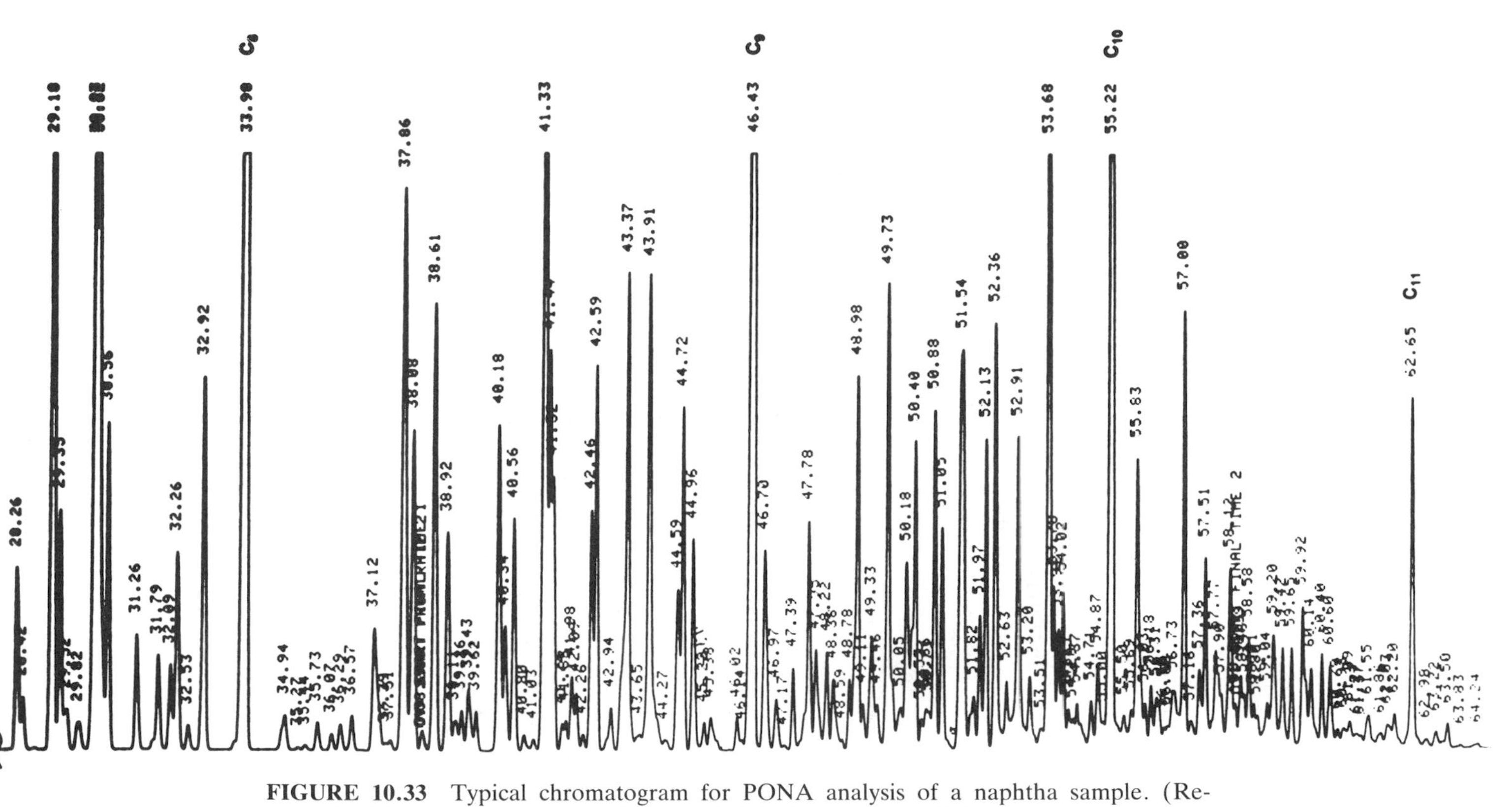

FIGURE 10.33 Typical chromatogram for PONA analysis of a naphtha sample. (Reprinted with permission from Reference 63, Hewlett–Packard Co.)

amount of solid support impregnated with sulfuric acid followed by an alkaline trap. This trapping allows for easy discrimination of saturated and unsaturated hydrocarbons.

A comparison of the PONA techniques including the classical FIA method, the multicolumn/trap/GC method, and the high-resolution capillary GC method was made by Kosal et al. (65). Capillary GC has the advantage in that it gives information on individual components. However, peak identification must be checked carefully, especially for different types of samples. It is best suited for laboratories dealing with the same type of samples that have slight variations in individual components. Separation of higher boiling components can also be a problem. The multicolumn/trap GC, on the other hand, is much more complicated, especially with maintenance of the switching valves. Finally, the classical FIA method provides only limited information with no individual component analysis or carbon number distribution. It also requires longer analysis time and more operator attention.

To monitor the catalytic conversion of methanol into gasoline, a PONA type analysis was developed by Bloch et al. (66). The catalytic process to convert methanol into gasoline produces a mixture of hydrocarbons with a maximum carbon number of 11. A complete analysis of these hydrocarbons was accomplished by using high-resolution columns along with aromatic precutting and olefin adsorption. A computer was used for programming of column temperature, flow, and valve switching. The computer also handled the data from three FIDs. An OV-275 WCOT column was used for precutting of the aromatics that were resolved on a squalene SCOT column. The paraffins, naphthenes, and olefins were separated on a squalene SCOT column. The effluent from this column was then split, with part going directly to a FID. The remainder was passed through a mercuric sulfate–sulfuric acid adsorber to remove the olefins and then onto the third FID. With this rather complex system, approximately 200 individual or combinations of compounds were identified and quantified.

Table 10.2 is a comparison of petroleum reformates and methanol-derived gasolines as determined by this analysis. The unleaded octane numbers for these samples are approximately equal. It can be seen from these data that the gasohol has higher olefin and naphthene content as well as a higher ratio of iso-/*n*-paraffins. Comparison of these data with FIA analysis (ASTM D-1319) revealed higher aromatic and lower saturate values along with excellent agreement on the olefins.

10.3.4 Sulfur and Nitrogen Compounds

Because of the unavailability of low-sulfur crude oils, heavy asphaltic crudes and synthetic crudes are now being utilized. These less desirable crudes contain heterocyclic sulfur and nitrogen compounds. The sulfur and nitrogen compounds, however, are catalyst poisons and must be removed before they

TABLE 10.2 Comparison of C_6^+ Hydrocarbon Types in Reformates and Methanol Derived Gasoline as Determined by Open Tubular Column Selective Olefin Absorption GC

Process Unit Charge Stock	C_6—360°F Mid-Continent Naphtha	C_6—365°F Nigerian Naphtha	Methanol
Octane number	97.7	99.2	98.2
Isohexanes	10.25	8.55	12.03
n-Hexane	4.61	3.86	0.60
Hexenes	0.10	0.10	1.98
Isoheptanes	7.37	5.89	5.14
n-Heptane	2.10	1.71	0.31
Heptenes	0.10	0.10	2.37
Isooctanes	2.57	2.50	1.05
n-Octane	0.64	0.50	0.06
Octenes	—	—	3.27
C_{9+} $(P + O + N_t)$[a]	0.46	0.91	2.20
Naphthenes (C_6–C_8)	1.23	1.67	9.10
Benzene	6.09	6.58	0.29
Toluene	22.04	20.40	2.96
C_8 Aromatics	25.24	23.43	18.40
C_9 Aromatics	14.20	17.72	25.26
C_{10} Aromatics	1.59	4.17	11.97
C_{11+} Aromatics	1.41	1.91	3.01
Paraffin total[b]	28.00	23.92	21.39
Olefin total	0.20	0.20	7.62
Naphthene total	1.23	−1.67	9.10
Aromatic total	70.57	74.21	61.89

Source. Reprinted with permission from Reference 66, *Journal of Chromatographic Science*, Copyright 1977, Preston Publications, Inc.

[a] N_t is total cyclopentanes and cyclohexanes.

[b] Includes C_{9+} $(P + O + N_t)$.

are refined. Their removal is also necessary for the performance, storage stability, and general acceptability of final petroleum products.

For monitoring of process streams and product quality, determination of total and individual sulfur compounds are required down to ppm levels. The characterization of individual compounds has primarily involved the combination of GC using packed columns and element-selective detectors. For sulfur compounds, Druschel (67) utilized a Carbowax 20M column with a microcoulometric detector. A combustion tube at 550°C converts the sulfur compounds into SO_2 as they elute from the column. The SO_2 is then titrated by electrogenerated iodine. A typical application for this analysis is shown in Figure 10.34. It was used to monitor the removal of condensed thiophenes

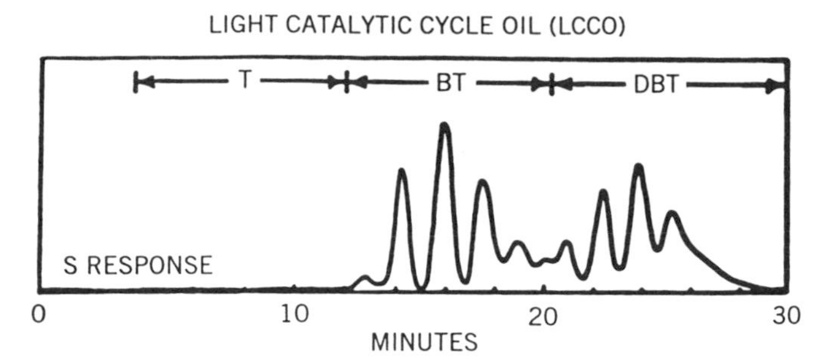

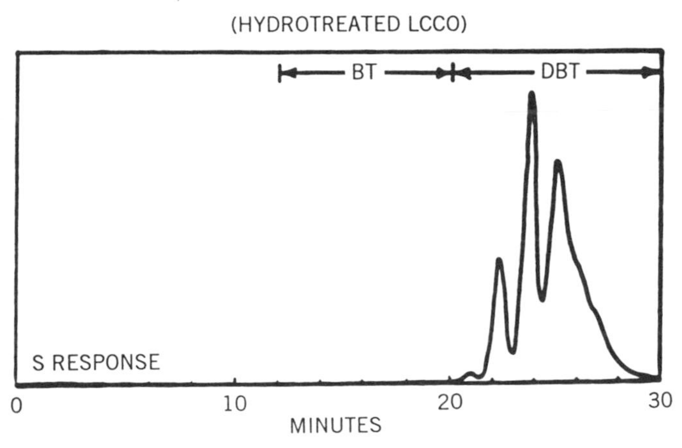

FIGURE 10.34 Gas chromatographic—microcoulometric chromatograms of light catalytic cycle oil (LCCO) demonstrating the removal of benzotheophenes by hydrotreating. (Reproduced with permission of the author, Reference 67.)

by hydrotreating of light catalytic cycle oil (LCCO). Although this analysis has low resolution, it is highly selective.

Nitrogen compounds have been similarly determined by Martin (68) and Albert (69). Hydrogenolysis is used to convert the nitrogen compounds to ammonia. The ammonia is then determined in a microcoulometric cell by generation of hydrogen ions. A typical chromatogram for LCCO using a column with a 12,000 molecular weight polyethylene liquid phase is shown in Figure 10.34. Like the sulfur analysis, this method has low resolution but high selectivity as long as a scrubber is used to remove the HCl from chlorine compounds.

A selective thermionic specific detector (TSD) has more recently been applied to the analysis of nitrogen compounds by Albert (70). This detector is commonly referred to as a *nitrogen phosphorous detector* (NPD). It is basically an alkali FID. Figure 10.35 compares the TSD and the microcoulometric detector. More resolution is obtained through the TSD with elimination of the mixing in the transfer line, reactor tube, and titration cell of the microcoulometer. A parallel FID can also be used to monitor the overall boiling point distribution of the sample.

To achieve high resolution of all the heterocyclic compounds, Druschel (71) has utilized fused-silica capillary columns with an SE-54 coating. A flame photometric detector (FPD) was used for sulfur and a TSD was used for nitrogen compounds. Figure 10.36 shows the effect of hydrotreating LCCO. As in Figure 10.34, the elimination of the peaks corresponding to the benzothiophenes is obvious. With the high resolution of the capillary column, many of the individual isomers are resolved. Comparison of the chromatograms in Figure 10.36 indicates that certain isomers of the remaining substituted dibenzothiophenes were easier to remove than others. This information allows for a more complete evaluation of catalyst activity.

High resolution of the nitrogen compounds was obtained with a TSD. Figure 10.37 shows the distribution of the individual two- and three-ring nitrogen compounds in LCCO. The resolution in these chromatograms is more obvious when they are compared to Figure 10.35. The effects of hydrotreating are illustrated in Figure 10.37. The two-ring compounds, primarily indoles, were denitrogenated. The high resolution allows identification of several hydrogenated intermediates that were formed without nitrogen removal. This is because hydrogenation is required before nitrogen removal can occur.

An effluent splitter allows for simultaneous detection with the FID and either the FPD or TSD. Besides determination of the overall hydrocarbon distribution, the FID is of value in monitoring for quenching of the FPD. The presence of large background hydrocarbon levels quench the emission from the sulfur compounds. The results are not quantitative if this occurs. Simultaneous detection with the FID also gives an indication of high levels of components that may cause a false response by the TSD.

A more recent development, the sulfur chemiluminescence detector (SCD), shows no interference from hydrocarbon quenching (72). It also provides part per billion sensitivity and response linearity over a large range. With its equimolar response to sulfur regardless of compound type, it allows accurate determination of total sulfur without the necessity for identification of all components. A further advantage is that the flame ionization detector acts as the sulfur converter. Thus, the FID remains functional as a simultaneous hydrocarbon detector. Figure 10.38 illustrates the determination of sulfur in gasoline. Although the current level of total sulfur in gasoline is 300–500 ppm, the California Air Resources Board (CARB) has called for a sulfur level below 40 ppm. The GC/SCD analysis in this figure indicated a sulfur content of 36 ppm versus the ASTM D4045 certified analysis of 28 ppm. The FID analysis reveals a different composition from the base naphtha feedstock from which it was formulated. Most obviously,

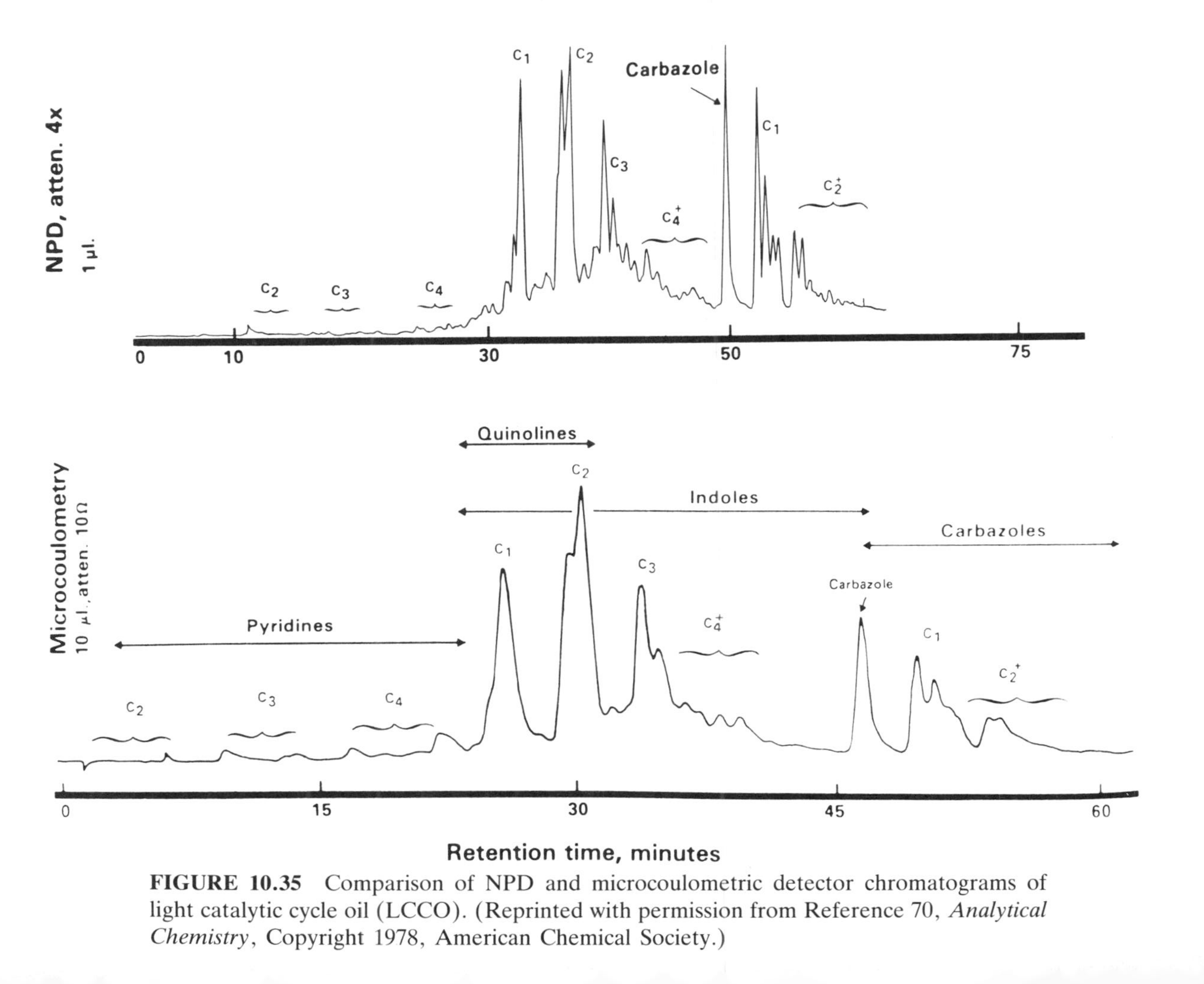

FIGURE 10.35 Comparison of NPD and microcoulometric detector chromatograms of light catalytic cycle oil (LCCO). (Reprinted with permission from Reference 70, *Analytical Chemistry*, Copyright 1978, American Chemical Society.)

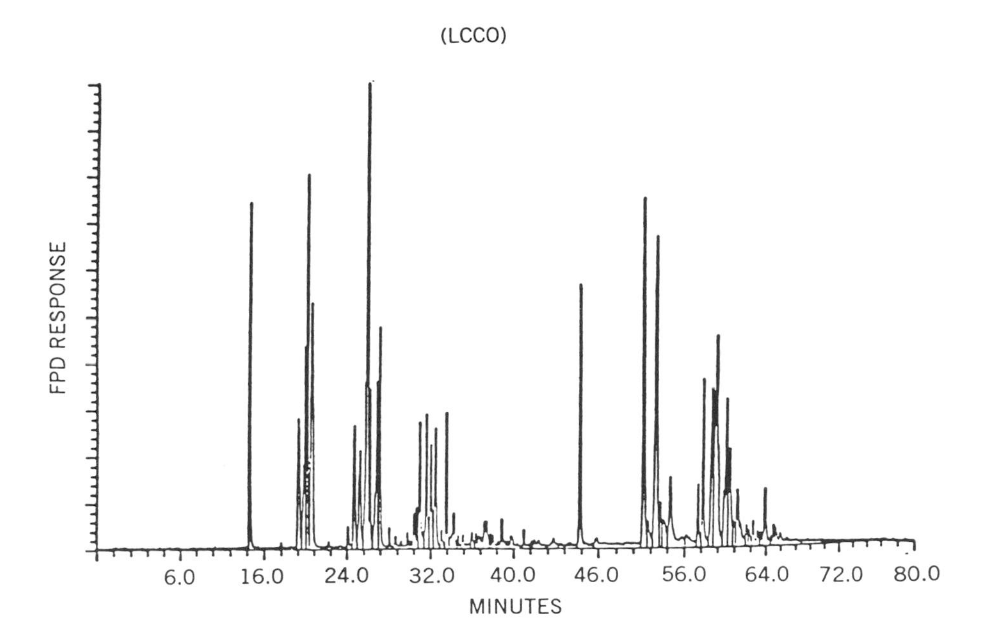

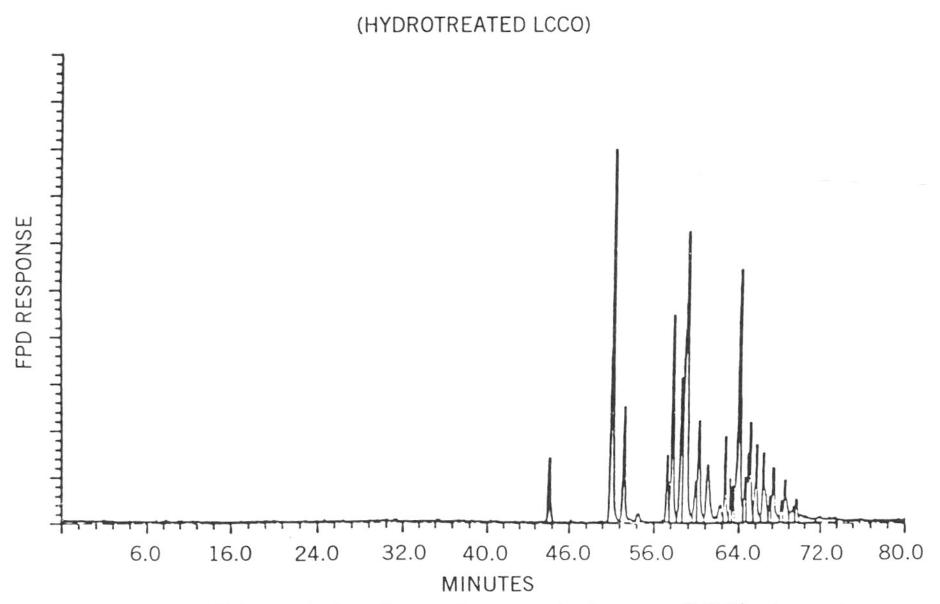

FIGURE 10.36 High-resolution flame photometric detector (FPD) chromatogram of light cycle gas oil (LCCO) demonstrating the effect of hydrotreating on condensed thiophenes. (Reproduced with permission of the author, Reference 71.)

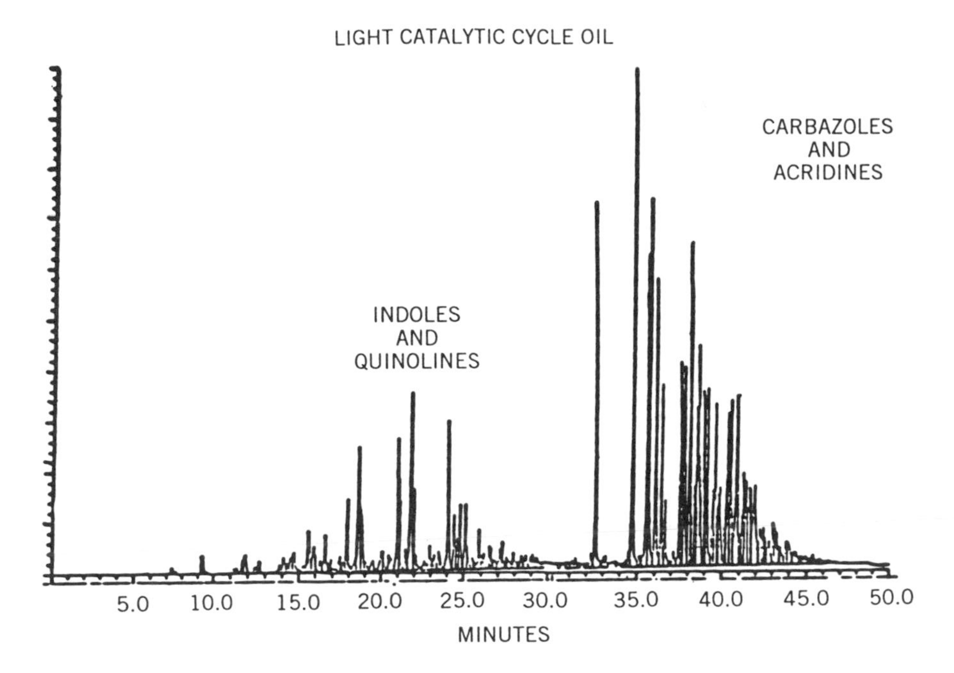

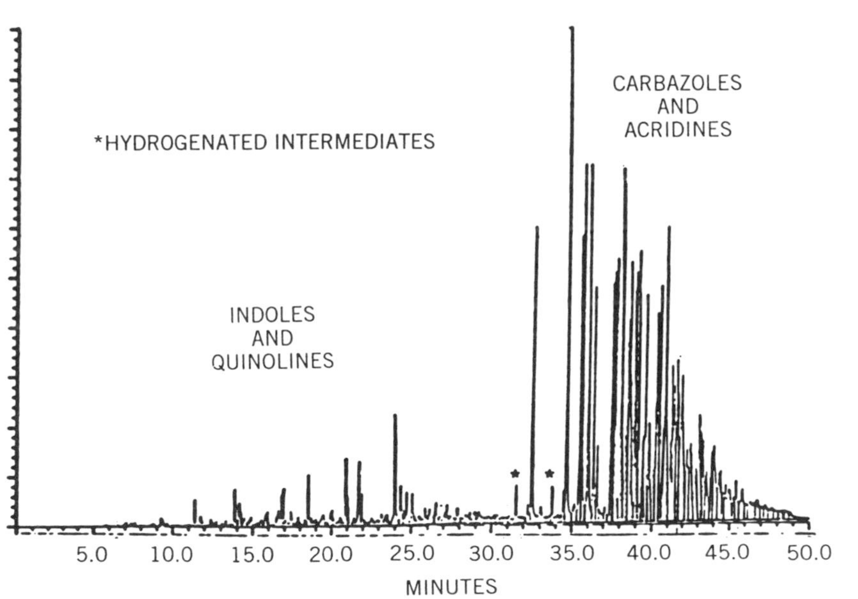

FIGURE 10.37 High-resolution thermionic specific detection (TSD) of light cycle gas oil (LCCO) demonstrating the removal of nitrogen compounds and the formation of hydrogenated intermediates due to hydrotreating. (Reproduced with permission of the author, Reference 71.)

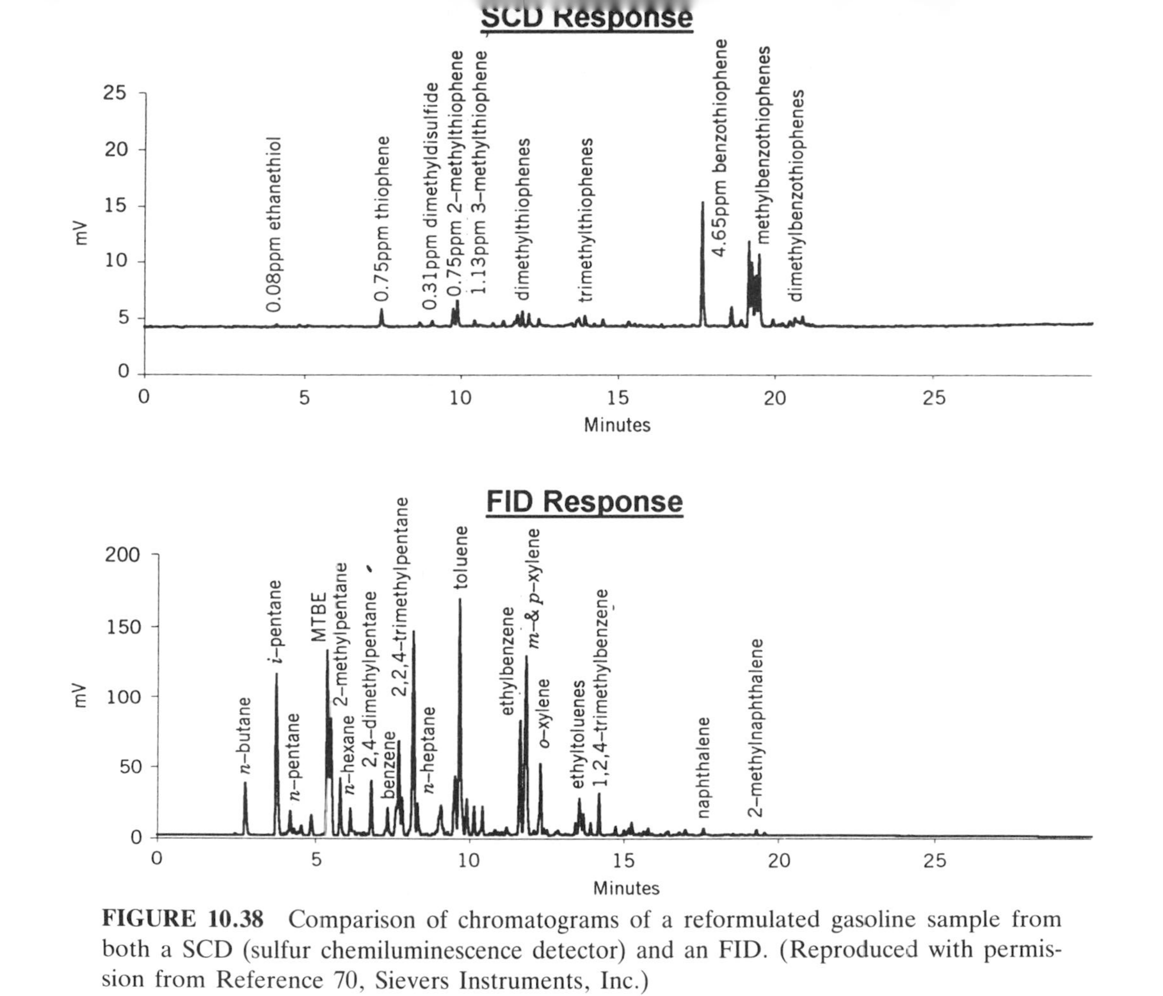

FIGURE 10.38 Comparison of chromatograms of a reformulated gasoline sample from both a SCD (sulfur chemiluminescence detector) and an FID. (Reproduced with permission from Reference 70, Sievers Instruments, Inc.)

methyl *tert*-butyl ether (MTBE) and 2,2,4-trimethyl pentane (isooctane) have been added to this commercial sample.

10.3.5 Gasoline Additives

In 1991, federal regulations mandated that oxygenated compounds such as alcohols and ethers be used as octane improvers for gasoline. They are intrinsically high-octane components, while they contain no environmentally objectionable heteroatoms like the previously used metal alkyls. Methyl *tert*-butyl ether (MTBE), ethyl *tert*-butyl ether (ETBE), and *tert*-amyl methyl ether (TAME) are now the most commonly used additives. Although ETBE and TAME have the advantage of lower volatility, they cost more to make than MTBE.

Due to the complexity of oxygenate containing samples, a multidimensional gas chromatograph was developed by Naizong and Green (73) for the analysis of these additives. Their approach has become the basis for the ASTM D-4815 method for C_1–C_4 alcohols and MTBE in gasoline (11). Unfortunately, this method determines a limited range of alcohols and ethers particularly with the *tert*-amyl alcohol internal standard. Work is currently in progress to expand the method to C_1–C_5 alcohols, MTBE, TAME, ETBE, and DIPE (di-isopropyl ether). Reproducibility is being improved and the range is being extended to 12 vol% for alcohols and 20% for ethers. The internal standard is also being changed to 1,2-dimethoxyethane (DME). A typical chromatogram is shown in Figure 10.39 (74).

The EPA has also endorsed the use of the oxygen-specific flame ionization detector (O-FID). The O-FID oxygenates analyzer (75) utilizes a single capillary column connected to two microreactors and an FID. A cracking reactor converts any oxygenate into carbon monoxide, which is then catalytically hydrogenated to methane for detection by the FID. The comparison in Figure 10.40 of the standard FID and O-FID for a gasoline sample demonstrates a very selective oxygenates analysis that is not quenched or suppressed by the sample matrix. The dynamic range of this analysis is 10^5 and 1-propanol is used as an internal standard.

For monitoring of process streams including MTBE process units, a single-column method is more desirable. Rather than use multiple injections, Spock (76) has proposed a 100-m Petrocol DH column. As shown in Figure 10.41, the oxygenates as well as the other hydrocarbons can be quantified using response factors rather than internal or external standards. This method also eliminates the risk of error due to leaking valves and inaccurate valve switch timing in multidimensional GC. However, multiple injections and standardization are required for levels below 100 ppm.

Prior to the establishment of recent environmental limitations, tetraethyl lead (TEL) was added to most gasolines as an antiknock agent. In addition to other alkyl leads, MMT (methylcyclopentadienyl manganese carbonyl) was also used. Methylcyclopentadienyl manganese carbonyl was more

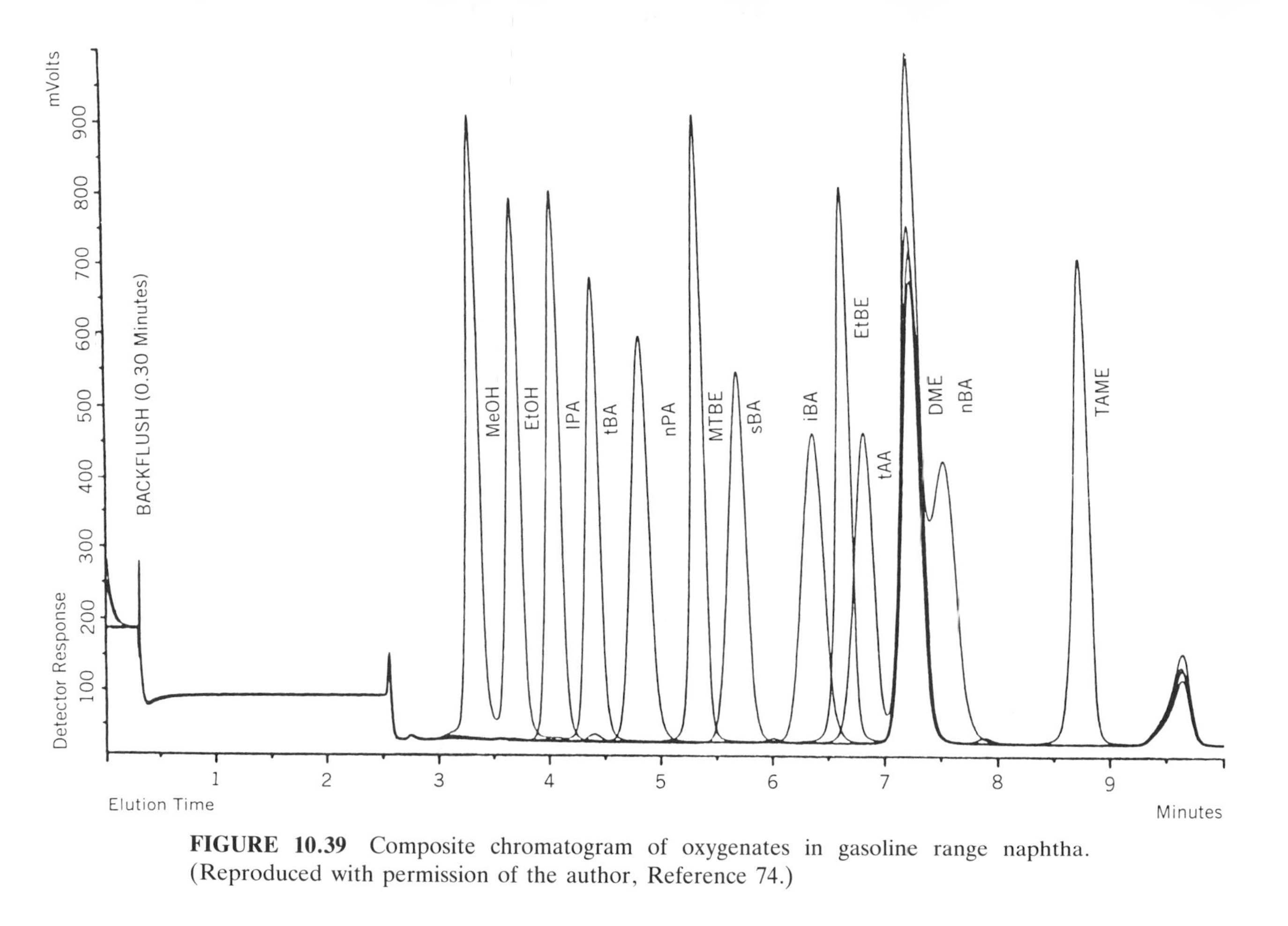

FIGURE 10.39 Composite chromatogram of oxygenates in gasoline range naphtha. (Reproduced with permission of the author, Reference 74.)

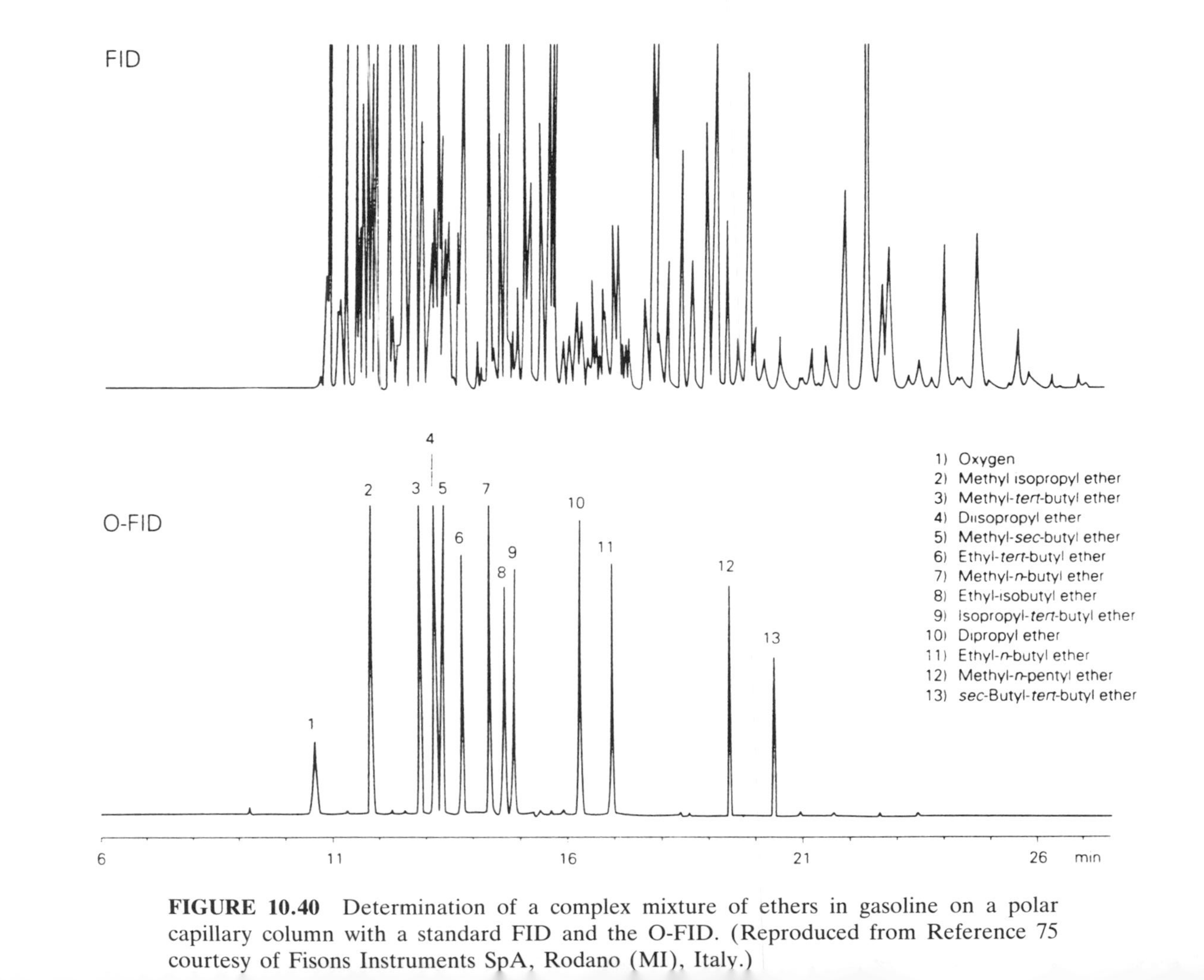

FIGURE 10.40 Determination of a complex mixture of ethers in gasoline on a polar capillary column with a standard FID and the O-FID. (Reproduced from Reference 75 courtesy of Fisons Instruments SpA, Rodano (MI), Italy.)

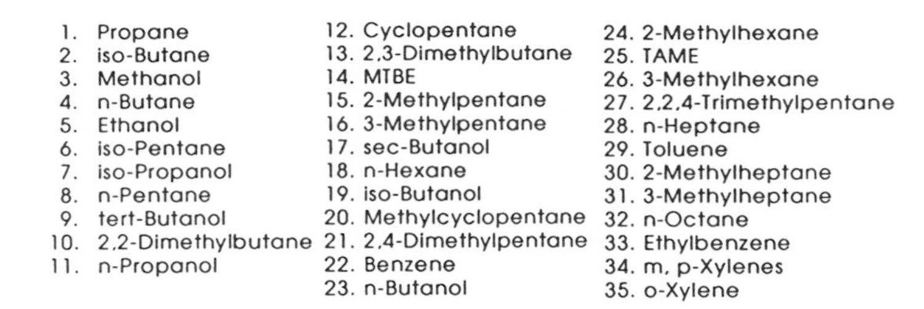

1. Propane
2. iso-Butane
3. Methanol
4. n-Butane
5. Ethanol
6. iso-Pentane
7. iso-Propanol
8. n-Pentane
9. tert-Butanol
10. 2,2-Dimethylbutane
11. n-Propanol
12. Cyclopentane
13. 2,3-Dimethylbutane
14. MTBE
15. 2-Methylpentane
16. 3-Methylpentane
17. sec-Butanol
18. n-Hexane
19. iso-Butanol
20. Methylcyclopentane
21. 2,4-Dimethylpentane
22. Benzene
23. n-Butanol
24. 2-Methylhexane
25. TAME
26. 3-Methylhexane
27. 2,2,4-Trimethylpentane
28. n-Heptane
29. Toluene
30. 2-Methylheptane
31. 3-Methylheptane
32. n-Octane
33. Ethylbenzene
34. m, p-Xylenes
35. o-Xylene

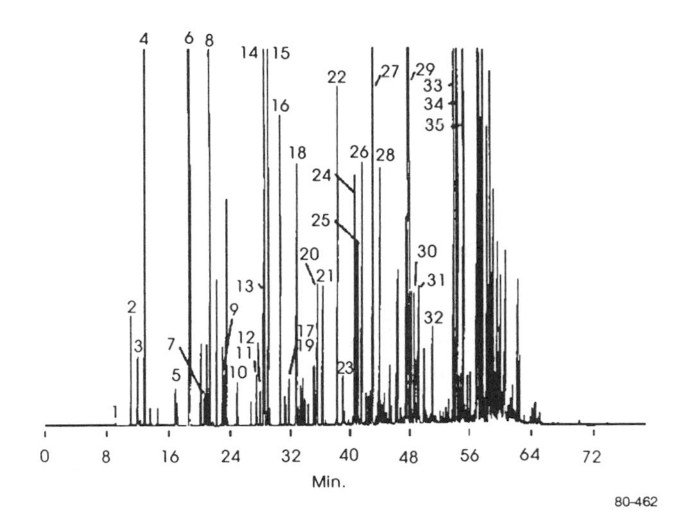

Petrocol DH column, 100m x 0.25mm, 0.50μm film, Col. Temp.: -20 °C/min., then to 225°C at 25°C/min., Injector Temp.: 225°C, Detector Temp: 275°C, Linear Velocity: 19cm/sec., He, Detector: FID (64 x 10^{-11} AFS), Injection Volume: 0.50μl, Split Ratio: 100:1, Sample: unleaded gasoline with 0.05 to 0.01% C1-C4 alcohols, 1.0% MTBE, and 0.5% TAME added.

FIGURE 10.41 Oxygenates in unleaded gasoline using a Petrocol DH column, 100 m × 0.25 mm, 0.5-μm film, column temperature: −20 to 65°C at 2°C/min, then to 225°C at 25°C/min. (Reprinted with permission from Reference 76, Supelco, Inc., Bellefonte, PA 16823.)

effective than lead, but was limited because of concern over manganese poisoning and its detrimental effects on automobile catalytic converters. Restricted amounts of these compounds are still used. X-ray fluorescence and atomic absorption spectroscopy are the most popular techniques for determination of these additives by the total metal content of gasolines.

For the determination of individual components, GC has been coupled with selective metal detectors. Coker (77) developed a gas chromatographic-atomic absorption technique for lead alkyls. A PEG 20M on Porasil C column provided the separation with a detection limit of 0.2 ppm. Uden et al. (78) utilized an argon plasma emission detector with a Dexsil 300 column detector for MMT with a detection limit of 3 ng of Mn. As for more typical gas chromatographic systems, several electron-capture detector (ECD) procedures have been developed (79–81). Analyses with an ECD are complicated by different sensitivities for each of the lead alkyl isomers. Also, there are interferences from the alkyl halides that are present as lead scavengers. Soulages (82) used a flame ionization detector following catalytic hydrogenolysis of methyl and ethyl lead alkyls.

A procedure was developed by DuPuis and Hill (83) for both lead and manganese antiknock compounds. They employed a hydrogen atmosphere flame ionization detector (HAFID) along with a conventional FID. The HAFID is a modified FID in which the flame burns in a hydrogen atmosphere that is doped with a small amount of silane. Typical chromatograms for regular and premium gasolines containing alkyl leads using this detector are given in Figure 10.42.

10.4 PETROCHEMICALS

The term "petrochemicals" refers to the basic chemicals that are derived from refinery petroleum cuts. They are produced by separation of the by-products from the cracking (pyrolysis) of hydrocarbon streams. These streams range from natural gas to the heavy distillate (gas oil) cuts from a refinery primary fractionator. Some chemicals, such as the aromatics, are separated from various refinery streams. Figure 10.43 is simplified schematic of a petrochemical process.

The basic petrochemicals, which are produced in the largest volumes, are separated into two classes: olefins and aromatics. The olefins include ethylene, propylene, and 1,3-butadiene. The aromatics are benzene and the xylenes. These chemicals are used primarily for the manufacturing of plastics, synthetic rubbers, and fibers. A wide range of other chemicals are also produced in somewhat lesser volumes, but with a variety of applications. This discussion is limited to the major chemicals identified previously.

Because of the large volumes involved and the interchange of products among companies, analyses of all the major petrochemicals are covered by ASTM standards (11). Although other analyses are often used for process control, the ASTM methods are used to resolve discrepancies between laboratories. Thus, the applicable ASTM procedures are discussed here along with each petrochemical.

10.4.1 Olefins

The primary purpose of the cracking of natural gas or a wide range of petroleum-derived streams is the production of ethylene. All the other olefins that form are considered to be by-products. Cracking feedstocks range from natural gas to heavy distillates, depending on their price and the

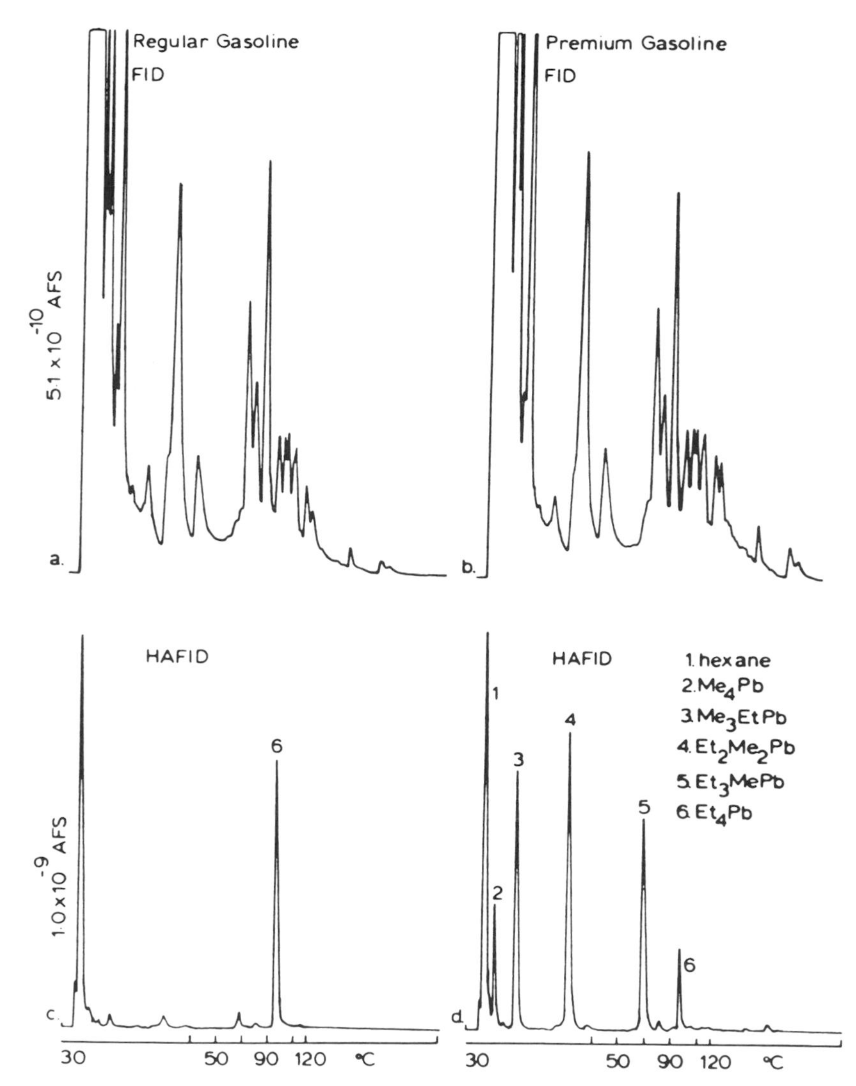

FIGURE 10.42 Hydrogen atmosphere FID and FID analysis of alkyl leads in gasoline: (a) FID tracing of a regular-grade gasoline; (b) FID tracing of a premium-grade gasoline; (c) HAFID tracing of a regular gasoline; (d) HAFID tracing of a premium gasoline. (Reprinted with permission from Reference 83, *Analytical Chemistry*, Copyright 1979, American Chemical Society.)

desired products. Originally, the low price of crude oil made petroleum streams, mainly the heavy distillate or gas oils, the most economical. Currently, this has shifted to the use of ethane from natural gas as the preferred feedstock. Besides profitability, the only other consideration in

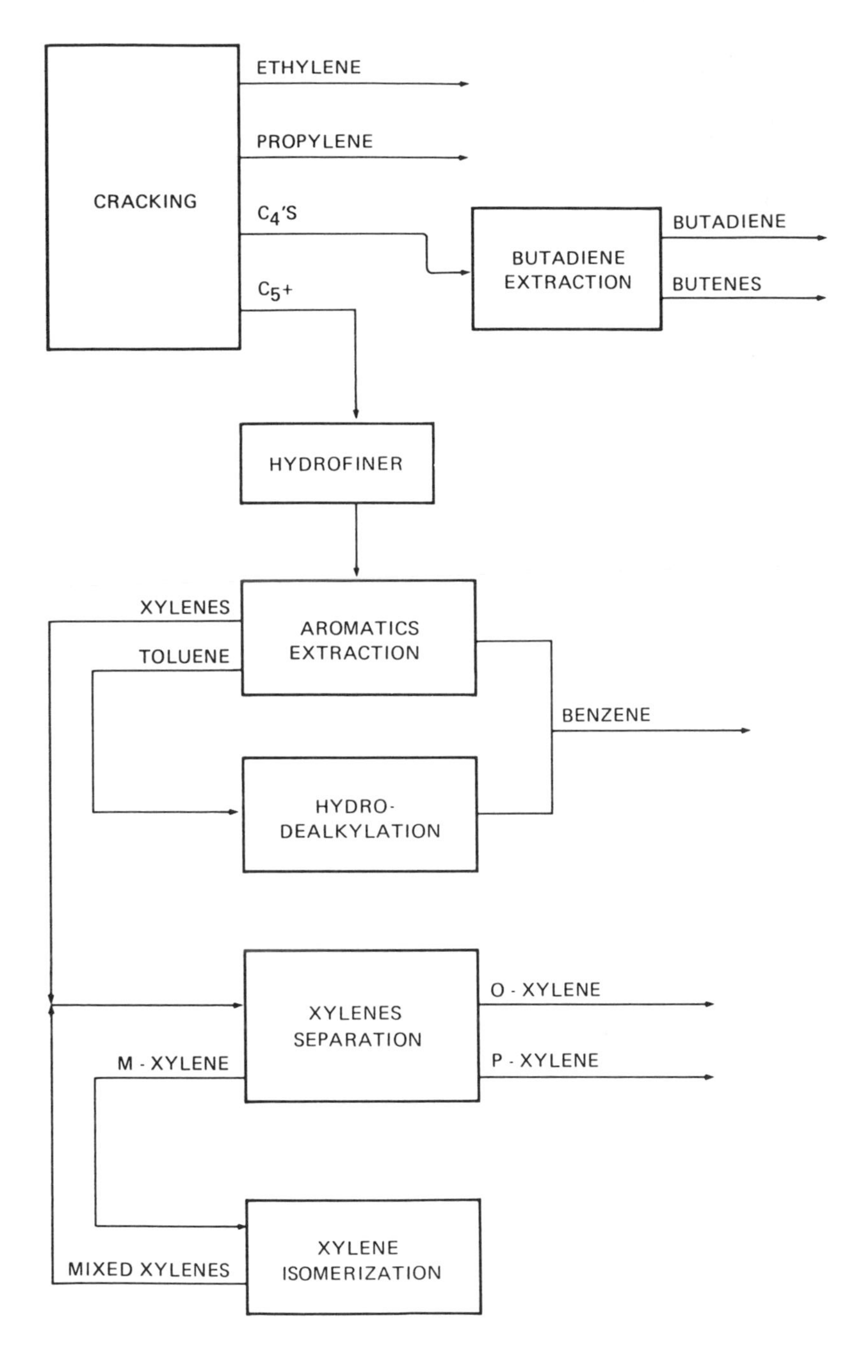

FIGURE 10.43 Simplified schematic of a petrochemical process.

choosing feedstocks is that the heavier feeds produce more of the heavier olefins, especially butylenes.

Cracking of hydrocarbons to produce olefins can be done thermally or catalytically. Thermal cracking is the most typical method. It is basically a pyrolysis step, as discussed in the section on refining. To avoid polymerization reactions with the olefins, steam can be injected to quench these side reactions. This process is commonly referred to as *steam cracking*.

With the variety of feedstocks involved, there is a need to optimize cracking conditions based on an analysis of the feedstock. For refinery distillate cuts, characterization is a difficult analytical problem. A hydrocarbon type analysis, as discussed in Section 10.3.4, can be correlated with cracking yields. Another technique was described by Greco (84) in which microscale pyrolysis was used to simulate cracking. A tube-type pyrolyzer with a quartz insert was coupled to a gas chromatograph. The pyrolysate was separated and determined with a Carbowax 20M on alumina column and an FID. With this system, a linear relationship was established between the microscale pyrolyzer and commercial cracker yields. Figure 10.44 is a

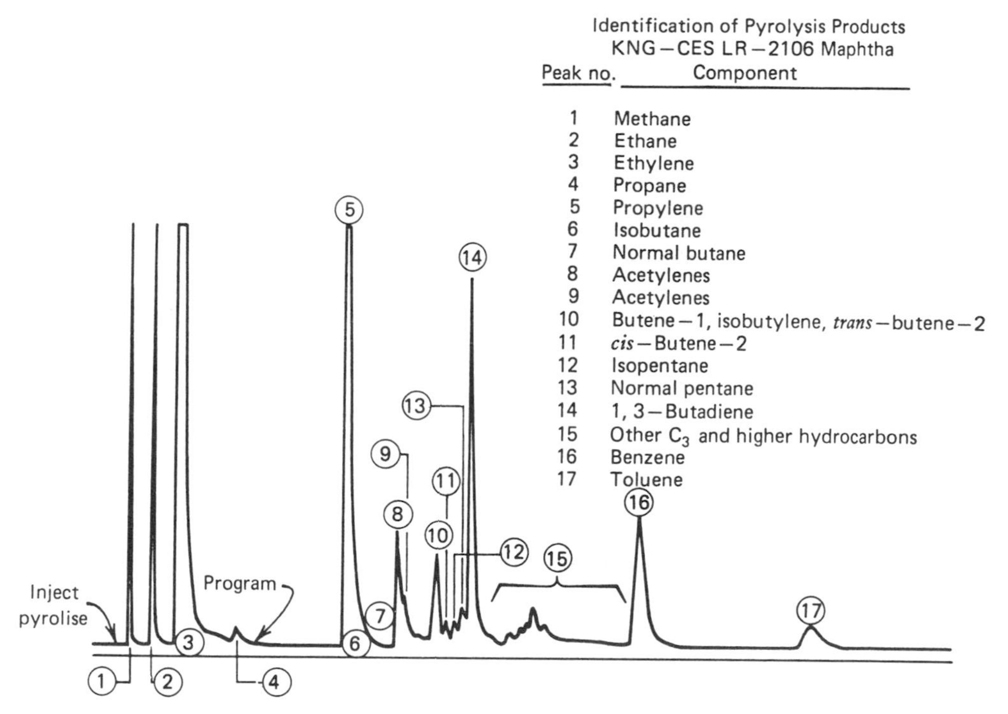

FIGURE 10.44 Chromatogram from analysis of bench-scale pyrolysis of a naphtha sample. (Reprinted with permission from Reference 84, *Journal of Chromatographic Science*, Copyright 1978, Preston Publications, A Division of Preston Industries, Inc.)

typical chromatogram for a light naphtha. A plot of product distribution as a function of temperature is shown in Figure 10.45. From this plot, an optimum operating temperature can be selected to achieve the desired cracking products. It should be noted, however, that regardless of the technique, exact prediction of cracking yields is extremely difficult.

Analysis of the effluent from a steam cracker is difficult in terms of both sampling and analysis. Because of the wide range of components present, condensation of the heavier compounds must be considered. Heated sample lines as well as heating of the sample bomb are required. Analysis of these components was accomplished by Jordan et al. (85), using a mixed liquid phase of SE-30 and tris-cyanopropane column with subambient temperature programming. Because of the similarity, refinery gas analyses are now typically utilized for this analysis. As mentioned in Section 10.3.1, the refinery gas analysis systems are well suited for this application. This

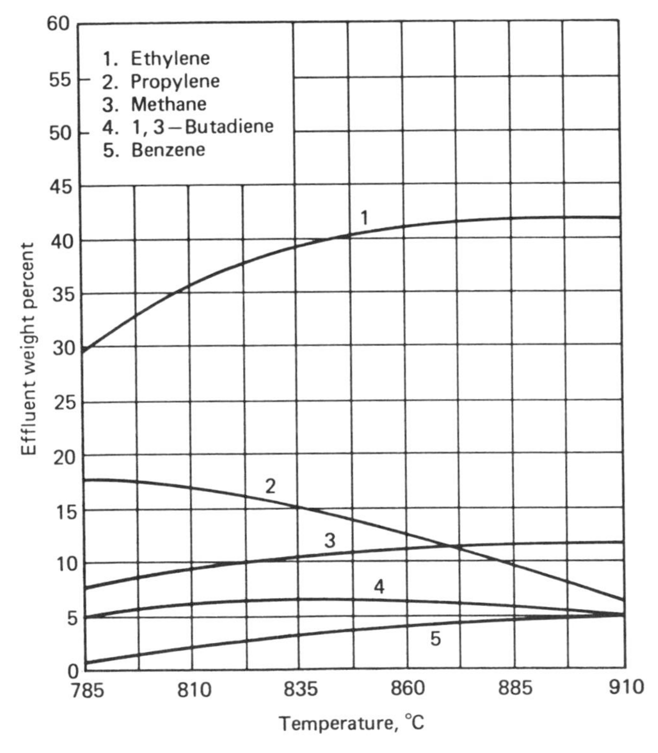

FIGURE 10.45 Plot of product distribution of selected hydrocarbons as a function of temperature for bench-scale pyrolysis of a naphtha sample. (Reprinted with permission from Reference 84, *Journal of Chromatographic Science*, Copyright 1978, Preston Publications, A Division of Preston Industries, Inc.)

includes the permanent gases, of which hydrogen is the most significant for analyzing cracker performance.

Ethylene. Ethylene is produced by fractionation of the steam cracker effluent. In order to monitor this process as well as product quality, many analyses have been developed. ASTM Method D-2504 for analysis of high-purity ethylene utilizes four different packed-column systems (11). Methane and ethane are determined by use of a silica gel column. Acetylene is analyzed by using a hexadecane column in series with a squalane column. A hexamethylphosphoramide column is used to determine propylene and heavier impurities. Carbon dioxide, the final component, is determined by using a column of activated charcoal that has been impregnated with a solution of silver nitrate and β,β'-oxydipropionitrile. Besides the difficulties associated with multiple analyses, this ASTM method does not achieve baseline resolution of the acetylene and propylene.

An analysis for all of the C_2 and lighter impurities in ethylene has been reported by Zlatkis and Kaufman (86). A Porapak Q column was used with a TC detector for the overall analysis (Figure 10.46). An FID was used for determination of trace acetylene levels. Those authors also used a carbon molecular sieve (Carbosieve) column for this analysis (87). With the use of short packed capillary columns, this analysis can be completed in less than 10 s. It was subsequently demonstrated by Supelco (88) that the C_3 hydrocarbons could also be separated with a Carbosieve column.

An analyzer has been developed by Wasson (89) for an extensive analysis of the impurities in ethylene. This is a dual automated multicolumn system

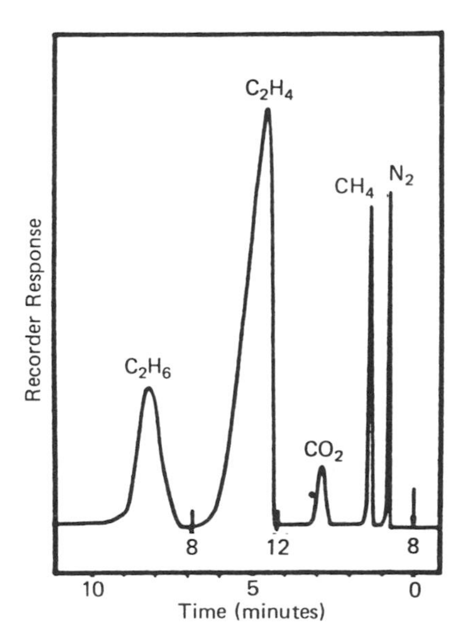

FIGURE 10.46 Chromatogram of ethylene and ppm levels of impurities. (Reprinted with permission from Reference 86, *Journal of Gas Chromatography*, Copyright 1966, Preston Publications, Inc.)

that utilizes parallel injections. An FID is used to determine trace levels of impurities, as shown in Figure 10.47. To monitor CO and CO_2, they are separated, catalyticly converted to methane, and then determined with the FID. Additionally, the analyzer is capable of monitoring methanol with a detection limit of 10 ppm. Methanol is important because it is used for deicing ethylene systems.

One interesting and very significant analytical problem in the operation of an ethylene plant is the control of acetylene along with the other impurities in the process. The tower system is set up with a deethanizer to first separate all the C_2 hydrocarbons from the C_3 and heavier components. The C_2 hydrocarbons taken overhead in this tower contain up to 2% acetylene. This stream is then passed through an acetylene converter that catalytically hydrogenates the acetylene. The C_2 splitter then separates ethylene overhead from the ethane that is also produced. Less than 1 ppm of acetylene should be in the product ethylene. Carson et al. (90) utilized a graphitized carbon black (Carbopack B) column for these analyzes. Figure 10.48 demonstrates the effectiveness of the acetylene converter. Table 10.3 gives the concentrations for each component. Unfortunately, the hydrogen required for this conversion had to be analyzed on a 5A molecular sieve column.

Propylene. Propylene is obtained as a coproduct in the production of ethylene. A propylene splitter tower recovers the propylene from the cracker effluent after the lighter components have been removed. The determination of trace levels of ethylene, total butylenes, acetylene, methyl acetylene, propadiene, and butadiene is covered in ASTM Method D-2712 (11). For this analysis, 11 systems using one or two packed columns are recommended (Table 10.4). Baseline resolution is not required in this method, but a resolution requirement is given.

A total analysis for propylene impuritics is very difficult. Wasson (91) has developed an automated multicolumn system to perform this analysis. One subsystem uses a switching valve and capillary columns with an FID to monitor the C_1–C_5 paraffins and the C_1–C_4 olefins. The second system uses packed columns with two switching valves and an FID for the analysis of trace methane, methanol and the individual C_2 hydrocarbons. Both CO and CO_2 are determined by means of a catalytic methanizer. Typical chromatograms from the two detectors are shown in Figure 10.49.

Butadiene. High-purity 1,3-butadiene is recovered from steam cracker product streams. By using dimethyl formamide as a solvent, butadiene can be extracted from the C_4 cuts. It is also recovered by catalytic dehydrogenation of mixed C_4 streams. Analysis of the main impurities resulting from the isolation of butadiene is covered in ASTM Method D-2593 (11). Ten possible columns or column combinations are suggested for this analysis (Table 10.5). Carson et al. (92) have indicated that a somewhat faster and

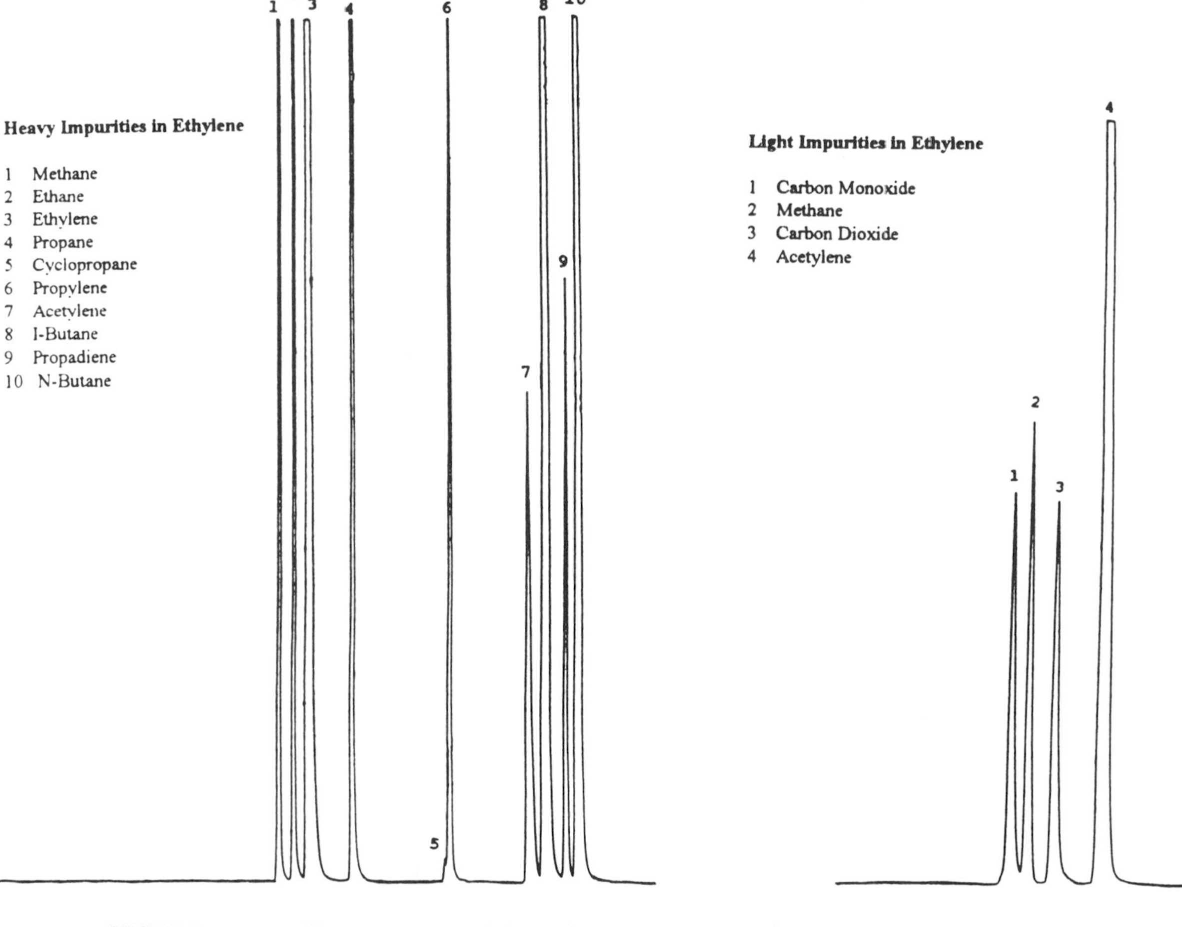

FIGURE 10.47 Chromatogram of impurities in ethylene. (Reprinted with permission from Reference 89, Wasson-ECE Instrumentation.)

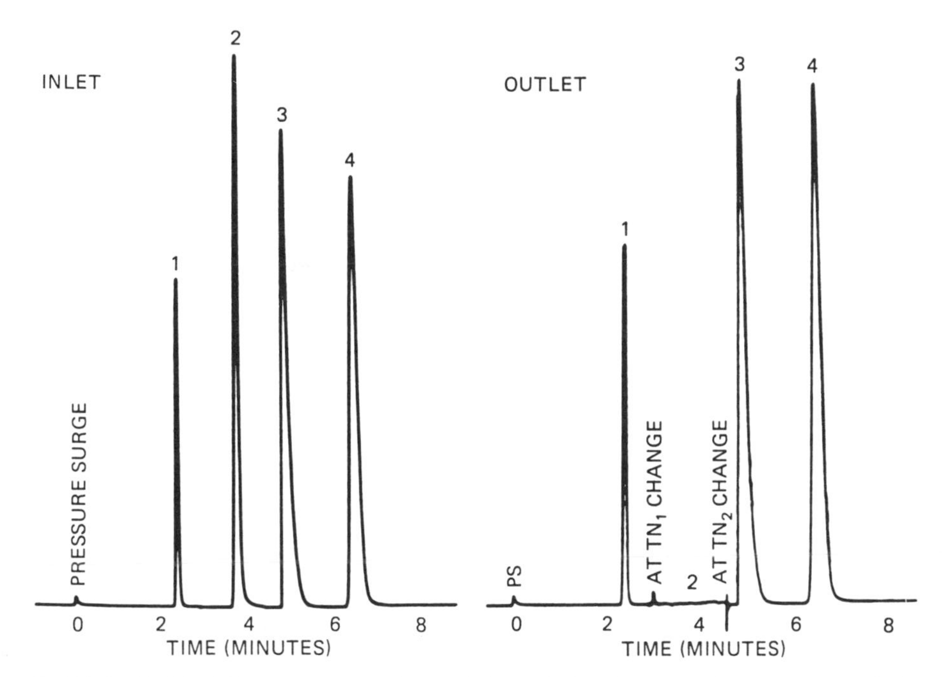

FIGURE 10.48 Chromatogram of a typical analysis of an acetylene converter inlet and outlet sample in an ethylene plant cracking naphtha. Peak identification is contained in Table 10.3. (Reprinted with permission from Reference 90, *Journal of Chromatographic Science*, Copyright 1975, Preston Publications, A Division of Preston Industries, Inc.)

TABLE 10.3 Acetylene Converter Analyses

Peak	Component	Inlet (mol%)	Outlet (mol%)
1	Methane	0.46	0.46
2	Acetylene	1.21	<1 ppm
3	Ethylene	87.13	89.71
4	Ethane	8.64	9.83
—[a]	Hydrogen	2.55	<1 ppm

Source. Reference 90.

[a] Analyzed on 5A molecular sieve column and normalized into analysis.

more complete analysis can be obtained through the use of another column system. They utilized a dibutyl maleate column followed by one with bis(2-methoxyethoxy)ethyl ether as the liquid phase. As shown in Figure 10.50, all the components listed in Table 10.6 can be determined. This

TABLE 10.4 ASTM D-2712 Column Systems for Propylene Analysis

1. 2,4-Dimethyl sulfolane (33% Chromosorb P, 0.19 in. × 4 ft); squalane (22%, Chromosorb P, 0.13 in. × 30 ft)
2. 2,4-Dimethyl sulfolane (Chromosorb P, 0.085 in. × 22 ft); β,β'-oxydipropionitrile (15% Chromosorb P, 0.085 in. × 20 ft); UCON (15%, Chromosorb, 0.085 in. × 8 ft)
3. 2,4-Dimethyl sulfolane (15%, Chromosorb P, 0.085 in. × 16 ft)
4. Silica gel (0.18 in. × 3.5 ft)
5. 1,2,3-Tris(2-cyano ethoxy) propane (20%) and SE-30 (25% Chromosorb P, 0.19 in. × 50 ft)
6. β,β'-Oxydipropionitrile (25%, Chromosorb P, 0.19 in. × 50 ft)
7. Normal hexadecane (20%, Chromosorb P, 0.085 in. × 20 ft)
8. Hexamethyl phosphoramide (30%, Chromosorb P, 0.085 in. × 20 ft)
9. Bis-2(methoxy ethoxy ethyl) ether (80%) and diisodecyl phthalate (20%, Chromosorb P, 0.085 in. × 25 ft)
10. Silica gel (modified with ferric chloride, 0.19 in. × 15 ft)
11. 2,4-Dimethyl sulfolane (33%, Chromosorb P, 0.085 in. × 8 ft); squalane (20%, Chromosorb P, 0.085 in. × 35 ft)

Source. Reference 11.

column was found to be useful in the analysis of butadiene plant streams for process control.

Graphitized carbon with a light loading of picric acid has been suggested for butadiene analyses. Figure 10.51 shows a chromatogram obtained by DiCorcia and Samperi (93). This column resolves butene-1 and isobutylene, which is one of the most difficult separations in the petrochemical laboratory. Finally, the refinery gas analyzers discussed in Section 10.3.2 can also be utilized for butadiene analyses.

Since one the major uses for butadiene is the production of styrene–butadiene rubber (SBR), analyses are also required for polymerization plant recycle streams. The determination of butadiene dimer and styrene is covered in ASTM Method D-2426 (11). A choice of several Carbowaxes and silicone oils is given in this method.

10.4.2 Aromatics

Aromatics (benzene, toluene, and xylenes (BTX)) are obtained from refinery and petrochemical light naphtha streams. Aromatics are produced in the reforming process and in steam cracking. Extraction or various extractive distillation processes are used to isolate and separate aromatics from the naphtha streams. Typical extraction processes are based on tetraethylene glycol, sulfolane, *N*,*N*′-methyl pyrolidene, or morpholine. They produce a mixture of aromatics that are subsequently separated by distillation, extractive distillation, or—in the case of xylene isomers—differential adsorption or fractional crystallization.

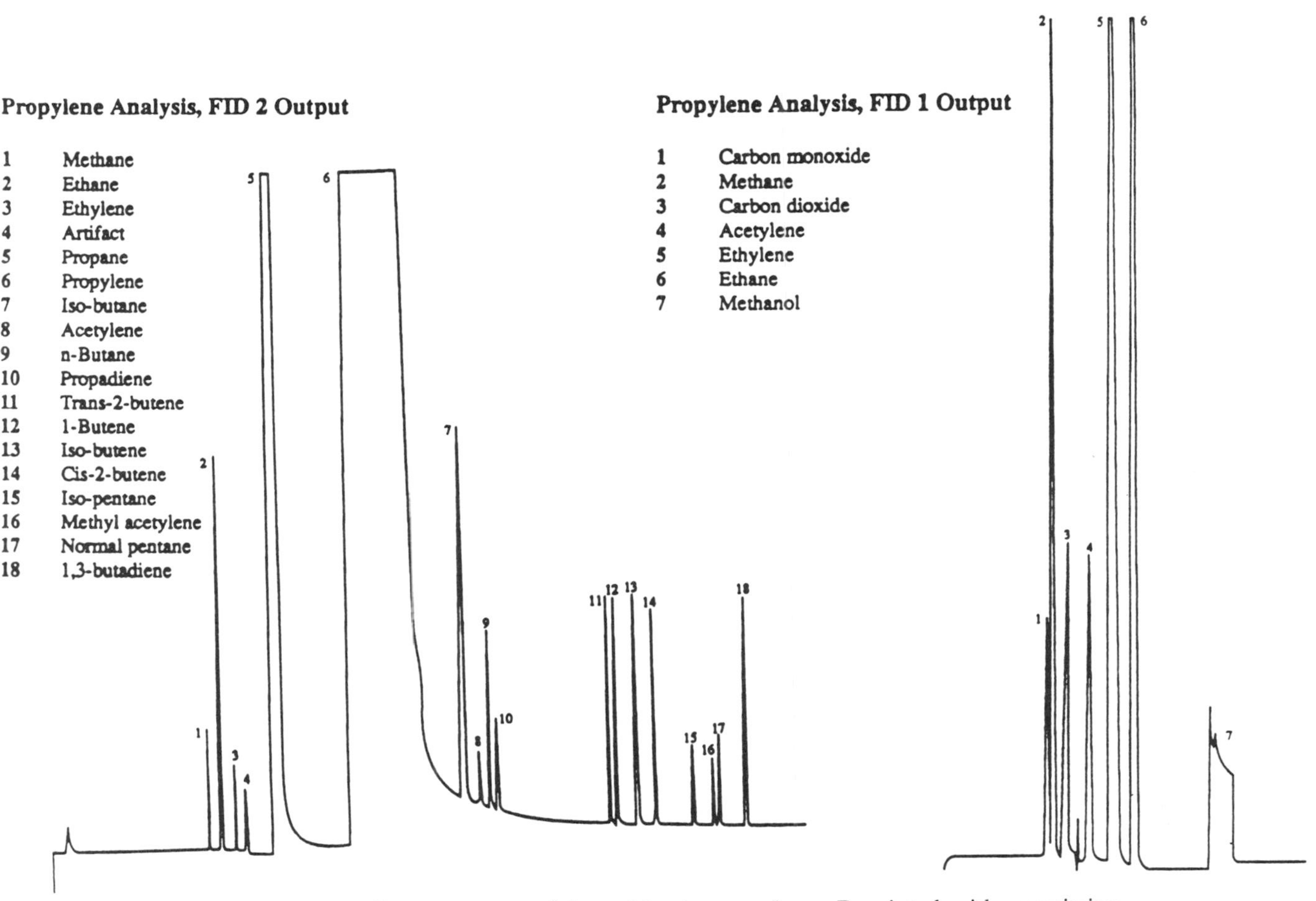

FIGURE 10.49 Chromatogram of impurities in propylene. Reprinted with permission from Reference 91, Wasson–ECE Instrumentation.)

TABLE 10.5 ASTM D-2593 Column Systems for Analysis of Butadiene

1. Bis-2(methoxy ethoxy ethyl) ether and diisodecyl phthalate (25%, Chromosorb P, $\frac{3}{16}$ in. × 20 ft)
2. Di-*n*-butyl maleate (15%, Chromosorb P, $\frac{3}{16}$ in. × 20 ft)
3. Sulfolane (30%, Chromosorb P, $\frac{1}{8}$ in. × 21 ft); didecyl phthalate (30%, Chromosorb P, $\frac{1}{8}$ in. × 3.5 ft)
4. Bis-2-methoxyethyl adipate (15%, Chromosorb P, $\frac{1}{4}$ in. × 24 ft); 1,2,3-tris(2-cyanoethyoxy) propane (15%, Chromosorb P, $\frac{1}{4}$ in. × 6 ft)
5. UCON LB-550X (20%, Chromosorb P, $\frac{1}{4}$ in. × 25 ft)
6. β,β'-Oxydipropionitrile (20%, Chromosorb P, $\frac{1}{4}$ in. × 25 ft)
7. Tributyl phosphate (15%, Chromosorb P, $\frac{1}{8}$ in. × 60 ft)
8. β,β'-Oxydipropionitrile (15%, Chromosorb P, $\frac{1}{8}$ in. × 20 ft)
9. Squalane (15%, Chromosorb P, $\frac{1}{4}$ in. × 5.67 ft); dimethyl sulfolane (20%, Chromosorb P, $\frac{1}{4}$ in. × 23 ft)
10. Propylene carbonate (30%, firebrick, $\frac{3}{16}$ in. × 16 ft)

Source. Reference 11.

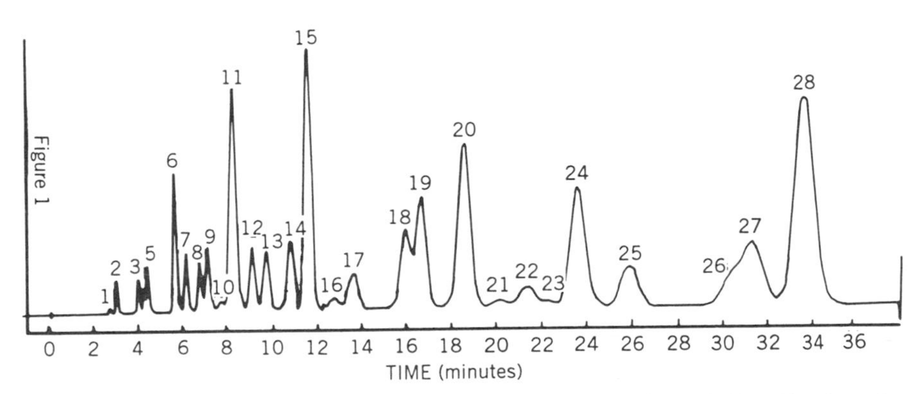

FIGURE 10.50 Analysis for butadiene process streams with peak identifications in Table 10.6 (Reprinted with permission from Reference 92, *Journal of Chromatographic Science*, Copyright 1972, Preston Publications, A Division of Preston Industries, Inc.)

When steam cracking or naphtha reforming produce an aromatics mixture short in benzene or *o*- and *p*-xylene, some interconversion is practiced. Toluene can be hydrodealkylated to benzene. Xylene can be isomerized to increase yields of *o*- and *p*-xylene. The analysis for aromatics thus falls into two general types to meet two different needs. Analysis for process optimization assists in obtaining the maximum product at the minimum unit cost. This involves analysis of feeds, products, and raffinate (purge) streams. These analyses must be tailored to the process and the plant streams involved. Generally, it is desirable to have one analytical procedure to apply to a variety of sample types. The final product specification analysis can also be used for process control. The ASTM

TABLE 10.6 Component Identification in Butadiene Analysis

Peak No.	Component
1	Methane
2	Ethane + ethylene
3	Propane
4	Acetylene
5	Propylene
6	Isobutane
7	Cyclopropane
8	Propadiene
9	*n*-Butane
10	Neopentane
11	Butene-1 + isobutylene
12	Methylacetylene
13	*trans*-Butene-2
14	*cis*-Butene-2
15	1,3-Butadiene
16	Isopentane
17	3-Methylbutene-1
18	*n*-Pentane
19	1,2-Butadiene
20	Ethylacetylene
21	2 Methylbutene-1
22	1,4-Pentadiene
23	*trans*-Pentene-2
24	Vinylacetylene
25	2 Methylbutene-2
26	2-Methylpentane
27	Isoprene
28	Dimethylacetylene

Source. Reprinted with permission from Reference 92, *Journal of Chromatographic Science*, Copyright 1972, Preston Publications, Inc.

standard tests for aromatics products include D-2360, D-2306, D-3797, and D-3798 (11). The latter three apply to xylene isomers.

The aromatic content of naphtha feeds to a BTX process can be measured by a procedure such as that described in ASTM D-2267 (11). Generally, a polar liquid phase is used on either an open tubular capillary or acid-washed Chromosorb P column. The aromatic content of the raffinate is used to determine the extraction efficiency and aromatic recovery of the process. A procedure similar to ASTM D-2600 measures the trace levels in the raffinate (11). Stationary phases used here are also polar, and, in

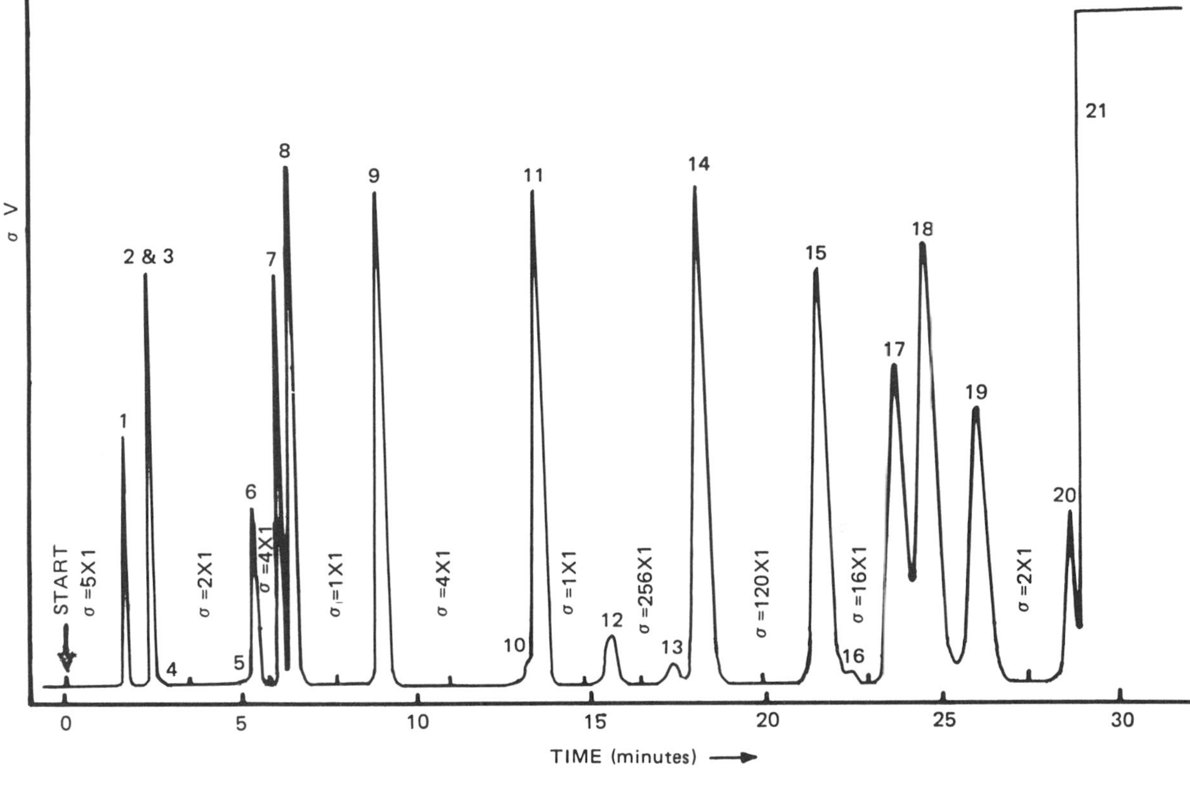

FIGURE 10.51 Analysis of trace light impurities in premium-grade 1,3-butadiene. Peak identifications are 1, methane (10 ppm); 2, ethylene (12 ppm); 3, ethane (7 ppm); 6, propane (5 ppm); 7, propene (46 ppm); 8, propadiene (70 ppm); 9, propyne (13 ppm); 11, isobutane (61 ppm); 14, 1-butene (5500 ppm); 15, isobutene (2500 ppm); 17, *cis*-2-butene (270 ppm); 18, *trans*-2-butene (390 ppm); 21, 1,3-butadiene. (Reprinted with permission from Reference 93, *Journal of Chromatography*, Copyright 1975.)

TABLE 10.7 ASTM Typical Columns for Aromatic Content Analysis

ASTM D-2267

1. Diethylene glycol succinate (20%, Chromosorb P, 25.6 mm × 2.4 m)
2. Polyethylene glycol-400 (0.25 mm × 30.5 m)
3. Tetracyano ethylated pentaerythritiol (10% Chromosorb PAW, 1.6 mm × 4.6 m)
4. 1,2,3-Tris(2-cyanoethoxy) propane (20% Chromosorb PAW, 4.6 mm × 6.1 m)

ASTM D-2600

1. *N*, *N*-Bis(2-cyanoethyl) formamide (17% Chromosorb PAW, 0.07 in. × 15 ft)
2. Polyethylene glycol-200 (30% Chromosorb PAW, 0.18 in. × 6 ft)
3. Tetracyano ethylated pentaerythritiol (0.01 in. × 100 ft)
4. Diethylene glycol succinate (20%, Chromosorb PAW, $\frac{1}{8}$ in. × 12 ft)

Source. Reference 11.

practice, the same chromatograph can be used for analyzing both the feed and raffinate if appropriate calibration procedures are used. Table 10.7 lists the stationary phases used in both ASTM D-2267 and D-2600.

An important process variable is the amount of extraction solvent leaving the process unit in either the raffinate or the aromatic stream. The analysis here will depend greatly on the extraction solvent used. One common feature of all of these solvents is their expense. Therefore, it is important to recover as much of the solvent as possible. Another common feature is the solvent polarity. Analysis on a polar column results in undesirably long analysis times. Analysis on a nonpolar column frequently results in a highly skewed peak and extensive tailing. For the analysis of sulfolane in BTX streams, Awwad (94) used a mixed stationary phase of 2% Carbowax 20M in SE-30. The resulting analysis is shown in Figure 10.52.

Analysis of product benzene and toluene is covered by ASTM D-2360. The method suggests Carbowax 1540 (25%) on Chromosorb P (60/80 mesh). Any stationary phase yielding the specified resolution can be used. An internal standard (*n*-butyl benzene) is used.

Analysis of mixed xylenes is by ASTM D-2306 with the use of a cross-linked polyethylene glycol (e.g. Carbowax 20M) stationary phase. Both capillary and packed columns are acceptable. Figure 10.53 illustrates this separation on a SP-1200/Bentone 34 column (95).

Analysis of product xylenes is by ASTM D-3797 (*o*-xylene) or D-3798 (*p*-xylene). These procedures are similar and satisfactory columns are given in Table 10.8. The concentration of impurities is measured by internal standard calibration and the purity of the *o*- or *p*-xylene is determined by subtracting the percent impurities from 100%.

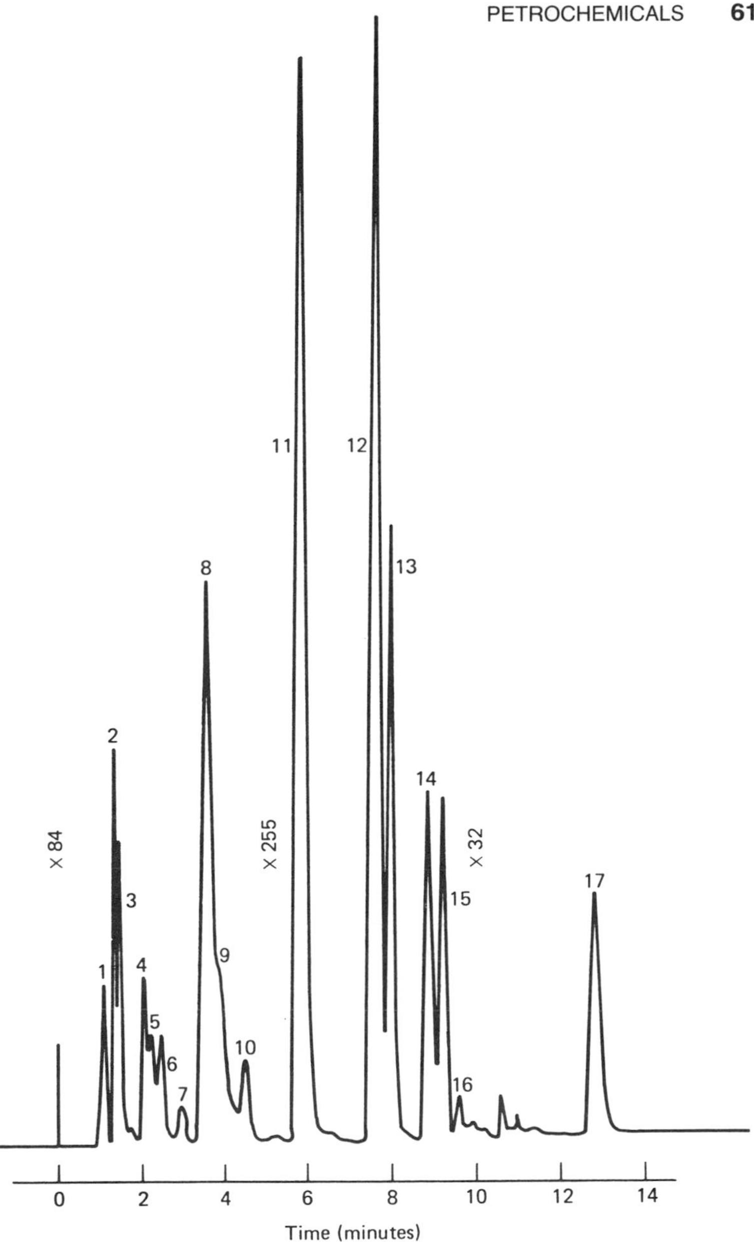

FIGURE 10.52 Typical chromatogram of sulfolane in BTX mixture extracted from Iragi Powerformate. Peak identifications are 1, *n*-pentane; 2, *cis*-2-pentene; 3, 2-methyl-2-butene; 7, *n*-hexane; 8, benzene; 9, cyclohexane; 10, 2,4-dimethylpentane; 11, toluene; 12, *m*- and *p*-xylene; 13, *o*-xylene; 14, *n*-propylbenzene; 15, *tert*-butylbenzene; 16, isobutylbenzene; 17, sulfolane. (Reprinted with permission from Reference 94, *Journal of Chromatographic Science*, Copyright 1979, Preston Publications, A Division of Preston Industries, Inc.)

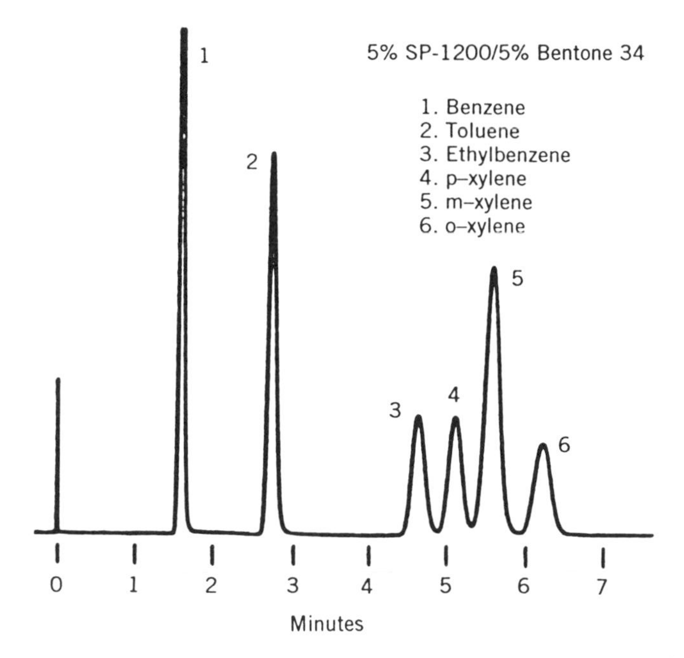

FIGURE 10.53 Chromatogram demonstrating the separation of aromatics on a 5% SP-1200/5% Bentone 34 on 100/120-mesh Supelcoport (6 ft × $\frac{1}{8}$ in.) at 100°C (82). (Reprinted from Reference 95 with permission of Supelco, Inc, Bellefonte, PA 16823.)

10.5 PROCESS CHROMATOGRAPHY

Process GC has developed into one of the most widely used on-line monitoring techniques in the petrochemical industry. Its popularity is due to the ability of GC to quickly analyze hydrocarbon streams for process control. As is the case for all on-line analyzers, process chromatographs are capable of continuous, unattended, in-plant operation. Because of the need for fast and specific analyses to provide feedback for process control, these chromatographs are usually designed for each specific application. The ideal process GC (PGC) has the following characteristics:

- Simple, reliable design to ensure a low failure rate
- Design that provides for simple, rapid repair
- Sensors strategically placed in the system to aid in diagnosis of problems
- Alarms to monitor proper operation

TABLE 10.8 Typical Columns for Analysis of Impurities of Xylenes

ASTM D-3797 o-Xylene

1. Butylbenzyl tetrachlorophthalate (SS 200 ft × 0.01 in.)
2. Dibutyl tetrachlorophthalate (SS 250 ft × 0.02 in.)
3. Di-*n*-propyl tetrachlorophthalate (SS 200 ft × 0.02 in.)
4. Carbowax 1540 (SS 300 ft × 0.01 in.)
5. Bentone 34 (5%)/OS 124 (5%) (Chromosorb WAW, SS 10 ft × $\frac{1}{8}$ in.); 1,2,3-tris(2-cyanoethoxy) propane (20%) (Chromosorb PAW, SS 18 ft × $\frac{1}{8}$ in.)

ASTM D-3798 p-Xylene

1. Polyethylene glycol or Carbowax 20M (fused silica, 50 m × 0.32 mm)
2. Diisodecylphthalate (3.5%) or Bentone 34 (Chromosorb W, 6.1 m × 3.2 mm)

Source. Reference 11.

The main components of the process gas chromatograph include a sample system, the gas chromatographic analyzer, and a programmer. A schematic of this system is shown in Figure 10.54. The sample system continuously circulates the process stream or streams of interest in the analyzer. Sample systems are designed to avoid the problems encountered in sampling for laboratory analyses. The analyzer is basically a gas chromatograph, but is designed for explosion-proof operation in the plant environment. The required separations are usually achieved with the use of valve switching and multiple columns. The programmer is a microprocessor that handles the operation of the chromatograph along with several sample streams. It also handles the data as well as any required computations, such as calorific (BTU) content. Currently, the chromatograph and programmer are integrated into a stand-alone unit. It can also be interfaced with other chromatographs as well as distributed process control systems.

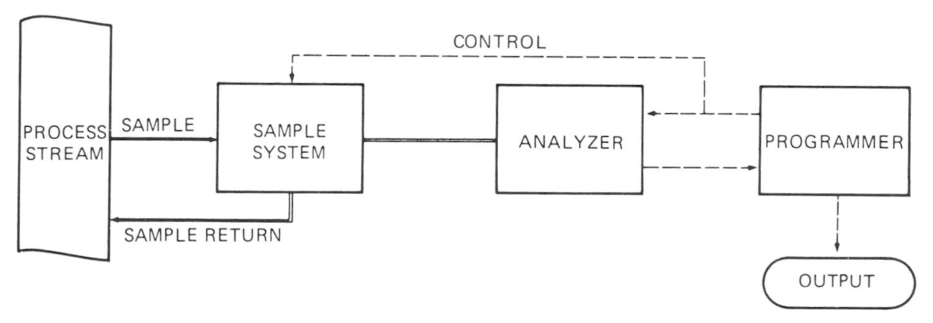

FIGURE 10.54 Simplified schematic of a process gas chromatographic analyzer system.

Overall, the trend is toward increased usage of process gas chromatographs in the future. This is based on the need for plant optimization as well as increased reliability and reduced maintenance of the microprocessor-controlled systems. Several reviews of this type of analyzer and their manufacturers have been recently published (96–100).

10.5.1 Process Chromatographs

Sample System. The sample system is the most critical part of the process analyzer. It is designed to provide a constant flow of sample to the analyzer. This includes conditioning of the sample stream so that a representative sample can be injected into the chromatograph. Thus the composition of each process stream must be carefully considered so that the sample system can be designed appropriately. Complete details of these considerations are discussed in the monograph by Houser (101) and by Cornish et al. (102). Unfortunately, the majority of the failures of process chromatographs can be attributed to the sample-handling system.

The sample stream originates from the process through a sample probe. The probe is inserted through a process valve into the center of the process stream. This provides a representative sample by minimizing the effects of laminar flow, gas bubbles, condensed liquids, and particulates.

The sample line itself is designed to provide a sample to the analyzer within the cycle time of the analysis. Typically, sample flow is obtained by the pressure drop between properly selected sample and return points. If necessary, a pump may be utilized, but this is expensive. The size of the sample line is a function of the required flow, the available pressure drop, and the length of the line. Relatively short and straight lines are the most desirable. Stainless-steel tubing is the most common choice for sample lines. The reactivity of the process stream, however, must be considered in choosing the proper materials. For example, the presence of hydrochloric acid requires the use of Monel. Insulation and heat tracing of the line may also be required to maintain the composition of the stream as well as to control the flow of viscous fluids. For gas streams, the possibility of condensation must be eliminated. The temperature and pressure of a liquid stream must be maintained to prevent it from reaching its bubble point. Tubing is now available that is manufactured with insulation and either steam or electrical tracing.

Most sample streams require filtration to remove particulates. Phase separation may also be required for removal of condensed liquid droplets from vapors or immiscible droplets from liquids. These are usually accomplished by slip-streaming the sample through a filter, as shown in Figure 10.55. Only periodic flow to the sample valve actually passes through the filter element. This arrangement greatly reduces the maintenance on the system. From the filter, automatic valve switching is used to provide flow to

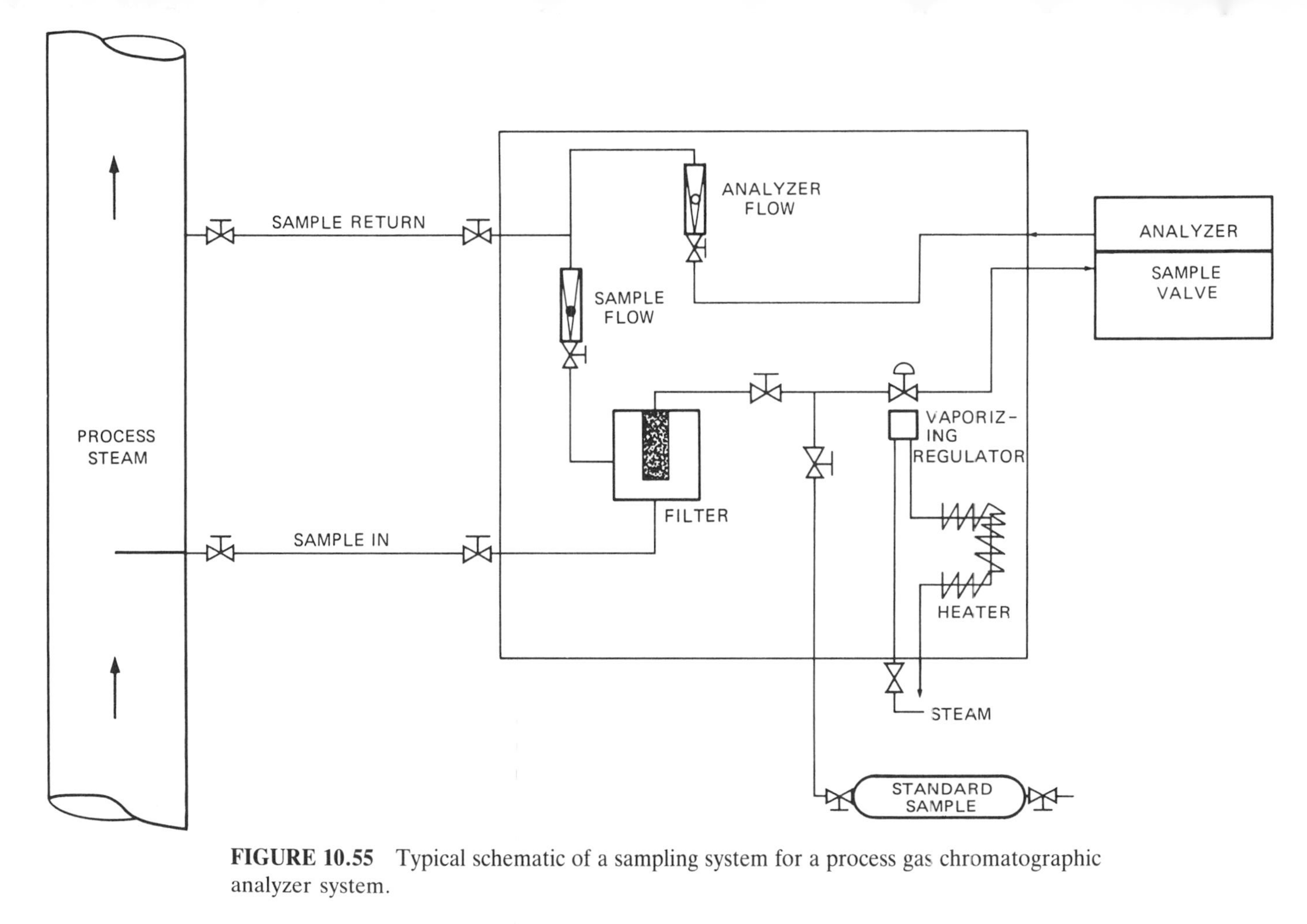

FIGURE 10.55 Typical schematic of a sampling system for a process gas chromatographic analyzer system.

the sample valve for injection. Valve switching is also used to sample multiple streams with a single analyzer. Flow controllers are installed on each line.

A vaporizer is utilized to convert mixed phase samples into gases prior to sampling. All of these components are contained in a temperature-controlled, air-purged cabinet. Valving and a vaporizer, if necessary, are provided for analysis of a calibration standard. Further details on calibration of the analyzer are discussed later.

Analyzer. The analyzer automatically performs the functions of a typical laboratory gas chromatograph. The key to the success of the process analyzer is the reliability of the sample valve. Multiport valves are used for sampling liquids or gases as well as column switching and backflushing. The most widely used valves include the diaphragm, sliding plate, O-ring, and rotary valves. A liquid injection valve with built-in vaporizer is also available. A sample loop is usually used to control the sample volume of gases. An internal hole or channel is used for liquids with pressures of up to 200 psig. All of these valves can be obtained with special metals or Teflon coating for reactive or corrosive streams.

The columns in the analyzer are arranged to provide repetitive analyses within the minimum amount of time possible. Analyses typically are performed within several minutes. To accomplish this, multiple columns with valve switching are used. The porous polymer column packings are widely used because of their high efficiency and stability. The term "stability" implies low bleed and resistance to normal and upset condition components in the sample. With the development of fused-silica columns, capillary columns are now being utilized. Most analyses are performed isothermally, although temperature-programmed analyzers are available.

Multicolumn techniques provide fast analyses by separating and purging the nonquantified components. The most widely used technique is backflushing. The backflush-to-measure method involves analysis of the initial components through the columns. Flow is then reversed through the column and the remaining components are measured as a single peak. The backflush-to-vent method uses two columns. The precolumn is used to separate the components of interest from the unwanted components. The components of interest pass on to the analysis column where they are further separated. While this is occurring, the unwanted components are backflushed from the precolumn and vented. Another technique utilizes dual analysis columns. Resolved components from the first column are sent to the detector while unresolved components are passed to the second column for further separation. The final technique is heartcutting. It is used mainly to measure a trace component that is not fully resolved from a major component. The method uses two columns, with the first column performing the primary separation. A narrow cut from the first column, which contains

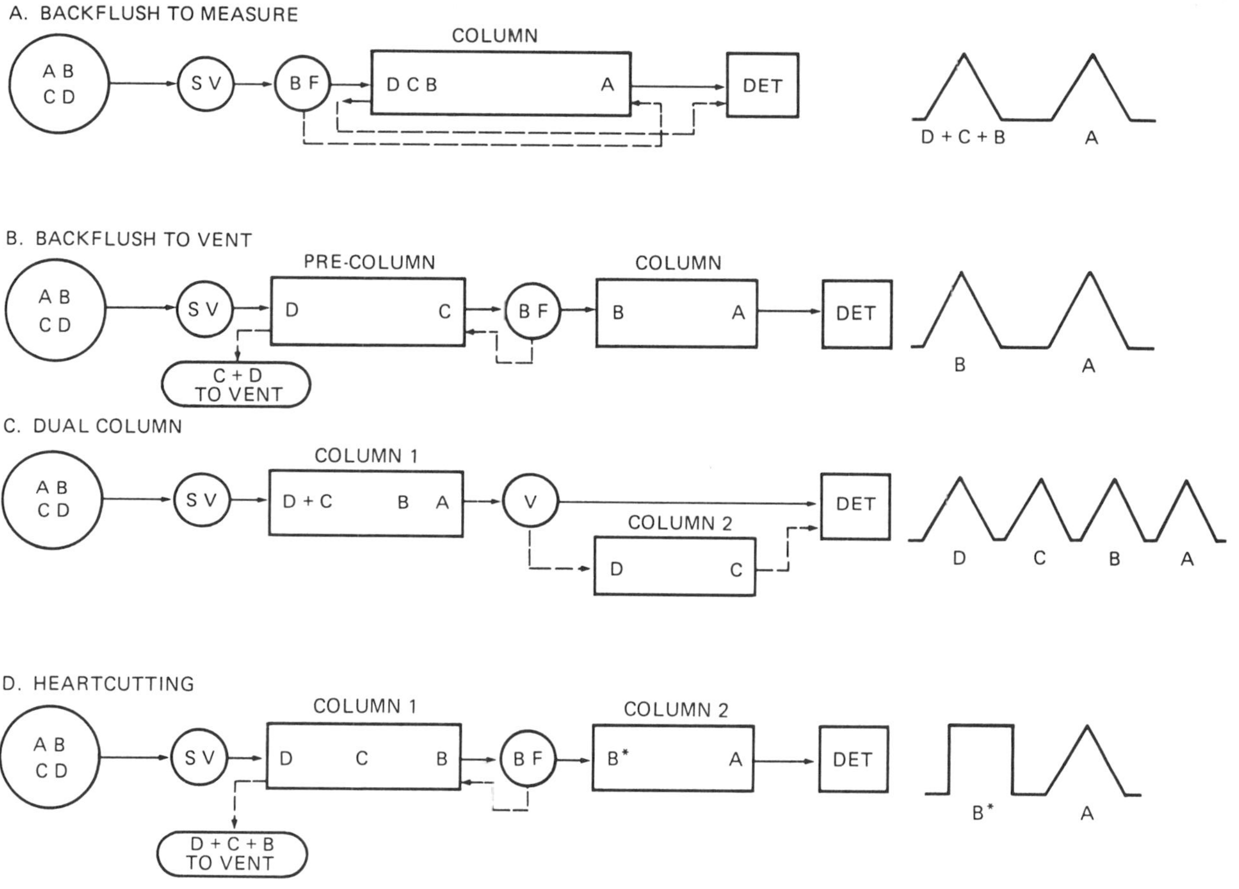

FIGURE 10.56 Simplified schematics and chromatograms for the multicolumn techniques used in process gas chromatographs (SV, sample valve; BF, blackflush valve). Dashed lines indicated flow pattern during backflush.

the component of interest along with a portion of the major component, is transferred to the second column for further separation. Each of these techniques is shown schematically in Figure 10.56.

The thermal conductivity detector is the most frequently used because of its simplicity and reliability. The FID is used when increased sensitivity is required. For hazardous areas, a pneumatic composition transmitter is used. It monitors the changes in differential pressure across an orifice for detection of the components as they elute (103). The FPD is available for determination of low levels of sulfur or phosphorous compounds. Oxygenated or chlorinated compounds can be monitored by use of an electron capture detector (ECD).

Programmer. The programmer controls the operation of the sample system and the analyzer. In addition to this, it handles collection and presentation of the chromatographic data. Most programmers are now microprocessor based. These systems have generally replaced the original cam timers and the solid-state timers. The basic function of the timer is to control valve switching for multiple streams and calibration standards in the sample system. For the analyzer, it controls the operation of the sample inject and column switching valves. With microprocessors, each stream can have different valve switching times and sequences. The detector can also be automatically zeroed.

In addition to timing, the programmer handles collection and reduction of the chromatographic data. Since only a few specific components are analyzed, the technique of gating is used. Gating integrates only the very small time band in which the component elutes. Quantification is provided by comparison to external standards. The newer systems can use an initial slope to detect the start of a peak. Tangent skimming and dropping of a perpendicular for unresolved peaks is also possible.

As mentioned earlier, quantification is usually achieved by comparison to external standards. However, internal normalization and comparison to laboratory analyses can also be used. External standards are blends of known composition or a process stream that has been analyzed. The process stream analyses are then correlated with the analysis of the calibration standard. An important consideration for these standards is that they must be stable over long periods of time. Care must be taken with components that may react with other components or the walls of the container. Loss of a very volatile component into the vapor space of a liquid standard must also be considered.

The programmer is typically interfaced with a process control computer network to provide the chromatographic data. Both system and data alarms are also transmitted. Another pathway is provided to direct maintenance data to an analyzer management station. This allows for statistical monitoring of the performance of the analyzer. Reprogramming can be done either through this station or at the analyzer itself.

10.5.2 Typical Applications

The most predominant application of the process gas chromatograph is to provide feedback for control of a fractionation tower. In these cases, one or more components are monitored, and the result then directly controls the operation of the tower. Griffen et al. (104) have summarized the considerations involved in analyzer control of fractionators. More complex analyses

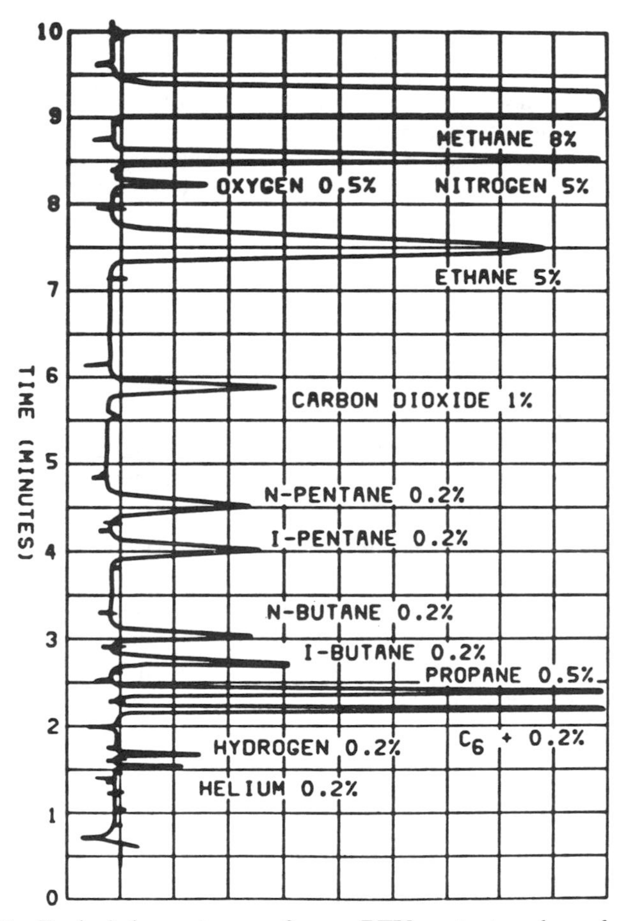

FIGURE 10.57 Typical chromatogram from a BTU content analyzer for natural gas streams. (Courtesy of Applied Automation/Hartman & Braun, Bartlesville, OK 74005.)

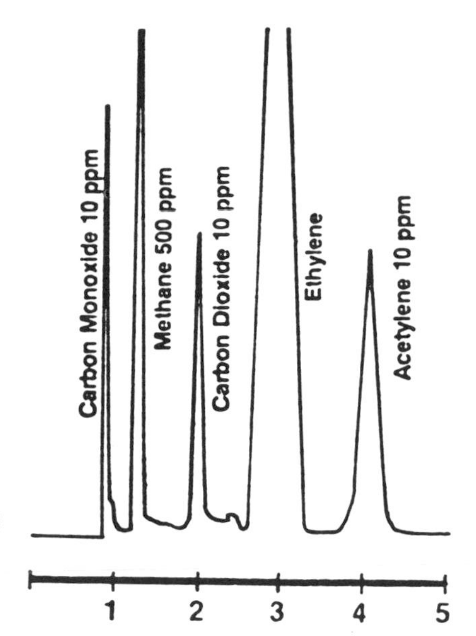

FIGURE 10.58 Typical ethylene product analysis from a process chromatograph. (Courtesy of ABB Process Analytics, Lewisburg, WV 24901.)

are required for applications such as cracking products, polymerization reactor feed impurities, and high-purity product quality control. Systems for determination of boiling-point distributions and octane ratings of refinery streams have also been developed.

Figure 10.57 is a typical chromatogram from an analyzer for BTU content on natural gas streams. In this analysis, the individual components are analyzed and then the BTU content and specific gravity are calculated. An ethylene product analysis is shown in Figure 10.58. For trace sulfur compounds, an FPD can be used. Figure 10.59 is a chromatogram from a natural gas sample. The speed and resolution achieved in these examples through column switching techniques should be noted. It is possible to obtain typical analyses in less than 30 sec through the use of microcolumns.

These analyses demonstrate the potential of the process chromatograph. Undoubtedly, they will continue to be used to improve the efficiency of refinery and chemical plant operations.

ACKNOWLEDGMENTS

The author gratefully acknowledges the contributions of the co-author for the previous edition, Kenneth E. Paulsen, particularly for the section on Exploration and Production. Dr. Paulsen is currently with Polaroid.

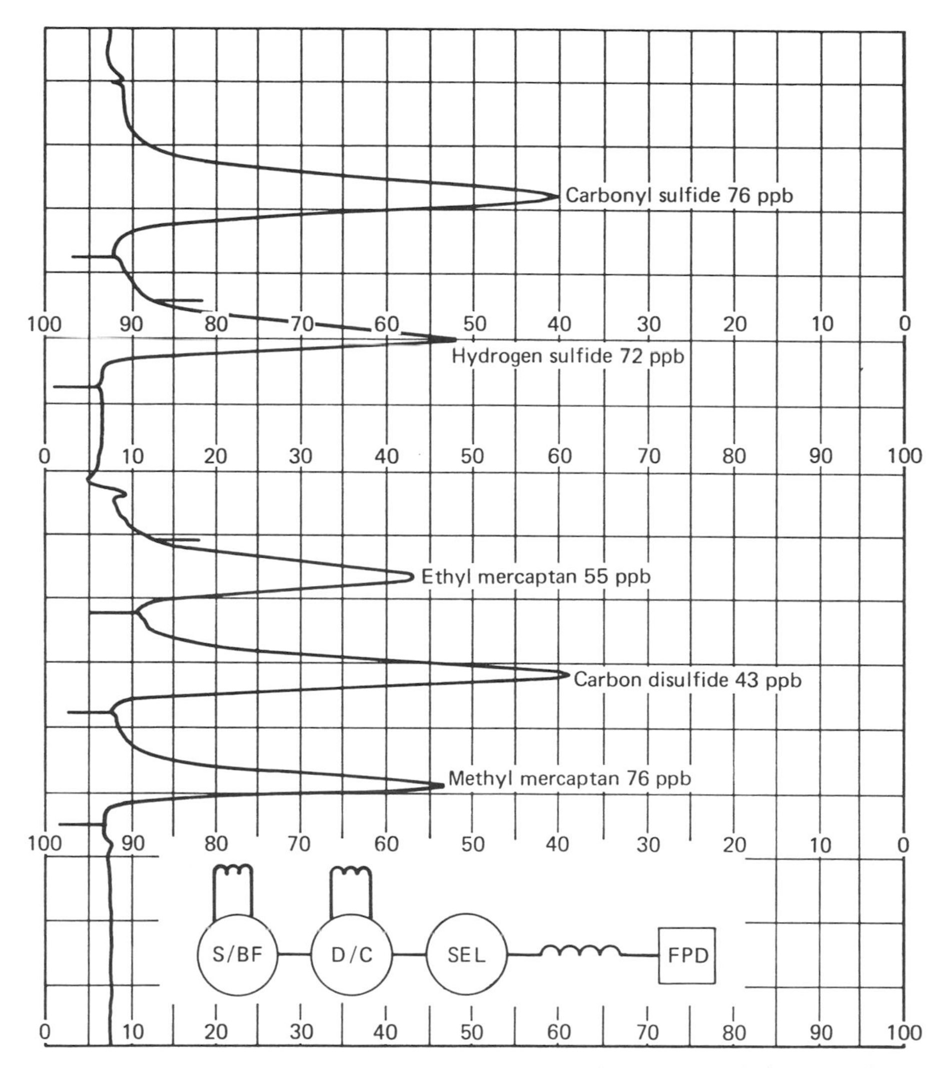

FIGURE 10.59 Typical chromatogram of trace sulfur compounds in natural gas obtained from a process gas chromatograph equipped with a flame photometric detector. (Courtesy of ABB Process Analytics, Lewisburg, WV 24901.)

REFERENCES

1. K. H. Altgelt and T. H. Gouw (Eds.), *Chromatography in Petroleum Analysis*, Chromatographic Science Series, Vol. 11, Dekker, New York, 1979.
2. J. G. Speight, *The Chemistry and Technology of Petroleum*, 2d ed., Dekker, New York, 1991.

3. W. J. Podbielniak, *Anal. Chem.*, **5**, 119–135 (1993).
4. H. H. Haussdorff, in *Vapor Phase Chromatography* (Proceedings, 1956 London Symposium), D. H. Desty (Ed.), Butterworths, London, 1957, p. 377.
5. F. T. Eggertsen, H. S. Knight, and S. Groennings, *Anal. Chem.*, **28**, 303 (1956).
6. D. H. Desty (Ed.), *Gas Chromatography* (Proceedings, 1958 Amsterdam Symposium), Butterworths, London, 1958.
7. D. H. Desty, A. Goldup, and W. T. Swanton, in *Gas Chromatography* (Proceedings, 1961 Lansing Symposium), N. Brenner, J. E. Callen, and M. D. Weiss (Eds.), Academic, New York, 1962, p. 105.
8. W. N. Sanders and J. B. Maynard, *Anal. Chem.*, **40**, 527 (19–68).
9. *Manual on Hydrocarbon Analysis*, 4th ed., American Society for Testing and Materials, 1916 Race Street, Philadelphia, PA 19103, 1989.
10. *ASTM Standards on Chromatography*, American Society for Testing and Materials, 1916 Race Street, Philadelphia, PA 19103, 1981.
11. *Annual Books of ASTM Standards*, Vols. 5.01, 5.02, 5.03, 6.03, and 14.01, American Society for Testing and Materials, 1916 Race Street, Philadelphia, PA 19103, 1990.
12. B. P. Tissot and D. H. Welte, *Petroleum Formation and Occurrence*, Springer, New York, 1978, pp. 10, 70; 2d ed., 1984.
13. J. M. Hunt, *Petroleum Geochemistry and Geology*, Freeman, San Francisco, 1979, p. 280.
14. C. R. Evans and F. L. Staplin, *Third International Geochemical Exploration Symposium Proceedings*, Canadian Institute of Mining and Metallurgy, Special Vol. 11, 1971, p. 517.
15. N. J. L. Bailey, C. R. Evans, and C. W. D. Milner, *Am. Assoc. Petrol. Geolog. Bull.*, **58**, 2284 (1974).
16. R. G. Schaefer, D. Leythaeuser, and B. Weiner, *J. Chromatogr.*, **167**, 355 (1978).
17. R. G. Schaefer, B. Weiner, and D. Leythaeuser, *Anal. Chem.*, **50**(13), 1848 (1978).
18. D. R. Dean, *Chromatographia*, **1**, 18 (1968).
19. K. F. M. Thompson, *Geochim. Cosmochim. Acta*, **43**, 657 (1978).
20. H. Wehner and M. Teschner, *J. Chromatogr.*, **204**, 481–490 (1981).
21. A. Jobson, F. D. Cook, and D. W. S. Westlak, *Appl. Microbiol.*, **23**(6), 1082 (1972).
22. P. Albrecht, M. Vanderbrouche, and M. Mandengue, *Geochim. Cosmochim. Acta*, **40**(7), 791 (1976).
23. W. K. Seifert and J. M. Moldowan, *Geochim. Cosmochim. Acta*, **42**, 77 (1978).
24. R. P. Philp and J. N. Oung, *Anal. Chem.*, **60** (15), 887A (1988).
25. R. J. Hwang, *J. Chromatogr. Sci.*, **28** (3), 109 (1990).
26. A. K. Burnham, J. E. Clarkson, M. F. Singleton, C. M. Wong, and N. W. Crawford, *Geochim. Cosmochim. Acta*, **46**, 1243 (1982).
27. A. J. Sood and R. B. Pannell, *J. Chromatogr. Sci.*, **20**, 39 (1982).
28. D. W. Later, M. L. Lee, K. D. Bartle, R. C. Kong, and D. L. Vassilaros, *Anal. Chem.*, **53**(11), 1612 (1981).

29. M. Novotny, J. W. Strand, S. L. Smith, D. Wiesler, and F. J. Schwende, *Fuel*, **60**, 213 (1981).
30. E. A. Harvey, in *Chromatography in Petroleum Analysis*, K. H. Altgelt and T. H. Gouw (Eds.), Chromatographic Science Series, Vol. 11, Dekker, New York, 1979.
31. C. J. Cowper and A. J. DeRose, *The Analysis of Gases by Chromatography*, Pergamon Series in Analytical Chemistry, Vol. 7, Pergamon, Elmsford, NY, 1983.
32. B. Thompson, *Fundamentals of Gas Analysis by Gas Chromatography*, Varian Associates, Palo Alto, CA, 1977.
33. R. Mindrup, *J. Chromatogr. Sci.*, **16**, 380 (1978).
34. R. Leibrand, *J. Gas Chromatogr.* **5**, 518–524 (1967).
35. J. S. Stufkens and H. J. Bogaard, *Anal. Chem.*, **47**(3), 383 (1975).
36. C. D. Pearson and W. J. Hines, *Anal. Chem.*, **49**, 123–126 (1977).
37. A. W. Drews and C. H. Pfeiffer, UOP Method 539-73, "Laboratory Test Methods for Petroleum and Its Products," Universal Oil Products Process Division, Des Plaines, IL 670016, 1973.
38. R. L. Firor and D. Kruppa, Application Note AN 228-126, Hewlett–Packard Co., Avondale, PA, 1991.
39. Application Note 383-00, Wasson–ECE Instrumentation, 1305 Duff Drive, Fort Collins, CO 80525.
40. F. T. Eggerston, S. Groennings, and J. J. Holst, *Anal. Chem.*, **32**, 904 (1960).
41. L. E. Green, L. J. Schmauch, and J. C. Worman, *Anal. Chem.*, **36**, 1512 (1964).
42. Bulletin 864, Supelco, Inc., Bellefonte, PA, 1990.
43. D. C. Ford, W. H. Miller, R. C. Thren, and R. Wertzler, ASTM Special Publication 577, American Society for Testing Materials, 1916 Race Street, Philadelphia, PA 19103, 1975, p. 20.
44. C. L. Stuckey, *J. Chromatogr. Sci.*, **16**(10), 482 (1978).
45. O. A. Houjan and K. M. Watson, *Chemical Process Principles*, Part 1, Wiley, New York, 1943, p. 329.
46. W. DeBruine and R. J. Ellison, *J. Petrol. Inst.*, **59**, 146 (1973).
47. J. C. Suatoni, *J. Chromatogr. Sci.*, **11**, 121–136 (1979).
48. J. C. Suatoni, H. R. Garber, and B. E. Davis, *J. Chromatogr. Sci.*, **13**, 367 (1975).
49. N. L. Soulages and A. M. Brieva, *J. Chromatogr. Sci.*, **9**, 492 (1971).
50. G. B. Ury, *Anal. Chem.*, **53**(3), 481–485 (1981).
51. L. L. Stavinoha, *J. Chromatogr. Sci.*, **11**, 515 (1973).
52. R. G. Mathews, J. Torres, and R. D. Schwartz, *J. Chromatogr. Sci.*, **20**(4), 160–164 (1982).
53. O. F. Folmer, Jr., *Anal. Chim. Acta*, **60**, 37 (1972).
54. C. L. Stuckey, *Anal. Chim. Acta*, **60**, 46 (1972).
55. J. V. Mortimer and L. A. Luke, *Anal. Chim. Acta*, **38**, 119 (1967).

56. F. Garilli, L. Fabiani, U. Filia, and V. Cusi, *J. Chromatogr.*, **77**, 3 (1973).
57. H. Boer and P. Van Arkel, *Hydrocarbon Process*, **51**(2), 80 (1972).
58. H. Boer, P. Van Arkel, and W. J. Boersma, *Chromatographia*, **13**(8), 500 (1980).
59. P. van Arkel, J. Beens, H. Spaans, D. Grutterink, and R. Verbeek, *J. Chromatogr. Sci.*, **25**, 141 (1987).
60. J. Curvers and P. van der Sluys, *J. Chromatogr. Sci.*, **26**, 267 (1988).
61. E. Boeren, R. Beyersbergen van Henegouwen, I. Bos, and T. Gerner, *J. Chromatogr.*, **349**, 377 (1985).
62. J. Curvers and P. van den Engel, *J. Chromatogr. Sci.*, **26**, 271 (1988).
63. L. E. Green and E. Matt, "PONA Analysis by High Resolution Fused Silica Gas Chromatography," paper presented at 33rd Pittsburgh Analytical Conference, 1982.
64. K. Shimoni, M. Shimono, H. Arimoto, and S. Takahashi, *J. High Resolut. Chromatogr.*, **14**(11), 729 (1991).
65. N. Kosal, A. Bhairi, and M. Ashraf Ali, *Fuel*, **69**(8), 1012 (1990).
66. M. G. Bloch, R. B. Callen, and J. H. Stockinger, *J. Chromatogr. Sci.*, **15**, 504 (1977).
67. H. V. Druschel, *Anal. Chem.*, **41**, 569 (1969).
68. R. L. Martin, *Anal. Chem.*, **38**, 1209 (1966).
69. D. K. Albert, *Anal. Chem.*, **39**, 1113 (1967).
70. D. K. Albert, *Anal. Chem.*, **50**(13), 1822 (1978).
71. H. V. Druschel, "High Resolution Simultaneous Determination of Sulfur Compounds, Nitrogen Compounds, and Hydrocarbons using Fused Silica Capillary Columns and Selective Detectors," paper presented at American Chemical Society Meeting, Las Vegas, NV, 1982.
72. Application Note No. 76, Sievers Instruments, Inc., 1930 Central Ave., Boulder, CO 80301, 1992.
73. Z. Naizhong and L. Green, *J. High Resolution Chromatogr.*, **9**, 400–404 (1986).
74. G. D. Dupre, "Oxygenates in Gasolines by Gas Chromatography: Progress and Status of ASTM Methodology," presented at the ASTM Committee D-2 Meeting, Boston, MA, 1993.
75. Application Note 611, Fisons Instruments, 32 Commerce Center, Cherry Hill Drive, Danvers, MA 01923, 1993.
76. P. S. Spock, Supelco Reporter Vol. XI (3), Supelco, Inc., Bellefonte, PA, 1992.
77. D. T. Coker, *Anal. Chem.*, **47**(3), 386 (1975).
78. P. C. Uden, R. M. Barnes, and F. P. DiSanzo, *Anal. Chem.*, **50**, 852 (1978).
79. H. J. Dawson, Jr., *Anal. Chem.*, **35**, 542 (1963).
80. E. J. Bonelli and H. Hartmann, *Anal. Chem.*, **35**, 1980 (1963).
81. E. A. Boettner and F. C. Dallos, *J. Gas Chromatogr.*, **3**(6), 190 (1965).
82. N. L. Soulages, *Anal. Chem.*, **38**, 28 (1966).
83. M. D. DuPuis and H. H. Hill, *Anal. Chem.*, **51**(2), 291 (1979).

84. M. Greco, *J. Chromatogr. Sci.*, **16**(4), 158 (1978).
85. J. H. Jordan, N. M. Broussard, and W. R. Holtby, *J. Chromatogr. Sci.*, **9**(6), 383 (1971).
86. A. Zlatkis and H. R. Kaufman, *J. Gas Chromatogr.*, **4**, 240 (1966).
87. A. Zlatkis, H. R. Kaufman, and D. E. Durbin, *J. Chromatogr. Sci.*, **8**, 416 (1970).
88. Bulletin 712, Supelco, Inc., Bellefonte, PA, 1971.
89. Application No. 260-0, Wasson–ECE Instrumentation, 1305 Duff Drive, Fort Collins, CO 80525.
90. J. W. Carson, G. Lege, and J. Irizarry, *J. Chromatogr. Sci.*, **13**(4), 168–172 (1975).
91. Application No. 261-00, Wasson-ECE Instrumentation, 1305 Duff Drive, Fort Collins, CO 80525.
92. J. W. Carson, J. D. Young, G. Lege, and F. Ewald, *J. Chromatogr. Sci.*, **10**, 737 (1972).
93. A. DiCorcia and R. Samperi, *J. Chromatogr.*, **107**(1), 99–105 (1975).
94. A. M. Awwad, *J. Chromatogr. Sci.*, **17**, 562 (1979).
95. Bulletin 740, Supelco, Inc. Bellefonte, PA, 1970.
96. R. Villaobos, *Anal. Chem.*, **47**(11), 983A (1975).
97. R. Annino, *Am. Lab*, **21**(10), 60 (1989).
98. K. J. Clevett, *Process Analyzer Technology*, Wiley, New York, 1986.
99. M. P. T. Bradley, *Chromatogr. Sci.*, **11**, 447 (1979).
100. F. D. Martin, *Instrum. Technol.*, **24**(1), 51 (1977).
101. E. A. Houser, "Principles of Sample Handling and Sample System Design," Instrument Society of America, Pittsburgh, PA, 1977.
102. D. C. Cornish, G. Jepson, and M. J. Smurthwaite, *Sampling Systems for Process Analyzers*, Butterworths, London, 1981.
103. R. Annino, J. Cruun, Jr., R. Kalinowski, E. Karas, R. Lindquist, and R. Prescott, *J. Chromatogr.*, **126**, 301 (1976).
104. D. E. Griffin, J. R. Parsons, and D. E. Smith, *ISA Trans.* **18**(1), 23 (1979).

CHAPTER ELEVEN

Polymer Analysis Using Gas Chromatography

D. J. SKAHAN

C. W. AMOSS

ARCO Chemical Company

Modern Practice of Gas Chromatography, Third Edition. Edited by Robert L. Grob.
ISBN 0-471-59700-7

11.1 INTRODUCTION

The major application of gas chromatography to the analysis of polymers is its use in the qualitative and quantitative determination of low molecular weight impurities and volatile oligomers in finished polymers. Gas chromatography continues to be one of the primary tools in the evaluation of monomer feedstocks, solvents, and catalysts employed in polymer production.

With the development and commercial availability of new pyrolysis techniques, inverse gas chromatography and the utilization of gas chromatographs coupled with mass and infrared spectrometers, polymer and copolymer structural elucidations are being reported in increasing numbers (1–6). Since many of the analytical techniques reported in the literature are dependent on sample pretreatment and auxiliary equipment, this chapter will identify some of these procedures as well as column technology and applications.

11.2 POLYMER TYPES

The increasing number and type of polymers and resins makes characterization difficult. However, for purposes of this chapter, the speciation of organic polymers listed below will provide some general characterization. Extensive information on the chemistry, structure, formation, and use of organic polymers is available in several texts (7–12).

11.2.1 Homopolymers

The least complex polymer is a chain of repeating single monomer units. The molecular weight of such polymers can vary from below one thousand to over one million. Polystyrene, polyethylene, polypropylene, polyvinyl acetate, and polyvinyl chloride are examples of homopolymers (13). The particular kind of polymerization undergone by ethylene in which many molecules of monomer are simply added together is called addition polymerization (14).

11.2.2 Copolymers

Copolymers are composed of two or more monomer units. The second monomer unit may alternate randomly with the first to form a chain as shown below:

A—A—B—A—B—B—A—A—A—B

Or it can be integrated in a fixed sequence of single units

A—B—A—B—A—B—A—B—A

or blocks

A—A—A—B—B—B—A—A—A

Each of the above copolymers is a linear chain of repeating monomer units. Each unit is bifunctional, that is, capable of forming only two bonds within the structural chain.

11.2.3 Cross-linked/Branched Copolymers

If the functional sites of the monomer unit is greater than two, more complex structures resulting in branching and linked branched chains can be produced (15). As illustrated in *Modern Practice of Gas Chromatography* (16), when vinyl acetate is condensed with divinyl adipate a cross-linked branched polymer is formed with the tetrafunctional adipate forming the base for the cross-linked pattern, as shown in Figure 11.1. A second example of a cross-linked polymer is shown in Figure 11.2.

11.2.4 Cyclolinear Polymers

Special linear polymers formed by linking ring systems can obtain heterocyclic and inorganic rings as well as benzene rings. Solubility of cyclolinear polymers is often very low and the tendency for crystallization is very high. When ring systems are linked together to form three-dimensional connecting units they are known as cyclomatrix polymers. Graphite is an example of a cyclomatrix polymer (15).

11.2.5 Graft Copolymers

When a polymer derived from one homopolymer is reacted with a second homopolymer a graft copolymer can result. Polymers of this type can be prepared by gamma or X-ray irradiation of a mixture of the two homopolymers or even by mechanical blending (15). Graft copolymers may also be prepared by polymerization of monomer B from initiation sites along the

```
         Ac           Ac                          Ac
         |            |                           |
         O            O                           O
         |            |                           |
- CH2 - CH - CH2 - CH - CH2 - CH - CH2 - CH -
                                   |
                                   O
                                   |
                                   C=O
                                   |
                                   (CH2)4
                                   |
                                   C=O
                                   |
                                   O
                                   |
- CH2 - CH - CH2 - CH - CH2 - CH - CH2 - CH -
         |            |            |           |
         O            O            O           O
         |            |            |           |
         Ac           Ac           Ac          Ac

            Note: O-Ac = CH3 - C=O
                               |
                               O
```

FIGURE 11.1 Branched polymer resulting from the condensation of vinyl acetate and divinyl adipate.

chain of polymer A:

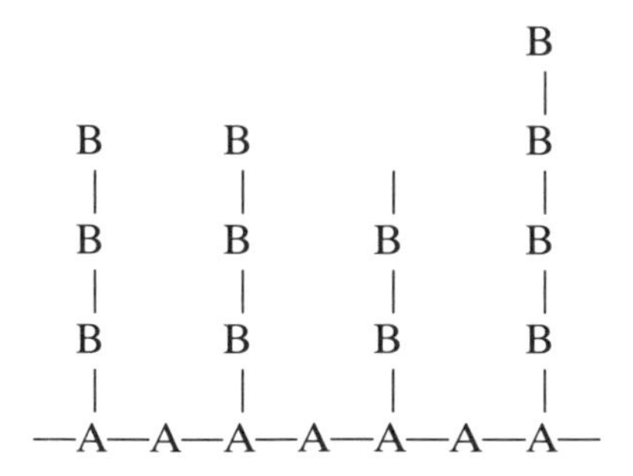

11.3 POLYMER FORMATION

The types of reactions used to synthesize polymers also provides a method of classification. Three such reactions are illustrated as examples of this classification: condensation, addition, and ring opening.

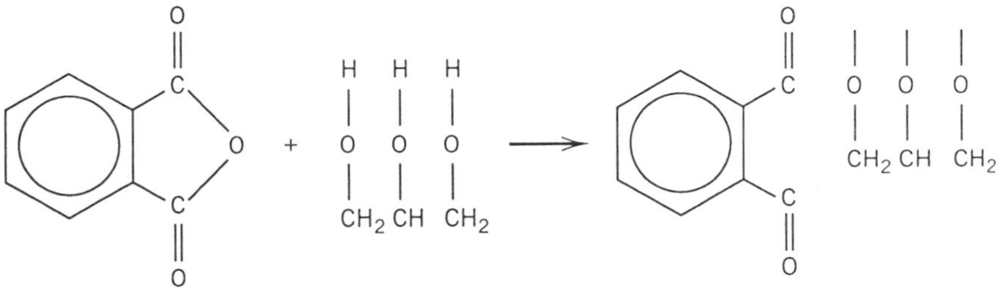

$$\text{Phthalic anhydride} + \underset{\text{Glycerine}}{\overset{\text{H}\quad\text{H}\quad\text{H}}{\underset{CH_2\;\;CH\;\;CH_2}{O\quad O\quad O}}} \longrightarrow C_6H_4\left(\overset{O}{\overset{\|}{C}}-\right)_2 \underset{CH_2\;\;CH\;\;CH_2}{O\quad O\quad O}$$

FIGURE 11.2 Glyptal resin formation form phthalic anhydride and glycerine polymerization.

11.3.1 Condensation Polymers

Polymers such as nylon, polyesters, and urea formaldehyde resins are formed by the reaction of two or more molecules with a concurrent loss of water or ammonia. As an example, Nylon 66 is made by the condensation of hexamethylene diamine with adipic acid:

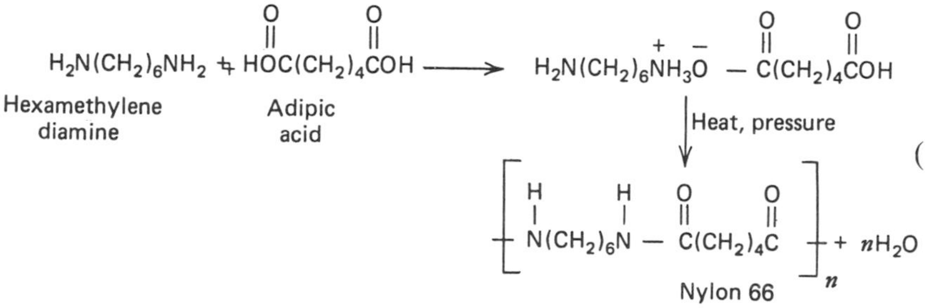

$$\underset{\text{Hexamethylene diamine}}{H_2N(CH_2)_6NH_2} + \underset{\text{Adipic acid}}{HO\overset{O}{\overset{\|}{C}}(CH_2)_4\overset{O}{\overset{\|}{C}}OH} \longrightarrow H_2N(CH_2)_6\overset{+}{N}H_3\overset{-}{O} - \overset{O}{\overset{\|}{C}}(CH_2)_4\overset{O}{\overset{\|}{C}}OH$$

$$\xrightarrow{\text{Heat, pressure}} \underset{\text{Nylon 66}}{\left[\overset{H}{\overset{|}{N}}(CH_2)_6\overset{H}{\overset{|}{N}} - \overset{O}{\overset{\|}{C}}(CH_2)_4\overset{O}{\overset{\|}{C}} \right]_n} + nH_2O \tag{11.1}$$

Polyesters such as Dacron or Mylar are made by the condensation of a dicarboxylic acid with a diol:

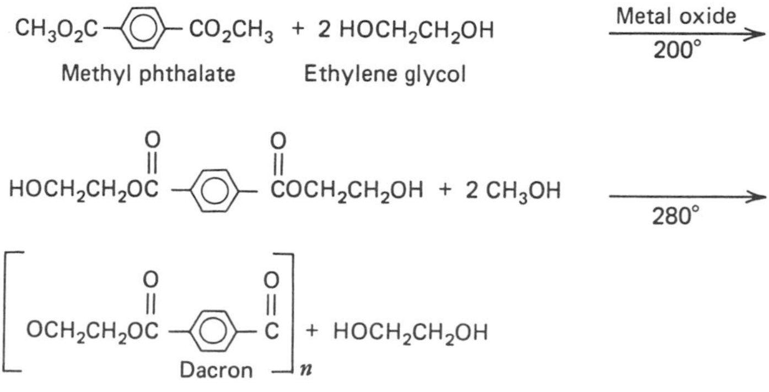

$$\underset{\text{Methyl phthalate}}{CH_3O_2C-C_6H_4-CO_2CH_3} + \underset{\text{Ethylene glycol}}{2\,HOCH_2CH_2OH} \xrightarrow[200^\circ]{\text{Metal oxide}}$$

$$HOCH_2CH_2O\overset{O}{\overset{\|}{C}}-C_6H_4-\overset{O}{\overset{\|}{C}}OCH_2CH_2OH + 2\,CH_3OH \xrightarrow[280^\circ]{} \tag{11.2}$$

$$\underset{\text{Dacron}}{\left[OCH_2CH_2O\overset{O}{\overset{\|}{C}}-C_6H_4-\overset{O}{\overset{\|}{C}} \right]_n} + HOCH_2CH_2OH$$

Inorganic condensation polymers, such as polysilicates and polyphosphates, are formed by the removal of water from the appropriate di-, tri-, or tetrahydroxy monomers:

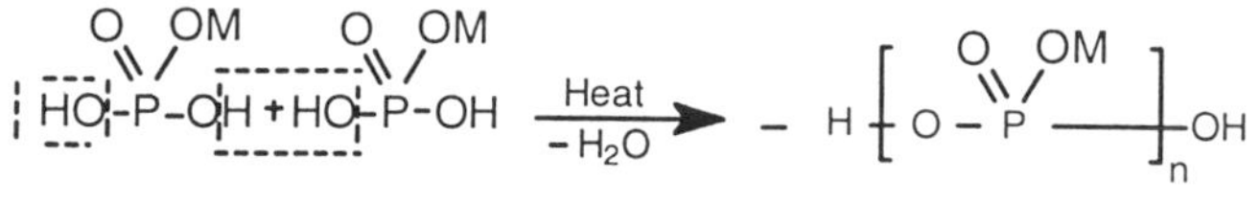

$$HO-P(=O)(OM)-OH + HO-P(=O)(OM)-OH \xrightarrow[-H_2O]{Heat} -H\left[O-P(=O)(OM)-\right]_n OH \qquad (11.3)$$

(M–alkali metal) Polyphosphate

Enzyme catalysis gives rise to the condensation polymerization of some amino acids to proteins, of sugars to polysaccharides, and the synthesis of nucleic acids such as DNA and RNA (15).

11.3.2 Addition Polymers

Addition polymers are formed by the addition reactions of olefins, acetylenes, aldehydes, or other compounds with "unsaturated" bonds. The reaction mechanism that proceeds from initiation through propagation to termination can be anionic, cationic, or free radical. Several well-known polymers are formed by the addition mechanisms, as illustrated in Table 11.1

11.3.3 Ring-Opening Polymerizations

These polymers are formed by catalytic cleavage of cyclic compounds followed by polymerization to form high molecular weight polymers. For example, caprolactam polymerizes to Nylon 6 and trioxane polymerizes to yield polyformaldehyde:

$$\overline{(CH_2)_5-C(=O)-NH} \longrightarrow \left[NH-(CH_2)_5-C(=O)\right]_n \qquad (11.4)$$

CAPRIOLACTAM NYLON 6

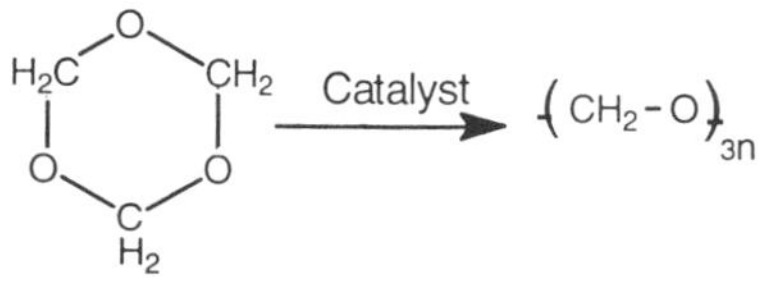

$$\text{cyclo-}(CH_2-O)_3 \xrightarrow{Catalyst} (CH_2-O)_{3n} \qquad (11.5)$$

TRIOXANE POLYFORMALDEHYDE

Inorganic ring systems are also polymerized by ring opening, such as

TABLE 11.1 Commercial Polymers Formed by Addition Reaction

Reaction	Polymer
$n\mathrm{CH_2{=}CH_2} \longrightarrow \mathrm{(CH_2{-}CH_2)}_n$	Polyethylene
$n\mathrm{CH_2{=}CH(Cl)} \longrightarrow \mathrm{(CH_2{-}CH(Cl))}_n$	Polyvinyl chloride
$n\mathrm{CH_2{=}CH(C{\equiv}N)} \longrightarrow \mathrm{(CH_2{-}CH(C{\equiv}N))}_n$	Poly(acrylonitrile)
$n\mathrm{CH_2{=}CH(C_6H_5)} \longrightarrow \mathrm{(CH_2{-}CH(C_6H_5))}_n$	Polystyrene
$n\mathrm{CH_2{=}C(CH_3)_2} \longrightarrow \mathrm{(CH_2{-}C(CH_3)_2)}_n$	Polyisobutylene
$n\mathrm{CH_2{=}CH{-}C(CH_3){=}CH_2} \longrightarrow \mathrm{(CH_2{-}CH{=}C(CH_3){-}CH_2)}_n$	*trans*-1,4-Polyisoprene
$n\mathrm{CH_2{=}CH{-}CH{=}CH_2} \longrightarrow \mathrm{(CH_2{-}CH{=}CH{-}CH_2)}_n$	*trans*-1,4-Polybutadiene
$n\mathrm{CF_2{=}CF_2} \longrightarrow \mathrm{(CF_2{-}CF_2)}_n$	Polytetrafluoroethylene (Teflon)

$$\underset{\text{Rhombic sulfur}}{S_8 \text{ (ring)}} \xrightarrow{\text{Heat}} \underset{\text{Plastic sulfur}}{(\mathrm{S{-}S})_{4n}} \tag{11.6}$$

$$n\,\underset{\text{Octamethylcyclotetrasiloxane}}{[\mathrm{(CH_3)_2SiO}]_4} \xrightarrow[\text{acid or base}]{\text{Trace of}} \underset{\text{Poly(dimethylsiloxane)}}{\left[\mathrm{O{-}Si(CH_3)_2}\right]_{4n}} \tag{11.7}$$

11.4 SAMPLE PREPARATION

11.4.1 Direct Sampling

When direct injection is used, the accuracy and amount of information obtained from gas chromatographic analysis of polymers is often dependent on the selection of relatively small volumes of sample from bulk quantities of resin. Statistical evaluation of numerous injections should be considered to improve the reliability of the analysis. Large sample volumes through a series of isolated valves have been used effectively to remove solvent and improve sample reproducibility (17).

11.4.2 Extraction or Dissolution

A common procedure used in the gas chromatographic analysis of polymers is extraction or dissolution. If the polymer is solid, a known amount is weighed into a suitable solvent. Low molecular weight oligomers and/or unreacted monomers can be chromatographically eluted. Autosampling systems or autoinjection systems that use serum-capped vials prevent loss of low boiling components such as blowing agents and degradation products (18).

If identification of trace levels of light components or selective impurities is desired, multiple extractions with both polar and nonpolar solvents is often used in conjunction with GC/MS or GC/IR techniques (19). If the entire polymer is dissolved, the higher molecular weight fraction is often precipitated by washing with a reagent in which the polymer is insoluble. For example, polystyrene can be dissolved in 10 parts by weight of tetrahydrofuran. The addition of 20 to 30 parts by weight of methanol will then precipitate the polystyrene. A portion of the liquid can then be filtered through a 0.2- to 0.45-μm syringe filter directly into the gas chromatographic sample vial. This technique is also used to prevent decomposition of the polymer in the injection system or columns of the gas chromatograph (20). As discussed in Chapter 7, internal standards can be added to the extraction of residual monomers such as acrylonitrile and styrene in polystyrene resins. In like manner, plasticizers and inhibitors in polymers are also quantitatiely determined by extraction procedures (21). Care must be taken to optimize the extraction procedures to prevent decomposition or loss of components.

Difficulties can occur when a polymer is neither dissolved nor swelled appreciably in typical solvents. Consider the case of extracting additivies such as hindered phenolic antioxidants from polyethylene. Even under strenuous Soxhlet extraction for 24 h, some additives may not be extracted to more than 50% recovery. Polyethylene is difficult to grind into small particles, particularly at room temperatue with a Wiley mill. Freeze grinding at dry ice or liquid nitrogen temperatures may yield finer, more easily extracted particles, but the procedure is laborious. Recently, several authors have reported success in extracting additives from polyethylene with supercritical fluid extraction.

11.4.3 Reaction Gas Chromatography

Polymer degradation by chemical reaction is primarily used to decouple or derivatize polymers in the gas chromatographic injection system or by reaction prior to injection (22). Reagents for these reactions, such as silanization, esterification, and deesterification agents, are readily available from commercial sources (23, 24). The determination of plasticizers in polyvinyl chloride polymers through hydrolysis was described by Krishen (19). The alcohols and methyl esters shown in Figure 11.3 were separated on a 6-ft 10% UCW98 packed column. Silanization of oligomers and low molecular weight hydroxy-terminated polymers, such as polyols, followed by GC/MS is a useful method for identification of impurities causing odor or color contamination of polymers (25, 26).

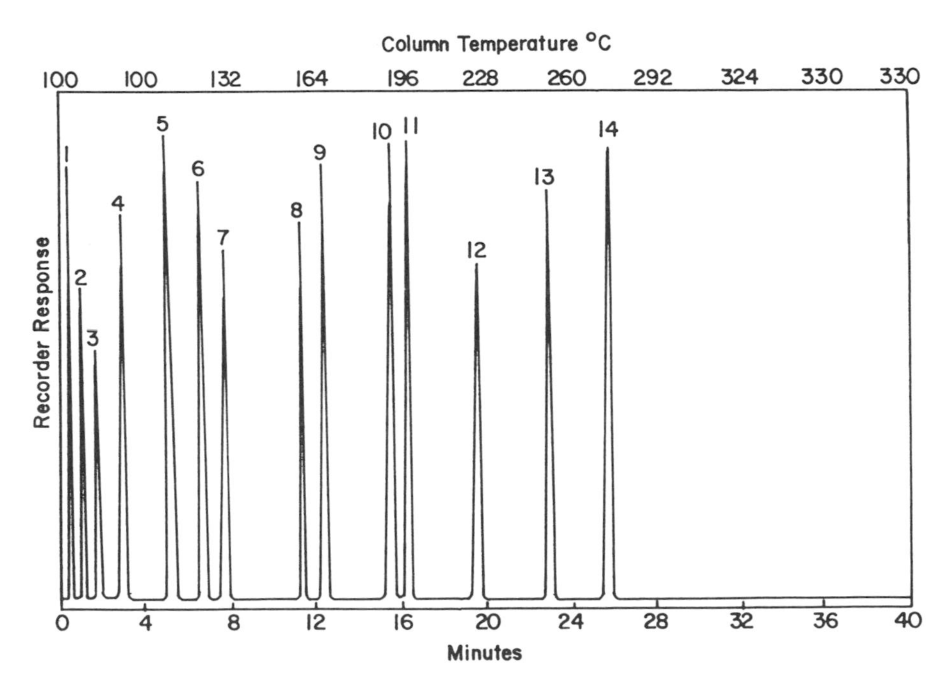

FIGURE 11.3 Gas chromatogram of alcohols and methyl esters of acids. Peaks: 1, methanol; 2, butanol; 3, pentanol; 4, hexanol; 5, heptanol; 6, 2-ethylhexanol; 7, octanol; 8, dimethyl adipate; 9, decanol; 10, dimethyl-*o*-phthalate; 11, dodecanol; 12, tetradecanol; 13, methyl palmitate; 14, methyl stearate. (With permission of the American Chemical Society.)

11.4.4 Fusion

Polymer decoupling by fusion of the material in a large excess of reagent followed by heating to above the melt temperature of the mix is a useful technique for generating low molecular weight products that can be separated by gas chromatography. Fusion is a convenient reaction technique that is often used when extraction, dissolution, or thermal decomposition cannot be used. Most compounds will undergo quantitative reactions at fusion temperatures of 300–350°C. Volatile products can be collected by cold trapping and analyzed by gas chromatography. Fusion reagents such as potassium hydroxide, alkali-metal hydroxides, sodium bisulfate, orthophosphoric acid, hydrazine, potassium metaperoxide, and sodium metaperiodate are readily available.

Techniques for both oxidative and reductive fusion are reported in the literature (27). The generation of volatile organic compounds such as alkanes and aromatics from the fusion of dimentylsiloxanes, phenyl-substituted polysiloxanes, polyamide, and polyimide polymers allows for direct quantification by in-line gas chromatography using both packed and fused-silica capillary column separation (28, 29).

11.4.5 Headspace Sampling

Gas chromatographic analyses employing headspace sampling techniques (see Chapter 14) are primarily used for the quantitative and/or qualitative determiation of volative impurities, residual monomer, or additives in finished polymers. The polymer is generally heated to 150°C or less for a short time in a septum-sealed container and the vapor in the headspace is subsequently injected into a gas chromatograph. Both static and dynamic headspace systems are commercially available as auxiliary equipment for gas chromatographs from manufacturers such as Hewlett Packard and Varian Associates. Quantification is achieved by internal or external standardization. Although headspace sampling is a common analytical technique for the determination of low molecular weight polymer impurities, it can also be used as an intermediate technique to collect decomposition products or monomers from the pyrolysis of polymers that can be easily decoupled, such as styrene–butadiene and vinyl acetate resins.

11.4.6 Pyrolysis

Pyrolysis or thermal decomposition has been an analytical technique for the identification of polymers for many years. Initially used in a static configuration under vacuum or an inert atmosphere, pyrolytic products were analyzed spectroscopically or by chemical or physical tests. Microfurnace pyrolyzers were readily available and the reproducibility of such systems was demonstrated on polyesters such as polyvinyl butyrate (30). Sequence distribution for copolymers such as butadiene–isoprene was also reported by pyrolytic gas chromatographic/mass spectrometry methods (31). Chemical derivatization followed by pyrolysis GC/MS was used to identify polar monomers at the 3% level in polyacrylate copolymers (32). To assure the reproducibility of products from polymer pyrolysis, several in-line gas chromatographic, spectroscopic, and classical chemical systems have been investigated, as illustrated in the study of vinyl chloride in vinyl chloride/vinyl acetate copolymers shown in Table 11.2.

Filament Pyrolysis. Commercially available pyrolysis probes using platinum or nichrome wire or ribbons provide in-line control of both temperature and time. Repeatability is often affected by recombination reactions occurring at the filament or ribbon surface. Wampler and Levy discussed some of the sources of nonrepoducibility in pyrolysis gas chromatography, such as sample size, nonhomogeneity, and pyrolizer design (33). Whiton and Morgan (34) described the modification of capillary injection ports to accommodate pyrolysis probes. A typical probe circuit is shown in Figure 11.4. Single-stage pyrolysis probes coupled to gas chromatographs have been in use for over 25 years. Filament and ribbon pyrolysis techniques have been used for the characterization of complex polymer mixtures, copolymer microstructure, and composition and the eludication of degradation mechanisms.

Curie-Point Pyrolysis. This technique utilizes the temperature achievable by transfer of energy from a radio-frequency field to ferromagnetic metals and alloys. A typical design of such a pyrolyzer is shown in Figure 11.5. The principal advantages of this type pyrolysis are rapid temperature rise and very good temperature control. The temperature rise time and Curie point for several alloy compositions is given in Table 11.3. Since no direct electrical contact between the power source and the pyrolysis wire occurs in this system, the wire can be coated with polymer, placed in the induction coil, and replaced after analysis.

Other pyrolysis methods for polymer degradation have been investigated. Fenter et al. (35) used a pulsed ruby laser to degrade a range of polymers.

TABLE 11.2 Percent Vinly Chloride in Vinyl Chloride/Vinyl Acetate Copolymers

Copolymer	By Pyrolysis Gas Chromatography (±2%)	By Infrared Analysis (±1%)	By Chlorine estimation (mean of 2 results)
A	55.8	54.7	60.8 ± 0.6
B	65.2	64.4	69.4 ± 0.1
C	67.8	66.7	69.1 ± 0.9
D	72.2	72.3	74.1 ± 0.3
E	83.9	84.8	81.8 ± 4.4
F	87.7	89.0	87.9 ± 1.0

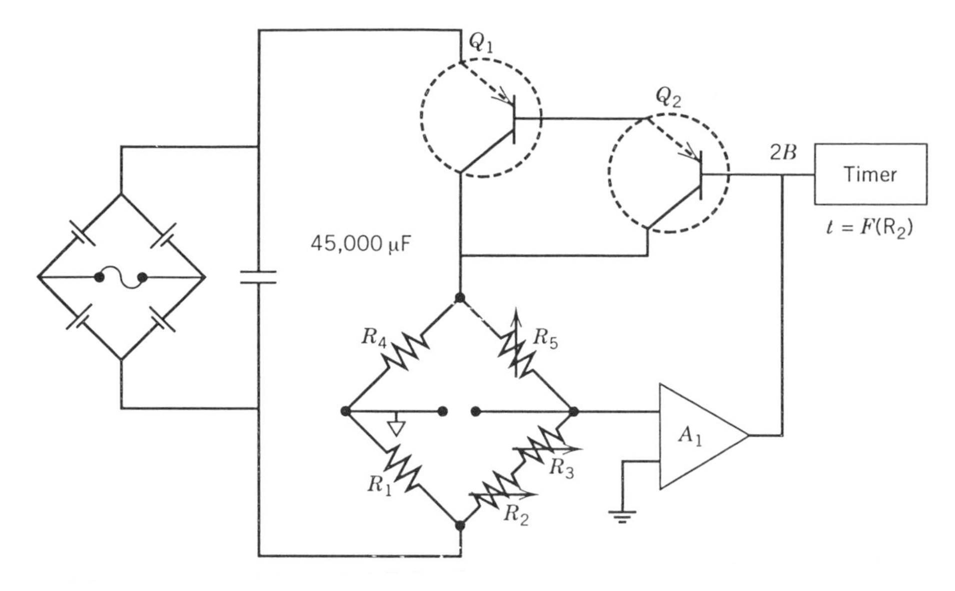

FIGURE 11.4 Schematic of pyrolysis probe circuit designed for fast temperature risetimes. (With permission of the American Chemical Society from *Anal. Chem.* **1980,** 52 (1983).)

A major drawback in laser pyrolysis is the intense local temperatures in a small portion of the polymer. The complex products that form a plasma recombine with acetylene, which is the major fragment. Further products due to these reactions make polymer characterization by this technique difficult. When defocused lasers are used, more characteristic degradation products are observed. Surface oxidation of weathered polymers was studied by Merrit et al. using a highly defocussed laser to strip the surface (36). Other pyrolytic methods, including dielectric breakdown in which the polymer sample is used as the dielectric of a high-voltage capacitor, produce severe degradation conditions, with most polymers yielding similar low molecular weight fragments and few products that are characteristic of the polymer structure.

In summary, the ideal pyrolysis system must provide a precise known temperature to give good reproducibility. It must reach the operating temperature very rapidly (Figure 11.6) and the products must be quickly passed into the gas chromatograph to reduce the possibility of secondary reactions. A comparison of the performance of various pyrolyzers is shown in Table 11.4.

11.5 QUALITATIVE ANALYSIS

The use of gas chromatography coupled with mass spectrometry and infrared spectrophotometry for qualitative analysis has been reviewed by

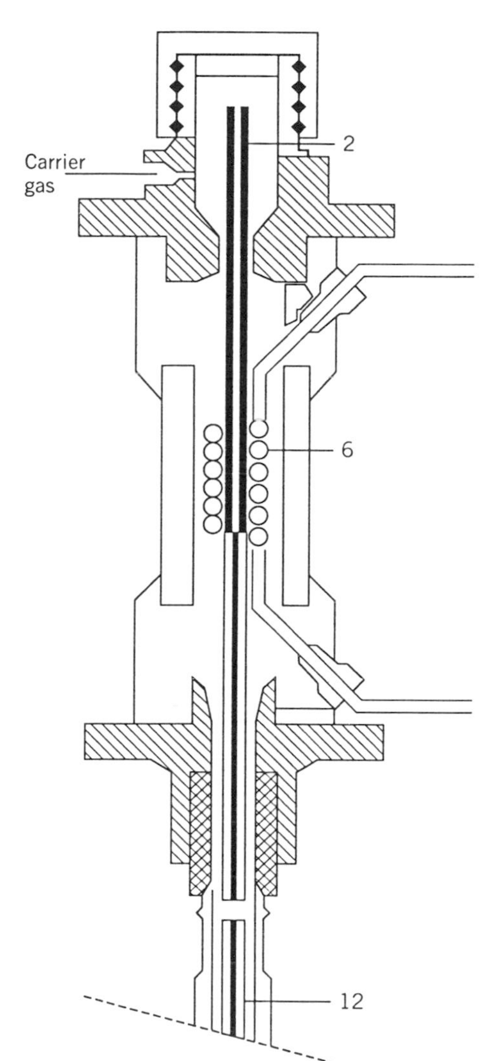

FIGURE 11.5 A Curie-point pyrolyzer design. The sample is coated onto ferromagnetic wire (not shown), which is lowered down a quartz tube (2) into the rf induction coil (6). The latter is a copper tube through which cooling water passes in order to prevent self-heating of the coil. After the rf current has pyrolyzed the sample, the products are swept by the carrier gas into the GLC column (12). (Reproduced by permission of Pergamon Press from *Chromatographica*, **12**, 59 (1976).)

Liebman and Wampler (37). This type instrumentation often provides the identity of trace impurities in polymers that cause odor, color, or reaction interference. GC/MS was used to identify trioxilane as the principal odor constituent in polyurethane foam at concentrations below 1 ppm (38). Less expensive detector systems such as flame ionization in conjunction with highly precise column temperature and carrier flow controls also provide an adequate means for component identity for both capillary and packed-column separations. This technique is widely used for the determination of trace monomers, plasticizers, inhibitors, and diluents in resins and polymers. Peak identification in systems using either flame ionization or thermal

TABLE 11.3 Temperature Rise Time and Curie-point Versus Alloy Composition

Alloy Composition (%)			Quoted Temperature Rise Times (ms)		
Fe	Ni	Co	Curie Point (°C)	Fisher-Varian (1500 W)	Phillips (30 W)
0	100	0	358	300	1300
61.7	0	38.3	400	40	500
50.6	49.4	0	510	150	700
42.0	41.0	16.0	600	70	500
29.2	70.8	0	610	130	1150
33.0	33.0	33.0	700	90	1350
100	0	0	770	110	2100

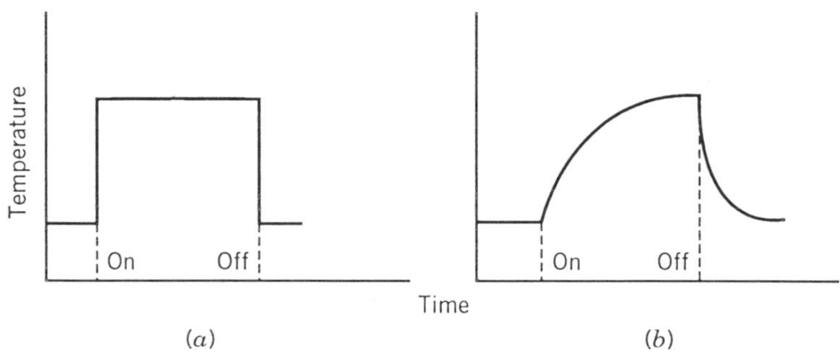

FIGURE 11.6 Temperature–time profiles for (a) an ideal pyrolyzer and (b) a more typical unit in which the filament is simply switched on and off.

conductivity detectors is based on external reference standards and retention indices using polar and nonpolar column substrates.

Recent advances in capillary column technology, including bonded coatings, allow the significant improvement in separations that can be achieved with capillary systems with the durability normally associated with packed columns. If, however, direct polymer injection is required, a precolumn usually packed with an inert substrate is recommended to prevent polymer deposition at the head of the capillary column. Installation of large-diameter capillary columns (0.52 mm) eliminates the need for sample splitting and provides increased sensitivity for trace components. Typical substrates that are routinely used for polymer analyses are shown in Table 11.5. Regardless of the column selection, if polymer analysis is carried out routinely, calibration standards should be scheduled for analysis to insure reproducibility of retention data for correct qualitative identification.

TABLE 11.4 Comparison of Performance of Various Pyrolyzers

Method	Temperature Rise Time to 600°C	Available on the Market	Repeat-ability	Disadvantages	Advantages
Boat	20–50 sec	Yes	Fair	1. Slow rise time 2. Requiring large samples 3. Does not lend itself to capillary columns	1. Any material, liquid or solid 2. Widest temperature range 3. Wide sample size range 4. Can be purchased
Conventional filament and ribbon	5–25 sec	Yes	Fair	1. Slow rise time 2. Aging of metal 3. Solid material hard to apply	1. Most widely used 2. Wide temperature range 3. Purchased or built 4. Simple to apply sample
Capacitive boosted filament	7–10 msec	No	Good	1. Solid material must be soluble 2. Must use small sample sizes 3. Cannot be purchased	1. Fast rise time 2. Wide temperature range 3. Good repeatability
Curie-point low-power (30–100 W)	1.5 sec	Yes	Good	1. Solid material hard to apply 2. Pyrolysis chamber must be heated 3. Slow rise time	1. Accurate temperature 2. Good repeatability 3. Can be purchased
Curie point high power (>200 W)	70 msec	Yes	Good	1. Solid material hard to apply 2. Pyrolysis chamber must be heated	1. Fast rise time 2. Accurate temperature 3. Good repeatability 4. Can be purchased

TABLE 11.4. *(Continued)*

Method	Temperature Rise Time to 600°C	Available on the Market	Repeat-ability	Disadvantages	Advantages
Vapor-phase gold tube reactor	N/A	Yes	Good	1. Only volatile materials	1. Good repeatability 2. Wide temperature range 3. Can be correlated with theory
Laser	Estimated 10 μ sec	No	Fair	1. Must use colored agent to pyrolyze 2. Complicated and expensive 3. Large sample size 4. Equilibrium temperature higher than other pyrolyzers 5. Laboratory hazards 6. Cannot be purchased 7. Complex fragmentation process	1. Pyrograms may be simple 2. Superior rise time 3. Not necessarily thermal degradation
Ribbon probe	10 msec	Yes	Good	1. Solid materials fall off 2. Must use small sample sizes	1. Excellent rise time 2. Wide temperature range 3. Can be purchased 4. Good repeatability

TABLE 11.4. (*Continued*)

Method	Sample Size	Sample Form Requirements and Limitations	Temperature Range	Pyrolysis Element Temperature Control	Catalytic Reactions Within Pyrolysis Chamber
Boat	1–5000 μg	All forms 1. Liquids 2. Solids	Up to 1500°C	Continuously variable	Slight
Conventional filament and ribbon	10–1000 μg	1. Soluble in a solvent	Up to 1200°C	Continuously variable	Yes
Capacitive boosted filament	5–10 μg	1. Soluble in a solvent	Up to 900°C	Continuously variable	No
Curie-point low-power (30–100 W)	10–50 μg	1. Soluble in a solvent 2. Some crystals and powders	Up to 980°C	Limited to 356, 480, 510, 600 770, 980°C	Yes
Curie-point high-power (>1000 W)	10–50 μg	1. Soluble in a solvent 2. Some crystals and powders	Up to 980°C	Limited to 356, 480, 510, 600, 770, 980°C	Yes
Vapor phase	0.001–10 μg	1. Volatile materials only	Up to 800°C	Continuously variable	No
Laser	500 μg or larger	1. Dark sample or media	Estimated 10^9 K	Not controlled	Slight
Ribbon probe	0.1–5 μg	1. Soluble in a solvent 2. Some powders	Up to 1000°C	Continuously variable	Slight

TABLE 11.5 Typical Gas Chromatographic Substrates Used for Analysis of Monomers and Polymers

Applications	Phases With Similar McReynolds Numbers	Composition	Polarity
Hydrocarbon Pesticides Phenols Aromtics	DB-1, SPB-1 OV-1, RTx-1 SE-30, GB-1	100% Dimethyl polysiloxane	Nonpolar
Esters Halogenated Compounds	DB-5, SPB-5 RTx-5, RSL-200 OV-73, SE-54	5% Diphenyl 95% Dimethyl polysiloxane	Nonpolar
Acids Alcohols Aldehydes Acrylates Ketones Nitriles Pyrolysate	SP-1000 Stabilwax DB-FFAP OV-351	Polyethylene glycol- TPA modified	Intermediate
Alcohols Free acids Ethers Glycols Solvents	Carbowax 20M DB wax, 007-CW Supelcowax-10 Stabilwax CP wax 52CB	Polyethylene glycol	Polar

11.6 QUANTITATIVE ANALYSIS

There are many different types of gas chromatographic analyses that require the use of quantitative analysis. Most frequently such analyses are used to insure the quality of a given polymer or to solve a problem related to a real or alleged undesirable characteristic associated with a particular batch. Typical quantitative applications include the determination of the amount of residual monomers, additives, or trace impurities. Each of these analysis types may require the use of specialized techniques to prepare the sample for analysis or to introduce the sample into the gas chromatograph. This section describes a variety of techniques and procedures that can be used for the analysis of polymer samples.

To determine volatile components trapped in a nonvolatile polymer matrix, the most common means of sample introduction involves some sort of dissolution or extraction of the volatile component. Pyrolysis and reaction gas chromatography are often used to break the polymer into small volatile fragments that can subsequently be determined quantitatively. Another popular technique uses headspace analysis to concentrate volatile components prior to introduction into the gas chromatograph. Regardless of the technique employed, a key objective is to avoid fouling the injection port with high molecular weight nonvolatile polymer.

The simplest case for extracting a substance from a polymer occurs when a component can be quantitatively extracted with a suitable solvent. This technique may be applied to polymer beads or to a cut up piece of finished product. The sample is weighed and extracted with a known volume of solvent. External standards can be made from a known concentration of the component of interest in the extracting solvent. Alternatively, an internal standard can be added to the extracting solvent. In some cases it is also desirable to "spike-in" a known quantity of the component of interest. This technique, also known as standard addition, is particularly useful for verifying the mount of trace impurities. For routine applications it is important to determine the extraction efficiency or percent recovery of the extraction procedure. Significant errors can occur if one assumes that the extraction efficiency is 100%. Likewise, it is necessary to run a solvent blank in addition to the samples and standards. Many solvents contain minor impurities or stabilizers that could add spurious peaks or peak areas to the analysis. Fortunately many vendors now market solvents that minimize such effects. Solvents that are sold as gas chromatographic or pesticide grade are usually pure enough to be used for such analyses.

Another amenable case occurs when the entire polymer can be dissolved in a solvent. The polymer can then be selectively precipitated with a nonsolvent, leaving the components of interest in solution. Combinations of solvent/nonsolvent pairs for many polymers have been published (39). The advantages of this technique are that it is rapid and generally gives quantitative recovery of the components of interest. A serious disadvantage can sometimes occur for trace compoments, since the 40-fold dilution may put some compoments near or below the limit of detection.

11.7 APPLICATIONS

The number of references covering the applications of gas chromatograpy to the analysis of polymers is quite large. The applications that are discussed in this chapter are representative of procedures and techniques that might be used to solve real problems facing contemporary scientists. Such applications often require very sophisticated procedures in order to sort out complex information.

11.7.1 Monomer Analysis

Zlatkis et al. have used capillary GC to determine trace impurities in styrene monomer feed (40). To concentrate the impurities, a 10-μL injection of styrene was first made onto a semipreparative packed column. The effluent from this column was passed into a tube containing the porous polymer Tenax. Collection of effluent was suspended while the main styrene peak eluted from the column. The concentrated impurities were then thermally desorbed into a capillary GC–mass spectrometer. More than 100 peaks were found in the capillary chromatogram, and of these, 60 peaks were identified. A list of these impurities can be found in Table 11.6.

Hydrogels are an important class of materials used in biomedical applications. They consist of a hydrophilic network polymer containing a significant amount of adsorbed water. Many of the applications of such polymers involve in vivo uses such as synthetic cartilage, wound coverings, or drug delivery systems. Since small changes in the monomer composition may have a large impact on the properties of the final polymer, it is important to carefully analyze the monomer before polymerization.

Roorda et al. have studied the purity of hydroxyethyl methacrylate (HEMA), used to make poly(hydroxyethyl methacylate) hydrogels, by capillary GC (41). Injections of 2.5% 2-hydroxyethyl methacrylate (HEMA) in acetone solution containing dimethylsulfoxide internal standard were made onto a Hewlett Packard fused-silica capillary column (25-m × 0.2-mm i.d.) coated with a 0.10-μm film thickness of Carbowax 20M. The column was maintained at 113°C while hydrogen carrier flowed at 2 mL/min. Under these conditions, concentrations of methacrylic acid, dimethyacrylethane, and hydroxyethoxyethyl methacrylate were determined in hydroxyethyl methacrylate in the range of 5×10^{-3} to 2%. A fourth impurity, hydroxyethyl acetate, was not quantified, since it normally is not incorporated into the polymer and can be washed out afterward. The method can also be used to monitor for residual monomer in polymer, an important consideration in biocompatibility. A chromatogram of HEMA and related impurities determined under the conditions described above is shown in Figure 11.7.

11.7.2 Structural Characterization

The physical properties of polymers are generally determined by variables such as molecular weight and molecular weight distribution. Other variables such as short chain branching can also play an important role. Pyrolysis of branched polymers followed by gas chromatography of hydrogenated fragments is an effective way to determine the length and frequency of polymer branches.

Pyrolysis–hydrogenation is a commonly used technique to study branching in low-density polyethylene (LDPE) and high-density polyethylene (HDPE). Due to the complex nature of the pyrograms, the best results have been achieved on high-resolution fused-silica capillary columns. In many cases, the gas chromatographic information can be compared with results from ^{13}C nuclear magnetic resonance spectroscopy in order to confirm structural assignments. Haney et al. have used relatively low pyrolysis temperatures (360°C) and have observed fragments suggesting pairing of

TABLE 11.6 Trace Impurities Present in Styrene Monomer

Diethylether	Allylbenzene	Undecane	Pentadecane
Acetone	Methylstyrene	C_5-alkylbenzene	C_9-alkylbenzene
Hexane	Methylstyrene (isomer)	Tridecene	Diethyl phthalate
Methanol	C_3-alkylbenzene (isomer)	Tridecane	Hexadecene
Ethanol	Dichlorobenzene	C_5-alkylbenzene (isomer)	Hexadecane
Methylene chloride	Indane	C_6-alkylbenzene	Phenyl benzoate
Benzene	C_4-alkylbenzene	C_7-alkylbenzene	Dipropyl phthalate
Chloroform	C_4-alkylbenzene (isomer)	Biphenyl	1,3-Diphenylpropane
Toluene	Dimethylstyrene	Tetradecene	C_{10}-alkylbenzene
Ethylbenzene	Decahydronaphthalene	Tetradecane	Stilbene
m,*p*-Xylene	Limonene	Diphenylmethane	Heptadecane
Nonene	C_4-alkylbenzene (isomer)	Dimethyl phthalate	Diphenylbutene
o-Xylene	Acetophenone	C_8-alkylbenzene	Diphenylbutadiene
n-Propylbenzene	C_4-alkylbenzene (isomer)	Pentadecene	Diphenylbutadiene (isomer)
C_3-alkylbenzene	Methyl benzoate	1,1-Diphenylethylene	Dibutyl phthalate

Source: From Reference 40 with permission of Elsevier Science Publishers B.V.

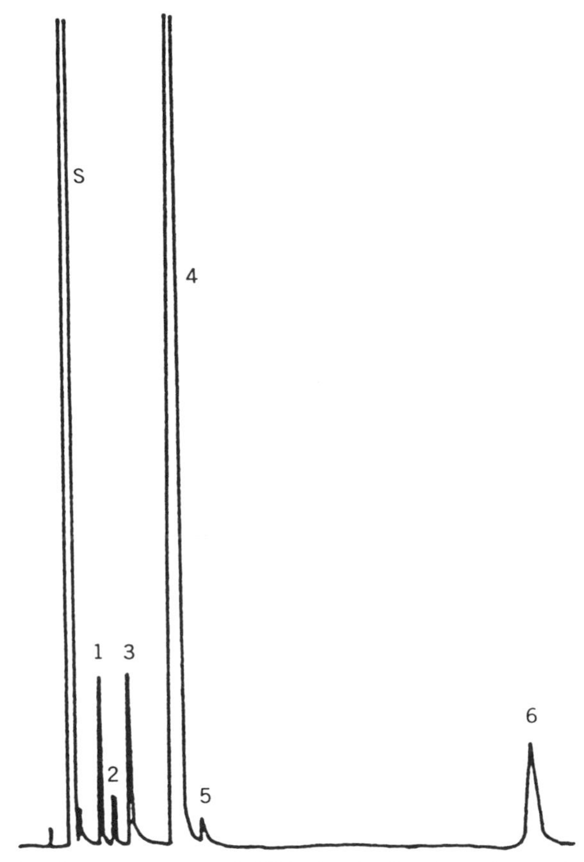

FIGURE 11.7 A chromatogram of HEMA. Peaks: S, solvent (retention time, 36 sec); 1, ODMSO (72 sec); 2, HEA (97 sec); 3, MAA (120 sec); 4, HEMA (174 sec); 5, DME (228 sec); 6, HEEMA (670 sec). (From Reference 41 with permission of Elsevier Science Publishers B.V.)

branches and branching of branches (42). Ohtani et al. used higher temperatures (650°C) to demonstrate the existence of amyl, hexyl, and larger branches (43). Samples of about 500 μg were pyrolyzed in a vertical microfurnace-type pyrolyzer maintained at 650°C. The pyrolysis occurred in the presence of hydrogen carrier gas, which then passed into a 5-cm OV-101 precolumn and a 10-cm hydrogenation catalyst column packed with 5% platinum. Both precolumns are maintained at 200°C. The first precolumn serves as a trap to prevent relatively nonvolatile components from fouling the catalyst and separation column. The hydrogenated pyrolysate passes into a 40:1 splitter and then onto the Hewlett Packard OV-1 WCOT fused-silica capillary column (0.2-mm i.d. $\times$ 50 m long).

A typical high-resolution hydrogenated product pyrogram from LDPE is

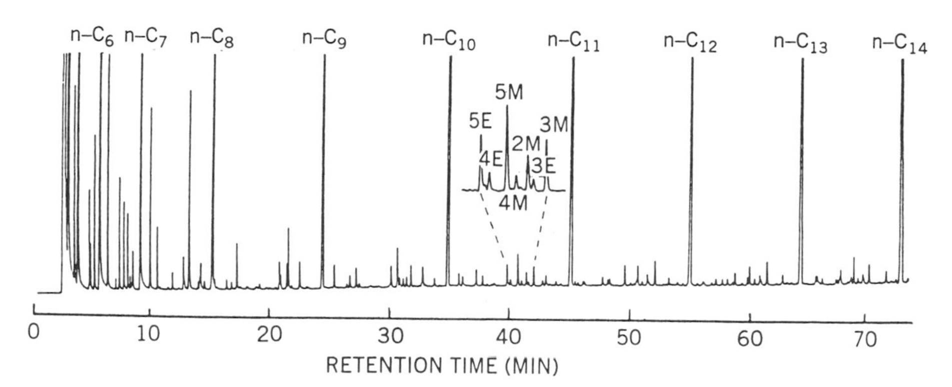

FIGURE 11.8 A high-resolution pyrogram of LDPE-B at 650°C: 2M, 2-methyldecane; 3M, 3-methyldecane, 4M, 4-methyldecane; 5M, 5-methyldecane; 3E, 3-ethylnonane; 4E, 4-ethylnonane; 5E, 5-ethylnonane. (From Reference 43 with permission of The American Chemical Society.)

illustrated in Figure 11.8. Using the relative intensity of each fragment in a given region of the chromatogram (e.g., all C_{11} peaks), the authors were able to determine the relative amounts of each branch size from methyl to hexyl. They reported a relative branch abundance in the order butyl > ethyl > amyl > methyl > hexyl. Usami et al. also used this technique to determine the degree of branching as a function of molecular weight (44). In this case the LDPE was first fractionated into 16 moleular weight cuts by depositing the polymer on Celite and then eluting the polymer with a solvent gradient of xylene in ethylcellosolve at 127°C. In general, ethyl and butyl branches were found to be more abundant in low molecular weight fractions, while some branches, such as methyl, were found to be relatively constant regardless of molecular weight.

11.7.3 Trace Analysis

Trace analyses are often necessary due to toxic properties of the material to be measured. Vinyl chloride monomer is a carcinogen found in poly-(vinylchloride) (PVC) at the part-per-billion (ppb) level. Poy et al. have reported an automated procedure using purge-and-trap and dynamic head-space of PVC solutions (45). Solutions containing 10% PVC in dimethyl-acetamide (DMA) are equilibrated in sealed vials for 1 h in a constant-temperature bath maintained at 90°C. The vial is then sparged with nitrogen into a cold trap filled with Tenax TA. At the completion of the purge-and-trap procedure, the trap is then placed in the carrier-gas stream and rapidly heated to inject the vinyl chloride monomer onto the chromatographic column. Most of the DMA is backflushed from the cold trap during the last portion of the purge-and-trap cycle. The analysis is performed on a 3-m × 3-

mm-i.d. column packed with 25% free fatty acid phase (FFAP) on Chromosorb P. The advantage of this column is that all of the DMA can be eluted from the column in 45 min, which is a shorter time than is possible with other packings. An overall advantage of this method using headspace of polymer solutions is that it can be applied to polymer samples that are not easily analyzed in the solid form by headspace gas chromatography.

Successful trace analyses are frequently dependent on the use of very careful techniques. Clean glassware, syringes, and vials are needed to prevent contaminants from giving spurious results. A thorough treatment of all the factors involved in good trace analysis is beyond the scope of this chapter. Examples of some of the factors to be considered can often be found in compendial methods published by major scientific societies. For example, The American Society for Testing and Materials (ASTM) publishes Standard Test Method F-1308-90 for Quantifying Volatile Extractables in Microwave Susceptors Used for Food Products (46). In this case a microwave susceptor is defined as "a packaging material which, when placed in a microwave field, interacts with the field and provides heating for the products the package contains." Susceptors are frequently made from polymeric materials. The precision of the method was tested on a susceptor of metallized polyethyleneterephthalate bonded to paperboard with ethylene vinyl acetate adhesive. Toluene and furfural were two volatile materials found in the susceptors tested.

ASTM Method F-1308-90 simulates the conditions to which a susceptor would be exposed in a microwave oven. The sample is exposed to microwave heating followed by headspace sampling and gas chromatography. Unknown extractables may be identified by gas chromatography with a mass spectrometer as a detector. Known extractables can be detected at a sensitivity level of 0.025 $\mu g/in^2$. A list of typical analytes is given in Table 11.7.

11.7.4 Copolymer Composition

Chemists are frequently called upon to determine the composition of unknown polymers. For example, polyurethane polymers can be made in a number of ways. One typical type of polyurethane is made from a polyether polyol and a diisocyanate. Depending on the starting alcohol (e.g., glycerol or pentaerythritol), the polyether polyol may be a diol, triol, or polyfunctional up to hexol or higher. The isocyanate portion could be from toluenediisocyanate, diphenylmethane diisocyanate, or other type of isocyanate. Haken has reviewed an analysis scheme for various condensation polymers such as polyurethanes (47). The chemical degradation followed by gas chromatographic analysis of volatile reaction products was originally developed by Whitlock and Siggia (27).

In the procedure outlined by Haken, rapid hydrolysis of the condensation polymer is achieved using high-temperature molten alkali fusion. An

TABLE 11.7 Analyte Recovery Without Microwaving

Compound	n^a	Recovery Recovery Mean %	Variablity % Within Laboratory	Overall Variability %
Benzene	5	97.7	7.8	9.0
2-Butoxy-ethanol	4	98.7	6.7	8.4
Dibutyl ether	5	109.7	16.5	23.7
Dodecane	3	101.1	10.7	10.7
2-Furfural	4	99.7	11.7	12.0
Furan-2-methanol	3	100.0	14.1	16.4
Isobutyl alcohol	4	98.0	7.1	7.9
Methylene chloride	5	103.5	16.7	22.6
2-Propanol	3	99.9	11.4	12.0
Styrene	5	100.8	8.5	9.3
Toluene	4	102.7	9.9	10.9
Overall		101.1	11.6	14.4

[a] n = number of laboratories submitting data on compound.

Source: From Reference 46 with permission of the American Society for Testing and Materials.

advantage of this procedure is that hydrolysis can be completed in a few hours at 250°C instead of a period of days required for refluxing methanolic caustic. Following the alkali fusion, dichloromethane is added to the reaction mixture and the resulting solution is extracted with hydrochloric acid. The acid extract contains diamine hydrochlorides derived from the original isocyanates. The acid solution is then made basic and the free diamines are recovered by extraction into dichloromethane. The diamines may be analyzed directly by gas chromatography or reacted with trifluoroacetic anhydride to form the trifluoroacetamide derivatives. The polyether portion can be also derivatized with a mixed anhydride reagent made from acetic anhydride and *p*-toluene sulfonic acid. The resulting polyol acetates may be too large in molecular weight to analyze by gas chromatography, unless the polyol was very small initially. Haken et al. have used the alkalai fusion technique to qualitatively analyze thermoplastic medical grade, polyether-based polyurethane elastomers (48).

11.7.5 Additives and Impurities

A very common problem related to polymers is that of determining additives and impurities in the monomer feed or the finished polymer. Additives in polymers may be determined by a variety of chromatographic techniques including GC, HPLC, thin-layer chromatography (TLC), and supercritcal-fluid chromatography (SFC). For some analyses, GC may offer significant advantages over other techniques. For example, long-chain thioester antioxidants such as dilauryl-thio-diproprionate (DLTDP) or long-chain amide slip agents such as oleamide are not easily analyzd by HPLC because they do not have UV chromophores and they cannot easily be derivatized to compounds that can be determined by UV or fluorescence detection. These compounds are readily determined by gas chromatography using a flame ionization detector. Another advantage of GC is that the high resolving power of GC can be used to screen for many types of additives in a single chromatographic run.

Nagata and Kishioka have reported the analysis of additives in polyolefins and petroleum resin by capillary GC (49). As shown in Figure 11.9, 22 additivies were separated in less than 40 min on a 15-m × 0.53-mm-i.d. × 1.0-μm fused-silica capillary column coated with 14% cyanopropylphenylmethylpolysiloxane stationary phase. Additives were extracted from a polyolefin film sample with refluxing chloroform for 2 h. The crude extract contained a significant amount of polyolefin oligomers (wax) that overlapped the additives of interest in the gas chromatographic trace. To overcome this problem, the extract was evaporated to near dryness, then applied as a hexane solution to a Florisil column. The waxes were eluted from Florisil with hexane and the additives were recovered by elution with chloroform. Using this procedure, additives were obtained with at least 92% recovery from the Florisil, and many of the recoveries were in the range of 96–100%. Poorer recoveries were achieved when the extracts were applied to silica gel.

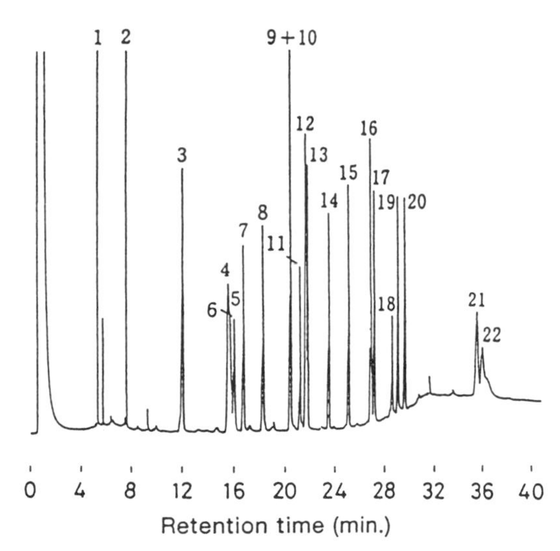

FIGURE 11.9 Gas chromatogram of standard solution of polymer additives: 1, BHT; 2, Sanol LS744; 3, palmitic acid amide; 4, Yoshinox 2246R; 5, oleic amide; 6, stearic acid amide; 7, Tinuvin 326; 8, Tinuvin 327; 9, Yoshinox BB; 10, UV 531; 11, erucic amide; 12, Tinuvin 120; 13, Yoshinox SR; 14, Sanol LS770; 15, Irgafos 168; 16, Irganox 1076; 17, DLTDP; 18, Ultranox 626; 19, Topanol CA; 20, DMTDP; 21, DSTDP; 22, Irganox 1330. (From Reference 49 with permission of Dr. Alfred Huethig Verlag GmbH.)

Castle et al. have used GC to determine epoxidized soya bean oil (ESBO) additives in various plastics (50). Epoxidized seed and vegetable oils are used as plasticizers, secondary heat stabilizers, and lubricants in such polymers as polystyrene, poly(vinyl chloride) (PVC), and poly(vinylidene chloride) (PVDC). Determination of the level of these additives is important in order to assess the level of protection they provide to the finished plastic, as well as to detemine their potential for migration when plastics containing epoxidized oil additives are used in food contact applications.

The first step in determining ESBO in plastics involves a simultaneous extraction and transmethylation. This was accomplished by adding sodium methoxide in methanol to the plastic sample and heating at 60°C for 2 h with occasional shaking. A dioxolane derivative was then formed by adding trimethylpentane, cyclopentanone, and boron trifluoride etherate and allowing the reagents to briefly react. The reaction was quenched by addition of aqueous sodium chloride. The gas chromatographic analysis of the organic phase was carried out using split (30:1) injection. Injections were made onto a 25-m × 0.22-mm-i.d. CP SIL 5CB fused-silica column (Chrompack) using a hydrogen carrier-gas flowrate of 2 mL/min. The temperature was programmed from 140°C at 20°C/min to a final temperature of 300°C, where it was held for 5 min. Typically, four key dioxolane derivatives were found in the chromatograms, resulting from monoepoxidized methyl octandecanoate, monoepoxidized methyl eicosanoate, and two isomers of diepoxidized methyl octadecanoate. Based on information provided by the manufacturers, the authors found 60–92% retention of the original level of ESBOs in various finished plastics. ESBO contents covered a wide range, varying from as low as <0.1% in PVC cooking oil bottles to 27.6% in a PVC gasket used in a fish paste jar.

11.7.6 Other Techniques

In the previous section of this chapter the problem of migration of monomers or additives into food was briefly considered. It is evident that polymers and plastics are used in a wide variety of forms for food contact applications. These applications frequently involve food or other products in which flavor or odor is a key property of the product. In many cases, the polymeric packaging has the potential to adsorb key flavors or aromas. The adsorptive properties of polymers can be studied by gas chromatography.

An example of this type of application involves measurement of the composition and quantity of flavor absorbed from dentifrices in contact with polymeric packaging. Tavss et al. have used headspace GC to measure the absorption of flavor from dentifrices (51). Disks of the material to be tested were punched out of a sheet and inserted into the vial caps for 9-mL headsapce vials so that they formed a septum in contact with a solution of flavor oil. The vials were inverted and kept at 22° or 32°C for a period of 4 weeks. The disks were then removed and thoroughly cleaned of any oil adhering to the surface. A smaller disk was then punched out of the sample and placed in a second sealed headspace vial. Following a 30-min equilibration at 90°C, a headspace aliquot was injected onto the capillary column. The conditions for this analysis included a 30-m × 0.32-mm-i.d. × 0.25-μm film thickness Suplecowax fused-silica column with helium carrier gas, 20:1 split ratio, and a temperature program holding at 50°C for 2 min followed by a 10°C/min program up to a final temperature of 150°C.

Using the conditions described above, the authors were able to demonstrate that the desorption of flavor from the packaging material follows first-order kinetics. Some samples were observed to lose up to 28% of their original flavor components after contact with a polymeric film. The advantage of this technique is that it can be applied quickly to a variety of potential packaging materials to determine which will have the lowest lost of flavor components.

Marin et al. have studied the effects of plastic polymers on the aroma character of orange juice (52). The containers studied were made from six or seven-layer laminates made of polyolefins, paper, paperboard, and aluminum foil. *d*-Limonene is a well-known aroma constituent of orange juice that was measured by GC with flame ionization detection. Other aroma compoments were measured by olfactometry using a method termed CharmAnalysis (53, 54). In this technique, a trained observer sniffed the effluent from the gas chromatograph and recorded a response to each aroma as it eluted from the column. These responses were recorded as a GC-olfactometry response curve. Comparisons were made of three-fold serial dilutions of each orange juice sample to produce a Charm chromatogram. To enhance the chromatograms for volatile components, orange juice samples were extracted into Freon 13 and ethyl acetate.

Two polymeric materials were tested for absorption of aroma components: a low-density polyethylene resin and a commercial resin, Surlyn (Dupont). The resins were fabricated into flat, round disks that were then placed in an incubated sample of orange juice. Samples of the juice were periodically taken and measured for limonene by GC-FID and for other volatiles by GC-olfactometry. Approximately 70–80% of the limonene was absorbed by the polymers in a 24-h period. Fifteen aroma volatiles were studied by olfactometry, and only two components were found to have significant decreases after exposure to the plastics.

The above application is typical of the problems encountered when one is trying to evaluate the components associated with an odor or aroma. It is often the case that components found in a chromatographic trace from a flame ionization detector may have no correlation to the actual odor of the sample. Therefore, the only sure way to determine which components actually contribute to the aroma is to sniff each component as it elutes from the chromatographic column.

REFERENCES

1. T. Visser and M. J. Vredenbregt, *Vib. Spectrosc.*, **1**, 205 (1990).
2. D. T. Williams, Q. Tran, P. Fellin, and D. A. Brice, *J. Chromatog.*, **549**, 297 (1991).
3. S. J. Stout and A. R. DaCunha, *J. Chromatog.*, **504**, 429 (1991).
4. J. V. Johnson, R. A. Yost, P. E. Kelly, and D. C. Bradford, *Anal. Chem.*, **62**, 2162 (1990).
5. W. V. Ligon, Jr., and H. Grade, *Anal. Chem.*, **63**, 255 (1991).
6. D. R. Lloyd, T. C. Ward, H. P. Schreiber, *Inverse Gas Chromatography, ACS Symposium Series 391*, Toronto, Canada, 1988.
7. P. C. Hiemenz, *Polymer Chemistry*, Dekker, New York, 1984.
8. H. S. Kaufman, *Introduction to Polymer Science and Technology*, Wiley, New York, 1977.
9. J. A. Manson and L. H. Sperling, *Polymer Blends and Composites*, Plenum, New York, 1976.
10. B. Ke (Ed.), *Newer Methods of Polymer Characterization*, Interscience, New York, 1964.
11. G. E. Ham, *Copolymerization*, Interscience, New York, 1964.
12. P. J. Flory, *Principles of Polymer Chemistry*, Cornell University Press, Ithaca, NY, 1953.
13. R. T. Morrison and R. N. Boyd, *Organic Chemistry*, 6th ed., Prentice Hall, Englewood Cliffs, NJ, 1992, p. 185.
14. R. T. Morrison and R. N. Boyd, *Organic Chemistry*, 6th ed., Prentice Hall, Englewood Cliffs, NJ, 1992, pp. 125–128.
15. H. R. Allcock and F. W. Lampe, *Contemporary Polymer Chemistry*, Prentice Hall, Englewood Cliffs, NJ, 1981.
16. J. F. Sullivan, in *Modern Practice of Gas Chromatography*, 2d ed., R. L. Grob (Ed.), Wiley, New York, 1985, pp. 724–728.
17. C. G. Smith, P. B. Smith, A. J. Pasztor, Jr., M. L. McKelvey, D. M. Meunier, S. W. Froelicher, and A. S. Ellaboudy, *Anal. Chem.*, **65**, 217 (1993).
18. Hewlett Packard Users Catlog, 1993, pp. 23–24.
19. A. Krishen, *Anal. Chem.*, **43**, 1130 (1991).
20. V. Rochelli and A. Razzini, *Proc. Int. Symp. Capillary Chrom.*, **6**, 419 (1985).
21. B. Kolb, P. Popisil, and M. Auer, *Chromatographia*, **19**, 113 (1984).
22. H. Hayashi and S. Matsuzawa, *J. Appl. Polym. Sci.*, **31**, 1709 (1986).
23. Supelco, Supelco Park, Bellefonte, PA 16823-0048.
24. Alltech Associates, 205 Waukegan Road, Deerfield, IL 60015.
25. A. Horna and J. Taborsky, *J. Chromatog.*, **325**, 367 (1985).
26. A. Horna and J. Taborsky, *J. Chroamtog.*, **360**, 89 (1986).
27. L. R. Whitlock and S. Siggia, *Sep. Purif. Methods.*, **3**, 299 (1974).
28. P. C. Rahn and S. Siggia, *Anal. Chem.*, **45**, 2336 (1973) 29.

29. P. A. Vimalasiri, J. K. Haken, and R. P. Burford, *J. Chromatog.*, **355**, 141 (1986).
30. P. A. Vimalasiri, J. K. Haken, and R. P. Burford, *J. Chromatog.*, **362**, 391 (1986).
31. G. Allen and J. Bevington, *Comprehensive Polym. Sci.*, **1**, 589 (1989).
32. M. R. Rao, T. V. Sebastian, T. S. Radhakrishnan, and P. V. Ravindran, *J. Appl. Polym. Sci.*, **42**, 753 (1991).
33. T. P. Wampler and E. J. Levy, *J. Anal. Appl. Pyrolysis*, **12**, 75 (1987).
34. R. S. Whiton and S. L. Morgan, *Anal. Chem.*, **57**, 778 (1985).
35. D. L. Fenter, R. L. Levy, and C. J. Wolf, *Anal. Chem.*, **44**, 43 (1972).
36. C. Meritt, R. E. Sacher, and B. A. Peterson, *J. Chromatog.*, **99**, 301 (1974).
37. S. A. Liebman and T. P. Wampler, *J. Chromatogr. Sci.*, **29**, 53 (1985).
38. S. H. Harris, P. E. Kreter, and C. W. Polley, *J. Cellular Plastics*, **24**, 486 (1988).
39. O. Fuchs, in *Polymer Handbook*, 3d ed., J. Brandrup and E. H. Immergut (Eds.), Wiley, New York, 1989, p. 379.
40. A. Zlatkis, J. W. Anderson, and G. Holzer, *J. Chromatog.*, **142**, 127 (1977).
41. W. Roorda, J. Verbeek, and H. Junginger, *J. Chromatog.*, **403**, 355 (1987).
42. M. A. Haney, D. W. Johnston, and B. H. Clampitt, *Macromolecules*, **16**, 1775 (1983).
43. H. Ohtani, S. Tsuge, and T. Usami, *Macromolecules*, **17**, 2557 (1984).
44. T. Usami, Y. Gotoh, S. Takayama, H. Ohtani, and S. Tsuge, *Macromolecules*, **20**, 1557 (1987).
45. F. Poy, L. Cobelli, S. Banfi, and F. Fossati, *J. Chromatog.*, **395**, 281 (1987).
46. ASTM Method F 1308-90, *Annual Book of ASTM Standards*, American Society for Testing and Materials, Philadelphia, 1990.
47. J. K. Haken, *Trends in Analytical Chemistry*, **9**, 14 (1990).
48. J. K. Haken, R. P. Burford, and P. A. D. T. Vimalasiri, *J. Chromatog.*, **349**, 347 (1985).
49. M. Nagata and Y. Kishioka, *HRC*, **14**, 639 (1991).
50. L. Castle, M. Sharman, and J. Gilbert, *J. Chromatog.*, **437**, 274 (1988).
51. E. Tavss, J. Santalucia, R. Robinson, and D. Carroll, *J. Chromatog.*, **438**, 281 (1988).
52. A. Marin, T. Acree, J. Hotchkiss, and S. Nagy, *J. Agric. Food Chem.*, **40**, 650 (1992).
53. T. Acree, J. Barnard, and D. Cunningham, *Food Chem.*, **14**, 273 (1984).
54. D. Cunningham, T. Acree, J. Barnard, R. Butts, and P. Braell, *Food Chem.*, **19**, 137 (1986).

CHAPTER TWELVE

Clinical Applications of Gas Chromatography

JUAN G. ALVAREZ

Beth Israel Hospital
Harvard Medical School

Modern Practice of Gas Chromatography, Third Edition. Edited by Robert L. Grob.
ISBN 0-471-59700-7

12.1 INTRODUCTION

In this chapter, the current clinical applications of gas chromatography are described. Rather than providing a compilation of all the existing methodologies for the gas chromatographic analysis of the various drugs described herein, the author has selected those procedures that are more frequently used in clinical laboratories and that provide both high resolution and high speed of analysis. The pharmacological properties, including pharmacological effects, mechanism of action, and absorption and elimination, corresponding to the drugs under analysis are described, followed by a discussion of the procedures used for their respective gas chromatographic analysis. In this way, the reader will have the opportunity to become familiar with the chemical properties, pharmacodynamics, and pharmacokinetics of each class of compounds before a detailed description of the analytical procedure is given.

12.2 AMPHETAMINES

Amphetamines comprise a group of adrenergic agonists that includes amphetamine, methamphetamine, methylphenidate, pemoline, ephedrine, ethylnorepinephrine, phenmetrazine, benzphetamine, phendimetrazine, phenmetrazine, diethylpropion, mazindol, fenfluramine, and phenylpropanolamine. Under federal law regulation amphetamines are considered schedule II drugs. Classical therapeutic applications of amphetamines include the treatment of obesity and narcolepsy. These drugs were found to produce weight loss by suppressing appetite (anorexia) rather than by increasing energy expenditure. Adverse effects of treatment include the potential for drug abuse and habituation, serious worsening of hypertension, sleep disturbances, palpitations, dry mouth, and depression.

In this section, the generic concept of sympathomimetic drugs and their pharmacological behavior is introduced, followed by the gas chromatographic analysis of amphetamines.

12.2.1 Pharmacological Considerations

Chemistry and Structure. β-Phenylethylamine can be viewed as the parent compound of the sympathomimetic amines, consisting of a benzene ring and an ethylamine side chain. The structure allows substitutions to be made on the aromatic ring, the α and β carbons, and the terminal amino group, to yield a variety of compounds with sympathomimetic activity. Norepinephrine, epinephrine, isopropterenol, and a few other agents have OH groups substituted in the 3 and 4 positions of the benzene ring. Since O-dihydroxybenzene is also known as catechol, sympathomimetic amines with these OH substitutions in the aromatic ring are designated catecholamines. Since substitution of polar groups on the phenylethylamine structure makes the resultant compound less lipophilic, alkyl-substituted compounds, including the amphetamines, cross the blood–brain barrier more readily and have more activity in the central nervous system.

Pharmacological Effects. The sympathetic nervous system is involved in the homeostatic regulation of a wide variety of functions, including heart rate, force of cardiac contraction, vasomotor tone, blood pressure, bronchial airway tone, and carbohydrate and fatty acid metabolism. Stimulation of the sympathetic nervous system normally occurs in response to physical activity, psychological stress, and generalized allergic reactions. As part of the response to stress, the adrenal medulla is also stimulated, resulting in elevation of the concentrations of epinephrine and norepinephrine in the circulation. As might be expected, naturally occurring sympathomimetic amines and drugs that mimic their actions comprise one of the more extensively studied groups of pharmacological agents. Some of the functions ascribed to catecholamines are summarized in Table 12.1.

Mechanism of Action. In 1948 Ahlquist (1) first proposed that there was more than one adrenergic receptor. He proposed the designations α and β for receptors in smooth muscle where catecholamines produce excitatory and inhibitory responses, respectively. Stimulation of α_1 receptors results in activation of phospholipase C mediated by a G protein, and the hydrolysis of membrane-bound polyphosphoinositides with the generation of the second messengers, diacylglycerol and inositol-1,4,5-triphosphate, which stimulate the release of Ca^{2+} from intracellular stores (2). Sympathomimetic

TABLE 12.1 Pharmacological Actions of Sympathomimetic Drugs

Peripheral excitatory	*Contraction* of smooth muscle in vessels supplying skin and mucous membranes, gland cells
Peripheral inhibitory	*Relaxation* of smooth muscle in the wall of the gastrointestinal tract, bronchial tree, and blood vessel supplying the skeletal muscle
Cardiac excitatory	*Increase* in the heart rate and force of contraction
Metabolism	*Increase* in rate of glycogenolysis in liver and muscle; liberation of fatty acids from adipose tissue
Endocrine system	*Modulation* of secretion of insulin, renin, and pituitary hormones
Central nervous system	*Stimulation* of respiration and increase in wakefulness, psychomotor activity, and a reduction of appetite

drugs influence both α and β receptors in the target tissue, but the ratio of the α and β activity varies widely between drugs in a continuous spectrum, from an almost pure α activity (phenylephrine) to an almost pure β activity (isopropterenol). An important factor in the response of any cell or organ to sympathomimetic amines is its density and proportion of α and β receptors. For example, norepinephrine has relatively little capacity to increase bronchial air flow since the receptors in bronchial smooth muscle are largely of the β_2 type. In contrast, isopropterenol and epinephrine are potent bronchiodilators. Cutaneous blood vessels possess α receptors almost exclusively; thus epinephrine and norepinephrine cause marked constriction.

Absorption and Elimination. Epinephrine and norepinephrine are not effective by oral administration because of their rapid conjugation and oxidation in the gastrointestinal mucosa and liver. Absorption from subcutaneous tissues occurs slowly because of local vasoconstriction. Absorption is more rapid after intramuscular than after subcutaneous injection. Epinephrine is rapidly inactivated in the body. The liver, which is rich in both of the enzymes responsible for the metabolization of circulating epinephrine (catecholamine methyl transferase and monoamino oxidase), is particularly important in this regard. Isopropterenol is also given parenterally or as an aerosol and it is rapidly metabolized in the liver, but unlike epinephrine and norepinephrine, it is not inactivated by the enzyme monoamino oxidase. Amphetamines in the form of sulfate salts are well absorbed orally.

12.2.2 Gas Chromatographic Analysis

Sample Preparation. The procedure described in this section is a modification of that reported by Lillsunde and Korte (3). Aliquots of 1 mL of serum or blood and control solutions are mixed for 30 sec with 1 mL of NaOH and 5 mL of toluene containing 30 ng/mL of 4-chloramphetamine as an internal standard. Following centrifugation, the organic layer is transferred to a glass conical tube, 1 mL of 0.2 M H_2SO_4 added, Vortex mixed for 30 sec, and centrifuged at 600 *g* for 5 min, and the organic layer is discarded. Aliquots of 1 mL of 0.5 M KOH are added to the aqueous layer and extracted with 1 mL of toluene. The toluene layer is transferred to a glass tube and 5 μL of the derivatizing agent heptafluorobutyric anhydride (HFBA) is added. Immediately 1 mL of 10% $NaHCO_3$ is added and centrifuged at 600 *g* for 5 min, and the upper layer is transferred to a glass conical tube.

Analytical Procedure. A Hewlett–Packard GC/MS 5990 equipped with a packed-column, 2% SP-2110/1% SP-2510 on 100/120 Supelcoport, 1 m, is used for identification. The oven temperature is held initially at 120°C for 1 min, and increased thereafter up to 210°C at a rate of 10°C/min. A Hewlett–Packard capillary gas chromatograph 5880 with an EC and an NP detector is used for quantification with a fused-silica capillary column, SE-54 (25 m × 0.3 mm, film thickness 0.17 μm). The oven temperature is programmed from 120°C (held for 1 min) and increased to 280°C at a rate of 10°C/min.

Quantification of Amphetamines. The retention times of nine different amphetamines, are shown in Table 12.2. The coefficient of variation for the derivatization procedure is approximately 6.5%, as calculated using repeated serum analysis at the level of 0.5 μg/mL. The lower limit of detection is 5 ng/mL with linear detector response extended to 2 μg/mL. The mass spectra of the HFB derivatives allow identification of the various amphetamines. Selective base peak ion monitoring is used for identification purposes (Table 12.2). The base peak of alkyl-substituted aromatic compounds contains a benzyl radical at m/z 91 for amphetamines. Another fragmentation ion derives from the cleavage of the C–N bond with m/z at 118 for the ring-unsubstituted amphetamines, amphetamine and methamphetamine.

A new procedure has been recently reported by Gjerde et al. (4) where amphetamine and methamphetamine in blood are analyzed as the perfluorooctanyl derivatives by GC/MS. Aliquots of 3 μL of the derivatized samples dissolved in butylacetate are injected into a Hewlett–Packard 5890 gas chromatograph equipped with a 5970A mass selective detector. The column used is a 12-m × 0.2-mm-I.D. methylsilicone, 0.33 μm thickness from Hewlett–Packard. The selected ions are monitored in the electron impact mode at m/z 118 and 440 for the amphetamine PFOC derivative, at an m/z of 118 and 454 for the methamphetamine PFOC derivative, and at an m/z of 121 and 443 for the amphetamine-d_3–PFOC derivative. Amphetamine-d_3 was used as the internal standard. The concentration of the various amphetamines in blood was calculated using amphetamine-d_3 as an internal standard. The ion ratios 118/121 and 440/443 are used to compute amphetamine concentrations, and the ion ratios 118/121 and

TABLE 12.2 Retention Times and Base Peaks of Amphetamines

Substance	Retention Time (min)	Base Peak (m/z)
Amphetamine	3.2	118
Methamphetamine	3.6	118
N-Ethylamphetamine	4.1	91
4-Methoxyamphetamine	5.8	121
3,4-Methylenedioxyamphetamine	7.5	135
2,5-Dimethoxyamphetamine	6.9	178
2,5-Dimethoxy-4-ethylamphetamine	7.8	179
3,4,5-Trimethoxyamphetamine	8.8	181
4-Bromo-2,5-dimethoxyamphetamine	10.2	229

TABLE 12.3 Accuracy and Precision Measurements of Amphetamines

Substance	Accuracy	Precision (RSD)
Amphetamine	0.49	1.5
Methamphetamine	0.50	2.2

454/443 are used to compute methamphetamine concentrations. The lower limit of detection is 0.081 μmol/L (11 μg/L) for amphetamine and 0.085 μmol/L (13 μg/L) for methamphetamine. The limits of quantification are 22 μg/L for amphetamine and 34 μg/L for methamphetamine. Linear detector response is obtained between 0.1 and 20.0 μmol/L for both amphetamine and methamphetamine. The accuracy and precision of the analysis are shown in Table 12.3. Precision is expressed as the relative standard deviation.

12.3 INHALATIONAL ANESTHETICS

12.3.1 Pharmacological Considerations

Chemistry and Structure. Halothane (Fluothane) is 2-bromo-2-chlor-1,1,1-trifluoroethane. Mixtures of halothane with air or oxygen are not flammable nor explosive. Enflurane is 2-chloro-1,1,2-trifluoroethyl difluoromethyl ether. It is a clear, colorless, nonflammable liquid with a mild, sweet odor. It is extremely stable chemically. It does not attack aluminum, tin, brass, iron, or copper. Enflurane is soluble in rubber and this property may prolong induction and recovery. Isoflurane is 1-chloro-2,2,2-trifluoroethyl difluoromethyl ether. The chemical and physical properties of isoflurane are similar to those of its isomer enflurane. It is not flammable in air or oxygen. Its vapor pressure is high, and delivery of safe concentrations necessitates the use of a precise vaporizer. Methoxyflurane is 2,2-dichloro-1,1-difluoroethyl methyl ether. It is a clear, colorless liquid with a sweet, fruity odor. It is stable in the presence of soda lime and is nonflammable and nonexplosive in air or oxygen, in anesthetic concentrations. It is very soluble in rubber. Sevoflurane, fluromethyl-1,1,1,3,3,3-hexafluoroisopropyl ether, is a relatively new, noninflammable inhalation anesthetic agent. Nitrous oxide, N_2O, is a colorless gas without appreciable odor or taste. Nitrous oxide has relatively low solubility in blood, the blood/gas partition coefficient at 37°C being 0.47 (Table 12.4).

Pharmacological Effects. Inhalational anesthetics currently used for general anesthesia include halothane, enflurane, isoflurane, methoxyflurane, and nitrous oxide. Nitrous oxide is a gas at normal ambient temperature and pressure, whereas the other four agents are volatile organic liquids. These

TABLE 12.4 Partition Coefficients of Inhalational Anesthetic Agents

Anesthetic	Oil/Gas, λ	Blood/Gas, λ
Methoxyfluorane	970	12.0
Halothane	224	2.3
Enflurane	98	1.9
Isoflurane	99	1.4
Nitrous oxide	1.4	0.47

agents share the common denominator of inducing rapid loss of consciousness that progresses to the absence of perception of all sensations, or anesthesia. During general anesthesia produced with an inhalational agent, the depth of anesthesia varies directly with the partial pressure of anesthetic agent in the brain, and the rates of induction and recovery depend on the rate of change of partial pressure in this tissue. The partial pressure of the anesthetic agent in the brain is always approaching that in arterial blood. The factors that determine the partial pressure of the anesthetic agent in the arterial blood and brain are (1) the concentration of the anesthetic agent in the inspired gas; (2) pulmonary ventilation delivering the anesthetic to the lungs; (3) transfer of the gas from the alveoli to the blood flowing through the lungs; and (4) loss of the agent from the arterial blood to all the tissues of the body.

Mechanism of Action. The molecular mechanism responsible for the anesthetic effect of inhalational anesthetics is related to their lipid solubility or hydrophobicity. NMR and EPR studies indicate that inhalational anesthetic agents act by causing a local disordering of the lipid matrix (5–7). It is hypothesized that fluctuations of volume in biological membranes are sufficiently large to be important in the regulation of the structural state of membrane-bound proteins, i.e., their state of aggregation, and therefore of their functional properties (8, 9). As inhibitors of such fluctuations, anesthetics could readily influence the fluxes of ions, which are crucial determinants of neuronal excitability or other functions of membranes that are determined by the proteins that function in the milieu of a dynamic lipid matrix (10).

Absorption and Elimination. When a constant partial pressure of anesthetic gas is inhaled, the corresponding partial pressure in arterial blood approaches that of the agent in the inspired mixture. For nitrous oxide the arterial partial pressure reaches 90% of the inspired pressure in about 20 min. With diethyl ether the same level is attained only after several hours. The solubility of the agent in blood is expressed as the blood/gas partition coefficient λ, which represents the ratio of anesthetic concentration in blood to anesthetic concentration in a gas phase when the two are in equilibrium. The λ value is as high as 12 for very soluble agents such as methoxyfluorane or diethyl ether, and as low as 0.47 for relatively insoluble agents such as nitrous oxide. The blood/gas partition coefficients for the commonly used inhalational anesthetics are given in Table 12.4.

The inhalational anesthetic agents are metabolized in the body to a variable extent. With most agents the amount metabolized is small. However, up to 15% of halothane and 70% of methylflurane is metabolized to various intermediate compounds, and in some cases to ionized halogens (11). The importance of the metabolism of anesthetic agents is not in the termination of their action but in that the metabolites produced may be responsible for their toxic aftereffects. Additional small losses of anesthetic gases from the body occur by diffusion across skin and mucous membranes, and by means of urinary excretion of the agent or its breakdown products (11).

12.3.2 Gas Chromatographic Analysis

Sample Preparation. Aliquots of 1 mL of the plasma sample, standards, or quality-control material are quickly transferred to an empty 10-mL red-top headspace tube and capped. A 19-gauge 1-in. needle (Becton Dickinson) is then inserted through the stopper. Another needle is inserted through the stopper of the stock standard. These serve as cannulas for the finer-gauged needle of a 50-μL fixed needle syringe (Hamilton, Reno, NV). The syringe needle is passed through the cannula into the liquid phase of the stock standard. A 40-μL aliquot of the stock standard is then quickly transferred to the working standard headspace tube. This cannula system allows rapid transfer without syringe needle damage. A 20-μL saturated solution of sevoflurane and halothane are added to each standard, sample, or quality control tube using the cannula system described above. All tubes are Vortex mixed for 15 sec and placed in a 25°C water bath. Saturated stock standards of sevoflurane and halothane are prepared by adding 100 μL of each to separate 10-mL draw Vautainer red-top evacuated blood collection tubes containing about 10 mL deionized water. These are then capped, shaken vigorously, placed in a 25°C water bath for 30 min, and centrifuged at 2000 *g* for 2 min at 4°C.

Analytical Procedure. A Varian 2100 gas chromatograph with flame ionization detection is used for analysis. The oven is operated isothermically at 165°C, and the detector and injector ports are set at 200°C. A 6-f, 0.25-in. o.d., 2-mm-i.d. glass column, packed with 80- to 100-mesh Porapak S (Supelco, Bellefonte, PA), is used for analysis. Prepurified nitrogen is used as the carrier gas at a flowrate of 45 mL/min, and compressed air (300 mL/min) and prepurified hydrogen (45 mL/min) are used as flame gases. All headspace equilibrations are performed at 25°C. Peak-height ratios equilibrate rapidly after the vortex mixing step. For routine analysis, samples and

TABLE 12.5 Retention Times of Inhalational Anesthetics

Substance	Retention Time (min)
Sevoflurane	4.0
Halothane	6.1

standards are allowed to equilibrate for 1 h. Aliquots of 1 mL of headspace vapor are injected with the use of a Pressure-Lok series A 2-mL glass syringe (Dynatech Precision Sampling, Baton Rouge, LA). Direct injections are also performed to compare results with headspace analysis. A solvent flush technique is used whereby 0.5 μL of water is introduced into a 5-μL Hamilton syringe, followed by 0.5 μL of air, 2.0 μL of sample, and finally 1 μL of air. These contents are emptied directly onto the gas chromatograph. Quality-control specimens are prepared by adding 50–500 μL of liquid sevoflurane or halothane to about 1 L of either deionized water or outdated blood bank plasma in a 1-L plastic bottle, mixed by inversion, and refrigerated.

Quantification of Inhalational Anesthetics. The retention times of sevoflurane and halothane following gas chromatographic analysis are shown in Table 12.5. Neither ethanol nor acetone volatiles, which are sometimes present in blood specimens, are found to interfere with the assay, since they have much shorter retention times. The within-run precision for sevoflurane and halothane as measured by the coefficient of variarion is about 2–3%. There is linear detector response between 12 and 944 μg/mL, with a corrclation coefficient of .9996. The use of saturated aqueous solutions of sevoflurane and halothane greatly enhances the precision of internal and external standardization. Reproducible stock standards are easily prepared without precise gravimetric or volumetric manipulations. Headspace analysis is favored by the low liquid/vapor partition coefficients and the large injection volume. This improves precision and decreases interfering peaks compared with direct injection.

12.4 TRICYCLIC ANTIDEPRESSANTS

Imipramine, amitriptyline, their *N*-demethyl derivatives, and other closely related compounds are the drugs currently most widely used for the treatment of major depression. Because of their structure, they are often referred to as "tricyclic" antidepressants. Their efficacy in alleviating major depression is well established, and support for their use in other psychiatric disorders is growing.

12.4.1 Pharmacological Considerations

Chemistry and Structure. In addition to the diabenzazepines, imipramine, and desipramine, there are amitriptyline and its *N*-demethylated metabolite nortriptyline (dibenzocycloheptadienes) as well as doxepin, a dibenzoxepine, and potrityline. Additional structurally related agents approved for general use in the United States are trimipramine, a benzodiazepine; maptroline, containing an additional methylene bridge across the central six-carbon ring; and amoxapine, a dibenzoxazepine with mixed neuroleptic and antidepressant properties. Since these agents all have a three-ring molecular core and produce therapeutic responses in most patients with major depression, the trivial name tricyclic antidepressants is used for this group.

Pharmacological Effects. Administration of therapeutic doses of a tricyclic antidepressant to depressed patients results in an elevation of the mood. About 2–3 weeks should be allowed for the antidepressant to exert its effect. For this reason, the tricyclic antidepressants cannot be prescribed on an as-needed basis. With some antidepressants, sedative or antianxiety effects may appear within a few days of treatment. The manner in which these agents relieve the signs of depression is not clear. However, maniac excitement as well as euphoria and insomnia can be induced in some patients, which contributes to the conclusion that antidepressant agents have clinically important mood-elevating actions.

Mechanism of Action. The administration of a tricyclic antidepressant produces an immediate reduction in the firing rate of neurons containing norepinephrine and a decrease in their turnover rate. These changes are thought to be a consequence of blockade of the uptake of monoamines by neurons with a resultant increase in their action upon presynaptic α_2-adrenergic receptors that serve to regulate the excitability of and transmitter release from monoaminergic neurons. Tricyclic antidepressants also act as antagonists at receptors for various neurohormones; these include moderate to high affinity at muscarinic cholinergic (12), α_1-adrenergic (13), and both H_1- and H_2-histaminergic receptors (14).

Absorption and Elimination. Imipramine and other tricyclic antidepressants are well absorbed after oral administration. Once absorbed, these lipophilic drugs are widely distributed. They are strongly bound to plasma protein and to constituents of tissues. The concentration of these drugs in plasma, that have been suggested to correlate best with satisfactory antidepressant responses, range between 50 and 300 ng/mL. Toxic effects can be expected when their concentrations in plasma rise to 1 μg/mL or even less (15). The tricyclic antidepressants are oxidized by hepatic microsomal enzymes followed by conjugation with glucoronic acid. The major route of metabolism of imipramine is to the active product desipramine. Biotransformation of either compound occurs largely by oxidation to 2-hydroxy metabolites, which retain some ability to block the uptake of amines and may have particularly prominent cardiac depressant actions (16). In contrast, amitriptyline and nortriptyline undergo preferred oxidation at the 10 position. The conjugation of hydroxylated metabolites with glucoronic acid extinguishes any remaining biological activity.

12.4.2 Gas Chromatographic Analysis

Sample Preparation. Aliquots of 1 mL of urine, plasma, or blood containing tricyclic antidepressants are mixed with 1 mL of 1 M sodium bicarbonate and 2 mL of water, and loaded onto a Sep-Pak cartridge. When blood is used, the mixture is centrifuged at 600 *g* for 8 min and the supernatant is used for the next step. The sample solution is poured into the preconditioned cartridges at a flowrate of 5 mL/min. Sep-Pak cartridges were preconditioned with 10 mL of chloroform-2-propanol (9:1, v/v), 10 mL of acetonitrile, and 10 mL of distilled water. Then 10 mL of water is added followed by 3 mL of chloroform-2-propanol (9:1, v/v) to elute the antidepressants. The eluate consists of an organic lower phase and an aqueous upper phase. The latter is discarded and the organic phase is evaporated to dryness. The residue is dissolved in 100 μL of methanol.

Analytical Procedure. Aliquots of 1 μL of the sample extract dissolved in methanol are injected in the splitless mode at 100°C into a Shimadzu GC-15A instrument equipped with SID system with a fused-silica SPB-1 capillary column (30 m × 0.32 mm i.d., film thickness 0.25 μm, Supelco, Bellefonte, PA) and a split splitless injector. Column temperature is 100–280°C at a rate of 6°C/min and the injection temperature is 200°C. SID conditions included a platinum emitter current of 2.2 A, an emitter temperature of 600°C, and a ring electrode bias voltage of +200 V with respect to the collector electrode.

Quantification of Tricyclic Antidepressants. The retention times corresponding to the antidepressants imipramine, amitriptyline, trimipramine, and chlorimipramine are shown in Table 12.6. Five nanograms of each were added to 1 mL of either urine, plasma, or blood and extracted with a

TABLE 12.6 Retention Times of Tricyclic Antidepressants

Substance	Retention Time (min)
Amitriptyline	8.90
Imipramine	25.3
Trimipramine	26.1
Chlorimipramine	32.2

Sep-Pak cartridge as indicated above. Recoveries are above 60%. Linear detector response is obtained between 10 and 80 pg in the injected volume (10–80 ng/mL). The lower limit of detection is 0.5 to 1.0 ng/mL of sample. SID detection provides an extremely sensitive alternative to the standard thermoionic ionization detector (TID) (17).

12.5 ANTIEPILEPTIC DRUGS

12.5.1 Pharmacological Considerations

Chemistry and Structure. The useful antiepileptic agents belong to several chemical classes. Most of the drugs introduced before 1965 are closely related in structure to phenobarbital, the oldest member of this therapeutic class. These include the hydantoins: phenytoin, mephenytoin, and ethotoin; the deoxybarbiturate primidone; the oxazolidinediones trimethadione and paramethadione; and the succinimides ethosuximide, methsuximide, and phensuximide. The agents introduced after 1965 include the benzodiazepines clonazepan and clorazepate, an iminostilbene, carbamazepine, and a branched-chain carboxylic acid, valproic acid.

Pharmacological Effects. There are two general ways in which drugs might abolish or attenuate seizures: through effects on pathologically altered neurons of seizure foci to prevent or reduce their excessive discharge, and through effects that would reduce the spread of excitation from seizure foci and prevent detonation and disruption of function of normal aggregates of neurons. Most of the antiepileptic drugs available exert their effect through the second mechanism, since all modify the ability of the brain to respond to various seizure-evoking stimuli.

Mechanism of Action. Antiepileptic drugs exert a stabilizing effect on excitable membranes of a variety of cells, including neurons and cardiac myocytes. They can decrease resting fluxes of Na^+ as well as Na^+ currents that flow during action potentials or chemically induced depolarization (18). As a result, antiepileptic drugs suppress episodes of repetitive neuronal firing that are induced by passage of intracellular current. Such effects can be achieved at concentrations of the drug below 10 μM. At concentrations in excess of 10 μM, phenytoin delays the activation of outward K^+ currents during action potentials in nerves, leading to an increased refractory period (19). Phenytoin can also reduce the size and duration of Ca^{2+}-dependent action potentials in cultured neurones at about 20 μM (20). On the other hand, the ability of the barbiturate phenobarbital to exert its anticonvulsant effect is thought to be mediated by producing a reduction in γ-aminobutyric acid (GABA) and Ca^{2+}-dependent release of neurotransmitters.

Absorption and Elimination. The pharmacokinetics of hydantoins are markedly influenced by their limited aqueous solubility and by dose-dependent elimination. Phenytoin is a weak acid with a pK_a of about 8.3; its aqueous solubility is limited, even in the intestine. Absorption of phenytoin after oral ingestion is slow, and significant differences in bioavailability of oral pharmaceutical preparations have been detected. Phenytoin is extensively bound to plasma proteins, mainly albumin. Less than 5% is excreted unchanged in urine. The major metabolite, the parahydroxyphenyl derivative, is inactive. Oral absorption of phenobarbital is complete but somewhat slow. Peak concentrations in plasma occur several hours after a single dose. It is 40–60% bound to plasma proteins and bound to a similar extent in tissues, including brain. The pK_a of phenobarbital is 7.3, and up to 25% of a dose is eliminated by pH-dependent renal excretion of the unchanged drug. The remainder is inactivated by hepatic microsomal enzymes.

The deoxybarbiturate primidone is rapidly and almost completely absorbed after oral administration. Peak plasma concentrations are reached within 3 h after ingestion. Primidone is converted to two active metabolites, phenobarbital and phenylethylmalonamide (PEMA). Primidone and PEMA are bound to plasma proteins to a much lesser extent than phenobarbital. Approximately 40% of primidone is excreted unchanged in the urine. The succinimide ethosuximide is also rapidly and completely absorbed after oral administration, reaching plasma peak concentrations after 3 h of administration. It is not significantly bound to plasma proteins. Valproic acid is rapidly and completely absorbed after oral administration. Peak concentration in plasma is observed in 1 to 4 h. Its extent of binding to plasma proteins is about 90%. The majority of the drug is converted to the conjugate ester of glucoronic acid, while β-oxidation in the mitochondria accounts for the remainder. Some of these metabolites, notably 2-propyl-2-pentenoic acid and 2-propyl-4-pentenoic acid, are nearly as potent anticonvulsant agents as the parent compound. However, only 2-en-valproic acid accumulates in plasma and brain to a significant extent.

12.5.2 Gas Chromatographic Analysis

Sample Preparation. The methodology herein described is a modification of that reported by Volmut et al. (21). A 500-μL aliquot of plasma or urine is thoroughly mixed with 100 μL of 0.5 M HCl in a 1.5-mL polypropylene microvial, and 10 μL of an internal standard solution in methanol containing a mixture of valproic acid, ethosuximide, phenobarbital, primidone, carbamazepine, and phenytoin. A 2-mL disposable polyethylene syringe (9 mm i.d.) is packed with 200 mg of Silipor C18 and the column is preconditioned with 1 mL of methanol followed by 1 mL of distilled water or 0.5 M HCl. The pretreated sample is poured onto the column and allowed to flow through. Then the column is rinsed with two 1-mL portions of water and dried under vacuum. The drugs are eluted with 1 mL of methanol. A 50-μL volume of a 0.05 M KOH solution in methanol is added to 1 mL of eluate, and the solvent is evaporated to dryness at 40°C. Addition of KOH prior to

TABLE 12.7 Recoveries for Antiepileptic Drugs

Substance	% Recovery
Valproic acid	85.1
Caprilic acid	78.0
Ethosuximide	31.5
Phenobarbital	96.0
Primidone	75.0
Carbamazepine	88.0
Phenytoin	100.0

the evaporation step prevents the loss of volatile antiepileptic drugs from the sample. The residue is dissolved in 50 μL of a 0.05 M solution of HCl in methanol. The internal standard caprilic acid or 5-phenylhidantoin is added to the samples at a concentration of 20 μg/mL.

Analytical Procedure. A HP-5790A gas chromatograph instrument equipped with a split/splitless capillary inlet system and a flame ionization detector are used. A fused-silica capillary column with cross-linked 5% phenylmethylsilicone gum phase HP-5, 25 m × 0.20 mm i.d., 0.33 μm film thickness, is used. Nitrogen is used as the carrier gas at an inlet pressure of 100 kPa. The oven is operated isothermally at 60°C for 0.5 min after injection, heated at 30°C/min to 200°C, then at 10°C/min to 250°C, and then held at 250°C for 10 min. Aliquots of 2 μL are injected in the split mode at a split ratio of 1:20 and a septum purge rate of 1 mL/min. The temperatures of injector and detector are 240 and 300°C, respectively. Hexobarbital was used as an internal standard during the chromatographic run, but it is not added to serum samples because of the presence of a peak in the biological matrix with a close retention time.

Quantification of Antiepileptic Drugs. The percent recoveries relative to 5-(4-methylphenyl)-5-phenylhydantoin for the various antiepileptic drugs using this methodology are shown in Table 12.7. The retention times for the various antiepileptic drugs are shown in Table 12.8. The chromatographic run-to-run reproducibility for various concentrations of antiepileptic drugs is calculated by the relative standard method using MPPH as the internal standard and ranges from 7 to 9.9%. Analysis of the various antiepileptic drugs can be accomplished in approximately 30 min.

12.6 BLOOD ALCOHOL

Alcoholic beverages have been used since the dawn of history, beginning with fermented beverages of relatively low alcohol content. When the Arabs introduced the alambique in Europe in the Middle Ages as a means of

TABLE 12.8 Retention Times for Antiepileptic Drugs

Substance	Retention Time (min)
Valproic acid	6.1
Caprilic acid	6.4
Ethosuximide	7.1
Hexobarbital	13.1
Phenobarbital	14.2
Primidone	20.2
Carbamazepine	23.1
Phenytoin	23.5
MMPH	27.6

distilling alcohol, alchemists believed that alcohol was the long-sought elixir of life. Alcohol was therefore held to be a remedy for practically all diseases, as indicated by the term whisky (Gaelic: *usquebaugh*, meaning "water of life"). It is now recognized that the therapeutic value of ethanol is extremely limited and that chronic ingestion of excessive amounts is a major social and medical problem. Methanol (methyl alcohol or wood alcohol) is a common industrial solvent. It is also used as an antifreeze fluid, a solvent for shellac and some paints and varnishes, and a component of paint removers. As an adulterant, it renders unpotable and tax-free the ethanol that is used for cleaning, paint removal, and other purposes. Isopropanol, used for rubbing alcohol, in hand lotions, and in deicing and antifreeze preparations, is occasionally the cause of accidental poisoning.

12.6.1 Pharmacological Considerations

Pharmacological Effects. Alcohol is primarily a continuous depressant of the central nervous system. The first mental processes to be affected are those that depend on training and previous experience. Memory, concentration, and insight are dulled and then lost. The psychic changes are accompanied by sensory and motor disturbances. As intoxication becomes more advanced, a general impairment of nervous function occurs and a condition of general anesthesia ultimately prevails. Methanol causes less inebriation than ethanol. Symptoms of methanol poisoning include headache, vertigo, vomiting, severe upper abdominal pain, blurring of vision, and hyperemia of the optic disk. The most pronounced laboratory finding is severe metabolic acidosis as a result of the oxidation of methanol to formic acid (22, 23). Visual disturbances, the most distinctive aspect of methanol poisoning in humans, become evident soon after the onset of acidosis. The final result is bilateral blindness, which is usually permanent. Like ethanol and methanol, isopropanol is a CNS depressant, but it does not produce retinal damage or acidosis as does methanol. Isopropanol produces a more prominent gastritis, with pain, vomiting, and hemorrhage. As with the other alcohols, hemodialysis is useful for removing isopropanol from the body (24).

Mechanism of Action. Since Chin and Goldstein in 1981 (25) reported the membrane fluidizing effects of ethanol, a number of investigators have shown a correlation between the degree of intoxication and the extent of ethanol-induced disordering of membranes (26). These disordering effects, however, are effected in regions or domains of biological membranes reflecting the nonuniform distribution of various phospholipids and cholesterol within the lipid bilayer. All of these properties closely resemble those of the anesthetic barbiturates, and they are shared by other aliphatic alcohols and a variety of anesthetic agents.

Absorption and Elimination. Ethanol is rapidly absorbed from the stomach, small intestine, and colon. The time from the last drink to maximal concentration in plasma usually ranges from 30 to 90 min. Vaporized ethanol can be absorbed through the lungs, and fatal intoxication has occurred as a result of its inhalation. Between 90 and 98% of the ethanol that enters the body is completely oxidized. In the adult, the average rate at which ethanol can be metabolized is about 30 mL in 3 h. The oxidation of ethanol occurs chiefly in the liver initiated by alcohol dehydrogenase. The product, acetaldehyde, is converted to acetyl-CoA, which is then oxidized through the Kreb's cycle or utilized in the synthesis of cholesterol, fatty acids, or other tissue constituents. It is generally agreed that threshold effects of intoxication appear when the concentration in plasma is 20–30 mg/100 mL (0.02–0.03%). More than 50% of persons are grossly intoxicated when the concentration is 150 mg/100 mL. The average concentration in fatal cases is about 400 mg/100 mL. The absorption of methanol and ethanol are similar. In addition, methanol is metabolized in humans by the same enzymes that metabolize ethanol:—alcohol dehydrogenase and aldehyde dehydrogenase—to form formaldehyde and formic acid (27). Oxidation of methanol, like that of ethanol, proceeds at a rate that is independent of its concentration in plasma. However, this rate is only one-seventh that of ethanol, and complete oxidation and excretion usually require several days.

12.6.2 Gas Chromatographic/Mass Spectrometric Analysis

Sample Preparation. The method described herein for the analysis of volatile organics is a modified procedure of that described by Schuberth (28). Ten to twenty milliliters of blood are obtained from the test subjects with the use of a Vacutainer and collected into 10-mL tubes containing 15 mg of ethylenediaminotetracetic acid (EDTA) and 100 mg of sodium fluoride as anticoagulate and preservative, respectively. The samples are then stored at

TABLE 12.9 Retention Times of Volatile Organics

Substance	Retention Time (min)	Limit of Detection (nmol/L)
Acetaldehyde	1:37	0.15
2-Propanone	2:04	0.015
Ethyl acetate	2:39	0.0005
2-Butanone	2:48	0.006
Methanol	2:49	1.5
2-Propanol	3:16	0.06
Ethanol	3:21	0.7
2-Butanol	5:13	0.03
1-Propanol	5:31	0.03

4°C. Aliquots of 1.5 mL of blood are then added to a headspace vial containing 1.8 g of sodium chloride. Headspace extraction is done at a bath temperature of 50°C and an equilibration time of 30 min.

Analytical Procedure. A 30-m × 0.25-mm DB-WAX capillary, coated with 0.25 μm of polyethylene glycol, is used. The valve/loop temperature is 54°C. Injection time is 1 sec and injection volume is 1 mL. The temperature program is of an initial temperature of 40°C held for 4 min and then increased at a rate of 10°C/min to a final temperature of 150°C. An ion trap detection system is utilized, with electron impact as the ionization mode (50–80 eV).

Quantification of Volatile Organics. The retention times for the various volatile organics in plasma are shown in Table 12.9 along with estimated detection limits for the various volatiles, as defined by the analyte concentration that gives a signal equal to three times the standard deviation of the baseline noise. No volatile compounds, with the exception of acetone, are normally found in blood. Ingestion of alcoholic beverages with more than 10 mmol of ethanol per liter results in the detection of ethanol, acetaldehyde, 2-propanone, ketones, and esters.

12.7 DRUGS OF ABUSE

Every society in recorded history has used drugs that produce effects on mood, thought, and feeling. Moreover, there were always a few individuals who digressed from custom with respect to the time, amount, and the situation in which these drugs were to be used. Thus, both the nonmedical use of drugs and the problem of drug abuse are as old as civilization itself.

Drugs of abuse include alcohol, cocaine, amphetamines, nicotine, tobacco, cannabinoids (marijuana), lysergic acid diethylamide (LSD), arylcyclohexylamines, and barbiturates. In the United States two-thirds of all adults use alcohol occasionally, and at least 12% of the users can be considered "heavy drinkers." The lifetime dependence or abuse is estimated at about 13%, with the risk for men far higher than women. It has been estimated that 20 million people in the United States have used cocaine. In 1988, 5% of young adults reported using cocaine and 2% reported using a stimulant other than cocaine during the 30 days prior to the survey. In recent years, increased use of cocaine by injection of its salts and by inhalation of the free alkaloid base ("crack") has been responsible for many serious toxic reactions and escalating crime rates. Per-capita consumption of cigarrettes in the United States has been declining since 1973. In 1988, 27% of adults were still smokers, but only 19% of high school seniors were regular smokers. Smokeless tobacco (snuff and chewing tobacco) is now used by 8% of young men. The rationale for considering use of tobacco as a form of drug dependence is presented in a report by Jaffe in 1990 to the Surgeon General. Marijuana, also known as "grass," "weed," "pot," and "reefer," is still by far the most commonly used illicit drug in the United States. About 55% of young adults report some lifetime experience with the drug. There is, however, a downward trend. Among high-school seniors, the use of marijuana has declined steady from 37% in 1978 to 18% in 1988. The incidence of daily use among high school seniors is currently reported to be 2.7%.

12.7.1 Pharmacological Considerations

Pharmacological Effects. The most relevant pharmacological effect induced by drugs of abuse, in addition to their particular effects on the target tissue, is the physical dependence. Physical dependence has been studied after long-term administration of opioids, ethanol, barbiturates, related hypnotics, benzodiazepines, amphetamines, cocaine, cannabinoids, phencyclidine, and nicotine. The withdrawal symptoms associated with many of these agents are generally characterized by rebound effects in those physiological systems that were initially modified by the drug. For example, amphetamines and cocaine alleviate fatigue, suppress appetite, and elevate mood; withdrawal from these drugs is characterized by lack of energy, increased appetite, and depression.

Mechanism of Action. A number of mechanisms have been proposed to explain the changes induced by drugs of abuse, some of which help to account for the observation that physical dependence is generally accompanied by tolerance and that the two phenomena develop and decay at about the same rate. However, there is growing evidence that for some drugs, notably ethanol, it is possible to distinguish the mechanisms responsible for tolerance from those responsible for physical dependence. The mechanisms responsible for opioid-induced physical dependence are among the most thoroughly studied. Although an increase in the number of opioid receptors follows a long-term administration of opioid antagonists, a continuous administration of opioid agonists does not change the number or affinity of such receptors in the CNS. However, adaptive changes in the second messenger systems that are altered by stimulation of opioid receptors can be detected. For example, in some areas of the brain the effects of opioids include inhibition of adenyl cyclase, an action mediated by the inhibitory guanine nucleotide-binding regulatory protein Gi; this effect is shared by α_2-adrenergic agonists. The long-term administration of morphine causes a compensatory increase in adenylate cyclase activity, which may be partially responsible for the rebound excitability of neurons in those areas of the brain that typically occurs during opioid withdrawal. The common intracellular mechanism helps to explain the utility of clonidine and other α_2-adrenergic agonists in suppressing some elements of the opioid withdrawal syndrome.

12.7.2 Gas Chromatographic Analysis

Sample Preparation. Dilute 1 mL of serum or plasma with 1 mL 0.1 M Na_2CO_3 buffer (pH 9). Force the mixture dropwise through a 1-mL Superclean LC-18 SPE tube, preconditioned with 1 mL of methanol followed by 1 mL of water. Wash the tube packing with 0.5-mL aliquots of water, evaporate to dryness, and elute the drugs with 0.1-mL aliquots of ethanol–ether (90:10 v/v).

Analytical Procedure. Nonpolar 15-m × 0.53-mm i.d. SPB-1 capillary columns offer a fast, effective, and economical sample screening for drugs of abuse at therapeutic or toxic levels. SPB-1 columns are bonded phase equivalents of SE-30 columns and are listed for in vitro diagnostic use with the U.S. Food and Drug Administration. For analysis of acid-neutral drugs, the column temperature program used is from 130 to 290°C to increase at a rate of 8°C/min, a flowrate of 3 mL/min, and FID detection. Complete resolution can be obtained in less than 20 min. The same column is used for analysis of basic drugs. The temperature program is 115°C held for 2 min, and then increased to 290°C at a rate of 7°C/min and held there for 10 min, a flowrate of 2.5 mL/min, and NPD detection. Resolution of the various components can be achieved in less than 25 min.

Quantification of Drugs of Abuse. The retention times for the acidic–neutral and basic drugs of abuse are shown in Tables 12.10 and 12.11, respectively. Complete resolution can be obtained for the acidic–neutral

TABLE 12.10 Retention Times for Acidic–Neutral Drugs

Substance	Retention Time (min)
Barbital	6.01
Apobarbital	7.85
Butabarbital	8.25
Amobarbital	9.34
Pentobarbital	9.66
Meperidine	9.89
Meprobamate	10.01
Secobarbital	10.12
Caffeine	10.20
Glutethimide	10.55
Hexobarbital	11.54
Theophylline	11.95
Phenobarbital	12.07
Primidone	14.67
Phenytoin	16.01
Diazepan	17.34

TABLE 12.11 Retention Times for Basic Drugs

Substance	Retention Time (min)
Amphetamine	3.05
Methamphetamine	3.67
Nicotine	5.31
Ephedrine	5.60
Benzocaine	8.02
Lidocaine	13.42
Procaine	15.01
Methadone	16.81
Imipramine	17.21
Cocaine	17.60
Codeine	19.23
Morphine	19.73
Heroin	21.86
Flurazepan	23.46

drugs in less than 20 min. Resolution of the basic components can be achieved in less than 25 min.

12.8 PROSTAGLANDINS

Prostaglandins are membrane-derived lipids that are formed from certain polyunsaturated fatty acids in response to diverse stimuli. Since prostaglandins principally derive from arachidonic acid and therefore have a 20-carbon backbone; they are also designated eicosanoids (from the Greek *eicosa*, which means twenty).

12.8.1 Pharmacological Considerations

Chemistry and Structure. Prostaglandins are derived from 20-carbon essential fatty acids that contain three, four, or five double bonds: 8,11,14-eicosatrienoic acid (dihomo-γ-linolenic acid); 5,8,11,14-eicosatetraenoic acid (arachidonic acid), and 5,8,11,14,17-eicosapentaenoic acid. Structurally, all prostaglandins have a "hairpin" configuration and are composed of a cyclopentanone nucleus with two side chains. They are derived from the hypothetical structure prostanoic acid. Primary prostaglandins contain a 15-hydroxy group with a double bond at carbon 13. Each group of protaglandins is allocated a letter (A, B, C, D, E, F, G, H, or I) that denotes particular functional groups in the cyclopentane ring. The degree of unsaturation of the side chains is indicated by the subscript numeral after the letter; thus PGE_1, PGE_2, and PGE_3, have one, two, and three double bonds, respectively. A description of the stereochemistry at position 9 in the cyclopentanone ring is denoted by the subscript α or β; thus the configuration of $PGF_{2\alpha}$ has the orientation of the 9-hydroxyl moiety oriented below the plane of the ring. $PGF_{2\beta}$ (the inactive isomer of $PGF_{2\alpha}$) has the 9-hydroxyl group oriented above the plane of the ring.

Pharmacological Effects. In most species, including humans, and in most vascular beds, the PGEs are potent vasodilators. The dilatation appears to involve arterioles, precapillaries, sphincters, and postcapillary venules. Similarly, PGD_2 causes also vasodilation in most vascular beds. An exception is the pulmonary circulation in which PGD_2 causes only vasoconstriction. Responses to $PGF_{2\alpha}$ vary with species and vascular bed. It is a potent vasoconstrictor of both pulmonary arteries and veins in humans (29). Systemic blood pressure generally falls in response to PGEs, and blood flow to most organs, including the heart, mesentery, and kidney, is increased. Prostaglandins contract or relax many smooth muscles besides those of the vasculature. In general, PGFs and PGD_2 contract and PGEs relax bronchial and tracheal muscle. Asthmatic individuals are particularly sensitive to $PGF_{2\alpha}$, which causes intense bronchospasm. PGEs relax the uterine smooth muscle, while PGFs induce contraction. The contractile response is most prominent before menstruation, whereas relaxation is greatest at midcycle. Uterine strips from pregnant women are uniformly contracted by PGFs and by low concentrations of PGE_2. Prostaglandins are widely used to induce midtrimester abortion.

Mechanism of Action. The diversity of the effects of prostaglandins is explained by the existence of a number of distinct receptors that mediate their actions. The receptors have been named for the natural prostaglandin for which they have the greatest apparent affinity and have been divided in five main types: DP (PGD), FP (PGF), IP (PGI_2), and EP (PGE), which has been subdivided into EP_2 (smooth muscle contraction) and EP_2 (smooth muscle relaxation). As with many other receptors, the prostaglandin receptors are coupled to effector mechanisms through G proteins (30). The second messenger systems adenyl cyclase and protein kinase C have been implicated in the action of prostaglandins. PGE antagonizes the lipolytic actions of epinephrine and the effects of antidiuretic hormone at least in part by inhibition of adenyl cyclase.

Absorption and Elimination. About 95% of infused PGE_2 is inactivated during one passage through the pulmonary circulation. Because of the unique position of the lungs between the venous and the arterial circulation, the pulmonary vascular bed constitutes an important filter for prostaglandins that act locally prior to their release into the venous circulation from endogenous sources. Several enzymatic catabolic reactions are responsible for the metabolization of prostaglandins. These involve the oxidation of the prostaglandin molecule by prostaglandin 15-OH dehydrogenase, reduction of the 15-keto group to produce the 13,14-dihydro derivative by prostaglandin Δ^{13}-reductase, and β and ω oxidation of the side chains. The degradation of PGI_2 apparently begins with its spontaneous hydrolysis in blood to 6-keto-$PGF_{1\alpha}$.

12.8.2 Gas Chromatographic Analysis

Sample Preparation. The procedure described herein is a modification of that described by Weber et al. (31). Methoxyamine hydrochloride (0.5 g) and 1.5 mL of 2.4 M phosphate buffer (pH 8.6) are added to 5 mL of urine. The samples are mixed and placed in a heating block at 35°C for 30 min. The reaction mixture is applied to a bonded-phase phenylboronic acid (0.1 g) cartridge prewashed with 3 mL of methanol and 3 mL sodium hydroxide (pH 9). Prostaglandins are eluted with 3 mL of water–methanol (60:40 v/v) adjusted to pH 9 with sodium hydroxide. The eluate is diluted with water to a water–methanol ratio of 85:15 v/v and adjusted to pH 3.5 using formic acid. The two solutions are applied to RP-18 cartridges preconditioned with 10 mL of methanol, 10 mL of water, and 5 mL of formic acid-acidified water (pH 3.5). The cartridges are flushed with 5 mL of methanol–water (15:85 v/v) (pH 3.5), and the prostaglandins are eluted with 3 mL of ethyl acetate from each cartridge and evaporated to dryness. The residues are repeatedly dried with ethanol and then redissolved in 25 μL of ethanol. The fractions are applied to silica gel TLC plates prewashed overnight in acetonitrile and dried at 60°C. The plates are developed in ethyl acetate–acetic acid–hexane–water (54:12:25:100 v/v/v/v). The corresponding zones of the eicosanoid-methoximes are localized by parallel development of 4 μg of a standard methoxime mixture following reaction with a 10% of copper sulfate in 8% phosphoric acid and heating in an oven with initial and final temperatures of 24 and 120°C, respectively. The corresponding bands are scraped from the plate, 0.5 mL of water is added, and the bands are extracted twice with a mixture of ethyl acetate–acetic acid–hexane (54:12:25 v/v/v). The organic layers are combined and evaporated to dryness. The dry residues of the methoxime derivatives are converted into their pentafluorobenzyl ester trimethylsilyl ether derivatives. Samples are dissolved in 20 μL of bis(trimethylsilyl)trifluoroacetamide. Recoveries for the various eicosanoids were determined using [^{3}H]-labeled PGE_2 and 6-keto-PGF1α.

Analytical Procedure. A Hewlett–Packard 5890 gas chromatograph equipped with a cooled injection system is used. Helium is used as the carrier gas and 0.5-μL aliquots of the samples in bis(trimethylsilyl)trifluoroacetamide are injected in the splitless mode. The injection port temperature is programmed from 60–250°C at 10°C/sec and held at 250°C for 1.5 min. Eicosanoids are chromatographed using an Ultra 2 (12-m × 0.2-mm i.d., 0.33-μm film thickness) column. The oven temperature is programmed from 140 to 290°C at a rate of increase of 25°C/min. Determination of $PGF_{2\alpha}$ is achieved using two coupled columns: an OV-17 column (30-m × 0.2-mm i.d., 0.33-μm thickness) and a DB-5 (30-m × 0.25-mm i.d., 0.25-μm film thickness) and a temperature program from 140 to 280°C at 25°C/min. The column is directly connected to the ion source of the mass spectrometer (MS 8230, Finnigan MAT). Analysis is carried out in the negative chemical ionization mode using ammonia as the reagent gas. MS parameters are 0.2 Pa, ion source pressure; 92 eV, electron energy; 200°C, ion source temperature; 0.1 mA, emission current.

Quantification of Prostaglandins. The recoveries obtained for the various eicosanoids are shown in Table 12.12. The various m/z ions obtained following GC/MS are summarized in Tables 12.13., 12.14., and 12.15.

12.9 STEROIDS

Steroids are a subclass of lipids that contain a basic skeletal structure of four fused rings referred to as perhydrocyclopentanophenanthrene. Steroids

TABLE 12.12 Recoveries of ^{3}H-Labeled Eicosanoids

Method	PGE_2 (%)	6-Keto-$PGF1_\alpha$ (%)
Bond Elut PBA	97	93
Bond Elut RP-18	97	93
TLC	77	77

TABLE 12.13 Retention Times and *m/z* Ions of Eicosanoids Obtained With an Ultra 2 Column

Eicosanoid	*m/z*	Retention Time (min)
D_4-6-Keto-$PGE_{1\alpha}$	590	8:25
	618	8:25
6-Keto-$PGE_{1\alpha}$	586	8:25
	614	8:25
D_4-PGD_2	528	9:20
PGD2	524	9:20
D_4-PGE_2	528	9:30
PGE_2	524	9:30

TABLE 12.14 Retention Times and *m/z* ions of Eicosanoids Obtained With an Ultra 2 Column

Eicosanoid	*m/z*	Retention Time (min)
D_4-$PGF_{2\alpha}$	573	8:55
$PGE_{2\alpha}$	569	8:55

TABLE 12.15 Retention Times and *m/z* Ions of Eicosanoids Obtained With an OV-17 Column Connected to a DB-5 Column

Eicosanoid	*m/z*	Retention Time (min)
D_4-$PGF_{2\alpha}$	573	35:10
$PGE_{2\alpha}$	569	36:00

comprise a subcategory of chemical compounds that form part of a large family of substances that include rubber, guttapercha, the phytol side chain of chlorophyll, numerous fragant oils, turpentine hydrocarbons, carotenoids, vitamins A, E, and K, and cholesterol. Cholesterol has the distinction of being the first isopentenoid isolated in pure form, and from it the generic term "steroid" is derived.

12.9.1 Pharmacological Considerations

Chemistry and Structure. The most common natural occurring steroids are listed in Table 12.16.

Mechanism of Action. The classic steroid hormones are estrogens, progesterone, androgens, glucocorticoids, mineralocorticoids, and vitamin D. These are potent hormones that regulate the developmental and physiologic functions of female phenotype (estrogen), pregnancy (progesterone), male phenotype (androgens), metabolism and stress responses (glucocorticoids), salt and water balance (mineralocorticoid), and calcium metabolism (vitamin D). To accomplish this task, the steroid hormones must bind and activate a group of specific gene-regulatory molecules called receptors. These receptors are proteins that are present in cells in low amounts but bind steroid hormones specifically and very tightly. The hormones are secreted from their respective endocrine glands into the blood stream, where they circulate, mostly bound (95%) to plasma transport proteins, which provide a reservoir for steroid supply to cells. Free steroid enters the cell and binds to inactive receptors in either the cytoplasmic or nuclear compartments. Upon complexing with hormone, the receptor undergoes an allosteric conforma-

TABLE 12.16 Trivial Names of Steroids

Trivial Name	Systematic Name
Cholesterol	5-Cholesten-3β-ol
Androstenedione	4-Androstene-3,17-dione
Testosterone	17β-Hydroxy-4-androsten-3-one
Androsterone	3α-Hydroxy-5α-androstan-17-one
Etiocholanolone	3α-Hydroxy-5β-androstan-17-one
Estrone	3-Hydroxy-1,3,5(10)-estratrien-17-one
Estradiol	1,3,5,(10)-Estratriene-3,17β-diol
Estriol	1,3,5,(10)-Estratriene-3,16α,17β,-triol
Pregnenolone	3β-Hydroxy-5-pregnen-20-one
Progesterone	4-Pregnene-3,20-dione
Pregnanediol	5β-Pregnane-3α,20α-diol
Cortisone	17α-21-Dihydroxy-4-pregnene-3,11,20-trione
Cortisol	11β-17α-21-Trihydroxy-4-pregnene-3,20-dione
Aldosterone	11β,21-Dihydroxy-3,20-dioxo-4-pregnen-18-al

tional change into an active form capable of affecting nuclear gene transcription.

Absorption and Elimination. When a steroid enters the blood compartment and flows through each tissue in the body, a certain amount of the steroid will be removed or extracted. The metabolic clearance rate (MCR) has been defined as the volume of blood that has been completely cleared of a substance per unit time (L/day). The rate of clearance of a steroid from the blood of the whole body is the sum of the clearance rates for each tissue or organ, and it is this overall value that is termed MCR. The liver is the principal tissue for removing steroids from blood, and it is possible to directly determine the hepatic clearance rate. The MCR minus the hepatic clearance rate indicates how much of the total clearance has occurred in extrahepatic tissue. Since hepatic blood flow is about 1500 L/day, a steroid with an MCR in excess of this value indicates that tissue other than the liver is extracting the steroid. The binding of steroids to specific plasma proteins such as testosterone–estradiol-binding globulin and cortisol-binding globulin or transcortin will suppress peripheral metabolism. This is reflected in relatively low MCRs for testosterone and cortisol; an increase in the level of a specific steroid-binding plasma protein can further decrease the MCR of the specifically bound steroid. Steroids are inactivated primarily in the liver, and the inactive metabolites excreted in the urine are conjugated generally as sulfate esters or β-glucoronates.

12.9.2 Gas Chromatographic Analysis

Sample Preparation The procedure herein described corresponds to that reported by Hamalainen et al. (32). The pH of the urine sample is adjusted to pH 3 with 1 M acetate buffer (pH 3; final concentration, 0.15 M) and the steroids are extracted with Sep-Pack C18 cartridges. The steroids are eluted with 5 mL of methanol, water is added to the eluate to obtain a 70% water–methanol solution (v/v), and the sample is added to a DEAE–Sephadex anion exchange column. Steroid conjugates are separated into free, monoglucoronide, monosulfate, and double-conjugate fractions on DEAE–Sephadex anion exchange columns (5 mm × 3 cm, packed in 70% methanol v/v). The first 10-mL fraction of the 70% methanol eluate contains the free steroids. The weak organic acids and colored substances are eluted with 10 mL of 0.2 M acetic acid, monoglucoronides with 15 mL of 0.4 M formic acid, monosulfates with 15 mL of formic acid–potassium formate, and double conjugates with 15 mL of 0.3 M lithium chloride plus 0.1 M formic acid in 70% methanol. The glucoronide conjugates are hydrolyzed overnight at 39°C in 5 mL of 0.1 M acetate buffer (pH 6.8) containing 5 U of *E. coli* β-glucoronidase, and the free steroids released are extracted with Sep-Pack C18 cartridges as described above. Steroid mono- and disulfate conjugates are desalted with Sep-Pack C18 cartridges and the solvolysis of sulfate moieties are allowed to proceed in a mixture of 300 μL of dimethylformamide and 5 μL of 6 N HCl in 3 mL dichloromethane overnight at 39°C. After solvolysis, the free steroids are purified on a DEAE–Sephadex anion exchange column in the acetate form as described above for the free steroid fractions.

After hydrolysis or solvolysis, the carbonyl groups of neutral steroids are derivatized with 4% methoxyamine hydrochloride in pyridine for 2 h at 80°C. This protection reaction is necessary for quantitative recovery of the polar corticosteroid metabolites, especially for 3α-, 17α-, 21-trihydroxy-5β-pregnan-11,20-dione. After methoximation, pyridine is evaporated to dryness, 0.5 mL of water is added, and steroids are extracted twice with 3 mL of ethylacetate. The combined organic phases are evaporated to dryness, dissolved in 0.5 mL of methanol, and applied to a 0.5 × 1.5-cm DEAE–Sephadex A-25 anion exchange column, in the free base form. Neutral steroids are eluted with 3 mL of methanol and separated from estrogens and other phenolic steroids. One-tenth of the DEAE-OH-column eluate is taken

TABLE 12.17 Neutral Steroid Conjugates Identified in Male Urine

Steroid	Relative Retention Time
3α-Hydroxy-5α-androsten-17-one	0.39
3α-Hydroxy-5β-androsten-17-one	0.40
5-Androstene-3β,17α-diol	0.41
3β-Hydroxy-5α-androsten-17-one	0.44
3β-Hydroxy-5α-androstan-17-one	0.44
5-Androstene-3β,17β-diol	0.44
3α-Hydroxy-5α-androstane-11,17-dione	0.46
3α-Hydroxy-5β-androstane-11,17-dione	0.46
3β,7α-Dihydroxy-5-androsten-17-one	0.46
3α,11β-Dihydroxy-5α-androstan-17-one	0.52
3α,11β-Dihydroxy-5β-androstan-17-one	0.54
3β,16α-Dihydroxy-5b-androstan-17-one	0.55
5β-Pregnane-3α,20α-diol	0.57
5β-Pregnane-3α,17α,20α-triol	0.59
3β,16β-Dihydroxy-5-androstan-16-one	0.62
5-Androstene-3β,16α,17β-triol	0.63
5-Pregnene-3β,17α,20α-triol	0.73
3α,17α,21-Trihydroxy-5β-pregnane-11,20-dione	0.73
3α,21-Dihydroxy-5β-pregnane-11,20-dione	0.74
3α,11β,17α,21-Tetrahydroxy-5β-pregnan-20-one	0.79
3α,11β,17α,21-Tetrahydroxy-5α-pregnan-20-one	0.80
3α,17α,20α,21-Tetrahydroxy-5β-pregnan-11-one	0.81
5β-Pregnane-3α,11β,17α,20β,21-pentol	0.84
3α,17α,20β,21-Tetrahydroxy-5β-pregnan-11-one	0.84
5β-Pregnane-3α,11β,17α,20α,21-pentol	0.88

for recovery determination. Stigmasterol is added as an internal standard to each fraction and the steroids are converted to their TMS and *O*-methoxime–TMS derivatives using overnight silylation with trimethylsilylimidazole at 80°C as the silylating reagent. The derivatives are purified by Lipidex 5000 microcolumns (5 mm × 2.5 cm) using hexane–pyridine–HMDS (98:1:1 v/v/v) as eluent. The eluent is then evaporated to dryness and redissolved in hexane and subsequently analyzed by GC or GC-MS.

Analytical Procedure. A Perkin-Elmer Sigma 1 gas-chromatographic system is used, which is equipped with a 25-m BP-1 bonded phase column utilizing flame ionization detector and hydrogen as the carrier gas (flow 2 mL/min). Analyses are carried out in the splitless injection mode utilizing a two-stage program from 100 to 220°C (25°C/min) and 220 to 290°C (1°C/min). Injector and detector temperatures are 300°C. GC-MS analyses are carried out with a Hewlett–Packard 5995B quadrupole gas chromatograph–mass spectrometer equipped with a 12-m BP-1 column. The steroids are identified by their relative retention times to stigmasterol, by ion chromatograms, and by complete mass spectra when compared to authentic standards.

Quantification of Steroids. The retention times of neutral steroid conjugates from male urine relative to stigmasterol are shown in Table 12.17.

REFERENCES

1. R. P. Ahlquist, *Am. J. Physiol.*, **153**, 586 (1948).
2. M. J. Berridge, *Proc. R. Soc. Lond.* [*Biol.*], **234**, 359 (1988).
3. P. Lillsunde, and T. Korte, *Forensic Sci. Int.*, **49**, 205 (1991).
4. H. Gjerde, I. Hasvold, G. Peterson, and A. Christophersen, *J. Anal. Toxicol.*, **17**, 65 (1993).
5. J. R. Trudell, W. L. Hubbell, and E. N. Cohen, *Biochim. Biophys. Acta*, **291**, 321 (1973).
6. J. R. Trudell, W. L. Hubbell, and E. N. Cohen, *Biochim. Biophys. Acta*, **291**, 328 (1973).
7. J. R. Trudell and W. L. Hubbell, *Anesthesiology*, **44**, 202 (1976).
8. M. J. Halsey, in *Anesthesic Uptake and Action* E. I. Eger II (Ed.), Williams & Wilkins, Baltimore, 1974, pp. 45–76.
9. M. J. Halsey, B. Warden-Smith, and S. Wood, *Br. J. Pharmacol.*, **89**, 299 (1986).
10. J. J. Richter, in *Clinical Anesthesia*, P. G. Barasch, B. F. Cullen, and R. K. Stoelting (Eds.), Lippincott, Philadelphia, 1989, pp. 281–291.
11. E. N. Cohen, *Anesthesiology*, **35**, 193 (1971).
12. S. H. Snyder and H. Yamamura, *Arch. Gen. Psychiatry*, **34**, 236 (1977).

13. D. C. U'Prichard, D. A. Greenberg, P. P. Sheeham, and S. H. Snyder, *Science*, **199**, 197 (1978).
14. E. Richelson, *Mayo Clin. Proc.*, **54**, 669 (1979).
15. R. J. Baldessarini, B. M. Cohen, and M. H. Teicher, *Arch. Gen. Psychiatry*, **45**, 79 (1988).
16. S. P. Kutcher, K. Reid, J. D. Dubben, and K. I. Shulman, *Br. J. Psychiatry*, **148**, 676 (1986).
17. H. Fugii and T. Arimoto, *Anal. Chem.*, **57**, 2625 (1985).
18. G. L. Jones, and G. H. Wimbish, in *Handbook of Experimental Pharmacology*, Vol. 74, Antiepileptic Drugs, H. H. Frey and D. Janz (Eds.), Springer, Berlin, 1985, pp. 725–765.
19. Y. Yaari, M. E. Selzer, and J. H. Pincus, *Ann. Neurol.*, **20**, 171 (1986).
20. M. J. McLean and R. L. MacDonald, *J. Pharmacol. Exp. Ther.*, **227**, 779 (1983).
21. J. Volmut, E. Matisova, and Pham Ti Ha, *J. Chromatogr.*, **527**, 428 (1990).
22. K. E. McMartin, G. Martin-Amat, A. B. Makar, and T. R. Tephly, *J. Pharmacol. Exp Ther.*, **201**, 564 (1977).
23. D. Jacobson and K. E. McMartin, *Med. Toxicol.*, **1**, 309 (1986).
24. L. King, K. P. Bradley, and D. L. Shires, *JAMA*, **1970**, 211 (1985).
25. J. H. Chin and D. B. Goldstein, *Mol. Pharmacol.*, **19**, 425 (1981).
26. W. G. Wood and F. Schroeder, *Life Sci.*, **43**, 467 (1988).
27. T. R. Tephly, A. B. Makar, K. E. McMartin, S. S. Hayreh, and G. Martin-Amat, in *Biochemistry and Pharmacology of Ethanol*, Vol 1. E. Majchrowicz and E. P. Noble, (Eds.), Plenum, New York, 1979, pp. 145–164.
28. T. Schuberth, *Biol. Mass Spectrometry*, **20**, 699 (1991).
29. H. Giles and P. Leff, *Prostaglandins*, **35**, 277 (1988).
30. P. V. Halushka, D. E. Mais, P. R. Mayeux, and T. A. Morinelli, *Annu. Rev. Pharmacol. Toxicol.*, **29**, 213 (1989).
31. C. Weber, M. Holler, J. Beetens, F. Clerck, and F. Tegtmeier. *J. Chromatogr.*, **562**, 599 (1991).
32. E. Hamalainen, T. Fotsis, and H. Adlercreutz, *Clin. Chim. Acta*, **199**, 205 (1991).

CHAPTER THIRTEEN

Forensic Science Applications of Gas Chromatography

THOMAS A. BRETTELL
New Jersey State Police

Modern Practice of Gas Chromatography, Third Edition. Edited by Robert L. Grob.
ISBN 0-471-59700-7

Part 1 Introduction

13.1 INTRODUCTION

Since its development at the beginning of this century, chromatography has become an important tool in many fields, including forensic science. Chromatographic techniques are widely used in this field because of their versatility, sensitivity, speed, and reliability. Gas chromatography (GC) in particular has been successfully applied to a number of specific and unique problems encountered in the crime laboratory. There is good reason for this success. Over the last two decades chromatographic instrumentation has become affordable to the routine crime laboratory. In addition, the basic principles and theory are readily understood, and the instrumentation has become reliable and simple enough for the novice to operate.

In this chapter gas chromatographic applications to problem samples, both routine and nonroutine, encountered in the crime laboratory are discussed. The material is covered in sufficient detail to enable the reader to understand each specific application and to render an appreciation for the type of work in which GC is applied in forensic science.

13.1.1 Definition and Scope of Forensic Science

Forensic science can be broadly defined as the application of science to law. It is more commonly applied to those laws (criminal and civil) that are enforced by police agencies in the criminal justice system. Specifically, it is the application of the principles of chemistry and related sciences to the examination of physical evidence collected at the scene of a crime, and the interpretation of the results of that examination in a court of law by an expert.

In the general field of Forensic Science many disciplines of science have been applied. Some do not use GC in their particular application and are beyond the scope of this chapter. These would include such areas as forensic geology, forensic archeology, forensic engineering, forensic anthropology, fingerprint examination, forensic odontology, and forensic psychiatry.

There are, however, a number of different types of forensic science that do apply GC to solve some unique problems. These are the areas that are addressed in this chapter. Most have to do with the area of drugs of abuse and toxicological analyses, including blood alcohol analysis. There are also some very unique applications in criminalistics, such as analysis of debris from fire scenes for accelerants, explosive analysis, and the examination of trace evidence, such as paints, fibers, and other polymers. Gas chromatographic applications dealing with these specific areas are detailed in the chapter.

13.1.2 Functions of the Forensic Scientist

The forensic scientist's responsibility in all of these areas is to analyze the evidential material with the best analytical techniques available and report the findings to the requesting authority. It is also the responsibility of the scientist to testify in court about the results of the tests performed, the scientific techniques used, and, if applicable, the meaning and interpretation of the data presented. The scientist must be able to articulate and explain to lay people, such as lawyers, police officers, and general citizens with no scientific background, the technical nature of the work in terms that nontechnically trained people can understand. The court testimony of a forensic scientist may or may not be accepted. His or her familiarity with the topic may be questioned and any failure to respond effectively creates uncertainty in the minds of the judge and jury. The forensic scientist can perform the most sophisticated analytical chemistry in the laboratory, but if the results are not communicated properly, the value of the analysis may be lost.

During testimony, the scientist must also be aware of the demands of the courts. The procedures used in the laboratory must not only rest on a firm scientific foundation, but must also satisfy the criteria of admissability (1–3). GC is a technique that is generally accepted by the scientific community as a reliable procedure in the analysis of physical evidence and has met all of the requirements imposed by the judicial system.

Another function performed by the scientist is the training of other law enforcement personnel regarding the capabilities of the laboratory. Prosecutors and investigators must understand the capability of the forensic laboratory and the value of the analyses that may be performed there. They need not know the theory and operation of the gas chromatograph, but a general knowledge of how evidence is examined can be helpful to determine what type of evidence to collect and submit to the laboratory.

This training provided by the forensic scientist may involve not only the analysis of evidence, but also the proper recognition and collection of physical evidence. It is important for the investigator to recognize evidence that is pertinent and valuable to the case while sorting through other items at the scene that may not be evidential. The investigator does not want to burden the laboratory with unnecessary pieces of evidence.

The forensic scientist may also provide technical support to prosecutors and investigators during several stages of the investigation and trial. The investigator must be trained to preserve the evidence and protect the chain of custody. The evidence must be submitted to the laboratory intact whenever possible. This means the evidence must be preserved from contamination, breakage, evaporation, accidental scratching, bending, or other types of loss. The forensic scientist's analysis can only be as good as the evidence that is submitted. The continuity of possession and transfer of evidence must be documented and the investigator must be trained to perform this properly.

13.2 PHYSICAL EVIDENCE

13.2.1 Types of Evidence

Physical evidence encompasses any and all objects that can establish that a crime has been committed or can provide a link between a crime and its victim or a crime and its perpetrator. The ultimate goal in examining physical evidence is to help determine or reconstruct the events of the crime and if possible the order of events.

Physical evidence can be any type of material, small or large, that can help in linking the victim or suspect to the scene(s) or to each other. Table 13.1 lists some common types of physical evidence. The evidence with asterisks are commonly analyzed by gas chromatographic techniques and will be discussed in more detail later in the chapter.

13.2.2 Identification versus Comparison

In the analysis of physical evidence the significance placed on that evidence depends on how narrowly the evidence can be related to the source. This is where forensic science differs from most other types of science. Generally, other sciences are satisified when an object can be placed into a specific class of the discipline. Criminalistics, or more generally, forensic science strives to relate the object to a particular source. This can be accomplished in one of two ways.

The evidence may need an identification of the particular substance or it may need to be compared to a standard or comparison sample. If an identification is needed, then a determination of the physical or chemical identity of a substance with as near absolute certainty as existing analytical techniques will permit must be performed. For example, the identification of a particular drug, an accelerant, or a type of explosive may be needed. GC can play a key role in this identification, especially when it is interfaced with highly specific detectors such as mass spectrometers or infrared spectrophotometers.

TABLE 13.1 Common Types of Physical Evidence

Blood*	Fibers*	Organs*
Semen	Fingerprints	Other physiological fluids*
Saliva*	Firearms & ammunition	Petroleum products*
Documents	Glass	Powder residues
Drugs*	Hair	Serial numbers
Explosives*	Impressions	Solids & minerals
Toolmarks	Polymers*	Wood & vegetation
Paint*	Soil	

* Commonly analyzed by GC.

When a comparison analysis is needed, the suspect evidence is subjected to the same set of tests as a comparison piece of evidence for the ultimate purpose of determining whether or not they have a common origin. GC is also used effectively in this type of analysis. For example, paint and synthetic fibers are routinely compared in the forensic laboratory by pyrolysis GC (PGC) to determine the source of origin.

13.2.3 Class versus Individual Characteristics

When a comparison analysis is undertaken in the forensic laboratory a two-step process must be performed. First, the best possible distinguishing properties must be selected for comparison of both the suspect and control evidence specimens. Second, and more important, it must be decided whether a conclusion can be drawn from the data as to the origin of the suspect evidence and control evidence. In other words, do both pieces of evidence come from the same source?

When considering the conclusion as to the origin of physical evidence the significance of the data and the power of differentiation both contribute to the value of the physical evidence. For the most part, physical evidence falls into two classifications: physical evidence with class characteristics only and physical evidence with individual identifying characteristics. When evidence is classified as having class characteristics only, such evidence, no matter how thoroughly examined, can only be placed into a class. A definite conclusion as to common origin can never be made since there is a possibility of more than one source of the evidence. An example of this type of evidence would be a single-layered paint chip or a footprint from a new pair of sneakers.

When evidence is classified as having individual identifying characteristics, this evidence can be definitely attributed to a person or source. The classic example of this type of evidence would be a set of fingerprints, since no two people have the same set of prints. GC can also be valuable in obtaining individual identifying characteristics, for example in cases such as the establishment of the origin of two drug specimens.

Of course, it is always desirable to have evidence that can be positively individualized, but the value of class evidence only should not be minimized. Class evidence can be valuable in an investigation where there is a preponderance of such evidence.

13.3 MODERN CRIME LABORATORY

13.3.1 Types of Forensic Laboratories

At present, there are hundreds of public crime laboratories operating in the United States at various levels of government—federal, state, county, and municipal. At the federal level, the Federal Bureau of Investigation (FBI) operates the largest state-of-the-art crime laboratory in the world. The Drug Enforcement Administration Laboratories (DEA) are responsible for the analysis of drugs seized in violation of federal laws and regulating the production, sale, and transportation of drugs. The Bureau of Alcohol, Tobacco, and Firearms (ATF) laboratories analyze alcoholic beverages and explosive residues. The U.S. Customs Service operates its own forensic laboratory in regulating goods entering the country. In July 1989, the National Fish and Wildlife Forensics Laboratory was established to aid wildlife enforcers in identifying the species victimized, determining cause of death, and linking a suspect to the crime.

Most state governments now have their own forensic laboratories, which service state and local law enforcement agencies. Many local governments maintain a crime laboratory, which provide services to county and municipal agencies. These are more frequently found in the highly populated urban areas of the large cities where the crime rate is high.

Most countries, including several third-world countries, have created and now maintain forensic science laboratories. Great Britain and Canada have two of the larger and better-known forensic institutions. In Canada, forensic services are provided by three government-funded institutes.

13.3.2 Crime Laboratory Organization

Taking into consideration the wide variety of crime laboratories at different levels of government, it is not surprising to find different services being offered at a particular forensic laboratory. There are many reasons for these differences, such as variations in local laws, budgetary restrictions, different capabilities and purposes of the governmental branch that the laboratory serves, and the historical evolution of the laboratory itself.

The basic forensic science laboratory usually consists of a chemistry section, biology unit, firearms unit, document examination unit, and a photography unit. There may be other services offered at different laboratories, such as a latent fingerprint examination unit, a polygraph unit, a voiceprint analysis unit, and a crime scene or evidence collection unit. The forensic science laboratory may consist of all or a combination of these units, depending on its needs.

The chemistry section is the unit that predominantly, if not solely, uses GC as a tool for physical evidence analysis. This section may be divided into a drug unit, for the analysis of drugs of abuse, a toxicology unit for the analysis of biological fluids for drugs and poisons, including alcohol, and a trace evidence unit for the analysis of small pieces of evidence, such as glass, paint, fibers, and explosives. The applications discussed in this chapter focus on these analyses. For further reading about forensic science laboratories and the analysis of physical evidence the reader is directed to general texts on forensic science (4–7).

Part 2 Drug Analysis by Gas Chromatography

13.4 CONSIDERATIONS IN FORENSIC DRUG ANALYSIS WITH GAS CHROMATOGRAPHY

13.4.1 Introduction to the Analysis of Drugs of Abuse

Drug abuse has become a major problem for the United States and many other countries around the world. The analysis of drugs of abuse now accounts for a major proportion of the workload of state and local forensic science laboratories. The most commonly encountered drugs of abuse at the present time are cannabis (marijuana), cocaine (crack), heroin, phencyclidine (PCP), lysergic acid diethylamide (LSD), amphetamines, including 3,4-methylenedioxyamphetamine (MDA) and 3,4-methylenedioxy-methamphetamine (MDMA), and some designer drugs.

The use of GC in the analysis of drugs of abuse is well established (7–9). A large number of methods with varying conditions are used for different drugs. The most common detector employed for the gas chromatographic analysis of drugs of abuse is the flame ionization detector (FID), although the mass-selective detector (MSD) and mass spectrometers, in general, are used extensively for confirmation. For nitrogen-containing drugs, alkali flame ionization detectors (AFID), or flame-photometric detectors are often used in detection for added sensitivity and selectivity. In addition, electron capture detectors (ECD) have been used after derivatization of the drug for added sensitivity and selectivity.

13.4.2 Controlled Dangerous Substance (CDS) Laws and Schedules

There are several ways to classify drugs of abuse. The most relevant classification to forensic science is the legal classification whereby the federal and state codes classify drugs of abuse as "controlled substances" by schedule. For the purposes of this chapter the discussion of the application of GC to drugs of abuse is limited to controlled substances. For the application of GC to other drugs the reader is referred to Chapter 12 on clinical applications of GC.

For practical law enforcement purposes the legal community has outlined the drug classifications and definitions in the drug laws. These laws are of particular interest to the forensic scientist, since they may impose certain requirements as to the analytical protocol for drug analysis. For example, the severity of a penalty associated with the manufacture, distribution, possession, and use of a drug may depend on the identification of a particular active compound, the weight of a drug, or the concentration of the drug. In these particular cases, the appropriate analytical approach must be performed and the drug analysis report must contain the pertinent information.

The Controlled Substance Act regulates the handling of drugs of abuse. There are five schedules, or lists, that classify drugs according to medical use and degree of abuse:

- Schedule I drugs have no accepted medical use in the United States but a high rate of abuse and/or lack accepted safety for use in treatment under medical supervision. Drugs controlled under this schedule include heroin, marijuana, methaqualone, and LSD.
- Schedule II drugs have an accepted medical use in the United States and a high rate of abuse, with either severe psychological or physical dependence potential. These drugs include morphine, codeine, cocaine, amphetamine, and most barbiturate preparations containing amobarbital, secobarbital, and pentobarbital.
- Schedule III drugs have an accepted medical use, but a lower potential for abuse than Schedules I or II, and have a potential for low or moderate physical dependency or high psychological dependency. Examples are all barbiturate preparations (except phenobarbital) not covered under Schedule II, some codeine preparations, and steroid preparations, such as testosterone and its esters.
- Schedule IV drugs have an accepted medical use and generally have a low potential for abuse relative to Schedule III. Drugs controlled under Schedule IV are generally the long-acting barbiturates, hypnotics, and minor tranquilizers, such as meprobamate, phenobarbital, diazepam, and dextropropoxyphene.
- Schedule V drugs have medical use, have low abuse potential, and have less potential for producing dependency than Schedule IV drugs. These drugs may be any of the drugs in the above schedules, usually in solution, at low concentration of controlled substance, which also may contain noncontrolled ingredients in sufficient quantities to effect qualities other than those possessed by the controlled substance alone; cough syrup is an example.

Under the Controlled Substance Act, a drug must be specifically classified in one of the five schedules if its use is to be considered illegal. This requirement has given rise to the existence of the so-called "designer drugs" as a means of circumventing this law. Designer drugs are substances that are chemically related to that of a controlled drug in Schedule I or II and are pharmacologically very potent. In response to the legal problems with scheduling designer drugs, the "Controlled Substance Analogue Act" was introduced to control these drugs except for legitimate use for medicine or research.

Controlled drugs can be procured only from licensed sources and accurate records of inventory must be kept as well as amount of drug used. Dilute standard solutions can be purchased, in limited quantities, usually without difficulty or license.

13.4.3 Types of Physical Evidence: Sample Preparation

A classification scheme for drugs of abuse that is of more practical interest is classification by the form(s) in which the drug substance is most often found when submitted as evidence to the laboratory.

Controlled substances are found in three major forms. The first type is plant or vegetative material. Marijuana, peyote, khat (*Catha edulis*) and mushrooms that contain psilocybin fall into this category. Substances in this form usually require some botanical examination as well as chemical analyses. To prepare samples for analysis by GC a specialized extraction procedure is normally required to separate the naturally occurring compounds from the drug to be chromatographed.

Another form of drug submissions is labeled tablets and capsules. For the most part, these are legitimately manufactured and generally bear identification marks that can be searched in the *Physician's Desk Reference* (*PDR*) for the identification of its contents.

The third form in which drugs of abuse are submitted to the laboratory is in the powder form. This is usually classified as a general unknown and grouped with other types of submissions that are in the liquid form. Drugs of abuse of this type vary, such as cocaine HCl, cocaine free base (crack), heroin, PCP, and methamphetamine. Many times this form of drug is adulterated, which has important implications in sample preparation and analytical methodology. Other unmarked tablets and capsules and clandestinely manufactured LSD, as well as steroid preparations, can be classified into this category.

Since illicit drugs are rarely encountered in their pure form, it is usually necessary to extract the drug of interest from any interfering compounds or adulterants before the gas chromatographic analysis. While dry extractions can be performed, one step liquid–liquid extractions are usually more effective. The analyte drug is extracted from an acidic or basic solution into an organic solvent such as methylene chloride or ether. Drugs of abuse may be classified as either acidic, basic, or neutral, depending on their pK_a. This classification is extremely useful to the forensic drug examiner in deciding on the appropriate methodology. Examples of basic drugs are heroin, cocaine, PCP, and amphetamines. Barbiturates are examples of acidic drugs.

Solid-phase extractions (SPE) may be used to separate drugs from interfering materials; however, these are usually not used for bulk formula-

TABLE 13.2 General Scheme of Analysis of Drugs of Abuse

1. Preliminary visual examination of all specimens
2. Weights of all exhibits; volume (if liquid is present)
3. Selection of representative samples
4. Microscopic examination (if vegetation)
5. Screening tests (usually spot tests, may include UV spectrophotometry and/or thin-layer chromatography)
6. Microcrystalline tests (optional)
7. Extraction
8. Confirmatory test (GC/MS or IR spectrometry)
9. Quantitative analysis (if needed)

tions on a routine basis but rather for toxicological analyses. Liquid–liquid extraction schemes can also require several steps, including extraction, filtration, centrifugation, and evaporation stages, which increase the analysis time of a single sample. However, when the analyst has completely prepared the sample for analysis, the sample is relatively clean, with the drug remaining in a few microliters (~50+) for analysis by GC.

Preparing the sample for GC is only one step in the examination of controlled dangerous substances. Several steps must be taken in the scheme of analysis of an unknown drug submission. Table 13.2 lists the steps in a general scheme of analysis for a typical unknown drug submission.

There is, of course, no one scheme of analysis for all drugs of abuse. Different laboratories may use different schemes for the same drugs. The methodology depends on a number of different factors, which include but are not limited to the particular drug, the laws that govern the particular locale or state, the instruments available in the laboratory, the number of cases submitted annually, the number of personnel available to analyze the case, and the training and background of the scientist.

For further reading on the analysis of drugs of abuse the reader is referred to an excellent chapter written by Siegel (10) and a comprehensive book on the subject written by Gough (8). For more detailed information on analytical data, References 11–20 are recommended.

13.5 QUALITATIVE ANALYSIS OF DRUGS OF ABUSE

Controlled drugs of abuse can generally be classified into five broad areas based on the drugs effects. Table 13.3 lists some examples of commonly abused drugs according to these five classes. GC is routinely used in the forensic laboratory to separate drugs in each of the classes. The following discussion illustrates some examples of the analysis of drugs of abuse by GC.

TABLE 13.3 Classification of Some Commonly Abused Drugs

Class	Drug
Narcotic	Morphine
	Heroin
	Codeine
	Methadone
	Fentanyl
Stimulant	Amphetamines
	Cocaine
Depressant	Barbiturates (both short- and long-acting)
	Ethanol
	Methaqualone
	Meprobamate
	Diazepam
	Chlordiazepoxide
Hallucinogens	Marijuana
	PCP
	LSD
	3,4-Methylenedioxyamphetamine
	3,4-Methylenedioxymethamphetamine
Steroids	Testosterone
	Stanzolone
	Testosterone esters

13.5.1 Narcotics

The class of narcotic drugs encompasses the opium-derived drugs of morphine, heroin, and codeine (Figure 13.1) as well as other narcotics, such as meperidine, hydromorphone, hydrocodone, and the fentanyl compounds. Because these compounds are highly polar and often require high temperatures for elution, GC is difficult and often demands derivatization. Morphine, for example, because of its amphoteric nature, is not only difficult to extract but must be derivatized to obtain good quantitative data.

Heroin (Figure 13.1A) is the most widely abused semisynthetic opiate. It was first synthesized in 1874 by acetylation of morphine (Figure 13.1B). Heroin is usually seen as a white crystalline powder and is mixed with diluents in illicit samples, such as sugar, quinine, caffeine, and even strychnine. Therefore it is necessary for chromatographic systems to resolve heroin and its commonly encountered diluents. In street samples heroin may comprise as low as 2% by weight of the total sample, which causes difficulty with separation when it is present in very complex mixtures. Heroin at its

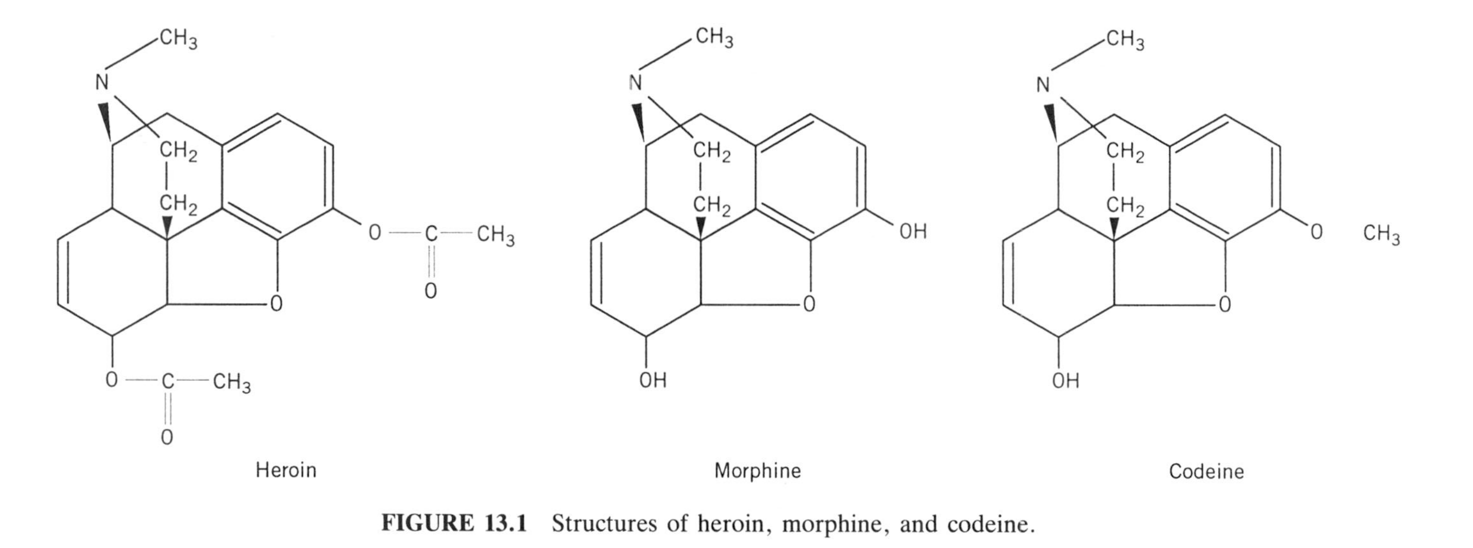

FIGURE 13.1 Structures of heroin, morphine, and codeine.

point of origin is not very pure, since it usually contains some acetylcodeine, morphine, and monoacetylmorphine. Analysis of illicit heroin and its impurities can be performed by GC/MS. A method has been developed for the simultaneous determination of heroin along with some of the commonly occurring adulterants utilizing simple dissolution of the sample along with an internal standard followed by capillary GC on a nonpolar methylsilicone column attached to a FID (21).

Heroin historically has been introduced into the body by intravenous syringe. More recently, however, heroin has shown signs of increasing purity to 30–40% or higher, which some experts believe indicates a user shift to nasal introduction ("snorting") (22). Wyatt and Grady (23) have reviewed the physical properties, synthesis, stability, metabolism, and analysis of heroin. A review of laboratory methods for the analysis of opiates and diluents in illicit drugs has been reported (24).

13.5.2 Stimulants

The most commonly encountered compounds in this class are the phenethylamines. These include amphetamine, methamphetamine, phentermine, and many other structurally related compounds that are controlled according to the Controlled Substance Act. Several factors make this class of drugs the most difficult to analyze. Because the phenethylamines are the most commonly clandestinely manufactured class of drug, the variety of closely related structural compounds can make the analysis a complicated process, requiring the GC to have high resolution, selective liquid phases, and good sensitivity. In addition, other structurally related compounds that are not controlled by law but are legitimately made are often found in these samples. These include ephedrine, phenylpropanolamine, and caffeine.

Amphetamine salts produce poor results during gas chromatographic analyses. The compounds must be prepared by extracting a basic solution of amphetamines into an organic solvent such as methylene chloride. This will produce a sample of free-base amphetamine which can be chromatographed easily. The use of methanol or ethanol as the injection solvent for gas chromatographic analysis of amphetamines is not recommended. The primary amines, such as amphetamine, MDA, and phenethylamine, yield imines upon injection of methanol or ethanol solutions (25). For this reason and because free-base amphetamines generally yield nondiscriminating mass spectra, several different derivatization techniques have been employed for the separation of these compounds (26). The formation of Schiff bases or conversion to amides are procedures that have been used, but halogenated derivatives account for many techniques that have sought better selectivity and sensitivity with the use of an electron capture detector (27, 28).

Tricholoracetyl and 4-carbethoxyhexafluorobutyryl chloride derivatives have been used to increase the molecular weight of the amphetamines. These particular derivatives consequently lengthen the elution times and help separate these compounds from potential interferences (29, 30). *N*-Monotrifluoroacetylated (TFA) derivatives of amphetamine analogues have been prepared by on-column derivatization with *N*-methylbis(trifluoroacetamide) (MBTFA) and analyzed by GC/MS on a 10% Carbowax 20M–2% KOH packed column (31). A separation on a 12-m methylsilicone capillary column of four phenethylamines with closely related structures is shown in Figure 13.2 using this procedure.

Most of the phases used in chromatographic analysis of the amine-type amphetamines have been polar, although with the advent of inert capillary columns, these compounds can be chromatographed with good success on nonpolar phases, as shown in Figure 13.2. The *AOAC Official Methods of Analysis*, 1990 edition, still lists a 1% Carbowax liquid phase for the analysis of amphetamine, although other phases have been used.

Chiral separation of amphetamines has also been performed. The enantiomeric composition of amphetamine samples can provide information about the synthesis and origin of drugs. For example, Liu et al. (32) used *N*-trifluoroacetyl-L-prolychloride (TPC) to separate isomers of methamphetamine on two different columns.

13.5.3 Cocaine

Cocaine, a benzoic acid ester of ecgonine (Figure 13.3), is a naturally occurring alkaloid from the plant *Erythroxylon coca*, grown in South America. Cocaine is a nervous system stimulant and local anesthetic drug

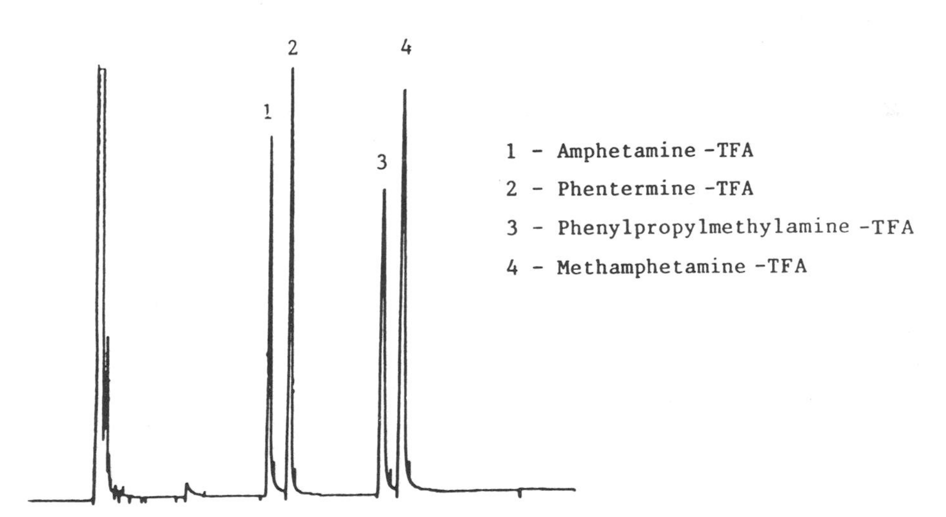

FIGURE 13.2 Separation of the TFA derivatives of amphetamine, methamphetamine, phentermine, and phenylpropylmethylamine using on-column derivatization with MBTFA. GC/FID conditions: 12-m HP-1 × 0.20-mm i.d. × 0.33-μm film; 120–180°C at 30°C/min; split ratio = 20/1. Courtesy of New Jersey State Police.

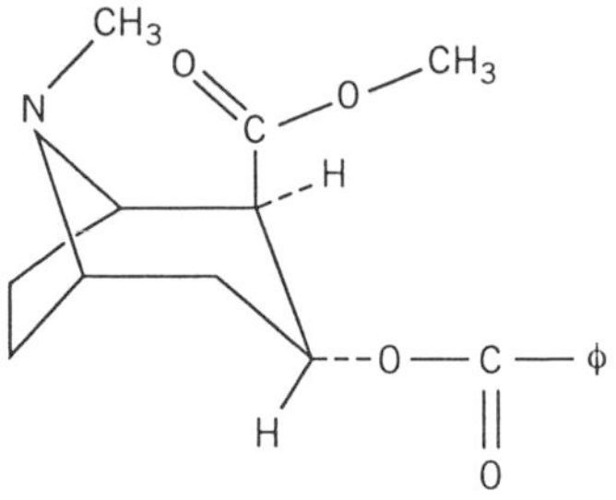

l-Cocaine

FIGURE 13.3 Structure of *l*-cocaine.

and is one of the most abused drugs in the United States. The most common route of administration is intranasal (IN) by insufflation or snorting. A more popular form of cocaine is the free-base form, known as "crack". Crack is made by alkalinizing the salt and extracting into the nonpolar solvents. The drug in this form can be smoked by the abuser, which produces a quicker and more intense euphoria.

Illicit cocaine can be cut with a number of different diluents, such as tetracaine, lidocaine, benzocaine, and procaine, to mimic the numbing sensation of cocaine when taste-tested. Other additives are mixed with cocaine to add bulk, such as sodium bicarbonate, starch, talcum powder, boric acid, and sugars.

Cocaine can be separated from the other caine diluents on a 12-m × 0.20-mm i.d., 0.33-μm film thickness capillary column using an isothermal temperature of 220°C in a very short run time (<5 min). Figure 13.4 shows the separation of a mixture. Previously the AOAC had adapted a gas chromatographic procedure that utilized a 3% OV-1 column for preliminary identification of cocaine (33). Packed-column gas chromatographic methods such as the AOAC method have been used to analyze cocaine samples, but capillary columns provide the best resolution in the shortest analysis time. Capillary GC coupled to a selective detector such as an NPD or a mass spectrometer has become the method of choice for analyzing illicit cocaine samples.

More recently, coca paste and several fractions from smoking products were analyzed by GC-FID and GC/MS (34). The inhalation efficiency and pyrolysis products of cocaine by the pyrolysis of crack and cocaine hydrochloride were studied by GC and GC/MS (35). GC/MS has also been used to study the injection-port-produced artifacts from cocaine (crack) exhibits (36).

13.5.4 Barbiturates

A variety of gas chromatographic methods have been reported for the analysis of barbiturates. The first chromatographic separation of barbiturates was described in the 1960s. Since that time most of the work

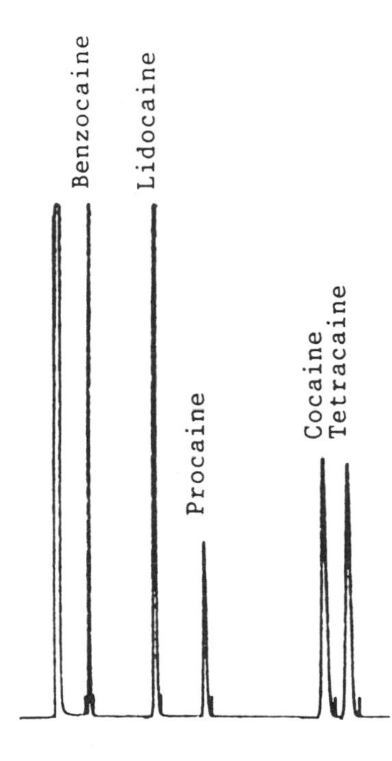

FIGURE 13.4 Separation of benzocaine, lidocaine, procaine, cocaine, and tetracaine. GC/FID conditions: 12-m HP-1 × 0.2-mm i.d. × 0.33-μm film; oven = 220°C, injection = 250°C, detector = 300°C; split ratio = 20/1. Courtesy of New Jersey State Police.

performed has been with derivatization techniques because free barbituric acids exhibit considerable adsorption and peak tailing. Jain and Cravey (37) have reviewed the literature up to 1974, and Pillai and Dilli (38) have reviewed the analysis of barbiturates by GC up to 1980. With the advent of modern fused-silica capillary columns the need for derivatization has lessened; however, some barbiturates give very similar mass spectra and diligent use of retention times is needed for additional identification. For this reason most analysts still use derivatization. Many reagents have been used, including BSTFA for silanization, trimethylanilinium (TMAH) for methylated compounds, and pentaflourobenzyl bromide (PFBB) for halogenation. Alkyl derivatives seem to enjoy the greatest success, especially when trimethylphenylammonium hydroxide (TMPAH) is used.

Sample preparation requires a simple acid/organic solvent extraction or a direct extraction into ethanol being dried down and then subsequently injected into the chromatographic column. Methyl- or phenylmethylsilicone columns have been used with most success.

Recently, a simple and reproducible method for the analysis of barbiturates by GC/MS after derivatization with DMF dipropylacetal has been reported (39).

13.5.5 Benzodiazepines

Benzodiazepine drugs are used as muscle relaxants and were introduced in the 1960s. These drugs have characteristically shown long retention times, but with better stationary phases and the use of stable fused-silica capillary columns, this has been improved.

Some compounds in this particular class of drugs tend to break down during gas chromatographic analysis. For example, chlordiazepoxide (Librium) breaks down to *N*-desmethyldiazepam, a noncontrolled substance. Clorazepate presents several problems in identification. In addition to rapid acid decarboxylation to *N*-desmethyldiazepam, extracts of the pharmaceutical forms of clorazepate contain substances that interfere with isolation of intact and unaltered clorazepate. The analyst should be cautioned about this particular problem. Many procedures use HPLC for the analysis of benzodiazipines for these reasons. However, procedures involving solvent extractions, GC, and detection via FID, NPD, or MS (MSD) give good detection limits (nanograms) for easy identification.

OV-1 and SE-30 phases have been used for packed columns and retention indices have been listed for this class of compounds (40). Plotczyk and Larson (41) have also studied the behavior of this class of drugs on 5% phenylmethylsilicone (DB-5, SE-52) fused-silica columns.

13.5.6 Cannabinoids

Cannabis usually comes in three forms: (1) cannabis (marijuana), (2) cannabis resin (hashish), and (3) extracts of cannabis resin (hashish oil). Most laboratories use a color test (modified Duquenois–Levine), a microscopical examination, and TLC to identify cannabis. Consequently, GC has not been widely used for the forensic identification of Cannabis samples.

Novotny et al. (42) were the first to indicate that there might be a correlation between the capillary gas chromatographic profiles and the country of origin. Brenneisen and ElSohly (43) later were able to differentiate samples of different origins using high-resolution capillary columns. Programmed temperature gas chromatographic analysis with high-resolution capillary columns can give information about batch origin, geographical origin, or identification of cannabinoids.

GC/MS analyses of cannabis have identified over 400 compounds in the plant with 61 of these being cannabinoids (44–46). The major cannabinoids—cannabidiol (CBD), tetrahydrocannabinol (THC), and cannabinol (CBN)—can be readily separated on a 12-m methylsilicone column and identified using a mass spectrometer. Sample preparation simply consists of an organic solvent wash with petroleum ether.

One of the major applications of GC in the forensic analysis of cannabis has been the determination of total THC content (including decarboxylated THC acid), which is used to determine the quality of the cannabis product (47). Although gas chromatographic analysis using different temperature programs can give information about the geographical origin, little success has been achieved in this area by comparing cannabinoids due to the decarboxylation on injection of the sample. Although a wide range of stationary phases and supports has been studied, most recent work has been conducted on fused-silica capillary columns. For example, the cannabinoid acid pattern of plant preparations from *Cannabis sativa* (hashish, marijuana) has been determined by a hexane extraction and by analysis of their methyl-TMS derivatives with high-resolution GC and GC/MS (48).

Some work has been done with headspace GC of cannabis extracts. Headspace analysis was performed on samples of marijuana and hashish and the results did not show a correlation between a geographical origin, but the method could differentiate common origin (49).

13.5.7 Hallucinogens

The most commonly abused hallucinogens are phencyclidine (PCP), lysergic acid diethylamide (LSD), psilocybin, mescaline, and 3,4-methylenedioxyamphetamine (MDA) and its analogues 3,4-methylenedioxymethamphetamine (MDMA) and its analogues, and 3,4-methylenedioxyethylamphetamine (MDEA). These compounds vary in structure and require different chromatographic conditions for analysis. Psilocybin for example, requires derivatization before chromatographic analysis and many extractions cleave the phosphoryl group from psilocybin converting it to psilocin. HPLC is an alternative technique for separation of this particular drug from other impurities.

LSD has isomers that complicate its chromatography. LSD, iso-LSD, and LAMPA, the methyl-propyl analog of LSD, all have molecular weights of 323 amu and similar mass spectra. The resolution of LSD and LAMPA is difficult to achieve since LSD suffers thermal degradation. Symmetrical peaks and a linear response can be achieved with glass columns and no derivatization, but glass columns are a prerequisite even with silyl derivatives. The separation of LSD and LAMPA is best achieved with a short, nonpolar fused-silica capillary column. Nichols et al. (50) used a 5-m SE-30 fused-silica capillary column and hydrogen as the carrier gas to obtain baseline resolution of these two compounds within 3 min. Figure 13.5 shows the baseline separation of LSD and LAMPA on a 12-m HP-1 fused-silica capillary column. It is essential that confirmation of the compounds in this class be accomplished by mass spectrometry because of the closely related isomers.

Phencyclidine (PCP), first sold as an animal tranquilizer, has been abused for years. PCP is submitted to the laboratory as either a powder or a liquid or may be found on vegetation such as marijuana or cigarettes in either a crystalline form or sprayed on as a liquid. The sole source of PCP is from clandestine laboratories, which means that the illicit samples are almost never pure but are mixed with precursors and by-products of the synthesis. The contamination of illicit PCP by the carbonitrile precursor piperidinocyclohexanecarbonitrile (PCC) is a likely possibility. In addition, several other PCP derivatives have been identified in street samples. Table 13.4 lists some homologs and analogs to PCP that can be found in illicit samples. Any

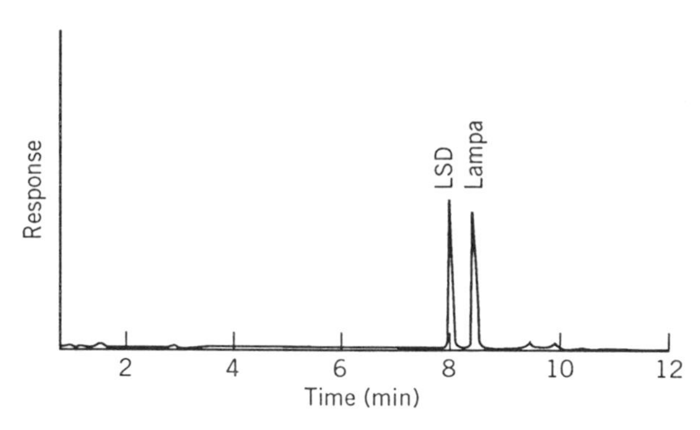

FIGURE 13.5 Separation of LSD and LAMPA. GC/MSD conditions: 12-m HP-1 × 0.2-mm i.d. × 0.33-μm film; 240–280°C at 10°C/min to 300°C at 5°C/min, injection = 270°C, split ratio = 20/1. Courtesy of New Jersey State Police.

TABLE 13.4 Homologs and Analogs of Phencyclidine

Abbreviation	Drug	Analog/Homolog
PCP	1-(1-Phenylcyclohexyl)piperidine	Phencyclidine
TCP	1-[1-(2-Thienyl)cyclohexyl]piperidine	Analog
PHP	1-(1-Phenylcyclohexyl)pyrrolidine	Homolog
PPP	1-(1-Phenylcyclopentyl)piperidine	Homolog
PCM	1-(1-Phenylcyclohexyl)morpholine	Analog
TCM	1-[1-(2-Thienyl)cyclohexyl]morpholine	Analog

or all of these compounds could be misidentified for PCP if the proper analytical conditions are not controlled. For the reasons above, high resolution is required for separation and identification.

Separation of PCP analogs is accomplished quite easily, with medium to nonpolar phases offering the best chromatography (51). PCP has often been observed to decompose to 1-phenylcyclohexane (PCH) upon gas chromatographic analysis due to thermal degradation and the presence of acidic sites (52). This can be eliminated by the use of fused-silica capillary columns and relatively low temperatures.

13.5.8 Anabolic Steroids

The federal Anabolic Steroids Control Act of 1990 classified all compounds of anabolic activity as Schedule III controlled dangerous substances effective 27 February, 1991 (53). Most state legislatures signed similar laws into effect shortly after this date, regulating the manufacture, distribution, or dispensing of anabolic steroids. The potential for abuse of anabolic steroids or androgens lies in the fact that they are responsible for many of the characteristics associated with male development. Athletes have used anabolic steroids since the 1950s to enhance athletic performance. In 1974, steroids were added to the list of barred substances by the International Olympic Committee (IOC). Today, it is not just the competitive athletes who are abusing steroids, but also those who want to improve their own physical appearance.

All anabolic steroids contain a five-membered cyclopentane ring fused to a fully reduced phenanthene ring system, similar to testosterone (Figure 13.6). Substitutions at the 3, 5, 9, and 17 positions give a variety of steroids with different properties. The variety and number of structurally related steroids makes the analysis of these compounds extremely complex for both the clinical samples and bulk formulations. Anabolic steroids can be analyzed by a variety of chromatographic techniques, including GC, HPLC, TLC, and supercritical fluid chromatography (SFC). However, the most reliable technique for specificity and sensitivity is GC/MS. Many GC/MS methods exist for the determination of steroids (54, 55). Most of these procedures were designed to detect steroids in biological fluids. Because metabolites and indogenous materials are present in biological samples and steroids are present at very low levels, sensitivity and specificity are absolute necessities for these procedures. To meet these requirements derivatization procedures are universally used to screen, detect, and identify anabolic steroids by GC/MS (56, 57). Most of these derivatization procedures are discussed in the *Handbook of Chromatography: Steroids* (58). The methyloxime-trimethylsilyl ether derivative has become the method of choice for derivatizing steroids. The success of this derivatization procedure is due to the ease by which both carbonyl and hydroxyl groups can be protected (58).

In contrast to clinical samples, bulk formulations provide more than an adequate amount of sample for analysis. Derivatization procedures are really unnecessary for unequivocal identification if adequate resolution can be achieved. Figure 13.7 shows the analysis of a mixture of 19 anabolic steroids on a 30-m × 0.25-mm-i.d. Rtx-5 (0.1-μm film thickness) fused-silica capillary column. Table 13.5 lists the corresponding compounds in this mixture. An analysis time of less than 18 min can be achieved using this column. Film thicknesses greater than 0.1 μm cause longer retention times and a corresponding deterioration in peak shape (59).

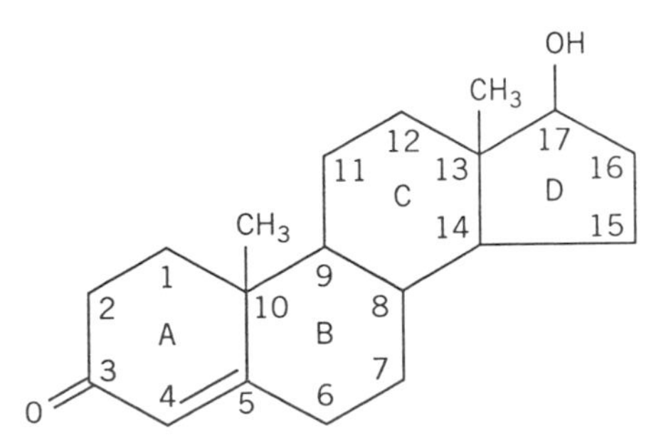

FIGURE 13.6 Structure of testosterone.

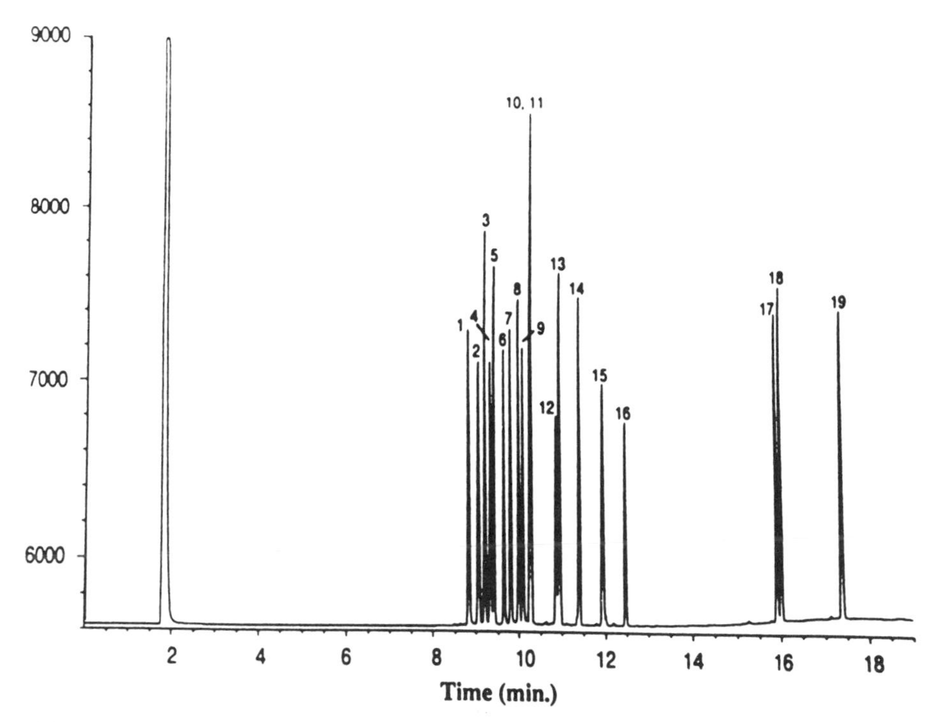

FIGURE 13.7 Separation of anabolic steroids. GC/FID conditions: 30-m Rtx -5×0.25-mm i.d. $\times$ 0.10-μm film; 180–340°C at 10°C/min (hold 3 min); injection = 280°C, detector = 340°C; split ratio = 50/1, 1-μl split injection, concentration = 1000 ng/μL. (Reprinted with permission from Reference 59).

Retention time is also strongly influenced by the choice of stationary phase. Several factors must be taken into account when selecting a column for separating androgens. Retention time and resolution will be affected by the choice of column length, stationary film thickness and polarity, and the temperature program rate. The programmed temperature optimization of a mixture of anabolic steroids found that a program rate of 9°C/min would satisfactorily separate a large number of anabolic steroids using a 12-m $\times$ 0.2-mm-i.d. (0.33-μm film thickness) HP-1 fused-silica capillary column (60). Table 13.6 lists the retention times for a number of steroids using this program and column conditions on both methylsilicone and 5% phenyl-methylsilicone fused-silica capillary columns.

Steroids are usually found in three dosage forms: (1) tablets and capsules, (2) aqueous suspensions (injectables), and (3) oil solutions (injectables). A variety of oils, vitamins, plant sterols, and plant extracts have been found in exhibits suspected of containing anabolic steroids. In addition, fillers and/or caffeine have been substituted in tablets. Mixtures of anabolic steroids are also found in some samples, none of which may be the steroid(s) listed on

TABLE 13.5 Anabolic Steroids in the Mixture of the Chromatogram from Figure 13.7

Peak No.	Anabolic Steroid
1	5-Androstene-3β,17β-diol
2	17α-Methyl-5-androstene-3β,17β-diol
3	5α-Androstan-17β-ol-3-one
4	19-Nortestosterone
5	17α-Methylandrostan-17β-ol-3-one
6	Mesterolone
7	Testosterone
8	17α-Methyltestosterone
9	1-Dehydrotestosterone
10	1-Dehydro-17α-methyltestosterone
11	Bolasterone
12	Oxymethalone
13	19-Nortestosterone-17-propionate
14	Testosterone propionate
15	Fluoxymesterone
16	4-Chlorotestosterone-17-acetate
17	Testosterone-17β-cypionate
18	1-Dehydrotestosterone benzoate
19	1-Dehydrotestosterone undecylenate

the label. For these reasons sample preparation of the steroid products is made somewhat complex because of the various ways in which these samples are found. Sometimes it may be necessary to refrigerate the sample first to allow the oil phase and aqueous phase to separate, subsequently providing a cleaner and more efficient extraction. An analytical method consisting of extraction, TLC, UV spectra, and GC/MS was devised for 13 commonly abused anabolic steroids (61). This extraction is outlined in Table 13.7.

A recently published paper outlined the capillary column GC/IR conditions for the separation and detection of all but one pair of testosterone and its 11 esters (62). Included in this paper are the analysis preparations for these particular compounds. For more information on the analysis and identification of anabolic steroids, the reader is referred to References 16, 63, and 64.

13.6 QUANTITATIVE ANALYSIS OF DRUGS OF ABUSE

The use of GC and GC/MS for quantitative analysis of drugs of abuse is now a routine procedure performed in nearly every forensic drug laboratory. The chromatographic conditions are critical to the success of the method. The major characteristics of the chromatographic system are that

TABLE 13.6 Retention Times of Anabolic Steroids

Peak No.	Steroid	Retention Times (min)	
		HP-1	HP-5
1	Androsterone	8.73	9.84
2	19-Nortestosterone	9.23	10.38
3	Testosterone	9.79	10.98
4	Methyltestosterone	10.06	11.20
5	Norethandrolone	10.61	11.72
6	Testosterone acetate	10.82	11.93
7	19-Nortestosterone 17-propionate	11.17	12.27
8	Testosterone proionate	11.69	12.78
9	Testosterone isobutyrate	12.10	13.21
10	Clostebol	12.76	13.98
11	Stanozolol	13.32	14.81
12	Testosterone enanthate	15.10	16.51
13	19-Nortestosterone benzoate	16.11	18.23
14	Testosterone-3-benzoate	16.85	19.13
15	Testosterone 17β-cypionate	17.28	19.57
16	19-Nortestosterone-17-decanoate	17.80	20.14
17	19-Nortestosterone-17-phenylpropionate	18.07	20.96
18	Testosterone undecanoate	20.30	23.50

Source: Reference 60.

TABLE 13.7 Extraction for Anabolic Steroids

1. Two tablets of the solid dosage are crushed to a fine powder and added to a 10 × 75-mm test tube and 1 mL of methanol[a] is added.
2. If the dosage form is a liquid injectable, 1 mL is placed in the test tube and 1 mL of methanol is added.
3. The mixture is shaken vigorously and vortexed for approximately 30 sec.
4. The emulsion that may form is broken up by centrifuging the mixture at high speed for 2 min.
5. If the top supernatant layer is not clear, it should be filtered through qualitative filter paper.
6. The clear filtrate or supernatant liquid is now ready for UV, TLC, and GC/MS analysis.

[a] Due to the low solubility in methanol, stanozolol and oxandrolone are extracted in dimethyl formamide (DMF) and methylene chloride, respectively.

the gas chromatographic column be thermally stable (low bleed) and inert, provide good peak shape, and provide satisfactory resolution of the analyte and any interfering compound. Fused-silica capillary columns generally fulfill all of these requirements for quantitative analysis of illicit drugs and the excellent resolution achievable with capillary columns contributes to improved sensitivity and specificity.

The first step in the development of a quantitative gas chromatographic procedure is to establish the performance characteristics of the drug in the absence of any matrix. Precision (repeatability), linearity, limit of detection (LOD), and reproducibility all must be demonstrated. Purity of the standard should be established by an independent technique to ensure structural integrity. The linearity range should be compatible with sample and standard availability and should bracket the concentration range of the analyte. The point should be to demonstrate that the chromatographic system is linear, reproducible, and compatible with the desired drug concentration range. Peak heights or areas can be used, depending on the detector. The solvent used should be compatible with both drug analyte and internal standard (if used) and should be one that will serve as the final extraction solvent in the sample preparation. This is important to avoid possible drug–solvent interactions that could result in abnormal detector responses and/or "ghost" peaks.

A "resolution standard" should be used before any quantitative data are generated to determine whether the chromatographic system is performing acceptably. A standard should be selected that closely resembles and behaves in a chromatographically similar way to the analyte drug. Fortunately, a variety of structural analogues is usually available in drug analysis, so that an appropriate selection can be made. A standard should be selected that will assure that the desired separation is reproducible on a day-to-day, basis. For example, Figure 13.4 shows the separation of cocaine and four analogues. This mixture could be used for this purpose.

An internal standard should be used for quantification of illicit drugs. There are two purposes for choosing an internal standard. The first is to compensate for any variations in the injection of the sample and standards. The second function of the internal standard is to improve the reproducibility and buffer against any chromatographic changes that might take place during the analysis. For best results, the internal standard should be structurally similar to the drug analyte such that they have similar polarity and volatility. This will result in optimum partitioning and reproducibility.

The discussion has so far assumed that no derivatization has been necessary for achieving good quantitative data. Derivatization can serve a variety of purposes. Most commonly it is used to improve the chromatographic characteristics of the analyte and hence improve resolution and sensitivity. The ideal derivative should

1. Be easily prepared
2. Be prepared in high yield
3. Have good chromatographic characteristics
4. Be chemically stable

5. Be thermally stable
6. Improve the detector response
7. Be efficiently ionized (if MS is used)
8. Give an abundant ion current at a structurally characteristic mass that is free from ions generated by coextractants (if MS is used) (65)

The derivatization step is an added complication, which can make quantification more difficult and lessen reproducability in some cases. It should be avoided if at all possible. Chemical derivatization of drugs for chromatographic and related analyses have been extensively reviewed (66, 67). Readers are referred to these sources for complete information. Specific applications of derivatization can be found in the separate sections of the particular drug classes covered in this chapter.

GC/MS is used routinely for quantitative analysis of drugs of abuse and quite often is used in the selective ion monitoring mode. Once the fragmentation pattern is known for a particular drug and a suitable internal standard is chosen, a quantitative method using GC/MS can be developed. Generally, the most abundant *m/e* fragments yield the best linearity and reproducibility. It is important to use the sample matrix as a blank, if possible (whether in the form of powders, tablets, or biological fluids, such as blood, serum, or urine), to assure that the analyte drug can be separated from potential interferences. Area or peak height of the selected ions are measured and provide the basis for quantification of the samples. Quantification based on isotopically labeled internal standards yield the best results, since the standards closely mimic the analyte drug chromatographically and give very similar detector responses. Isotopially labeled drug standards are available commercially and are easily obtained in small quantities without a DEA license.

13.7 SOURCE DISCRIMINATION AND IDENTIFICATION

Often the forensic scientist is asked to compare two drug samples to show that they may have come from the same source or that they originated from the same batch of drug. Unlike legitimate pharmaceutical preparations, illicit drug samples are often contaminated with impurities and adulterants. The impurities can originate from a variety of sources, such as the precursors and chemicals used to synthesize the drug, the synthetic procedure, including incomplete reactions and side reactions, decomposition, and handling and packaging of the drug. Taking this information into consideration, it is understandable that chromatographic patterns resulting from illicit drug samples can be used to compare samples. A great deal of information can be learned about the history of the sample and the route of synthesis by these "chemical signature" analyses.

Initially, crime laboratories attempted to accomplish this type of analysis by determining the ratio of the concentrations of the parent drug to that of any adulterant or diluent present in the sample. A more successful technique, however, has been the qualitative and quantitative determination of the impurities in the illicit sample. This has been done for heroin (68), amphetamine (69), methamphetamine (70), hashish (71), and phenmetrazine and morphine (72). The ratio of the impurity to the main drug, rather than its absolute concentration, is used. This is done in order to eliminate the effect of added adulterants and diluents (69). This procedure has been used for the comparison of methamphetamine, heroin, and cocaine specimens.

Cocaine, for example, is a naturally occurring alkaloid that can be extracted from the leaves of *Erythoxylon coca*. Cocaine purity can vary, depending on the extraction and purification process. The amounts and varieties of related alkaloids available for sample comparisons are also dependent on the source of the leaves and the extraction and purification process.

The amount of sample used in the analysis should be as large as the methodology will permit. The following procedure can be used as an example (73):

1. Accurately weigh an amount of sample equivalent to 50 mg of cocaine into a glass-stoppered test tube.
2. Add 1.0 mL of $CHCl_3$ containing 0.5 mg/mL of octacontane and octacosane.
3. Add 1 mL of BSA (bistrimethylsilyacetamide).
4. Heat at 60°C for 15 min.
5. Inject sample.

Figure 13.8 shows the separation of a typical uncut cocaine sample as a result of this procedure. Some of the components present, besides cocaine, are methylecgonine, ecgonine, and benzoylecgonine. These compounds can be used for sample comparisons, but one must be cautious, since they can be formed from decomposition. Also present are *cis*- and *trans*-cinnamoylcocaine. These particular compounds can be used to significant advantage in comparisons, since their ratio and concentrations can vary significantly with geographical origins of the coca plant. Cocaine sample differentiation requires the determination of synthetic or natural origin. Synthetic samples are characterized by the presence of optical isomers, certain diastereoisomers and other by-products, and chemical residues. Samples derived from the coca plant are characterized by the presence of certain natural products and their derivatives, and residual chemicals. The approaches to this differentiation, including chromatography, have been reviewed elsewhere (74).

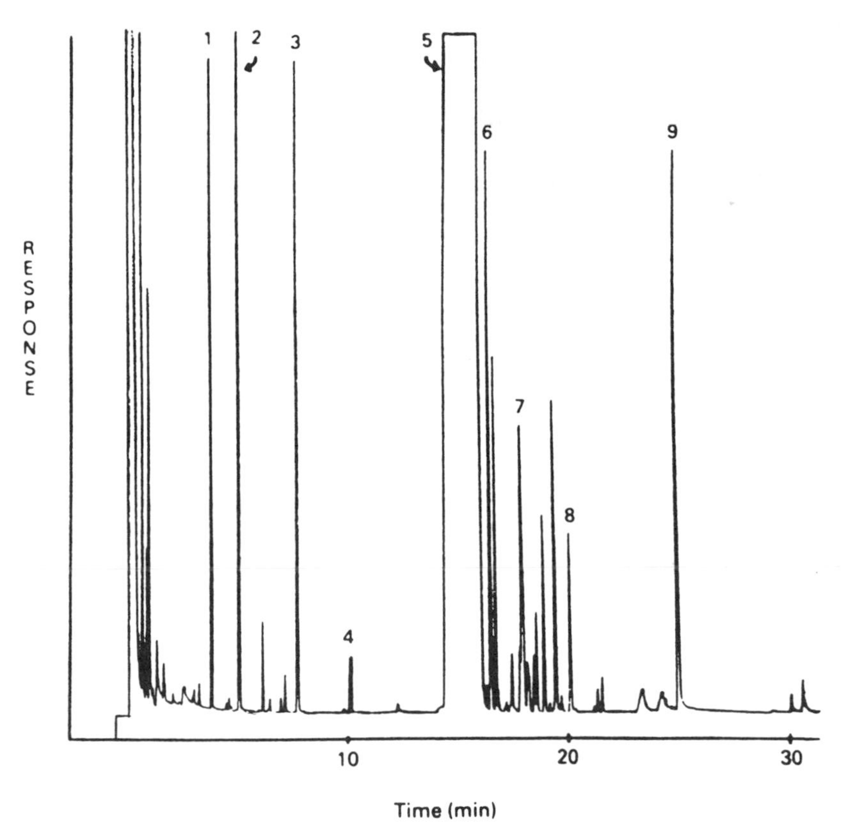

FIGURE 13.8 Separation of impurities in an illicit cocaine sample. 1, ecgonine methyl ester TMS; 2, ecgonine di-TMS; 3, n-C_{18} internal standard; 4, tropacocaine; 5, cocaine; 6, benzoylecgonine TMS; 7, *cis*-cinnamoylcocaine; 8, *trans*-cinnamoylcocaine; 9, n-C_{28} internal standard. Oven temperature program: initial temperature 170°C, initial hold 1 min, heating rate 5°C/min, final temperature 270°C, final hold 10 min. (Reprinted with permission from Reference 7).

13.8 CLANDESTINE LABORATORY ANALYSIS

The analysis of materials seized from clandestine drug laboratories falls into the realm of the responsibilities of the forensic scientist. Because this is essentially a "chemical investigation," the forensic scientist plays a major role in all phases of the investigation. Based on the information gathered the scientist's responsibilities may include the following:

1. Formulation of an opinion as to what drug is being synthesized
2. Determination of synthesis route

3. Estimation or determination of production capability
4. Projection of synthesis time
5. Determination of the degree of hazard to be encountered
6. Determination of the function of laboratory apparatus
7. Preservation and collection of evidence
8. Analysis of evidence submitted to laboratory
9. Aiding clean-up, removal, and destruction of chemicals
10. Acting as scientific advisor to prosecutors and investigators
11. Offering expert testimony

Laboratory analysis in this type of crime normally focuses on the identification of drugs and precursors to determine the synthesis route(s) and to estimate the production capability. GC is a highly effective tool in these analyses. Since many samples may contain complex mixtures of precursors, impurities, and by-products, high resolution is essential and the use of capillary columns are recommended.

The most common drugs clandestinely manufactured in the United States are methamphetamine, amphetamine, MDA and its analogues, PCP, LSD, and methaqualone. The Leuckart reaction has been the most popular method for synthesizing illicit amphetamine in the United States, while illicit methamphetamine has been produced primarily by reductive amination using benzylmethylketone and methylamine. PCP is commonly prepared using precursors such as piperidine, cyclohexanone, and phenyl magnesium bromide. All of these clandestinely manufactured drugs have several different synthesis routes that use different reagents and precursors.

GC has been applied in the analysis of clandestine samples by separating the different components and identifying the precursors and chemicals that have been used in the synthesis. Mass spectrometry has been almost universally employed as the detector in this type of analysis, since unequivocal identification of the components is essential. As an example, Figure 13.9 shows the total ion chromatogram (TIC) from a 2-mg sample of vegetation (mint) adulterated with PCP thermally desorbed directly at 85°C for 5 min. In addition to PCP, the precursors used in the synthesis, cylcohexanone and piperidine, are easily detected. This is a rather unique application involving thermal desorption of the chemicals from the vegetation directly into the injection port of the gas chromatograph. More commonly, a solvent extraction is performed and a sample of the liquid extract is injected into the gas chromatographic column for separation and detection of the precursors.

In contrast to legitimate drug formulations, illicit drug samples are often contaminated with impurities as a result of inadequate purification procedures. As previously discussed, gas chromatographic patterns originating

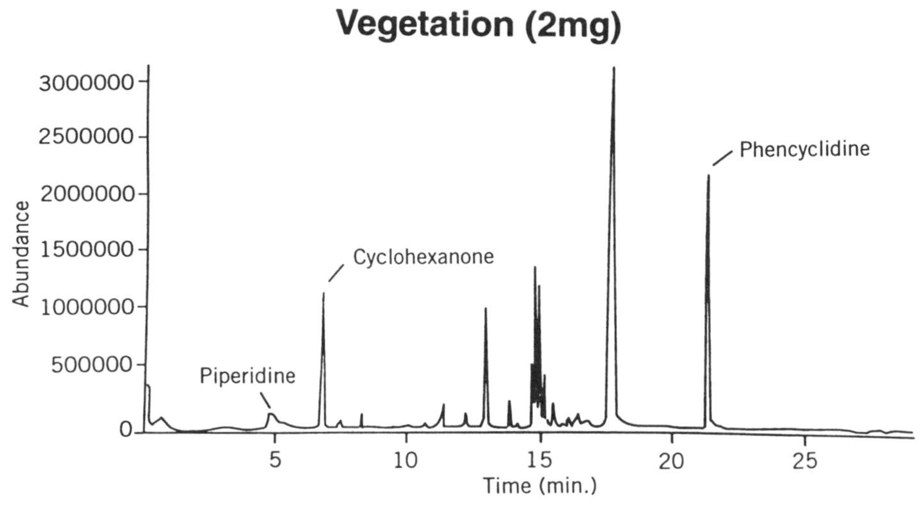

FIGURE 13.9 Total ion chromatogram resulting from volatiles of a 2-mg sample of vegetation (mint) adulterated with phencyclidine (PCP) desorbed directly into the gas chromatographic injection port with a Short Path Thermal Desorption Unit at 85°C for 5 min. GC/MSD conditions: 12-m HP-1 × 0.20-mm i.d. × 0.33-μm film; −10–250°C at 10°C/min; injection = 250°C. Courtesy of New Jersey State Police.

from these drugs contain valuable information about the drug and its synthesis route. In recent years some basic work has been done regarding the nature of contaminants encountered in the different synthesis of drugs of abuse (70).

Part 3 Gas Chromatography in Forensic Toxicology

13.9 APPLICATIONS OF GAS CHROMATOGRAPHY IN FORENSIC TOXICOLOGY

13.9.1 Drug Analysis in Biological Fluids and Tissues

Forensic toxicology is the application of toxicology for legal purposes. The classic example is postmortem toxicology, where specimens from deceased individuals are analyzed to determine whether compounds that were found were a cause of or a contributing factor to the death of the victim. This type of analysis involves detection, identification, and quantification of an array of toxic chemicals and drugs (and metabolites), including alcohol, poisons (and metabolites), and other chemicals, such as solvents and gases. Forensic toxicology also includes the screening of drugs, including alcohol, for the determination of whether someone is under the influence of a particular drug while driving a motor vehicle. Some forensic laboratories are also asked to perform workplace testing of employees and police officers.

Samples submitted to the forensic toxicology laboratory for analysis are either removed by the pathologist at autopsy or are gathered by the arresting officer in the case of a person suspected of driving under the influence (DUI) (at the police station or hospital). These samples normally include blood, urine, brain, kidney, and bile. Other sample types may include stomach contents, foodstuffs or other drug tablets and capsules found near the victim. A vast variety of compounds may be encountered in the analyses, but the majority are drugs and volatiles, such as alcohol.

GC is applied in a variety of ways and is one of the most important techniques in this particular area. GC provides the retention time or retention index (RI) of an unknown substance that can be used for its identification. GC is routinely utilized to separate the analyte from indogenous interferences for more specific identification via mass spectrometry and can also be used to provide quantitative information about the drugs present. The following applications focus on the identification and quantification of drugs and volatiles in biological fluids by GC.

Sample Preparation. The process of screening for drugs of abuse can be divided into two stages: sample preparation and analysis of the sample. The initial step in screening for drugs of abuse is to separate the drug of interest from the biological matrix. This first involves a sample pretreatment step commonly involving dilution of samples such as plasma, serum, and urine. Whole blood can be sonicated and diluted, while tissues are usually treated by either protein precipitation or enzymic digestion. When analyte drugs are present in a conjugated form, deconjugation is required. The use of β-glucoronidase for enzymic hydrolysis of samples is the preferred procedure for the analysis of morphine (75). The main purposes for sample pretreatment are the following (76):

1. Release of drugs from the biological matrix
2. Removal of proteins and particulate matter, which would interfere with further analysis
3. Adjustment of the pH, ionic strength, and concentration of the sample to allow optimum extraction efficiencies

After proper pretreatment of the sample, extraction of the drug from the matrix must be completed. Liquid–liquid extraction procedures have been the most commonly used methods for extraction of drugs of abuse. However, solid-phase extraction (SPE) has gained popularity in recent years because it uses fewer solvents and more samples can be extracted at one time. Presently, many types of SPE materials are commercially available for extraction of drugs. Some contain as many as three different solid phases for extracting acidic, basic, and neutral drugs. When developing an SPE procedure for drug screening each step must be carefully optimized to gain maximum recovery of the particular drug. Following are some of the many factors that affect the recovery of a drug during an SPE:

1. Selection of sorbent
2. pH of sample
3. Flowrate of the sample and eluent passing through the column or disk
4. Properties and volume of solvent wash
5. Properties and volume of solvent eluent
6. Proper pH and type of buffer

Many SPE methods and procedures for the extraction of drugs of abuse have been published in the literature and by manufacturers. These have included automated methods that have used robotic systems. Comprehensive reviews have appeared recently that specifically discuss the SPE of abused drugs in toxicological samples (76, 77).

Screening for Drugs of Abuse. Analysis for drugs of abuse in forensic toxicology is similar to the methodology used in the clinical laboratory, except the drug levels encountered in forensic samples are usually higher. The National Institute on Drug Abuse (NIDA) guidelines (78) have become the standard for drug testing in laboratories that conduct workplace testing for federal agencies. This standard is increasingly being demanded of private laboratories as well and is serving as a model for forensic laboratories performing analyses for drugs of abuse in biological fluids. Analysis generally begins with a screening test. This may be an immunoassay, such as enzyme multiplied immunoassay technique (EMIT) or radioimmunoassay (RIA), or may encompass an array of chromatographic screening techniques, such as GC, TLC, or HPLC. The confirmation test is almost always GC/MS.

Most toxicological drug screening has been done on conventional packed columns with the stationary phase coated onto inert supports (40, 79). In more recent years fused-silica capillary columns have grown in use as forensic toxicologists have become more aware of the higher separation efficiency, resolution, and sensitivity of these columns. In 1981, the Committee for Systematic Toxicological Analysis of the International Association of Forensic Toxicologists recommended SE-30 or OV-1 as the stationary phase of choice on the basis of discriminating power, proven reliability, and general availability (80). Extensive evaluations in different laboratories have shown that dimethylsilicone is the preferred phase when using capillary columns for screening drugs in forensic toxicology. Most laboratories use a combination of dimethylsilicone and phenylmethylsilicone (5–50%) capillary columns for the screening of drugs and poisons. The added polarity of the phenylmethylsilicone phase can help solve some separation problems not feasible with short (12–15 m) methylsilicone capillary columns.

Various drug screening procedures have been described using capillary columns with detectors such as the FID, nitrogen phosphorus detector (NPD), and mass spectrometric detectors (81–83). Figure 13.10 shows the separation of a drug standard mixture on a fused-silica capillary column (12-m × 0.2-mm i.d.) consisting of cross-linked dimethylsilicone with helium as the carrier gas and an FID. The temperature program and chromatographic conditions are shown in the figure. Extracts from urine, blood, or other samples can be screened on this column and unknown drugs can be

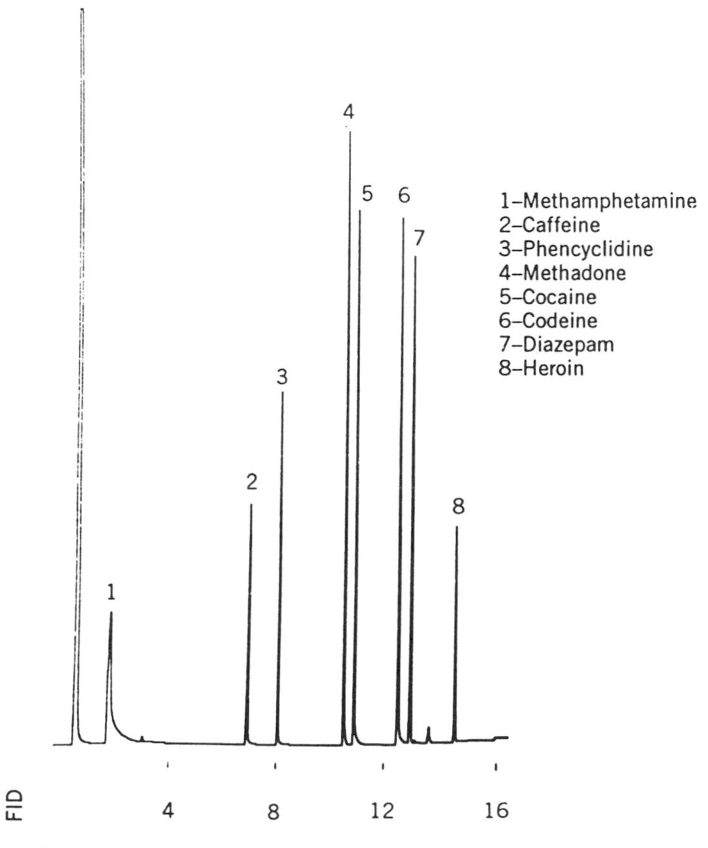

FIGURE 13.10 Separation of methamphetamine, caffeine, PCP, methadone, cocaine, codeine, diazepam and heroin. GC/FID conditions: 12-m HP-1 × 0.2-mm i.d. × 0.33-μm film; 140°C (1 min)–260°C at 10°C/min; injection = 270°C, detector = 300°C; split ratio = 20/1. Courtesy of New Jersey State Police.

tentatively identified by comparing the retention times. The same standard mixture (diluted 1000-fold) is shown in Figure 13.11 separated on a similar column but using an NPD. The responses are different for the compounds on both detectors, but the retention data can be used for preliminary identification. Once the sample is screened, the extract can then be confirmed by using similar chromatographic conditions with GC/MS.

Lillsunde and Korte (84) reported a screening procedure covering 300 substances, including drugs of abuse and metabolites, in which they used a combination of packed and capillary columns. Wide-bore capillary columns have also been used successfully for the screening and confirmation of drugs in forensic toxicological samples (85).

In 1981, the Committee for Systematic Toxicological Analysis of the

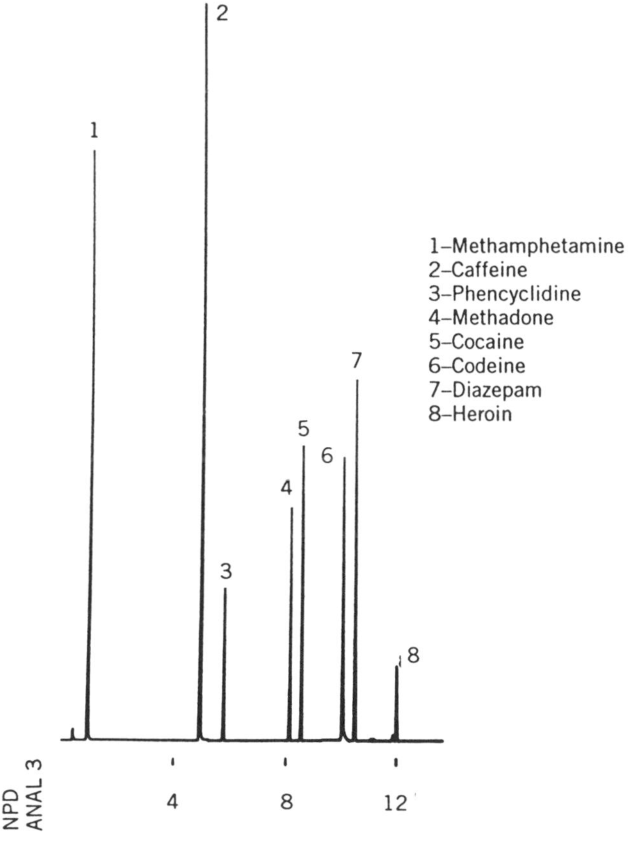

FIGURE 13.11 Separation of methamphetamine, caffeine, PCP, methadone, cocaine, codeine, diazepam and heroin. GC/NPD conditions, 12-m HP-1 × 0.2-mm i.d. × 033-μm film; 140°C (1 min)–260°C at 10°C/min; injection = 270°C, detector = 300°C; split ratio = 20/1. Courtesy of New Jersey State Police.

International Association of Forensic Toxicologists also recommended that retention behavior be expressed in terms of Kovat's retention indices (RI) (80). The interlaboratory standard deviation of RIs at that time on SE-30 or OV-1 packed columns was about 15–20 RI units, so that a search window of about 50 RI units had to be taken into account when trying to identify an unknown compound (40). More recently, J.-P. Franke et al. (86) have suggested using a carefully selected secondary standard drug mixture to improve the interlaboratory reproducibility of RI values, which even under vastly different operational conditions allows a much better search window than that which must be applied when using alkane or substituted alkane homologues. Table 13.8 lists two test mixtures suggested for use with capillary columns. Before starting a gas chromatographic screen for drugs or a study to collect retention indices, the chromatographic system should be checked by means of one of these test mixtures with regard to quality of the column separation efficiency and detection sensitivity.

These test mixtures cover a broad range of RI values and the chromatographic system should be able to detect 100 ng of each component with a good separation of all substances with acceptable peak shape. An example of a separation of mixture B is given in Figure 13.12 (amphetamine, trimipramine, and haloperidol are not shown). The calculation of RIs for individual unknown substances can be accomplished by comparing retention times of the unknown substance to the retention times and retention indices of the "bracketing" standards. The use of nitrogen-containing substances for calibration and for generation of RIs for use with a NPD can also be used

TABLE 13.8 Reference Drug Mixtures for Acidic, Basic, and Neutral Drugs for the Determination of Retention Indices

Mixture A Acidic and Neutral Drugs		Mixture B Basic and Neutral Drugs	
Ethosuximide	RI = 1205	Amphetamine	RI = 1125
Ethinamate	1365	Ephedrine	1365
Barbital	1489	Benzocaine	1545
Aprobarbital	1618	Methylphenidate	1725
Secobarbital	1786	Diphenhydramine	1870
Phenobarbital	1953	Tripelenamine	1976
Heptabarbital	2055	Methaqualone	2135
Primidone	2250	Trimipramine	2215
Phenylbutazone	2367	Codeine	2375
Bis(2-ethylhexyl)-phthalate	2507	Nordiazepam	2490
Prazepam	2648	Prazepam	2648
Clonazepam	2823	Papaverine	2825
		Haloperidol	2930
		Strychnine	3116

Source. Reprinted with permission from Reference 87.

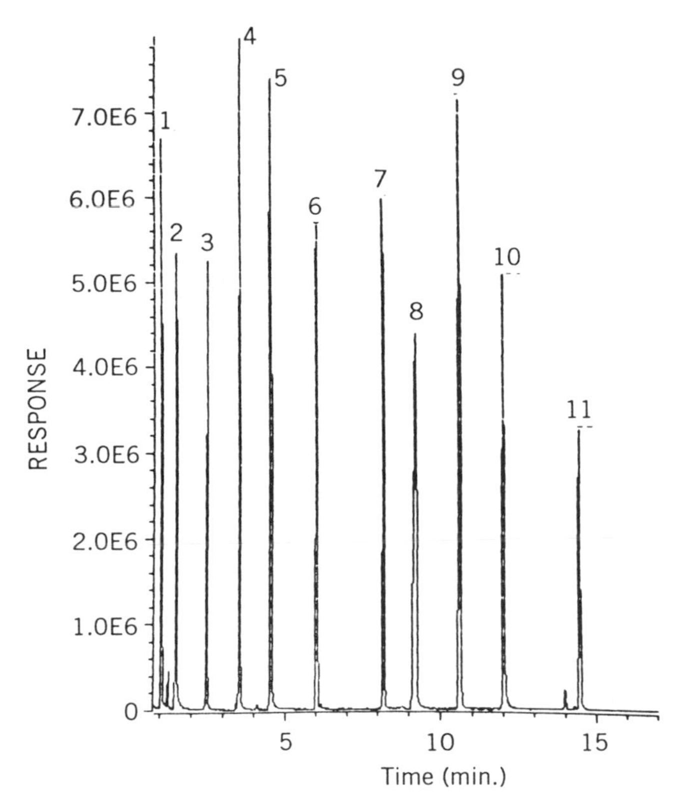

FIGURE 13.12 Separation of reference mixture B (amphetamine, trimipramine, and haloperidol are not shown). 1, ephedrine; 2, benzocaine; 3, methylphenidate; 4, diphenhydramine; 5, tripelenamne, 6, methaqualone; 7, codeine; 8, nordiazepam; 9, prazepam; 10, papaverine; 11, strychnine. GC/MSD conditions: 12-m HP-1 fused-silica capillary column × 0.2-mm i.d. × 0.33-μm film; oven = 190°C (2 min)–300°C at 9°C/min, injection = 270°C; split ratio = 20/1.

successfully (88, 89). An overview on the standardization of chromatographic methods for screening drugs in toxicology by means of retention indices and secondary standards has been reported (90).

Opiates are a major group of abused drugs for which the forensic toxicology laboratory routinely tests. Morphine, codeine, and 6-monoacetylmorphine (6-MAM) are common drugs and metabolites that are routinely identified in drug overdose specimens. Morphine is present in many prescriptions for treatment of pain and cough suppression and is also a metabolite of codeine and ethylmorphine. Because of this, morphine's presence cannot be used solely for identifying heroin use. It has been reported that 6-MAM can be used as an indicator of heroin use, since heroin is metabolized in the body first as 6-MAM and then to morphine (91).

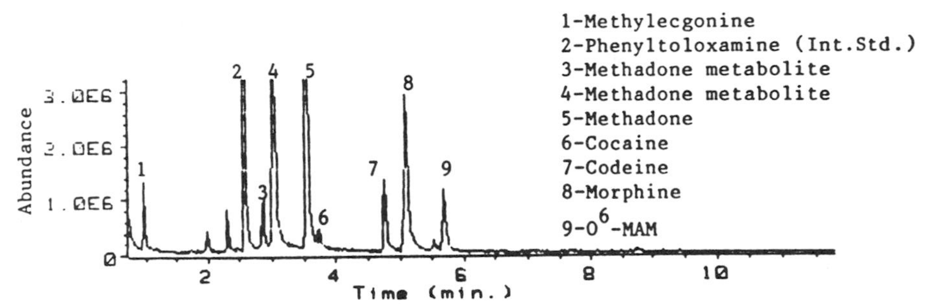

FIGURE 13.13 Total ion chromatogram of a urine extract, GC/MSD conditions: 12-m HP-1 fused-silica capillary column (0.2-mm i.d. × 0.33-μm film); oven = 200–260°C at 15°C/min, injection = 270°C; split ratio = 20/1. Courtesy of New Jersey State Police.

Figure 13.13 shows the total ion chromatogram (TIC) of an extract of urine (5 mL) from a driver arrested for being under the influence of drugs. This particular sample is a good example of a urine containing multiple drugs and metabolites. The presence of methadone, cocaine, codeine, and morphine, and several metabolites, including 6-MAM are shown. Gas chromatographic methods for the determination of 6-MAM in various body fluids have been reported (see Table 13.9). These procedures have used a variety of methods and have included derivatization and GC/MS.

Cocaine is metabolized into two major metabolites, benzoylecgonine (BZE) and ecgonine methyl ester (EME), and to a lesser extent into other metabolites such as norcocaine, and ecgonine. In addition, cocaine is unstable in aqueous solutions, including urine, above pH 5, and also in blood. All of these factors make the interpretation of cocaine concentrations difficult. The normal procedure for analysis of biological fluids generally involves the identification of cocaine in addition to its metabolites. Gas chromatographic procedures for the detection of cocaine and its metabolites in biological fluids are numerous and diverse. Cocaine and EME can be identified by GC/MS without the need for derivatization; however, methods for the determination of BZE and ecgonine usually require some type of derivatization. Table 13.9 lists some examples of derivatization used to chromatograph these metabolites. The majority of the methods for the identification of cocaine and its metabolites are done on short, narrow-bore, fused-silica capillary columns coated with methylsilicone or phenylmethylsilicone and chromatographed using programmed-temperature runs. GC/MS is universally used for the confirmation of the drugs.

A new metabolite has been detected recently in cocaine cases where ethanol has been present. This metabolite, cocaethylene, appears in concentrations in the blood similar to cocaine and can be chromatographed using the same chromatographic conditions as cocaine with no derivatization necessary (105).

Amphetamine and methamphetamine are the most commonly abused

TABLE 13.9 Gas Chromatographic Methods for the Determination of Drugs and Metabolites in Biological Fluids

Drug	Derivative	Reference
6-MAM	Propionyl ester	92
	MBTFA	93
Morphine, codeine, and 6-MAM	Acetic anhydride (deuteratcd)	94
Codeine and morphine	Silylation	95
	Pentafluoropropionylation	96
	Trifluoroacetylation	91, 97
	Heptafluorobutyrylation	98
	Acetylation	99
Cocaine and BZE	Butyl ester	100
	Trimethylsilyl	101
	Propyl ester	102
	Pentafluoropropionylation	103
	Hexafluoroisopropylation	104
Amphetamine and methamphetamine	Heptafluoro	84
	Heptafluorobutylation	107
	Trifluoracetic anhydride	109
	Trichloroacetylation	110
	N-Trifluoroacethyl-L-prolyl chlorides	111
	MTBSTFA	112
	4-Carbethoxyhexafluorobutyryl chloride	113
	Perfluorooctanoyl chloride	114
Barbiturates	DMF dipropylacetal	117
	TMAH	118
Phencyclidine	Heptafluorobutyric anhydride	123
	BSTFA/TCMS	124
THC and metabolites	3-Pyridine diazonium chloride	125
	Hexafluroisopropyl/ penta fluoropropionyl	126
	Pentafluorobenzyl bromide	127
	TMAH and DMSO	116
	MSTFA	128
	tert-Butyldimethylsilyl	129

central nervous stimulants (CNS) and consequently are often found in toxicological specimens. Diethylpropion, phentermine, and phendimetrazine, as well as MDA and MDMA, are also found in biological samples. Methodology for the determination of therapeutic levels of underivatized amphetamine have been reported for whole blood (106), urine (107), and plasma and urine (108). Generally, these compounds can be screened using capillary columns with relatively low oven temperatures (150°C) and FID.

Nitrogen–phosporus detectors can be used used to increase the sensitivity and reduce background peaks when screening for low levels of amphetamines in biological samples (see Figures 13.10 and 13.11). Like most other drug analyses, confirmation of abused CNS drugs is accomplished by GC/MS.

When screening for CNS type drugs the free base is normally chromatographed and detected with the use of the FID and/or the NPD. When analyzed without derivatization, however, peak tailing and resulting sensitivity problems are often encountered. To correct these problems, a variety of derivatization reagents have been employed. Derivatization is almost always used to confirm the identity of the amphetamines with GC/MS because the free-base amphetamines do not give high molecular weight ions, which results in mass spectra that are not very discriminating. The most common derivatizing agent for amphetamine and methamphetamine is trifluorocetic anhydride (109); however, Hornbeck and Czarny examined several different derivatizing agents and found that the trichloroacetyl derivatives are the best for the analysis of amphetamines (110). Table 13.9 list some derivatization procedures that have been used to chromatograph these compounds, including chiral reagents.

New methods for identification of amphetamine and methamphetamine in urine have been employed using GC/Fourier transform IR (GC/FTIR) spectroscopy (115). These methods have provided identification of the amphetamines and metabolites at the low picogram levels. Developments in cryogenic sample deposition for GC/FTIR spectroscopy have allowed the highly selective ability of IR spectroscopy to be used for identification and quantification of these drugs.

Barbiturates are commonly detected in forensic specimens from both overdose cases and driving while intoxicated (DWI) cases. Butalbital and phenobarbital are two of the most commonly abused barbiturates. The determination of barbiturates in biological fluids has been routinely accomplished with both packed columns and wide-bore capillary columns. The formation of the *N*,*N*′-dimethylderivatives of the 5,5′-disubstituted barbiturates is a common procedure used in many clinical and forensic laboratories to reduce the adsorption on the column and peak tailing. Mule and Casella (116) reported a detection limit of 20 ng/mL in human urine using GC/MS with this procedure. The barbiturates have been detected using other derivatization techniques, including the use of on-column derivatization with trimethylanilinium hydroxide (TMAH) (see Table 13.9).

The 1,4 and 1,5 benzodiazepines are among the most prescribed tranquilizers, hypnotics, and muscle relaxant drugs available today. Hence, they are frequently abused and often found in DWI or drug overdose cases in combination with other drugs or alcohol. Benzodiazepine concentrations are much higher in the urine than the blood so most procedures are designed to detect these compounds in the urine. Benzodiazepines and their metabolites are normally excreted as the glucuronide conjugates and require either acid or enzyme hydrolysis for good recovery. Hydrolysis of the benzodiazepines yields the corresponding benzophenone, which can be identified by GC/MS and related back to the parent benzodiazepine. In some cases, however, the specific benzodiazepine cannot be identified because some benzodiazepines yield the same benzophenone after acid hydrolysis. In addition, some benzodiazepines yield the same metabolites. For example, diazepam and chlordiazepoxide both metabolize to desmethyldiazepam and oxazepam.

To eliminate this problem, it is possible to derivatize the benzodiazepines using BSTFA to form their trimethylsilyl derivatives (116). This method has been reported to give a limit of detection of 50 ng/mL for all of the dicyclobenzodiazepines on two different capillary columns. ECD has gained popularity in recent years because of the increased sensitivity over the FID or NPD and also due in part to the lower therapeutic doses of the newer, more potent benzodizepines, such as triazolam and alprazolam (119). Triazolam and alprazolam give well-defined peaks as underivatized drugs, while anhydrides (trifluoroacetic, pentafluoropropionic, and acetic) have been used to form the esters of the metabolites. The acetic anhydride derivatives produce the greatest yields with few apparent decomposition products (120).

The identification and quantification of lorazepam in blood is of forensic interest. A specific and sensitive analytical method is required because of the low concentrations normally detected in blood. Lorazepam has been quantified by GC/ECD without derivatization and after hydrolysis of the parent drug to the benzophenone. Trimethylsilyl and heptafluorobutyryl derivatives have also been used for quantification of lorazepam in biological fluids. A successful application of GC/negative ion chemical ionization mass spectrometry (NICIMS) for the analysis of lorazepam and triazolam in postmortem blood has reported a detection limit of 0.5 ng/mL (121). This method used a fused-silica capillary column (DB-1, 15-m $\times$ 0.25-mm i.d., 0.1-μm film thickness) coupled directly to the ion source of a mass spectrometer set up in the negative chemical ionization mode with methane as the reagent gas.

Phencyclidine (PCP) is rapidly metabolized in the body and excreted in the urine as several hydroxy metabolites and the parent drug. Many gas chromatographic methods have been developed for the detection of PCP. Several of these methods have incorporated the identification of PCP with a general gas chromatographic screening of different drugs, such as that shown in Figure 13.10. Vereby and DePace described a method for the confirmation of PCP that used a 15-m capillary column and an FID that gave a detection limit of 5–10 ng of extracted PCP (122). The detection of such low levels was previously hard to achieve using packed columns and FID without derivatization due to the adsorption on these columns. Now most procedures use short capillary columns and either FID or NPD for screening and mass spectrometric methods for confirmation. Other procedures are referenced in Table 13.9 including Holsztynska and Domino's (123), which used conventional packed column, derivatization, and NPD to achieve a detection level of 5 pmol per injection.

There are numerous reports for the gas chromatographic determination of THC and its metabolites, 11-nor-Δ-9-tetrahydrocannibinol-9-carboxylic acid (THC-COOH) and 11-hydroxy-Δ-9-tetrahydrocannabinol (11-OH-THC) in urine and blood. THC is not normally found in urine, so it must be determined in blood at levels around 2–4 ng/mL. The TMS derivative is the most widely used derivatization procedure with GC/MS for the determination of cannabinoids. In addition to the obvious advantages of derivatizing the THC metabolites, the acidic constituents of cannabis must be derivatized because they can easily decarboxylate above 80°C. Most gas chromatographic procedures today use fused-silica capillary columns for this analysis. Determination of THC in blood is routinely done in forensic toxicological samples, and the detection and quantification of the two THC metabolites in urine is a routine procedure for proof of cannabis use in workplace testing. Several of the procedures used for this type of analysis are listed in Table 13.9.

Lysergic acid diethylamide (LSD) is rapidly metabolized in the body and hence less than 1% is excreted unchanged. This makes LSD very difficult to identify since dosages are in the 100-μg range. In addition, LSD must be derivatized, usually as the trimethylsilyl derivative, which degrades very rapidly in the presence of water. Also samples must be stored away from sunlight, since this also adds to the degradation. Short fused-silica capillary columns (12–15 m) GC/MS, and derivatization with BSTFA have been successful in detecting LSD in biological fluids (130). Using the trimethlsilyl derivative, a detection limit of 10 pg/mL in urine has been reported (131). Even with these procedures the column must be conditioned to neutralize excess silicic acid, which reduces sensitivity. The *N*-trifluoroacetyl derivative of LSD has been used along with GC/negative chemical ionization mass spectrometry to measure less than 100 pg/mL of LSD in plasma (132).

Two other groups of drugs encountered in casework samples are the antipsychotics and anti-inflammatory drugs. The phenothiazines, such as chloropromazine and its analogues, and the tricyclic antidepressants, such as amitriptyline, nortriptyline and imipramine, account for the majority of antipsychotic drugs that are normally detected. The most frequently used procedures for the detection of these drugs are GC/FID, GC/NPD, and GC/ECD, with the NPD being the method of choice. Orsulak et al. (133) have reviewed the procedures for antidepressants.

Aspirin, acetaminophen, ibuprofen, ketoprofen, and indomethacin are the most widely detected anti-inflammatory drugs detected in the forensic toxicology laboratory. Generally, the gas chromatographic analysis of these particular drugs involves the formation of the methyl derivative with iodomethane and potassium carbonate, not with TMAH.

Analysis of Unconventional Samples. Historically, blood, urine, organs, and other tissues have been the common forensic specimens chosen for analysis of drugs of abuse and poisons. More recently however, interest in unconventional samples such as hair, nail, saliva, and sweat has increased in the forensic field. This interest has largely been due to the several potential advantages over current drug methodologies that employ body fluids. These samples are noninvasive and samples such as hair and nails retain drugs over long periods of time, providing valuable information on the degree and pattern of drug use.

The analysis of hair for drugs of abuse has received considerable attention recently. Many laboratories are now reporting the ability to confirm the presence of drugs of abuse and drug metabolites in human hair. Among the drugs confirmed are cocaine, opiates, amphetamines, and PCP using a combination of techniques such as immunoassay, GC/MS, and MS/MS. For quantitative analysis, it is difficult to obtain a representative sample because the drugs are not uniformly distributed along the shaft of the hair or between hairs. In addition, other considerations complicate the analysis, such as extraction of the drug and environmental contamination.

Despite these difficulties, laboratories have been successful in detecting and quantifying drugs of abuse in hair. Welch et al. (134) have described the development of a standard hair reference material and a method for quantifying cocaine, benzoylecgonine, codeine, and morphine using GC/MSD. The method used a 1- to 5-μL splitless injection onto a DB-5, 12-m $\times$ 0.20-mm-i.d. fused-silica capillary column (0.33-μm film thickness). Selected ion monitoring with isotopic dilution was used for quantification. The limit of detection was reported to be at 0.5 ng/mg of hair. Extraction of hair from a cocaine user with 0.1 N HCl at 45°C overnight gave a high recovery of both cocaine and benzoylecgonine.

Since saliva levels of many drugs correspond to the concentration in plasma, interest has grown in this specimen for use in forensic investigations. One of the advantages of saliva as a sample is the minimal requirements in sample preparation. Drugs in saliva are usually extracted by liquid–liquid extraction or by SPE. Deproteinization with methanol and perchloric acid prior to extraction may be necessary. The use of saliva in the forensic detection of drugs has been reviewed by Caddy (135). Included in this report are a number of systems and references for the gas chromatographic analysis of drugs in saliva.

Nail analysis has not been widely utilized in forensic toxicology, but some methods have been reported using GC (136). Nail samples receive similar treatment to hair samples for extraction of the drug.

The analysis of sweat for detecting drugs is rarely performed because it is extremely difficult to estimate drug levels. Sweat samples are collected on gauze or cotton by wiping the surface of the skin, eluting with water, and extracting by liquid–liquid extraction. The detection of cocaine, morphine, and amphetamine have been reported in sweat as well as drugs in perspiration stains (137).

A novel "patch" approach of sweat collection has been introduced recently for testing of drugs. Though not yet cleared for drug-testing purposes, the adhesive patch is approved for collecting perspiration, and may eventually prove to be a viable technique. The device consists of an adhesive layer on a 2 × 3-in. transparent film that adheres to the skin. The patch contains an adsorbent pad in its center that collects sweat as it exits the body. The device is intended to be worn for up to two weeks at a time, after which the collected sweat residue is removed by a simple extraction procedure. So far, the sweat patch has been effective in monitoring low levels of cocaine and heroin.

Insect larvae have also been a source for drug detection in death investigations. For example, cocaine and benzoylecgonine have been determined in insect larvae found on a decomposed body using GC/MS (138).

13.9.2 Analysis of Ethanol and Other Volatiles

GC is the most widely used technique for identification and quantification of ethanol in biological fluids. The fact that ethanol has a low molecular weight and high vapor pressure and can be chromatographed easily on polar liquid phases makes GC the technique of choice. In addition, ethanol and other volatiles can be quantified and identified simultaneously. Blood alcohol analysis in driving while intoxicated (DWI) cases is the most often requested analysis in the forensic toxicology laboratory. Since the relationship between blood alcohol concentration (BAC) and driving impairment is well established, laws have mandated a BAC level above which driving is considered unsafe and prohibited.

BAC levels are also necessary for death investigations, since fatal accidents may involve alcohol as a contributing factor. Deaths due to acute or chronic effects of alcohol alone or in combination with other drugs may also require determination of BAC levels. In addition, blood or urine determination may also be part of the protocol for many workplace drug testing programs.

Many other toxic volatile liquids and gases in addition to ethanol are abused or are involved in death investigations, including fatal accidents. GC is also used for their determination. For example, methanol (wood alcohol), isopropanol (rubbing alcohol), ethylene glycol (antifreeze), solvents such as toluene (glue), acetone, 1,1,1-trichloroethane, Freon, volatiles from commercial products, particularly butane, and nitrous oxide have all been determined by GC in forensic investigations. In addition, fire-related deaths require the analysis of body fluids for carbon monoxide, hydrogen cyanide, and nitriles. Gas chromatographic procedures for all these compounds and others have been described. Table 13.10 lists some volatile liquids and gases with corresponding references to specific gas chromatographic methods used for their determination in forensic applications.

GC of the nonhalogenated volatiles is typically done on the same system

TABLE 13.10 Examples of Toxic Volatiles and Gases Determined by GC in Forensic Investigations

Volatile	Reference
Acetonitile	141
Acrylonitile and acetonitrile	142
Butanol	143
Carbon monoxide	144
Cyanide	145
Cyclopropane	146
Kerosene	147
C_2ClF_5 and $CHClF_2$	148
Phenol	149
Enflurane	150
Methanol	151
Propane and ethyl mercaptan	152
Paint thinner	153
Toluene	154

as ethanol. This procedure normally uses an FID and polar liquid phase such as Porapak Q or Carbowax 20M. The temperature may be modified from the routine blood alcohol procedure and a thick-film capillary column is recommended for best resolution of volatiles. The analysis of chlorinated and fluorinated volatiles requires the use of ECD since the sensitivity of FID does not meet the requirements of forensic samples. GC/MS procedures have also been applied to the screening of volatiles in biological specimens but even these procedures are sometimes lacking necessary sensitivity and may require specialized techniques, such as enhanced mass resolution. For further information on the analysis of volatile substances in toxicology the reader is referred to Reference 139. A book has been recently published which provides reliable gas chromatographic retention indices for volatile substances frequently encountered in analytical toxicology (140).

Determination of Ethanol in Biological Fluids. GC is used to determine ethanol in blood from DWI suspects and in autopsy specimens, which may include blood, urine, vitreous humor, spinal fluid, and organs such as brain tissue. Urine samples are often analyzed for ethanol, but the variation in conversion of urine to blood ethanol values leaves the result of little forensic value.

Blood samples are normally collected into an evacuated tube containing preservatives and anticoagulants such as sodium fluoride and potassium oxalate. The collection tube should be 75% or more filled with blood to reduce the risk of any loss of volatiles. It should also be properly sealed, labeled, and stored under refrigeration. If the tube is improperly resealed after analysis or if it is not refrigerated for extended periods of time, volatiles may be lost.

The gas chromatographic determination of ethanol is a well-established procedure. FID has been the universal detector, but TCD and MS have been used for some applications. Polar liquid phases are normally employed with packed columns. The phases currently on the market that are used for this purpose include Porapak Q/S, Carbowax 20M, and Carbowax 20M on Carbopack B. Most analyses are performed using packed columns, but more laboratories are finding the use of megabore columns useful for this procedure.

Specimens may be subjected to distillation, protein precipitation, or solvent extraction for the separation of ethanol from the biological fluid. Most recently, the determination of ethanol and other volatiles is accompanied by introducing the sample into the gas chromatograph either directly as a liquid or as a gas from the headspace.

Direct Injection Technique. When using the direct injection technique, the liquid blood sample can be either injected directly or diluted prior to injection. The sample may also be extracted prior to the injection. Sample volume is typically 1–3 μL when using this procedure and it is not uncommon for the needle to clog. Also, many nonvolatile components of the blood and other samples are injected into the injection port and consequently lodge onto the column requiring frequent maintenance. Many of the earlier methods used direct injection and these have been thoroughly reviewed (155). The recommended procedure for direct injection is as follows:

1. Homogenize clotted or inhomogeneous samples.
2. Dilute samples (1:10) with aqueous internal standard to minimize maintenance.
3. Mix samples thoroughly.
4. Cap and analyze immediately (inject 1–3 μL).

A diluter/pipetter will give better reproducibility and allow more sample to be handled in a shorter amount of time than diluting by hand. The method described by Jain (156) is a typical example of a direct injection procedure. This method simply mixes 0.5 mL of blood and 0.5 mL of internal standard solution (50 mg/100 mL isobutanol) with no extraction. Then 0.1–0.5 μL is injected onto a column of Carbowax 20M on acid washed 60/80-mesh Chromosorb W. The oven temperature is usually 100–130°C, with an injection port temperature of 160°C and an FID temperature of 200°C.

Static Headspace Procedure. According to Henry's law the concentration of ethanol in the headspace of a blood sample in a closed vial is directly proportional to the concentration of ethanol in the blood solution when the system is in equilibrium. Thus the concentration of ethanol in the blood can be determined by measuring the peak area, or height, of a chromatographic peak resulting from a static headspace sample. The principles of Henry's law are described in Chapters 1 and 14 of this book.

Static headspace GC was originally developed for the determination of ethanol in blood, and today it is the method of choice for this application. Use of this technique offers distinct advantages over direct injection methods. Most importantly is prevention of contamination of the column and syringe. The literature has numerous publications in this area describing various methods and studies of different factors affecting the determination of ethanol in blood. The most recent and perhaps state-of-the-art methods for the determination of ethanol involve automated headspace analysis using chromatographic systems that are capable of analyzing 30 samples in sequence. The recommended procedure for static headspace sampling is as follows:

1. Homogenize clotted or inhomogeneous samples.
2. Dilute samples (1:10) with aqueous internal standard (i.e., *n*-propanol) (may contain sodium chloride or sodium sulfate).
3. Mix samples thoroughly.
4. Cap and place in constant temperature bath.
5. Equilibrate and inject (25 μL–10 mL).

A diluter/pipetter will give better reproducibility and allow more samples to be handled in a shorter amount of time. The use of a high dilution factor and the addition of salt to the samples eliminates any differences between blood/air and water/air partition ratios, therefore allowing the use of aqueous standards for calibration. The addition of salt increases the volatilization of the ethanol and the internal standard, lowering the liquid/air partition ratio and improving sensitivity. This technique is called "salting out".

It should also be noted that at elevated temperatures ethanol can become oxidized to acetaldehyde, so the gas chromatographic procedure must separate ethanol from acetaldehyde. The author's laboratory uses a 6-ft glass column (2 mm i.d.) packed with 5% Carbowax 20M on Carbotrap B 60/80 mesh. When the oven temperature is held at 75°C isothermally, with a nitrogen flow of 30 mL/min, this column will separate ethanol from acetaldehyde with baseline resolution. Figure 13.14 shows the separation of a headspace injection of a mixture of volatiles including ethanol, acetaldehyde, and the internal standard, *n*-propanol. This separation is completed within 4 min. Figure 13.15 shows the separation of a headspace injection of a typical case sample. Dubowski has detailed a headspace procedure that uses an internal standard (acetonitrile) in the *Manual for the Analysis of Ethanol in Biological Liquids* (157).

The practice of breath collection onto a silica adsorbent for later analysis

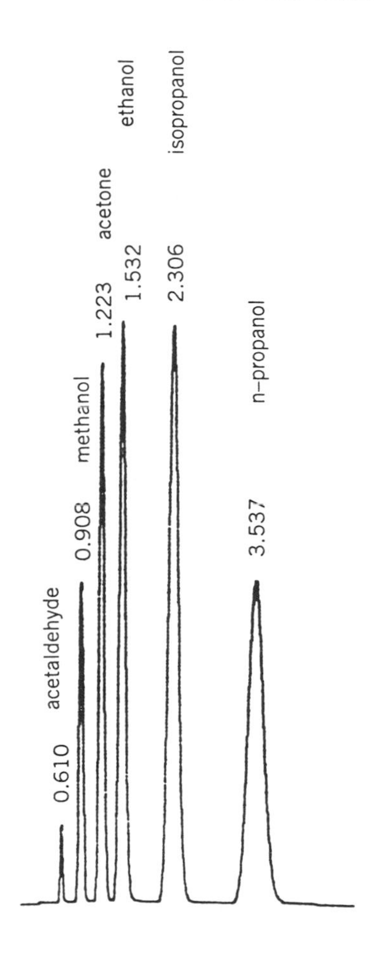

FIGURE 13.14 Gas chromatogram of a 1-mL headspace sample of a volatile mixture containing acetaldehyde, methanol, acetone, ethanol, isopropanol, and *n*-propanol. GC/FID conditions: 6-ft glass column packed with 5% Carbowax 20M on 60/80 Carbopack B; oven = 75°C; injection = 150°C, detector = 200°C; loop = 50°C. Courtesy of New Jersey State Police.

to compare to the results of a breath-testing devise is currently being performed in some states in the United States. The contents are emptied into a vial, diluted with an aqueous internal standard solution (*n*-propanol) and analyzed by headspace GC using procedures similar to those for blood alcohol analysis, but adjusted for sensitivity differences. Reanalysis of breath samples collected in this manner is not recommended, however, due to factors other than instrument performance, such as sample collection and operator errors.

Miscellaneous. GC is commonly used to perform the above procedures routinely in many forensic toxicology laboratories around the world. The precision of the blood alcohol procedure should show a coefficient of variation of 3% or less on replicate analyses, and should be accurate to 5% compared to primary standards. Samples should be run in duplicate and

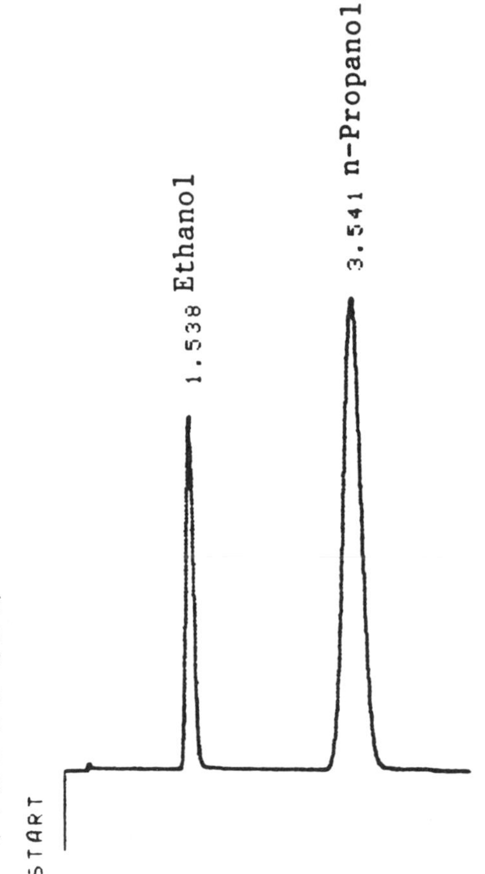

FIGURE 13.15 Gas chromatogram of a 1-mL headspace sample of blood containing 0.100% (w/v) ethanol. GC/FID conditions: 6-ft glass column packed with 5% Carbowax 20M on 60/80 Carbopack B; oven = 75°C; injection = 150°C, detector = 200°C; loop = 50°C; internal standard: *n*-propanol (0.100% w/v). Courtesy of New Jersey State Police.

blanks should be run periodically to demonstrate that the system has no carryover. Quality assurance primary standards should be run periodically throughout the analyses to ensure linearity of the method. Documentation of the results of the standards, as well as instrument certification, maintenance history, and proficiency testing, is mandatory for good quality assurance. Chain of custody of samples and good record keeping are also mandatory in this area.

Forensic laboratories typically report BAC values differently than clinical laboratories. Forensic BAC values are reported consistent with the state statute, such as grams per 100 mL of blood (g/100 mL). Most clinical values are reported in mg% or mg/dL. In addition, most hospital laboratories report the results of serum alcohol, which is approximately 1.16 times higher than a whole blood reading.

Part 4 Applications of Gas Chromatographic Analysis of Trace Evidence

13.10 DETECTION OF ARSON ACCELERANTS WITH GAS CHROMATOGRAPHY

13.10.1 Introduction

Among the various responsibilities of the forensic science laboratory is the examination of evidence from the scenes of suspicious fires. GC has universally been the method of choice for analysis of fire debris, with the analysis normally centering around the detection of accelerants. An accelerant can be defined as any material that insures ignition, accelerates the rate of fire spread, or promotes the spread of fire (158). For a comprehensive listing of terms used in fire debris analysis the reader is referred to a published glossary of terms (159).

The most commonly used accelerants are commercial flammable or combustible liquids, such as gasoline, kerosene, paint thinners, charcoal lighter fluids, alcohols, mineral spirits, and fuel oils. In the investigation of a suspicious fire, fire investigators first identify the location of the origin(s) of the fire and then the source of ignition. An obvious indication that a fire has been deliberately set is the severity of damage or unusual burn patterns indicative of the presence of an accelerant. Detection and identification of accelerants provides the investigator with scientific proof that the fire was incendiary and may help link the suspect to the crime.

Debris recovered from the fire scene is often wet and burned, and may consist of material such as wood, carpet, carpet padding, tile, and other synthetic materials. All of these can contribute interfering volatile pyrolysis products that can make the identification of the accelerants difficult. The loss of accelerants through adsorption into the debris, evaporation from the heat of the blaze, and the presence of water all contribute to make the identification of accelerants a challenging task. GC can be a powerful tool in the analysis to separate and identify the accelerant in the presence of these interferences. Several reviews have been published on various aspects of fire investigation in the laboratory, including those by Midkiff (160), Willson (161), Camp (162), The Forensic Science and Engineering Committee (163), and Stafford (164). An excellent chapter has also been written by Fultz and DeHaan on GC in arson and explosive analysis (165).

13.10.2 Collection and Packaging of Evidence

Because of the high vapor pressure of accelerants, the debris from the fire scene must be packaged in vapor-tight, clean, unused containers for transmittal to the laboratory. It is important for the laboratory to periodically check containers used by investigators for any contamination that could interfere with the chromatography. Clean, new, unused metal paint cans with friction lids are the most commonly used container to package fire debris. These cans come in a variety of sizes and are available lined to prevent rusting. Glass jars can also be used to package fire debris but have the disadvantage of being breakable. The rubber sealant in the lids may also be destroyed by liquid solvent or vapors in the sample.

Plastic bags have been investigated for packaging fire debris evidence. Polyethylene bags are permeable to hydrocarbon vapors and not suitable for arson evidence. Nylon film bags have been effective in retaining volatiles but can be difficult to seal. These bags have gained some popularity, especially in Great Britain, but are not as popular in the United States, although one U.S. manufacturer has released nylon bags that are heat sealable, overcoming the sealing problem. DeHaan and Skalsky (166) in 1981 evaluated heat-sealable polyester/polyolefin bags sold by KAPAK and found that they were effective for packaging fire debris. In 1985, however, the filmstock changed and the bags were found to contain traces of a medium petroleum distillate (167). In 1990, a new source of filmstock was used and the bags were found to be free of contamination (168). This example stresses the importance for the forensic science laboratory to periodically test the containers that are being submitted to their laboratory for background contamination.

13.10.3 Chromatographic Characterization of Accelerants

GC is the method of choice for the detection and characterization of accelerants from fire debris. Since petroleum products are by far the most common types of accelerants and because GC is used to characterize the type of accelerant, the forensic scientist must have a basic understanding of petroleum products and their manufacturing process.

The refining and manufacturing of petroleum products is basically a distillation procedure, with the commercial products being distributed accordingly. Fortunately for the analyst, only those products with high vapor pressure and low flash point are used as accelerants. GC can easily separate these compounds and, in fact, when using temperature programming of the gas chromatographic oven, the accelerants can be placed into a relatively simple classification scheme based on overall chromatographic retention patterns and refining processes.

Table 13.11 shows the classification scheme adapted by the American Society of Testing and Materials (ASTM) in 1991 (169). This is similar to a scheme originally proposed from studies by the National Bureau of Standards and Bureau of Alcohol, Tobacco and Firearms in 1982. This classification is based on the elution of specific compounds within a retention time window defined by *n*-alkane carbon number. For example, for an accelerant

TABLE 13.11 Accelerant Classification Scheme

Class No.	Class Name	Retention Time Windows[a]	Examples
1	Light petroleum distillates (LPD)	C_4–C_8	Petroleum ethers, pocket lighter fuel, some rubber cement solvents
2	Gasoline	C_4–C_{12}	Automotive gasoline
3	Medium petroleum distillates (MPD)	C_8–C_{12}	Paint thinner, mineral spirits, some charcoal starters
4	Kerosene	C_9–C_{16}	No. 1 fuel oil, Jet-A (aviation) fuel, insect spray, some charcoal starters
5	Heavy petroleum distillates (HPD)	C_{10}–C_{23}	No. 2 fuel oil, diesel fuel
0	Unclassified	Varies	Single compounds, such as alcohols, acetone, or toluene; xylenes, isoparaffinic mixtures, some lamp oils, camping fuels, lacquer thinners, duplicating fluids, others

[a] Based on *n*-alkane carbon numbers. (From Reference 169.)

to be classified as a light petroleum distillate, the chromatogram must have at least four major peaks in the n-C_4 to n-C_8 range with no major peak above n-C_8. Examples of light petroleum distillates are listed in Table 13.11. Figure 13.16 shows the chromatogram for a headspace sample of CAM 2 racing fuel, a light petroleum distillate. Similarly, for an accelerant to be classified as a medium petroleum distillate the chromatogram must contain at least three major peaks between n-C_8 and n-C_{12}. Figure 13.17 shows the chromatogram for a headspace sample of MAB paint thinner, a medium petroleum distillate. This classification scheme divides the accelerants into five classes of petroleum distillates. There is a separate class of unclassified single components that also includes nonpetroleum compounds. Examples in this class include alcohols, acetone, toluene, and speciality products such as isoparaffinic mixtures that do not qualify for any of the other five classes of accelerants. Gasoline and kerosene produce chromatographic patterns distinctive enough to be placed into separate classes.

Chromatographic characterization of accelerants based on this classifica-

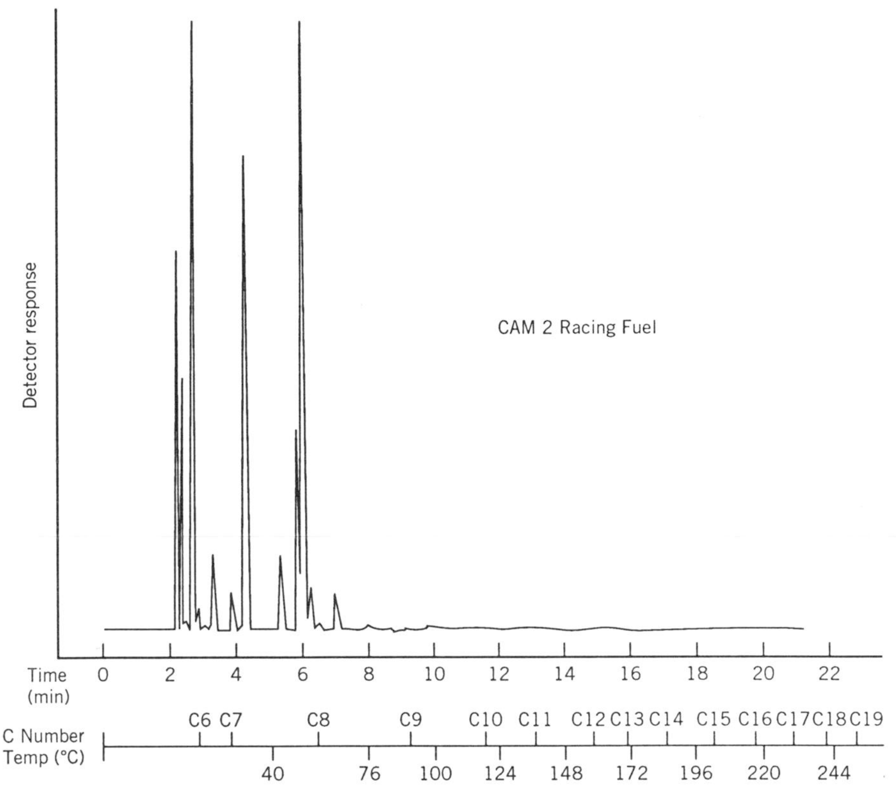

FIGURE 13.16 Gas chromatogram of a 1-mL headspace sample of CAM 2 Racing Gasoline (25-μL/qt can) heated for 20 min at 90°C. GC/FID conditions: 30-m SPB-1 × 0.75-mm i.d. × 1.0-μm film; 40°C (5 min)–250°C at 12°C/min; injection = 260°C, detector = 280°C. (Courtesy of New Jersey State Police.)

tion requires enough column efficiency to separate the *n*-alkanes from butane up to tricosane (n-C_{23}) in such a manner as to be able to evaluate unknown complex mixtures. Figure 13.18 shows the separation of a standard mixture of normal alkanes (n-C_6 to n-C_{28}) that can be used for evaluation of accelerants in this scheme. Under these conditions, sufficient separation and resolution is provided for identification of accelerants listed in this classification scheme. The chromatographic conditions for this evaluation usually require the use of a nonpolar liquid phase on which the elution order of most compounds can be directly related to boiling point. Methylsilicone liquid phases are the best stationary phases for this application and are widely used for accelerant identification.

Today, most laboratories use temperature programming and fused-silica capillary columns to separate the wide-boiling point range of petroleum distillates that are found as accelerants. The particular column and chro-

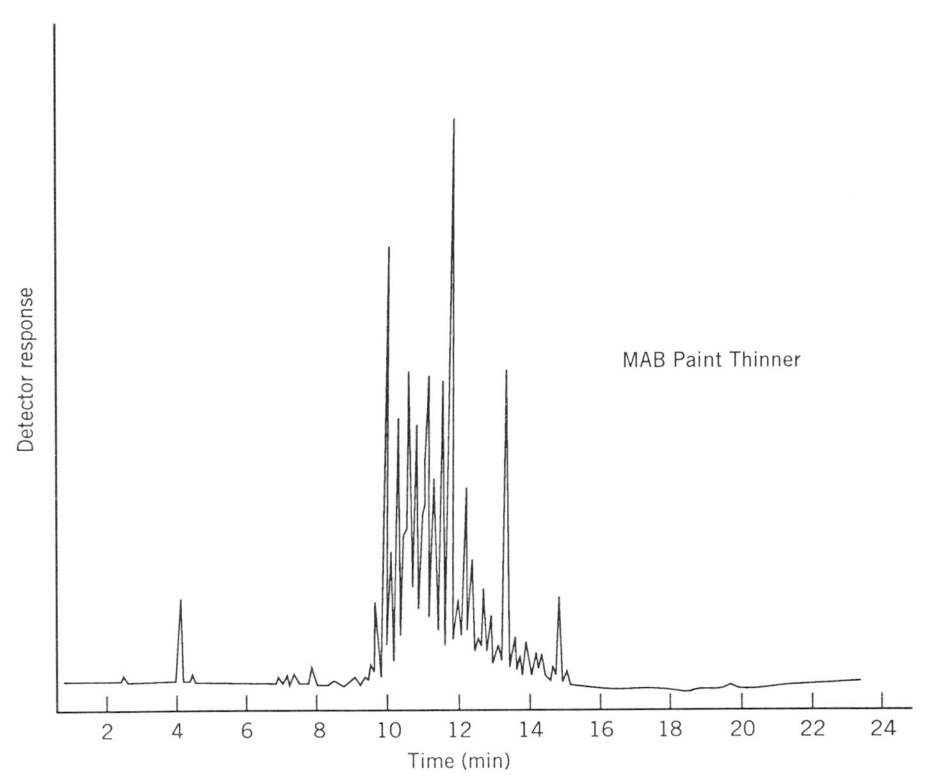

FIGURE 13.17 Gas chromatogram of a 1-mL headspace sample of MAB paint thinner (10-μL/qt can) heated for 20 min at 90°C. GC/FID conditions: 30-m SPB-1 × 0.75-mm i.d. × 1.0-μm film; 40°C (5 min)–250°C at 12°C/min; injection = 260°C, detector = 280°C. (Courtesy of New Jersey State Police.)

matographic conditions used are not as important as long as the column can provide enough resolution to effectively separate and identify the complete range of accelerants (165). ASTM Standard Test Method E-1387-90 (169) recommends that a test mixture of equal parts by weight of even-numbered normal alkanes ranging from *n*-hexane through *n*-eicosane plus toluene, *p*-xylene, *o*-ethyltoluene, *m*-ethyltoluene, and 1,2,4-trimethylbenzene must be resolved for the column to be considered adequate for identification of petroleum products.

Wide-bore columns have been shown to be advantageous in arson analysis, especially when headspace sampling is used (170). They offer high capacity and provide high enough efficiency to separate the components of most accelerants and to differentiate between the classes of accelerants. Wide-bore columns also accommodate a variety of injection methods, some of which are not compatible with narrow-bore columns without accessories such as cryofocusing. The high capacity of wide-core columns is a great

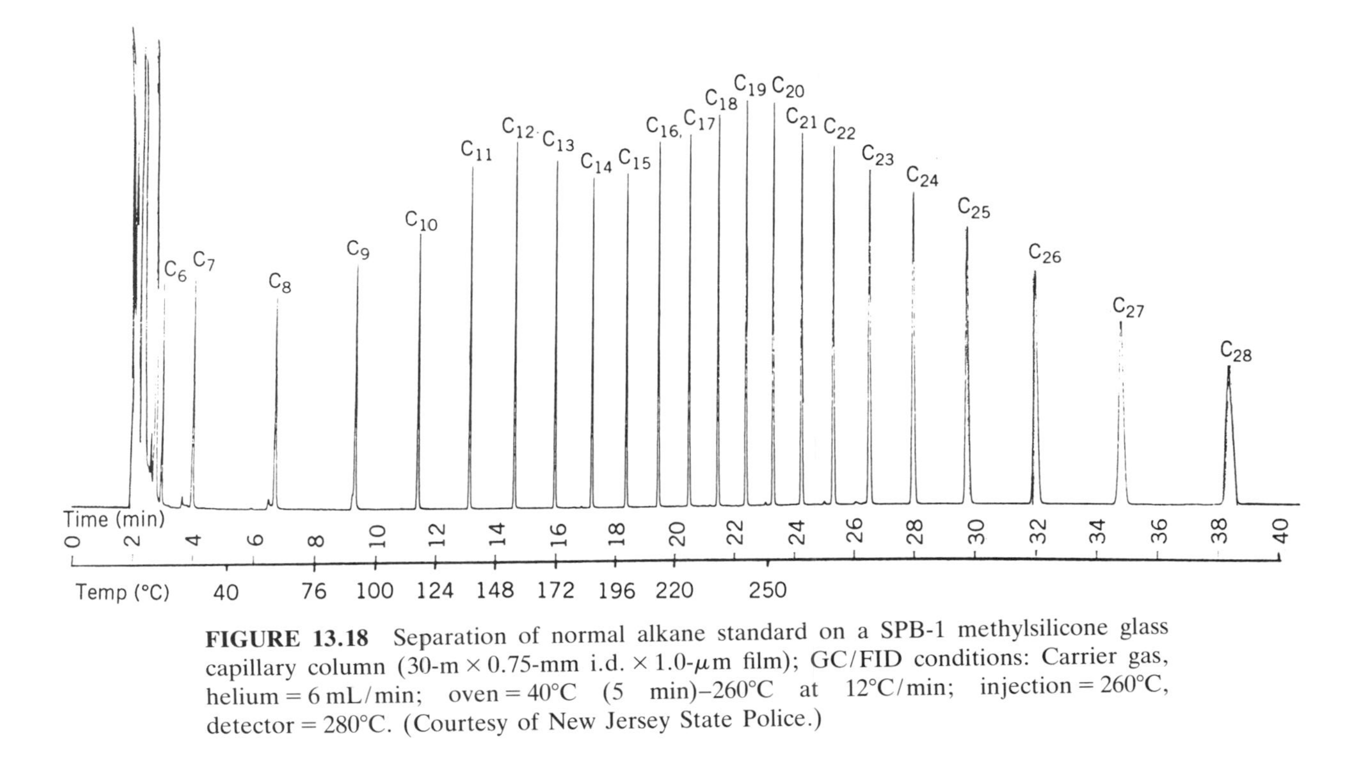

FIGURE 13.18 Separation of normal alkane standard on a SPB-1 methylsilicone glass capillary column (30-m × 0.75-mm i.d. × 1.0-μm film); GC/FID conditions: Carrier gas, helium = 6 mL/min; oven = 40°C (5 min)–260°C at 12°C/min; injection = 260°C, detector = 280°C. (Courtesy of New Jersey State Police.)

advantage, especially when headspace sampling is used, since most times the analyst has little control of the amount of volatiles being sampled from fire debris. Narrow-bore columns can be easily overloaded in this case scenario.

Because of the large number of components present in petroleum products, they produce complex characteristic patterns that can be used to identify different accelerants by pattern-recognition techniques. Most laboratories build a library of chromatograms using the columns, conditions, and sample preparation technique that is most frequently used for casework. Chromatograms obtained by headspace sampling most often differ enough from chromatograms obtained by liquid injections to justify building a separate headspace chromatogram library. The library of standards should also include chromatograms from common accelerants at various stages of evaporation.

Accelerants recovered from fire debris generally have been exposed to extreme heat and therefore have lost some or most of the volatile components through evaporation. This is commonly refered to as "weathering." Because of this evaporation, the patterns of the accelerants will differ; therefore they must become part of the library. Figures 13.19 and 13.20 show the headspace chromatograms of gasoline and 75% evaporated

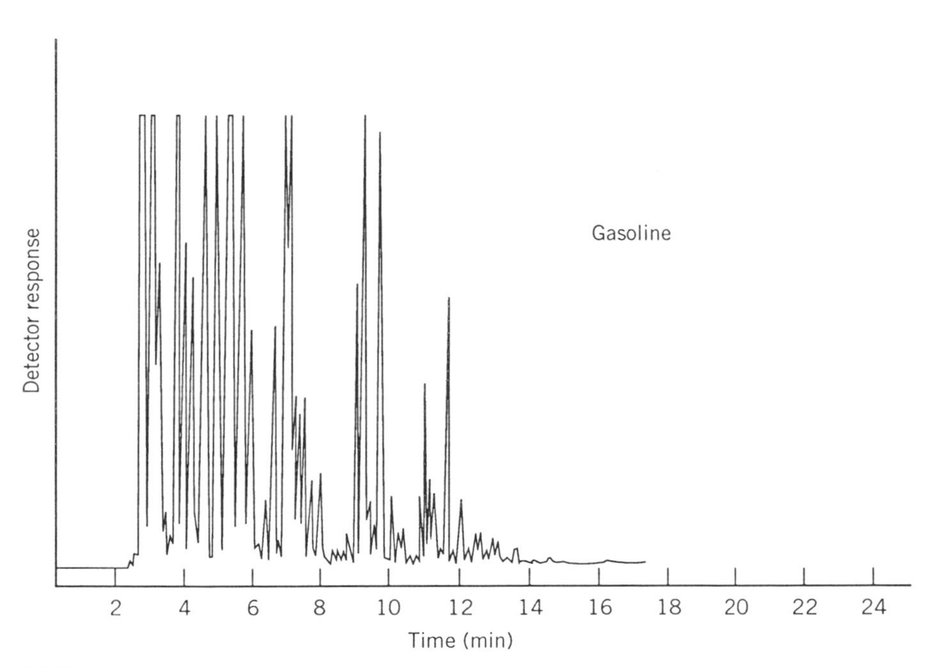

FIGURE 13.19 Gas chromatogram of a 1-mL headspace sample of gasoline (10-μL/qt can) heated for 20 min at 90°C. GC/FID conditions: 30-m SPB-1 × 0.75-mm i.d. × 1.0-μm film; 40°C (5 min)–250°C at 12°C/min; injection = 260°C, detector = 280°C. (Courtesy of New Jersey State Police.)

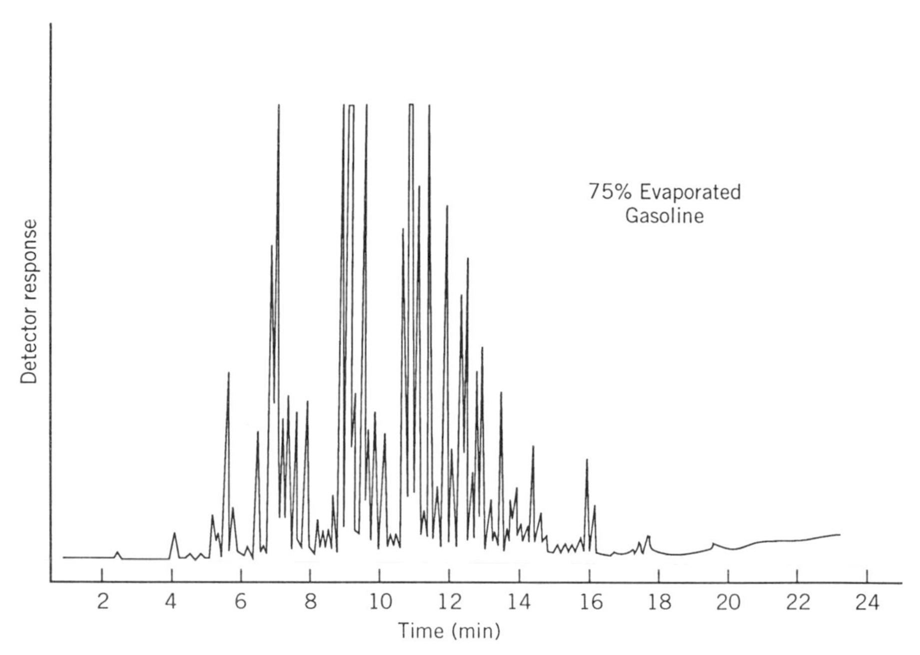

FIGURE 13.20 Gas chromatogram of a 1-mL headspace sample of 75% evaporated gasoline (10-μL/qt can) heated for 20 min at 90°C. GC/FID condition: 30-m SPB-1 × 0.75-mm i.d. × 1.0-μm film; 40°C (5 min)–250°C at 12°C/min; injection = 260°C, detector = 280°C. (Courtesy of New Jersey State Police.)

gasoline. The obvious loss of volatiles and pattern change can be seen easily in these chromatograms. It is recommended that several different standards of gasoline be run unevaporated to at least 90% evaporated and placed in the library.

Pattern recognition and chromatogram interpretation can be a very difficult task in arson analysis. An excellent checklist that the analyst can use in obtaining a chromatogram suitable for pattern recognition interpretation and steps used to interpret the chromatogram are given in Reference 165.

13.10.4 Sample Preparation

Several sample preparation techniques have been used to recover accelerants from fire debris. Not all are in use today, nor is there one technique that is universally applied to all types of flammable or combustible liquids. The sample preparation technique must be simple, efficient, rapid, free of contamination, and able to recover the sample in identifiable quantities. The analyst must be aware of the advantages and pitfalls of each method and must also recognize how the sample preparation technique will effect the chromatogram of the accelerant. Caddy et al. (171) recently reviewed the sample preparation methods used for the detection of petroleum products in fire debris. In addition, a proposed analysis scheme has been presented which shows how these techniques fit into an overall analysis scheme for the identification of flammable and combustible liquids (165).

Distillation. Distillation methods were among the first techniques used to isolate petroleum products from fire debris samples. Distillation methods include simple distillation, steam distillation, and vacuum distillation. Ethylene glycol and vacuum distillations offer the highest recovery of petroleum products of the three distillation methods. These methods are time-consuming and cumbersome, however, and the complications involved in these methods make their usefulness for fire debris preparation uncommon. The procedure of separation and concentration of accelerants in fire debris samples containing visible amounts of water has been outlined (172). These procedures are recommended only for samples that have a detectable odor of petroleum distillate at room temperatures. The advantage of this technique is that the distilled liquid can be harvested for courtroom presentation, the odor being recognizable by the jury.

Solvent Extraction. Solvent extraction is one of the original techniques used to recover accelerants from fire debris (173, 174). The method is based on the solubility of the accelerant, generally hydrocarbons, in the extraction solvent and has the advantage of recovering nonvolatile materials that are not recoverable by other methods, generally because of their vapor pressure. The technique usually involves soaking the fire debris in a suitable solvent for a length of time, decanting the solvent, filtering, and then evaporating to a small concentrated volume for analysis. In this process, volatile accelerants may be lost and nonsoluble accelerants may not be recovered. In addition, a major disadvantage of solvent extraction is that contaminants and pyrolysis products may also be extracted, complicating the chromatogram and the analysis.

Solvent extraction may be necessary to distinguish between kerosene and diesel fuel, since headspace methods do not have good recovery of hydrocarbons greater than C_{18}. Several extraction solvents have been studied. Table 13.12 lists some solvents that have been used in the recovery of fire debris. Practical guidelines have been published for the solvent extraction of flammable and combustible liquids from fire debris, which is useful for all classes of petroleum distillates (182).

Static Headspace. Static headspace sampling (SHS) followed by GC has become one of the more common techniques for initial screening of fire debris evidence. The ease of sample preparation, the speed with which a large number of samples can be analyzed, and the versatility of handling

TABLE 13.12 Solvents Used in the Recovery of Petroleum Distillates From Fire Debris

Solvent	References
Carbon tetrachloride	173, 174, 175,177
Acetone	173
Carbon disulfide	173, 174, 179
Hexane	161, 173, 174
Ethyl ether	173, 174, 176, 177
n-Pentane	173, 174, 178, 180
n-Heptane	161
Petroleum ether	161
Dodecane	173, 174
Hexadecane	173, 174
Methylene chloride	173, 174, 181
Chloroform	173, 174
Benzene	173, 174

different types of evidence, permit SHS followed by GC to be a very attractive and practical technique for accelerant detection.

When there is a sufficient amount of any flammable or combustible liquid present in fire debris with enough vapor pressure to be present in the headspace of the container, SHS followed by GC will yield reproducible and meaningful results. Heating the sample container will increase the vapor pressure, yielding a higher concentration of accelerant vapors and improving sensitivity. The limit of detection has been reported to be in the vicinity of not less than 5–10 μL of petroleum product in a gallon can (183).

As previously mentioned, the most common container for fire debris is an unused metal paint can. Normally a hole is punched in the lid of the can and sealed with tape before the can is sampled. Samples are typically heated in an oven or on a hot plate from 10 to 30 min at temperatures from 50 to 100°C. If enough accelerant is present to cause an odor then the analyst may elect to sample the container at room temperature. Heating samples that contain water above 100°C may cause the container to vent or burst. In addition, over heating the sample may cause pyrolysis of debris and complicate the interpretation of an already complex chromatogram. Typically, 0.5–3 mL of headspace vapor is removed from the container and injected directly into the injection port of the gas chromatograph. The amount depends on the diameter and film thickness of the column. Procedures for sampling of headspace vapors from fire debris have been outlined (184).

The first preliminary work in which vapor examination of headspace samples was used for the analysis of traces of accelerants was reported by Midkiff and Washington in 1972 (185). Since then many studies have been performed in regard to heating conditions, container size, sample size, the effect of water, the degree of interference from pyrolysis products, and the chromatographic conditions, including the type and dimension of column. A comparative study on the use of headspace and steam distillation has been performed (186) as well as a direct comparison of the headspace technique and the charcoal adsorption technique (187).

Generally, SHS followed by GC is used as a screening technique, especially when an odor of petroleum distillate is found. The technique is particularly good for accelerants with high vapor pressure or when single-component solvents are present. Chromatograms resulting from the headspace of accelerant vapors may differ from chromatograms of liquid flammables and the analyst must be wary of these differences in the interpretation of the chromatogram. A good review of the methods published in the literature has been presented recently and the reader is referred to this publication (171).

Passive Headspace. Passive diffusion of accelerant vapors onto an adsorbent placed inside the container of fire debris has recently gained acceptance in the United States because it is nonlaborious and takes little time to perform. Dietz (188) has reported a procedure that uses carbon-coated Teflon strips, similar to devices used in environmental monitoring badges, to recover as little as 0.2 μL of an equal mixture of gasoline, kerosene, and diesel fuel. In this procedure, samples are heated at 60°C for one hour or left at room temperature overnight with little interference of water. The accelerants are desorbed with carbon disulfide, which can be concentrated to improve sensitivity and shows little response, but may cause a pressure disturbance to the FID. Another advantage to this technique is the fact that multiple analyses may be performed from one sample. The carbon strip can be easily cut into smaller pieces, placed into vials and frozen for later analysis. For more information about the actual conditions to use with this technique, see Reference 189.

Earlier attempts at passive diffusion were not quite as successful. Variations of this technique included the use of Curie-point pyrolysis wires coated with finely divided activated charcoal and pyrolysis GC (190, 191). Other workers reported using charcoal-coated wires and elution with various solvents (187, 192, 193). Most of these procedures did not gain general acceptance and most laboratories now use the carbon strip approach with carbon disulfide elution.

A relatively new technique, Solid-Phase Microextraction (SPME), shows promise as an alternative passive headspace sampling technique for gasoline in fire debris (193A). SPME is a simple, solventless extraction procedure in which a phase-coated fused silica fiber is exposed to the headspace above the fire debris packaged in a closed container. The investigators in this initial report used a 100 μm polydimethylsiloxane-coated SPME fiber to detect 0.04 μL of gasoline in a quart can with a capillary column and FID.

Dynamic Headspace. Dynamic headspace sampling (DHS) is a nonequilibrium process in which air or an inert gas such as nitrogen is passed over the sample (in the case of a solid) or through the sample (in case of a liquid). In the case of a liquid sample this is more commonly referred to as the purge and trap technique, which is used widely in the environmental field.

DHS methods are more sensitive than SHS methods are not as cumbersome or time-consuming as distillation methods. However, like other headspace techniques, DHS procedures are best suited for light petroleum distillates and medium petroleum distillates and do not give good recoveries of high petroleum distillates, such as diesel fuel.

Samples collected on adsorbents can be desorbed by heat (thermal desorption) or by solvent extraction. The use of various forms of activated carbon is the most widely used procedure for the collection and adsorption of accelerants in forensic science laboratories. Thermal desorption of samples from charcoal is not efficient because of the high temperature needed (950°C) to remove hydrocarbons from the charcoal (194). For this reason, most procedures use carbon disulfide to extract the adsorbed accelerants. In 1967 Jennings and Nursten (195) reported concentrating analytes from a large volume of aqueous solution using activated charcoal as the adsorbent and extracting with carbon disulfide. Since then many adaptations of this method have been used to detect accelerants in fire debris. Chrostowski and Holmes (196) used a procedure to collect accelerants from fire debris that used activated charcoal packed into a Pasteur pipet. The can containing the fire debris was purged with preheated (110°C) nitrogen. The nitrogen passed out of the top of the can through the pipet containing the activated charcoal. Carbon disulfide was then used to extract the accelerants from the charcoal. Kubler et al. (197) also used a similar method. Tontarski and Strobel (198) used activated charcoal and allowed heated room air to flush the debris onto the charcoal tube via vacuum. The disadvantage to this procedure was the possibility of contamination from room air. Crandall and Pennie (199) improved on this method by using charcoal to filter the air entering the can. Juhala and Beever (200) reported using charcoal packed in a disposable hypodermic needle and pulling a vacuum through the needle. Although different variations in these techniques were common in forensic laboratories more recently passive headspace procedures using charcoal strips are replacing these methods.

Thermal desorption from porous polymers is theoretically the most sensitive method for detecting volatile accelerants from fire debris, since the entire sample that has been trapped is injected into the gas chromatograph, rather than a portion of diluted sample. Since desorption efficiency of charcoal is poor, thermal desorption methods that have been developed have used other sorbents. Tenax GC, a porous polymer capable of trapping a wide range of hydrocarbons, has been particularly useful to concentrate petroleum distillates from fire debris (201, 202). DHS methods using this type of sorbent have shown increased sensitivity over static headspace methods of several orders of magnitude. One of the major advantages of Tenax GC over other adsorbents such as charcoal is the fact that it does not adsorb water. The use of Tenax GC and DHS has been extended to the fire scene to recover accelerants (201). This method has been successfully used to recover charcoal lighter fluid, gasoline, and kerosene at scenes up to 15 h after the fire has been extinguished.

Andrasko (203) compared Porapak Q, Tenax GC, and Chromosorb 102 for effectiveness of trapping hydrocarbon vapors from fire debris. The study found that although Porapak Q and Chromsorb 102 seemed to trap the vapors more strongly, Tenax GC had the best desorption efficiency and therefore was more suitable for analysis of fire debris.

While there seems to be no one universal method for collecting, trapping, and analyzing fire debris for accelerants, the majority of laboratories in this field use either Tenax GC or activated carbon for collecting vapors from fire debris. Thermal desorption systems, especially those using Tenax GC, readily lend themselves to automation, but the equipment can be expensive. Activated charcoal and carbon disulfide extraction systems, while either passive or dynamic, are cheaper and require less expensive equipment, but demand the handling of carbon disulfide, a toxic solvent. Cryogenic focusing has been shown to improve chromatographic resolution although it has not received wide acceptance in the field (204). Practical guidelines for using dynamic headspace concentration methods have been outlined (205).

The forensic science laboratory should be prepared to examine fire debris with all of these methods and use the appropriate method for the particular sample, which would be dependent on the presence of either LPD, MPD, HPD, or single nonpetroleum solvents. Headspace analysis lends itself well to light and medium petroleum distillates and solvent extraction results in the best recoveries of HPD.

Detection. By far the most popular detector used in the gas chromatographic detection of accelerants in fire debris is the FID. It offers adequate sensitivity and because of the complex chromatogram that is generated with FID (pattern recognition), it is used in most circumstances for class identification of petroleum products. The photoionization detector (PID) and the thermionic ionization detector (TID) have both found applications in the analysis of petroleum products and hydrocarbons. However, neither are used very often in the detection of arson accelerants.

GC/MS has been used in the past for the detection and identification of single components, such as solvents, or accelerants with few peaks. Forensic scientists recognized the value of the mass spectrometer for identification of compounds in arson debris as early as 1976 (206). It has not been used routinely in the crime laboratory, however, because until recently many forensic laboratories lacked the funds to purchase and maintain mass spectrometers. In addition, most crime laboratories are so overburdened with drug submissions that even if they do have mass spectrometers at their facility they usually are dedicated to drug analysis so there is little time or few personnel to dedicate to the development of GC/MS methods for the analysis of arson debris.

In the last few years the cost of the mass spectrometers has dropped into a range that has made them affordable to the local forensic laboratory. Until recently even those laboratories that operated mass spectrometers lacked the data-handling capabilities to analyze the number of components in a complex chromatogram, such as those generated from a petroleum product in fire debris. Today, mass spectrometers are equipped with low-cost data-handling systems, allowing for the analysis of the complex mixtures from petroleum distillates. Because it is not feasible to identify each peak from the chromatogram, especially when using capillary columns, most recent applications have concentrated on the use of mass chromatography to identify accelerants and differentiate petroleum products from background and pyrolysis products. Smith (207) reported a GC/MS method using mass chromatography for the identification of petroleum products. This method used several characteristic ions for each of the major classes of compounds present in petroleum distillates. These ions and corresponding compounds are listed in Table 13.13. Mass chromatograms corresponding to each class of compounds are calculated by summing the intensities of the characteristic ions for each group. Comparison of mass chromatograms of unknown samples to known standards will identify the particular class of accelerant. Figure 13.21 shows the mass chromatograms that result from an evaporated gasoline sample.

Mass chromatography is not needed for every fire debris sample analyzed in the forensic science laboratory; however, for those samples that are inconclusive or when accelerant identification cannot be made from the GC/FID pattern, GC/MS can be of great value (208). For a comprehensive discussion on mass chromatography the reader is referred to Reference 209 and Chapter 6 in this book. Other mass chromatographic techniques have been developed for the identification of accelerants, which have included the identification of additives such as methyl-*tert*-butyl ether (MTBE) and lead alkyl compounds (210), the identification and quantification of target compounds (211), and the use of GC/MS identification as a basis of an accelerant classification scheme (212).

TABLE 13.13 Characteristic Ions and Compounds in Petroleum Distillates

Types of Compounds	Characteristic Ions (m/e)
Saturated aliphatic hydrocarbon	57, 71, 85, 99 (fragments) 100, 114, 128 (molecular)
Alicyclic and unsaturated aliphatic hydrocarbons	55, 69, 83, 97
Alkylbenzenes	91, 105, 119, 133 (fragment) 92, 106, 120 (molecular)
Alkyl-substituted condensed polynuclear aromatics	142, 178 (molecular)
Monoterpenes	93 (fragment) 136 (molecular)

Source. From Reference 207 with permission.

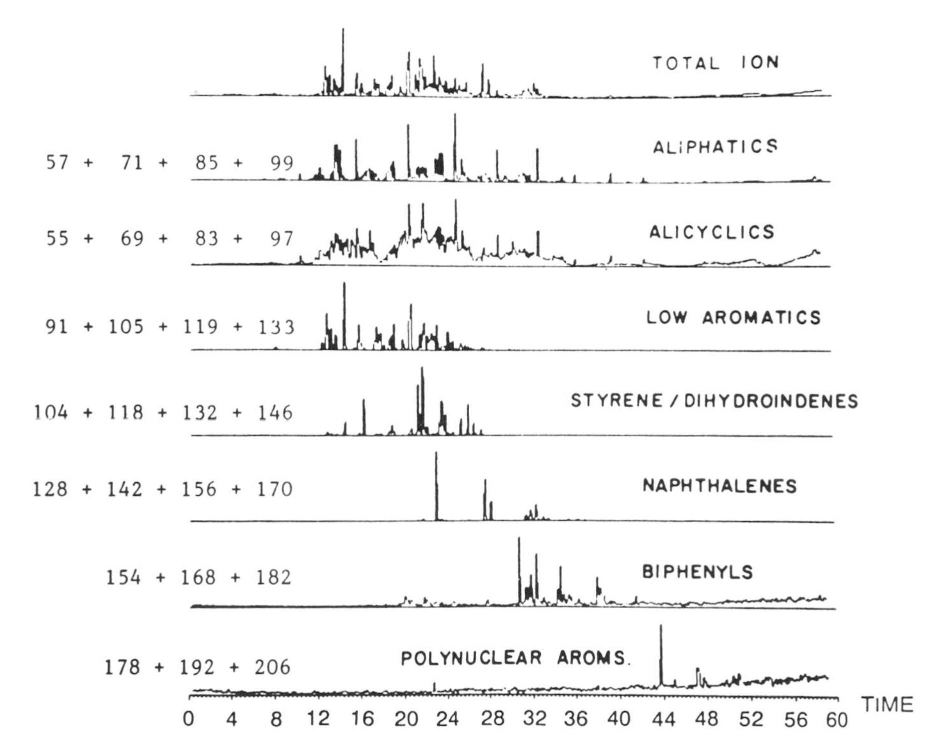

FIGURE 13.21 Family mass chromatogram from evaporated gasoline. (Reprinted with permission from Reference 207.)

One application of GC/Fourier transform infrared spectroscopy (GC/FTIR) to the identification of petroleum products in fire debris samples has been reported (213). Various accelerants were analyzed to determine if characteristic IR absorption bands could be used to identify each type of product. In a comparison to FID and MS detectors the FTIR was the least sensitive and did not discriminate accelerants from common pyrolysis products as well as GC/MS.

13.10.5 Comparison of Gasoline Samples

The forensic scientist is occasionally asked to compare brands of gasoline for the determination of common source or for the identification of the brand used in the arson. Unfortunately, brand identification is extremely difficult and, depending on the sample, it may be impossible (214). The main reasons for this difficulty are the marketing practices of the refinery industry and the changes the product undergoes in storage. Despite these limitations there has been some published work in this area and most of the publications have involved GC in the methodology. GC/MS has been used to determine the presence or absence of organolead compounds (209), which can help distinguish between leaded and unleaded neat liquid samples of gasoline. Mann (215) reported a comparison method to differentiate between unevaporated liquid gasoline samples, which involved measuring the quantitative differences of compounds eluting in thc *n*-butane to *n*-octane region of the chromatogram when using a capillary column yielding baseline resolution of the compounds. Mann (216) further extended this work to fire debris samples. In both reports the method was abel to differentiate between samples not having a common source, but a conclusive determination of common origin was not possible.

13.11 EXPLOSIVES ANALYSIS WITH GAS CHROMATOGRAPHY

13.11.1 Introduction

The forensic identification of bulk explosives and postexplosion residues is important in bombing investigations. The information can be used to determine the type of explosive, to link the suspect to the explosive, and ultimately to provide evidence in court. Many analytical techniques have been applied to the identification of explosives and explosive residues and several texts have been written on the subject (217–220).

The unambiguous identification of an explosive is complicated by the interfering compounds present in the postblast debris and the small amount recovered at the scene. GC can be a valuable tool in the separation of the explosive from the interfering substance and when coupled with a specific and sensitive detector can give the unambiguous identification of an explosive. GC is widely used for the analysis of explosives and explosive residues, despite the fact that many explosive compounds are inherently thermally unstable.

Basically there are two types of explosives, high- and low-order explosives, which are primarily differentiated by the burning rate and the manner of initiation. Low-order explosives burn more slowly and are usually initiated with a burning fuse or other heat source. High-order explosives are initiated by shock, usually through a booster charge or another explosive, such as a blasting cap. Examples of low-order explosives are black powder, pyrodex, single- and double-based smokeless powders, and flash powders. Analytical methods for examining this type of explosive include techniques other than GC, such as microchemical spot tests and thin-layer chromatography.

High-order explosives generally fall into three categories: commercial, military, or improvised, depending on their intended use and manufacturing process. Table 13.14 lists some explosives that have been analyzed by GC. Gas chromatographic methods for the examination of a variety of explosives have been reported (221). Prior to the introduction of fused-silica capillary columns, gas chromatographic analysis was limited to bulk explosives of

TABLE 13.14 Commercial and Military Explosives Analyzed by GC

Name(s)	Abbreviation
Dinitrotoluene isomers	DNT
Trinitroluene	TNT
Ethylene glycol dinitrate	EGDN
Nitroglycerin (glycerol trinitrate)	NG
Nitrocellulose (cellulose nitrate)	NC
Isosorbide dinitrate	ISDN
Pentaerythritol tetranitrate	PETN
1,3,5,7-Tetranitro-1,3,5,7-tetrazacyclooctane	HMX
1,3,5-Trinitro-1,3,5-triazacyclohexane	RDX
2,4,6-*N*-Tetranitro-*N*-methylaniline	Tetryl

nitroaromatics, because of the thermal lability of these compounds (221). The polarity of this class of compounds also made analysis on packed columns difficult because of irreversible adsorption on most liquid phases. Glass columns, on-column injection, and low-temperatures were common techniques used to prevent decomposition.

Trace analysis of explosives in postblast debris samples requires a detector considerably more sensitive than an FID. ECD, nitric oxide detection (TEA), and GC/MS have all been used for the chromatographic detection of explosives. Gas chromatographic analysis with specific detection is normally used for screening purposes. Other methods, such as TLC, microscopy, XRD, and SEM are often used to confirm the presence of trace amounts of explosives. Comprehensive reviews of the analysis of explosive residues using a variety of analytical methods have been reported (222, 223).

13.11.2 Electron-Capture Detection of Explosives

The electron-capture detector (ECD) was the first detector available that had the necessary selectivity and sensitivity to detect trace amounts of explosives (224). Douse used silica capillary columns coated with OV-101 to separate picogram quantities of explosives (225). Jane et al. used this GC/ECD method to identify nitroglycerin in gunshot residue (226). NG, TNT and RDX in the low-nanogram range were detected in handswab extracts using a 12-m $\times$ 0.25-mm BP-1 fused-silica capillary column and a ^{63}Ni ECD (227). Twibell et al. (228) also analyzed organic explosives from handswabs and their work indicated that GC/ECD was the most sensitive technique for this kind of work.

In 1982, Yip used short, mixed liquid phases and combined packed-capillary columns to separate EGDN, EGMN (ethylene glycol mononitrate), and NG at levels of 10^{-12} g/mL (229). A method has been reported using a packed column with dual detection using an ECD and photoionization detector (PID) and response ratios for the identification of TNT, RDX, Tetryl, and NG (230).

Belkin et al. detected ppb levels of TNT-type explosives by capillary column GC/ECD (231). Hobbs and Conde developed a headspace technique utilizing capillary column GC/ECD to detect vapors of several types of explosives (232).

13.11.3 Thermal Energy Analyzers

The thermal energy analyzer (TEA) is a specific detector for the measurement of *N*-nitroso compounds. The TEA is based on chemiluminescence and was used to detect explosives as early as 1978 (233). The principle of the detector involves the pyrolysis of *N*-nitroso compounds to release NO_2 that emits a characteristic infrared chemiluminescence, which is detected by a photomultiplier tube. The TEA is linear over six orders of magnitude and sensitive to *N*-nitroso compounds in the picogram range. Fine et al. used the TEA to examine postblast debris and handswabs from volunteers handling gelatin dynamite or the military explosive, C-4 (234). On-column injection into a 30-m × 0.32-mm DB-5 fused-silica capillary column with TEA detection was used to detect picogram amounts of NG, PETN, ISDN, EGDN, 2,4-DNT, TNT, RDX, and Tetryl (235).

For the analysis of explosives the TEA has been found to be superior in sensitivity and selectivity over the FID, ECD, and thermionic specific detector (TSD). Douse reported the TEA detection of explosives from handswabs in the low-picogram range using heated splitless injection with fused-silica capillary columns (236). Several modifications have been made to improve the sensitivity of the TEA detector. One variation involved the development of a method to trap high-performance liquid chromatographic eluent from a microcolumn and inject the eluent directly onto a gas chromatographic retention gap of unmodified silica, eliminating the need for evaporation and concentration prior to the analysis by GC/TEA (237). Kolla used GC/TEA to analyze trace amounts of explosives that were extracted from debris using solid-phase extraction (238).

13.11.4 Gas Chromatography/Mass Spectrometry

Gas chromatography/mass spectrometry (GC/MS) has been used only to a limited extent for the analysis of explosives due to the thermal instability of the explosives in the heated gas chromatograph (see Chapter 6). However, Zitrin recommended the use of GC/MS identification of explosive residue (239). Fused-silica capillary columns have increased the sensitivity of the technique and chemical ionization (CI), a soft ionization technique that yields molecular weight information, has helped renew interest in the method. The combination of CI and EI mass spectra complement each other and provide a reliable identification of explosive compounds. Negative ion mass chromatography (NIMS), has also been used for explosive analysis, which has led to increased sensitivity and selectivity. Using this technique with a fused-silica capillary column, TNT was analyzed with a detection limit of 50 pg (240). An excellent text has been published on the forensic analysis of explosives by mass spectrometry, which provides an extensive review of the literature (241). Pate and Mach reported the separation and identification of EGDN, NG, 2–4 DNT, and TNT at the nanogram level on glass packed columns using CIMS (242). Tamiri and Zitrin have described the application of GC/MS to confirm the results of a TLC analysis of postblast explosive residues (243).

GC/MS has been used for the analysis of smokeless gunpowders. Thirty-three commercial smokeless powders were compared using GC/MS with EI and CI (methane) (244). The same investigators unsuccessfully attempted to use this procedure to test for the presence of gunshot residue (GSR) found on hands by measuring either traces of the original volatilizable organic constituents present in the smokeless powder or characteristic organic compounds formed during the firing process (245). GC/EIMS was later used to analyze recovered unreacted GSR particles from double-base powder (246). Martz et al. developed a library of reloader smokeless powders based on electron impact mass spectral data of the volatile components of the smokeless powders (247). Selavka et al. replaced high-performance liquid chromatographic methods of analysis for smokeless powder additives by GC/FID (248). GC/FID was also used to characterize smokeless powder flakes from fired cartridge cases and from discharge patterns on clothing (249).

For further review of the application of GC/MS and MS/MS to the forensic identification of explosives refer to Reference 250.

13.12 FORENSIC SCIENCE APPLICATIONS OF PYROLYSIS GAS CHROMATOGRAPHY

13.12.1 Introduction

Pyrolysis consists of the thermal transformation of a compound into another compound or compounds, usually in the absence of air. The usefulness of pyrolysis is the technique's ability to convert a nonvolatile organic material into a number of volatile organic compounds that can be separated by GC, identified, and related to the chemical composition of the original material.

The combination of pyrolysis and GC (PGC) is a powerful and sensitive method for discriminating certain types of physical evidence. PGC has been applied to a wide variety of sample types, but by far the major application of PGC has been in polymer analysis (see Chapter 11). In the crime laboratory, PGC provides one of the most discriminating tests for forensic paint comparisons.

Dr. Paul Kirk and his coworkers were first to apply and develop many of the applications and techniques of PGC to various types of physical evidence (251), such as commercial plastics, paints, and drugs. The application of PGC to forensic samples was reviewed by Wheals (252) in 1981, by Saferstein (253) in 1985, and by Blackledge (254) in 1992. A bibliography of analytical pyrolysis applications, which includes a forensic section, was published by Wampler (225) in 1989.

13.12.2 Pyrolysis Gas Chromatographic Methods

PGC can be classified into two distinct types, depending on the method in which heat is applied to the sample: static-mode (furnace) reactors and dynamic (filament, pulse-mode) reactors. Furnace-type pyrolysis systems are seldom used in forensic laboratories. The most common pyrolysis systems used are the Curie-point (inductive heating) and the ribbon or filament-type pyrolyzers (resistive heating). A brief description of each pyrolysis system is provided below. For more detailed information the reader is referred to Chapter 11 of this book and to other texts on the subject (253, 256). In addition a good introductory article written by Munson (257) provides a basic understanding for a laboratory initially setting up a pyrolysis system.

Curie-Point Pyrolysis (Inductive Heating). Curie-point pyrolysis uses a high-frequency induction coil to heat a ferromagnetic wire containing the sample to the wire's Curie point (the temperature at which the ferromagnetic wire becomes paramagnetic). The sample is centered in a glass or quartz tube, connected to the inlet of the gas chromatograph in a position to be in the flow of the carrier gas. Proper control of the pyrolysis conditions (temperature and flow), including wire composition, is required to obtain repeatable pyrolysis data. The major advantage of the use of a pulse-mode heating arrangement is that the sample is heated quickly and the products are removed from the hot zone before any significant secondary reactions occur.

Curie-point pyrolysis is widely used by forensic laboratories in Europe and Asia. It is seldom used by forensic scientists in the United States, basically because of the lack of marketing and available service to these units.

Filament and Ribbon Pyrolysis (Resistive Heating). In this type of system a resistively heated platinum or nichrome wire coil or ribbon is used to rapidly heat the sample. The wire is continuously swept with carrier gas, whereupon the pyrolysis vapors are transported into the chromatographic column. Heating times are relatively large (up to 20 sec) for this system, which may lead to nonrepeatable pyrograms and secondary reactions. The pyrolysis conditions, sample size, and location must be carefully controlled to obtain repeatable data. Two possible heating modes are available for this system: pulse mode or programmed mode. For most forensic applications the pulse mode has been used.

Various commercial pyrolysis systems are available that offer filament or ribbon resistive heating. Forensic laboratories in the United States almost exclusively use various models of the Pyroprobe (Chemical Data Systems, Oxford, PA). The Pyroprobe systems use either a self-sensing resistivity platinum wire coil or ribbon to heat the sample. The coil probe is used for solid samples, viscous liquids, and semisolids that are not soluble in a volatile solvent. Samples are normally placed in a quartz boat or between quartz wool plugs in a quartz tube. The sample should be centered and placed in nearly the exact spot of the tube or boat for good repeatability. Care must also be taken not to contaminate either the sample or holder when preparing for analysis.

13.12.3 Applications

Paint. Paint is one of the more common types of physical evidence that is submitted to the crime laboratory. Paint evidence may originate from a number of different sources, such as tools, household items (windows and doors), and automobiles. The forensic examination of paint evidence involves a scheme of analysis with a variety of analytical techniques, such as microscopy, solvent tests, scanning electron microscopy with energy-dispersive x-ray spectrometry (SEM-EDAX), infrared spectroscopy (IR), neutron activation analysis (NAA), x-ray diffraction (XRD), and PGC.

PGC is one of the most powerful techniques available to the forensic scientist for examining paint specimens (258). Forensic PGC of paint samples has been extensively reported in the literature. A wide range of pyrolysis operating conditions have been used and the difficulties in reproducibility have caused a lack of standardization in this particular area. However, PGC has been shown to be sufficiently characteristic and reproducible to differentiate different manufacturers of similar paint (259, 260).

The discrimination power of PGC in a paint comparison has been shown to be markedly linked to the gas chromatographic stationary phase employed (253). Over the years many different types of phases of various polarities have been evaluated for the analysis of paint samples by PGC. Carbowax 20M and Porapak Q have been used for discriminating paints by PGC. Fekuda (261) used a packed-column PGC system with a Curie-point pyrolyzer to differentiate 78 automobile topcoats obtained from Japanese automobile manufacturers.

Stanley and Peterson (262) were first to connect a pyrolysis unit to a capillary gas chromatographic column. They used isothermal PGC to separate the pyrolysis products and, as expected, much greater resolution was achieved over that obtained with packed columns. A few years later

Barbour demonstrated the advantages of using cryogenic temperature programming in PGC (263).

The use of capillary columns has evolved with PGC to improve the resolution and in turn the discriminating power. Twibell et al. designed a splitless injection system with a Curie-point pyrolyzer and a capillary chromatographic column (264). Another group used a split injection with a 25-m × 0.2-mm-i.d. SP-2100 fused-silica capillary column and achieved sufficient baseline resolution and high reproducibility to make a quantitative comparison of pyrograms meaningful (265). Wampler et al. used cryofocusing and splitless injection to analyze pyrolyzate vapors with capillary GC (266). They showed improved chromatographic resolution and sensitivity over methods that utilize packed columns or split capillary systems.

While capillary columns have improved the resolution of pyrolyzate compounds, the type of stationary phase is still important in the discriminatory power of PGC. Bates et al. (267) were not able to improve the discrimination of a test group of 25 alkyd resins when they compared a capillary column and packed column of the same phase. Nowicki and Vanderwerff (268) used two packed columns of differing polarity in sequence with PGC to improve the discriminating power in their scheme of analysis for paints. A dual-column method has been reported in an effort to further improve the discrimination of PGC of paint samples (269). This method uses a polar and a nonpolar capillary column connected to the same injection port of a gas chromatograph via a two-holed ferrule. The pyrolyzate vapors are split between the two columns and a separate, different pyrogram is generated simultaneously for the same sample. Figure 13.22 shows the corresponding pyrograms from a paint sample using this configuration. The repeatability of this system is good and allows the scientist to gather much more information with only one sample. Table 13.15 lists the chromatographic and pyrolysis conditions that were used to generate this sample. These conditions are typical of a PGC analysis of a paint sample using this configuration.

In 1978, Audette and Percy introduced the idea of first examining a paint chip by infrared spectroscopy (IR) using a KBr pellet and then pyrolyzing the same pellet for gas chromatographic analysis (270). This method offers high discriminative power with the use of two distinctly different techniques and a very small amount of sample (3–5 μg). At the present time there have been no other reports applying this method of analysis to paint evidence.

Fibers. The identification and comparison of fiber evidence in the forensic laboratory can encompass a number of different techniques. Nondestructive examination of fibers by polarized light microscopy, microspectrophotometry, and FTIR spectroscopy very often provide a high enough discrimination in most casework comparisons that other analytical tests are not needed. When a sufficient amount of fiber evidence exists, a comparison of fiber dyes by thin-layer chromatography (TLC) will often further the

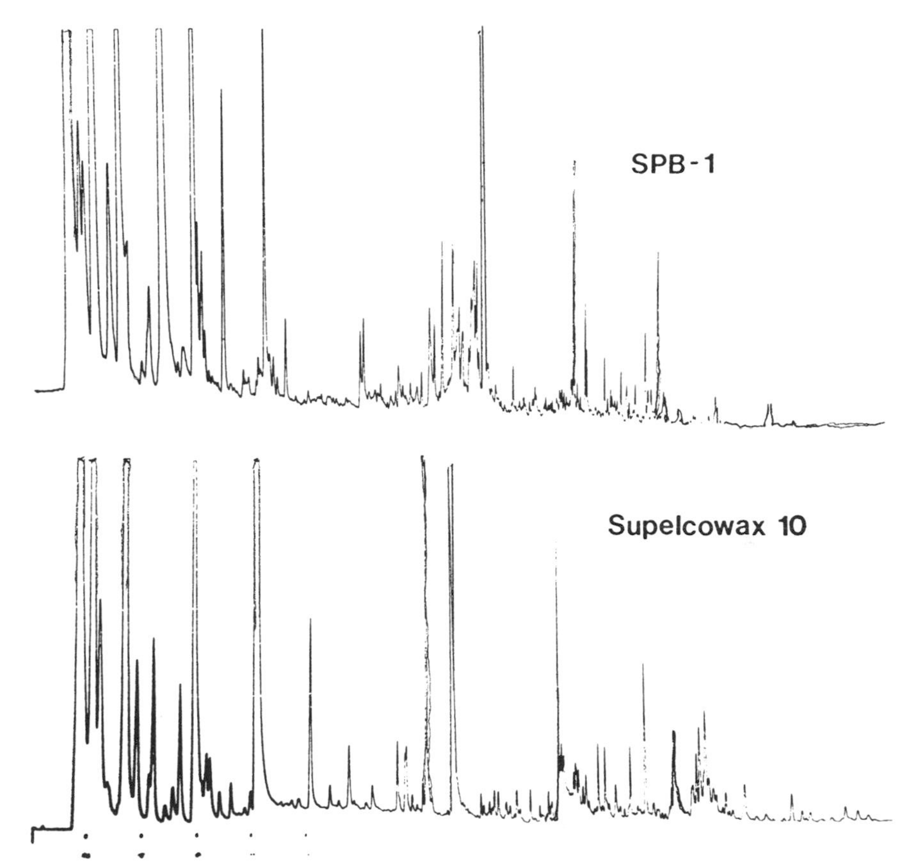

FIGURE 13.22 Pyrograms of a red acrylic enamel automobile paint resulting from the pyrolzolate being split onto a 25-m SPB-1 capillary column and a 25-m Supelcowax 10 capillary column; GC/FID conditions: See Table 13.15. (Courtesy of New Jersey State Police.)

discrimination. Auxiliary techniques such as PGC, pyrolysis mass spectrometry (PMS), and SEM-EDAX are only sporadically used when the evidence or situation is required.

When considering this scheme of fiber comparisons it is not surprising to find that the literature is not as abundant with PGC applications of fiber identification as is the case for paint evidence. This is disappointing, since there is ample evidence that PGC can distinguish between various types of fibers, such as polyesters, acrylics, nylons, cellulose, and acetates (271–274). The discrimination power of PGC for distinguishing various types belonging to a single fiber class, however, is a subject of question. Bortnial et al. in 1971 examined a number of acrylics and modacrylics and were able to differentiate various types of nylon fibers, distinguishing, for example, nylon 66, 11, 610, 6, and Nomex (275). Almer recently analyzed polyacrylonitrile

TABLE 13.15 Conditions for PGC of Paints (see Figure 13.22)

Sample	Red acrylic enamel automobile paint (1 mm^3)
Column	SPB-1 25-m × 0.25-mm i.d. (0.11 μm)
	Supelcowax-10 25-m × 0.25-mm i.d. (0.2 μm)
Oven temperature	40–100°C at 4°C/min
	100–180°C at 10°C/min
Injection temperature	225°C
Detector temperature	250°C
Detector	FID
Pyrolysis temperature	550°C
Pyrolysis duration	10°C
Interface temperature	200°C
Linear gas velocity	26.62 cm/sec, helium
Split ratio	10/1

fibers, 63 acrylic and 22 modacyrlic, with a Pyroprobe and capillary column using as little as 10 μg of sample (276). In a review, Wheals concluded that PGC was slightly less effective than IR spectroscopy for discriminating fibers and that the conditions normally used for the analysis of paint chips was unsuitable for fiber comparisons (277).

Figure 13.23 shows a pyrogram resulting from the pyrolysis of a polypropylene fiber. The conditions used to analyze this particular fiber (listed in Table 13.16) are typical of conditions used to analyze a variety of different fibers and should be contrasted to the conditions used for PGC of paints (Table 13.15).

There have been some other forensic applications of PGC of fibers. Levy and Wampler used a Pyroprobe and a capillary column to examine various synthetic fibers (278). Hardin and Wang used a Pyroprobe with quartz sample tubes and packed-column GC to examine several types of textile fibers (279). Wright et al. used a microfurnace pyrolysis unit (MP) interfaced

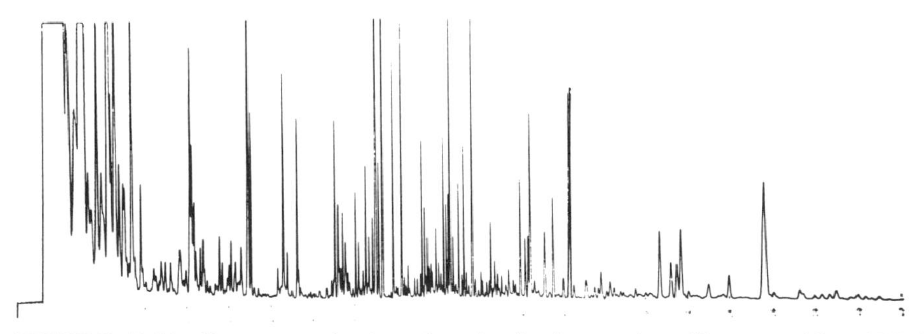

FIGURE 13.23 Pyrogram of a 1-cm length of polypropylene fiber on a 25-m SPB-1 fused-silica capillary column (0.25-mm i.d. × 0.11-μm film); GC/FID conditions: See Table 13.16. (Courtesy of New Jersey State Police.)

TABLE 13.16 Conditions for PGC of Fibers (see Figure 13.23)

Sample	Polypropylene fiber (1 cm)
Column	SPB-1 25-m × 0.25-mm i.d. (0.11 μm film thickness)
Oven temperature	50–100°C at 4°C/min
	100–180°C at 10°C/min
Injection temperature	225°C
Detector temperature	250°C
Detector	FID
Pyrolysis temperature	770°C
Pyrolysis duration	10 sec
Interface temperature	200°C
Linear gas velocity	26.23 cm/sec, helium
Split ratio	8/1

to a capillary column gas chromatograph to pyrolyze various synthetic fibers (280). They used multidimensional techniques, including heartcutting and cryofocusing, to obtain better resolution and discrimination of the fibers. A study of the effects of a soil environment on the biodeterioration of manmade textiles showed that fibers could still be identified by IR spectroscopy and PGC as to fiber type after 12 months of exposure, even though the solubility behavior and optical properties (polarized light microscopy and dispersion stains) both changed (281).

Other Polymers. PGC has been applied to the characterization of a number of different types of polymers besides the traditional paint chips and fiber evidence submitted to the crime laboratory. Synthetic plastics have been readily identified and differentiated by PGC (271). Hume et al. (282) applied PGC to the characterization of motor-vehicle body fillers utilizing a 15-m Carbowax 20M gas chromatographic column and a computer-based system to compare pyrograms.

Davis (283) used a Pyroprobe with quartz tubes to examine rubber from a truck tire, the handle of a hammer, and an automobile bumper guard. Blackledge (284) used PGC to compare a rubber bumper guard removed from a suspect car to one recovered at a hit-and-run scene. Both investigators used a different approach than the traditional high-temperature PGC. The polymer samples in both cases were first heated by the injection port of the gas chromatograph to produce a chromatogram of the rubber's volatile components and then the sample was pyrolyzed. This approach first reported by Chih-an Hu (285, 286) is really a combination of thermal desorption and PGC and has the potential for further increasing the power of discrimination of both techniques.

Lennard and Margot (287) examined and compared traces of synthetic material from shoe soles by FTIR microspectrometry and packed-column PGC. Ding and Liu (288) in a similar analysis examined 10 different rubber-soled shoes and 12 different automobile tires by Curie-point PGC using two separate packed columns. They used varying ratios of butadiene–styrene, natural rubber, and butadiene–acrylonitrile to differentiate the samples.

Miscellaneous Applications of Pyrolysis Gas Chromatography. Beyond paints, fibers, and other polymers, PGC has been applied in the forensic science laboratory to characterize and compare a number of different types of material submitted as evidence in criminal casework. The utility of PGC for the characterization of adhesives has been described (271, 289), as well as various methods for the comparison of tapes with adhesive backings (290). Vinyl tile with an asphaltic-type glue from a safe-cracking case was analyzed by PGC (291). Williams and Munson (292) used capillary column PGC to examine 30 black PVC tapes, distinguishing 26, and even partially burned tapes could be examined. The analysis of photocopy toners by PGC has been reviewed (293).

Newlon and Booker (294) applied PGC to the identification of smokeless powders. They were able to differentiate 40 samples on the basis of their chemical composition. However, Keto (295) examined smokeless powders by capillary column PGC and concluded that PGC has only limited value for the source identification of smokeless powders. A comparative study of samples of chewing gum bases has been conducted using PGC (296). Criminalists at the police crime laboratory in Osaka, Japan, used capillary PGC/MS to "fingerprint" Japanese cedar, Japanese cypress, and American pine (297). PGC has been used to differentiate fetal hemoglobin (Hbf) from adult hemoglobin (HbA) and sickle cell hemoglobin (SbS) in blood (298) and also to differentiate fetal and adult bloodstains (299).

PGC was explored as a technique for the individualization of hair as early as 1960 (300), and in 1987 Munson reviewed the status of hair comparisons by PGC (301). As recently as 1990, Ishizawa and Misawa examined head hairs by PGC (302). PGC has not been shown to be a viable technique for this purpose.

13.13 MISCELLANEOUS FORENSIC APPLICATIONS OF GAS CHROMATOGRAPHY

The aforementioned applications of GC have been from several major areas of forensic science, specifically drug analysis and toxicology and other areas of trace evidence, such as pyrolysis, explosives, and the detection of accelerants. There are, however, a few other forensic analyses by GC that have been reported.

In 1985, a gas chromatographic procedure for comparing the relative ages of ballpoint inks was reported (303). This procedure is based on the quantitative analysis of solvents contained in the inks, which are reported to remain in the paper for up to one year. The technique involves extracting the dried ink on paper with strong solvents, such as pyridine, and then performing a quantitative analysis by GC.

The quantitative determination of oxalic acid in biological fluids is critical for the effective diagnosis of fatal ethylene glycol or oxalic acid intoxications. In 1980, a procedure was described that is a specific and quantitative micromethod for the determination of oxalic acid in forensic specimens as its 2,3-trimethylsiloxyquinoxaline derivative using GC/MS (304).

In 1981, GC was used to determine the racemization of aspartic acid in dentin for the estimation of age. The *N*-trifluoroacetyl isopropyl esters of amino acids in dentin from teeth were quantitatively compared to estimate the age (305).

A thermal desorption capillary gas chromatographic method has been use to analyze volatiles from clingfilms (306). This procedure was able to discriminate fourteen brands of polyvinylchloride film and seven brands of polyethylene film. The forensic examination of clingfilms is often requested and this gas chromatographic procedure offers an alternative to physical methods for comparing control and recovered clingfilms.

These few examples should give the readers some idea of how GC is used in forensic science. For more examples or a thorough review of the topic the reader is referred to the literature on forensic science (307).

REFERENCES

1. *Frye vs. United States*, 293 Fed. 1013, 1014 (D.C. Cir. 1923).
2. *Coppolino vs. State*, 223 So. 2nd 68 (Fl. App. 1968).
3. *Daubert vs. Merrell Dow Pharmaceutical, Inc.*, 113 S. Ct. 2786 (1993).
4. *Criminalistics: An Introduction to Forensic Science*, 4th ed., R. Saferstein (Ed.), Prentice-Hall, Englewood Cliffs, NJ, 1990.
5. *Forensic Science Handbook*, Vols. 1–3, R. Saferstein (Ed.), Prentice-Hall, Englewood Cliffs, NJ, 1992.
6. P. R. DeForest, R. E. Gaensslen, and H. C. Lee, *Forensic Science: An Introduction to Criminalistics*, McGraw-Hill, New York, 1983.
7. P. A. McDonald and T. A. Gough, *Anal. Methods in Forensic Chemistry*, Matt Ho (Ed.), Ellis Horwood, Chichester, UK, 1990, pp. 149–172.
8. T. Gough, *The Analysis of Drugs of Abuse*, Wiley–Interscience, New York, 1992.
9. *Toxicological and Forensic Applictions of Chromatography*, E. J. Cone and Z. Deyl (Eds.), Elsevier, New York, 1992.
10. J. Siegel, *Forensic Science Handbook*, Vol. 2, R. Saferstein (Ed.), Prentice-Hall, Englewood Cliffs, NJ, 1988, pp. 68–160.
11. A. C. Moffat, J. V. Jackson, M. S. Moss, and B. Widdop, *Clarke's Isolation and Identification of Drugs*, The Pharmaceutical Press, King of Prussia, PA, 1986.

12. T. A. Mills III and J. C. Roberson, *Instrumental Data for Drug Analysis*, Vols. 1–5, Elsevier, New York, 1987.
13. *AOAC Official Methods of Analysis*, William Horowitz (Ed.), AOAC, Washington, DC, 1990.
14. *Analytical Profiles of Cocaine, Local Anesthetics and Common Diluents Found With Cocaine*, CND Analytical, Auburn, AL, 1988.
15. *Analytical Profiles of Substituted 3,4-Methylenedioxyamphetamines: Designer Drugs Related to MDA*, CND Analytical, Auburn, AL, 1988.
16. *Analytical Profiles of the Anabolic Steroids and Related Substances*, CND Analytical, Auburn, AL, Vol. 1, 1989; Vol. 2, 1991.
17. *Analytical Profiles of Amphetamines and Related Phenethylamines*, CND Analytical, Auburn, AL, 1989.
18. *Analytical Profiles of Designer Drugs Related to 3,4-Methylenedioxyamphetamines*, Vol. 2, CND Analytical, AL, 1993.
19. *Analytical Profiles of Benzodiazepines*, CND Analytical, AL, 1993.
20. *Analytical Profiles of Precursors and Essential Chemicals*, CND Analytical, AL, 1993.
21. A. Sperling, *J. Chromatogr.*, **538**(2), 269 (1991).
22. E. J. Cone, B. A. Holicky, T. M. Grant, W. D. Darwin, and B. A. Goldberger, *J. Anal. Toxicol.*, **17**(10), 327 (1993).
23. D. K. Wyatt and L. T. Grady, *Anal. Profiles Drug Subst.* **10**, 357 (1981).
24. V. Navaratnam and N. K. Fei, *Bull. Narc.*, **36**(1), 15 (1984).
25. C. C. Clark, J. DeRuiter, and F. T. Noggle, *J. Chromatogr. Sci.*, **30**(10), 399 (1992).
26. N. Narasimhachari and R. O. Friedel, *J. Chromatogr.*, **164**, 386 (1979).
27. R. D. Budd and W. J. Leung, *J. Chromatogr.*, **179**, 355 (1979).
28. N. C. Jain, T. C. Sneath, R. D. Budd, and B. A. Olson, *J. Anal. Toxicol.*, **1**, 233 (1977).
29. C. L. Hornbeck and R. J. Czarny, *J. Anal. Toxicol.*, **13**, 144 (1989).
30. R. J. Czarny and C. L. Hornbeck, *J. Anal. Toxicol.*, **13**, 257 (1989).
31. T. A. Brettell, *J. Chromatogr.*, **257**, 45 (1983).
32. J. H. Liu, W. W. Ku, J. T. Tsay, M. P. Fitzgerald, and S. Kim, *J. Forensic Sci.*, **27**(1), 39 (1982).
33. C. C. Clark, *J. AOAC*, **69**(5), 814 (1986).
34. M. A. ElSohly, R. Brenneisen, and A. B. Jones, *J. Forensic Sci.*, **36**(1), 93 (1991).
35. Y. Nakahara and A. Ishigami, *J. Anal. Toxicol.*, **15**(3), 105 (1991).
36. J. F. Casale, *J. Forensic Sci.*, **37**(5), 1295 (1992).
37. N. C. Jain and R. G. Cravey, *J. Chromatogr. Sci.*, **1**, 587 (1978).
38. D. N. Pillai and S. Dilli, *J. Chromatogr.*, **220**, 253 (1981).
39. A. D. Barour, *J. Anal. Toxicol.*, **15**(4), 214 (1991).
40. R. E. Ardery and A. C. Moffat, *J. Chromatogr.*, **220**, 195 (1981).
41. L. L. Plotczyk and P. Larson, *J. Chromatogr.*, **257**, 211 (1983).
42. M. Novotny, M. L. Lee, C. E. Low, and A. Raymond, *Anal. Chem.*, **48**, 24 (1976).

43. R. Brenneisen and M. A. ElSohly, *J. Forensic Sci.*, **33**, 1385 (1985).
44. R. J. H. Liu, M. P. Fitzgerald, and G. V. Smith, *Anal. Chem.*, **51**, 1875 (1979).
45. R. J. H. Liu and M. P. Fitzgerald, *J. Forensic Sci.*, **25**, 815 (1980).
46. T. B. Vree, *J. Pharmacol. Sci.*, **66**, 144 (1977).
47. P. B. Baker, T. A. Gough, S. J. M. Johncock, B. J. Taylor, and L. T. Wyles, *Bull. Narc.*, **34**(3), 101 (1982).
48. G. Lercker, F. Bocchi, N. Frega, and R. Bartolomeazzi, *Farmaco*, **47**(3), 367 (1992).
49. H. V. S. Hood and G. T. Barry, *J. Chromatogr.*, **166**, 499 (1978).
50. H. S. Nichols, W. H. Anderson, and D. T. Stafford, *J. High Res. Chromatogr.*, **6**, 101 (1983).
51. R. L. Epstein, P. Lorimer, and E. J. Sloma, *J. Forensic Sci.*, **22**(1), 61 (1977).
52. D. Legault, *J. Chromatogr. Sci.*, **20**, 228 (1982).
53. *Federal Register*, **56**(30), 98 (1991).
54. J. J. Vrbanac, W. E. Braselton, J. F. Holland, and C. C. Sweeney, *J. Chromatogr.*, **239**, 265 (1982).
55. G. P. Cartoni, M. Ciardi, A. Giarusso, and F. Rosati, *J. Chromatogr.*, **239**, 515 (1983).
56. B. C. Chung, H.-Y. P. Choo, T. W. Kim, K. D. Eom, O. S. Kwon, J. Suh, J. Yang, and J. Park, *J. Anal. Toxicol.*, **14**, 91 (1990).
57. R. Masse, H. Bi, and P. Du, *Anal. Chem. Acta*, **247**(2), 211 (1991).
58. H. Lamparczyk, *Handbook of Chromatography*, J. Sherma (Ed.), CRC Press, Boca Raton, FL, 1992.
59. R. Morehead, *The Restek Advantage*, 8 November, 1992.
60. T. Brettell and L. Manley, 45th Annual Meeting of American Academy of Forensic Sciences, Boston, MA, February 1993.
61. G. Koverman, *Southwetern Assoc. Forensic Sci. J.*, **15**(1), 16 (1993).
62. J. Hugel, *Can. Soc. Forensic Sci. J.*, **24**(3), 147 (1991).
63. D. M. Chiong, E. Consuegra-Rodriquez, and J. R. Almirall, *J. Forensic Sci.*, **37**(2), 488 (1992).
64. P. D. Coleman, E. A'Hearn, R. W. Taylor, and S. D. Le, *J. Forensic Sci.*, **36**(4), 1079 (1991).
65. R. L. Foltz, *Advances in Analytical Toxicology*, Vol. 1, R. C. Baselt (Ed.), Biomedical Publications, Foster City, CA, 1984, p. 125.
66. D. R. Knapp, *Handbook of Analytical Derivatization Reactions*, Wiley, New York, 1979.
67. J. W. Brooks, C. G. Edmonds, S. J. Gaskell and A. G. Smith, *Chem. Phys. Lip.*, **21**, 403 (1978).
68. E. P. J. Van der Slooten and H. T. Van der Helm, *Forensic Sci.*, **16**, 83 (1975).
69. L. Stromberg, *J. Chromatogr.*, **106**, 355 (1975).
70. K. Tanaka, T. Ohmori, and T. Inoue, *Forensic Sci. Int.*, **56**(2), 157 (1992).
71. L. Stromberg, *J. Chromatogr.*, **68**, 253 (1972).
72. L. Stromberg and A. C. Machly, *J. Chromatogr.*, **109**, 67 (1975).
73. C. C. Clark, *Analytical Methods in Forensic Chemistry*, Matt Ho (Ed.), Ellis Horwood, Chichester, UK, 1990, p. 173.

74. L. D. Baugh and R. H. Liu, *Forensic Sci. Rev.*, **3**(2), 101 (1991).
75. T. A. Jennison, E. Wozniak, G. Nelson, and F. M. Urny, *J. Anal. Toxicol.* **17**(4), 208 (1993).
76. X-H. Chen, J.-P. Franke, and R. A. DeZeeuw, *Forensic Sci. Rev.*, **4**(2), 147 (1992).
77. G. E. Platoff, Jr., and J. A. Gere, *Forensic Sci. Rev.* **3**(2), 117 (1991).
78. *Federal Register*, **53**, 1, 1979 (1988).
79. R. A. DeZeeuw, M. Bogusz, J.-P. Franke, and J. Wijsbeek, *Advances in Analytical Toxicology*, Vol. 1, R. C. Baselt (Ed.), Biomedical Publications, Foster City, CA, 1984, p. 41.
80. A. C. Mofatt, B. S. Finkle, R. K. Muller, and R. A. DeZeeuw, *Bull. Int. Assoc. Forensic Toxicol.*, Special Issue (1981).
81. R. A. Cox, J. A. Crifasi, R. E. Dickey, S. C. Ketzler, and G. L. Pshak, *J. Anal. Toxicol.*, **13**, 224 (1989).
82. J. Park, S. Park, D. Lho, H. P. Choo, B. Chung, C. Yoon, H. Min, and M. J. Choi, *J. Anal. Toxicol.*, **14**, 66 (1990).
83. S. J. Mule and G. A. Casella, *J. Anal. Toxicol.*, **12**, 102 (1988).
84. P. Lillsunde and T. Korte, *J. Anal. Toxicol.*, **15**, 17 (1991).
85. X. Chen, J. Wijsheek, J. Van Ween, J.-P. Franke, and R. A. DeZeeuw, *J. Chromatogr. (Biomed. Applic.)*, **529**, 161 (1990).
86. J.-P. Franke, J. Wijsbeek, and R. A. DeZeeuw, *J. Forensic Sci.*, **35**(4), 813 (1990).
87. *Gas Chromatographic Retention Indices of Toxicologically Relevant Substances on Packed or Capillary Columns With Dimethyl-silicone Stationary Phases*, R. A. DeZeeuw, J.-P. Franke, H. H. Maurer, and K. Pfleger (Eds.), VCH, Weinheim, Germany, 1992.
88. D. Manca, L. Ferron, and J.-P. Weber, *Clin. Chem.*, **35**(4), 601 (1989).
89. V. W. Watts and T. F. Simmonick, *J. Anal. Toxicol.*, **11**, 210 (1987).
90. J.-P. Franke, M. Bogusz, and R. A. DeZeeuw, *Fresenius' J. Anal. Chem.*, **347**(3–4), 67 (1993).
91. P. Kintz, P. Mangin, A. A. Lugnier, and A. J. Chaumont, *Eur. J. Clin. Pharmacol.*, **37**, 531 (1987).
92. B. D. Paul, J. M. Mitchell, L. D. Mell, and J. Irvins, *J. Anal. Toxicol.*, **13**, 2 (1989).
93. B. A. Goldberger, W. D. Darwin, T. M. Grant, A. C. Allen, Y. H. Caplan, and E. J. Cone, *Clin. Chem.*, **39**(4), 670 (1993).
94. L. J. Bowie and P. B. Kirkpatrick, *J. Anal. Toxicol.*, **13**, 326 (1989).
95. F. Medzihradsky and P. J. Dahlstrom, *Pharmacol. Res. Commun.*, **7**, 55 (1975).
96. J. Combie, J. W. Blake, T. E. Nugent, and T. Tobin, *Clin. Chem.*, **28**, 83 (1982).
97. L. W. Hayes, W. G. Krasselt, and P. A. Meuggler, *Clin. Chem.*, **33**(6), 806 (1987).

98. J. M. Moore, *J. Chromatogr.*, **147**, 327 (1987).
99. D. F. Mathis and R. D. Budd, *Clin. Toxicol.*, **16**, 181 (1980).
100. P. Jacob, B. A. Elias-Baker, R. T. Jones, and N. L. Benowitz, *J. Chromatogr.*, **417**(2), 277 (1987).
101. R. W. Taylor, N. C. Jain, and M. P. George, *J. Anal. Toxicol.*, **11**(5), 233, (1987).
102. V. R. Spiehler and D. Reed, *J. Forensic Sci.*, **30**(4), 1003.
103. K. Vereby and A. DePace, *J. Forensic Sci.*, **34**(1), 46 (1989).
104. J. Y. Zhang and R. L. Foltz, *J. Anal. Toxicol.*, **14**, 201 (1990).
105. S. Rose, J. Cofino, W. Lee-Hearn, and G. W. Hime, 43rd Annual Meeting of the American Academy of Forensic Sciences, Anaheim, California, 1991.
106. A. S. Christopherson, E. Dahlin, and G. Petersen, *J. Chromatogr.* (Biomed. Applic.), **432**, 290 (1988).
107. R. W. Taylor, S. D. Le, S. Philip, and N. C. Jain, *J. Anal. Toxicol.*, **13**(5), 293 (1989).
108. P. Kintz, A. Tracqui, P. Manqin, A. A. Lugnier, and A. J. Chaumont, *Forensic Sci. Int.*, **40**(2), 153 (1989).
109. S. J. Mule and G. A. Casella, *J. Anal. Toxicol.*, **12**, 102 (1988).
110. C. L. Hornbeck and R. J. Czarny, *J. Anal. Toxicol.*, **13**, 144 (1989).
111. R. L. Fitzgerald, R. V. Blanke, R. A. Glennon, M. Y. Yousif, J. A. Rosecrans, P. Francom, D. Andrenyak, H.-K. Lim, R. R. Bridges, R. L. Foltz, and R. T. Jones, *J. Anal. Toxicol.*, **12**, 1 (1988).
112. R. Melgar and R. C. Kelly, *J. Anal. Toxicol.*, **17**(7), 399 (1993).
113. R. J. Czarny and C. L. Hornbeck, *J. Anal. Toxicol.*, **13**, 257 (1989).
114. H. Gjerde, I. Hasvold, G. Petterson, and A. Christophersen, *J. Anal. Toxicol.*, **17**(2), 65 (1993).
115. K. Kalasinsky, B. Levine, M. L. Smith, J. Magluilo, Jr., and T. Schaefer, *J. Anal. Toxicol.*, **17**(6), 359 (1993).
116. S. J. Mule and G. A. Casella, *J. Anal. Toxicol.*, **13**(3), 179 (1989).
117. B. Rollmann, B. Lefebure, A. Goetemans, and B. Tilquin, *J. Pharm. Belg.* **45**(4), 245 (1990).
118. R. D. Budd, *J. Chromatogr.*, **237**(1), 155 (1982).
119. H. Seno, O. Suzuki, T. Kumazawa, and H. Hattoria, *J. Anal. Toxicol.*, **15**, 21 (1991).
120. W. A. Joern and A. B. Joern, *J. Anal. Toxicol.*, **11**(6), 247 (1987).
121. E. M. Koves and B. Yen, *J. Anal. Toxicol.*, **13**, 69 (1989).
122. K. Vereby and A. DePace, *J. Forensic Sci.*, **34**(1), 46 (1989).
123. D. J. Holsztynska and E. G. Domino, *J. Anal. Toxicol.*, **10**, 107 (1986).
124. J. R. Woodworth, M. Mayerson, and S. M. Owens, *J. Anal. Toxicol.*, **8**, 2 (1984).
125. L. K. Ritchie, Y. H. Caplan, and J. Park, *J. Anal. Toxicol.*, **11**, 205 (1987).
126. R. L. Foltz and I. Sunshine, *J. Anal. Toxicol.*, **14**(6), 374 (1990).
127. M. A. ElSohoy, E. S. Arafat, and A. B. Jones, *J. Anal. Toxicol.*, **8**, 7 (1984).
128. A. H. B. Wu, N. Liu, Y. J. Cho, K. G. Johnson, and S. S. Wong, *J. Anal. Toxicol.*, **17**(4), 215 (1993).

129. R. Cloutte, M. Jacob, P. Koteel, and M. Spain, *J. Anal. Toxicol.*, **17**(1), 1 (1993).
130. B. D. Paul, J. M. Mitchell, R. Burbage, M. May, and R. Sroka, *J. Chromatogr.* (Biomed. Applic.), **529**, 103 (1990).
131. N. Bukowski and A. N. Eaton, *Rapid Commun. Mass Spectrom.*, **7**(1), 106 (1993).
132. D. I. Papac and R. L. Foltz, *J. Anal. Toxicol.*, **14**(3), 189 (1990).
133. P. J. Orsulak, M. C. Haven, M. E. Burton, and L. C. Akers, *Clin. Chem.*, **35**(7), 1318 (1989).
134. M. J. Welch, L. T. Sniegoski, C. C. Allgood, and M. Habram, *J. Anal. Toxicol.*, **17**(7), 389 (1993).
135. B. Caddy, *Advances in Analytical Toxicology*, Vol. 1, R. C. Baselt (Ed.), Biomedical Publications, Foster City, CA, 1984, p. 198.
136. O. Suzuki, H. Hattori, and M. Adano, *Forensic Sci., Int.*, **24**, 9 (1984).
137. O. Suzuki, T. Inoue, H. Hori, and S. Inayaama, *J. Anal. Toxicol.*, **13**, 176 (1989).
138. K. B. Nolte, R. D. Pinder, and W. D. Lord, *J. Forensic Sci.*, **37**(4), 1179 (1992).
139. R. J. Flanagan, M. Ruprah, R. J. Meredith, and J. D. Ramsey, *Drug Saf.*, **5**(5), 359 (1990).
140. *Gas Chromatographic Retention Indices of Solvents and Other Volatile Substances for Use in Toxicological Analysis*, R. A. DeZeeuw, J.-P. Franke, G. Machata, M. R. Muller, R. K. Muller, A. Graefe, D. Tiess, K. Pfleger, and M. G. Von Mallinckrodt (Eds.), VCH, Weinheim, Germany, 1992.
141. A. W. Jones, A. Lofgren, A. Eklund, and R. Grundin, *J. Anal. Toxicol.*, **16**(2), 104 (1992).
142. R. V. Babakhanyan, A. Kirsanov, and L. V. Petrov, *Sud.-Med. Ekspert*, **32**(2), 28 (1990).
143. W. Gubala, *Forensic Sci. Int.* **46**(1–2), 127 (1990).
144. R. Iffland, P. Balling, and G. Eiling, *Beitr. Gerichtl. Med.*, **47**, 87 (1989).
145. Y. Seto, T. Shinohara, and N. Tsunoda, *Hochudoku*, **8**(2), 56 (1990).
146. J. G. Krause and W. B. McCarthy, *J. Forensic Sci.*, **34**(4), 1011 (1989).
147. K. Kimura, T. Negata, K. Kudo, T. Imamura, and K. Hara, *Biol. Mass Spectrom.*, **20**(8), 493 (1991).
148. R. L. Fitzgerald, C. E. Fishel, and L. L. E. Bush, *J. Forensic Sci.*, **38**(2), 476 (1993).
149. C. LoDico, Y. H. Caplan, B. Levine, D. F. Smyth, and J. E. Smialek, *J. Forensic Sci.*, **34**(4), 1013 (1989).
150. B. Jacob, C. Heller, T. Daldrup, K. F. Burrig, J. Barz, and W. Bonte, *J. Forensic Sci.* **34**(6), 1408 (1989).
151. A. Pla, A. F. Hernanadez, F. Gil, M. Garcia-Alonzo, and E. Villanueva, *Forensic Sci. Int.*, **49**, 193 (1991).
152. W. T. Lowry, B. Gamse, A. T. Armstrong, J. M. Corn, L. Juarez, J. L. McDowell, and R. Owens, *J. Forensic Sci.*, **36**(2), 386 (1991).
153. K. Kato, T. Nagata, K. Kimura, K. Kudo, T. Imamura, and M. Noda, *Forensic Sci. Int.*, **44**(1), 5 (1990).

154. H. Gjerde, A. Smith-Kielland, P. T. Normann, and J. Morgland, *Forensic Sci. Int.*, **44**, 77 (1990).
155. N. C. Jain and R. H. Cravey, *J. Chromatogr. Sci.*, **10**, 263 (1972).
156. N. C. Jain, *Clin. Chem.* **17**(2), 82 (1971).
157. K. M. Dubowski, *Manual for Analysis of Ethanol in Biological Liquids*, Dept. of Transportation Report No., DOT-TSC-NHTSA-76-4, January 1977.
158. J. J. Lentini, M. L. Fultz, A. Armstrong, B. Davie, J. DeHaan, R. Henderson, J. O'Donnell, B. J. Rogers, and J. Small, *Fire and Arson Investigator*, **41**(2), 50 (1990).
159. *Fire and Arson Investigator*, **40**(2), 25 (1989).
160. C. R. Midkiff, *Arson Analysis Newsletter*, **2**, 8 (1978).
161. D. Willson, *Forensic Sci. Int.*, **10**, 243 (1977).
162. M. Camp, *Anal. Chem.*, **52**, 422A (1980).
163. Forensic Science and Engineering Committee of the International Associations of Arson Investigators, March 1988, *The Fire and Arson Investigator*, **38**, 45 (1988).
164. D. T. Stafford, *Crime Lab. Dig.*, **14**, 7 (1987).
165. M. L. Fultz and John D. DeHaan, *Gas Chromatography in Forensic Science*, Ian Tebbett (Ed.), Ellis Horwood, Chichester, UK, 1992.
166. J. D. DeHaan and F. A. Skalsky, *Arson Analysis Newsletter*, **5**(1), 6 (1981).
167. W. R. Dietz and D. D. Mann, *Scientific Sleuthing Newsletter*, **12**(3), 5 (1988).
168. W. D. Kinard and C. R. Midkiff, 42 Annual Meeting, American Academy of Forensic Sciences, Cincinnati, OH, February 1990.
169. ASTM E-1387-90, *Standard Test Method for Flammable or Combustible Liquid Residues in Extracts From Samples of Fire Debris by Gas Chrmatography*, ASTM, Philadelphia, PA, 1991.
170. T. A. Brettell, P. A. Moore, and R. L. Grob, *J. Chromatogr.*, **358**, 423 (1986).
171. B. Caddy, F. P. Smith, and J. Macy, *Forensic Sci. Rev.*, **3**(1), 58 (1991).
172. ASTM E-1385-90, *Standard Practice for Separation and Concentration of Flammable or Combustible Liquid Residues From Fire Debris Samples by Steam Distillation*, ASTM, Philadelphia, PA, 1991.
173. B. V. Ettling and M. F. Adams, *J. Forensic Sci.*, **13**(1), 76 (1968).
174. C. R. Midkiff, *Arson Analysis Newsletter*, **2**(6), 8 (1978).
175. D. L. Adams, *J. Crim. Law, Criminol. Police Sci.*, **47**, 593 (1957).
176. R. N. Thaman, *Arson Analysis Newsletter*, **3**, 9 (1979).
177. R. L. Harmer, R. D. Moss, T. E. Nolan, and R. N. Thaman, *The Fire and Arson Investigator*, **33**, 3 (1983).
178. J. F. O'Donnell, *The Fire and Arson Investigator*, **41**, 4 (1990).
179. A. Claussen, *Arson Analysis Newsletter*, **6**, 105 (1982).
180. J. Nowicki and C. Strock, *Arson Analysis Newsletter*, **7**, 98 (1983).
181. B. V. Ettling, *J. Forensic Sci.*, **8**, 261 (1963).
182. ASTM E-1386-90, *Standard Practice for Separation and Concentration of Flammable Liquid Residues From Fire Debris Samples by Solvent Extraction*, ASTM, Philadelphia, PA, 1991.

183. T. A. Brettell, P. Moore, M. LaMachia, and R. L. Grob, *Arson Analysis Newsletter*, **8**(4), 86 (1984).

184. ASTM E-1388-90, *Standard Practice for Sampling of Headspace Vapors From Fire Debris Samples*, ASTM, Philadelphia, PA, 1991.

185. C. R. Midkiff, Jr., and W. D. Washington, *J. Assoc. Off. Anal. Chem.*, **55**(4), 840 (1972).

186. D. G. Kubler, D. Greene, C. Stackhouse, and T. Stoudemeyer, *Arson Analysis Newsletter*, **5**, 82 (1981).

187. V. Reeve, J. Jeffrey, D. Weihs, and W. Jennings, *J. Forensic Sci.*, **31**, 479 (1986).

188. W. R. Dietz, *J. Forensic Sci.*, **35**(1), 111 (1991).

189. ASTM E:1413-91, *Standard Practice for Separation and Concentration of Flammable or Combustible Liquid Residues From Fire Debris Samples by Passive Headspace Concentration*, Philadelphia, PA, 1991.

190. J. D. Twibell and J. M. Home, *Nature*, **268**(5622), 711 (1977).

191. J. Andrasko, *J. Forensic Sci.*, **28**(2), 330 (1983).

192. D. J. Tranthim-Fryer, *J. Forensic Sci.*, **35**(2), 271 (1990).

193. J. A. Juhala, *Arson Analysis Newsletter*, **6**(2), 32 (1982).

193A. R. F. Mindrup, Supelco Reporter, **14**(1), 2 (1995).

194. M. Frenkel, S. Tsaroom, Z. Aizenshtat, S. Kruas, and D. Daphna, *J. Forensic Sci.*, **29**(3), 723 (1984).

195. W. Jennings and H. E. Nursten, *Anal. Chem.*, **39**, 521 (1967).

196. J. E. Chrostowski and R. N. Holmes, *Arson Analysis Newsletter*, **3**, 1 (1979).

197. D. G. Kubler, D. Greene, C. Stackhouse, and T. Stoudemeyer, *Arson Analysis Newsletter*, **5**, 82 (1981).

198. R. E. Tontarski and R. A. Strobel, *J. Forensic Sci.*, **27**(3), 710 (1982).

199. D. S. Crandall and J. T. Pennie, *Arson Analysis Newsletter*, **8**(2), 47 (1984).

200. J. A. Juhala and F. K. Beever, *Arson Analysis Newsletter*, **9**(1), 1 (1986).

201. R. Saferstein and S. A. Park, *J. Forensic Sci.*, **27**(3), 484 (1982).

202. D. Willson, International Association of Forensic Sciences Meeting, Oxford, UK, Abstract No. 83, 1984.

203. J. Andrasko, *J. Forensic Sci.*, **28**(2), 330 (1983).

204. H. J. Kobus, K. P. Kirkbride, and A. Maehly, *J. Forensic Sci. Soc.*, **27**, 307 (1987).

205. ASTM E:1412-91, *Standard Practice for Separation and Concentration of Flammable or Combustible Liquid Residues From Fire Debris Samples by Dynamic Headspace Concentration*, Philadelphia, PA, 1991.

206. J. A. Zoro and K. Hadley, *J. Forensic Sci. Soc.*, **16**, 103 (1976).

207. R. M. Smith, *J. Forensic Sci.*, **28**(2), 318 (1983).

208. J. Nowicki, *J. Forensic Sci.*, **36**(5), 1536 (1991).

209. R. M. Smith, *Forensic Mass Spectrometry*, J. Yinon (Ed.), CRC Press, Boca Raton, FL, 1987, p. 131.

210. R. L. Kelly and R. M. Martz, *J. Forensic Sci.*, **29**(3), 714 (1984).
211. R. O. Keto and P. L. Wineman, *Anal. Chem.*, **63**, 1964 (1991).
212. J. F. Nowicki, *J. Forensic Sci.*, **35**(5), 1064 (1990).
213. S. E. Hipes, J. W. Witherspoon, G. A. Bertleson, P. A. Beyer, and M. E. Kurtz, *MAAFS Newsletter*, **20**(1), 48 (1991).
214. C. R. Midkiff, *Fire and Arson Investigator*, **26**(2), 18 (1975).
215. D. Mann, *J. Forensic Sci.*, **32**(3), 606 (1987).
216. D. Mann, *J. Forensic Sci.*, **32**(3), 616 (1987).
217. T. Urbanski, *Chemistry and Technology of Explosives*, Pergamon, Oxford, UK, Vol. 1, 1964; Vol. 2, Vol. 3, 1967.
218. I. Lurie and J. Wittwer (Eds.), *HPLC in Forensic Chemistry*, Dekker, New York, 1983.
219. J. Yinon and S. Zitrin, *The Analysis of Explosives*, Pergamon, Oxford, UK, 1981.
220. J. Yinon and S. Zitrin, *Modern Methods and Applications in Analysis of Explosives*, Wiley, New York, 1993.
221. J. Yinon, *Analysis of Explosives: CRC Critical Reviews in Analytical Chemistry*, CRC Press, Cleveland, OH, **7**(1), 1 (1977).
222. A. D. Beveridge, *J. Energetic Mater.*, **4**, 29 (1986).
223. A. D. Beveridge, *Forensic Sci. Rev.*, **4**, 17 (1992).
224. L. Elias, *Proceedings*, New Concepts Symposium and Workshop on Detection and Identification of Explosives, 30 October–1 November, 1978, NTIS: Springfield, VA, 1978, p. 265.
225. J. M. F. Douse, *J. Chromatogr.*, **234**, 415 (1982).
226. I. Jane, P. G. Brookes, J. M. F. Douse and K. A. O'Callaghan, *Proceedings*, International Symposium on the Analysis and Detection of Explosives, 29 March, US Department of Justice, FBI, 1983, p. 475.
227. J. M. F. Douse and R. N. Smith, *J. Energetic Mater.*, **4**, 169 (1986).
228. J. D. Twibell, J. M. Howe, K. W. Smalldon, and D. G. Higgs, *J. Forensic Sci.*, **27**, 783 (1982).
229. I. H. L. Yip, *Can. Soc. Forensic Sci. J.*, **15**(2), 87 (1982).
230. I. S. Krull, M. Swartz, K. H. Xie, and J. H. Driscoll, *Proceedings*, the International Symposium on the Analysis and Detection of Explosives, March, FBI, 1983, p. 107.
231. F. Belkin, R. W. Bishop, and M. V. Sheely, *J. Chromatogr. Sci.*, **24**, 532 (1985).
232. J. R. Hobbs and E. Conde, in *Proceedings*, the Third Symposium on Analysis and Detection of Explosives, Mannheim-Neuostheim, Germany, 1989, p. 41.
233. A. L. Lafleur, B. D. Morrison, and D. H. Fine, *Proceedings*, New Concepts Symposium and Workshop on Detection and Identification of Explosives, 30 October–1 November, NTIS, Springfield, VA, 1978, p. 597.
234. D. H. Fine, W. C. Yu, E. U. Goff, E. C. Bender and D. Reutter, *J. Forensic Sci.*, **29**(3), 732 (1984).
235. A. L. Lafleur and K. M. Mills, *Anal. Chem.*, **53**(8), 1202 (1981).

236. J. M. F. Douse, *J. Chromatogr.*, **256**, 359 (1983).
237. J. B. F. Lloyd, *J. Energetic Mater.*, **9**, 1 (1991).
238. P. Kolla, *J. Forensic Sci.*, **36**(5), 1342 (1991).
239. S. Zitrin, *J. Energetic Mater.*, **4**, 199 (1986).
240. A. S. Cummings and K. P. Park, *Proceedings*, International Symposium on the Analysis and Detection of Explosives, 29–31 March, U.S. Department of Justice, FBI, 1983, p. 259.
241. J. Yinon, in *Forensic Mass Spectrometry*, Jehuda Yinon (Ed.), CRC Press, Boca Raton, FL, 1987, p. 106.
242. C. T. Pate and M. H. Mach, *Int. J. Mass Spectrom., Ion Phys.*, **26**, 267 (1978).
243. T. Tamiri and S. Zitrin, *J. Energetic Mater.*, **4**, 215 (1986).
244. M. H. Mach, A. Pallos, and P. F. Jones, *J. Forensic Sci.*, **23**, 433 (1978).
245. M. H. Mach, A. Pallos, and P. F. Jones, *J. Forensic Sci.*, **23**, 446 (1978).
246. T. G. Kee, D. M. Holmes, K. Doolan, J. A. Hamill, and R. M. E. Griffin, *J. Forensic Sci. Soc.*, **30**, 285 (1990).
247. R. M. Martz, T. O. Munson, and L. D. Laswell, *Proceedings*, International Symposium on the Analysis and Detection of Explosives, 29–31 March, U.S. Dept. of Justice, FBI, 1983, p. 245.
248. C. M. Selavka, R. A. Stroebel, and R. E. Tontorski, *Proceedings*, 3rd Symposium on Analysis and Detection of Explosives, Fraunhofer-Institut fur Chemische Technologie (ICT), Berghausen, Germany, 1989, p. 3.
249. J. Andrasko, *J. Forensic Sci.*, **37**(4), 1030 (1992).
250. J. Yinon, *Forensic Sci. Rev.*, **3**(1), 17 (1991).
251. P. L. Kirk, *J. Gas Chromatogr.*, **1**, 11 (1967).
252. B. B. Wheals, *J. Anal. Appl. Pyrolysis*, **2**, 277 (1981).
253. R. Saferstein, *Pyrolysis and Gas Chromatography in Polymer Analysis*, S. A. Liebman, E. J. Levy (Eds.), Dekker, New York, 1985, Chapter 7.
254. R. D. Blackledge, *Forensic Sci. Rev.*, **4**, 1 (1992).
255. T. P. Wampler, *J. Anal. Appl. Pyrolysis*, **16**, 291 (1989).
256. D. J. Freed and S. A. Liebman, *Pyrolysis and Gas Chromatography in Polymer Analysis*, S. A. Liebman and E. J. Levy (Eds.), Dekker, New York, 1985, Chapter 2.
257. T. O. Munson, *Crime Lab. Digest*, **13**, 82 (1986).
258. P. R. DeForest, *J. Forensic Sci.*, **19**, 113 (1974).
259. J. M. Challinor, *J. Anal. Appl. Pyrolysis*, **18**(3–4), 233 (1991).
260. B. B. Wheals and W. Noble, *J. Forensic Sci. Soc.*, **14**, 23 (1974).
261. K. Fekuda, *Forensic Sci. Int.*, **29**, 227 (1985).
262. C. W. Stanley and W. R. Peterson, *Soc. Plastics Engineers Trans.*, **2**, 298 (1962).
263. W. M. Barbour, *J. Gas Chromatogr.*, **3**, 228 (1965).
264. J. D. Twibell, J. M. Home, and K. W. Smalldon, *Chromatographia*, **14**, 366 (1981).
265. B. M. Dixon, J. R. Lalhey, and L. M. Powell, Presented at the 20th Eastern Analytical Symposium, New York, 1981.

266. T. P. Wampler, W. A. Bowe, J. Higgins, and E. J. Levy, *Am. Lab.*, **17**(8), 82 (1985).
267. J. W. Bates, T. Allison, and T. S. Bal, *Forensic Sci., Int.*, **40**(25), (1989).
268. J. Nowicki and K. Vanderwerff, *MAFS News*, **16**, 1 (1987).
269. R. Saferstein and E. C. Ostberg, *Crime Lab. Digest*, **15**, 39 (1988).
270. R. F. Audette and R. F. E. Percy, *J. Forensic Sci.*, **23**, 672 (1978).
271. B. B. Wheals and W. Noble, *Chromatographia*, **5**, 553 (1972).
272. R. A. Janiak and K. A. Damerau, *J. Crim. Law, Criminol. Pol. Sci.*, **59**, 434 (1968).
273. *Identification of Textile Materials*, 7th Ed., The Textile Institute, Manchester, UK, 1975.
274. J. S. Crighton, in *Analytical Pyrolysis*, C. E. R. Jones and C. A. Cramers (Eds.), Elsevier Scientific, Amsterdam, 1977, p. 337.
275. J. P. Bortniak, S. E. Brown, and E. H. Sild, *J. Forensic Sci.*, **16**, 380 (1971).
276. J. Almer, *Can. Soc. Forensic. Sci. J.*, **24**, 51 (1991).
277. B. B. Wheals, in *Analytical Pyrolysis*, C. E. R. Jones and C. A. Cramers (Eds.), Elsevier Scientific, Amsterdam, 1977, p. 89.
278. E. J. Levy and T. P. Wampler, *J. Chem. Educ.*, **63**, A64 (1986).
279. J. R. Hardin and X. Q. Wang, *Textile Chem. Colorist*, **21**, 29 (1989).
280. D. W. Wright, K. O. Mahler, L. B. Ballard, and E. Dawes, *J. Chromatogr. Sci.*, **24**, 13 (1986).
281. D. M. Northrop and W. F. Rowe, *Biodeterioration Res.*, **1**, 7 (1986).
282. J. M. Hume, J. D. Twibell, and K. W. Smalldon, *Med. Sci. Law*, **20**, 163 (1980).
283. R. J. Davis, *Pyrolysis GC Analysis of Rubber: "Tieline"*, California Criminalistics Institute, Sacramento, CA, **5**(1), 24 (1978).
284. R. D. Blackledge, *J. Forensic Sci.*, **26**, 557 (1981).
285. J. Chih-An Hu, *Anal. Chem.*, **49**, 537 (1977).
286. J. Chih-An Hu, *Anal. Chem.*, **53**: 331A (1981).
287. C. J. Lennard and P. A. Margot, *J. Forensic Ident.*, **39**, 329 (1989).
288. J.-K. Ding and H.-S. Liu, *Forensic Sci. Int.*, **43**, 45 (1989).
289. C. J. Lennard and W. D. Mazella, *J. Forensic Sci. Soc.*, **31**, 365 (1991).
290. R. D. Blackledge, *Applied Polymer Analysis and Characterization*, Hanser, Munich, Germany, 1987, Chapter III-F.
291. R. D. Blackledge and J. A. Blackledge, *The Examination and Comparison of Floor Tile Glues*, 1991 Joint Meeting of the Federation of Analytical Chemistry and Spectroscopy Societies and the Pacific Conference on Chemistry and Spectroscopy, Anaheim, CA, 1991.
292. E. R. Williams and T. O. Munson, *J. Forensic Sci.*, **33**, 1163 (1988).
293. R. N. Totly, *Forensic Sci. Rev.*, **2**, 1 (1990).
294. N. A. Newlon and J. L. Booker, *J. Forensic. Sci.*, **24**, 87 (1979).
295. R. O. Keto, *Forensic Sci.*, **34**, 74 (1989).
296. F. H. Cassidy, *Chewing Gum Analysis by Pyrolysis Gas Chromatography: Tieline*, Calfornia Criminalistics Institute, Sacramento, CA, **5**(4), 59 (1979).

297. *Anrn: Cedar, Cypress or Pine?* PEAK, Hewlett–Packard, Palo Alto, CA, Summer 1988, p. 2.
298. F. L. Bayer, T. J. Hopkins, and F. M. Menzer, in *Analytical Pyrolysis*, C. E. R. Jones and C. A. Cramers (Eds.), Elsevier Scientific, Amsterdam, The Netherlands, 1977, p. 217.
299. P. K. Clausen and W. F. Rowe, *Forensic Sci.*, **25**, 765 (1980).
300. P. R. DeForest, *Individualization of Human Hair: Pyrolysis Gas Chromatography*, D. Crim. Dissertation, University of California, Berkeley (University Microfilm Order #70-6029).
301. T. O. Munson, *Crime Lab Digest*, **14**, 112 (1987).
302. F. Ishizawa and S. Misawa, *J. Forensic Sci. Soc.*, **30**, 201 (1990).
303. L. F. Stewart, *J. Forensic Sci.*, **30**(2), 405 (1985).
304. G. L. Gauthier, J. J. Rousseau, and M. J. Bertrand, *Can. Soc. Forensic Sci. J.*, **13**(4), 1 (1980).
305. S. Ohtani and K. Yamamoto, *J. Forensic Sci.*, **36**(3), 792 (1991).
306. J. Gilbert, J. M. Ingram, M. P. Scott, and M. Underhill, *J. Forensic Sci. Soc.* **31**, 331 (1991).
307. T. A. Brettell and R. Saferstein, *Anal. Chem.*, **65**(12), 293R (1993).

CHAPTER FOURTEEN

Environmental Applications of Gas Chromatography

JOSEPH M. LOEPER[☆]
Roy F. Weston, Inc.

[☆] Current Address: Environmental Resources Management, Inc., Exton, PA. 19341

Modern Practice of Gas Chromatography, Third Edition. Edited by Robert L. Grob.
ISBN 0-471-59700-7

14.1 INTRODUCTION

People have always been aware of the environment and its importance to their well being. As early as 1871, only four years after acquiring all of the land that now comprises the continental United States, the federal government set aside Yellowstone National Park to preserve some of the natural wonders of the country (1). These early attempts at environmental protection in the United States are best classified as "conservational measures." But an unprecedented period of prosperity in the United States, which began in the 1950s, brought to the forefront a different set of problems. Technical advancements associated with many fields created new products and processes, but they also generated a wide variety of chemical wastes and by-products. At the time, the benefits obtained from the advancements were truly remarkable and there was no real focus on environmental impact. As a result, ever increasing amounts of chemical substances were introduced into the land, water, and atmosphere through poor disposal and storage techniques and inefficient processes. Chemical pollution of natural resources was silently evolving into a significant problem.

By the 1960s, the environmental dilemma created by this wave of technology began to reveal itself in different ways. Visible problems like polluted air and water, chemical spills, dump sites, and dying wildlife were apparent either on a first-hand basis or as presented by expanded media coverage. Well-publicized personal tragedies like those encountered at Love Canal (2), Times Beach, and Bhopal (3) also contributed to an increased public awareness and led to a firm public demand for a resolution of environmental problems from chemical contamination.

The analysis of air, soil, and water samples has been and still is an important component of the overall effort to resolve the environmental problems associated with undesirable and unhealthy chemicals. The introduction of gas–liquid chromatography by James and Martin in 1952 (4) and its subsequent enhancements coincided roughly with the increased demand for quick, reliable, and sensitive analytical methods for the determination of the many chemical compounds that contaminated the environment. Gas chromatography (GC) proved invaluable for many of these applications and continues to play a major role in this effort.

In this chapter, a general overview of the applications of GC to environmental analysis is provided. Topics are segregated into the various sample preparation techniques and methods of analysis, according to the sample type and the chemical class of the analytes, since these are the critical parameters that must be evaluated when devising a scheme of analysis for chemical compounds in environmental samples. There is usually more than one way to perform an analysis for a given compound, but the sample matrix and ultimate requirements for the analysis will dictate the most practical approach.

14.2 REGULATIONS AND GOVERNMENT AGENCIES

Environmental laws and regulations and government agencies are obviously major participants in the realm of activities related to the protection of the environment, and an effective plan for the analysis of environmental samples must be designed within the framework of these laws and the agencies that administer them. While these concept would appear to be quite logical, in reality it has been a problem that both chemists and regulators have grappled with for quite some time. An American Chemical Society (ACS) panel provided the following assessment on the status of environmental analysis in 1977:

> Analysis too often is a secondary activity; unwarranted assumptions are made of the simplicity of the measurement and the reliability of the method . . . The result in part has been masses of data that are frequently useless for the intended purposes. (5)

This appraisal reinforces one of the basic requirements for the analyst in the design phase of any type of analysis, and that is the importance of knowing the proper background information on the project. This provides the foundation required to make the appropriate technical decisions. A report by the ACS Committee on Environmental Improvement in 1983 reiterated the importance of including the analyst in the planning stages of a project:

> It cannot be assumed that the person requesting an analysis will also be able to define the objectives of the analysis properly. Numerous discussions between the analyst and those who will use the results may be necessary until there is agreement on what is required of the analysis, how the results will be used, and what the expected results may be. The analytical methodology must meet realistic expectations regarding sensitivity, accuracy, reliability, precision, interferences, matrix effects, limitations, cost, and the time required for the analysis. (6)

Obviously, then, any analytical work performed in conjunction with a federal or state regulation or government agency must conform to the guidelines that they provide, and a working knowledge of both the guidelines and the capabilities of the various analytical techniques is essential for any analyst involved in the planning process. A detailed description of the many laws and government agencies is not within the scope of this chapter (it is estimated that there are 80 federal departments or agencies that have responsibilities in the area of environmental affairs) (8). The information that is provided can serve as an introduction for anyone who is not familiar with these topics. Detailed information on a particular government agency or regulation can be obtained directly from the agency.

14.2.1 Government Agencies

Government agencies that participate in environmental monitoring and remediation projects function at the federal, state, and municipal levels. Frequently, one (or more) of these agencies will provide some type of guidance, general oversight, and final approval for projects and laboratory analysis at every level of complexity. The interaction between an agency overseeing a project and the laboratory providing support for that project can be significant. At a minimum, most organizations will require that a laboratory is approved through an on-site audit prior to the initiation of work for their project. The laboratories may also be subject to subsequent audits on an annual or biannual basis. Agencies with a sizable technical staff and laboratory facility may develop their own methods and mandate that the support laboratory utilize these methods for their projects. Other agencies allow the laboratory to exercise its own judgment and simply require that the work conform to their quality assurance (QA) plan.

At the federal level, the United States Environmental Protection Agency (USEPA or EPA) is probably the most widely recognized organization. The EPA was formed in 1970 as a means to combine federal authority and know-how for the regulation and reduction of pollution and to address any other environmental problems the nation faces. From the regulatory perspective, the primary function of the EPA is to interpret a law, as outlined by the Congress, and formulate an operating practice. Table 14.1 lists several prominent federal agencies charged with the protection of the environment and pollution control. In addition to the federal agencies, there are also state and municipal agencies responsible for regulation and environmental protection within their region.

Environmental legislation is enacted at all three levels of government. The actions of various agencies at the different levels and the restrictions introduced by their regulations result in some degree of overlap and duplication of effort or even conflicts, and concessions are required. Federal regulations give the individual states the latitude of managing their own environmental affairs, provided that the state desires to do so and conforms to the boundaries outlined by the federal law. Failure to enforce these laws

TABLE 14.1 Federal Agencies Charged with Environmental Control

U.S. Environmental Protection Agency (USEPA)
U.S. Army Corps of Engineers (COE)
Department of Energy (DOE)
• Federal Energy Regulatory Commission
Department of the Interior (DOI)
• Bureau of Mines
• Bureau of Reclamation
• U.S. Geological Survey
• Office of Surface Mining
Department of Justice—Land and Natural Resources Division
Department of Labor (DOL)
• Occupation Safety and Health Administration (OSHA)
National Institute of Occupational Safety and Healthy (NIOSH)

Source. Reference 1.

at the state level, however, will lead to federal regulation within the state. Similarly, environmental regulations enacted at the state level must not impede the mission of any federal regulation.

14.2.2 Environmental Laws and Regulations

The early 1970s marked the beginning of a period in which the federal government enacted a significant number of new regulations designed to control pollution. The legislation included federal policies concerning air and water pollution, hazardous waste, radioactive waste, toxic materials, controls on mining activities and the protection of wildlife and wilderness. Table 14.2 lists the major federal environmental legislation enacted by the government for the purpose of controlling air pollution, water pollution, solid waste, and hazardous materials in the United States. A review of the information listed in Table 14.2 reveals that 16 out of the total number of 28 acts and their subsequent amendments were implemented on or after 1970, a clear indication of the heightened awareness in this time frame. More recent legislation has yielded numerous amendments to the original laws as well as new guidance focused on the control of international and global problems such as ozone depletion and acid rain.

14.3 OVERVIEW OF SAMPLE PREPARATION TECHNIQUES

With the sophisticated instrumentation available to the analytical chemist today, it is easy to lose sight of the importance of sample preparation techniques. While the advancements in both chemistry and engineering have

TABLE 14.2 Major Federal Environmental Legislation

Water Pollution Control

1899 **Rivers and Harbors Act** Required permit from chief of engineers for discharge of refuse into navigable waters.

1948 **Federal Water Pollution Control Act** Gave the federal government authority for investigations, research, and surveys. Left primary responsibility for pollution control with the states. Amended in 1956, 1961, 1972, and 1977.

1987 **Federal Water Pollution Control Act Amendments** Requires states to identify nonpoint sources and develop plans for a comprehensive water pollution control program.

1965 **Water Quality Act** Created Federal Water Pollution Control Administration.

1966 **Clean Water Restoration Act** Increased grant authorizations.

1970 **Water Quality Improvement Act** Established liability for oil spills, created new rules regarding thermal pollution.

1974 **Safe Drinking Water Act** Directed EPA to set standards for all public water systems.

1986 **Safe Drinking Water Act** Established primary (enforceable) and secondary (advisory) national drinking water regulations.

Air Pollution Control

1955 **Air Pollution Control Act** Authorized a federal program for research, training, and demonstrations relating to air pollution control (extended for four more years in 1959).

1963 **Clean Air Act** Gave the federal government enforcement powers through enforcement conferences similar to the 1956 approach to water pollution. Amended in 1970, 1974, and 1977.

1990 **Clean Air Act Amendments** Required EPA to categorize sources, regulate emissions of 189 hazardous air pollutants, evaluate risks, and address sudden accidental releases.

1965 **Motor Vehicle Air Pollution Control Act** Added new authority to 1963 act, giving HEW power to prescribe emission standards for automobiles as soon as practicable.

1967 **Air Quality Act** (1) Authorized HEW to oversee state standards for ambient air quality and state implementation plans; (2) set national standards for auto emissions.

Solid Waste and Resource Recovery

1965 **Solid Waste Disposal Act** Promoted the application of solid waste management and resource recovery systems.

1976 **Resource Conservation and Recovery Act** Promoted the development of management plans and facilities for the recovery of energy and other resources from discarded materials. Regulated management of hazardous wastes.

1984 **Hazardous and Solid Waste Amendments** Requirements added to regulate small-quantity waste generators and to ban land disposal of hazardous wastes.

TABLE 14.2. (*Continued*)

1980	**Comprehensive Environmental Response, Compensation and Liability Act** Established "superfund" from selected taxes to clean up abandonded hazardous waste sites. Mandated the establishment of National Hazardous Substance Response Plan.
1986	**Superfund Amendments and Reauthorization Act** Provided $8.5 billion for development of solutions for hazardous waste sites. Title III established Emergency Planning and Community Right-to-Know Act requiring states to develop emergency response plans for toxic releases.
	Chemicals
1972	**Federal Insecticide, Fungicide and Rodenticide Act** Authorized federal regulation of pesticides and related chemicals including banning, manufacture, sale, and use.
1976	**Toxic Substance Control Act** Required testing and necessary use restriction on certain chemical substances.
	Comprehensive Environmental Acts
1969	**The National Environmental Policy Act** Established national environmental policy and the Council on Environmental Quality. Coordinated federal projects and programs.
1970	**Environmental Quality Improvement Act** Required federal departments involved with work that affected the environment to implement the policies established under existing law.

Source. Adapted from References 1 and 28. Only the final amendment is listed for those acts with multiple amendments.

provided the chemist with some truly amazing instruments, even the most powerful of these will be useless with a sample that was improperly or inadequately prepared.

The particular type of preparation procedures required is dependent on several factors, related to both the analytes and the sample. Preliminary consideration must be given to the following:

1. The physical properties of the analytes, which include their volatility, solubility (in both water and organic solvents), stability, and acidic or basic qualities.
2. The physical nature of the sample matrix.
3. The ultimate purpose for the analysis, which, in turn, dictates the requirements for the data. An analysis meant to determine if a water supply is safe for public consumption will certainly have different requirements than a screening analysis intended to identify "hot spots" or areas of high contamination at a known hazardous waste site.

14.3.1 Categories of Analysis

Analysis schemes are designed to perform measurements in pre-ordained concentration ranges established in the project planning stage. The sensitivity of the instrument and the manner in which the sample/extract is handled are the factors that will ultimately determine whether the measurement can be performed in the desired concentration range. Umbreit (9) has put forth the classification of analysis categories listed in Table 14.3, and it is useful to review this information to obtain some perspective for the various analysis requirements. At one end of the spectrum is the major component analysis, a category that is rarely required for environmental analysis. The remaining two categories, characterized as minor component and trace analysis, are routinely required for environmental analysis. If it is assumed that the sensitivity of the instrument has already been optimized through the selection of the proper column, detector, and operating conditions (the reader is referred to the chapters on columns and detectors in this book as well as Section 14.7 in this chapter for specific guidance and examples), then the analyst is challenged to select those sample preparation steps that will also provide the conditions required to measure the analytes at the requested concentration levels. The remainder of this section considers the various aspects of sample preparation procedures in general terms. In this way, the analyst can evaluate practical applications provided within and outside of this book and identify the logic or reasoning behind the various modifications with the many procedures that will be encountered. This same logic can also be applied to the development of new methods designed to solve analytical problems that are not addressed by established techniques.

14.3.2 Trace Analysis

Methods of trace analysis require the highest level of instrument sensitivity and sample preparation steps that can enhance the concentration level of the analytes of interest prior to the analysis. The sample preparation techniques for trace level analysis can be broken down into three major stages: extraction, concentration, and clean-up.

TABLE 14.3 Analysis Categories and Concentration Ranges

Categories	Component Concentration	Range	Matrix Concentration
Trace	1 ppb–100 ppm	10^5	>99.999%
Minor component	100 ppm–5%	500	>95%
Major component or purity assay	5–100%	20	0–95%

Source. Reference 9.

Extraction Stage. The analytes of interest are transferred from the sample matrix into a medium that can be introduced into the instrument. In the case of a gas chromatograph, an appropriately medium is either a liquid or a gas. Under ideal circumstances, the analytes of interest would be selectively transferred or extracted from the sample matrix and any undesirable sample components that might interfere during the analysis would remain unextracted. The extraction usually effects some preliminary enhancement of the concentration levels of the analytes in the extraction medium relative to their original concentration level in the sample matrix.

Concentration Stage. This step serves to enhance or increase the concentration levels of the analytes in the extract, thereby improving the overall sensitivity of the analysis. The concentration of analytes contained in a solvent can be achieved through various evaporative techniques. A Kuderna–Danish (KD) concentration device or a rotary evaporator can be utilized for the selective volatilization of 200–300 mL of an extraction solvent. The evaporative flask containing the solvent is placed in a heated bath maintained at a temperature slightly above the boiling point of the extraction solvent, and a condenser column is attached to the flask. The excess extraction solvent passes through the condenser column, but the analytes are recondensed and carried back down to the solvent remaining in the flask. Small volumes of volatile solvents can be evaporated by directing a gentle stram of an inert gas at the surface of the solvent. Gentle heating of the solvent/extract will facilitate evaporation.

Several commercially marketed devices that provide for the automation of the concentration step have also become available recently. A combination of gentle heating and a gas flow that creates a vortex effect provides for the evaporation of the extraction solvent. An electronic sensing device automatically interrupts the gas flow when the solvent volume reaches approximately one milliliter.

Analyte concentration for gas-phase extracts or gaseous samples is achieved by adsorption onto a solid adsorbent, by partitioning into a liquid phase coated onto a support material, or by cryogenic trapping. The gaseous sample or extract is forced through the device responsible for analyte collection. The analytes can be thermally desorbed from the material and introduced directly into the gas chromatograph. Analytes collected on a solid adsorbent can also be desorbed with an organic solvent, which is subsequently injected into the instrument. Impingers are also utilized to concentrate analytes in a gaseous matrix. The gaseous sample flows into the impinger and through a collection solvent, which retains the analytes but allows the gas to escape. The solvent can then be further concentrated or directly injected into the gas chromatograph.

Cleanup Step. Various cleanup procedures may help to isolate the analytes of interest from coextracted sample material that may interfere with the analysis. Most cleanup techniques utilize classical gravity-fed liquid chromatography columns (typically 30-cm × 1-cm i.d.) packed with an appropriate adsorbent. The extract is transferred to the column and rinsed through the adsorbent with the solvents that will effect the separation of the analytes and the interferences. Alternative cleanup techniques utilize chemical reactions to destroy undesirable compounds and/or a secondary extraction designed to isolate the analytes of interest from interferants. Table 14.4 lists some of the common clean-up techniques outlined in various EPA procedures.

Derivatization Step. Derivatization is an additional preparation technique that is used less frequently than the three previously described processes, but one that can be invaluable in selected situations. Derivatization techniques chemically convert an analyte into another chemical species to improve sensitivity or instrument performance. One popular application of derivatization techniques is to convert a compound that does not respond to an electron capture detector (ECD) into a product with electronegative properties and a favorable ECD response. Chemical species with acidic or basic properties are also derivatized to neutral products, which are more easily chromatographed than the original polar species. (Also see Chapter 7, Section 7.5.4.)

Most methods for trace analysis require all three of the processing steps (the extraction, concentration, and cleanup) described here. This represents a lengthy preparation procedure with the potential for error and partial losses of analytes with each step. The more volatile analytes are particularly susceptible to losses during the concentration step. Sensitive compounds may also undergo degradation during the prolonged exposure to heat encountered with certain evaporative processes. These are important points to consider when selecting the best sample preparation technique for a particular application. Simplified processes should be utilized whenever possible.

14.3.3 Minor Component Analysis

A minor component analysis encompasses a higher range of analyte concentrations than the trace analysis. The preparation steps (extraction, concentration, cleanup, and occasional derivatization) are the same as those outlined for the trace analysis, but since the end requirements for the minor component analysis are not as demanding, the procedure can usually exclude one or more of the manipulations required for a lower-level determination. Under certain circumstances, a sample may be analyzed directly as received in the laboratory without any preparation if the sensitivity of the instrument is sufficient to detect the analytes at the required concentration level. Or a simple extraction, dilution, and/or

TABLE 14.4 Common Sample Cleanup Methods

Method	Description	Materials Removed/Separated
SW3650A	Acid–base partition	Long-chain hydrocarbons fats, lipids, and phthalates
SW3620A	Florisil column cleanup	Nitrogen compounds from hydrocarbons; aromatic compounds from aliphatic–aromatic mixtures; steroids esters, ketones, glycerides, alkaloids, and certain carbohydrates
SW3610/3611A	Alumina column cleanup	*Basic* (pH 9–10); basic and neutral compounds stable to alkali, alcohols, hydrocarbons, steroid alkaloids, natural pigments; polymerization, condensation can occur *Acidic* (pH 4–5); acidic pigments, strong acids *Neutral* aldehydes, ketones, quinones, esters, lactones glycoside (less active than basic form)
SW3640A	Gel permeation cleanup	Lipids, polymers, protein, natural resins, cellular components, viruses, steroids, and dispersed high molecular weight compounds
SW3630A	Silica gel cleanup	*Activated form*: hydrocarbons
SW3660A	Sulfur clean-up with copper powder, mercury, or tetrabutyl ammonium sulfite	Sulfur

Source. Reference 7.

cleanup prior to the analysis may be the only processing required. The sample should be manipulated as little as possible, since every additional step introduces another source of error.

14.4 PREPARATION OF AQUEOUS SAMPLES

In a limited number of circumstances it is feasible to analyze an aqueous sample by injecting it directly into the gas chromatograph, but the vast majority of environmental applications require extraction or removal of the analytes from the water matrix prior to analysis. The sample preparation techniques for water take advantage of the different physical properties of the analytes and the matrix. Conditions that favor the extraction of an analyte from water are those that create an environment for which the analyte has a greater affinity.

Systems consisting of trace concentrations of an organic compound in a water matrix have been evaluated quite extensively, and it is not difficult to design an appropriate extraction method for a noncomplex water sample. The extremely polar characteristics of the water molecules create an environment that tends to repel nonpolar organic compounds. This results in very low aqueous solubilities and favorable conditions for the removal of the analyte by extraction into a nonpolar organic solvent, displacement by a gaseous phase, or adsorption onto a solid phase.

Drinking water is probably the best example of an ideal water matrix, because it will contain minimal amounts of extraneous chemicals. Environmental water samples, on the other hand, can be quite complex and contain many chemical species at elevated levels. Many ionic compounds are native to water obtained from the environment, and their presence may actually improve the extraction conditions by "salting out" the nonpolar analytes. In fact, salts may be added to the water sample just before extraction to improve extraction efficiency. Various types of organic matter contained in the water can create the opposite effect and impede the extraction of organic analytes. Surfactants and water-miscible organics are two types of compounds that can increase the aqueous solubility of organic compounds and create problems with the extraction. Suspended organic matter in the water can bind organic compounds to its surface and prevent the analytes from dispersing into the extractant.

The pH of the sample is another parameter that can affect aqueous solubilities of organic compounds and extraction performance; thus the sample pH must be controlled before or during the extraction to ensure that the appropriate analytes are removed from the sample. Compounds affected by the pH are those that can exist in water as ionic species due to their acidic or basic characteristics. Equation 14.1 describes the equilibrium of an acidic compound in water:

$$\text{(Basic) } H^+ + A^- = HA \text{ (Acidic)} \tag{14.1}$$

Under basic conditions, compound HA will dissociate into H^+ and A^-. These ionic species are soluble in water and insoluble in nonpolar organic solvents, so compound HA would essentially remain in the aqueous phase during the extraction. Under acidic conditions, however, HA will be undissociated and have a low solubility in water and a high solubility in an organic solvent. These conditions (acidic pH) would be favorable for the extraction of compound HA. Analogous situations exist for compounds that are weak bases (i.e., extractable only under basic conditions).

These acid/base properties may be utilized to effect a preliminary cleanup for complex samples during the extraction process. A sample requiring extraction for phenols (weakly acidic compounds) can be preextracted under basic conditions. This will remove the neutral compounds (organics that are extracted under any pH conditions) and basic compounds, with the acids remaining in the aqueous phase. After preextraction, the organic solvent can then be discarded, the sample pH adjusted to acidic conditions, and extraction performed for the removal of the phenols and any other acidic compounds. Acidic or basic conditions can also cause certain compounds to breakdown in water. If an analyte of interest does decompose under the pH conditions required for the extraction of other compounds of interest, the pH-sensitive compounds can be extracted first, followed by a pH adjustment and a second extraction for the remaining compounds.

14.4.1 Liquid–Liquid Extraction

This mode of extraction depends on the distribution of analytes between two immiscible liquid phases. The extraction solvent utilized for the preparation requires the following characteristics;

1. A low solubility in water
2. The physical properties that will induce the analytes to have a greater affinity for the extraction solvent than the water

Consideration must also be given to the volatility of the solvent if an additional concentration step is required and the compatibility of the solvent with the column/detector configured in the gas chromatograph, although it is always possible to "exchange" the analytes into a more appropriate solvent before the analysis. Several solvents perform suitably for these purposes, and specifics are discussed with the particular methods in subsequent sections in this chapter.

Several useful equations describe the distribution of analytes in an extraction. The distribution law, as described in Chapter 1 for liquid–liquid extractions, states that the distribution of an analyte between two phases is based on the solubility of the analyte in each of the two phases. To describe the extraction of environmental contaminants, the presentation of Equation

1.23 can be modified as follows:

$$K = \frac{C_o}{C_{aq}} \tag{14.2}$$

where c_o is the concentration of the analyte in the organic phase or extraction solvent, C_{aq} is the concentration of the analyte in the aqueous phase or sample, and K is the distribution coefficient describing the equilibrium between the two phases. Clearly, then, if the distribution coefficient is large (which is the case with organic contaminants), Equation 14.2 will describe a system that favors the extraction of the analyte from the sample into the organic solvent.

The fraction or percentage of an analyte that will be removed from an aqueous sample by an extraction can be calculated by the following equation (10):

$$\%E = \frac{C_o V'_o}{C_o V_o + C_{aq} V_{aq}} \times 100\% \tag{14.3}$$

where $\%E$ is the percentage of the analyte extracted, V_o is the volume of the organic phase, and V_{aq} is the volume of the aqueous phase. Using Equation 14.2, Equation 14.3 can be rearranged to

$$\%E = \frac{100K}{K + (V_{aq}/V_o)} \tag{14.4}$$

The actual extraction can be performed in either macro or micro proportions, according to the ratio of the extraction solvent volume to the sample volume. Macro techniques, which can probably be best classified as the classical method, utilize solvent-to-sample ratios on the order of 0.05 to 0.2. The extractions are usually performed with either a separatory funnel or a continuous liquid–liquid extraction apparatus with some type of solvent concentration step following the extraction process. With most separatory funnel extraction procedures, quantitative extraction (i.e., 99% removal of an analyte) is achieved only through multiple (typically three) extractions of the sample. This can be illustrated by a simple calculation with Equation 14.3. A single extraction of a 1000-mL sample of water with 30 mL of an organic solvent would remove 85.7% of an analyte with a distribution coefficient K of 200. Two additional extractions of the sample under the same conditions would remove a combined total of 99.7% of the analyte from the water sample. The continuous liquid–liquid extraction device constantly recycles pure solvent through the aqueous sample. These extractions are allowed to run for extended periods (18 h), and quantitative extraction is achieved in a single extended extraction process.

In contrast to the macro techniques, microextractions are performed with extraction solvent volume to sample volume ratios on the order of 0.001 to 0.01. Ratios of this order of magnitude yield a nonquantitative analyte extraction like the macro separatory funnel techniques, but they also yield an extract from a single extraction step with an analyte concentration that is elevated relative to that which would be obtained with larger solvent to sample ratios. Consequently, there is no need for the additional concentration step required with the macro techniques. Table 14.5 provides a comparison of the performance characteristics of these two techniques through a hypothetical extraction situation and calculations based on Equation 14.4. These calculations demonstrate that the macro technique does extract a larger percentage of the analyte than the micro technique, but the net effect is a lower concentration for the macro technique relative to the micro technique.

The microextraction is usually performed in a volumetric flask using an extraction solvent with a density less than that of water. In this way, the small volume of extraction solvent can accumulate in the narrow neck of the volumetric flask for easy removal at the completion of the extraction process. A measured amount of salt can also be added to the water to force a larger proportion of the organic compounds into the organic phase by decreasing their solubility in the water. One example system evaluated for a number of organic compounds on the priority pollution list combined a 90-mL sample volume, 30 g of salt, and 1 mL of extraction solvent in a 100-mL volumetric flask (11). The contents of the flask were agitated in a manner that would ensure adequate mixing of the organic and aqueous phases. This was followed by an equilibration period for the contents of the flask and, finally, the removal of the extract for immediate analysis or storage.

Both the macro and the microextraction techniques remain popular alternatives for the preparation of aqueous samples for gas chromatographic analysis. The classical macro techniques require some type of concentration step to increase analyte concentrations to a certain level. The microextrac-

TABLE 14.5 Comparison of Macro and Micro Liquid–Liquid Extraction Characteristics

	Macro	Micro
Solvent volume:	30 mL	1 mL
Sample volume:	1000 mL	100 mL
% of analyte extracted in a single step	85.7%	66.7%
Concentration of analyte in extract	28.6 mg/mL	66.7 mg/mL

Note. Analyte concentration = 1000 mg/L
Distribution coefficient = 200.

tions eliminate the requirement for this preparation step and reduce the total volume of sample and solvent required for the analysis. While it is true that the quantitative extraction and concentration step associated with the macro technique can theoretically provide higher analyte levels in an extract than a microextraction, it is important to realize that the concentration step will usually result in some volatilization and subsequent loss of certain analytes, sometimes in excess of 50% of the total amount extracted. The concentration phase is also an additional processing step for the extract, which means increased handling and probability for error as well as longer processing time. The microextraction, on the other hand, requires fewer processing steps, but the nonquantitative extraction of analytes requires that appropriate internal standards be used or that calibration standards are extracted in the same manner as the samples.

The macro techniques are the most prevalent in the various EPA methods, but there are selected methods for drinking water analysis that now include the option to use the microextraction. These applications are described in Section 14.7.

14.4.2 Gas–Liquid Extraction

An alternative method for removing analytes from a liquid matrix is the use of a gas-phase extraction technique. This approach, which is also known as headspace or vapor equilibration analysis, is particularly useful with those organic compounds that can be classified as volatiles. Microextractions have been used to some extent to handle this class of compounds, but the classical macro techniques are essentially useless, because most if not all of the analyte could easily be lost by volatilization during the concentration and/or extraction step.

The basis for a gas–liquid extraction process is the same as that of the liquid–liquid extraction process: An analyte distributes itself between two immiscible phases at some constant ratio once an equilibrium has been achieved. With the gas–liquid system, the immiscible phases are now the gas and liquid phase or aqueous phase. Once again, the distribution coefficient K describes the partitioning of the analyte:

$$K = \frac{C_{\mathrm{g}}}{C_{\mathrm{l}}} \tag{14.5}$$

where C_{g} is the concentration of the analyte in the gaseous phase, C_{l} is the concentration of the analyte in the liquid phase or sample, and K is the distribution coefficient describing the equilibrium between the two phases (Chapter 1 provides additional information on these systems).

A large numerical value for K will describe a system that favors partitioning of the analyte into the gaseous phase, thereby creating a situation that favors the detection of very low concentration levels. The two physical properties of a compound or class of compounds that influence the distribution coefficient are the solubility in water and the vapor pressure. A low solubility in water and a high vapor pressure (which equates to increased volatility) favors distribution into the gas phase. This would include compounds that are classified as aromatics, halocarbons, and hydrocarbons (both cyclic and noncyclic), a grouping that happens to include many compounds that can be considered environmental pollutants. Compounds that would exhibit lower distribution coefficients are polar compounds like the short-chain alcohols and glycols. Additional gas–liquid system parameters that can affect an analyte's distribution coefficient are the temperature and the ionic strength. Elevated temperatures and ionic strength increase volatility and decrease water solubility, respectively.

There are two modes of gas–liquid extraction: the static and the dynamic modes. In the static mode the analysis begins with the addition of an accurate volume of sample to a vial. The sample volume must be selected so as to allow for the appropriate volume of gas phase or headspace in the remaining volume of the vial. Once the sample has been added, the vial is sealed with a septum cap, placed in a temperature-controlled environment and allowed to equilibrate for a certain period of time. During the equilibration period, the analytes will distribute themselves between the two phases in the vial, according to the system parameters. After equilibrium has been achieved, a sample of the gaseous phase in the vial is removed by a gas-sampling syringe and injected into the gas chromatograph.

In the dynamic or purge-and-trap mode, extraction of the analytes is effected by passing a constant flow of gas through the aqueous sample. The analytes are continuously purged from the sample and carried by the gas flow to a medium that can collect (and concentrate) the analytes. The most common type of collection medium is a solid adsorbent contained within a metallic tube and enclosed within a heating jacket. Upon completion of the purging sequence, the trap is connected directly into the carrier-gas line flowing to the gas chromatographic column. The trap temperature is rapidly increased and the analytes are desorbed from the trap and transferred to the chromatographic column by the flow of carrier gas, whereupon the chromatographic analysis is initiated. Alternative trapping media to the solid adsorbents are solvents that can be directly injected into the gas chromatograph and cold traps that function in a manner similar to the solid-adsorbent traps.

Conceptually each of the vapor-phase techniques is somewhat analogous to the liquid–liquid extraction techniques. The static or headspace analysis technique can be compared with the microextraction technique for liquid–liquid systems. The extraction is usually performed in a single step with incomplete removal of the analytes from the sample, but the appropriate selection of the various parameters (temperature and gas-to-liquid phase volume ratio) will yield a system with favorable sensitivity for a great many compounds. The dynamic mode is similar to the continuous liquid–liquid extraction method, with the extractant, the gas, being continuously passed through the sample for the quantitative removal and subsequent concentration of the analytes.

Instrument calibration for the static mode is accomplished with standard solutions prepared in water and maintained under the same conditions as the samples. Accurate quantification of analytes in the sample will require the analyst to make some attempt at preparing standards in a matrix that is similar to that of the actual samples being analyzed, because standards and samples of different ionic strength will exhibit different distribution coefficients. This may be done with the technique of standard additions or by adding a large amount of salt to both standards and samples to compensate for any minor differences in the two matrices. Instrument calibration in the dynamic mode is also performed with standards prepared in an aqueous solution and processed in a manner that is identical to that in which samples are processed. In this way, if the transfer of analyte from the aqueous phase to the chromatographic system is not quantitative, it will be consistent and yield accurate quantification. With a complex sample, however, the possibility of a poor purging efficiency always exists, and caution should be exercised.

The hardware required for the headspace or static system is relatively simple and usually available in any laboratory. The important components are the vials with septum-sealed caps, a stable water bath or comparable thermostatted environment, and a gas-tight syringe. There are also several commercially available units marketed by various manufacturers of gas chromatographs. These systems are attached directly to the instrument and provide consistent and continuously automated performance. Before beginning any type of headspace analysis, particularly with new analytes, some preliminary investigation into the optimization of the system parameters is required for both a manual injection system and fully automated system. The accessories and techniques required for a dynamic gas–liquid extraction tend to be a little more complicated than the static systems, because many elements of the purging system can adversely affect the analysis. These many components make it impractical for almost any laboratory to set up its own system, so they are usually purchased from any of a number of manufacturers. Figure 14.1 illustrates the various components of a typical purge-and-trap system.

In general, the dynamic techniques usually provide more sensitivity than the static techniques, and they are the techniques specified in many of the EPA methods for volatile compounds, as described in Section 14.7. The hardware required for the dynamic technique is more complex than that required for the static technique, and an experienced operator must be available to troubleshoot any problems. The static techniques are popular as screening tools both in the laboratory and in the field. They are also advantageous with complex samples that contain high levels of extraneous organic material, which can contaminate hardware required for dynamic processes.

14.4.3 Solid-Phase Extraction

Solid-phase adsorbents have provided analysts with an alternative technique for the extraction of organic contaminants from aqueous samples. This technique reduces the extraction time and the amount of hazardous solvent required for the liquid–liquid extraction methods, making solid-phase extraction (SPE) an attractive alternative to the classical approaches.

The fundamental principles describing the solid-phase extraction process are the same as those that apply to liquid chromatography (LC) and ion-exchange separations. In 1974, Kirkland (12) discussed how analytes in an aqueous sample could be focused and enriched on liquid chromatographic columns prior to analysis, since many organic compounds will collect

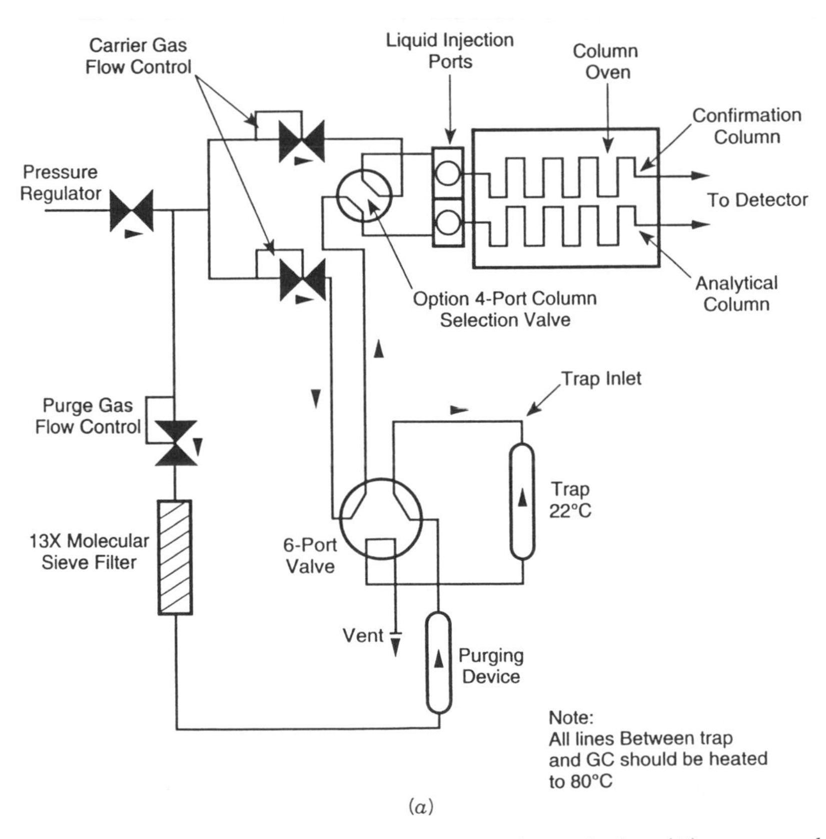

FIGURE 14.1 Schematic diagram of a purge-and-trap device: (A) purge mode; (B) desorb mode. (From Reference 7.)

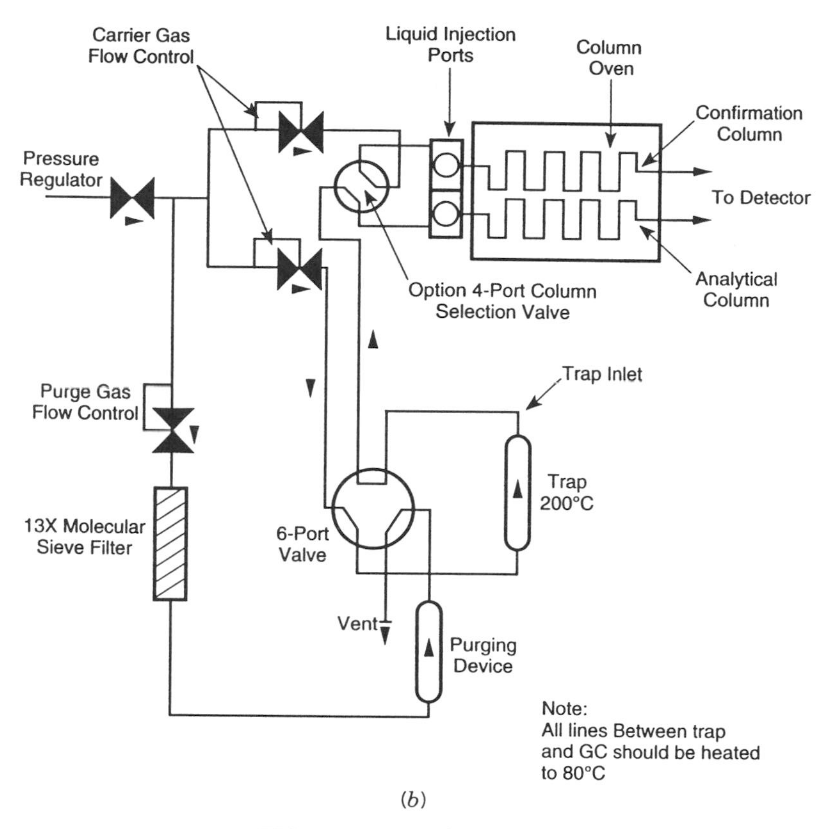

(*b*)

FIGURE 14.1. (*Continued*)

on the front portion of the chromatographic column and remain stationary as a water matrix continues to pass through it. When the composition of the liquid passing through the column is changed to a certain percentage of an organic solvent, the organic compounds are then carried through the column. The same principles are applied on a larger scale with SPE to prepare samples for gas chromatographic analysis.

Many types of solid adsorbent materials have been evaluated for SPE applications. They include polymers and/or copolymers of styrene/divinylbenzene, Tenax-GC, acrylic ester polymers, and both octyl (C_8) and octadecyl (C_{18}) reversed-phase liquid chromatographic packings (13). Many of these adsorbents are commercially available as prepacked columns or cartridges for SPE applications. Extractions disks or membranes consisting of octyl- or octadecyl-bonded silica particles enmeshed in a network of PTFE fibrils were introduced in 1990 (14) as a substitute for the SPE columns and cartridges. Kraut-Vass and Thoma (15) evaluated the performance of the extraction disk for 43 semivolatile organic compounds and compared the results with the extraction cartridge results reported in EPA

Method 525. The extraction disks offered nearly the same reliability as the cartridges and a dramatically decreased extraction time.

Solid-phase extraction technology is generally limited to compounds of low-volatility and compounds that are extracted by separatory funnel or continuous liquid–liquid extraction methods. The extraction of most nonpolar compounds from the aqueous samples is attributed to the attraction that these compounds exhibit toward the hydrophobic (nonpolar) moieties that exist on the solid-phase material and their low solubilities in water. The physical interaction between the analytes and the stationary phase takes place as the aqueous sample passes through the cartridge or membrane. Analytes that are within close proximity to the hydrophobic phase will be adsorbed to the material. An efficient extraction process requires a large surface area for the hydrophobic material in order to maximize the possibilities for analyte interaction with the stationary phase. When the sample has completely passed through the SPE material, the retained analytes can then be eluted from the adsorbent with a small amount (ca. 30 mL) of organic solvents like methylene chloride and ethyl acetate. This solvent is collected and concentrated to an appropriate volume with a gentle stream of dry nitrogen.

The extraction of acidic or basic compounds with the hydrophobic solid-phase materials requires that the sample pH be adjusted to provide the conditions under which these compounds will remain undissociated (see Equation 14.1). Dissociated compounds will be more water soluble and less hydrophobic, and will not be retained by the SPE material. The retained acids or bases are then eluted with an organic solvent. Ionizable species can also be extracted on an ion-exchange basis with stationary ionizable compounds attached to the solid-phase material. Changing pH levels of the water will affect the retention characteristics for the ionized analytes.

Solid-phase extraction materials must be properly prepared before the process is started. The adsorbents and associated hardware must be rinsed with a solvent to remove residual materials that may appear as an interference during the chromatographic analysis. This is especially important with plastic cartridges, because they can contain significant levels of plasticizers. The use of disposable glass cartridges or the extraction disks with a glass filtration apparatus can eliminate some of these problems. After the rinsing step, the solid-phase adsorbent must be rinsed with a wetting solvent like methanol to ensure effective contact between the hydrophobic material, the aqueous sample matrix, and analytes. The sample is then immediately introduced to the cartridge or disk, before the wetting agent drys.

A limited amount of water solubility is required with the first solvent used for analyte elution. Solvents that are not miscible with water may not be able to penetrate a micro layer of water that remains on the solid-phase adsorbent. An initial rinse with acetone will remove the residual water and some of the analytes, and follow-up rinses with nonpolar solvents will remove the remainder of the analytes. Ethyl acetate, a solvent with a moderate solubility in water, is also effective.

In general, the SPE technique provides a viable alternative to the liquid–liquid extraction techniques required for most EPA methods. It has already been approved as an option or replacement for many EPA drinking water methods, and is currently under evaluation for wastewater applications with EPA methods. The process is faster than the conventional methods and eliminates a significant amount of hazardous solvent (several hundred milliliters) for each extraction, and it will eventually be the method of choice for the preparation of all aqueous samples. The SPE technique has also been used to extract analytes from an aqueous sample in the field. In this way, only the extraction cartridge or disk must be sent to the laboratory for subsequent elution and analysis.

14.5 PREPARATION OF SOIL AND SEDIMENT SAMPLES

Soil and sediment materials play a complex role in the interplay of chemical pollutants and the environment. The chemical makeup of these materials, the surrounding population of microorganisms and animal life, and the hydrology of the area all combine to determine the fate of the pollutants that reside in these materials. Ultimately, a pollutant will remain stationary or continuously migrate, according to the characteristics of the groundwater in the area, the solubility of the pollutants in this water and the forces that might cause a compound to adhere to the soil or sediment particles. At the same time, chemical pollutants can be modified or degraded by biological or chemical processes or even volatilized and eventually expelled from the soil. Effective sample preparation techniques and analytical procedures are an important tool for the evaluation of these complex mechanisms.

There are many ways in which chemical pollutants initially make their way into a soil or sediment. Accidental spills and leaks along with both proper and improper disposal procedures contribute to the problem. Pesticides and herbicides are applied directly to the land to help with agricultural production and aesthetic appeal, but they can accumulate in certain areas by numerous pathways. Air pollutants attached to particulate matter in the atmosphere can eventually make their way to solid land and become integrated into the top soil. Water flow can also transport chemical pollution and deposit it in a certain type of sediment. The mode of transport can be either as a solute dissolved in the water or as the chemical adhering to a suspended particle which gradually settles into the sediment layer. Measures have been implemented to control or eliminate some of the avenues for the generation of soil pollution, but there are still many that remain.

Soil and sediment samples present the analyst with some unusual problems that are not encountered with aqueous samples. Organic pollutants are not usually homogeneously integrated into a soil or sediment in the same way that they can be solubilized and evenly distributed in water. They can be bound or adsorbed to the exterior surface of an individual soil or sediment particle, according to the adsorptive properties of the individual particle. Strong adsorption forces and a large surface area for a given type of solid material will bind a greater amount of a specific chemical compound to that material. The effective removal of an analyte requires that the adsorption process be reversed, resulting in the dissolution of the analytes into the extraction solvent.

The adsorption properties of a particular type of sediment or soil are governed by the composition of the material. Table 14.6 lists some of the specific organic and inorganic components of soils and sediments. Manahan (16) offers a more generalized classification of these two types of materials:

> *Sediments* typically consist of mixtures of clay, silt, sand, organic matter, and various minerals. Their composition may range from pure mineral matter to predominantly organic matter.
>
> *Soil* is a variable mixture of minerals, organic matter, and water capable of supporting plant life on the Earth's surface. The solid fraction of a typical productive soil is approximately 5% organic matter and 95% inorganic matter. Some soils, such as peat soils, may contain as much as 95% organic material. Other soils contain as little as 1% organic matter.

The materials that contribute most significantly to the adsorptive forces of soils and sediments are the organic (humic) substances and the secondary

TABLE 14.6 Components of Soil and Sediment

	Organic Components
Humus	Most abundant organic component, a residue from plant decay. Consists primarily of C, H, and O.
Fats, resins, and waxes	Comprise only several percent of organic soil matter.
Saccharides	Major food source for microorganisms. Consist of cellulose, starches, hemicellulose and gums.
Nitrogen and phosphorus—organics	Primarily amino acids, amino sugars, phosphate esters, inositol phosphates, and phospholipids.
	Inorganic Components
May consist of quartz, silicates, iron oxides, manganese oxides, titanium oxides, and calcium carbonate. The clay or secondary minerals (hydrated aluminum and iron silicates) also occur widely.	

Source. Adapted from Reference 16.

minerals or clays. Adsorption can occur through ion-exchange interactions with ionizable organic compounds. Complex pesticides can act as ligands and coordinate with various metals in soil mineral matter. Neutral species may be bound by van der Waals forces or induced dipole–dipole interaction. Grathwohl (17) has investigated the adsorption of chlorinated aliphatic hydrocarbons on soils of varied origin and geological history. He postulates that the elemental composition of the natural organic matter, which is a functions of its age, plays a more significant role in adsorption processes than was initially expected.

Ash or fly ash, the solid debris that remains after incineration processes, is another class of solid sample material that is of environmental concern. The ash material is not a soil per se, but it is extracted by the same processes used for soils and sediments and is introduced in this section for that reason. One type of ash material that requires chemical analysis is the particulate matter that escapes into the atmosphere from certain high-temperature manufacturing processes. Numerous organic compounds formed during the process may be adsorbed to the particles as they enter the atmosphere and contribute to air, soil, and water pollution. Extraction and analysis of the ash is a means of monitoring the pollution that occurs from these sources.

Another type of ash that must be analyzed for chemical contamination is that generated from the incineration of hazardous waste. The use of this technology has increased significantly in recent years as a consequence of the 1984 amendment to the Resource Conservation and Recovery Act (RCRA), which prohibited the landfilling of many types of hazardous waste. This resulted in an additional 7 million tons of hazardous waste a year that required treatment (18). Waste material or contaminated soil that has been characterized as hazardous can be treated by the incineration process for the destruction of the chemical contamination. The remaining ash material must be analyzed to verify that the chemical contaminants have been destroyed.

Adsorption processes encountered with ash materials are those that are attributable to inorganic materials, like ion exchange phenomena or complexes formed between the organic contaminants and the inorganic components. Fisher et al. (19) has also identified the formation of spherical particles contained within larger spherical particles in fly ash generated from certain operations. Contaminants adsorbed to the ash surface of the enclosed spherical particles can be trapped inside of the larger particles. Both of these phenomena contribute to difficulties with the extraction of organic compounds from ash materials.

Three types of preparation methods can be employed for soil, sediment, and ash sample matrices: extraction by organic solvent (liquid–solid), a gas (gas–solid), and supercritical fluids. The solvent and gaseous extraction techniques are the conventional methods that have been in place for many years and are currently the designated procedure in the various EPA methods. Supercritical fluid extraction (SFE) has been receiving a significant amount of attention more recently as an alternative for the organic solvent extraction procedures. It eliminates the need for large volumes of hazardous solvents and has been proven effective for environmental applications. Consequently, it is under evaluation as a replacement for certain procedures and will most likely replace the current organic solvent extraction methods.

14.5.1 Solvent Extraction

Techniques that utilize organic solvents for the extraction of organic pollutants from soils and sediments are currently the most widely used methods. Removal of the analytes from the sample material is achieved via the simple process of contacting the solvent with the entire surface area of the sample material. Analytes that are soluble in the extraction solvent will be distributed between the sample matrix and the solvent.

Many factors govern the effectiveness with which a solvent can remove an analyte from a solid material. At the molecular level, the appropriate solvent competes with the analytes for the adsorption sites and gradually displaces the analytes from these sites. Proper selection of the extraction solvent will facilitate the process.

Junk and Richard (20) performed a rather extensive evaluation of extraction efficiencies obtained with different solvents. Their evaluation indicated that one technique or solvent was not consistently superior and that the best alternative was to evaluate different systems for a given application. In addition to their examinations of solvent combinations, they also pretreated fly-ash samples with different complexing agents to facilitate the extraction of PAHs from this material. But all attempts with the added complexing agents were unsuccessful and PAH extraction efficiency remained poor. They concluded that the PAHs did not form a complex or that the wrong complexing agent was selected.

Cooke et al. (21) investigated the addition of dilute HCl to assist in the solvent extraction of organic compounds from fly ash. Improved recoveries were observed with the addition of the acid, and this was attributed to the breakdown of the spherical particles identified by Fisher et al. (19), thus enabling the solvent to contact particles that had previously been enclosed within larger particles. The acid is also believed to assist in the displacement of analytes from the ion exchange sites.

In another report, Freeman and Cheung (22) identified additional aspects of liquid solvent extraction of pond sediments that could impact on the efficiency with which an organic contaminant was removed. They discussed swelling effects for the humin–kerogen polymeric structure of the sediment material which was created by certain solvents. Partial swelling impeded the diffusion of compounds contained within the polymeric structure and reduced extraction rates. Solvents that generated the maximum amount of swelling for the structure allowed compounds to diffuse freely from within the structure and exhibited maximum rates of extraction. Investigations by Haddock et al. (23) and Fowlie and Bulman (24) described reduced extraction efficiencies for PAH compounds in soils with increased time after spiking. The reduced efficiency was attributed to increased partitioning or movement of the compounds into the interclay structure of the sample matrix.

Others factors that can influence the extraction efficiency are the ability of the solvent to physically contact the surface of all sample particles, the length of time solvent contact occurs, and the amount of moisture contained in the sample. Portions of a sample with a hardened crust could inhibit complete solvent–particle surface contact. Physically breaking apart hardened materials and large sample pieces will promote thorough dispersion of the solvent throughout these subsample portions. Water contained in a soil or sediment sample will curtail the effectiveness with which a nonpolar solvent can contact the individual particles, because the solvent will not be able to penetrate the aqueous layer that surrounds them. To avoid this type of problem the sample can be mixed with an anhydrous salt like sodium sulfate to remove the water. Nonpolar extraction solvents can also be combined with miscible, polar organic solvent like methanol or acetone to yield a solvent system that is more readily dispersed through soils and sediments containing a moderate level of moisture.

Once the soil–solvent mixture has been prepared, it must be circulated or agitated by some means. This will improve the kinetics of the extraction process by promoting solvent–sample contact and eliminating high analyte concentrations in the solvent localized around contaminated particles. Mixing by manual handshaking is the simplest method, but it is useful only with analytes that are easily removed from the sample matrix under investigation, because the process can be continued only for several minutes. Mechanical wrist action shakers are more practical for procedures requiring extended mixing intervals. At the conclusion of the initial mixing sequence, the extraction solvent can be removed and the process repeated two more times if a single extraction does not prove to be quantitative. The total volume of extraction solvent is combined and concentrated to the final volume required for the analysis.

The Soxhlet extraction device is an apparatus designed to continuously reflux the extraction solvent through the sample material. The extraction is allowed to continue for an extended period of time, typically 18 h. At the completion of the extraction, the solvent is removed, filtered, and concentrated appropriately. This technique does expose the extracted analytes to elevated temperatures for an extended period of time and may lead to the degradation of thermally labile compounds. Sonication is another means of promoting solvent–sample contact. This can be performed in an ultrasonic bath or with an ultrasonic probe. The energy of the bath or the probe keeps the soil or sediment particles in constant motion throughout the extraction process. Extraction in an ultrasonic bath offers an advantage in some instances, since the bath can be cooled, an option that can prove useful with compounds that are sensitive to exposure to an elevated temperature for an extended period of time. Extraction by the ultrasonic bath or the probe may be completed in a single step or may require a three-step process similar to the mechanical shaking procedure described previously.

14.5.2 Gas-Phase Extraction

Gas-phase extraction is another means of extracting analytes from a solid (soil or sediment) matrix. The application is analogous to the gas–liquid extraction technique, with the analytes partitioning from the solid phase or sample into the gas phase for subsequent introduction into the gas chromatograph. Equilibration of the gas phase and the sample once again takes place in either the static mode (headspace analysis) or the dynamic mode. The gas-phase extraction techniques for soils and sediments are typically utilized for volatile compounds, since their high vapor pressure promotes the partitioning into the vapor phase.

The vapor-phase equilibration can be performed directly on an aliquot of the soil or sediment sample, with a measured volume of water added to the container to disperse the sample evenly. If the sample material is not well dispersed and free flowing, equilibration may take an inordinately long time, since analytes migrate very slowly from soil particles that do not have exposed surfaces. The application of heat is also important under these conditions. The solubility of most organic compounds in water is very low, and this does not favor the partitioning of the analytes from the particle surface into the water. When the water–sample mixture is heated, analytes will volatilize more readily, and they are forced from the particle surface for subsequent gas–liquid partitioning.

This approach can be applied to both the static and dynamic techniques. With the static techniques the soil will affect the partitioning to some degree, so it is important to use a soil–water mixture with calibration standards in order to obtain accurate analyte quantification. This is best performed with a clean sample matrix whose composition is similar to that of the samples under investigation or by means of standard additions. Matrix matching is usually not so critical with the dynamic techniques because the analyte extraction is considered quantitative. For analyses that do not require the highest level of sensitivity, an alternative procedure can be utilized. An aliquot of the soil or sediment sample can be extracted with methanol or another suitable solvent. A volume of the methanolic extract can then be combined with a measured volume of water, and this water–methanol can be analyzed by either the static or dynamic vapor-phase equilibration techniques.

14.5.3 Supercritical Fluid Extraction

Supercritical fluid extraction (SFE) is a relatively new technique in comparison to the soil extraction methods discussed previously. Analytical chemists utilized the diverse properties of supercritical fluids with a technique known as supercritical fluid chromatography, but by the mid 1980s these same fluids were being used for purposes of extraction as well. Since that time, it has been viewed as a likely technique to replace the classical organic solvent extraction methods for soil and sediment samples. Some of the advantages that make SFE an attractive alternative are the speed of the extraction process, the low extraction temperatures and ability to handle thermally sensitive compounds, the elimination of hazardous solvents, and the ability to forgo the concentration step.

A supercritical fluid forms when a material is maintained at a temperature and pressure that are above its critical level. Under these conditions, the material will exhibit the rapid diffusion characteristics and viscosities of a gas and the solvating characteristics of a solvent. The gas-like properties are particularly important, because they allow the fluid to permeate the sample matrix and contact the analytes much more rapidly than the solvent. Once the contact is initiated, the process depends on the solvating ability of the fluid to displace the analytes from the adsorption sites on the sample particles.

One of the most widely used supercritical fluids for the extraction of organic compounds from solid matrices is CO_2. It has a relatively low critical temperature (31°C) and pressure (73 atm), it is not toxic or reactive, and it is available at a high purity for a low price. Variations in the temperature and pressure at which supercritical CO_2 is maintained will increase or decrease the solvent strength and the selectivity of the extraction. At a constant temperature, supercritical CO_2 will extract analytes of low polarity at low pressure and analytes of high polarity at high pressures.

In practice, SFE with CO_2 is performed at pressures that are not high enough to prove effective with polar compounds. Hawthorne (25) has classified supercritical CO_2 as an excellent extraction medium for nonpolar species, such as alkanes and terpenes; reasonably good for moderately polar species, including polycyclic aromatic hydrocarbons (PAHs), polychlorinated biphenyls (PCBs), aldehydes, esters, alcohols, and organochlorine pesticides; but less useful for more polar compounds. The effectiveness of supercritical CO_2 as an extractant can be further enhanced through the addition of a small amount of a modifier designed to increase solvating power and polarity. Some common organic solvents that are popular modifiers are methanol, isopropanol, acetonitrile, and benzene (26). Supercritical N_2O (with and without modifiers) has also been useful as an alternative to CO_2 for certain applications (25).

The various parameters that control the extraction process can also prove to be advantageous for sample preparation in a quite different sense. A programmed increase or decrease in one or more of these parameters during the sample extraction can provide sequential extraction of certain classes of compounds. This means that the analyst can perform an extraction under more than one set of "optimized" conditions. The same principle can be used as a means to fractionate analytes of interest from sample matrix components that might pose an interference during the chromatographic analysis. In this sense, the programming capabilities can be used essentially as a cleanup technique during the sample preparation process.

The extraction process can be performed in either a static mode or a dynamic mode. With a static extraction, the cell containing the sample is pressurized with the supercritical fluid and sealed. The fluid and sample remain in contact for a set period of time, after which the fluid is allowed to flow from the cell and is directed to the analyte collection device. In the dynamic mode, a constant flow of supercritical fluid passes through the cell, contacts the sample, and is then directed to the analyte collection device. Analytes are continuously extracted by the constant flow of supercritical fluid and carried to the collection medium by the fluid. Both the static and dynamic mode of SFE offer complete sample preparation in a time frame that is usually less than 2 h, a significant reduction in comparison with the conventional solvent extraction processes.

Collection of the analytes after their removal by the supercritical fluid extraction can be performed in a manner that provides the analyst with an extract much like the solvent extraction methods. One way of executing this is to direct the flow of the supercritical fluid into an appropriate solvent. The fluid usually exits through a piece of fused-silica tubing, and this tubing is placed into the solvent. As the supercritical fluid escapes from the tubing it becomes a gas. The majority of this gas passes through the solvent, but the analytes are retained. A side benefit of the expansion of the escaping supercritical fluid is that it reduces the temperature of the solvent. This cooling effect reduces or eliminates evaporation of the solvent and also promotes effective collection of the extracted analytes. Once the extraction is completed, the volume of the collection solvent can be adjusted and the extract can be analyzed.

The fluid flow can also be directed to a solid sorbent material for the collection of the analytes. At the completion of the extraction, the analytes can be removed from the adsorbent material with a solvent, which can then be injected into the gas chromatograph. Both of these techniques provide for the concentration of the analytes, but without the use of the conventional evaporative techniques required with solvent extraction process. This is one of the major advantages of SFE, because the analytes are no longer subject to heat and evaporative losses encountered during conventional solvent concentration procedures. Substantial volumes of hazardous solvents are also eliminated from the process.

The effluent flow from the SFE cell can also be transferred directly to a capillary column gas chromatograph to provide a technique that is known as "on-line" or "coupled" SFE. This modification provides an analysis scheme with the highest level of sensitivity for limited sample sizes, with the total mass of analyte contained in the sample aliquot being transferred completely to the instrument. The supercritical extraction fluid will be deposited inside of the instrument's injection port where the analytes can then be cryogenically focused on the chromatographic column. Analysis proceeds as usual once the extraction has been completed.

Supercritical fluid extraction devices are commercially available from a number of sources. They are marketed as both manually operated units, which are capable of handling only one sample at a time, and as computer-controlled units with the ability to process multiple samples automatically. The minimum requirements for the hardware are a syringe pump to deliver the supercritical fluid, a temperature-controlled extraction cell designed to withstand elevated levels of pressure, and a means of restricting the outflow of supercritical fluid to maintain the required extraction pressures. Automated units will require switching devices to properly divert the ingoing and outgoing flow of supercritical fluid to the proper extraction cell and the computer hardware and software to control the process. The costs are obviously proportional to the level of sophistication of the device.

14.6 PREPARATION TECHNIQUES FOR AIR OR GASEOUS SAMPLES

Organic compounds present as air pollutants can exist either as a gas or as particulate matter. These compounds can originate directly from various manufacturing and mechanical processes as a result of the incomplete combustion of the fuel material or as a result of the spontaneous volatilization of material from a chemical spill, uncontained waste, or uncontained feedstocks. Vegetation sources also release a significant amount of organic compounds into the atmosphere.

Pollutants that exist primarily in the gaseous state are described as volatile organic compounds (VOCs). These are compounds with boiling points typically around or below 100°C and vapor pressures of $>10^{-1}$ Torr (27). The semivolatile organic compounds (SVOCs) have vapor pressures in the range of 10^{-1} to 10^{-7} Torr and can exist in the gaseous phase or in the particle bound phase, either as an adsorbed material or as a liquid droplet or mist (27). These adsorbed compounds can remain attached to the particles as they are dispersed through the atmosphere or they can undergo chemical reaction in the atmosphere. Alcohols, aldehydes, ketones, organic acids, esters, and organic nitrates have been identified as components of organic smog aerosols (16), which are essentially liquids.

The preparation steps required for the trace analysis of most air or gaseous samples differ in one sense from those required for solid or aqueous samples in that they are performed partially during the actual sample collection process in the field, so the gaseous sample itself is never really transported to the laboratory. The organic contaminants contained in these samples are usually extracted or removed from the gaseous matrix and concentrated during the sampling procedure. A measured volume of the sample is drawn into or through the sampling system by some type of pump.

In its simplest form, the sampling system consists only of a collection medium. The analytes are retained on or in the collection medium, and it is the medium that is actually transported to the laboratory for preparation and analysis. But the same basic phases of the sample preparation remain, the extraction and concentration of the analytes in the gaseous matrix, followed by preparation of the collection medium and extract cleanup (when required) prior to the gas chromatographic analysis. Under certain circumstances, the gaseous sample may be collected in a Tedlar bag or an appropriate container (glass or inert metal) and transported to a laboratory for direct gas chromatographic analysis with no intermediate preparation steps. Transportation of a gaseous sample in this manner is useful only when the analytes of interest are stable and low detection limits are not required.

Analytical procedures for air or gaseous sample matrices are required for three types of analysis: (1) air emissions methods for stationary or point sources like incinerators, power plants, and various industrial processes; (2) ambient air monitoring methods; and (3) methods for monitoring indoor air contamination in the workplace. Air emissions and ambient air monitoring fall within the jurisdiction of the state agencies and several offices of the EPA, most notably the EPA's Atmospheric Research and Exposure Assessment Laboratory. Indoor air monitoring falls into the realm of industrial hygiene and the regulatory effort provided by OSHA. The basic principles for the sample collection, preparation, and analysis of industrial hygiene samples are the same as those applied to outdoor air pollution. Many of the compounds that are of concern environmentally are also encountered in this area and it is not unusual to find these procedures associated with environmental analysis.

14.6.1 Filter Collection Procedures

Various types of filters can be used to collect particulate matter in air samples. Analyses are performed on the particulate to measure the concentrations of those organic compounds that are adsorbed on the surface or trapped within the particles. As the air sample flows through the filter, the solid particles are blocked or retained by it. The entire filter is shipped to the laboratory, where it is extracted by conventional techniques utilized for solid sample matrices. The extract of the filter can also be subject to appropriate cleanup procedures and concentration steps. The filtering media typically employed for air sampling are quartz or glass filters.

14.6.2. Sorbent Collection Procedures

Collection of analytes on sorbent materials is designed to trap or retain those analytes that exist in the vapor phase in the sample. The sorbent material selected for the sampling/analysis is one that efficiently retains the analytes of interest as the sample passes through it. Popular sorbent materials for the collection of organic compounds include charcoal, alumina, silica gel, the various types of porous polymers (including Tenax, the Porapaks, the Chromosorbs, and the XAD materials), polyurethane foam (PUF) cartridges, and liquid stationary phases coated onto a support phase.

Any sorbent material can retain only a finite amount of an analyte, and when the sorbent's capacity has been exceeded the analyte will no longer be retained by the material. For this reason, two or more beds of sorbent are usually connected in series. If the first bed of sorbent is overloaded by a high level of analyte, analysis of the subsequent bed will indicate that breakthrough has occurred. If the amount collected on the beds that follow is not excessive, the analyst can assume that the analyte was quantitatively removed from the sample. But when the amount of analyte collected on the last bed is significant, the integrity of the sample must be questioned, since some amount of analyte may have escaped from the last bed of sorbent. Consideration must also be given to the other components of the sample matrix and their effect on analyte retention. High levels of water or other organic compounds can saturate the sorbent material and reduce the efficiency of the sorbent materials.

The volatility of the analytes collected on the sorbent will dictate which technique(s) can be employed for their extraction or removal. Compounds that are extracted from soil samples by solvent extraction techniques (i.e., Soxhlet extraction) are extracted from sorbents by the same process. This assumes that the sorbent will not dissolve or degrade upon exposure to the solvent/temperature conditions for the extraction. Extracts of the sorbents can also be subject to the appropriate concentration and cleanup steps.

Compounds of high volatility can be thermally desorbed from the adsorbent material or extracted by solvent extraction techniques. Solvent extraction must be performed under mild conditions without the use of heat. This precludes any type of concentration step, thereby limiting the lower level at which analytes can be detected. Thermal desorption techniques allow for the concentration of analytes and provide lower detection limits. They are usually performed on an instrument configured for the purge-and-trap or dynamic gas–liquid extraction technique. The sorbent cartridge or tube is connected in series to the purge-and-trap device with the carrier gas flowing through the sorbent material. As the temperature of the sorbent is elevated, the analytes are desorbed and carried by the carrier gas to the trap employed for the purge-and-trap analysis. The desorbed analytes are then measured by the same process utilized for the purge-and-trap analysis of soil and water samples. Volatile analytes can also be thermally desorbed directly into the gas chromatographic column where they can be cryogenically focused or trapped on the front of the column. This process can introduce a significant amount of water into the chromatigraphic system, creating problems at low temperatures.

14.6.3 Liquid or Impinger Collection Procedures

Airborne organic contaminants can be removed from a gaseous sample by passing the sample through an appropriate solvent contained in an impinger. The impinger is designed to distribute the total gas flow in a manner that yields a stream of finely divided bubbles through the solvent. The dispersion of the gas sample as it flows through the solvent increases the effectiveness with which the solvent can solubilize the analytes. Impinger solvents are selected on the basis of their vapor pressure and the solubility of the analytes. Chemical reagents can also be added to the solvent to convert the analyte to a new compound that is easier to analyze. The impinger solvent can be analyzed directly for the analytes of interest, or it can be extracted and/or concentrated before analysis (see Chapter 7).

Under certain conditions, an aqueous condensate phase is generated during the collection of an air sample. The condensation is usually created by a condenser that precedes the adsorbent traps. The primary purpose of the condenser is to remove the water vapor from the gaseous sample before it passes through the adsorbent. This aqueous condensate can be extracted by the typical extraction methods employed for water samples. Very little, if any of the nonpolar organic compounds present in the gaseous sample will be retained by an aqueous condensate, but high levels of polar organics contained in the gaseous sample can condense as well and increase the solubility of certain analytes in the condensate.

14.6.4 Miscellaneous Collection Procedures

Other techniques that are available for gaseous samples include sample collection in a Tedlar bag (or other inert container) at atmospheric pressure, cryogenic trapping, and collection in a stainless steel canister under both subatmospheric and pressurized conditions (27). Collection in the Tedlar bags at atmospheric pressure is a straightforward process, but it is done without the benefits of analyte enhancement, yielding higher detection levels than most other air-sampling techniques. Cryogenic or cold-trap techniques rely on the condensation of analytes as the gaseous sample passes through a low-temperature zone. Analytes can be thermally desorbed from this collection device for analysis. The stainless steel canister provides a mechanism for collecting an ambient air sample over an extended time period. The canister also provides a convenient means of sample transportation. Preevacuated canisters allow a controlled flow of the air sample to enter the system over a set period of time. The pressurized sampling mode requires the use of a pump to create a positive pressure inside the canister. The sample canister can be connected directly to the gas chromatograph for analysis. The analytes are usually cryogenically focused in a trap that precedes the gas chromatograph. After a set time, the trap temperature is elevated and the volatilized analytes are transferred to the chromatographic column for a routine analysis (27).

14.7 APPLICATIONS WITH SPECIFIC COMPOUND CLASSES

The sections dealing with the various sample preparation techniques for environmental analysis focused on the physical properties of the different classes of chemical compounds. These properties dictated which techniques could be employed as sample preparation steps. These same properties influence the selection of the chromatographic system (i.e., columns and detectors) required for the analysis and the applications with actual environmental samples.

The various listings of EPA methods provide a convenient grouping for many of the classes of compounds that are of environmental concern, and the material on applications will be organized similarly. The techniques included in certain EPA methods are described, along with alternative approaches referenced in the literature. This material by no means addresses every chemical compound that has had to be measured in an environmental sample, or even every class of compound for that matter. But these methods do address those compounds and instrumental techniques for which the vast majority of environmental analyses are performed. When confronted with the need for an analysis that is not specifically covered by one of these methods, the analyst can determine if the compound in question falls within one of the classes for which a method does exist and attempt to determine this compound by the method for that class of compounds.

The two sets of EPA method compendiums considered in this section include

1. "Test Methods for Evaluating Solid Waste, Physical/Chemical Methods," 3d ed., EPA Publication SW-846 (7)
2. "Methods for the Determination of Organic Compounds in Drinking Water" (29)

The first group of methods are frequently referred to as the 8000 series methods, since each is numerically identified with a number in the 8000 s. These are methods that the EPA has evaluated and identified as appropriate techniques of analysis under Subtitle C of the Resource Conservation and Recovery Act of 1976 (RCRA). Several regulations under Subtitle C of RCRA specifically require the use of SW-846 methods for testing. Under other circumstances, the methods are intended to provide guidance for the analyst. They apply to all types of environmental matrices, including groundwater, wastewater, soils, sediments, air samples, and liquid organic wastes.

The third edition of SW-846 and its first update were approved for use in August 1993. Surprisingly, many of the columns recommended in these methods are packed columns, even though capillary columns are now considered superior for most environmental applications. Exceptions to this would be screening applications, analysis of clean samples exhibiting simple chromatograms, and the analysis of samples for one or two analytes at high concentration levels. For some classes of compounds there are separate capillary- and packed-column methods in SW-846. Only the capillary-column methods are summarized in this chapter when two options exist. There is also a proposed Update II for SW-846, which includes new or revised capillary-column methods for gas chromatographic analysis. The proposed capillary-column methods are summarized along with the currently approved packed-column methods to provide a comprehensive listing.

The second group of methods are known as the drinking water methods, and they are identified as the 500 series methods. These methods were developed as a result of the Safe Drinking Water Act (SDWA). Some are required test methods under the National Primary Drinking Water Regulations, while others have been proposed for future regulations under the SDWA. They can be applied to fairly clean water matrices like drinking water and some ground and surface waters. Most of the drinking water methods recommend capillary columns for analysis, but when both a packed- and capillary-column method are provided, only the capillary column methods is summarized.

Both the 500 and 8000 series methods require a confirmation analysis for methods that do not use a mass spectrometer for analyte detection, and alternative columns are referenced. All columns in the methods are recommended columns only, and alternatives can be used if they perform satisfactorily.

Volatile Compounds. The compounds that are classified as volatile organic compounds (VOCs) have a low solubility in water and a high vapor pressure. When they are present in a water or solid matrix, these compounds will readily partition themselves into the gaseous phase. Dynamic gas extraction (purge-and-trap techniques) and, to a lesser extent, microextraction techniques are methods that are commonly utilized to determine these types of compounds with EPA methods. The SW-846 purge-and-trap procedure (Method 5030A) is listed separately from the gas chromatographic methods for the individual compound classes. The static gas extraction (headspace techniques) and solid-phase extraction are additional approaches that are used in certain circumstances for volatile compounds.

Semivolatile Compounds. Compounds that are classified as semivolatiles have higher boiling points and lower vapor pressures than the volatile class of compounds. Lower vapor pressures mean that analytes can be exposed to temperatures above ambient conditions, conditions that are associated with evaporative concentration steps, and liquid–liquid or liquid–solid extraction procedures.

The SW-846 extraction and cleanup procedures are listed as separate methods from the determinative methods. These extraction and cleanup

TABLE 14.7 SW-846 Extraction Methods for Semivolatiles

Technique	Method Number
Separatory funnel	3510A
Continuous liquid–liquid extraction	3520A
Soxhlet extraction	3540A
Sonication extraction	3550A
Waste dilution	3580A

Source. Reference 7.

methods are given in Table 14.7 along with their corresponding method numbers.

14.7.1 Halogenated, Aromatic, and Nonhalogenated Volatile Compounds

The volatile compounds can be divided into three major groups: halogenated, aromatic, and nonhalogenated volatile compounds. The group of compounds included with the halogenated volatiles represents an important set of industrial chemicals. These compounds are used as refrigerants and degreasing and general purpose solvents. They are also important components in the manufacture of products such as polymers, plastics, and chlorinated pesticides.

The aromatic compounds all contain the basic structure of the benzene molecule. The variations to the basic structure are provided by the different types of chemical groups that can be substituted around the benzene ring. These substituents range in complexity from the single methyl group characteristic of toluene to the fused ring structures of the polynuclear aromatic hydrocarbons like naphthalene.

Many of the basic aromatic volatile compounds are obtained from petrochemical processes. They are a major component of the C_5–C_{12} range of petroleum products, which includes gasoline. The basic aromatic compounds and the more complex substituted aromatic compounds are important in industrial processes as chemical intermediates in the manufacturing of more complex products.

A small group of nonhalogenated compounds are included in various purge-and-trap methodologies. The group consists primarily of some of the oxygenated classes (i.e., alcohols, ethers, aldehydes, and ketones).

EPA Methods for Halogenated Volatile Compounds. Halogenated volatile compounds in water, soil, and waste samples can be determined by Method 8010A. Method 502.1 provides an identical procedure for the same class of compounds in drinking water. The analysis is performed by the purge-and-trap technique or by direct injection for elevated concentration levels encountered in soil or waste samples. The analytical columns listed in the methods are an 8-ft × 0.1-in. stainless steel or glass column packed with 1% SP-1000 on Carbopack-B (60/80 mesh) and a 6-ft × 0.1-in. stainless steel or glass column packed with chemically bonded *n*-octane on Porasil-C (100/120 mesh). The electrolytic conductivity detector (ELCD) is used for detection.

With the purge-and-trap technique, the analytes are removed from an aqueous sample by purging with an inert gas (nitrogen or helium) for 11 min. They are collected on a trap consisting of 2,6-diphenylene oxide polymer, silica gel, and coconut charcoal. When purging is completed, the trap is coupled to a gas chromatograph and the carrier-gas flow is directed through the trap in the opposite direction of the purge flow. The trap is rapidly heated to a temperature of 180°C and the trapped analytes are desorbed and transferred onto the chromatographic column.

Low concentrations (<0.1 mg/kg) of volatile compounds in soil are measured by a similar technique. A 5-g aliquot of the soil is transferred to the purging chamber along with 5 mL of water. The soil/water mix is heated to a temperature of 40°C, and then purged and desorbed as aqueous samples. High concentrations of volatiles in soil can be determined after a methanolic extraction of the soil. A 4-g aliquot of the soil is extracted with 10.0 mL of methanol, after which the appropriate volume of the extract is added to 5.0 mL of water for the purge-and-trap process. Liquid waste samples can also be handled by the high concentration soil technique. The methanolic extracts of both soil and solid wastes can also be directly injected into the gas chromatograph if concentration levels of analytes are high enough to be detected by this technique.

A microextraction technique for the determination of 1,2-dibromoethane (EDB) and 1,2-dibromo-3-chloropropane (DBCP) in drinking water and groundwater is described in Methods 504 and 8011. Six grams of NaCl are added to 35 mL of sample or aqueous standard solution. The samples and standards are then extracted with 2 mL of hexane. Each of the extracts is analyzed by capillary column GC with an electron capture detector (ECD). Recommended columns for the analysis include Durawax-DX3, DB-1, and Rt_x-volatiles capillary columns. Microextraction provides greater sensitivity than the purge-and-trap technique for these particular compounds.

EPA Methods for Aromatic Volatile Compounds. Techniques for the determination of volatile aromatic compounds are provided in Methods 8020 (for soil, water, and waste matrices) and 503.1 (for drinking water). Both methods utilize the purge-and-trap technique for all matrices, with Method 8020 also providing the option for direct injection of methanolic extractions of high level soil and waste samples as in Method 8010A. Method 503.1 also includes a few unsaturated compounds, but the aromatic compounds make up the majority of the listed analytes.

The columns specified in both methods are a 6-ft × 0.082-in. stainless steel or glass column packed with 5% SP-1200 and 1.75% Bentone-34 on

Supelcoport (100/120 mesh) and an 8-ft × 0.1-in. stainless steel or glass column packed with 5% 1,2,3-Tris-(2-cyanoethoxy)propane on Chromosorb W-AW (60/80 mesh). The photoionization detector (PID) is used for detection of the analytes. Method 8020 designates the same type of trap material and purging procedure as Methods 8010A and 502.1 described previously. The drinking water method's trap is 25 cm × 0.105 in. packed with 1.0 cm of methyl silicone packing at the inlet and the remainder of the trap filled with 2,6-diphenylene oxide polymer. After the 11-min sample purge, the trap is dried by directing a purge gas flow directly through the trap for 4 min. The analytes are then desorbed at 180°C for the chromatographic analysis.

EPA Methods for the Simultaneous Determination of Halogenated and Aromatic Compounds. Two additional EPA methods for volatile compounds combine the analysis for halogenated compounds with the aromatic compounds: 8021 and 502.2. The purge-and-trap portion of the analysis is the same as described previously. Detection of analytes is achieved with a PID and an ELCD connected in series, with the nondestructive PID preceding the ELCD. The connection is made with a short piece of fused-silica capillary tubing which has no stationary phase coating. Method 502.2 lists three capillary columns for the analysis: a 60-m × 0.75-mm VOCOL column (1.5-μm film thickness), a 105-m × 0.53-mm RTX-502.2 column (3.0-μm film thickness), and a 30-m × 0.53-mm DB-624 column (3.0-μm film thickness). Method 8021 lists only the VOCOL column for the analysis.

Each of these methods provides for a subambient column temperature of 10°C at the start of the chromatographic analysis with a VOCOL column. This lower temperature allows the analytes to be focused at the front end of the column, which provides improved peak shape and separations for the early eluting compounds. The subambient temperatures are achieved by introducing either liquid carbon dioxide or liquid nitrogen into the gas chromatograph's oven. Most instruments are readily configured to handle these cryogenic liquids to achieve subambient temperatures.

EPA Methods for Nonhalogenated Volatile Organic Compounds. Acrolein and acrylonitrile are the compounds included in Method 8030A. The method can be applied to water, soil, or waste sample matrices with either purge and trap for low concentrations or direct injection of methanolic extracts for elevated concentrations. The only packing material required for the trap is Tenax (2,6-diphenylene oxide polymer).

The analytes are purged from the sample for 15 min, with the sample maintained at a temperature of 85°C during the purge. Analytes are desorbed from the trap at 180°C for 1.5 min. The columns designated in the method are a 10-ft × 2-mm stainless steel or glass column packed with Porapak-QS (80/100 mesh) and a 6-ft × 0.1-in. stainless steel or glass column packed with Chromosorb 101 (60/80 mesh). A flame ionization detector (FID) is used for analyte detection.

Method 8015A provides a procedure for the determination of diethyl ether, ethanol, methyl ethyl ketone, and methyl isobutyl ketone. The purge-and-trap portion of the analysis is identical to that listed for Method 8030A. The analysis is peformed with an FID and the same chromatographic columns that are used in Method 8010A.

Alternative Methods for Volatile Compounds. Louch et al. (30) have described a system for the determination of volatile compounds in water which is based on solid-phase extraction principles. Their extraction device consisted of a poly(dimethylsiloxane)-coated fused-silica optical fiber mounted in a syringe needle. The fiber is placed in an aqueous solution containing volatile compounds for a controlled period of time. The analytes are partitioned from the water onto the fiber at a rate that is dependent on their affinity for the solid phase and their rate of diffusion. The fiber is then removed from the water and withdrawn into the needle of the syringe, which serves as a protective covering. The needle is introduced into the heated injection port of a gas chromatograph. The fiber is once again pushed from the needle and the analytes are thermally desorbed and carried to the chromatographic column for separation and measurement. The authors reported sub-ppm detection of volatile compounds with analysis by capillary column GC-FID.

Pratt and Pawliszyn (31) used an 8-cm-long hollow fiber membrane to extract 1,1,1-trichloroethane, trichloroethane, and tetrachloroethane from aqueous samples. The water was pumped through the fiber while a carrier-gas flow was directed around the perimeter of the fiber. The analytes diffused across the membrane and into the gas flow where they were transferred and cryofocused at the front of a chromatographic column. The authors evaluated both a nonporous silicone rubber and microporous polypropylene membrane and concluded that each performed satisfactorily.

Adsorption/thermal desorption (ATD) was used for the in situ collection of volatiles by Rosen and Pankow (32). The trap, constructed of Pyrex glass, was packed with 0.13 g of 60/80 mesh Tenax-TA. Aqueous samples were passed through the Tenax trap to adsorb the volatile analytes and extract them from the water. The trap was then connected to a gas chromatograph for thermal desorption. A water-removal trap was placed between the analyte trap and the gas chromatograph. The volatiles were separated on a 30-m $\times$ 0.53-mm DB-624 fused-silica column (3-μm film thickness) and detected by a mass spectrometer. Performance was comparable to that provided by the purge-and-trap method.

Two groups reported on an evaluation of GC/FTIR as a means of determining volatile compounds. Phillipparearts and coworkers (33) compared purge-and-trap GC/FTIR with purge-and-trap GC/MS for the analysis of volatile compounds. Data obtained by GC/FTIR were complimentary to the GC/MS data in the area of identifying isomeric forms of compounds. Most compounds could be detected at a concentration of 10 ppb. Gurka et al. (34) compared direct aqueous injection with ion trap GC/MS and GC/FTIR for the determination of nonhalogenated organic compounds. Analytes were separated on a 30-m × 0.32-mm fused-silica capillary column coated with a 10-μm film of Poraplot Q or a 15-m × 0.53-mm PTE-5 column (0.5-μm film thickness). The detection limits provided by GC/FTIR were in the low ppm range, while those provided by ion trap GC/MS were in the low to mid ppb range.

Dingyvan and Jianfei (35) reported on the determination of tri-halomethanes by direct aqueous injection and packed-column GC/ECD. The column was packed with a porous polymer support (GDX-103) coated with 1% SE-30. This provided adequate separation of solvent (water) and target analytes and yielded excellent performance in terms of peak shape and detector response. Detection limits were reported at the 1-ppb concentration level.

The technique of static headspace extraction and gas chromatographic analysis was evaluated as a field screening method for soil samples (36). Soil samples were transferred to an appropriate size vial and 30 mL of deionized water was also added to facilitate extraction of the analytes from the soil. The vial was sealed and agitated to disperse the soil and extract the analytes. An aliquot of the vial's vapor phase was then analyzed by packed-column GC to measure the concentration of volatiles.

When compared to a methanolic extraction followed by purge-and-trap analysis, the aqueous extraction/headspace gas chromatographic technique performed adequately as a screening method for four test compounds. Soils with a high organic carbon content yielded the largest discrepancies, with headspace/GC results that were 30% of those obtained by purge and trap. The gas chromatographic analysis was performed with a completely portable field instrument equipped with a packed column and a PID.

A simple modification to the headspace technique, known as multiple headspace extractions (MHE), has also served as an alternative method for the determination of volatile compounds in soil samples. This approach provides a certain amount of qualitative information for sample analytes in the form of a constant that is proportional to the partition coefficient for an analyte. It also eliminates the need to matrix match standards in a material that provides analyte partition coefficients that are the same as the sample. The technique of MHE was initially introduced by McAuliffe (37) for aqueous samples and it was later adapted to handle solid sample materials as well (38).

Application of the MHE technique begins with the analysis of the vapor phase or headspace equilibrated with the sample in a sealed vial. After the analysis, a given volume of the vapor phase is removed and replaced with an equivalent volume of a clean gas. The sample and vapor phase are again allowed to equilibrate and the headspace of the vial is once again analyzed for the compounds of interest. A plot of the natural log of the peak area for an analyte obtained from the analysis vs. the equilibration number yields a straight line which can be described by Equation 14.6:

$$\ln \text{peak area} = -kN + q \tag{14.6}$$

where N = the equilibration or injection number
ln peak area = natural log of the peak area for injection N
k = slope of the line, which is a constant related to instrument response and partition coefficient for an analyte
q = intercept

The total mass of analyte in the sample can then be determine by the following relationship:

$$A_{\text{tot}} = \frac{A_1}{1 - e^{-k}} \tag{14.7}$$

where A_{tot} = total peak area for an analyte that would be theoretically obtained from an infinite number of equilibrations that would exhaustively extract the analytes from the sample
A_1 = peak area obtained from the first equilibration
k = constant obtained from Equation 14.6

The total mass of the analyte in a sample is then determined from its response factor and A_{tot}.

Maggio and coworkers (39, 40) described an evaluation of the MHE technique for the analysis of soil samples for volatile compounds. Determinations were based on nine equilibrations of each of the samples for the MHE measurement. A comparison was made to a single headspace equilibration for the same samples using the technique of standard additions. Analytes of interest were determined by capillary column GC with an FID. Results obtained by MHE provided better accuracy and precision than those obtained by the single equilibration method.

14.7.2 Phenols

The phenols are characterized as a class of compounds with the basic structure of a benzene ring and an attached hydoxyl group. Variations will include any of a number of organic substituents to the basic structure. The hydroxyl group imparts acidic characteristics to the phenols, and requires the use of certain procedures and precautions during preparation and analysis. The acidic strength of a particular phenol is dictated by the other substituents of the base phenol molecule.

In water, the hydroxyl group can lose a hydrogen ion and become ionized under basic conditions. Therefore, all extractions of water samples must be performed under acidic conditions to ensure that the phenols do not dissociate and remain in the polar aqueous phase. Glassware used for sample extraction should be acid washed to neutralize any basic sites on the glassware and prevent losses of the phenols by irreversible adsorption at these basic sites.

The extremely polar nature of the hydroxyl group also creates adsorption problems and peak tailing during the chromatographic analysis. Active sites in any of the components of the instrumentation (i.e., injection port liners, high boiling residues deposited at the front end of the chromatographic column) will provide sites for hydroxyl interaction and lead to peak tailing. Silanization of injection port liners or inserts and frequent column maintenance can reduce these occurrences. Derivatization techniques can also be used to generate a product that is easier to chromatograph than the phenol and, in some cases, provide a product that yields a greater detector response than the original parent phenol compound.

EPA Methods for Phenols. Various phenols can be determined in soil, water, and nonaqueous samples by EPA Method 8040A. Water samples are adjusted to pH 2 and extracted with methylene chloride using either a separatory funnel or a continuous liquid–liquid extractor. Solid samples are extracted as received (i.e., no pH adjustment) with 1:1 methylene chloride:acetone. The extracts of both solid and aqueous samples are exchanged to 2-propanol, concentrated to a final volume of 1.0 mL, and analyzed on a 1.8-m × 2-mm glass column packed with 1% SP-1240DA on Supelcoport (80/100 mesh) and measured with an FID. A GPC or acid–base partition cleanup procedure can be included for extracts containing high levels of interferences. Pentafluorobenzylbromide derivatives of the phenols can also be prepared for analysis with an ECD to obtain greater sensitivity. The column recommended for the derivatized phenols is a 1.8-m × 2-mm glass column packed with 5% OV-17 on Chromosorb W-AW-DMCS (80/100 mesh).

Several phenolic compounds are also included in Method 515.2, a procedure for the determination of chlorinated acids in water, and Method 8150A, a procedure for the determination of chlorinated herbicides in soil and water samples. Each of these methods are discussed in more detail in Section 14.7.4.

Alternative Methods for Phenols. A technique utilizing solid-phase extraction followed by supercritical fluid extraction and derivatization for phenols was reported by Hawthorne and coworkers (41). Aqueous samples were acidified and passed through a C_{18} extraction disk. The phenols were then removed from the disk and derivatized by supercritical fluid extraction with CO_2 and a modifier of 1% trimethyl phenyl ammonium hydroxide (TMPA) in methanol. The TMPA generated methyl ether derivatives of the phenols during the SFE process. The SFE was performed with an initial static phase for analyte derivatization, followed by a dynamic phase to complete the extraction process. The derivatized phenols were analyzed by capillary column GC and detected by an ECD, FID, or MS.

14.7.3 Phthalate Esters

Various phthalate esters are used as plasticizers to improve the quality of plastic materials, sometimes making up almost 40% of a given product. In the environment, soil or water that contacts a plastic material will become contaminated with a phthalate. The popularity and widespread use of plastic materials provides countless pathways for the entry of phthalate esters into the environment.

The ability to perform trace analysis for phthalates can be hampered by elevated background levels of these compounds, since any phthalate containing plastic in the laboratory can directly or indirectly contaminate a sample or extract during preparation or handling. Extra caution is always required to minimize the occurrence of cross-contamination when working with samples requiring an analysis for this class of compounds.

EPA Methods for Phthalates. Method 8060 describes procedures for the determination of phthalate esters in soil, water, or sludge samples. The analytes are extracted from soil and sediment samples by either the Soxhlet or the ultrasonic method. Acceptable extraction solvents include methylene chloride/acetone or hexane/acetone, each at a 1:1 ratio. Aqueous samples can be prepared by liquid–liquid extraction with a separatory funnel. The continuous liquid–liquid extractor is not recommended since the longer chain esters can adsorb to the glassware during the extended extraction times encountered with this technique. Methylene chloride is used for the separatory funnel extraction with the sample maintained at pH 5–7.

All extracts are exchanged into hexane and analyzed by packed-column GC with either an FID or an ECD, with the latter providing greater sensitivity. Recommended columns for the analysis are $1.8\,m \times 4\,mm$ packed with 1.5% SP-2250/1.95% SP-2401 on Supelcoport (100/120 mesh) or 3% OV-1 on Supelcoport (100/120 mesh). Florisil, alumina, sulfur (for ECD analysis), and/or GPC cleanup techniques can be used to remove interferences prior to analysis.

A proposed update to this method is described in Method 8061. This update includes an expanded compound list, an option for solid-phase extraction of aqueous samples and analysis with wide-bore capillary columns. The SPE method prescribes analyte retention on a C_{18} extraction disk and then elution with approximately 10 mL of acetonitrile. All extracts are exchanged into hexane prior to analysis with an ECD. Recommended columns for the analysis are a DB-5 (1.5-μm film thickness) and a DB-1701 (1.0-μm film thickness), each with dimensions of $30\,m \times 0.53\,mm$.

Phthalates in drinking water can also be determined by Method 506. The analytes can be extracted by separatory funnel (neutral pH) with three separate 60-mL portions of methylene chloride followed by 40 mL of hexane. The procedure requires the addition of 50 g of NaCl to the 1-L water sample prior to the extraction. Alternatively, extraction can be achieved by solid-phase techniques using either a C_{18} extraction disk or cartridge followed by elution with methylene chloride and/or acetonitrile. Chromatographic columns are DB-5 and DB-1 fused-silica capillary columns, each 30 m × 0.32 mm with a 0.25-μm film thickness. Analytes are detected by a PID.

Alternative Methods for Phthalates. Ritsema et al. (42) reported on the determination of phthalate esters in surface water and suspended particulate matter. A 500-mL surface water sample was passed through a C_8 extraction cartridge. The analytes were then eluted with a 3:1 mixture of hexane:ether. A 2-g sample of suspended particulate was extracted with 15 mL of a 1:1:1 mixture of acetone:water:hexane. The hexane layer was removed, washed, and then analyzed by capillary-column GC. An electron-capture detector and a mass spectrometer were used for the analysis. Detection limits of 0.01 ppb were reported for aqueous samples and 0.01 to 1 ppm for the suspended particulates. The mass spectrometer was the preferred means of detection because it provided the selectivity required for the samples.

14.7.4 Chlorinated Herbicides and Acids

The compounds included in this class behave in a manner that is similar to the phenols (some of the phenols are actually a subset of the list of analytes included in these methods). The analytes must be extracted at a pH of 2 or less and are then derivatized prior to chromatographic analysis.

Sample preparation steps for these analytes must include a hydrolysis step prior to the extraction. Many of these compounds exist as salts or some type of an ester (i.e., methyl, ethyl, or propyl ester) of the acid. The hydrolysis under basic conditions converts all forms of the esters to their corresponding acid.

EPA Methods for Chlorinated Herbicides and Acids. Method 8150A provides a procedure for the determination of chlorinated herbicides in solid and aqueous samples. An aqueous sample is adjusted to a pH of less than 2. The analytes are extracted from the water by separatory funnel, using diethyl either as extraction solvent. Soil samples are extracted with a combination of acetone and diethyl ether, with a wrist-action shaker providing the means of solvent/soil contact. Concentrated HCl is added to the soil before the solvent to acidify the sample. The acetone/diethyl ether extract of the soil is then washed with an aqueous solution of 5% acidified sodium sulfate.

The organic extracts of soils and waters are then combined with a measured volume of 37% KOH and heated to complete the hydrolysis step.

The ether is driven off during this process, and the dissociated form of the herbicide acid is generated. This form remains soluble in the basic aqueous phase. After cooling, the aqueous phase is washed with diethyl ether, adjusted to a pH of 2, and reextracted with diethyl ether for partitioning of the herbicides from the water into the organic phase. The diethyl ether containing the herbicides is dried with approximately 7 g of acidified sodium sulfate and concentrated to a volume of approximately 1–2 mL. Diazomethane is then introduced into the solvent to generate the methyl ester forms of the chlorinated herbicides. Following derivatization, the excess diazomethane is destroyed with silicic acid, and the extract volume is adjusted to 10.0 mL for analysis.

The herbicide extracts are separated and analyzed on a gas chromatograph equipped with packed columns and either an ECD or an ELCD. The method includes a total of three chromatographic columns for the separation of all of the target analytes. The first two columns are 1.8-m × 4-mm glass, packed with 1.5% SP-2250/1.95% SP-2401 on Supelcoport (100/120 mesh) and 5% OV-210 on Gas Chrom Q (100/120 mesh). The third column in the method, 1.98-m × 2-mm glass packed with 0.1% SP-1000 on Carbopak C (80/100 mesh), is suggested for the analysis of one of the more volatile analytes included in the method.

The herbicide compounds in Method 8150A are reported as their acid equivalents. Standards for analysis should be prepared from the acid form of the herbicide in diethyl ether, and then derivatized with diazomethane to generate the methyl ester form of the herbicides. Standards can also be prepared directly from the methyl ester forms of the herbicides, but the final calculation of herbicide concentrations must include a correction for the molecular weight of the methyl ester versus the acid herbicide.

Method 8151 is a proposed update to Method 8150A, which include capillary columns and an expanded compound list. This method provides an optional hydrolysis step, sonication extraction for soils and sediments, and herbicide derivatization by either diazomethane or pentafluorobenzyl bromide. Recommended capillary columns include a DB-5, DB-1701, and DB-608.

Method 515.2 is applied to the determination of chlorinated acids (which includes many of the Method 8150A herbicides) in drinking water. The procedure is similar in many respects to the technique provided in Method 8150A for aqueous samples. The sample is transferred to a separatory funnel, and 250 g of NaCl is dissolved in the sample. The sample pH is then adjusted to 12 with 6 N NaOH for the hydrolysis step, which is allowed to proceed for 1 h at room temperature.

After the hydrolysis, the basic water phase is washed 3 times with methylene chloride. After the third wash, the sample pH is adjusted to 2 or less with 12 N H_2SO_4 and extracted with ethyl ether to remove the analytes from the aqueous phase. The analytes can also be extracted from the acidified aqueous phase by the solid-phase extraction technique. The acidified sample is passed through a 47-mm poly(styrene-divinylbenzene) extraction disk. Analytes are then removed from the disk with 10% methanol in methyl *tert*-butyl ether. The extracts obtained from either method are dried with acidified sodium sulfate for at least 2 h, concentrated, and then treated with diazomethane to generate the methyl ester form of the chlorinated acids.

The sample extracts and standards are measured with a gas chromatograph equipped with 30-m $\times$ 0.25-mm capillary columns and an ECD. Stationary phases for the columns are a DB-5 and a DB-1701, each with a 0.25-μm film thickness. The chlorinated esters are reported as their acid equivalents.

Alternative Methods for Chlorinated Herbicides and Acids. Most of the method-development/improvement work related to the determination of herbicides focuses on new derivatization techniques that are faster and more reliable than the conventional approach. Propionic acid herbicides in water can be determined by trifluoroanilide (TFA) derivatization with EC detection through a one-step process (43). The derivatization can be performed directly in the water under acidic conditions with the addition of a solution of ethyl acetate containing TFA and dicyclohexylcarbodimide. The derivatized herbicides are then extracted from the aqueous phase with ethyl acetate, processed through an alumina column cleanup, and determined by capillary-column GC with either an ECD or MS for detection. Tetrapion and Dalapon were each measured at the low ppb level by this method.

Phenoxy acetic acid and phenoxy propionic acid herbicides can be determined in water as the related 2,2,2-trifluoroethyl ester by derivatization with 2,2,2-trifluoroethanol (44). The herbicides are extracted from the water under basic basic conditions with an ion exchange resin. After sufficient time has elapsed for interaction of the herbicides with the resin, most of the water is removed. The water that remains is adjusted to an acidic pH and the resin is extracted with diethyl ether. The ether is blown to dryness and the residue is treated with a mixture of 2,2,4-trimethyl pentane, 2,2,2-trifluoroethanol, and a small amount of H_2SO_4. The mixture is maintained at 70°C to complete the derivatization. An aqueous buffer at pH 7 is added and the herbicides are reextracted with 2,2,4-trimethyl pentane. The solvent is adjusted to the appropriate volume and the herbicides are determined by capillary-column GC with an ECD. Herbicides evaluated were MCPA, MCPP, 2,4-D, 2,4-DP, 2,4,5-T, and 2,4,5-TP. The derivatization enhanced the electron-capture response of these herbicides through the addition of electronegative fluoro groups.

Herbicides in soil can be extracted and converted to their corresponding methyl ester through a simple one-step process (45). A 5-g soil sample is equilibrated with water and H_2SO_4. A 40% solution of benzyltrimethylammonium chloride in methanol is added and the soil/solvent mixture is shaken occasionally for 10 min to extract and derivatize the herbicides. The solvent is then removed by filtration, processed through a Florisil cleanup, and analyzed by capillary-column GC with an ECD. Herbicides that have been determined by this technique include 2,3-D, 2,4-D, 2,4,5-T, and MCPP.

A variation of the static mode SFE/derivatization technique described previously for phenols (41) has been applied to the determination of 2,4-D and Dicamba in soils. Preliminary extraction/derivatization was performed with static mode SFE and supercritical CO_2 with a methanol/TMPA modifier. The herbicide methyl ester extraction was then completed with SFE in the dynamic mode. A methanol/BF_3 modifier was also used to selectively derivatize only the 2,4-D. Analysis was performed by capillary-column GC/ECD.

14.7.5 Pesticides and PCBs

Pesticides and PCBs are an important group of compounds in a commercial sense. They are often linked together in discussions of analytical methods for the simple reason that PCBs can be determined by the same techniques that are used for chlorinated pesticides, but they represent two different types of pollution problems. Samples that are obtained from a site with an unknown history will usually be analyzed for both chlorinated pesticides and PCBs at the same time. The analysis of samples obtained from a known pesticide or PCB contamination site, required for the purpose of determining the extent of contamination for a known group of chlorinated pesticides or PCBs, can be optimized specifically for that determination.

Numerous compounds are classified as pesticides. The pesticides that can be determined by gas chromatographic methods are divided into the chlorinated pesticides and the nitrogen–phosphorus pesticides. Several pesticides included in EPA methods have been banned in this country for health reasons, but their use is still permitted in foreign countries. Some of the pesticides are extremely sensitive and degrade in the chromatographic system, which can complicate the analysis. Endrin and DDT are two chlorinated pesticides that can break down when they contact hot metal surfaces, active sites on injection port liners and the walls of glass columns, and residues of nonvolatilized sample components. This requires extreme care in the preparation and handling of certain instrument components before and during the analysis of certain pesticides. Other pesticides like Toxaphene and Technical Chlordane are mixtures of several compounds that yield multipeak patterns in the associated chromatograms rather than single-peak responses.

The PCBs or polychlorinated biphenyls are a group of compounds that were used (and in some cases still are used) as hydraulic fluids, capacitor and transformer fluids, and in various other industrial applications. They are a very stable, inert, and non-volatile group of compounds, which adds to their persistence as pollutants in the environment. Their manufacture was banned in the United States in 1979.

The term polychlorinated biphenyl applies to any of 209 possible chlorinated biphenyl compounds ranging in level of chlorination from monochlorobiphenyl to decachlorobiphenyl. The PCBs were manufactured and distributed commercially as the Aroclor mixtures. These products consisted of certain percentages of the individual PCBs.

The chromatogram obtained from the gas chromatographic analysis of an Aroclor will exhibit a multipeak pattern or fingerprint. This fingerprint is created by the particular mixture of chlorinated biphenyls that are contained in that product, and the pattern will be unique for that product. Analytically, the Aroclors are usually identified by comparing a multipeak sample pattern to those exhibited by Aroclor standards.

EPA Methods for Pesticides and PCBs. Chlorinated pesticides and PCBs can be determined by Method 8080, while certain organophosphorus pesticides can be determined by Method 8041. Both methods include a methylene chloride extraction at a neutral pH for aqueous samples using a separatory funnel or a continuous liquid–liquid extractor. Soil and sediment samples are extracted by either the Soxhlet or ultrasonic techniques with 1:1 methylene chloride:acetone. All extracts are exchanged into hexane and adjusted to a final volume of 10.0 mL for analysis.

The chlorinated pesticide/PCB extracts can undergo a Florisil, GPC, or sulfur cleanup before analysis. The gas chromatographic analysis by Method 8080 is performed with a 1.8-m × 4-mm glass column packed with Supelcoport (100/120 mesh) coated with either 1.5% SP-2250/1.95% SP-2401 or 3% OV-1. Detectors are an ECD or an ELCD. The organophosphorus pesticide extracts can undergo a Florisil or sulfur cleanup. Method 8041 lists three megabore columns for the analysis, each with dimensions of 15 m × 0.53 mm. The three columns are a DB-210 and a DB-5 (each 1-μm film thickness) and an SPB-608 (1.5-μm film thickness). The nitrogen–phosphorus detector (NPD) or the flame photometric detector (FPD) can be used for detection, with each set in the phosphorus mode. Several of the organophosphorus pesticides are also halogenated, and the ELCD can be used specifically for the detection of these compounds.

Methods for the determination of chlorinated pesticides and PCBs as well as nitrogen- and phosphorus-containing pesticides in drinking water are provided by Methods 505, 507, and 508. Method 505 describes a microextraction for the analysis of chlorinated pesticides and PCBs. Six grams of NaCl are added to a 35-mL aliquot of the sample or standard solution. The salt-water mixture is then extracted with 2.0 mL of hexane. After extraction, the solvent layer is removed and analyzed by capillary-column GC with an ECD. Three fused-silica capillary columns are listed in the method, a 30-m × 0.32-mm DB-1 (1-μm thickness), a 30-m × 0.32-mm

Durawax-DX3 (0.25-μm thickness), and a 25-m $\times$ 0.32-mm OV-17 (1.5-μm thickness).

The PCBs and a larger list of chlorinated pesticides are also included in Method 508. One hundred grams of NaCl is added to a 1-L aliquot of the aqueous sample. The sample is then extracted with methylene chloride at a neutral pH. After extraction, the solvent is exchanged to methyl *tert*-butyl ether and adjusted to a final volume of 5.0 mL for analysis. The chromatographic columns listed in the method are a DB-5 and a DB-1701 fused-silica capillary column, each 30 m $\times$ 0.25 mm with a 0.25-μm film thickness. Analyte detection is by ECD. Method 507 includes an extraction procedure and chromatographic columns that are identical to Method 508 for the determination of nitrogen- and phosphorus-containing pesticides. The analysis is performed with an NPD.

Method 8081 is a new method for chlorinated pesticides and PCBs in soils and waters that has been proposed for addition to SW-846. This method lists both wide- and narrow-bore capillary columns for the analysis, including the DB-608, DB-5, and DB-1701 columns. The compound list has been expanded to include more pesticides that are listed in the current Method 8080. The halowaxes have also been added to this method. Extraction techniques are the same as those listed for Method 8080.

Alternative Methods for Pesticides. Benfenati et al. (46) extracted 50 organochlorine and organophosphate pesticides from water by solid-phase extraction and then determined the compounds by GC/MS analysis. The extraction was performed by passing a 10-L water sample through a 15-cm $\times$ 5-mm column packed with 0.8 g of C_{18} and 0.4 g of pH silica bonded phases. The pesticides were eluted from the column with 4 mL of methylene chloride. The analysis was performed with a 25-m $\times$ 0.32-mm CP Sil 8 CB capillary column (0.12-μm film thickness). Sensitivity was at least 50 ppt for the pesticides.

A chromic acid digestion–extraction technique has been described for pesticides and PCBs (47). This preparation technique yielded recoveries that were better than those obtained by conventional separatory funnel extraction methods. River water samples containing a high level or organic material were transferred to a roundbottom flask containing 5 mL of chromic acid and 200 mL of hexane. The contents of the flask were refluxed for 2 h, cooled, and poured into a separatory funnel. After phase separation, the hexane layer was removed, dried, and concentrated to an appropriate volume for analysis.

The digestion–extraction technique exhibited recoveries for organochlorine pesticides that were better than those obtained by separatory funnel extraction of the river water sample. The improvement was attributed to the complete extraction of analytes adsorbed to the organic matter in the aqueous sample, which was aided by the chromic acid digestion. This provided a total sample concentration for the analytes, while the separatory funnel extraction provided a concentration only for the dissolved pesticides. The digestion process also eliminated the occurrence of emulsions and similar phase separation problems that have been encountered with this sample matrix.

Alternative Methods for PCBs. On-line SFE/GC has been used for the determination of PCBs in sediments (48). The SFE extraction device was connected directly to the gas chromatograph by a precolumn consisting of a 2-m × 0.32-mm of fused-silica column that contained a bonded phase of SE-54 (1.2-μm film). The precolumn was enclosed in a temperature-controlled environment.

The SFE of the PCBs was performed in the static mode with CO_2 and a modifier of 2% methanol. The analytes emerged from the SFE device and were trapped on the precolumn, which was maintained at 5°C. At the completion of the extraction/collection step, the analytes were transferred to the gas chromatograph by heating the precolumn and directing the carrier-gas flow through the pre-column and into the analytical column.

The separation was performed with a 30-m × 0.25-mm SE-52 fused-silica column, while an ECD was used for detection. The technique provided quantitative extraction of PCBs from sediments with contamination levels that ranged from 20 to 200 ppm. Impurities in the CO_2 created interferences with determinations below 20 ppm, but the elimination of these impurities would lead to improved sensitivity with this technique.

Improved chromatographic separations, required for the determination of individual PCB congeners in the presence of other congeners, were obtained with multidimensional GC (49). Sixty-meter capillary columns are usually used for the analysis of individual congeners, but many will coelute with one or more of the other congeners, even with a column of this length. The multidimensional system was evaluated as an alternative means of achieving this difficult separation.

The system consisted of a preliminary 30-m × 0.25-mm DB-5 capillary column with a film thickness of 0.25 μm and a second column which was 20 m × 0.20 mm and coated with a chemically bonded liquid crystalline phase. Each of the columns were housed in separate instrument ovens to optimize temperature programming. The two columns were connected to an automatically controlled tee valve which was also attached to an FID. The effluent from the preliminary column was initially directed to the FID for a pre-determined period of time. The effluent flow was switched to the second column when the PCB congeners began to elute from the first column. The congener elution pattern for the preliminary column were based primarily on the analyte's boiling point, while the liquid-crystal column provided separations according to molecular shape. These characteristics provided the ability to separate certain congeners from other congeners that coeluted on single column systems.

Alternative cleanup techniques evaluated for PCBs include partitioning with dimethylsulfoxide (DMSO) (50) and fractionation with carbon cartridges (51). The DMSO partitioning process was useful for eliminating matrix interferences from aliphatic-based waste oils. Hexane extracts were mixed with DMSO and then water and hexane to back extract the PCBs into the hexane for analysis. Extracts were carried through a multilayer column cleanup after the partitioning step. The carbon cartridges were capable of separating planar and nonplanar PCB congeners from each other. Elution with 1:1 hexane:methylene chloride removed the nonplanar PCBs and DDE from the cartridge. Subsequent elution with toluene removed the planar PCBs, hexachlorobenzene, and polychlorinated naphthalenes from the carbon cartridge.

14.7.6 Polynuclear Aromatic Hydrocarbons

The polynuclear aromatic hydrocarbons are a class of compounds that consist of two or more aromatic rings that share a pair of carbon atoms. These compounds receive a considerable amount of attention, since many of the more complex PAHs have cancer-producing properties. They are associated with certain petroleum products and they are also by-products of many combustion processes.

EPA Methods for Polynuclear Aromatic Hydrocarbons (PAHs). Method 8100 can be used to determine certain PAHs in soil, water, or liquid waste samples. Aqueous samples are extracted at a neutral pH with methylene chloride using either a separatory funnel or a continuous liquid–liquid extractor. Soils and sediments are extracted with 1:1 methylene chloride:acetone using a Soxhlet or an ultrasonic extractor. Sample extracts can undergo a silica gel and a GPC cleanup to remove coextracted interferences. Acid–base partitioning and alumina column cleanup is also a useful technique with extracts of petroleum wastes. The extract volume is adjusted to 1.0 mL for greatest sensitivity.

The method lists both packed and capillary columns and an FID for analysis. The packed column is 1.8 m × 2 mm glass, packed with 3% OV-17 on Chromosorb W-AW-DCMS (100/120 mesh). A 30-m × 0.25-mm or 0.32-mm fused-silica capillary column coated with SE-54 (no film thickness provided) is also designated.

Several pairs of PAHs cannot be completely separated by the packed-column chromatographic conditions provided in Method 8100. The capillary columns may provide adequate resolution for these PAH pairs. If sufficient resolution cannot be achieved, the analytes in question must be reported as the quantitative sum of the unresolved pair.

Alternative Methods for PAHs. Polynuclear aromatic hydrocarbons in soil have been extracted and analyzed by on-line SFE (52). The effluent from the SFE device was introduced directly into the injection port of a gas chromatograph from the extraction cell. The cell was located just above the chromatograph with its exit line connected to the injection port.

To begin the extraction process, the sample aliquot is transferred to the cell and the carrier-gas flow to the instrument is interrupted. The column oven is cryogenically cooled to subambient temperature and sample extraction is then performed in the dynamic mode for 5–30 min. The extraction effluent enters the injection port and then the gas chromatographic column, with the PAH compounds trapped at the front end of the column under the subambient temperature conditions. Upon completion of the extraction process, the column temperature is elevated and programmed appropriately to initiate the analysis and separate the target compounds.

Both NO_2 and CO_2 were evaluated as supercritical extraction fluids, and it was determined that the NO_2 yielded a more rapid extraction of the PAHs. Overall, the method was faster than conventional extraction methods with excellent sensitivity.

A liquid chromatograph has been coupled directly to a gas chromatograph (53) to provide an automated cleanup procedure for PAH extracts prior to analysis. Extracts of crankcase oil and PUF cartridges were automatically processed and analyzed by capillary column GC with FID for the determination of PAHs.

14.7.7 Nitrosamines, Nitroaromatics, and Cyclic Ketones

EPA Methods for Nitrosamines, Nitroaromatics, and Cyclic Ketones. Method 8070 provides a method for determining certain nitrosamine compounds in soil and water. The continuous liquid–liquid and the separatory funnel techniques are included for aqueous samples, with extraction at a neutral pH by methylene chloride. Soil and sediment samples are extracted by the ultrasonic or Soxhlet techniques using 1:1 methylene chloride:acetone. All extracts are washed with dilute HCl to remove free amines, exchanged into methanol, and concentrated to a final volume of 10.0 mL for analysis. Extracts can be carried through a GPC, alumina, or Florisil cleanup procedure prior to analysis.

The extracts are analyzed by packed-column GC with an NPD, a thermal energy analyzer, or a reductive Hall detector. Chromatographic columns are 1.8 m × 4 mm glass packed with Chromosorb W AW (80/100 mesh) coated with 10% Carbowax 20M/2% KOH or Supelcoport (100/120 mesh) coated with 10% SP-2250. One of the analytes in this method, *N*-nitrosodiphenylamine, forms diphenylamine at the elevated temperatures encountered with the gas chromatograph, so this compound is actually chromatographed and measured as diphenylamine. This requires that any diphenylamine contained in the sample is removed before performing an analysis for *N*-nitrosodiphenylamine. The diphenylamine is removed by Florisil or alumina cleanup prior to analysis.

Nitroaromatics and cyclic ketones can be determined in the same matrices by Method 8090. Options for sample extraction are the same as those provided for the nitrosamines, without the acid-washing procedure. Extracts are exchanged into hexane for analysis. Florisil and/or GPC cleanup procedures can be used prior to analysis. The dinitrotoluenes included in this method are analyzed with an ECD while the remaining compounds are analyzed with a FID. The chromatographic columns described in the method are a 12.2-m × 2-mm or 4-mm glass column packed with 1.95% QF-1/1.5% OV-17 on Gas-Chrom Q (80/100 mesh) and a 3-m × 2-mm or 4-m glass column packed with 3% OV-101 on Gas-Chrom Q (80/100 mesh).

14.7.8 Haloethers and Chlorinated Hydrocarbons

EPA Methods for Haloethers and Chlorinated Hydrocarbons. Analytical methodology for the determination of haloethers is provided in Method 8110, while Method 8120 provides a procedure for chlorinated hydrocarbons. Extraction and cleanup procedures are the same for both methods. Continuous liquid–liquid and separatory funnel techniques are acceptable for aqueous samples, and Soxhlet and ultrasonic techniques are acceptable for solid matrices. Aqueous samples are extracted at a neutral pH with methylene chloride. Solid samples are extracted with 1:1 methylene chloride:acetone. Extracts are exchanged into hexane and concentrated to a volume of 10.0 mL for analysis. Extracts for both sets of analytes can be carried through Florisil and/or GPC cleanup prior to analysis to remove coextracted interferences.

The analysis for haloethers is performed by packed-column GC with either an ECD or the more selective ELCD. The columns listed in the method are both 1.8 m × 2 mm. Column 1 is packed with Supelcoport (100/120 mesh) coated with 3% SP-1000. Column 2 is packed with 2,6-diphenylene oxide polymer (Tenax-GC 60/80 mesh).

Chlorinated hydrocarbons are also analyzed by packed-column GC with column 1 described previously or a 1.8-m × 2-mm glass column packed with 1.5% OV-1/2.4% OV-225 on Supelcoport (80/100 mesh). The ECD is the detector listed in Method 8120.

Method 8121 is a capillary column procedure for the determination of chlorinated hydrocarbons, which has been proposed for addition to SW-846. Sample extraction and cleanup procedures are the same as those outlined in method 8120. The recommended capillary columns are a DB-5 (1.5-μm film thickness), a DB-1701 (1.0-μm film thickness), a DB-WAX, and a DB-210 (no film thicknesses designated), all 30 m × 0.53 mm.

14.7.9 Analysis for Volatile and Semivolatile Compounds by GC/MS

The analyses described in the previous sections for the various classes of chemical compounds can also be performed by GC/MS techniques. The sample preparation procedures and the chromatographic conditions for the GC/MS methods are very similar to the methods already described. The major differences with these methods is with the mass spectrometer and the information it can provide to the analyst.

The most significant advantage provided by the mass spectrometer is the qualitative data it provides in the form of a mass spectrum for every detected peak. This provides much more conclusive data on compound identification than those obtained by a retention time match. Spectra that are not similar to any of the compounds included in the calibration standards can be compared to library spectra to tentatively identify compounds. Compounds that elute at similar retention times can also be detected and measured simultaneously if their spectra are not similar.

The EPA-GC/MS methods listed in the following section are, in general, less sensitive than the previously listed EPA methods for the specific compound classes. Selected ion monitoring (SIM) techniques can be used to enhance sensitivity for a GC/MS analysis, but this approach will limit the analysis to a certain class (e.g., chlorinated dioxins and furans, PCBs, or PAHs) or number of compounds.

EPA Methods for Volatile Compounds by GC/MS. Methods 8260 (soils, waters, and waste material) and 524.2 (drinking water and groundwater) are GC/MS methods for the determination of volatile compounds. The purge-and-trap procedure, the chromatographic analysis, and the compounds determined by these methods are similar to Methods 8021 and 502.2 described in Section 14.7.2. Analyte quantification is performed by internal standardization. The GC/MS methods provide the option for a larger aliquot for aqueous samples (25 mL rather than 5 mL) if greater sensitivity is required. A narrow-bore 30-m × 0.32-mm DB-5 fused-silica capillary column (1-μm film thickness) is listed in the GC/MS methods in addition to the wide-bore VOCOL and DB-624 columns.

The narrow-bore column requires the use of a capillary precolumn interface that is cryogenically cooled. The narrow-bore column cannot handle the high carrier-gas flowrate required for efficient desorption of the analytes from the trap. The device is located between the trap and the chromatographic column. Analytes desorbed from the trap are condensed on a short piece of uncoated fused-silica capillary tubing which is maintained at −150°C with liquid nitrogen. The tubing is then heated to 250°C in 15 sec or less to transfer the analytes to the column.

EPA Methods for Semivolatile Compounds by GC/MS. All classes of semivolatile compounds (including those in Sections 14.7.4 through 14.7.10) that are amenable to conventional extraction processes and gas chromatographic analysis can be determined by GC/MS techniques. Two GC/MS methods for semivolatiles are Methods 8270 (soils, waters, and wastes) and 525.1 (drinking water and groundwater). Each method recommends a 30-m × 0.25-mm DB-5 fused-silica capillary column for the separation of analytes. The mass spectrometer is operated in the electron impact ionization mode for analyte identification and quantification.

The Method 8270 extraction of aqueous samples is performed by separatory funnel or continuous liquid–liquid extraction at two pH values to extract the acidic, basic, and neutral analytes. The sample pH is initially adjusted to <2, and extraction is performed with methylene chloride. Following the acid extraction, the pH is adjusted to >11, and the sample is extracted again with methylene chloride. The extracts are then concentrated to a volume of 1.0 mL for analysis. The acid and base extracts for a particular sample can be combined prior to the concentration step, concentrated and combined immediately before analysis, or analyzed separately if desired. Soil and sediment extracts are extracted by the Soxhlet technique or the ultrasonic technique using 1:1 methylene chloride: acetone or methylene chloride. These extracts are concentrated to a final volume of 1.0 mL for analysis.

Cleanup options for the entire range of compounds in Method 8270 are limited, since there are multiple classes of compounds with varying characteristics. Gel permeation chromatography can be used as a general cleanup if all analytes are to be determined. When a single class of compounds must be determined, the various cleanup techniques described in the previous sections can be employed.

The only extraction technique provided in Method 525.1 is the solid-phase extraction technique. The procedure is performed using a C_{18} phase contained in either a cartridge or a 47-mm filter disk. Extraction is performed at a $pH < 2$, with analyte elution by methylene chloride. Extracts are concentrated to 1.0 mL for analysis.

Both methods use isotopically labeled internal standards for analyte quantification. The Method 8270 internal standards are added directly to the extract just before the analysis. The internal standards for Method 525.1 are added to the sample before extraction.

14.7.10 Polychlorinated Dibenzo-*p*-Dioxins and Polychlorinated Dibenzofurans

Dioxin is probably the term that is most readily recognized by the general public as an environmental pollutant. The dioxins that are of concern are really the class of compounds known as the chlorodibenzodioxins. They are inadvertently generated during the manufacture of chlorinated aromatic compounds.

The base structure of the chlorodibenzodioxins is dibenzodioxin, and it can accept from one to eight chlorine atoms in various substitution patterns. Toxicological data indicate that the tetrachloro isomers are perhaps the most dangerous of the entire group (54). One specific tetra isomer, 2,3,7,8-tetrachlorodibenzodioxin (TCDD) appears more frequently than the other isomers and therefore receives the most attention. The well known "Agent

Orange," which was widely used by the military in Vietnam, contains TCDD as a contaminant.

The dibenzofurans or chlorinated dibenzofurans are related to the dioxins. Structurally, they differ only in the atoms that form the bridge between the two benzo structures: Furans are joined by one oxygen and one carbon atom, while dioxins are joined by two oxygen atoms. The furans can also accept from one to eight chlorine atoms as substituents. Biologically they can interact in a manner that is similar to the dioxins, but they are believed to be less toxic. They are found as trace impurities in PCB formulations.

EPA Methods for Dioxins and Furans. The tetra through octachlorinated dibenzo-*p*-dioxin and dibenzofuran isomers can be determined by Method 8280. Matrix specific extraction and cleanup procedures are provided in the method. The analysis is performed by capillary column GC with low resolution mass spectrometry. The mass spectrometer is operated in the electron impact ionization mode with selected ion monitoring. Instrument operating parameters must be adjusted to provide maximum resolution of the 2,3,7,8-tetrachlorodibenzo-*p*-dioxin from the other tetrachlorinated isomers. Three fused-silica capillary columns are recommended in the method: a 50-m CP-Sil-88, a 30-m × 0.25-mm DB-5 (0.25-μm film thickness), and a 30-m SP-2250.

Aqueous sludge samples are refluxed with 50-mL of toluene in a device fitted with a Dean–Stark separator for the removal of water. The solvent is then evaporated to near dryness, reconstituted with hexane, transferred to a separatory funnel, and washed with a 5% NaCl solution. Still bottom samples are simply mixed with toluene, filtered, concentrated, and reconstituted with hexane, and then washed with the NaCl solution. A fly-ash sample is mixed with anhydrous sodium sulfate and Soxhlet extracted with toluene for 16 h. The extract is then handled in the same manner as the sludge and still bottom samples.

Soil samples are combined with anhydrous sodium sulfate and then mixed with a 1:4 solution of methanol:petroleum ether on a wrist-action shaker for 2 h. The soil extract is filtered and exchanged into hexane. Aqueous samples are extracted with methylene chloride (with no pH adjustment) using either a separatory funnel or a continuous liquid–liquid extractor. The extract is then dried, concentrated, and exchanged into hexane. The hexane extracts obtained from all sample matrices are then carried through the following series of washes and column cleanups:

- Wash with 20% KOH solution
- Wash with 5% NaCl solution
- Wash with concentrated H_2SO_4
- Wash with 5% NaCl and concentrate to 2 mL
- Alumina column cleanup

- Carbon/silica gel column cleanup

The final elution solvent for the carbon/silica gel column, toluene, is concentrated to a final volume of 100 μL for soil samples and 500 μL for all other matrices to provide maximum sensitivity.

Quantification of the analytes is performed by the internal standard technique. The internal standards are isotopically labeled analytes which are added to the sample prior to the extraction process.

Drinking water and groundwater samples can be analyzed for 2,3,7,8-tetrachlorodibenzo-*p*-dioxin by Method 513. Instrumental analysis is performed by capillary column GC with high-resolution mass spectrometry and electron impact ionization. The aqueous samples are extracted as in Method 8280 or by a 47-mm C_{18} extraction disk with elution of analytes by a benzene wash. Extracts are processed through a series of silica gel, alumina, and carbon column cleanups and concentrated to a volume of 500 μL for analysis.

The chromatographic column selected for the analysis must separate the 2,3,7,8 dioxin isomer from all other tetraclorodioxin isomers. Recommended columns are the previously referenced CP-Sil 88 column and a 60-m SP-2330. Isotopically labeled analogues of the 2,3,7,8 dioxin isomer added prior to the sample extraction are used as internal standards for quantification of the target analyte.

Alternative Methods for Chlorinated Dioxins and Furans. Bicking and Wilson described a high-performance size exclusion chromatography (HPSEC) process as a cleanup procedure for chlorinated dioxins and furans (55). Two different HPSEC columns were connected in series, a 100-Å pore size followed by a 50-Å pore size. Elution patterns were monitored with a diode array detector. Motor oil and sediment extracts were fractionated with chloroform and recoveries for chlorinated dioxins and furans were quantitative. Dioxin/furan-contaminated fat extracts, which were fractionated with THF, yielded lower recovery data. A secondary sulfonic acid ion exchange cleanup was also utilized and, when combined with the HPSEC, yielded extracts that were free of interference when analyzed by GC/MS.

14.7.11 Analysis of Airborne Pollutants

Many of the same compounds that must be determined in soil and water samples must also be determined in air samples. The major procedural difference required for the analysis of air samples is with the collection of the sample itself. The gas chromatographic portion of the analysis for a given class or group of compounds is essentially the same as that which is described for the analysis of extracts prepared from soil and water samples.

EPA Methods for Air Samples. Samples required to monitor the stack emissions associated with various industrial processes can be collected by SW-846 Methods 0010 and 0030 (7), which were described in Section 14.6. Method 0010, which is designed to collect semivolatile analytes, requires a combination of filters, XAD resin, solvent rinses, and aqueous impinger solutions for sampling. Figure 14.2A illustrates the components and configuration of the Method 0010 sampling train. The filters and resins are extracted by the Soxhlet technique, with the solvent rinse from the probe usually added to the corresponding filter sample. The aqueous impinger samples can be extracted by separatory funnel or a continuous liquid–liquid extractor. All extracts can then be processed through the required cleanups and concentrated to the appropriate volume for analysis. Analysis can be performed by GC/MS using Method 8270 or by an individual 8000 series GC method designed for a specific class of compound.

Method 0030 is designed to collect the more volatile analytes which are not retained effectively by the Method 0010 sampling technique. The major sampling components are sorbent beds consisting of Tenax and charcoal and a condensate sample. The Method 0030 sampling train components are illustrated in Figure 14.2B. Prior to analysis, the analytes are thermally desorbed from the sorbent bed using a procedure that is outlined in SW-846 Method 5040 (packed column) and 5041 (capillary column). The sorbent bed is maintained at a temperature of 180°C while an inert gas flow is directed through the heated sorbent bed and then directed to a purge-and-trap device. The inert gas flow exiting from the sorbent bed is bubbled through 5 mL of water in the purge-and-trap device with the remainder of the analysis proceeding as with a normal purge-and-trap procedure. The analytes are subsequently determined by one of the 8000 series volatiles methods, usually a GC/MS technique. The aqueous condensate samples are analyzed by the same purge-and-trap procedures routinely used for water samples.

Additional EPA methods for the determination of toxic organic compounds in ambient air by gas chromatographic techniques are summarized briefly in Table 14.8. This group of methods is commonly identified as the TO methods.

Alternative Methods for Air Samples. Numerous articles outlining alternative techniques for the determination of air pollutants by GC are proved in the literature. Several references are provided in Table 14.9 with comments on gas chromatographic detectors and collection/concentration techniques.

14.8 CHEMICAL STANDARDS

Chemical standards are required for both instrument calibration and the preparation of sample spikes and method control samples. They are an

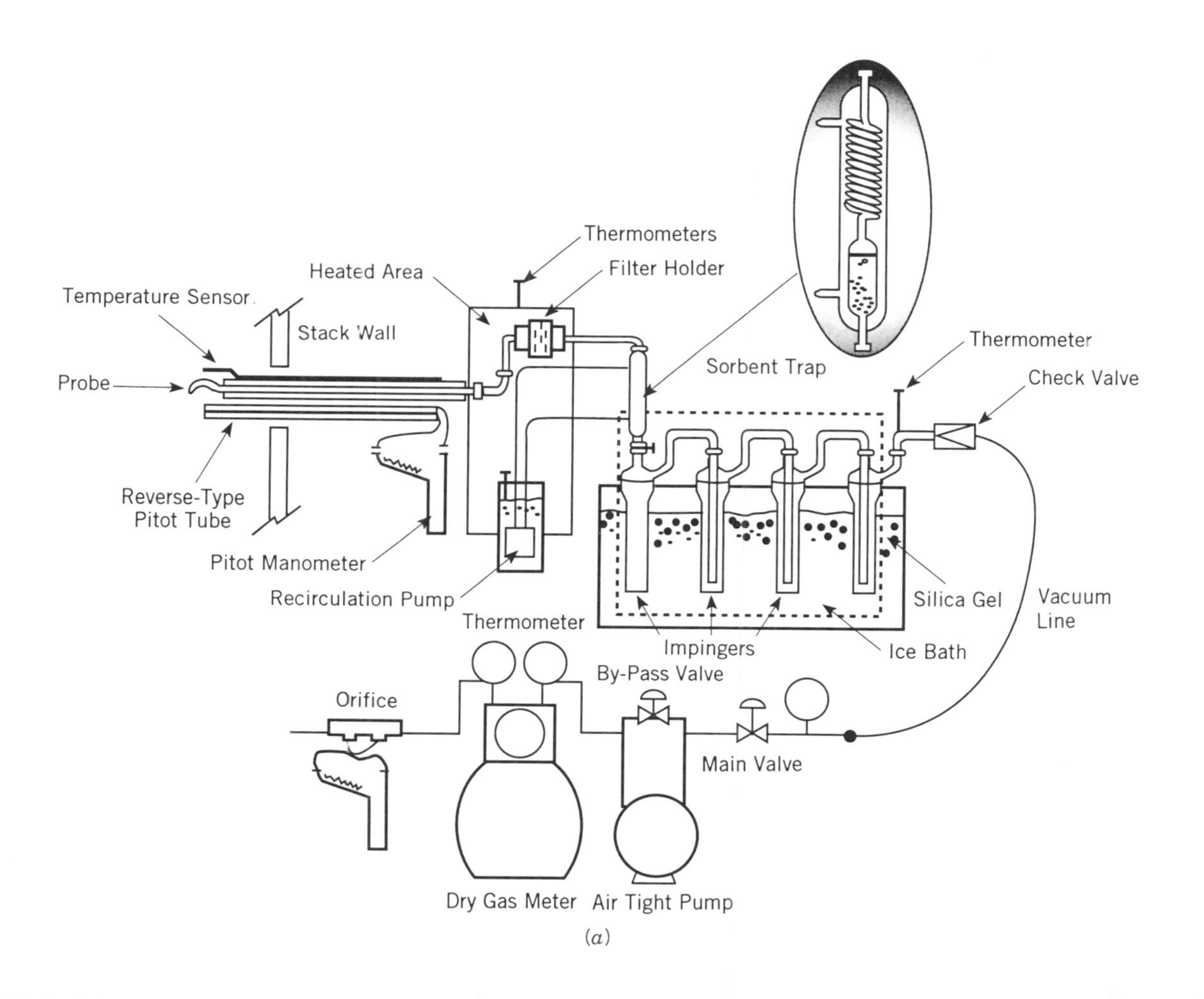
Thermometers
Heated Area
Filter Holder
Temperature Sensor
Stack Wall
Probe
Sorbent Trap
Thermometer
Check Valve
Reverse-Type Pitot Tube
Pitot Manometer
Recirculation Pump
Thermometer
Silica Gel
Vacuum Line
Impingers
Ice Bath
By-Pass Valve
Orifice
Main Valve
Dry Gas Meter
Air Tight Pump

(a)

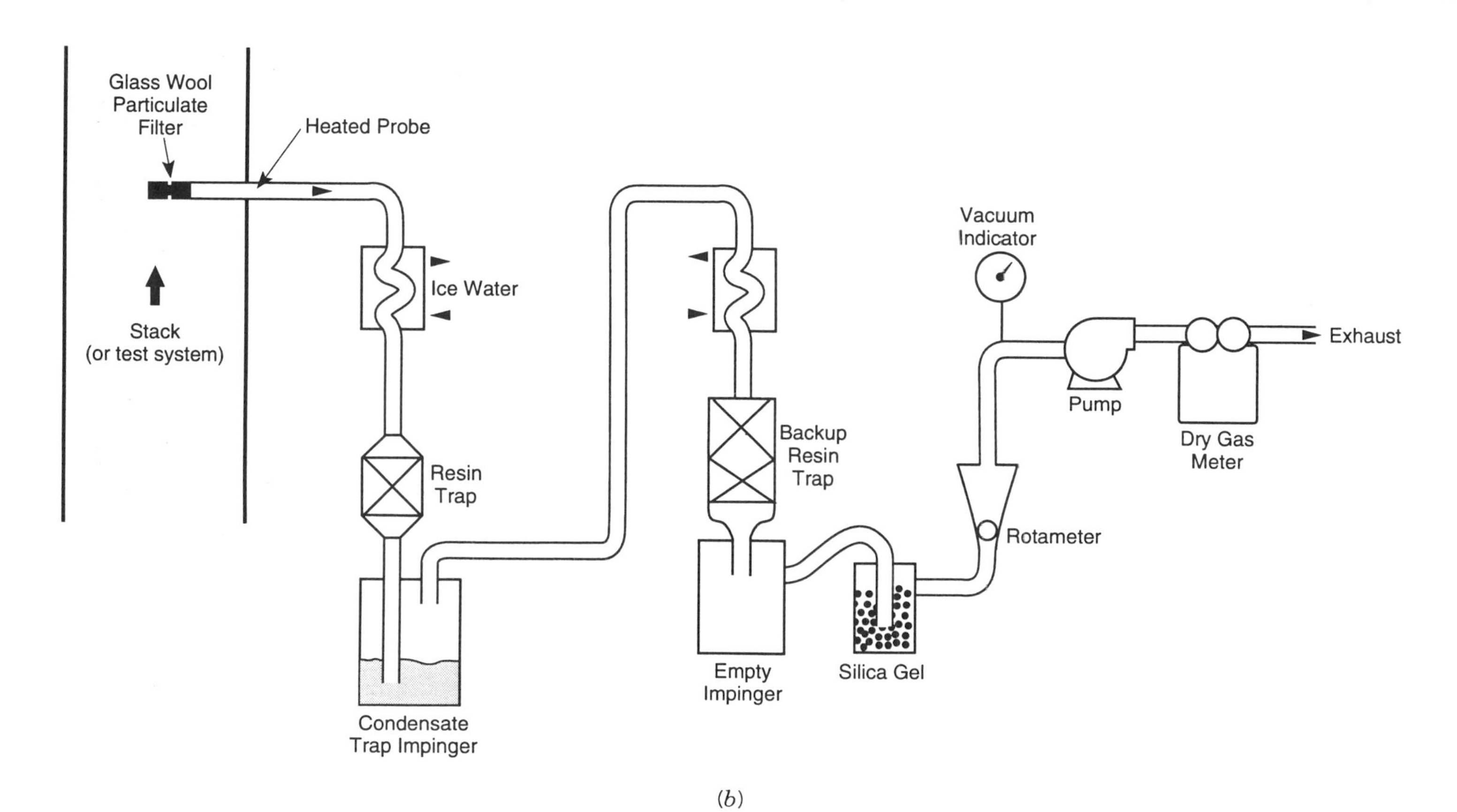

(*b*)

FIGURE 14.2 (a) Method 0010 sampling train; (b) Method 0030 sampling train. (From Reference 7.)

TABLE 14.8 TO Methods for the Determination of Air Toxics in Ambient Air

Method Number	Method Description
TO-1	Determination of Volatile Organic Compounds in Ambient Air Using Tenax Adsorption and GC/MS
TO-2	Determination of Volatile Organic Compounds in Ambient Air by Carbon Molecular Sieve Adsorption and Gas Chromatography (GC/MS)
TO-3	Determination of Volatile Organic Compounds in Ambient Air Using Cryogenic Preconcentration Techniques and Gas Chromatography with FID and ECD
TO-4	Determination of Organochlorine Pesticides and PCBs in Ambient Air
TO-7	Determination of *N*-Nitrosodimethylamine in Ambient Air Using Gas Chromatography
TO-9	Determination of Polychlorinated Dibenzo-*p*-Dioxins (PCDDs) in Ambient Air Using High-Resolution GC/High-Resolution MS
TO-10	Determination of Organochlorine Pesticides in Ambient Air Using Low-Volume Polyurethane Foam (PUF) Sampling with GC/ECD
TO-13	Determination of Polynuclear Aromatic Hydrocarbons in Ambient Air Using High-Volume Sampling with GC/MS and HPLC Analysis
TO-14	Determination of Volatile Organic Compounds in Ambient Air Using SUMMA Polished Canister Sampling and Gas Chromatographic Analysis

Source. Adapted from Reference 27.

integral part of the analysis because they provide the data required for the identification and quantification of the chemical contaminants. Many chemical standards are readily available from any of a number of chemical supply houses as either a neat material or as a solution prepared at a specified concentration. For nonroutine analysis, it is always important to determine if the standard materials are available since vendors do not usually stock materials that are not in great demand.

14.8.1 Second Source Verification

Verification of a chemical standard against a second source is a sound practice for both technical and legal purposes. It provides confirmation of compound identity, stability, and purity. At a minimum, the verification should be performed at the time that the working standard is prepared, but a more frequent check will provide assurances that the working standards are still valid.

TABLE 14.9 Additional Reference—Air Sample Analysis

Compound	Detector	Reference	Comments
CO	Flame ionization	56	Converted to CH_4
Atmospheric NH_3	Flame thermionic	57	
N_2O	Flame photometric	58	
N_2O	Electron capture	59	
Alkyl nitrate Peroxypropionyl nitrate Peroxyacetyl nitrate	Electron capture	60, 61	
$(CH_3)_2SO_4$	Flame photometric	62	Florisil adsorption tube collection
H_2S, $(CH_3)_2S$	Flame photometric	63	
CS_2	Electron capture	64	
CH_2O	Flame ionization	65	Chemically modified XAD-2 collection
C_2–C_5 aldehydes	FID, ECD, FPD, FTD	66	Liquid O_2 cold trap collection
Carboxylic acids	Mass spectrometer	67	Ion exchange resin collection
CH_3I	Electron capture	68	
C_1–C_2 halocarbons	Electron capture	69	Stainless steel cannister collection
Organobromines	Mass spectrometer	70	Tenax-GC trap
C_4–C_{14} hydrocarbons	FID and MS	71	Carbon adsorption trap
C_3–C_5 hydrocarbons	FID and PID	72	

The second source material may be a certified reference standard or a material that has been compared with a certified reference material. When a certified material is not available, the analyst should attempt to obtain the chemical standard from two separate sources. If the material is available from only one source, the purity and composition should be confirmed by appropriate analytical techniques. Two separate stock standards should be prepared independently by separate analysts for the verification.

The comparison or verification of standards is performed under the conditions used for the calibration of the instrument. A qualitative comparison is provided by the retention time data and any other information (i.e., spectra) provided by the various detectors. Quantitative comparison is provided by the responses provided by the two standard materials.

14.8.2 Documentation of Standard Preparation

The proper documentation of standard preparation provides a permanent record of the source of all standard materials. This should include the organization or vendor that supplied the material, the lot number of the material, and any information that documents purity and identity. The documentation system should also include a labeling convention that uniquely identifies each solution used in the standard preparation. This system will provide a means of tracing all standard components back to their original source material. The documentation system can also provide a place to record other pertinent standard preparation information. This would include expiration dates, solvents, and any special requirements for the dissolution of materials.

14.9 QUALITY ASSURANCE AND QUALITY CONTROL

The concepts of quality assurance (QA) and quality control (QC) are important components of any organization. In general terms, QA and QC can be described as a system that is developed and implemented to ensure that an organization's product meets a certain standard or level of quality. Techniques in analytical chemistry have always played a major role in this endeavor, particularly in manufacturing operations where chemical analyses are used to monitor the quality of incoming raw materials, intermediate production steps and the final product. In a laboratory performing environmental analyses, the data or analytical results are the final product, and an appropriate QA/QC program provides concrete guidelines for the generation and acceptability of this product.

The quality of a product, service, or analysis must not be confused with the complexity of the product. As an example, two environmental labs will most likely perform pH measurements as part of their services. In addition to pH measurements, one of the labs may also perform more technically complicated analyses that the other lab is not equipped to perform. This does not mean that one lab is of a higher quality than the other lab. The quality is reflected in the caliber of the measurements, on their own, and in the context of the requirements for a particular project.

Quality assurance and quality control are frequently considered as a single entity, but they can be discussed as separate elements. The Quality Assurance Program of the U.S. Army Toxic and Hazardous Materials Agency (73) provides a concise explanation of QA and QC as distinct operations:

> Quality Assurance refers to the system whereby an organization provides assurance that monitoring of quality-related activities has occurred. Quality Control refers to specific actions taken to ensure that system performance is consistent with established limits . . . . Implementation of the QA Program in the laboratory is designed to ensure that data are collected under in-control conditions rather than simply to ensure documentation of poorly conducted analyses.

14.9.1 Quality Assurance

Quality assurance is probably best classified as a management function in the sense that it provides direction and guidance for the proper completion of the analysis. To be effective, QA programs must be developed and maintained through a continuous interaction of those individuals charged with the QA responsibilities and the analysts at the bench level. This will lead to meaningful QA activities and realistic quality goals for the analysts and their work.

Quality Assurance Plans. The specific guidelines for a QA program or system are documented within a QA plan, or QAP. The plan will include a general discussion of the QA measures associated with the sample analysis. The various components of the QC system are also outlined within the plan for objective performance evaluations of analytical methods.

Quality assurance plans are developed from both a project perspective and a laboratory operations perspective. Project-specific QAPs, prepared during the initial planning process, outline the analytical requirements for a project. Objectives for the data must be defined in terms of realistic and meaningful goals during this phase. Once the objectives are defined, appropriate sampling techniques and analytical methods must be identified. Method selection obviously involves finding the proper analytical technique for the measurement of the contaminants of interest. But it also involves the selection of techniques suitable for the sample matrices under investigation.

The laboratory QAP describes the QA program that is in place within a laboratory. This laboratory QAP is available for evaluation by any party obtaining analytical results from the laboratory. Any discrepancies between a project QAP and lab QAP must be resolved and documented prior to the initiation of the analytical phase of the project.

Data Quality Indicators. Five parameters are evaluated in order to obtain a tangible measure of data quality. These five parameters—precision, accuracy, representativeness, completeness, and comparability—are usually identified as the PARCC parameters.

Quatifiable values and/or descriptive values for each of these parameters can be translated into a meaningful measurement of performance. In the context of a QAP, the parameters can be used to establish data quality objectives (DQOs) for specific projects or to define QA objectives for laboratory performance.

Table 14.10 provides a brief description of each of the PARCC parameters. When evaluating sample results and the corresponding quality indicators, it is important to realize that the data can be influenced by sample collection techniques, laboratory proficiency, and unexpected interferences inherent to the sample matrix. Judicious application of laboratory

TABLE 14.10 Data Quality Indicators

Data Quality Indicator	Description
Precision	An indication of the variability of a series of measurements. Determined by the analysis of duplicate samples collected in the field or through the preparation and analysis of two separate aliquots of a single sample within the laboratory. Also determined through the addition of target compound spikes to duplicate samples. Usually expressed as a relative percent difference for two measurements and a standard deviation or relative standard deviation for three or more measurements.
Accuracy	A measure of the bias of sample results. Determined through the analysis of artificially prepared samples containing known levels of analytes or through the addition of known amounts of target compounds to site samples. Expressed as a Percent Recovery.
Representativeness	Describes how effectively sample data reflect the desired characteristics of a site. This is dependent on the proficiency of the analysis (i.e., how well the data describe or represent the samples) and the design and implementation of the sampling plan (i.e., how well the proposed set of samples and accompanying data represent the site characteristics). Representativeness, as it pertains to laboratory work, is expressed through the accuracy, precision, and completeness parameters.
Completeness	Represents the number of measurements out of the total that are deemed valid. Valid measurement implies that results were obtained within specified method criteria and that the sample matrix had no adverse effects upon the analysis. Expressed as a percentage of the total number of analyses.
Comparability	Describes the level of confidence with which various data sets can be compared. An appropriate level of comparability for data obtained from different sources is achieved through the proficient use of standard methods of analysis

and field quality control measures will usually provide the information required to isolate sampling errors from laboratory errors.

Additional Quality Assurance Components. Some additional aspects of QA are the fundamental practices that are associated with any type of analysis. The first of these would be the thorough documentation of all aspects of the analysis. This would begin with the standard operating procedures (SOPs) or documented methods that were used for the sample analysis. Appropriate measures must also be available for the management of samples. These management practices must ensure that samples are labeled and stored properly and analyzed within the prescribed period of time. Laboratory notebooks, raw data (i.e., chromatograms) and other pertinent records of the sample preparation and analysis must provide all information required to follow the sample history within the laboratory.

The data reporting format is another facet of the analysis that is influenced by the QA program. The results must be presented in a manner that accurately reflects the capabilities of the method. The primary concern is that the numbers that are reported actually reflect the level of confidence that is attained by the measurement. This takes on added importance in light of the fact that results from the analysis of environmental samples may be interpreted by individuals unaccustomed to evaluating technical data. An earlier publication on QA for environmental analysis (74) presents just such a dilemma: "Lawyers usually attempt to dispense with uncertainty and try to obtain unequivocal statements; . . . a value of 1.001 without a specified uncertainty, for example, may be viewed as legally exceeding a permissible level of 1."

14.9.2 Quality Control

The QC system is established as a means to control errors and generate reproducible results for the laboratory analysis. There are various QC checks designed to for this purpose, and they were implemented at various stages of the analysis. The data obtained from these check parameters can be evaluated by any of a number of mathematical calculations to provide an indication of instrument stability and/or method performance.

Method Blanks. Various types of external contaminants can be introduced into any sample during its collection, preparation, or analysis. Non-target-compound contaminants can elute in the characteristic retention time region of an analyte of interest and create an interference that renders the analysis useless. Cross contamination can also occur whereby a "clean" sample is contaminated with target compounds from another sample that contains high levels of those target compounds. Laboratory or method blanks, prepared in conjunction with a set of samples, provide the analyst with a means to monitor and isolate contamination that may be introduced during the sample preparation and analysis.

The method blanks are generated at a predefined rate, like 1 per day or 1 per every 20 samples. For aqueous samples, a measured volume of "clean" laboratory water equivalent to the typical sample volume is extracted as a blank. For soil or sediment samples, all chemicals and reagents used for the preparation are carried through the procedure. If a clean soil or sediment matrix is available, a sample of this material can be substituted as a blank. The blank accompanies the samples through the entire preparation and analysis. Upon completion of the analysis, final results are calculated and reported for the blank in the same way that they are reported for the samples. Should a target compound or artifact be identified in the blank, the potential impact on the sample data must be determined.

Matrix Spike Samples. Method performance data for a specific site matrix is obtained from target compound spikes added directly to a sample of the site matrix (the term "spike" refers to the act of spiking or dosing a sample with a compound or set of compounds). The matrix spike samples are carried through the entire extraction and analysis along with an unspiked portion of the same sample. Upon completion of the analysis, recovery data can be determined for the spike compounds. Results must be obtained for an unspiked portion of the sample as well, because the analyst must correct the matrix spike sample results for the background levels in the unspiked sample.

The recovery data obtained from the matrix spike samples is particularly important for the evaluation of method performance. Variations in the makeup of the sample matrix can have a profound effect on the efficiency with which certain compounds or classes of compounds are extracted from a particular sample type. Matrix spikes are performed on a predetermined percentage of samples collected at a site, usually 1 for every 20 samples collected. If all samples are obtained within close proximity, then all should exhibit the same matrix effects as the matrix spike sample.

Surrogates. A surrogate is a non-target compound that is added to every sample at a predefined concentration prior to the preparation and analysis. Upon completion of the analysis, recoveries are determined for the surrogates to provide an evaluation of method performance for each sample.

The compound selected as a surrogate must be readily detectable by the method and clearly separated from all other target compounds required for the analysis, and it should not be expected in the samples. Chemically, it should behave similarly to the target compounds. When more than one class of compounds is being determined, surrogates should be selected to represent each class of compounds.

Methods that use a mass spectrometer as a detector allow for a rather novel approach with surrogates—the use of isotopically labeled analogues of target compounds. These labeled compounds are identical to their non-labeled counterparts in terms of their chemical behavior, yet each can be detected and measured independently of the other with a mass spectrometer.

Duplicates. Method precision data for a specific sample group are obtained through the preparation and analysis of two separate aliquots of the same sample. Quite frequently, method precision data are obtained by the analysis of duplicate spike samples, which are identified as a matrix spike (MS) and matrix spike duplicate (MSD). This approach is useful when there are no target compounds anticipated in samples.

The precision obtained for a given sample group can be expressed in terms of the relative percent difference (RPD) by Equation 14.8:

$$\mathrm{RPD_A} = \frac{\mathrm{Conc_A} - \mathrm{Conc_{ADup}}}{\mathrm{Avg\ Conc_A}} \times 100\%$$

where $\mathrm{RPD_A}$ = RPD of compound A

$\mathrm{Conc_A}$ = concentration of compound A in sample

$\mathrm{Conc_{ADup}}$ = concentration of compound A in duplicate

$$\mathrm{Avg\ Conc_A} = \frac{\mathrm{Conc_A} + \mathrm{Conc_{ADup}}}{2}$$

Alternatively, the precision can also be expressed as the range or difference of the two results. With this approach, comparisons between multiple samples are valid only if the spike concentrations are identical.

Control Samples. A control sample is a synthetic laboratory matrix that is fortified with known levels of target compounds. This type of sample will allow for the evaluation of method performance in the absence of uncontrolled matrix effects created by site samples. The only influences on the recoveries will be the technique of the analyst, instrument performance, the integrity of the calibration and spike standards, and the effectiveness of the method.

The control samples can be generated through the addition of the target compound spike to an identical set of the reagents and chemicals used for the method blank. A thoroughly evaluated substitute matrix can also be utilized. Sand that has been pre-treated by baking at 700°C is a popular substitute for soils or sediments. ASTM Type II water, which contains 100 mg/L of both sulfate and chloride, can serve as an aqueous substitute.

Instrument Calibration. The control parameters pertaining to the calibration of a gas chromatograph focus on two areas: the linear response range for the detector and the verification of response stability during the analysis. The linear response range is established during the initial calibration through the analysis of standard solutions prepared at multiple concentration levels for each of the compounds of interest.

The data obtained from the initial calibration can be evaluated by one of a number of statistical approaches to verify that the detector provides acceptable response characteristics. One technique is to determine the % relative standard deviation (% RSD) of the response factors obtained for the multilevel standards. Smaller numerical values for the RSD indicate a more consistent detector response throughout the calibration range than larger RSDs. A common upper limit is 20% RSD for an acceptable calibration. Alternatively, the calibration can be evaluated by the method of least squares with the calculation of a correlation coefficient. A minimum value for the correlation coefficient can be established for an acceptable calibration.

Once an analytical sequence has been initiated, the stability of the detector response (relative to the calibration) must be monitored. Continuing calibration standards, which contain either all of the calibration compounds or a representative subset, are analyzed at prescribed intervals. The component responses must fall within a set limit, usually ±15% of that obtained from the calibration.

Retention Time Windows. Retention time stability for an instrument can be monitored by means of a retention time window (RTW). This is a predefined increment of time that brackets the established retention time by a set interval. This window defines the expected amount of variability that should be encountered with a compound in the ensuing analysis. The windows are also used by the analyst as a guide for compound identification.

The RTWs can be established experimentally for each column/instrument system on which an analysis will be performed. Component retention time data obtained from the analysis of at least three standards on the respective instrument are tabulated. The standard deviation (SD) of the retention times is determined, and the RTW is then defined as $\pm(3 \times \mathrm{SD})$.

The individual component retention time data can be obtained either from standards analyzed at time intervals that approximate the intervals at which standards will be analyzed during a run sequence or from the multilevel standards analyzed during the initial calibration.

Performance Evaluation Standards. Certain instrument performance checks may be required for specific compounds or classes of compounds that are sensitive to variations in the chromatographic system. The performance evaluation standards contain a representative number of compounds that serve as an indicator of acceptable instrument performance. Some of the commonly evaluated operating parameters include the following:

1. Resolution of two closely eluting compounds
2. The column efficiency (HETP)

3. Degradation of unstable compounds
4. Degree of tailing encountered with polar compounds
5. Minimum detector response

A method might require that any number of these or other parameters be evaluated prior to the calibration and during the analysis. The minimum/maximum limits must also be outlined.

Control Charts. The control chart provides a graphical depiction of data that is arranged in chronological order. It is an extremely useful tool for the evaluation of method proficiency for analytes, verification of results obtained for method QC indicators, and the identification of trends or biases that may indicate potential problems with the analysis. Graphic control charts are most frequently used to monitor recovery data for spikes and surrogates as well as reproducibility as either an RPD or a range. Figure 14.3 provides an example of a control chart.

The fundamental features of the chart or graph are the warning and control limits for the particular parameter as well as its average value, as calculated from historical data. The warning limits are $\pm(2 \times SD)$ of the data points, centered around the average value. Experimental values are expected to fall within these warning limits 95% of the time. The control limits are $\pm(3 \times SD)$ of the data points, which is the 99.7% confidence limits. Data points that fall outside of the warning limits but within the control limits provide the analyst with a warning of possible problems. Data points that appear outside of the control limits represent an out-of-control event or measurement, and require investigation. Another control chart

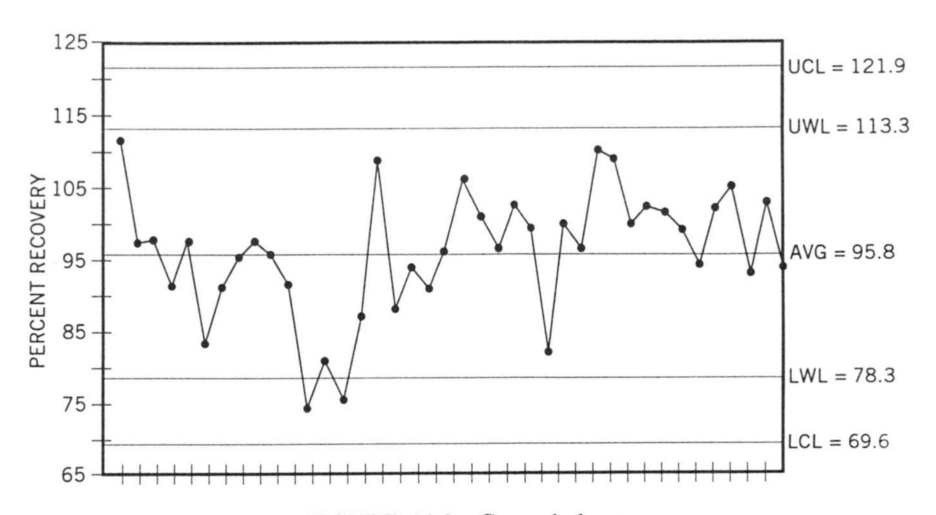

FIGURE 14.3 Control chart.

feature that can be evaluated is trends consisting of multiple points (usually seven or more in a row) that appear on one side of the average.

REFERENCES

1. N. Stoloff, *Regulating The Environment: An Overview of Federal Environmental Law*, Oceana, Dobbs Ferry, NY, 1991.
2. A. G. Levine, *Love Canal: Science, Politics and People*, Heath, Lexington, MA, 1982.
3. A. deGrazia, *Cloud Over Bhopal*, Lalit Dalal, Bombay, India, 1985.
4. A. T. James and A. J. P. Martin, *Biochem. J.*, **50**, 679 (1952).
5. *Cleaning Our Environment: A Chemical Perspective*, 2d ed., prepared by a task force of the ACS Committee on Environmental Improvement, American Chemical Society, Washington, DC, 1977.
6. ACS Committee on Environmental Improvements, *Anal. Chem.*, **55**, 2210 (1983).
7. *Test Methods for Evaluating Solid Waste:Physical/Chemical Methods*, EPA-SW-846, 3d ed., U.S. Environmental Protection Agency, Washington, DC, 1992.
8. D. H. Henning and W. R. Mangun, *Managing the Environmental Crisis*, Duke University Press, Durham, NC, 1989.
9. G. R. Umbreit, in *Trace Analysis by Gas Chromatography*, R. L. Grob (ed.), *Modern Practice of Gas Chromatography*, 2d ed., Wiley–Interscience, New York, 1985, p. 425.
10. G. D. Christian, *Analytical Chemistry*, 4th ed., Wiley, New York, 1986, p. 428.
11. J. W. Rhoades and C. P. Nulton, *J. Environ. Sci. Health, Part A, Environ. Sci. Eng.*, **A15**, 467 (1980).
12. J. Kirkland, *Analyst*, **99**, 859 (1974).
13. W. A. Saner, J. R. Jadamec, R. W. Sager, and T. J. Killeen, *Anal. Chem.*, **51**, 2180 (1979).
14. D. F. Hagen, C. G. Markell, G. A. Schmitt, and D. D. Blevins, *Anal. Chim. Acta*, **236**, 157 (1990).
15. A. Kraut-Vass and J. Thoma, *J. Chromatogr.*, **214**, 335 (1981).
16. S. E. Manahan, *Environmental Chemistry*, 5th ed., Lewis, Chelsea, MI, 1991.
17. P. Grathwohl, *Environ. Sci. Technol.*, **24**, 1687 (1990).
18. D. J. Hanson, *Chem. Eng. News*, **71** (13), 7, 1993.
19. G. L. Fisher, D. P. Y. Chang, and M. Brummer, *Science*, **192**, 553 (1976).
20. G. A. Junk and J. J. Richard, *Anal. Chem.*, **58**, 962 (1986).
21. R. M. M. Cooke, J. W. A. Lustenhouwer, K. Olie, and O. Hutzinger, *Anal. Chem.*, **53**, 461 (1981).
22. D. H. Freeman and L. S. Cheung, *Science*, **214**, 790 (1981).
23. J. D. Haddock, P. F. Landrum, and J. P. Glesy, *Anal. Chem.*, **55**, 1197 (1983).
24. P. J. A. Fowlie and T. L. Bulman, *Anal. Chem.*, **58**, 721 (1986).
25. S. B. Hawthorne, *Anal. Chem.*, **62**, 633A (1990).

26. L. Myer, J. Tehrani, C. Thrall, and M. Gurkin, *Supercritical Fluid Extraction (SFE): Advantages, Applications and Instrumentation for Sample Preparation*, Isco Applications Bulletin 69, 1991.
27. W. T. Winberry, Jr., *Environ. Lab*, **5**(3), 46 (1993).
28. W. T. Winberry, Jr., *Environ. Lab*, **5**(4), 52 (1993).
29. *Methods for the Determination of Organic Compounds in Drinking Water*, EPA/600/4-88/039, United States Environmental Protection Agency, December 1988 (revised July 1991).
30. D. Louch, S. Motlagh, and J. Pawliszyn, *Anal. Chem.*, **64**, 1187 (1992).
31. K. F. Pratt and J. Pawliszyn, *Anal. Chem.*, **64**, 2107 (1992).
32. M. E. Rosen and J. F. Pankow, *J. Chromatogr.*, **537**, 321 (1991).
33. J. Phillippaerts, C. Vanhoof, and E. F. Vansant, *Talanta*, **39**, 681 (1992).
34. D. F. Guka, S. M. Pyle, and R. Titus, *Anal. Chem.*, **64**, 1749 (1992).
35. H. Dingyvan and T. Jianfei, *Anal. Chem.*, **63**, 2078 (1991).
36. A. D. Hewitt, P. H. Miyares, D. C. Leggett, and T. F. Jenkins, *Environ. Sci. Technol.*, **26**, 1932 (1992).
37. C. D. McAuliffe, *Chem. Technol.*, **46**, 46 (1971).
38. B. Kolb, *Chromatographia*, **15**, 587 (1982).
39. A. Maggio, M. R. Milanai, M. Denaro, R. Feliciani, and L. Gramiccioni, *J. High Resolut. Chromatogr.*, **14**, 618 (1991).
40. M. R. Milana, A. Maggio, M. Denaro, R. Feliciani, and L. Gramiccioni, *J. Chromatogr.*, **552**, 205 (1991).
41. S. B. Hawthorne, D. J. Miller, D. E. Nivens, and D. C. White, *Anal. Chem.*, **64**, 405 (1992).
42. R. Ritsema, W. P. Cofino, P. C. M. Frintrop, and U. A. T. Brinkman, *Chemosphere*, **18**, 2161 (1989).
43. H. Ozawa and T. Tsukioka, *Anal. Chim. Acta*, **267**, 25 (1992).
44. M. Adolfsson-Erici and L. Renberg, *Chemosphere*, **23**, 845 (1991).
45. L. Chiang, R. J. Magee, and B. D. James, *Anal. Chim. Acta*, **255**, 187 (1991).
46. E. Benfenati, P. Tremolada, P. Chiappetta, L. Frassanito, R. Bassi, N. DiTorro, R. Fanelli, and G. Stella, *Chemosphere*, **21**, 1411 (1990).
47. M. S. Driscoll, J. P. Hassett, C. L. Fish, and S. Litten, *Environ. Sci. Technol.*, **25**, 1432 (1991).
48. F. I. Gnuska and K. A. Terry, *J. High Resolut. Chromatogr.*, **12**, 527 (1989).
49. F. R. Guenther, S. N. Chesler, and R. E. Rebbert, *J. High Resolut. Chromatogr.*, **12**, 821 (1989).
50. B. Larsew, R. Tilio, and S. Kapila, *Chemosphere*, **23**, 1077 (1991).
51. L. Zupancic-Kralj, J. Jan, and J. Marsel, *Chemosphere*, **23**, 841 (1991).
52. S. B. Hawthorne, D. J. Miller, and J. J. Langenfeld, *J. Chromatogr. Sci.*, **28**, 2 (1990).
53. L. Ostman, A. Bemgard, and A. Colmsjo, *J. High Resolut. Chromatogr.*, **15**, 437 (1992).
54. M. A. Ottoboni, *The Dose Makes the Poison*, Vincente Books, Berkley, CA, 1984.

55. M. K. Bicking and R. L. Wilson, *Chemosphere*, **22**, 437 (1991).
56. K. Lee, Y. Yanagisawa, M. Hishinuma, J. D. Spengler, and I. Billick, *Environ. Sci. Technol.*, **26**, 697 (1992).
57. N. Yamamoto, H. Nishiura, T. Honjo, and H. Inoue, *Anal. Sci.*, **7**, 1041 (1991).
58. G. Gassmann and S. Dahlke, *J. Chromatogr.*, **598**, 313 (1992).
59. O. Furukawa, *Nippon Kankyo Eisel Senta Soho*, **17**, 50 (1990).
60. K. P. Mueller, J. Rudolph, and K. Wohlfart, *Phys.-Chem. Behav. Atmos. Pollut.* [*Proc. Eur. Symp.*], **5**, 705 (1989).
61. G. Mineshos, N. Roumells, and S. Glavas, *J. Chromatogr.*, **541**, 99 (1991).
62. S. Fukui, M. Morishima, S. Ogawa, and Y. Hanazaki, *J. Chromatogr.*, **541**, 459 (1991).
63. L. Lukacovic and S. Vankova, *Petrochemia*, **30**, 154 (1990).
64. M. V. Russo, *Ann. Chim.* (*Rome*), **82**, 397 (1992).
65. C. Muntuta-Kinyanta and J. K. Hardy, *Talanta*, **38**, 1381 (1991).
66. M. Hoshika and G. Muto, *J. High Resolut. Chromatogr.*, **14**, 330 (1991).
67. S. Sollinger, K. Levsen, and M. Emmrich, *J. Chromatogr.*, **608**, 297 (1992).
68. M. Tsetsi, F. Petitet, P. Carlier, and G. Mouvier, *Phys.-Chem. Behav. Atmos. Pollut.* [*Proc. Eur. Symp.*], **5**, 32, (1989).
69. O. Furukawa and T. Nezu, *Nippon Kankyo Eisei Senta Shoho*, **16**, 60 (1989).
70. G. J. Sharp, Y. Yokouchi, and H. Akimoto, *Environ. Sci. Technol.*, **26**, 815 (1992).

Appendices

APPENDIX A

Effect of Detector Attenuation Change and Chart Speed on Peak Height, Peak Width, and Peak Area

I

1. Attenuation Change: Inverse proportional change in peak height h and peak area A; that is, increase in attenuation (sensitivity decrease) decreases peak height h and peak area A.

2. Chart Speed Change: Proportional change in peak width W and peak area A, i.e., increase in chart speed increases peak width W, and peak area A.

Example A:

Doubling Chart Speed

Initial conditions: $A_1 = h_1 \times W_1$

Final conditions: $A_2 = h_2 \times W_2$

Note: These conditions will be the same for all examples.

$h_1 = h_2$ but $W_2 = 2W_1$

$A_2 = h^2 \times 2W_1$

Thus, $A_2 = 2A_1$

Example B:

Halving Chart Speed

$h_1 = h_2$ but $W_2 = 0.5W_1$

$A_2 = h_2 \times 0.5W_1$

Thus, $A_2 = 0.5A_1$

Example C:

Doubling Attenuation (Halving Sensitivity)

$W_1 = W_2$ but $h_2 = 0.5h_1$

$A_2 = 0.5h_1 \times W_2$

Thus, $A_2 = 0.5A_1$

Example D:

Halving Attentuation (Doubling Sensitivity)

$W_1 = W_2$ but $h_2 = 2h_1$

$A_2 = 2h_1 \times W_2$

Thus, $A_2 = 2A_1$

Example E:

Doubling Chart Speed and Attenuation

$W_2 = 2W_1$ and $h_2 = 0.5h_1$

$A_2 = 0.5h_1 \times 2W_1 = h_1 \times W_1$

Thus, $A_1 = A_2$

Example F:

Halving Chart Speed and Attenuation

$W_2 = 0.5W_1$ and $h_2 = 2h_1$

$A_2 = 2h_1 \times 0.5W_1 = h_1 \times W_1$

Thus, $A_1 = A_2$

Example G:

Double Chart Speed and Halving Attenuation

$W_2 = 2W_1$ and $h_2 = 2h_1$

$A_2 = 2h_1 \times 2W_1 = 4h_1W_1$

Thus, $A_1 = 0.25A_2$

Example H:

Halving Chart Speed and Doubling Attenuation

$h_2 = 0.5h_1$ and $W_2 = 0.5W_1$

$A_2 = 0.5h_1 \times 0.5W_1 = 0.25h_1W_1$

Thus $A_1 = 4A_2$

II

1. In general, then, we know that
   a. Each increase in attenuation (decrease in sensitivity) *halves* the previous h or A, that is, 1, 0.5, 0.25, 0.125, 0.0625, and so on.
   b. Each decrease in attenuation (increase in sensitivity) *doubles* the previous h or A, that is, 1, 2, 4, 8, 16, and so forth.
   c. Each increase or decrease in chart speed causes a proportional change in W and A.
2. Let x = initial attenuation setting. If we *increase* the attenuation n times, the change in h or A will be

$$x(0.5)^n \times h \text{ or } A$$

If we *decrease* the attenuation n times, the change in h or A will be

$$x(2)^2 \times h \text{ or } A$$

3. Let Y = initial chart speed and N = final chart speed. If we *increase* the chart speed, the change in W or A will be

$$\frac{N}{Y} \times W \text{ or } A$$

If we *decrease* the chart speed, the change in W or A will be

$$\frac{1}{Y/N} \times W \text{ or } A$$

APPENDIX B

Gas Chromatographic Acronyms and Symbols and Their Definitions

A	Peak area; surface area of solid granular adsorbent; van Deemter equation eddy diffusion term
A_c	Cross-sectional area of a column (internal)
AAD	Atomic absorption detector
AFID	Alkali flame-ionization detector
AN	Area normalization
ANRF	Area normalization with response factors
API	Atmosphere pressure ionization
ARF	Absolute response factor
B	van Deemter equation molecular diffusion term; second virial coefficient
B_0	Specific permeability
C	van Deemter equation mass-transfer term
C_G	Concentration of solute component in gas phase
C_i	Concentration of a test substance in the mobile phase at the detector
C_M	Concentration of solute component in mobile phase
C_{p1}	Gram-specific heat ratio of carrier gas at constant pressure
C_{p2}	Gram-specific heat ratio of sample at constant pressure
CRF	Chromatographic response function
C_S	Concentration of solute component in stationary phase
CI	Chemical ionization
D_A	Density of absorbent
D_c	Concentration distribution ratio
D_g	Distribution coefficient
D_m	Mass distribution ratio
D_s	Distribution coefficient
D_v	Distribution coefficient
D	Minimum detectability of a detector
D	Diffusion coefficient in general; density; distribution ratio
D_G	Diffusion coefficient in the gas phase

D_L	Diffusion coefficient in the liquid stationary phase
D_M	Diffusion coefficient in the mobile phase
D_S	Diffusion coefficient in the stationary phase
E^*	Activation energy
EA	Electron affinity
ECD	Electron-capture detector
EI	Electron impact
EST	External standard technique
F; F	Frequency; Faraday
F_a	Mobile phase flowrate at ambient temperature
F_c	Mobile phase flowrate corrected to column temperature
$\bar{\bar{F}}_c$	Average flowrate of mobile phase in column
F_0	Initial flowrate of mobile phase into the column
FD	Field desorption
FFF	Field flow fractionation
FID	Flame ionization detector
FPD	Flame photometric detector
FSOT	Fused-silica open tubular column
FTIR	Fourier transform infrared
ΔG°	Free energy of adsorption
GC	Gas chromatography (noun)
GLC	Gas–liquid chromatography (noun)
GSC	Gas–solid chromatography (noun)
H	Plate height (height equivalent to one theoretical plate); McReynolds constant for 2-methylpentanol-2
H_{eff}	Effective plate height (height equivalent to one effective plate)
ΔH°_{ST}	Isoteric heat of adsorption
HAFID	Hydrogen atmosphere flame-ionization detector (*Anal. Chem.*, **51**(2), 291 (1979))
HCOT	Helically coiled open tube (column)
HDPE	High-density polyethylene
HECD	Hall electrolytic conducivity cell
HETP	Height equivalent to a theoretical plate
HTS	Hydrogen-transfer system
I	Retention index
$\Sigma\,\Delta I$	Sum of McReynolds numbers; used for stationary phase characterization
I^0	Initial photon flux
IR	Infrared; infrared detector
IST	Internal standard technique
I^T	Retention index obtained in programmed temperature analysis
J	McReynold's constant for iodobutane
K	Absolute temperature; distribution constant in general; McReynold's constant for 2-octyne

K_c	Distribution constant in which the concentration in the stationary phase is expressed as weight of substance per volume of the phase
$\tilde{K}_c$	Equilibrium distribution constant
K_D	Distribution constant
$K^0_{D(R,S)}$	Distribution coefficients on pure-phase R or S
K^0	Thermodynamic distribution constant, Equation 1.24
K_g	Distribution constant in which the concentration in the stationary phase is expressed as weight of substance per weight of the dry solid phase
K_s	Distribution constant in which the concentration in the stationary phase is expressed as weight of substance per surface area of the solid phase
L	Column length; McReynold's constant for 1,4-dioxane
LPDE	Low-density polyethylene
LPG	Liquefied petroleum gas
M	Molecular weight
M	McReynold's constant for *cis*-hydrindane
M_i	Mass rate of the test substance entering the detector
MAOT	Maximum allowable operating temperature for stationary phases
MDL	Minimum detectable level (detector)
MPD	Microwave plasma detector
MS	Mass spectrometry
N	Noise of a detector; Avogadro's number; plate number (number of theoretical plates)
N_{eff}	Effective theoretical plate number
NPD	Nitrogen–phosphorus detector
OPGV	Optimum practical carrier-gas velocity
OTC	Open tubular columns (capillary columns)
P	Pressure in general; relative pressure
PGC	Pyrolysis gas chromatography
$PH(GC)^2$	Pyrolysis–hydrogenation with glass capillary gas chromatography
PI	Performance index
PID	Photoionization detector
PLOT	Porous-layer open tubular column
PTGC	Programmed-temperature gas chromatography
Q	Heat flow
R	Gas constant; resistance; retardation factor in column chromatography; fraction of a sample component in the mobile phase
RAN	Raw area normalization
1-R	Fraction of sample component in the stationary phase
RF	Response factor

RRF	Relative response factor
R_s	Peak resolution
S	Separation factor according to Purnell, Equation 2.82; surface area; detector sensitivity
$\overline{\Delta S^\circ}$	Entropy of adsorption
SCOT	Support-coated open tubular column
SIM	Single ion monitoring
SN	Separation number
STP	Standard temperature and pressure (25°C and 1 atm)
SVP	Saturation vapor pressure
T	Temperature in general
T_a	Ambient temperature
T_c	Column temperature
TCD	Thermal conductivity detector
TEA	Thermal energy analyzer
TID	Thermionic detector
TIM	Total ion mode
TSD	Thermionic specific detecor (name used for NPD detector)
TZ	Trennzahl number
V	Volt
V	Volume in general
V_A	True adsorbent volume
V_c	Column (tube) volume (cm^3); $V_c = A_c L$
V_{ext}	Extra-column volume
V_g	Specific retention volume at 0°C
V_g	Specific retention volume at column temperature
V_G	Interstitial mobile phase volume (interparticle volume) = V_I
V	Interstitial volume of mobile phase
V_L	Volume liquid stationary phase
V_M	Holdup volume; i.e., retention volume of nonretained peak; mobile phase holdup volume
V_M^0	Corrected holdup volume for nonretained peak
V_N	Net retention volume
V_0	STP volume of one mole of gas; interparticle volume of column
V_R	Absolute retention volume
V_R	Peak elution volume
V_R'	Adjusted retention volume
V_R^0	Corrected retention volume
V_S	Volume stationary phase
V_t	Total mobile phase in the column
V_S^T	Specific retention volume in gas-solid chromatography
W	Mass (weight) in general
WCOT	Wall-coated open tubular column
W_i	Mass (weight) of a test substance present
W_L	Mass (weight) of the liquid phase

W_S	Mass (weight) of the stationary phase
WWCOT	Whisker-wall-coated open tubular column
WWPLOT	Whisker-wall porous-layer open tubular column
WWSCOT	Whisker-wall support-coated open tubular column
X_s	Mole fraction in the stationary phase
Y	Pen reponse
Z	Area response factor
a	Uptake, in grams per gram of adsorbent; packing porosity
$a^u_{s(M)}$	Activity coefficient in the stationary phase (mobile)
c	Gas-phase concentration
c_a	Adsorbed-phase concentration
d	Tube diameter
d_c	Column inside diameter
d_f	Thickness of liquid-phase film
d_p	Diameter of support particle
f	Relative detector response factor; also frequency
h	Reduced plate height; Peak height
j	Compressibility correction factor
k	Retention factor (capacity factor)
n	Moles of a substance in a mixture; mole fraction
n_{ne}	Required plate number
p	Pressure in general
p_i	Inlet pressure
p_o	Outlet pressure
p°	Vapor pressure of a pure substance
p_w	Partial pressure of water at ambient temperature
p	Pressure drop
$r_{a/b}$	Relative retention
r_c	Column tubing radius, i.d.
r_G	Unadjusted relative retention
r_p	Pore radius
s	Rohrschneider constant for pyridine; sound pathlength
s'	McReynold's constant for pyridine
t	Time in general; analysis time, based on solute component more readily sorbed (see Equation 2.94)
t_M	Mobile phase holdup time; it is also equal to the retention time of an unretained compound; referred to as "air peak"
t_N	Net retention time
t_{ne}	Minimum analysis time
t_0	Injection point time
t_R	Total retention time; absolute retention time
t'_R	Adjusted retention time
t^0_R	Corrected retention time
t^T_R	Total retention time in temperature-programmed analysis

$\bar{t}_R$	Peak elution time
u	Linear velocity of mobile phase; interstitial velocity of mobile phase; Rohrschneider constant for nitromethane
$\bar{u}$	Mean interstitial velocity of mobile phase; i.e., average linear gas velocity
u_D	Diffusion velocity
u_o	Carrier-gas velocity at column outlet; $=F_c L/V_c$
u'	McReynolds constant for nitropropane
w_b	Peak width at the base
w_h	Peak width at half height
w_i	Peak width at the inflection points
x	Rohrschneider constant for benzene
x'	McReynolds constant for benzene
y	Rohrschneider constant for ethanol
y'	McReynolds constant for butanol-l
z	Rohrschneider constant for methyl ethyl ketone; number of carbon atoms of a *n*-paraffin eluting before the peak of interest
z'	McReynolds constant for methyl *n*-propyl ketone
$z+1$	Number of carbon atoms of a *n*-paraffin eluting after the peak of interest
α	Separation factor (relative retardation)
α_G	Unadjusted separation factor (relative retention)
β	Phase ratio
γ	Tortuosity factor, expressing uniformity of support particle size and shape; activity coefficient; specific heat ratio
ε	Interparticle porosity; $\varepsilon = V_o/V_c$
ε_I	Interstitial fraction
ε_S	Stationary phase fraction
λ	Packing term, expressing uniformity with which a packed column is filled; thermal conductivity
η	Mobile phase velocity; efficiency coefficient of ionization
π	Constant = 3.1416
κ	$\log k$
ν	Reduced mobile phase velocity
ρ	Hammett constant; density
ρ_L	Density of liquid phase at column temperature
σ	Standard deviation of a Gaussian peak; area occupied by one molecule; Hammett constant; absorption cross section
σ^2	Variance of a Gaussian peak
Φ	Fraction of total solute in a given phase; degrees phase change; flow resistance parameter
$\Phi_{A,S}$	Mole fraction of stationary phases A and S in binary mixture
Φ	Fraction remaining in original phase after extraction

INDEX

Where the statue stood
Of Newton, with his prism and silent face,
The marble index of a mind for ever
Voyaging through strange seas of though alone.

—William Wordsworth (1770–1850)
The Prelude, book iii, line 61